Handbook of Experimental Pharmacology

Volume 223

For further volumes:
http://www.springer.com/series/164

Bernd Nilius • Veit Flockerzi
Editors

Mammalian Transient Receptor Potential (TRP) Cation Channels

Volume II

Springer

Editors
Bernd Nilius
Laboratory Ion Channel Research
Campus Gasthuisberg
KU Leuven Department
Cell Mol Medicin
Leuven, Belgium

Veit Flockerzi
Institut für Experimentelle u. Klinische
Pharmakologie und Toxikologie
Universität des Saarlandes
Homburg, Germany

ISSN 0171-2004 ISSN 1865-0325 (electronic)
ISBN 978-3-319-05160-4 ISBN 978-3-319-05161-1 (eBook)
DOI 10.1007/978-3-319-05161-1
Springer Cham Heidelberg New York Dordrecht London

Library of Congress Control Number: 2014937453

Printed on acid-free paper

Springer is part of Springer Science+Business Media (www.springer.com)

Preface

When we, in 2007, edited the first issue on transient receptor potential channel in the *Handbook of Experimental Pharmacology*, we were all very excited by the progress in this field although only one decade after cloning the first TRP channel had passed. At this time, somewhat less than 5,000 papers were published on TRP channel (January 1, 1960, until December 31, 2006). If we check now the period (January 1, 2007, until January 13, 2014), additional 9,300 papers can be found in a PubMed search.[1] Needless to say, the general interest on these 28 members of the Trp gene family which encode ion channels is nearly exponentially growing. Therefore, it seemed to be indicated, although many excellent books on TRP channels have been published meanwhile, to jump into a new adventure editing a comprehensive source book in this successful Springer Handbook series again on the same topic. This is not only an update of the 2007 book but also an impressive introduction of novel areas which TRP channels have entered. The 2007 view that TRP channels are mainly cell sensors with an intriguing variability concerning the modes of activation has dramatically extended into the evolutionary field, the structural approach, and especially the advent of the important role of TRP channels in hereditary and acquired diseases. Important new data concerning the role of TRP channels in intracellular compartments are included. We also refer to the still controversial topic how TRP channels are involved in store-dependent Ca^{2+} entry. Indeed, the TRP field expansion did not lose the fast speed. It is extending into so far unexpected areas. The *gain of knowledge* has reached such an extent that we have not been able to restrict the source book into a single volume; rather, we had to agree on a two-volume publication. In the first volume, we go through all the known TRP channels. Leading experts in the field summarize features of individual TRP channels starting with the description of the gene, expression patters, associated proteins, biophysical and biochemical function properties, and transgenic animal models and closing with cellular TRP functions, dysfunctions, and their role in diseases. The second volume starts with a chapter on sensor properties and functions

[1] The used search string was ("transient receptor potential" OR trpa* OR trpc* OR trpm* OR trpp* OR trpv* OR PKD* OR stim1 OR stim2 OR orai1 OR orai2 OR orai3 OR trpa*). Note that this search included also the main players of store-operated Ca^{2+} entry, because of the still often reported links to TRP as also discussed in Volume 2.

of TRP channels. This was highlighted in the 2007 book but is not very much extended. Surprising new features are reported, e.g., new insights into thermo- and light-sensing, novel roles or TRPs in taste perception and chemesthesis, and especially their functional importance as chemosensors for gasotransmitters, including oxygen sensing, which was evidenced only in the last 5 years. In the second part, more general topics related to TRP functions and features are discussed such as channel structure; TRPs as targets of pharmacological modulation, including a wealth of natural compounds; and the exciting discovery of novel channel toxins. New aspects are discussed concerning the role of TRPs as important players in the physiology of reproduction and in neural networks which control reproductive behavior opening a *TRP window* into neuroendocrinology, i.e., their role in hormone-secreting cells. We finish this book with some critical remarks on the current state of TRP research, controversies, and surprises.

We hope that this book will guide a large reader community through the fascinating world of the TRP channel family from basic science to pathophysiology and disease. May this voluminous source/textbook also help to establish interactions between the fundamental and clinical research and the research in drug discovery and development! We are convinced that this book is "translational" in the best meaning of this word. Despite the many advances in the understanding of the molecular mechanisms and function features of TRP channel, there is still a tremendous need for more in-depth understanding of the structure of TRP channels, their implementation in diverse signal cascades, and more mechanistic insight into channel function at the molecular and systemic level, as well as the need for identifying selective pharmacological tools and new therapeutic targets and developing new treatment options. We hope this book stimulates further research. Finally, we may conclude that we might be still in a period of the end of the beginning rather than the beginning of the end! The editors wish to thank all authors for excellent contribution and also Wilma McHugh (Springer) for all expert support and very helpful editorial advice!

Leuven, Belgium — Bernd Nilius
Homburg, Germany — Veit Flockerzi

Contents for Volume II

Contents for Volume I

Part V

TRPs as Special Cell Sensors

TRP Channels and Thermosensation

Thomas Voets

Contents

Abstract

Several TRP channels exhibit highly temperature-dependent gating properties, which leads to steep changes in depolarising current upon either cooling or heating. Based on this characteristic feature, these so-called "thermoTRPs" have been widely studied with the aim to elucidate their potential key role as thermosensors in the somatosensory system and to understand the basis of their high thermal sensitivity. In this chapter, I provide a brief critical overview of current knowledge on the role of TRP channels in thermosensing and on the thermodynamic and molecular basis of their steep temperature dependence.

Keywords

Somatosensory system • Temperature dependence • Gating mechanisms • Channel biophysics

---

T. Voets (✉)
Laboratory of Ion Channel Research and TRP Research Platform Leuven (TRPLe), KU Leuven, Herestraat 49 bus 802, 3000 Leuven, Belgium
e-mail: thomas.voets@med.kuleuven.be

B. Nilius and V. Flockerzi (eds.), *Mammalian Transient Receptor Potential (TRP) Cation Channels*, Handbook of Experimental Pharmacology 223, DOI 10.1007/978-3-319-05161-1_1,

1 Cellular and Molecular Basis of Thermosensation

Temperature sensing is primarily mediated by sensory neurons that have their sensory endings in the skin and mucosa and their cell bodies in the trigeminal (TG) and dorsal root ganglia (DRG) (Damann et al. 2008; Julius and Nathans 2012). Under resting conditions, voltage-gated K^+ channels ensure a negative membrane potential in these cells (Damann et al. 2008). In response to changes in temperature, different types of depolarising channels open in the nerve endings, leading to membrane depolarisation and, when a certain threshold is crossed, action potential initiation via voltage-gated Na^+ channels. These action potentials are then propagated along the sensory neuron and lead to neurotransmitter release onto second-order neurons in the dorsal horn (DRG) or sensory nucleus in the brain (TG) (Damann et al. 2008). The depolarising channels that open in response to thermal stimulation can be considered as the primary molecular sensors of temperature, nature's thermometers, and understanding their molecular nature and modus operandi represents an important goal in sensory physiology. In the last 16 years, several members of the transient receptor potential (TRP) superfamily have been identified as thermosensitive (hence the name "thermoTRPs") and put forward as primary temperature sensors in the somatosensory system (Caterina et al. 1997; Voets et al. 2005; Talavera et al. 2008). In this chapter, I present a critical reappraisal of the role of TRP channels as thermosensors in the mammalian somatosensory system and bring up some considerations on the basis of their temperature sensitivity.

2 ThermoTRPs as Prime Molecular Thermometers(?)

Since all processes in nature exhibit some degree of temperature sensitivity, three discerning criteria have been put forward that may be used to identify "true" thermoTRPs (Voets 2012). As a first criterion, the candidate thermoTRP's biophysical properties must be such that it can produce robust depolarising currents in response to a thermal stimulus. The robustness of the thermal response can be quantified as the Q_{10} value, which is defined as the relative change in current amplitude upon a 10° increase in temperature. A Q_{10} value of more than 5 (or less than 0.2 for of a cold-activated channel) has been put forward as a minimal requirement, although it should be noted that experimental Q_{10} values are strongly dependent on the cellular context and experimental approach. As a second criterion, the candidate thermoTRP must be expressed at sites in the body that face significant variations in temperature, including the skin, mouth, upper airways and oesophagus. As a third and probably most important criterion, there must be in vivo evidence that the thermal sensitivity of a candidate thermoTRP confers thermosensitivity to a physiological process.

With respect to the first criterion, according to the literature, 11 mammalian TRP channels exhibit Q_{10} values >5 or <0.2 when heterologously expressed in cell lines or Xenopus laevis oocytes: currents mediated by TRPV1, TRPV2, TRPV3, TRPV4,

TRPM2, TRPM3, TRPM4 and TRPM5 increase upon warming (Caterina et al. 1997, 1999; Guler et al. 2002; Peier et al. 2002b; Smith et al. 2002; Xu et al. 2002; Talavera et al. 2005; Togashi et al. 2006; Vriens et al. 2011), whereas currents mediated by TRPM8, TRPA1 and TRPC5 increase upon cooling (McKemy et al. 2002; Peier et al. 2002a; Story et al. 2003; Zimmermann et al. 2011). It should be noted, however, that heat-induced activation is observed for mouse and rat TRPV2 but not for the human orthologue (Neeper et al. 2007) and that cold-induced activation of mammalian TRPA1 has been consistently observed by some laboratories (Story et al. 2003; Sawada et al. 2007; Fajardo et al. 2008; Karashima et al. 2009) but not by others (Jordt et al. 2004; Zurborg et al. 2007; Cordero-Morales et al. 2011).

With respect to the second and third criteria, the situation is even much more contentious. This is particularly the case when considering the role of TRP channels as thermosensors in the somatosensory system. Following the cloning of TRPV1 as the first heat-activated TRP channel and the subsequent identification of closely (TRPV2-V4) and more distally related (TRPM8, TRPA1) as differently tuned thermosensors, it was generally assumed for many years that these six channels would be sufficient to cover the entire range of temperatures that the mammalian somatosensory system can discriminate (Voets et al. 2005; Dhaka et al. 2006; Basbaum et al. 2009). Indeed, the TRP channel literature and physiology handbooks are replete with a variation of figures and schemes with these channels as sensors for noxious (TRPA1) and innocuous (TRPM8) cold, warmth (TRPV3 and TRPV4), heat (TRPV1) and extreme heat (TRPV2).

However, with hindsight, it appears that these schemes were somewhat overoptimistic. Actually, with respect to acute thermosensing, the only TRP channel for which there is general agreement in studies from multiple groups is TRPM8, which seems to be crucial for accurately discriminating temperatures between ~15 and ~30° (Bautista et al. 2007; Colburn et al. 2007; Dhaka et al. 2007). For TRPV1, there is consensus that it plays a crucial role in heat sensing under conditions of inflammation (Caterina et al. 2000; Davis et al. 2000). However, whereas some reports show that TRPV1-deficient mice have a deficit in acute heat sensing (Caterina et al. 2000), others studies have not been able to replicate this (Davis et al. 2000). For TRPV3 and TRPV4, initial studies reported clear behavioural deficits in warmth sensing at temperatures between ~25 and ~35° (Lee et al. 2005; Moqrich et al. 2005). However, more recent studies indicate that the alteration in temperature preference in TRPV3-deficient mice is highly dependent on the genetic background of the mice (Huang et al. 2011; Miyamoto et al. 2011). Moreover, TRPV3/TRPV4 double knockout mice on a C57BL6 background exhibited thermal preference and heat avoidance behaviour that was virtually indistinguishable from that of wild-type C57BL6 mice, even when TRPV1 function was eliminated (Huang et al. 2011). Although heat activation and expression in sensory neurons is widely documented for mouse TRPV2 (Caterina et al. 1999; Neeper et al. 2007), TRPV2-deficient mice show no indication of altered thermosensation (Park et al. 2011). In the case of TRPA1, there are unexplainable differences between labs, not only with respect to cold activation in vitro but also concerning its role in vivo. Indeed, whereas some laboratories report significant deficits in noxious cold sensing in TRPA1-deficient mice (Kwan et al. 2006; Karashima et al. 2009), others found no acute thermosensing phenotype whatsoever (Bautista et al. 2006; Knowlton et al. 2010).

More recently, evidence has been presented that TRPC5 (as a cold sensor) (Zimmermann et al. 2011) and TRPM3 (as a heat sensor) (Vriens et al. 2011) may play a role in acute thermosensing in mice, but it seems wise to await further independent confirmation of these interesting findings. TRPM2 and TRPM4 are expressed in sensory neurons (Vandewauw et al. 2013), but there is no evidence so far that they confer relevant thermosensitivity to somatosensation or any other physiological process. There is also strong evidence that TRPM5 confers thermosensitivity to taste (Talavera et al. 2005), but its expression level in somatosensory neurons is probably too low to play any significant role in (acute) thermosensation (Vandewauw et al. 2013).

Ongoing research in this field may cause the list of thermoTRPs to further grow (or shrink). It also seems warranted to envisage that important aspects of thermosensation may depend on thermosensitive conductances that are not mediated by TRP channels, such as K^+ channels (Noel et al. 2009), TMEM16A (Cho et al. 2012), Stim-Orai (Xiao et al. 2011) and probably more.

3 Temperature-Dependent Gating: Thermodynamic Considerations

Steep temperature-dependent activation of an ion channel can, in principle, be based on either intrinsic steep temperature dependence of the channel itself or, alternatively, on the steep temperature sensitivity of another cellular component (e.g., enzyme, membrane, cytoskeleton) that directly or indirectly regulates the activity of the channel. Without excluding a contribution of the latter mechanism, there is a growing body of strong evidence that the temperature sensitivity of at least some thermoTRPs, including TRPM8 and TRPV1, is largely preserved in cell-free patches and even upon purification and reconstitution in artificial lipid membranes (Zakharian et al. 2009; Cao et al. 2013). This indicates that in these channels temperature sensitivity is an intrinsic property of the channel's structures that determine gating and permeation.

In general, the ion flux through an open (TRP) channel pore exhibits mild thermal sensitivity, comparable to the temperature dependence of ionic diffusion in aqueous solution, with typical Q_{10} values <2 (Voets 2012). Consequently, research has focused on the origin of temperature-sensitive channel gating. However, despite significant efforts and a growing number of papers, we are still far from fully understanding the thermodynamic processes and structural rearrangements that underlie the highly temperature-sensitive gating of thermoTRPs.

In the TRP channel literature, there are widely diverging global views on thermosensitive gating of TRP channels. On the one hand, thermosensitivity of

TRP channels has been explained using a simple two-state model, where the actual channel gating step is temperature sensitive and thermosensitivity is not confined to a specific part of the channel (Voets et al. 2004, 2005; Clapham and Miller 2011). On the other hand, models have been put forward in which the origin of thermosensitivity is confined in one or few restricted domains of the channel, which act as thermosensor "modules" and are allosterically coupled to the channel gate (Brauchi et al. 2004; Latorre et al. 2007; Matta and Ahern 2007). Both extreme views can be easily criticised; for instance, a two-state model is obviously an (over)simplification, given the extensive evidence (e.g., from kinetic and single-channel analyses) for the existence of multiple open and closed states, whereas restricting thermosensitivity to delineated channel domains disregards the fact that every atom in the channel protein is affected by temperature.

Irrespective of the model, the behaviour of steeply temperature-dependent TRP channels implies that the equilibrium between closed and open states is highly thermosensitive. In thermodynamic terms, the equilibrium between two global states, for instance, the closed and open state of a channel or the inactive and active state of a thermosensor module, is given by

$$K_{\mathrm{eq}} = \exp\left(\frac{-\Delta G}{RT}\right),$$

where K_{eq} represents the ratio between the open and closed state of the channel (or between the active and inactive conformation of the thermosensor module), ΔG the free energy change, R the universal gas constant and T the temperature in Kelvin (Fig. 1a).

If we disregard effects of mechanical or electrical forces, ΔG is given by

$$\Delta G = \Delta H - T\Delta S,$$

where ΔH represents the difference in enthalpy (in J mol^{-1}) and ΔS the difference in entropy (in J mol^{-1} K^{-1}) between the two states.

When $\Delta G = 0$, it follows that $K_{\mathrm{eq}} = 1$, which means that 50 % of the channels (or thermosensor modules) are in the open state (active conformation). This occurs at the temperature for half-maximal activation (T_{50}), which is given by

$$T_{50} = \frac{\Delta H}{\Delta S}.$$

Assuming that T_{50} has a realistic (i.e., positive) value, it follows that ΔH and ΔS must have the same sign. When ΔH and ΔS are both positive, ΔG decreases upon heating. This leads to a heating-induced shift of the equilibrium towards the open state/active conformation, as would be the case in a heat-activated (TRP) channel. Oppositely, when ΔH and ΔS are both negative, ΔG decreases upon cooling. This leads to a cooling-induced shift of the equilibrium towards the open state/active conformation, as would be the case in a cold-activated (TRP) channel (Fig. 1b).

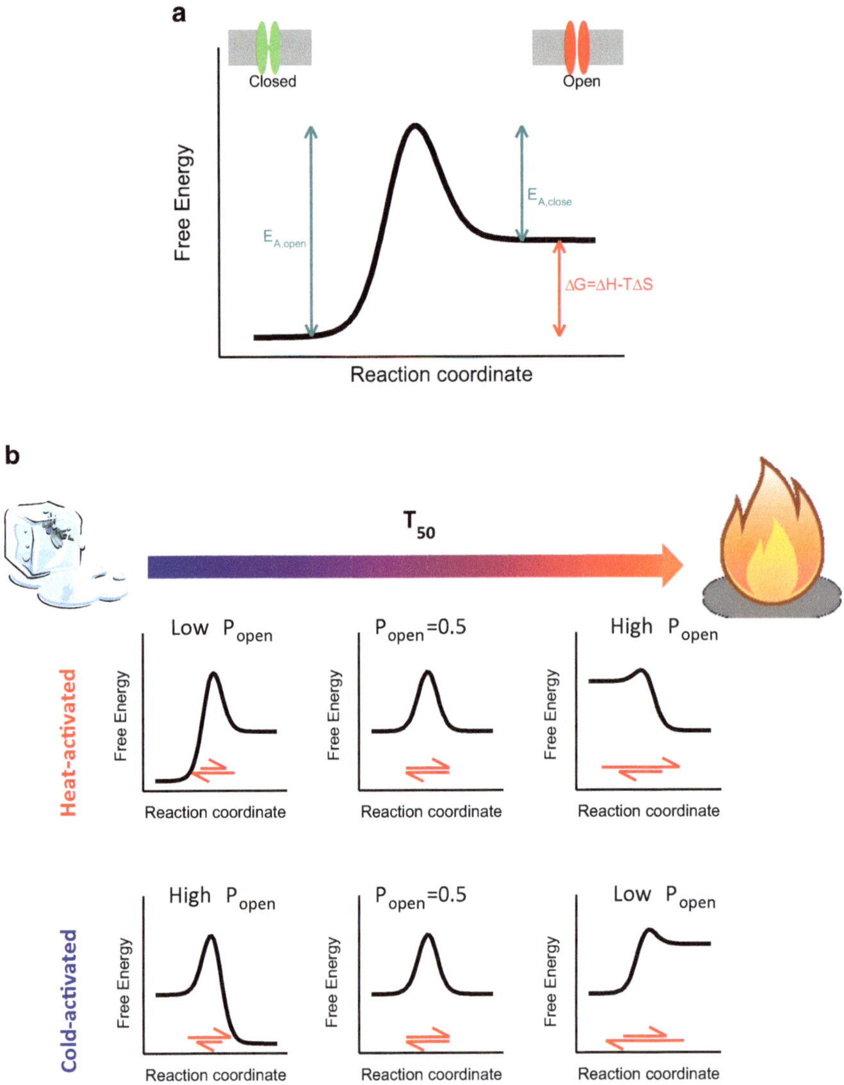

Fig. 1 Energy diagrams describing the gating process of temperature-sensitive channels. (**a**) General scheme. ΔG, the difference in Gibbs free energy between the close and open states determines the equilibrium, whereas the activation energies for opening ($E_{A,open}$) and closing ($E_{A,close}$) determine the opening and closing rates (indicated by the length of the *red arrows* in panel **b**). (**b**) Influence of temperature on the energy diagram in heat- and cold-activated channels

Energy diagrams such as those shown in Fig. 1 also allow understanding the effect of temperature on the gating kinetics of cold- and heat-activated TRP channels. In the case of a heat-activated channel, increasing temperature causes a destabilisation of the closed state. This reduces the activation energy for channel opening ($E_{A,open}$), which leads to a steep increase of the channel opening rate in response to heating, whereas channel closing rates are only mildly temperature dependent. In contrast, in the case of a cold-activated channel, decreasing temperature causes a stabilisation of the open state. This increases the activation energy for channel opening ($E_{A,close}$), which leads to a steep decrease of the channel closing rate in response to cooling, whereas in this case channel opening rates are only mildly temperature dependent (Voets et al. 2004).

At very low open probabilities, the temperature dependence of channel gating, expressed as $Q_{10,gating}$, is directly related to ΔH, according to

$$Q_{10,\text{gating}} \approx 10^{4.34 \times \frac{\Delta H}{RT^2}}.$$

It should be noted that the Q_{10} for current through a thermoTRP is determined not only by $Q_{10,gating}$ but also by $Q_{10,permeation}$, which quantifies the temperature dependence of ion flux through the open channel pore. Taking this into account, we arrive at

$$Q_{10} = Q_{10,\text{gating}} \times Q_{10,\text{permeation}}.$$

For ion diffusion through ion channels, $Q_{10,permeation}$ generally has values between 1.1 and 2, thus providing a minor contribution to the high Q_{10} of heat-activated TRP channels and counteracting the cold activation of cold-activated TRP channels.

Several studies have provided measurements of ΔH and ΔS for different temperature-sensitive TRP channels, based on steady-state current measurements at different temperatures and voltages or kinetic analyses of currents in response to heat and/or voltage jumps. These studies indeed consistently show $\Delta H > 0$ and $\Delta S > 0$ for the opening of heat-activated TRP channels and $\Delta H < 0$ and $\Delta S < 0$ for the opening of cold-activated TRP channels. Estimates for ΔH range between ± 150 and ± 400 kJ mol^{-1} and for ΔS between ± 500 and $\pm 1{,}200$ J mol^{-1} K^{-1}. (Note that I use the SI unit joules (J) rather than the commonly used calorie (cal) as unit for energy; 1 cal = 4.184 J.)

So how can one interpret ΔH and ΔS for a channel gating reaction? A positive value for ΔH, as found in heat-activated channels, indicates that the opening of the channel is an endothermic reaction, absorbing heat from the surroundings. This is, for instance, the case when gating involves the disruption of internal interactions in the channel, such as salt bridges, cation–π interactions or hydrogen bonds. Oppositely, opening of cold-activated channels has a negative ΔH, indicative of the formation of stabilising internal interactions. As a yardstick, typical ΔH values for disruption of a single hydrogen bond in a protein are in the order of ~5 kJ mol^{-1}, whereas ΔH values for disruption of cation–π interactions or salt bridges are typically 2–4-fold higher (Jackson 2006).

A positive value of ΔS, as found in heat-activated channels, indicates that gating leads to an increase in the degrees of freedom of the channel, for instance, due to exposure of amino acid side chains upon unfolding of a close-packed protein domain. Oppositely, opening of cold-activated channels has a negative ΔS, indicating a reduction in the degrees of freedom in the open state. Published estimates of ΔS for protein unfolding are in the order of 20 $\text{mol}^{-1}\ \text{K}^{-1}$ per amino acid residue (Jackson 2006).

Understanding the molecular basis of thermosensitivity in (TRP) channels comes down to understanding the (sub)molecular events that underlie these changes in enthalpy and entropy when the channel transits between closed and open states. Temperature-sensitive TRP channels are tetramers consisting of between ~3,500 and 4,500 amino acids. Moreover, ΔH and ΔS refer to the complete system, which includes not only the channel protein but also interacting water molecules and lipids. As such, there is ample "room" for local or global conformational changes to accommodate the changes in enthalpy and entropy. For instance, it has been estimated that membrane channels typically contain 0.29 hydrogen bonds per amino acid residue (Joh et al. 2008). If this holds through for TRP channels, a change in the number of hydrogen bonds during gating of only a few percent would suffice to explain the typical ΔH values.

In addition, it has been pointed out that ΔH and ΔS should not necessarily be viewed as constants, but vary with temperature whenever gating is accompanied by changes in heat capacity (ΔC_p, expressed in kJ $\text{mol}^{-1}\ \text{K}^{-1}$) (Clapham and Miller 2011):

$$\Delta H(T) = \Delta H_0 + \Delta C_\text{p}(T - T_0),$$

and

$$\Delta S(T) = \Delta S_0 + \Delta C_\text{p}\ln(T/T_0),$$

where ΔH_0 and ΔS_0 represent the change in enthalpy and entropy at a reference temperature T_0. If ΔC_p is large, ΔH and ΔS will change significantly with temperature and may even change sign, which implies that a channel is both cold and heat activated. Although this is an interesting and provocative idea, there is currently no evidence for such behaviour in TRP channels. In contrast, several studies indicate that ΔH remains fairly stable within the biologically relevant temperature range (Voets et al. 2004, 2007; Yang et al. 2010; Yao et al. 2010a, 2011).

4 (No) Lessons from Structure–Function Studies?

There is a steadily growing number of studies aimed at understanding the structural basis of temperature sensitivity in TRP channels. In most cases, these studies started from the assumption that TRP channels are modular machines that contain clearly delineated functional regions and even transplantable thermosensing modules. Several approaches were taken to pursue such modules, including "brute force" methods consisting of random mutagenesis followed by medium- to high-

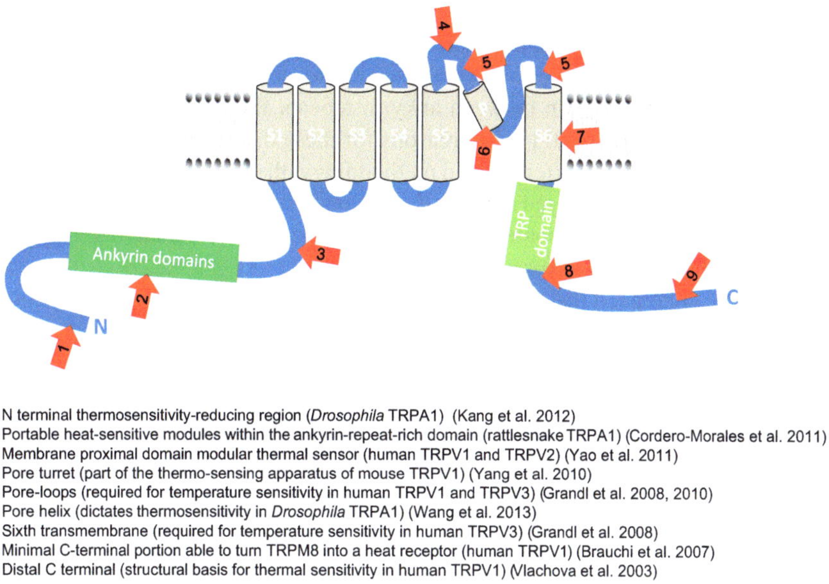

Fig. 2 Schematic overview of regions that have been implicated in thermosensitivity in the TRP channels TRPV1, TRPV2, TRPV3, TRPM8 and TRPA1

throughput functionality screens, site-directed mutagenesis aimed at disturbing the functionality of putative functional domains, characterisation of orthologues and splice variants with distinct and/or opposite thermosensitivities and analysis of chimeric channels obtained by swapping putative thermosensing modules between closely or distantly related TRP channels (Vlachova et al. 2003; Brauchi et al. 2006, 2007; Grandl et al. 2008, 2010; Yang et al. 2010; Cordero-Morales et al. 2011; Yao et al. 2011; Kang et al. 2012; Zhong et al. 2012; Wang et al. 2013). The main outcomes of such studies performed on mammalian isoforms of TRPV1, TRPV3 and TRPM8, as well as on TRPA1 orthologues from mammals, snakes and Drosophila, are schematically illustrated in Fig. 2.

Although most individual studies claim identification of specific domain required for thermosensation, the overall picture that emerges from these studies is fuzzy, with studies inculpating specific domains in the N- and C-terminal cytosolic loops, transmembrane domains and pore region. Moreover, at least in the case of TRPV1 and TRPA1, there are also some apparently conflicting data between studies. For instance, Cordero-Morales et al. (2011) reported that the N-terminal cytoplasmic domain of heat-sensitive TRPA1 from *Drosophila* and rattlesnake and specific ankyrin repeats therein contain transplantable heat-sensitive modules that can confer heat sensitivity to heat-insensitive human TRPA1. In contrast, Wang et al. (2013) reported that heat sensitivity was conferred to the human TRPA1 channel upon transplantation of the pore region, but not the N-terminal part of *Drosophila* TRPA1. Further complexity is added by the surprising finding that a specific part of Drosophila TRPA1 acts as a suppressor of heat sensitivity (Kang et al. 2012). Similarly, whereas some studies strongly implicate the pore region and pore turret in setting the thermosensitivity of TRPV1 (Grandl et al. 2010; Yang et al. 2010), other studies dismiss this possibility and rather implicate the domain between ankyrin repeats and the first transmembrane helix (Yao et al. 2010b, 2011). The reasons for these discrepancies are not always obvious, but may reflect differences in methodology, the lack of robustness of some of the parameters that are used to quantify thermal sensitivity and possibly artefacts caused by the harshness of some experimental protocols to evoke thermal responses from TRP channel-(over)expressing cells.

It may be concluded that, in contrast to voltage-gated Na^+, K^+ or Ca^{2+} channels, in which the voltage sensor is mainly contained within the fourth transmembrane (S4) domain, or to ligand-gated channels such as the cys-loop and ionotropic glutamate receptors, in which the ligand binding site is located in the extracellular domain, thermosensing does not seem to be generated in a clearly delineated and conserved domain of TRP channels. Possibly, the significant difference in enthalpy between closed and open states arises from multiple submolecular rearrangements occurring in different and diffuse areas of the channel.

Obviously, detailed understanding of the molecular basis of thermosensing in TRP channels is therefore still far ahead. Deeper insight may be obtained by meticulous analysis, using standardised patch-clamp assays and/or calorimetry, of the thermodynamic consequences of thousands of point mutations. Once TRP channel structures in closed and open configuration will be available, it may also become feasible to calculate the contribution of domains, residues and atoms to the thermodynamic properties of TRP channels.

Acknowledgements I wish to thank all present and former members of the Laboratory of Ion Channel Research for enlightening discussions. Our work on thermoTRPs was supported by grants from the Belgian Federal Government (IUAP P7/13), the Research Foundation-Flanders (FWO) (G.0565.07 and G.0825.11) and the Research Council of the KU Leuven (GOA 2009/07, EF/95/010 and PF-TRPLe).

References

Basbaum AI, Bautista DM, Scherrer G, Julius D (2009) Cellular and molecular mechanisms of pain. Cell 139:267–284

Bautista DM, Jordt SE, Nikai T, Tsuruda PR, Read AJ, Poblete J, Yamoah EN, Basbaum AI, Julius D (2006) TRPA1 mediates the inflammatory actions of environmental irritants and proalgesic agents. Cell 124:1269–1282

Bautista DM, Siemens J, Glazer JM, Tsuruda PR, Basbaum AI, Stucky CL, Jordt SE, Julius D (2007) The menthol receptor TRPM8 is the principal detector of environmental cold. Nature 448:204–208

Brauchi S, Orio P, Latorre R (2004) Clues to understanding cold sensation: thermodynamics and electrophysiological analysis of the cold receptor TRPM8. Proc Natl Acad Sci U S A 101:15494–15499

Brauchi S, Orta G, Salazar M, Rosenmann E, Latorre R (2006) A hot-sensing cold receptor: C-terminal domain determines thermosensation in transient receptor potential channels. J Neurosci 26:4835–4840

Brauchi S, Orta G, Mascayano C, Salazar M, Raddatz N, Urbina H, Rosenmann E, Gonzalez-Nilo F, Latorre R (2007) Dissection of the components for PIP2 activation and thermosensation in TRP channels. Proc Natl Acad Sci U S A 104:10246–10251

Cao E, Cordero-Morales JF, Liu B, Qin F, Julius D (2013) TRPV1 channels are intrinsically heat sensitive and negatively regulated by phosphoinositide lipids. Neuron 77:667–679

Caterina MJ, Schumacher MA, Tominaga M, Rosen TA, Levine JD, Julius D (1997) The capsaicin receptor: a heat-activated ion channel in the pain pathway. Nature 389:816–824

Caterina MJ, Rosen TA, Tominaga M, Brake AJ, Julius D (1999) A capsaicin-receptor homologue with a high threshold for noxious heat. Nature 398:436–441

Caterina MJ, Leffler A, Malmberg AB, Martin WJ, Trafton J, Petersen-Zeitz KR, Koltzenburg M, Basbaum AI, Julius D (2000) Impaired nociception and pain sensation in mice lacking the capsaicin receptor. Science 288:306–313

Cho H, Yang YD, Lee J, Lee B, Kim T, Jang Y, Back SK, Na HS, Harfe BD, Wang F, Raouf R, Wood JN, Oh U (2012) The calcium-activated chloride channel anoctamin 1 acts as a heat sensor in nociceptive neurons. Nat Neurosci 15:1015–1021

Clapham DE, Miller C (2011) A thermodynamic framework for understanding temperature sensing by transient receptor potential (TRP) channels. Proc Natl Acad Sci U S A 108:19492–19497

Colburn RW, Lubin ML, Stone DJ Jr, Wang Y, Lawrence D, D'Andrea MR, Brandt MR, Liu Y, Flores CM, Qin N (2007) Attenuated cold sensitivity in TRPM8 null mice. Neuron 54:379–386

Cordero-Morales JF, Gracheva EO, Julius D (2011) Cytoplasmic ankyrin repeats of transient receptor potential A1 (TRPA1) dictate sensitivity to thermal and chemical stimuli. Proc Natl Acad Sci U S A 108:E1184–E1191

Damann N, Voets T, Nilius B (2008) TRPs in our senses. Curr Biol 18:R880–R889

Davis JB, Gray J, Gunthorpe MJ, Hatcher JP, Davey PT, Overend P, Harries MH, Latcham J, Clapham C, Atkinson K, Hughes SA, Rance K, Grau E, Harper AJ, Pugh PL, Rogers DC, Bingham S, Randall A, Sheardown SA (2000) Vanilloid receptor-1 is essential for inflammatory thermal hyperalgesia. Nature 405:183–187

Dhaka A, Viswanath V, Patapoutian A (2006) Trp ion channels and temperature sensation. Annu Rev Neurosci 29:135–161

Dhaka A, Murray AN, Mathur J, Earley TJ, Petrus MJ, Patapoutian A (2007) TRPM8 is required for cold sensation in mice. Neuron 54:371–378

Fajardo O, Meseguer V, Belmonte C, Viana F (2008) TRPA1 channels mediate cold temperature sensing in mammalian vagal sensory neurons: pharmacological and genetic evidence. J Neurosci 28:7863–7875

Grandl J, Hu H, Bandell M, Bursulaya B, Schmidt M, Petrus M, Patapoutian A (2008) Pore region of TRPV3 ion channel is specifically required for heat activation. Nat Neurosci 11:1007–1013

Grandl J, Kim SE, Uzzell V, Bursulaya B, Petrus M, Bandell M, Patapoutian A (2010) Temperature-induced opening of TRPV1 ion channel is stabilized by the pore domain. Nat Neurosci 13:708–714

Guler AD, Lee H, Iida T, Shimizu I, Tominaga M, Caterina M (2002) Heat-evoked activation of the ion channel, TRPV4. J Neurosci 22:6408–6414

Huang SM, Li X, Yu Y, Wang J, Caterina MJ (2011) TRPV3 and TRPV4 ion channels are not major contributors to mouse heat sensation. Mol Pain 7:37

Jackson MB (2006) Molecular and cellular biophysics. Cambridge University Press, Cambridge, MA

Joh NH, Min A, Faham S, Whitelegge JP, Yang D, Woods VL, Bowie JU (2008) Modest stabilization by most hydrogen-bonded side-chain interactions in membrane proteins. Nature 453:1266–1270

Jordt SE, Bautista DM, Chuang HH, McKemy DD, Zygmunt PM, Hogestatt ED, Meng ID, Julius D (2004) Mustard oils and cannabinoids excite sensory nerve fibres through the TRP channel ANKTM1. Nature 427:260–265

Julius D, Nathans J (2012) Signaling by sensory receptors. Cold Spring Harb Perspect Biol 4: a005991
Kang K, Panzano VC, Chang EC, Ni L, Dainis AM, Jenkins AM, Regna K, Muskavitch MA, Garrity PA (2012) Modulation of TRPA1 thermal sensitivity enables sensory discrimination in Drosophila. Nature 481:76–80
Karashima Y, Talavera K, Everaerts W, Janssens A, Kwan KY, Vennekens R, Nilius B, Voets T (2009) TRPA1 acts as a cold sensor in vitro and in vivo. Proc Natl Acad Sci U S A 106:1273–1278
Knowlton WM, Bifolck-Fisher A, Bautista DM, McKemy DD (2010) TRPM8, but not TRPA1, is required for neural and behavioral responses to acute noxious cold temperatures and cold-mimetics in vivo. Pain 150:340–350
Kwan KY, Allchorne AJ, Vollrath MA, Christensen AP, Zhang DS, Woolf CJ, Corey DP (2006) TRPA1 contributes to cold, mechanical, and chemical nociception but is not essential for hair-cell transduction. Neuron 50:277–289
Latorre R, Brauchi S, Orta G, Zaelzer C, Vargas G (2007) ThermoTRP channels as modular proteins with allosteric gating. Cell Calcium 42:427–438
Lee H, Iida T, Mizuno A, Suzuki M, Caterina MJ (2005) Altered thermal selection behavior in mice lacking transient receptor potential vanilloid 4. J Neurosci 25:1304–1310
Matta JA, Ahern GP (2007) Voltage is a partial activator of rat thermosensitive TRP channels. J Physiol 585:469–482
McKemy DD, Neuhausser WM, Julius D (2002) Identification of a cold receptor reveals a general role for TRP channels in thermosensation. Nature 416:52–58
Miyamoto T, Petrus MJ, Dubin AE, Patapoutian A (2011) TRPV3 regulates nitric oxide synthase-independent nitric oxide synthesis in the skin. Nat Commun 2:369
Moqrich A, Hwang SW, Earley TJ, Petrus MJ, Murray AN, Spencer KS, Andahazy M, Story GM, Patapoutian A (2005) Impaired thermosensation in mice lacking TRPV3, a heat and camphor sensor in the skin. Science 307:1468–1472
Neeper MP, Liu Y, Hutchinson TL, Wang Y, Flores CM, Qin N (2007) Activation properties of heterologously expressed mammalian TRPV2: evidence for species dependence. J Biol Chem 282:15894–15902
Noel J, Zimmermann K, Busserolles J, Deval E, Alloui A, Diochot S, Guy N, Borsotto M, Reeh P, Eschalier A, Lazdunski M (2009) The mechano-activated K + channels TRAAK and TREK-1 control both warm and cold perception. EMBO J 28:1308–1318
Park U, Vastani N, Guan Y, Raja SN, Koltzenburg M, Caterina MJ (2011) TRP vanilloid 2 knock-out mice are susceptible to perinatal lethality but display normal thermal and mechanical nociception. J Neurosci 31:11425–11436
Peier AM, Moqrich A, Hergarden AC, Reeve AJ, Andersson DA, Story GM, Earley TJ, Dragoni I, McIntyre P, Bevan S, Patapoutian A (2002a) A TRP channel that senses cold stimuli and menthol. Cell 108:705–715
Peier AM, Reeve AJ, Andersson DA, Moqrich A, Earley TJ, Hergarden AC, Story GM, Colley S, Hogenesch JB, McIntyre P, Bevan S, Patapoutian A (2002b) A heat-sensitive TRP channel expressed in keratinocytes. Science 296:2046–2049
Sawada Y, Hosokawa H, Hori A, Matsumura K, Kobayashi S (2007) Cold sensitivity of recombinant TRPA1 channels. Brain Res 1160:39–46
Smith GD, Gunthorpe MJ, Kelsell RE, Hayes PD, Reilly P, Facer P, Wright JE, Jerman JC, Walhin JP, Ooi L, Egerton J, Charles KJ, Smart D, Randall AD, Anand P, Davis JB (2002) TRPV3 is a temperature-sensitive vanilloid receptor-like protein. Nature 418:186–190
Story GM, Peier AM, Reeve AJ, Eid SR, Mosbacher J, Hricik TR, Earley TJ, Hergarden AC, Andersson DA, Hwang SW, McIntyre P, Jegla T, Bevan S, Patapoutian A (2003) ANKTM1, a TRP-like channel expressed in nociceptive neurons, is activated by cold temperatures. Cell 112:819–829

Talavera K, Yasumatsu K, Voets T, Droogmans G, Shigemura N, Ninomiya Y, Margolskee RF, Nilius B (2005) Heat activation of TRPM5 underlies thermal sensitivity of sweet taste. Nature 438:1022–1025

Talavera K, Nilius B, Voets T (2008) Neuronal TRP channels: thermometers, pathfinders and life-savers. Trends Neurosci 31:287–295

Togashi K, Hara Y, Tominaga T, Higashi T, Konishi Y, Mori Y, Tominaga M (2006) TRPM2 activation by cyclic ADP-ribose at body temperature is involved in insulin secretion. EMBO J 25:1804–1815

Vandewauw I, Owsianik G, Voets T (2013) Systematic and quantitative mRNA expression analysis of TRP channel genes at the single trigeminal and dorsal root ganglion level in mouse. BMC Neurosci 14:21

Vlachova V, Teisinger J, Susankova K, Lyfenko A, Ettrich R, Vyklicky L (2003) Functional role of C-terminal cytoplasmic tail of rat vanilloid receptor 1. J Neurosci 23:1340–1350

Voets T (2012) Quantifying and modeling the temperature-dependent gating of TRP channels. Rev Physiol Biochem Pharmacol 16:91–119

Voets T, Droogmans G, Wissenbach U, Janssens A, Flockerzi V, Nilius B (2004) The principle of temperature-dependent gating in cold- and heat-sensitive TRP channels. Nature 430:748–754

Voets T, Talavera K, Owsianik G, Nilius B (2005) Sensing with TRP channels. Nat Chem Biol 1:85–92

Voets T, Owsianik G, Janssens A, Talavera K, Nilius B (2007) TRPM8 voltage sensor mutants reveal a mechanism for integrating thermal and chemical stimuli. Nat Chem Biol 3:174–182

Vriens J, Owsianik G, Hofmann T, Philipp SE, Stab J, Chen X, Benoit M, Xue F, Janssens A, Kerselaers S, Oberwinkler J, Vennekens R, Gudermann T, Nilius B, Voets T (2011) TRPM3 is a nociceptor channel involved in the detection of noxious heat. Neuron 70:482–494

Wang H, Schupp M, Zurborg S, Heppenstall PA (2013) Residues in the pore region of Drosophila transient receptor potential A1 dictate sensitivity to thermal stimuli. J Physiol 591:185–201

Xiao B, Coste B, Mathur J, Patapoutian A (2011) Temperature-dependent STIM1 activation induces Ca(2) + influx and modulates gene expression. Nat Chem Biol 7:351–358

Xu H, Ramsey IS, Kotecha SA, Moran MM, Chong JA, Lawson D, Ge P, Lilly J, Silos-Santiago I, Xie Y, DiStefano PS, Curtis R, Clapham DE (2002) TRPV3 is a calcium-permeable temperature-sensitive cation channel. Nature 418:181–186

Yang F, Cui Y, Wang K, Zheng J (2010) Thermosensitive TRP channel pore turret is part of the temperature activation pathway. Proc Natl Acad Sci U S A 107:7083–7088

Yao J, Liu B, Qin F (2010a) Kinetic and energetic analysis of thermally activated TRPV1 channels. Biophys J 99:1743–1753

Yao J, Liu B, Qin F (2010b) Pore turret of thermal TRP channels is not essential for temperature sensing. Proc Natl Acad Sci U S A 107:E125, author reply E126-127

Yao J, Liu B, Qin F (2011) Modular thermal sensors in temperature-gated transient receptor potential (TRP) channels. Proc Natl Acad Sci U S A 108:11109–11114

Zakharian E, Thyagarajan B, French RJ, Pavlov E, Rohacs T (2009) Inorganic polyphosphate modulates TRPM8 channels. PLoS One 4:e5404

Zhong L, Bellemer A, Yan H, Ken H, Jessica R, Hwang RY, Pitt GS, Tracey WD (2012) Thermosensory and nonthermosensory isoforms of Drosophila melanogaster TRPA1 reveal heat-sensor domains of a thermoTRP Channel. Cell Rep 1:43–55

Zimmermann K, Lennerz JK, Hein A, Link AS, Kaczmarek JS, Delling M, Uysal S, Pfeifer JD, Riccio A, Clapham DE (2011) Transient receptor potential cation channel, subfamily C, member 5 (TRPC5) is a cold-transducer in the peripheral nervous system. Proc Natl Acad Sci U S A 108:18114–18119

Zurborg S, Yurgionas B, Jira JA, Caspani O, Heppenstall PA (2007) Direct activation of the ion channel TRPA1 by Ca2+. Nat Neurosci 10:277–279

TRPs in Mechanosensing and Volume Regulation

Tim D. Plant

Contents

Abstract

Mechanosensitive channels allow cells to respond to changes in membrane stretch that occur due to external stimuli like pressure or flow or that occur because of osmotically induced cell swelling or shrinkage. Ion fluxes through the channels change the membrane potential and ion concentrations and link the stretch to cellular signalling. Changes in cellular activity evoked by mechanical stimuli can be used to elicit local tissue responses or can be transmitted further to generate more widespread responses. Channels can respond directly to membrane stress, can be conferred mechanosensitive by interaction with structural proteins, or can be activated by mechanosensitive signalling pathways. Because mechanosensitive channels are often nonselective cation channels, and invertebrate TRP isoforms are involved in mechanosensation, many of the mammalian TRP isoforms have been investigated with regard to their mechanosensitivity. There is evidence that members of the TRPC, TRPV, TRPM, TRPA and TRPP

T.D. Plant (✉)
Pharmakologisches Institut, BPC-Marburg, FB-Medizin, Philipps-Universität Marburg, Karl-von-Frisch-Straße 1, 35032 Marburg, Germany
e-mail: plant@staff.uni-marburg.de

B. Nilius and V. Flockerzi (eds.), *Mammalian Transient Receptor Potential (TRP) Cation Channels*, Handbook of Experimental Pharmacology 223, DOI 10.1007/978-3-319-05161-1_2,

subfamilies could be in some way mechanosensitive, and each of the activation mechanisms described above is used by a TRP channel. TRP channels may be involved in mechanosensitive processes ranging from flow and pressure sensing in the vasculature and other organs to mechanosensation in sensory neurones and sensory organs. There is also evidence for a role of mechano- or osmosensitive TRP isoforms in osmosensing and the regulation of cell volume. Often, a number of different TRP isoforms have been implicated in a single type of mechanosensitive response. In many cases, the involvement of the isoforms needs to be confirmed, and their exact role in the signalling process determined.

Keywords
Mechanosensitive • Cation channel • Volume regulation • TRPC • TRPV • TRPM • TRPA • TRPP

1 Mechanosensitive Channels

Most cell types are able to detect and respond to mechanical stimuli. These responses can be used by cells to modulate their activity in response to local changes in pressure, to stretch or to changes in cell volume. The change in cellular activity may then be used to initiate a local tissue response, or, in the case of sensory cells, can be used to transduce a mechanical stimulus into a signal that can be transmitted over longer distances, and elicit complex responses of the whole organism.

Although the responses to mechanical stimuli can differ greatly in their complexity, similar mechanisms are often involved in sensing the mechanical stimulus at the cellular level. Ion channels are capable of rapidly transducing mechanical stimuli into an ion flux resulting in an electrical signal and a change in intracellular ion concentration.

Channels that respond to mechanical stimuli have a number of different mechanisms of activation (Chalfie 2009; Gillespie and Walker 2001; Hamill 2006; Kung 2005; Kung et al. 2010; Nilius and Honoré 2012; Sharif-Naeini et al. 2008; Sukharev and Sachs 2012). One of the simplest, giving rise to a truly mechanosensitive channel, is when the channel itself responds to changes in forces in the lipid bilayer. In this case, the channel should be activated within milliseconds of applying the mechanical stimulus. Other mechanosensitive channels present in the lipid bilayer are conferred mechanosensitive by their interaction with other proteins, like channel-attached cytoskeletal proteins, the extracellular matrix or other structural proteins. A further group of channels are a distal component of a signalling cascade triggered by mechanical stimuli and are themselves not directly involved in detection of the mechanical stimulus. Here, because of the number of steps involved, a delay in the activation of the channel is likely.

2 Cell Volume Regulation

Changes in extracellular osmolarity lead to movement of water and to changes in cell volume. Mammalian cells generally respond to cell swelling in hypotonic solutions with a regulatory volume decrease (RVD) and to shrinkage in hypertonic solutions with a regulatory volume increase (RVI) (for a review, see Hoffmann et al. 2009). Mechanosensitive channels are sometimes involved in volume regulation, because changes in volume affect membrane tension. Because of the simplicity of applying solutions of different osmolarities to cells, compared to the application of direct mechanical stimuli, osmotic stimuli have often been used to study mechanosensitive channels. Mechanosensitive channels are, however, only one means of detecting changes in cell volume. Other systems for detecting volume changes use ion channels as part of the signalling cascade involved in volume regulation. During RVD or RVI, cells respond to alterations in volume by changing the concentration of intracellular electrolytes, often ions such as K^+ and Cl^-, or organic osmolytes, and, thus, modify the intracellular osmotic pressure. The fluxes of K^+ and Cl^- are regulated by the opening and closing of ion channels. Ca^{2+} entry though mechanosensitive channels or channels activated by a mechanosensitive signalling cascade is often an important trigger or component of processes involved in cell volume regulation. The change in intracellular Ca^{2+} concentration results in the activation of Ca^{2+}-activated channels or modification of Ca^{2+}-sensitive metabolic pathways.

3 TRP Channels as Mechanosensitive Channels and Their Role in Volume Regulation

The idea that mammalian TRP channels could be involved in mechano- or osmosensing came from studies on invertebrate isoforms. In *Drosophila* and *Caenorhabditis elegans*, it was found that mutation of genes for TRP isoforms leads to deficits in the responses to mechanical or osmotic stimulation, or in responsiveness to auditory stimuli (Barr and Sternberg 1999; Colbert et al. 1997; Kim et al. 2003; Tracey et al. 2003; Walker et al. 2000). Therefore, it was logical to investigate whether mammalian isoforms, and particularly those isoforms closely related to the invertebrate TRPs, are involved in similar processes in mammals.

In this chapter, I will review the evidence for a role of mammalian TRP channels in mechanosensation and cell volume regulation. A number of previous reviews have also addressed the role of TRP channel families in mechanosensitivity and osmosensitivity (Liedtke 2007; Nilius and Honoré 2012; Pedersen and Nilius 2007; Sharif-Naeini et al. 2008; Yin and Kuebler 2010). For details of the properties of the TRP channels discussed, readers are referred to the specific chapters in this handbook.

3.1 TRPC Subfamily

TRPC channels have mainly been characterised to be receptor-operated cation channels activated by intracellular messengers or, perhaps, by depletion of intracellular Ca^{2+} stores. Some TRPC channel isoforms have also been suggested to form mechanosensitive channels.

3.1.1 TRPC1

3.1.1.1 TRPC1 and Mechanosensitivity

TRPC1 was identified immunologically as the cation channel forming the mechanosensitive Ca^{2+}-permeable cation channel (MscCa) or stretch-activated channel (SAC) that is activated with a latency of a few milliseconds by stretch in patches from *Xenopus* oocytes (Maroto et al. 2005). Overexpression of TRPC1 caused a tenfold increase in the current in patches, and TRPC1 knockdown reduced the endogenous current. In the same study, expression of TRPC1 increased mechanosensitive currents in Chinese hamster ovary (CHO) cells. In contrast, a later study showed no significant increase in the mechanosensitive current in CHO and COS cells following overexpression of TRPC1 (Gottlieb et al. 2008).

Although TRPC1 is expressed in most tissues, deletion of the gene in $TRPC1^{-/-}$ mice had no obvious effect on the phenotype of the mice (Dietrich et al. 2007). Pressure-induced vasoconstriction in cerebral vessels of these animals was normal, and currents in response to cell swelling and cell inflation were similar in vascular smooth muscle cells from $TRPC1^{-/-}$ and $TRPC1^{+/+}$ mice (Dietrich et al. 2007). Recently, evidence has been presented that TRPC1 forms mechanosensitive channels in growth cones of *Xenopus* spinal neurones (Kerstein et al. 2013).

Transcripts for TRPC1 have been found in dorsal root ganglion neurones (DRGs Elg et al. 2007; Kress et al. 2008). A subpopulation of DRGs responded to hypotonic solutions with an increase in intracellular Ca^{2+} (Staaf et al. 2009). The proportion of responsive neurones was more than halved after downregulation of TRPC1 with a short hairpin RNA (shRNA Staaf et al. 2009). In a skin-nerve preparation from $TRPC1^{-/-}$ mice, the number of action potentials fired in response to mechanical stimulation was reduced in slowly adapting Aβ fibres (largely from Merkel cells) and in rapidly adapting Aδ down-hair fibres compared to control mice. No differences were found in the responses of nociceptive afferents. Consistent with these findings, behavioural responses to light stroking or light punctate stimuli were reduced in the $TRPC1^{-/-}$ mice, but responses to intense (painful) mechanical stimulation were unaffected (Garrison et al. 2012).

3.1.1.2 TRPC1 and Volume Regulation

TRPC1 may also influence the regulation of cell volume in response to hypotonic stimuli, but results from different cell types appear contradictory. In liver cells, knockdown of TRPC1 results in a larger increase in cell volume, but a more rapid RVD than in control cells (Chen and Barritt 2003). An involvement of TRPC1 in Ca^{2+} entry leading to RVD in a kidney cell line has also been suggested, because knockdown of TRPC1 in this case decreased the rate of the RVD (Madsen et al. 2012).

3.1.2 TRPC3 and TRPC6

TRPC3 and TRPC6 are both group 3 TRPCs and can form heteromultimers. Spassova et al. (2006) showed that TRPC6 expressed in human embryonic kidney (HEK293) or CHO cells can be activated by cell swelling and that a channel with a similar conductance to TRPC6 opens in response to application of pressures >84 mmHg to inside-out membrane patches. Channel activation by swelling or pressure application was phospholipase C independent and, thus, did not involve the signalling pathway that regulates channel activity after G protein-coupled receptor stimulation. Channel activation by stretch, receptor stimulation or the diacylglycerol (DAG) analogue 1-oleoyl-2-acetyl-*sn*-glycerol (OAG) was inhibited by the tarantula peptide GsMTx-4, an inhibitor of a number of stretch-activated channels that acts on the lipid bilayer. In contrast, other studies could not demonstrate direct mechanosensitivity of TRPC6 in patches (Gottlieb et al. 2008; Inoue et al. 2009; Mederos y Schnitzler et al. 2008), although the pressures applied were ≤ 80 mmHg, and below the average threshold reported by Spassova et al. (2006). In the study by Mederos y Schnitzler et al. (2008), it was shown that not the channel but G_q protein-coupled receptors in the membrane were mechanosensitive leading to phospholipase C (PLC)-dependent activation of TRPC6 and also the isoforms TRPC3 and TRPC7. Another group showed that TRPC6 was not activated by mechanical stimuli but that receptor-mediated activation was enhanced by various forms of mechanical stimuli (osmotic swelling, shear stress, a bulging agent and pipette pressure, Inoue et al. 2009). The enhancement mechanism involved the release of arachidonic acid from membrane phospholipids by phospholipase A_2 (PLA_2) and its subsequent hydroxylation to 20-hydroxyeicosatetraenoic acid (20-HETE).

Expression of TRPC6 did not confer sensitivity to mechanical stimulation with a glass pipette on cell lines (Quick et al. 2012). However, in the same study, expression of TRPC3 or, even more, coexpression of TRPC3 and TRPC6 resulted in currents in response to mechanical displacement in a cell line derived from sensory neurones, but not in HEK293 or CHO cells. This suggests that interaction partners required for this form of mechanosensitivity are not expressed in the latter cell lines.

3.1.2.1 TRPC6 and Mechanosensitivity in the Cardiovascular System

In intact cerebral arteries, small interfering RNA (siRNA)-mediated knockdown of TRPC6 led to a decrease in the vasoconstriction and smooth muscle cell depolarisation that occurs on increasing the intra-arterial pressure (Welsh et al. 2002). Currents activated by hypotonic solutions in isolated smooth muscle cells from siRNA-treated arteries were also reduced. Another study also indicated that TRPC6 is involved in the myogenic contraction, but deletion of the channel increased myogenic tone because of a compensatory upregulation of TRPC3 (Dietrich et al. 2005). In line with their reports of more indirect activation or sensitisation of TRPC6 by mechanical stimuli via mechanosensitive G_q-coupled receptors (Mederos y Schnitzler et al. 2008) or PLA_2-dependent production of 20-HETE (Inoue et al. 2009), respectively (Sect. 3.1), these groups also showed that similar mechanisms were involved in the regulation of myogenic tone in cerebral and mesenteric arteries (Inoue et al. 2009; Mederos y Schnitzler et al. 2008). However, contradictory to the findings of Inoue et al. (Inoue et al. 2009), another group found no effect of G_q-coupled receptor activation on a hypotonicity-activated TRPC-like current in myocytes from cerebral arteries (Anfinogenova et al. 2011).

In ventricular myocytes, mechanical stimulation of the membrane with a glass stylus activated nonselective cation channels (Dyachenko et al. 2009). These were inhibited by polyclonal antibodies against TRPC6 and by the toxin GsMTx-4. The stretch-sensitive cation channels and stretch-sensitive K^+ channels were found to be located in the membrane of the transverse tubuli, and it was proposed that TRPC6 may contribute to extrasystoles resulting from myocardial stretch.

3.1.2.2 TRPC6: Activation by Stretch in Podocytes

In the glomerulus of the kidney, podocytes may regulate filtration, and they respond to distension of glomerular capillaries with a Ca^{2+}-dependent contraction. Although other TRPCs are expressed in these cells, TRPC6 has gained particular interest because several mutations that result in a gain of function have been associated with autosomal dominant nephrotic syndrome (for review see e.g. Dryer and Reiser 2010). In a recent study, it has been shown that cells of podocyte cell line display a cation current in response to osmotic swelling, membrane indentation or extracellular fluid flow with some properties characteristic of TRPC6 (Anderson et al. 2013). Furthermore, the current activation was prevented by a TRPC6 siRNA. Responses to stretch were enhanced by disruption of the cytoskeleton and inhibited by GsMTx-4. The sensitivity of cells to stretch was reduced by the expression of podocin, a protein that interacts with TRPC6 (Anderson et al. 2013).

3.1.3 TRPC Subunit Combinations in Mechanosensation and Hearing

TRPC1, TRPC3 and TRPC6 are expressed in primary sensory neurones (Elg et al. 2007; Kress et al. 2008). Intrathecal injection of antisense oligonucleotides to TRPC1 or TRPC6 was shown not to influence baseline mechanical nociception, but did reduce mechanical hypersensitivity induced by inflammatory mediators (Alessandri-Haber et al. 2009). More recently, it was shown that mechanical nociception was unaffected in *TRPC3*$^{-/-}$, *TRPC6*$^{-/-}$ and double knockout mice (Quick et al. 2012). However, deletion of the genes, that are coexpressed in single sensory neurones, did influence non-nociceptive mechanosensation (Quick et al. 2012). Whilst the proportions of mechanosensitive and insensitive small-diameter sensory neurones were similar to wild-type mice, there was a shift in the adaptation properties of mechanosensitive currents from rapidly adapting to intermediately adapting in *TRPC3*$^{-/-}$ neurones. In double knockout mice, there was a loss of rapidly adapting currents and a reduction in the percentage of mechanoresponsive neurones. Mice with deletions of one gene did not display deficits in mechanosensation, but those with double deletions did, suggesting that one of the isoforms may compensate for the other. The double deletion, but not single deletions, also had deficits in hearing and vestibular function (Quick et al. 2012). TRPC3 and TRPC6 are coexpressed in cochlear outer hair cells. In organotypic cultures of the cochlea from double mutants, mechanoelectrical transduction currents of hair cells from the basal, but not the apical coil, were reduced by 75–80 % (Quick et al. 2012). For *TRPC3*$^{-/-}$ mice, another study has shown that they display disrupted Ca^{2+} homeostasis leading, in contrast to the double knockout described above, to hyperacusis (Wong et al. 2013).

3.1.4 TRPC5

HEK293 cells expressing TRPC5 respond to hypotonic solutions with a current that has typical characteristics of TRPC5 (Gomis et al. 2008). The response was blocked by GsMTx-4. In contrast to osmotically induced activation of TRPC6 (Mederos y Schnitzler et al. 2008), the activation of TRPC5 was not dependent on PLC, but required the presence of phosphatidylinositol-4,5-bisphosphate (PIP_2) in the cells (Gomis et al. 2008). TRPC5 was also activated by positive pressure applied to the patch pipette in whole-cell recordings (Gomis et al. 2008).

3.2 TRPV Subfamily

3.2.1 TRPV1

TRPV1 is a heat-, ligand- (e.g. capsaicin) and pH-sensitive, voltage-dependent cation channel involved in thermal nociception. A splice variant of TRPV1 has been implicated in osmosensing in neurones (Sect. 3.2.1.1), but there are few reports on the osmosensitivity of TRPV1 after heterologous expression. No Ca^{2+} increase was observed in cells expressing full-length TRPV1 in response to hypotonic solutions (Liedtke et al. 2000; Strotmann et al. 2000), and hypertonic solutions had no effect on currents at room temperature (Ahern et al. 2005). In another study, there was hardly any Ca^{2+} response to changes in osmolarity of the extracellular solution at room temperature. However, around body temperature, responses were observed (Nishihara et al. 2011). Cells expressing the channel responded to increases in osmolarity with increased Ca^{2+} influx, whereas hypotonic solutions caused small decreases in influx (Nishihara et al. 2011). Thus, in contrast to most of the other TRP channels described in this chapter, TRPV1 may be activated by shrinkage or a decrease in membrane stretch. Although, as indicated, this finding is based on one report and needs to be confirmed.

3.2.1.1 TRPV1 and Central Osmosensing

Central osmosensitive neurones are involved in the regulation of extracellular fluid osmolarity by a number of mechanisms including water and salt intake and excretion (Bourque 2008). The primary osmoreceptors are probably located in organum vasculosum laminae terminalis (OVLT), one of the circumventricular organs, but other neurones in the subfornical organ and supraoptic nucleus (SON) are osmosensitive. Most of these osmosensitive neurones respond to increases in extracellular osmolarity with a depolarisation and an increase in firing and to hypotonic solutions with a decrease in activity. Osmosensitive neurones of the OVLT and the SON express mechanosensitive cation channels that open in response to hypertonic stimulation and close under hypotonic conditions. The channels are activated by cell shrinkage. TRPV1 has been shown to be involved in the response of both sets of neurones (Ciura and Bourque 2006; Sharif Naeini et al. 2006). SON neurones express an N-terminally truncated TRPV1 splice variant that is capsaicin insensitive (Sharif Naeini et al. 2006). Neurones from *TRPV1*$^{-/-}$ animals failed to respond to hypertonic solutions with a cation current, depolarisation and an increase in firing (Ciura and Bourque 2006; Sharif Naeini et al. 2006). Responses to hypertonic stimuli were counteracted by cell inflation and similar responses evoked by cell deflation, indicating that the channels are regulated by a mechanical process (Zhang et al. 2007). The same study provided evidence that the responses to osmotic stimuli are dependent on the actin cytoskeleton. Despite the evidence for a role of TRPV1 in mechanosensing in osmosensitive neurones, the role of the channel in vivo remains unclear since *TRPV1*$^{-/-}$ mice showed only moderate differences (Ciura and Bourque 2006) or no difference (Taylor et al. 2008) compared to wild-type mice in the amount of water they drank in response to NaCl injection.

3.2.2 TRPV2

CHO cells, transiently transfected with TRPV2, have been shown to respond to hypotonic solutions with an increase in current. Stretch-activated nonselective cation channel activity in patches from these cells was also increased (Muraki et al. 2003). Furthermore, the Ca^{2+} response to stretch of CHO or HEK293 cells growing on an elastic substrate was increased after expression of TRPV2 (Iwata et al. 2003; Muraki et al. 2003; Shibasaki et al. 2010).

3.2.2.1 TRPV2 in Striated and Smooth Muscle

TRPV2 is expressed in myocytes from skeletal and cardiac muscle, and expression is increased in dystrophic muscle (Iwata et al. 2003). TRPV2 has been implicated in Ca^{2+}-induced damage in dystrophic muscle because its expression in CHO cells leads to an increase in stretch-activated Ca^{2+} influx (Iwata et al. 2003). In aortic myocytes, TRPV2 expression was detected, and cell swelling and Ca^{2+} influx in response to hypotonic solutions was reduced by an siRNA against TRPV2 (Muraki et al. 2003).

3.2.2.2 TRPV2 in Other Tissues

Expression of TRPV2 was found in developing motor neurones, DRG neurones and in the dorsal horn of the spinal cord (Shibasaki et al. 2010). In motor neurones and DRGs, protein expression was detected in the soma, axon shafts and growth cones. Stretch-induced Ca^{2+} influx in embryonic DRGs was reduced by a dominant-negative TRPV2. Stretch-induced axon outgrowth in vitro, and axon outgrowth in vivo, was increased in neurones transfected with TRPV2 and decreased in neurones transfected with the dominant-negative construct or a TRPV2 short hairpin RNA (Shibasaki et al. 2010).

Rapid strain applied to the membrane of alveolar type II cells with a spatula mounted on a piezo-driven manipulator induced extremely fast Ca^{2+} entry (within 30 ms) which stopped immediately upon removal of the stimulus, suggesting that the channel involved is directly mechanosensitive (Fois et al. 2012). The strongest mRNA expression for TRPV and TRPC channels in these cells was for TRPV2, and silencing of this subunit with siRNA reduced the Ca^{2+} response to strain. Evidence was also obtained for a role of focal adhesions and the cytoskeleton in the response to strain (Fois et al. 2012).

3.2.3 TRPV4

The mammalian TRPV channel most often connected with mechanosensitivity or responses to changes in osmolarity is TRPV4. A number of groups including our own showed that TRPV4 expressed in HEK293 or CHO-K1 cells was activated by the application of hypotonic solutions (Liedtke et al. 2000; Strotmann et al. 2000; Wissenbach et al. 2000). However, in preliminary experiments in HEK293 cells, we could not detect a change in activity in response to membrane stretch (Strotmann et al. 2000). Vriens et al. (2004) later showed that TRPV4 is at the end of a cascade activated by cell swelling that involves PLA_2-dependent generation of arachidonic acid and its cytochrome P450 epoxygenase-dependent metabolism to epoxyeicosatrienoic acid (EET). The ability of TRPV4 to respond to osmotically induced cell swelling is regulated by a protein–protein interaction. Sensitivity to swelling is reduced when the channel interacts with the protein PACSIN 3 (D'Hoedt et al. 2008), which prevents the N-terminal tail of the channel from interacting with PIP_2 (Garcia-Elias et al. 2013).

3.2.3.1 TRPV4 in the Kidney

Initial studies on TRPV4 showed that it is strongly expressed in the kidney (Liedtke et al. 2000; Strotmann et al. 2000; Wissenbach et al. 2000), an organ in which cells can be exposed to strong differences in osmolarity of the extracellular fluid. An older study indicated that TRPV4 expression may be localised to water-impermeable nephron segments and that expression was mainly basolateral, suggesting a possible role in sensing interstitial osmolarity (Tian et al. 2004). More recent studies, however, have shown that TRPV4 is expressed near the apical membrane of cells of the aquaporin 2-positive medullary and cortical collecting duct system (Berrout et al. 2012a). Cells in this region respond to hypotonic solutions with increases in Ca^{2+}, and endogenous, Ca^{2+}-permeable TRPV4 was activated by hypotonic solutions in a cortical collecting duct cell line (reviewed see Pochynyuk et al. 2013; Wu et al. 2007). In response to hypotonic stimuli, cells in this region of the nephron release ATP which then increases intracellular Ca^{2+} via metabotropic purinergic receptor (P2Y) signalling. TRPV4 was shown to be essential for the plateau phase of the Ca^{2+} signal (Mamenko et al. 2011). By inhibiting vasopressin-dependent water transport, purinergic signalling is part of a local control mechanism for water transport. The exact contribution of TRPV4 to this mechanism and the link between the P2Y receptor and TRPV4 activation is still unclear. TRPV4 also responds to the flow of tubular fluid. It was shown that flow-dependent increases in Na^+ absorption and K^+ secretion in cortical collecting ducts was dependent on the expression of TRPV4 (Taniguchi et al. 2007). TRPV4 is also necessary for high flow-induced increases in Ca^{2+} in the collecting duct system (Berrout et al. 2012a).

Because they are exposed to large osmotic gradients, kidney cells need to regulate their volume. TRPV4 is involved in cell volume regulation (the RVD) in a cortical collecting duct cell line (Galizia et al. 2012). This response required aquaporin 2 and was blocked by ruthenium red, which unspecifically blocks TRPVs, and stimulated by a 4α-phorbol ester activator of TRPV4. The RVD response involved the cytoskeleton and trafficking of TRPV4 to the cell membrane (Galizia et al. 2012).

3.2.3.2 TRPV4 in the Cardiovascular System

TRPV4 is also involved in flow-dependent signalling in endothelial cells. Shear stress on the endothelium stimulates the release of vasodilatory factors (nitric oxide, prostacyclin, H_2O_2) and the generation of endothelium-derived hyperpolarizing factor (EDHF) signalling from endothelial cells. This leads to smooth muscle relaxation. TRPV4 is expressed in large arteries and arterioles, and there is good evidence that it is involved in the response to shear stress. Shear-stress-induced relaxation is strongly reduced by inhibitors of TRPV4, by TRPV4 siRNAs, or by genetic deletion of the channel (Hartmannsgruber et al. 2007; Köhler et al. 2006; Loot et al. 2008; Mendoza et al. 2010). Furthermore, application of activators of TRPV4 also results in vasodilation (Köhler et al. 2006). Ca^{2+} entry through TRPV4 is thought to couple channel activation to the production of vasodilatory mediators like NO or prostaglandins (Köhler et al. 2006) or to membrane hyperpolarisation via activation of Ca^{2+}-activated K^+ channels (Köhler and Hoyer 2007; Sonkusare et al. 2012). The hyperpolarisation may be transmitted from the endothelium to smooth muscle cells by electrical coupling and may, indeed, be EDHF. Activation of TRPV4 by shear stress seems to be indirect and to involve PLA_2-dependent release of arachidonic acid and its subsequent metabolism to EETs (Hartmannsgruber et al. 2007; Köhler et al. 2006; Loot et al. 2008). A recent study has shown that TRPV4 located in microdomains of endothelial cells together with Ca^{2+}-activated K^+ channels at the myoendothelial junction can also be activated by low intravascular pressure (Bagher et al. 2012). In the lung, Ca^{2+} entry through endothelial TRPV4 has been implicated to play a role in high vascular pressure-induced lung injury, a process involving the production of EETs (Jian et al. 2008).

In addition to its presence in endothelial cells, TRPV4 is also expressed in vascular smooth muscle cells of some arteries. There it could mediate the response to EETs, which have also been proposed to be EDHF (Earley et al. 2005). Ca^{2+} entry through TRPV4 located close to ryanodine receptors induces sparks of Ca^{2+} release from the sarcoplasmic reticulum. This leads to the activation of Ca^{2+}-activated K^+ channels, hyperpolarisation of the membrane and smooth muscle relaxation. In addition to this inhibitory role on smooth muscle cells, TRPV4 has also been reported to increase smooth muscle tone. TRPV4 is upregulated in chronic hypoxic rats leading to an increase in myogenic tone and pulmonary hypertension (Yang et al. 2012).

3.2.3.3 TRPV4 and Mechanosensation in Sensory Neurones

In primary sensory neurones, TRPV4 was shown to be involved in the activation of nociceptors by hypotonic solutions, an effect enhanced by inflammation (Alessandri-Haber et al. 2003). Later, some of the first reports on *TRPV4*$^{-/-}$ mice described an increase in threshold for the responses to strong noxious mechanical stimulation in mice lacking the channel (Liedtke and Friedman 2003; Suzuki et al. 2003). Other studies, however, suggest that TRPV4 does not normally respond to painful stimuli, but plays a role in mechanical hyperalgesia induced by inflammatory mediators (Alessandri-Haber et al. 2006; Grant et al. 2007) or neuropathy (Alessandri-Haber et al. 2004). The channel has also been shown to be involved in tonicity-induced neurogenic inflammation (Vergnolle et al. 2010).

3.2.3.4 TRPV4: Involvement in Mechano- or Osmosensitivity in Other Tissues

TRPV4 expression was detected in airway smooth muscle cells, and the channel may be involved in the contraction caused by exposure to hypotonic solutions (Jia et al. 2004). TRPV4 was also reported to be involved in the response of the airway epithelium to shear stress caused by airflow (Sidhaye et al. 2008). Shear stress increased intracellular Ca^{2+} in epithelial cells and decreased paracellular epithelial permeability. In the signalling cascade, TRPV4 activation preceded the opening of voltage-gated Ca^{2+} channels, Ca^{2+} influx and a decrease in the expression of aquaporin 5 (Sidhaye et al. 2008).

In the bladder, TRPV4 is present in urothelial cells and plays an important role in the Ca^{2+}-induced release of ATP evoked by hypotonic solutions, stretch or increases in intravesicular pressure (Birder et al. 2007; Gevaert et al. 2007; Mochizuki et al. 2009). The activation of TRPV4 leads to the stimulation of mechanosensitive, capsaicin-insensitive bladder C fibre afferents that facilitate the micturition reflex (Aizawa et al. 2012). Indeed, *TRPV4*$^{-/-}$ mice show deficits in bladder voiding, confirming a possible role in sensing or transducing intravesicular pressure or volume (Gevaert et al. 2007).

TRPV4 is expressed in cells of the cochlea including outer and inner hair cells. Outer hair cells of *TRPV4*$^{+/+}$ mice respond to hypotonic solutions or TRPV4 agonists with an increase in Ca^{2+}, whereas cells from *TRPV4*$^{-/-}$ mice do not, indicative of an osmo- or mechanosensory role of the channel in these cells (Shen et al. 2006). In one study, TRPV4$^{-/-}$ mice were shown to respond like wild-type mice to acoustic stimulation (Liedtke and Friedman 2003), whereas in another deletion of TRPV4 was reported to cause a delayed loss of hearing (Tabuchi et al. 2005). These results, together with the fact that the channel properties do not fit with those of the hair cell mechanotransduction channel, suggest that TRPV4 is not the mechanotransduction channel but may be involved in the production of endolymph (for review see Christensen and Corey 2007).

3.2.3.5 TRPV4 and Systemic Osmosensing

Because TRPV4 is activated in cells exposed to hypertonic stimuli, it could potentially be involved not only in cellular but also in systemic osmosensing. One study showed that *TRPV4*$^{-/-}$ mice did not counterregulate systemic hypertonicity as well as *TRPV4*$^{+/+}$ mice. They drank less and released less antidiuretic hormone, probably as a result of a defect in osmotic sensing central nervous system (Liedtke and Friedman 2003). However, another showed increased ADH release in response to water deprivation (Mizuno et al. 2003). As discussed above (Sect. 3.2.1.1), neurons in the OVLT are the primary osmosensing neurons in mammals, but the cation channel in these neurones, unlike TRPV4, is activated in response to hyperosmotic stimuli. Furthermore, the response of OVLT neurones is affected by deletion of TRPV1, but not TRPV4 (Ciura et al. 2011).

3.2.3.6 TRPV4 and the RVD

In addition to a role in volume regulation in kidney cells (Sect. 3.2.3.1, Galizia et al. 2012), Ca^{2+} entry through TRPV4 has been reported to be involved in the RVD in a number of cell types. CHO cells do not express endogenous TRPV4 and normally do not display an RVD, but show an RVD when transfected with TRPV4 (Becker et al. 2005). Other cell types for which an involvement of TRPV4 in RVD has been described include airway epithelial cells (Arniges et al. 2004), keratinocytes (Becker et al. 2005), salivary gland cells (Liu et al. 2006), corneal epithelial cells (Pan et al. 2008), chondrocytes (Phan et al. 2009) and astrocytes (Benfenati et al. 2011).

Like in kidney cells (Sect. 3.2.3.1, Galizia et al. 2012), the responses in salivary gland cells (Liu et al. 2006) and astrocytes (Benfenati et al. 2011) required the concerted activity of TRPV4 and aquaporin isoforms. Furthermore, in a number of cell types, an intact actin cytoskeleton was shown to be required for the TRPV4-dependent RVD (Becker et al. 2009; Liu et al. 2006), and one study suggested that primary cilia and possibly TRPV4 therein are involved in the Ca^{2+} response to hypo-osmotic solutions (Phan et al. 2009).

3.3 TRPM Subfamily

Compared to the subfamilies of TRP channels described above, there are only few reports of sensitivity of TRPM channels to mechanical stimuli or changes in cell volume.

3.3.1 TRPM3

One splice variant of TRPM3 has been shown to be spontaneously active after heterologous expression, and the activity of this variant is reduced by hypertonic and increased by hypotonic solutions (Grimm et al. 2003). Nothing is known of the physiological importance of this responsiveness, although the channel is expressed in the kidney of some species (Grimm et al. 2003).

3.3.2 TRPM4 in Vascular Smooth Muscle

The Ca^{2+}-activated, Ca^{2+}-impermeable cation channel TRPM4 has been shown to be expressed in smooth muscle cells of cerebral arteries (Earley et al. 2004). Antisense oligonucleotides against TRPM4 decreased the depolarisation and the myogenic constriction in response to increased intra-arterial pressure (Earley et al. 2004).

3.3.3 TRPM7

A channel activated by membrane stretch (half maximal activation at −3 cm H_2O) in inside-out patches from HeLa epithelial cells was inhibited by external Mg^{2+} and channel activity abolished by knockdown of TRPM7 with siRNA (Numata et al. 2007b). In the same study, currents activated by cell swelling and the RVD in HeLa cells were strongly reduced by a TRPM7 siRNA. In a similar study by the same authors on TRPM7 expressed in HEK cells, increases in current were observed with shear stress or on cell swelling (Numata et al. 2007a). In excised patches, a channel was activated by stretch, with similar properties and a similar pressure sensitivity to that seen in HeLa cells. In contrast to these data, an endogenous TRPM7-like channel in microglia was insensitive to cell swelling (Jiang et al. 2003). Another study in HEK293 cells found that TRPM7 activity and endogenous TRPM7-like currents are sensitive to the extracellular osmolarity (Bessac and Fleig 2007). Hyperosmotic solutions decreased currents and hypo-osmotic solutions increased currents. Currents were insensitive to membrane stretch and there was no effect of TRPM7 expression on the RVD. The osmosensitivity was shown to arise in part from swelling- or shrinkage-induced changes in the intracellular Mg^{2+} concentration to which the channel is sensitive.

3.4 TRPA Subfamily

3.4.1 TRPA1

Heterologously expressed TRPA1 and TRPA1 in sensory neurons are activated by the application of hypertonic solutions, but not by hypotonic solutions (Zhang et al. 2008). The mechanism of activation by hypertonic solutions/cell shrinkage has not been reported.

TRPA isoforms are involved in mechanosensation in invertebrates. Mammalian TRPA1 is expressed in auditory hair cells and is located to the stereocilia. Knock-down experiments indicated that TRPA1 was a candidate for the auditory transduction channel (Corey et al. 2004). *TRPA1*$^{-/-}$ mice, however, did not have hearing defects (Bautista et al. 2006; Kwan et al. 2006), so that the channel's physiological role in hair cells remains unclear. One of these studies (Kwan et al. 2006), but not the other, found a reduced sensitivity to noxious mechanical stimuli in *TRPA1*$^{-/-}$ mice. In a more detailed investigation in a skin-nerve preparation, the same group confirmed deficits in mechanotransduction in *TRPA1*$^{-/-}$ mice (Kwan et al. 2009). These mice showed weaker C fibre responses to noxious mechanical stimulation over a wide range of stimulus intensities and reduced Aδ fibre responses to intense stimulation. In addition, adaptation of non-noxious mechanoresponses was modified. Slowly adapting mechanosensitive currents activated by indentation of the membrane were completely lost in isolectin B4-negative small-diameter dorsal root ganglion neurons (that give rise to small-diameter nociceptive C fibres) from *TRPA1*$^{-/-}$ mice (Vilceanu and Stucky 2010). TRPA1 is also expressed in mechanosensitive visceral afferents, and responses to noxious colonic distension were reduced in *TRPA1*$^{-/-}$ mice (Brierley et al. 2009). This group found a reduction in amplitude of intermediately adapting currents in response to neurite stimulation in small-diameter DRG neurones, but not in larger-diameter neurones (Brierley et al. 2011).

3.5 TRPP Subfamily

Polycystin 2 (referred to here as TRPP2, but also called polycystic kidney disease 2 (PKD2), PC2 and TRPP1 by IUPHAR) and the two polycystin 2-like proteins TRPP3 (polycystic kidney disease 2-like 1 (PKD2L1), IUPHAR: TRPP2) and TRPP5 (polycystic kidney disease 2-like 2 (PKD2L2), IUPHAR: TRPP3) belong to the TRP family, whereas polycystin 1 (PKD1, sometimes called TRPP1 in the nomenclature used here) and its relatives, because of their additional transmembrane domains and because they have little sequence similarity to TRP channels, do not. It is noteworthy that differences in nomenclature can cause considerable confusion when surveying the polycystin literature. Mutations in PKD1 and TRPP2 have been shown to cause polycystic kidney disease (ADPKD), and hence the name polycystin (Hofherr and Kottgen 2011). PKD1 and TRPP2 co-localise to nonmotile primary cilia of epithelial and endothelial cells where there is evidence that they confer responses to mechanical stimuli. In addition to forming a surface membrane channel, PKD1 and TRPP2 interact with and amplify Ca^{2+} release from inositol trisphosphate receptors in the endoplasmic reticulum (Koulen et al. 2002).

3.5.1 TRPP2

3.5.1.1 TRPP2 in the Kidney

Cultured kidney epithelial cells responded to shear stress with an increase in Ca^{2+} that was lost in cells from *PKD1*$^{-/-}$ mice or in cells treated with antibodies against PKD1 or the extracellular domain of TRPP2 (Nauli et al. 2003). The response was also lost in cells lacking fully developed primary cilia and depended on Ca^{2+} release through RyRs. TRPV4 has also been reported to interact with TRPP2 to form mechanosensitive channels in renal epithelial cells (Kottgen et al. 2008). After coexpression of TRPV4 and TRPP2 in *Xenopus* oocytes, swelling-activated currents were larger than when TRPV4 was expressed alone. TRPV4 knockdown abolished flow-activated currents in a cell line derived from the renal epithelium.

3.5.1.2 TRPP2 in the Vasculature

PKD1 and TRPP2 are also expressed in primary cilia of endothelial cells where they have been shown to be involved in the response to shear stress. Knockout of PKD1 or TRPP2 in mice resulted in a loss of the shear-stress response (AbouAlaiwi et al. 2009; Nauli et al. 2008). Similarly, endothelial cells from a ADPKD patient had no response to shear stress and lacked TRPP2 expression (AbouAlaiwi et al. 2009). In the endothelium, polycystin-dependent Ca^{2+} increases were linked to the production of the vasodilator NO (AbouAlaiwi et al. 2009; Nauli et al. 2008). The authors suggest that the loss of shear-stress responses may contribute to hypertension in ADAKD patients. If, as in kidney cells, TRPV4 in endothelial cells were to interact with polycystins, it may be possible to explain the results of TRPV4 (Sect. 3.2.3.2) or polycystin deletion on shear-stress-induced vasodilation.

In addition to an involvement in endothelial mechanosensation, TRPP2 and PKD1 influence a mechanosensitive channel in arterial smooth muscle cells. Overexpression of TRPP2 decreased stretch-activated channel activity in excised patches from COS cells (Sharif-Naeini et al. 2009), whereas expression of PKD1 alone had no effect. Coexpression of PKD1 with TRPP2 reversed SAC inhibition by TRPP2. Consistent with these effects, deletion of PKD1 in arterial myocytes reduced SAC activity and impaired the myogenic response, and depletion of TRPP2 in PKD1-deficient myocytes restored SAC activity and the myogenic response. The effect of TRPP2 on SAC activity was dependent on the actin cytoskeleton and the protein filamin A that interacts with TRPP2.

TRPP2 has also been shown to interact with TRPC1 (Tsiokas et al. 1999), and it has recently been suggested that these mediate Ca^{2+} influx in response to stretch-induced injury in blood–brain barrier endothelial cells (Berrout et al. 2012b).

3.5.1.3 TRPP2 and Embryonic Development

In addition to the mechanosensitivity in adult tissues described above, TRPP2 plays a role in embryonal development. Visceral organs are normally asymmetric with respect to their position and/or shape. *TRPP2*$^{-/-}$ mice display deficits in left–right (L–R) symmetry, e.g. in lobation of the lung and liver and in location of the heart stomach and spleen (Pennekamp et al. 2002). The breaking of L–R symmetry is mediated by unidirectional fluid flow in the ventral node of the embryo. TRPP2 expressed in the nonmotile cilia of perinodal crown cells is required for sensing of fluid flow and likely involves Ca^{2+} signalling (Yoshiba et al. 2012).

3.5.2 TRPP3 and PKD1/TRPP3 Heteromers

TRPP3 was first reported to be retained in the endoplasmic reticulum (ER) and trafficked to the cell surface when coexpressed with PKD1 or after removal of an ER retention signal (Murakami et al. 2005). In the same study, it was shown that HEK293 cells coexpressing PKD1 and TRPP3 or expressing TRPP3 mutants lacking the retention signal responded to hypo-osmotic stimulation with an increase in Ca^{2+}. More recently, it was shown that TRPP3 forms constitutively active channels after expression alone in HEK293 cells (Shimizu et al. 2009). Cell volume increases in response to hypotonic solutions led to an increase in channel activity, whereas cell shrinkage decreased channel activity (Shimizu et al. 2009). The changes in activity resulted from swelling-induced shifts in the potential dependence of the channel.

3.6 TRPML Subfamily

3.6.1 TRPML3

The mucolipidin TRP channel TRPML3 is expressed in various cells of the inner ear including outer and inner hair cells, where most expression is intracellular, but some is detected in stereocilia (Di Palma et al. 2002). Because of this, and because heterozygote mice carrying the varitint-waddler mutation in TRPML3 are deaf (Cable and Steel 1998; Di Palma et al. 2002), it was hypothesised that the channel may be involved in mechanosensation (Di Palma et al. 2002). The varitint-waddler mutation leads to a gain of function in TRPM3 and to degeneration of hair cells (for a review see e.g. Atiba-Davies and Noben-Trauth 2007), but the channel does not seem to form the mechanotransduction channel (van Aken et al. 2008), and hair cell-specific deletion of TRPML3 does not lead to hearing or balance impairments (Jörs et al. 2010).

Conclusions and Outlook

From the information presented in this chapter, it is clear that a number of TRP isoforms from different subfamilies have been proposed to be involved in responses to mechanical stimuli. However, for only few of these channels is the evidence supporting this role comprehensive and comprises evidence for mechanical activation in native cells and after heterologous expression, evidence at the biophysical (single channel conductance and channel properties) and molecular level (knockout, knockdown, expression) for channel involvement, knowledge of the activation mechanism (direct, indirect, pathway), and an indication of its possible physiological roles (response to knockout, knockdown). In my view, the channel that meets these criteria best amongst the TRP channels is TRPV4. For other channels (e.g. TRPP1), evidence is good at some levels (expression, physiological role), but lacking at others (activation mechanism, biophysical characterisation). For others, indeed for most, however, the evidence that the channel responds to mechanical stimuli is often based on a single report or work from one group and lacks independent confirmation, or there are conflicting reports on the same channel from different studies. In some cases, the evidence that the TRP isoform is involved in a response relies on the specificity of the technique used (e.g. the antibody or siRNA), and relatively indirect measurements of channel activity (e.g. changes in Ca^{2+} concentration or contraction), or behavioural assays. Since most TRP channels are Ca^{2+} permeable and often show some spontaneous activity, they can influence cellular Ca^{2+} homeostasis and the activity of other Ca^{2+}-sensitive channels.

Some TRP channels (there is at least some evidence for TRPC1, TRPC3/6, TRPV2, TRPM7, TRPA1, TRPP1) may respond rapidly and directly to changes in membrane tension as indicated by responses to pressure applied to membrane patches or to mechanical stimulation of cells using a probe and may thus fulfil a major criterium for a mechanosensitive channel. They may, however, particularly in native cells, influence other non-TRP cation channels present in the cells that respond to these stimuli. For some TRP channels (TRPC6, TRPV1, TRPV2, TRPV4, TRPP1), there are reports that the cytoskeleton is involved in the response to mechanical stimulation. Many of the TRP channels may not be mechanosensitive per se, but be activated by multistep mechanosensitive signalling pathways. A feature of many TRP channels is their complex regulation by the membrane potential and/or a variety of intracellular signals including G proteins, Ca^{2+}, PIP_2, DAG, EETs and phosphorylation. Thus, mechanosensitive pathways that influence these factors could influence TRP channel activity. An indirect activation pathway has clearly been shown for TRPV4 and may apply to a number of other TRP channels, particularly if there is a delay in activation. In some processes, such as the responses of vascular smooth muscle cells to pressure and mechanosensation in sensory neurones, there is evidence for the involvement of a number of TRP channel isoforms from different subfamilies. In these cases, it needs to be clarified how the channels interact in the response, which channel, if any, is the primary mechanosensor and which are activated subsequently by mechanosensitive signalling cascades. The evidence for an involvement of TRPs in volume regulation is again best for TRPV4, but TRPC1 may also play a role.

Future studies will hopefully pinpoint the role of individual TRPs in mechanosensing and volume regulation and determine their importance in comparison to other mechanosensitive non-TRP cation channels and mechanosensitive K^+ channels.

References

AbouAlaiwi WA, Takahashi M, Mell BR, Jones TJ, Ratnam S, Kolb RJ, Nauli SM (2009) Ciliary polycystin-2 is a mechanosensitive calcium channel involved in nitric oxide signaling cascades. Circ Res 104:860–869

Ahern GP, Brooks IM, Miyares RL, Wang XB (2005) Extracellular cations sensitize and gate capsaicin receptor TRPV1 modulating pain signaling. J Neurosci 25:5109–5116

Aizawa N, Wyndaele JJ, Homma Y, Igawa Y (2012) Effects of TRPV4 cation channel activation on the primary bladder afferent activities of the rat. Neurourol Urodyn 31:148–155

Alessandri-Haber N, Yeh JJ, Boyd AE, Parada CA, Chen X, Reichling DB, Levine JD (2003) Hypotonicity induces TRPV4-mediated nociception in rat. Neuron 39:497–511

Alessandri-Haber N, Dina OA, Yeh JJ, Parada CA, Reichling DB, Levine JD (2004) Transient receptor potential vanilloid 4 is essential in chemotherapy-induced neuropathic pain in the rat. J Neurosci 24:4444–4452

Alessandri-Haber N, Dina OA, Joseph EK, Reichling D, Levine JD (2006) A transient receptor potential vanilloid 4-dependent mechanism of hyperalgesia is engaged by concerted action of inflammatory mediators. J Neurosci 26:3864–3874

Alessandri-Haber N, Dina OA, Chen X, Levine JD (2009) TRPC1 and TRPC6 channels cooperate with TRPV4 to mediate mechanical hyperalgesia and nociceptor sensitization. J Neurosci 29:6217–6228

Anderson M, Kim EY, Hagmann H, Benzing T, Dryer SE (2013) Opposing effects of podocin on the gating of podocyte TRPC6 channels evoked by membrane stretch or diacylglycerol. Am J Physiol Cell Physiol 305(3):C276–C289

Anfinogenova Y, Brett SE, Walsh MP, Harraz OF, Welsh DG (2011) Do TRPC-like currents and G protein-coupled receptors interact to facilitate myogenic tone development? Am J Physiol Heart Circ Physiol 301:H1378–H1388

Arniges M, Vazquez E, Fernandez-Fernandez JM, Valverde MA (2004) Swelling-activated Ca^{2+} entry via TRPV4 channel is defective in cystic fibrosis airway epithelia. J Biol Chem 279:54062–54068

Atiba-Davies M, Noben-Trauth K (2007) TRPML3 and hearing loss in the varitint-waddler mouse. Biochim Biophys Acta 1772:1028–1031

Bagher P, Beleznai T, Kansui Y, Mitchell R, Garland CJ, Dora KA (2012) Low intravascular pressure activates endothelial cell TRPV4 channels, local Ca^{2+} events, and IK_{Ca} channels, reducing arteriolar tone. Proc Natl Acad Sci U S A 109:18174–18179

Barr MM, Sternberg PW (1999) A polycystic kidney-disease gene homologue required for male mating behaviour in C. elegans. Nature 401:386–389

Bautista DM, Jordt SE, Nikai T, Tsuruda PR, Read AJ, Poblete J, Yamoah EN, Basbaum AI, Julius D (2006) TRPA1 mediates the inflammatory actions of environmental irritants and proalgesic agents. Cell 124:1269–1282

Becker D, Blase C, Bereiter-Hahn J, Jendrach M (2005) TRPV4 exhibits a functional role in cell-volume regulation. J Cell Sci 118:2435–2440

Becker D, Bereiter-Hahn J, Jendrach M (2009) Functional interaction of the cation channel transient receptor potential vanilloid 4 (TRPV4) and actin in volume regulation. Eur J Cell Biol 88:141–152

Benfenati V, Caprini M, Dovizio M, Mylonakou MN, Ferroni S, Ottersen OP, Amiry-Moghaddam M (2011) An aquaporin-4/transient receptor potential vanilloid 4 (AQP4/TRPV4) complex is essential for cell-volume control in astrocytes. Proc Natl Acad Sci U S A 108:2563–2568

Berrout J, Jin M, Mamenko M, Zaika O, Pochynyuk O, O'Neil RG (2012a) Function of transient receptor potential cation channel subfamily V member 4 (TRPV4) as a mechanical transducer in flow-sensitive segments of renal collecting duct system. J Biol Chem 287:8782–8791

Berrout J, Jin M, O'Neil RG (2012b) Critical role of TRPP2 and TRPC1 channels in stretch-induced injury of blood-brain barrier endothelial cells. Brain Res 1436:1–12

Bessac BF, Fleig A (2007) TRPM7 channel is sensitive to osmotic gradients in human kidney cells. J Physiol 582:1073–1086

Birder L, Kullmann FA, Lee H, Barrick S, de Groat W, Kanai A, Caterina M (2007) Activation of urothelial transient receptor potential vanilloid 4 by 4alpha-phorbol 12,13-didecanoate contributes to altered bladder reflexes in the rat. J Pharmacol Exp Ther 323:227–235

Bourque CW (2008) Central mechanisms of osmosensation and systemic osmoregulation. Nat Rev Neurosci 9:519–531

Brierley SM, Hughes PA, Page AJ, Kwan KY, Martin CM, O'Donnell TA, Cooper NJ, Harrington AM, Adam B, Liebregts T, Holtmann G, Corey DP, Rychkov GY, Blackshaw LA (2009) The ion channel TRPA1 is required for normal mechanosensation and is modulated by algesic stimuli. Gastroenterology 137(2084–2095):e3

Brierley SM, Castro J, Harrington AM, Hughes PA, Page AJ, Rychkov GY, Blackshaw LA (2011) TRPA1 contributes to specific mechanically activated currents and sensory neuron mechanical hypersensitivity. J Physiol 589:3575–3593

Cable J, Steel KP (1998) Combined cochleo-saccular and neuroepithelial abnormalities in the Varitint-waddler-J (VaJ) mouse. Hear Res 123:125–136

Chalfie M (2009) Neurosensory mechanotransduction. Nat Rev Mol Cell Biol 10:44–52

Chen J, Barritt GJ (2003) Evidence that TRPC1 (transient receptor potential canonical 1) forms a Ca^{2+}-permeable channel linked to the regulation of cell volume in liver cells obtained using small interfering RNA targeted against TRPC1. Biochem J 373:327–336

Christensen AP, Corey DP (2007) TRP channels in mechanosensation: direct or indirect activation? Nat Rev Neurosci 8:510–521

Ciura S, Bourque CW (2006) Transient receptor potential vanilloid 1 is required for intrinsic osmoreception in organum vasculosum lamina terminalis neurons and for normal thirst responses to systemic hyperosmolality. J Neurosci 26:9069–9075

Ciura S, Liedtke W, Bourque CW (2011) Hypertonicity sensing in organum vasculosum lamina terminalis neurons: a mechanical process involving TRPV1 but not TRPV4. J Neurosci 31:14669–14676

Colbert HA, Smith TL, Bargmann CI (1997) Osm-9, a novel protein with structural similarity to channels, is required for olfaction, mechanosensation and olfactory adaptation in Caenorhabditis elegans. J Neurosci 17:8259–8269

Corey DP, Garcia-Anoveros J, Holt JR, Kwan KY, Lin SY, Vollrath MA, Amalfitano A, Cheung EL, Derfler BH, Duggan A, Geleoc GS, Gray PA, Hoffman MP, Rehm HL, Tamasauskas D, Zhang DS (2004) TRPA1 is a candidate for the mechanosensitive transduction channel of vertebrate hair cells. Nature 432:723–730

D'Hoedt D, Owsianik G, Prenen J, Cuajungco MP, Grimm C, Heller S, Voets T, Nilius B (2008) Stimulus-specific modulation of the cation channel TRPV4 by PACSIN 3. J Biol Chem 283:6272–6280

Di Palma F, Belyantseva IA, Kim HJ, Vogt TF, Kachar B, Noben-Trauth K (2002) Mutations in Mcoln3 associated with deafness and pigmentation defects in varitint-waddler (Va) mice. Proc Natl Acad Sci U S A 99:14994–14999

Dietrich A, Mederos YSM, Gollasch M, Gross V, Storch U, Dubrovska G, Obst M, Yildirim E, Salanova B, Kalwa H, Essin K, Pinkenburg O, Luft FC, Gudermann T, Birnbaumer L (2005) Increased vascular smooth muscle contractility in TRPC6$^{-/-}$ mice. Mol Cell Biol 25:6980–6989

Dietrich A, Kalwa H, Storch U, Mederos Y, Schnitzler M, Salanova B, Pinkenburg O, Dubrovska G, Essin K, Gollasch M, Birnbaumer L, Gudermann T (2007) Pressure-induced and store-operated cation influx in vascular smooth muscle cells is independent of TRPC1. Pflügers Archiv 455:465–477

Dryer SE, Reiser J (2010) TRPC6 channels and their binding partners in podocytes: role in glomerular filtration and pathophysiology. Am J Physiol Renal Physiol 299:F689–F701

Dyachenko V, Husse B, Rueckschloss U, Isenberg G (2009) Mechanical deformation of ventricular myocytes modulates both TRPC6 and Kir2.3 channels. Cell Calcium 45:38–54

Earley S, Waldron BJ, Brayden JE (2004) Critical role for transient receptor potential channel TRPM4 in myogenic constriction of cerebral arteries. Circ Res 95:922–929

Earley S, Heppner TJ, Nelson MT, Brayden JE (2005) TRPV4 forms a novel Ca^{2+} signaling complex with ryanodine receptors and BKCa channels. Circ Res 97:1270–1279

Elg S, Marmigere F, Mattsson JP, Ernfors P (2007) Cellular subtype distribution and developmental regulation of TRPC channel members in the mouse dorsal root ganglion. J Comp Neurol 503:35–46

Fois G, Wittekindt O, Zheng X, Felder ET, Miklavc P, Frick M, Dietl P, Felder E (2012) An ultra fast detection method reveals strain-induced Ca^{2+} entry via TRPV2 in alveolar type II cells. Biomech Model Mechanobiol 11:959–971

Galizia L, Pizzoni A, Fernandez J, Rivarola V, Capurro C, Ford P (2012) Functional interaction between AQP2 and TRPV4 in renal cells. J Cell Biochem 113:580–589

Garcia-Elias A, Mrkonjic S, Pardo-Pastor C, Inada H, Hellmich UA, Rubio-Moscardo F, Plata C, Gaudet R, Vicente R, Valverde MA (2013) Phosphatidylinositol-4,5-biphosphate-dependent rearrangement of TRPV4 cytosolic tails enables channel activation by physiological stimuli. Proc Natl Acad Sci U S A 110:9553–9558

Garrison SR, Dietrich A, Stucky CL (2012) TRPC1 contributes to light-touch sensation and mechanical responses in low-threshold cutaneous sensory neurons. J Neurophysiol 107:913–922

Gevaert T, Vriens J, Segal A, Everaerts W, Roskams T, Talavera K, Owsianik G, Liedtke W, Daelemans D, Dewachter I, Van Leuven F, Voets T, De Ridder D, Nilius B (2007) Deletion of

the transient receptor potential cation channel TRPV4 impairs murine bladder voiding. J Clin Invest 117:3453–3462
Gillespie PG, Walker RG (2001) Molecular basis of mechanosensory transduction. Nature 413:194–202
Gomis A, Soriano S, Belmonte C, Viana F (2008) Hypoosmotic- and pressure-induced membrane stretch activate TRPC5 channels. J Physiol 586:5633–5649
Gottlieb P, Folgering J, Maroto R, Raso A, Wood TG, Kurosky A, Bowman C, Bichet D, Patel A, Sachs F, Martinac B, Hamill OP, Honoré E (2008) Revisiting TRPC1 and TRPC6 mechanosensitivity. Pflügers Arch 455:1097–1103
Grant AD, Cottrell GS, Amadesi S, Trevisani M, Nicoletti P, Materazzi S, Altier C, Cenac N, Zamponi GW, Bautista-Cruz F, Lopez CB, Joseph EK, Levine JD, Liedtke W, Vanner S, Vergnolle N, Geppetti P, Bunnett NW (2007) Protease-activated receptor 2 sensitizes the transient receptor potential vanilloid 4 ion channel to cause mechanical hyperalgesia in mice. J Physiol 578:715–733
Grimm C, Kraft R, Sauerbruch S, Schultz G, Harteneck C (2003) Molecular and functional characterization of the melastatin-related cation channel TRPM3. J Biol Chem 278:21493–21501
Hamill OP (2006) Twenty odd years of stretch-sensitive channels. Pflügers Arch 453:333–351
Hartmannsgruber V, Heyken WT, Kacik M, Kaistha A, Grgic I, Harteneck C, Liedtke W, Hoyer J, Köhler R (2007) Arterial response to shear stress critically depends on endothelial TRPV4 expression. PLoS One 2:e827
Hoffmann EK, Lambert IH, Pedersen SF (2009) Physiology of cell volume regulation in vertebrates. Physiol Rev 89:193–277
Hofherr A, Kottgen M (2011) TRPP channels and polycystins. Adv Exp Med Biol 704:287–313
Inoue R, Jensen LJ, Jian Z, Shi J, Hai L, Lurie AI, Henriksen FH, Salomonsson M, Morita H, Kawarabayashi Y, Mori M, Mori Y, Ito Y (2009) Synergistic activation of vascular TRPC6 channel by receptor and mechanical stimulation via phospholipase C/diacylglycerol and phospholipase A2/omega-hydroxylase/20-HETE pathways. Circ Res 104:1399–1409
Iwata Y, Katanosaka Y, Arai Y, Komamura K, Miyatake K, Shigekawa M (2003) A novel mechanism of myocyte degeneration involving the Ca^{2+}-permeable growth factor-regulated channel. J Cell Biol 161:957–967
Jia Y, Wang X, Varty L, Rizzo CA, Yang R, Correll CC, Phelps PT, Egan RW, Hey JA (2004) Functional TRPV4 channels are expressed in human airway smooth muscle cells. Am J Physiol Lung Cell Mol Physiol 287:L272–L278
Jian MY, King JA, Al-Mehdi AB, Liedtke W, Townsley MI (2008) High vascular pressure-induced lung injury requires P450 epoxygenase-dependent activation of TRPV4. Am J Respir Cell Mol Biol 38:386–392
Jiang X, Newell EW, Schlichter LC (2003) Regulation of a TRPM7-like current in rat brain microglia. J Biol Chem 278:42867–42876
Jörs S, Grimm C, Becker L, Heller S (2010) Genetic inactivation of Trpml3 does not lead to hearing and vestibular impairment in mice. PLoS One 5:e14317
Kerstein PC, Jacques-Fricke BT, Rengifo J, Mogen BJ, Williams JC, Gottlieb PA, Sachs F, Gomez TM (2013) Mechanosensitive TRPC1 channels promote calpain proteolysis of talin to regulate spinal axon outgrowth. J Neurosci 33:273–285
Kim J, Chung YD, Park DY, Choi S, Shin DW, Soh H, Lee HW, Son W, Yim J, Park CS, Kernan MJ, Kim C (2003) A TRPV family ion channel required for hearing in Drosophila. Nature 424:81–84
Köhler R, Hoyer J (2007) Role of TRPV4 in the mechanotransduction of shear stress in endothelial cells. In: Liedtke WB, Heller S (eds) TRP ion channel function in sensory transduction and cellular signaling cascades. CRC Press, Boca Raton, FL
Köhler R, Heyken WT, Heinau P, Schubert R, Si H, Kacik M, Busch C, Grgic I, Maier T, Hoyer J (2006) Evidence for a functional role of endothelial transient receptor potential V4 in shear stress-induced vasodilatation. Arterioscler Thromb Vasc Biol 26:1495–1502

Kottgen M, Buchholz B, Garcia-Gonzalez MA, Kotsis F, Fu X, Doerken M, Boehlke C, Steffl D, Tauber R, Wegierski T, Nitschke R, Suzuki M, Kramer-Zucker A, Germino GG, Watnick T, Prenen J, Nilius B, Kuehn EW, Walz G (2008) TRPP2 and TRPV4 form a polymodal sensory channel complex. J Cell Biol 182:437–447

Koulen P, Cai Y, Geng L, Maeda Y, Nishimura S, Witzgall R, Ehrlich BE, Somlo S (2002) Polycystin-2 is an intracellular calcium release channel. Nat Cell Biol 4:191–197

Kress M, Karasek J, Ferrer-Montiel AV, Scherbakov N, Haberberger RV (2008) TRPC channels and diacylglycerol dependent calcium signaling in rat sensory neurons. Histochem Cell Biol 130:655–667

Kung C (2005) A possible unifying principle for mechanosensation. Nature 436:647–654

Kung C, Martinac B, Sukharev S (2010) Mechanosensitive channels in microbes. Annu Rev Microbiol 64:313–329

Kwan KY, Allchorne AJ, Vollrath MA, Christensen AP, Zhang DS, Woolf CJ, Corey DP (2006) TRPA1 contributes to cold, mechanical, and chemical nociception but is not essential for hair-cell transduction. Neuron 50:277–289

Kwan KY, Glazer JM, Corey DP, Rice FL, Stucky CL (2009) TRPA1 modulates mechanotransduction in cutaneous sensory neurons. J Neurosci 29:4808–4819

Liedtke WB (2007) TRPV channels' function in osmo- and mechanotransduction. In: Liedtke WB, Heller S (eds) TRP ion channel function in sensory transduction and cellular signaling cascades. CRC Press, Boca Raton, FL

Liedtke W, Friedman JM (2003) Abnormal osmotic regulation in trpv4-/- mice. Proc Natl Acad Sci U S A 100:13698–13703

Liedtke W, Choe Y, Marti-Renom MA, Bell AM, Denis CS, Sali A, Hudspeth AJ, Friedman JM, Heller S (2000) Vanilloid receptor-related osmotically activated channel (VR-OAC), a candidate vertebrate osmoreceptor. Cell 103:525–535

Liu X, Bandyopadhyay BC, Nakamoto T, Singh B, Liedtke W, Melvin JE, Ambudkar I (2006) A role for AQP5 in activation of TRPV4 by hypotonicity: concerted involvement of AQP5 and TRPV4 in regulation of cell volume recovery. J Biol Chem 281:15485–15495

Loot AE, Popp R, Fisslthaler B, Vriens J, Nilius B, Fleming I (2008) Role of cytochrome P450-dependent transient receptor potential V4 activation in flow-induced vasodilatation. Cardiovasc Res 80:445–452

Madsen CP, Klausen TK, Fabian A, Hansen BJ, Pedersen SF, Hoffmann EK (2012) On the role of TRPC1 in control of Ca^{2+} influx, cell volume, and cell cycle. Am J Physiol Cell Physiol 303: C625–C634

Mamenko M, Zaika O, Jin M, O'Neil RG, Pochynyuk O (2011) Purinergic activation of Ca^{2+}-permeable TRPV4 channels is essential for mechano-sensitivity in the aldosterone-sensitive distal nephron. PLoS One 6:e22824

Maroto R, Raso A, Wood TG, Kurosky A, Martinac B, Hamill OP (2005) TRPC1 forms the stretch-activated cation channel in vertebrate cells. Nat Cell Biol 7:179–185

Mederos y Schnitzler M, Storch U, Meibers S, Nurwakagari P, Breit A, Essin K, Gollasch M, Gudermann T (2008) G_q-coupled receptors as mechanosensors mediating myogenic vasoconstriction. EMBO J 27:3092–3103

Mendoza SA, Fang J, Gutterman DD, Wilcox DA, Bubolz AH, Li R, Suzuki M, Zhang DX (2010) TRPV4-mediated endothelial Ca^{2+} influx and vasodilation in response to shear stress. Am J Physiol Heart Circ Physiol 298:H466–H476

Mizuno A, Matsumoto N, Imai M, Suzuki M (2003) Impaired osmotic sensation in mice lacking TRPV4. Am J Physiol Cell Physiol 285:C96–C101

Mochizuki T, Sokabe T, Araki I, Fujishita K, Shibasaki K, Uchida K, Naruse K, Koizumi S, Takeda M, Tominaga M (2009) The TRPV4 cation channel mediates stretch-evoked Ca^{2+} influx and ATP release in primary urothelial cell cultures. J Biol Chem 284:21257–21264

Murakami M, Ohba T, Xu F, Shida S, Satoh E, Ono K, Miyoshi I, Watanabe H, Ito H, Iijima T (2005) Genomic organization and functional analysis of murine PKD2L1. J Biol Chem 280:5626–5635

Muraki K, Iwata Y, Katanosaka Y, Ito T, Ohya S, Shigekawa M, Imaizumi Y (2003) TRPV2 is a component of osmotically sensitive cation channels in murine aortic myocytes. Circ Res 93:829–838

Nauli SM, Alenghat FJ, Luo Y, Williams E, Vassilev P, Li X, Elia AE, Lu W, Brown EM, Quinn SJ, Ingber DE, Zhou J (2003) Polycystins 1 and 2 mediate mechanosensation in the primary cilium of kidney cells. Nat Genet 33:129–137

Nauli SM, Kawanabe Y, Kaminski JJ, Pearce WJ, Ingber DE, Zhou J (2008) Endothelial cilia are fluid shear sensors that regulate calcium signaling and nitric oxide production through polycystin-1. Circulation 117:1161–1171

Nilius B, Honoré E (2012) Sensing pressure with ion channels. Trends Neurosci 35:477–486

Nishihara E, Hiyama TY, Noda M (2011) Osmosensitivity of transient receptor potential vanilloid 1 is synergistically enhanced by distinct activating stimuli such as temperature and protons. PLoS One 6:e22246

Numata T, Shimizu T, Okada Y (2007a) Direct mechano-stress sensitivity of TRPM7 channel. Cell Physiol Biochem 19:1–8

Numata T, Shimizu T, Okada Y (2007b) TRPM7 is a stretch- and swelling-activated cation channel involved in volume regulation in human epithelial cells. Am J Physiol Cell Physiol 292:C460–C467

Pan Z, Yang H, Mergler S, Liu H, Tachado SD, Zhang F, Kao WW, Koziel H, Pleyer U, Reinach PS (2008) Dependence of regulatory volume decrease on transient receptor potential vanilloid 4 (TRPV4) expression in human corneal epithelial cells. Cell Calcium 44:374–385

Pedersen SF, Nilius B (2007) Transient receptor potential channels in mechanosensing and cell volume regulation. Methods Enzymol 428:183–207

Pennekamp P, Karcher C, Fischer A, Schweickert A, Skryabin B, Horst J, Blum M, Dworniczak B (2002) The ion channel polycystin-2 is required for left-right axis determination in mice. Curr Biol 12:938–943

Phan MN, Leddy HA, Votta BJ, Kumar S, Levy DS, Lipshutz DB, Lee SH, Liedtke W, Guilak F (2009) Functional characterization of TRPV4 as an osmotically sensitive ion channel in porcine articular chondrocytes. Arthritis Rheum 60:3028–3037

Pochynyuk O, Zaika O, O'Neil RG, Mamenko M (2013) Novel insights into TRPV4 function in the kidney. Pflügers Arch 465:177–186

Quick K, Zhao J, Eijkelkamp N, Linley JE, Rugiero F, Cox JJ, Raouf R, Gringhuis M, Sexton JE, Abramowitz J, Taylor R, Forge A, Ashmore J, Kirkwood N, Kros CJ, Richardson GP, Freichel M, Flockerzi V, Birnbaumer L, Wood JN (2012) TRPC3 and TRPC6 are essential for normal mechanotransduction in subsets of sensory neurons and cochlear hair cells. Open Biol 2:120068

Sharif Naeini R, Witty MF, Seguela P, Bourque CW (2006) An N-terminal variant of Trpv1 channel is required for osmosensory transduction. Nat Neurosci 9:93–98

Sharif-Naeini R, Dedman A, Folgering JH, Duprat F, Patel A, Nilius B, Honoré E (2008) TRP channels and mechanosensory transduction: insights into the arterial myogenic response. Pflügers Arch 456:529–540

Sharif-Naeini R, Folgering JH, Bichet D, Duprat F, Lauritzen I, Arhatte M, Jodar M, Dedman A, Chatelain FC, Schulte U, Retailleau K, Loufrani L, Patel A, Sachs F, Delmas P, Peters DJ, Honoré E (2009) Polycystin-1 and -2 dosage regulates pressure sensing. Cell 139:587–596

Shen J, Harada N, Kubo N, Liu B, Mizuno A, Suzuki M, Yamashita T (2006) Functional expression of transient receptor potential vanilloid 4 in the mouse cochlea. Neuroreport 17:135–139

Shibasaki K, Murayama N, Ono K, Ishizaki Y, Tominaga M (2010) TRPV2 enhances axon outgrowth through its activation by membrane stretch in developing sensory and motor neurons. J Neurosci 30:4601–4612

Shimizu T, Janssens A, Voets T, Nilius B (2009) Regulation of the murine TRPP3 channel by voltage, pH, and changes in cell volume. Pflügers Arch 457:795–807

Sidhaye VK, Schweitzer KS, Caterina MJ, Shimoda L, King LS (2008) Shear stress regulates aquaporin-5 and airway epithelial barrier function. Proc Natl Acad Sci U S A 105:3345–3350
Sonkusare SK, Bonev AD, Ledoux J, Liedtke W, Kotlikoff MI, Heppner TJ, Hill-Eubanks DC, Nelson MT (2012) Elementary Ca^{2+} signals through endothelial TRPV4 channels regulate vascular function. Science 336:597–601
Spassova MA, Hewavitharana T, Xu W, Soboloff J, Gill DL (2006) A common mechanism underlies stretch activation and receptor activation of TRPC6 channels. Proc Natl Acad Sci U S A 103:16586–16591
Staaf S, Maxvall I, Lind U, Husmark J, Mattsson JP, Ernfors P, Pierrou S (2009) Down regulation of TRPC1 by shRNA reduces mechanosensitivity in mouse dorsal root ganglion neurons in vitro. Neurosci Lett 457:3–7
Strotmann R, Harteneck C, Nunnenmacher K, Schultz G, Plant TD (2000) OTRPC4, a nonselective cation channel that confers sensitivity to extracellular osmolarity. Nat Cell Biol 2:695–702
Sukharev S, Sachs F (2012) Molecular force transduction by ion channels: diversity and unifying principles. J Cell Sci 125:3075–3083
Suzuki M, Mizuno A, Kodaira K, Imai M (2003) Impaired pressure sensation in mice lacking TRPV4. J Biol Chem 278:22664–22668
Tabuchi K, Suzuki M, Mizuno A, Hara A (2005) Hearing impairment in TRPV4 knockout mice. Neurosci Lett 382:304–308
Taniguchi J, Tsuruoka S, Mizuno A, Sato J, Fujimura A, Suzuki M (2007) TRPV4 as a flow sensor in flow-dependent K + secretion from the cortical collecting duct. Am J Physiol Renal Physiol 292:F667–F673
Taylor AC, McCarthy JJ, Stocker SD (2008) Mice lacking the transient receptor vanilloid potential 1 channel display normal thirst responses and central Fos activation to hypernatremia. Am J Physiol Regul Integr Comp Physiol 294:R1285–R1293
Tian W, Salanova M, Xu H, Lindsley JN, Oyama TT, Anderson S, Bachmann S, Cohen DM (2004) Renal expression of osmotically responsive cation channel TRPV4 is restricted to water-impermeant nephron segments. Am J Physiol Renal Physiol 287:F17–F24
Tracey WD Jr, Wilson RI, Laurent G, Benzer S (2003) Painless, a Drosophila gene essential for nociception. Cell 113:261–273
Tsiokas L, Arnould T, Zhu C, Kim E, Walz G, Sukhatme VP (1999) Specific association of the gene product of PKD2 with the TRPC1 channel. Proc Natl Acad Sci U S A 96:3934–3939
van Aken AF, Atiba-Davies M, Marcotti W, Goodyear RJ, Bryant JE, Richardson GP, Noben-Trauth K, Kros CJ (2008) TRPML3 mutations cause impaired mechano-electrical transduction and depolarization by an inward-rectifier cation current in auditory hair cells of varitint-waddler mice. J Physiol 586:5403–5418
Vergnolle N, Cenac N, Altier C, Cellars L, Chapman K, Zamponi GW, Materazzi S, Nassini R, Liedtke W, Cattaruzza F, Grady EF, Geppetti P, Bunnett NW (2010) A role for transient receptor potential vanilloid 4 in tonicity-induced neurogenic inflammation. Br J Pharmacol 159:1161–1173
Vilceanu D, Stucky CL (2010) TRPA1 mediates mechanical currents in the plasma membrane of mouse sensory neurons. PLoS One 5:e12177
Vriens J, Watanabe H, Janssens A, Droogmans G, Voets T, Nilius B (2004) Cell swelling, heat, and chemical agonists use distinct pathways for the activation of the cation channel TRPV4. Proc Natl Acad Sci U S A 101:396–401
Walker RG, Willingham AT, Zuker CS (2000) A Drosophila mechanosensory transduction channel. Science 287:2229–2234
Welsh DG, Morielli AD, Nelson MT, Brayden JE (2002) Transient receptor potential channels regulate myogenic tone of resistance arteries. Circ Res 90:248–250
Wissenbach U, Bödding M, Freichel M, Flockerzi V (2000) Trp12, a novel Trp related protein from kidney. FEBS Lett 485:127–134

Wong AC, Birnbaumer L, Housley GD (2013) Canonical transient receptor potential channel subtype 3-mediated hair cell Ca^{2+} entry regulates sound transduction and auditory neurotransmission. Eur J Neurosci 37:1478–1486

Wu L, Gao X, Brown RC, Heller S, O'Neil RG (2007) Dual role of the TRPV4 channel as a sensor of flow and osmolality in renal epithelial cells. Am J Physiol Renal Physiol 293:F1699–F1713

Yang XR, Lin AH, Hughes JM, Flavahan NA, Cao YN, Liedtke W, Sham JS (2012) Upregulation of osmo-mechanosensitive TRPV4 channel facilitates chronic hypoxia-induced myogenic tone and pulmonary hypertension. Am J Physiol Lung Cell Mol Physiol 302:L555–L568

Yin J, Kuebler WM (2010) Mechanotransduction by TRP channels: general concepts and specific role in the vasculature. Cell Biochem Biophys 56:1–18

Yoshiba S, Shiratori H, Kuo IY, Kawasumi A, Shinohara K, Nonaka S, Asai Y, Sasaki G, Belo JA, Sasaki H, Nakai J, Dworniczak B, Ehrlich BE, Pennekamp P, Hamada H (2012) Cilia at the node of mouse embryos sense fluid flow for left-right determination via Pkd2. Science 338:226–231

Zhang Z, Kindrat AN, Sharif-Naeini R, Bourque CW (2007) Actin filaments mediate mechanical gating during osmosensory transduction in rat supraoptic nucleus neurons. J Neurosci 27:4008–4013

Zhang XF, Chen J, Faltynek CR, Moreland RB, Neelands TR (2008) Transient receptor potential A1 mediates an osmotically activated ion channel. Eur J Neurosci 27:605–611

TRPs as Chemosensors (ROS, RNS, RCS, Gasotransmitters)

Shunichi Shimizu, Nobuaki Takahashi, and Yasuo Mori

Contents

S. Shimizu
Division of Physiology and Pathology, Department of Pharmacology, Toxicology and Therapeutics, Showa University School of Pharmacy, Tokyo, Japan

N. Takahashi
Laboratory of Molecular Biology, Department of Synthetic Chemistry and Biological Chemistry, Graduate School of Engineering, Kyoto University, Kyoto, Japan

Advanced Biomedical Engineering Research Unit, Kyoto University, Kyoto, Japan

Y. Mori (✉)
Laboratory of Molecular Biology, Department of Synthetic Chemistry and Biological Chemistry, Graduate School of Engineering, Kyoto University, Kyoto, Japan

CREST, JST, Chiyoda-ku, Japan

Laboratory of Environmental Systems Biology, Department of Technology and Ecology, Hall of Global Environmental Studies, Kyoto University, Kyoto 615-8510, Japan
e-mail: mori@sbchem.kyoto-u.ac.jp

B. Nilius and V. Flockerzi (eds.), *Mammalian Transient Receptor Potential (TRP) Cation Channels*, Handbook of Experimental Pharmacology 223, DOI 10.1007/978-3-319-05161-1_3,

Abstract

The transient receptor potential (*trp*) gene superfamily encodes TRP proteins that act as multimodal sensor cation channels for a wide variety of stimuli from outside and inside the cell. Upon chemical or physical stimulation of cells, TRP channels transduce electrical and/or Ca^{2+} signals via their cation channel activities. These functional features of TRP channels allow the body to react and adapt to different forms of environmental changes. Indeed, members of one class of TRP channels have emerged as sensors of reactive oxygen species (ROS), reactive nitrogen species (RNS), reactive carbonyl species (RCS), and gaseous messenger molecules including molecular oxygen (O_2), hydrogen sulfide (H_2S), and carbon dioxide (CO_2). Hydrogen peroxide (H_2O_2), an ROS, triggers the production of ADP-ribose, which binds and activates TRPM2. In addition to TRPM2, TRPC5, TRPV1, and TRPA1 are also activated by H_2O_2 via modification of cysteine (Cys) free sulfhydryl groups. Nitric oxide (NO), a vasoactive gaseous molecule, regulates TRP channels directly via Cys S-nitrosylation or indirectly via cyclic GMP (cGMP)/protein kinase G (PKG)-dependent phosphorylation. Anoxia induced by O_2-glucose deprivation and severe hypoxia activates TRPM7 and TRPC6, respectively, whereas TRPA1 serves as a sensor of mild hypoxia and hyperoxia in vagal and sensory neurons. TRPA1 also detects other gaseous molecules, such as hydrogen sulfide (H_2S) and carbon dioxide (CO_2). In this review, we highlight our current knowledge of TRP channels as chemosensors for ROS, RNS, RCS, and gaseous molecules and discuss their functional impacts on physiological and pathological events.

Keywords

TRP channels • Chemosensor • ROS • RNS • RCS • Oxygen

1 Sensing of ROS by TRP Channels

Oxidative stress is generally defined as an imbalance that favors the production of prooxidants [represented by reactive oxygen species (ROS)] over antioxidants (which detoxify the reactive intermediates in biological systems) (Fig. 1). ROS include superoxide anions ($O_2^{\cdot-}$) and hydrogen peroxide (H_2O_2), which can produce the severely damaging hydroxyl radical ($^{\cdot}OH$). These species are produced mainly by one-electron reduction of molecular oxygen (O_2), but are also generated as a byproduct of respiration in mitochondria or by specific enzymes of the NADPH oxidase (NOX) and dual oxidase (DUOX) family (Orrenius et al. 2003; Kulkarni et al. 2007). Because ROS are highly chemically reactive, they are commonly considered as nonspecific toxins that cause random damage to cellular components. However, accumulated evidence has shown that oxidative stress also controls the activation of a number of complex and interrelated signaling events (Chandra

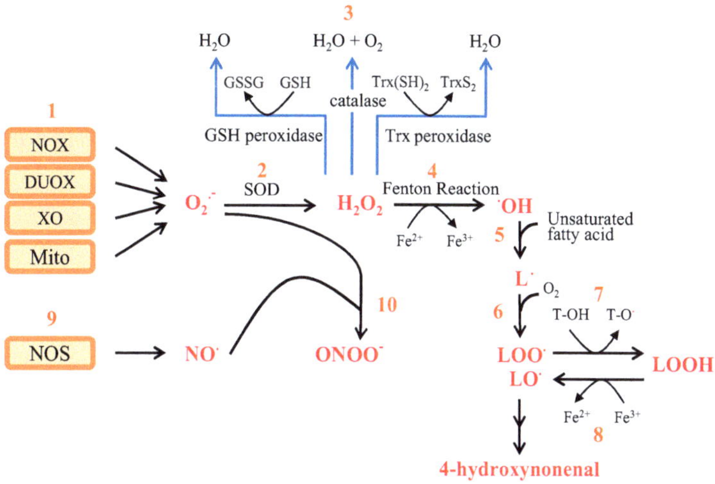

Fig. 1 Pathways of ROS formation. Reaction 1: the superoxide anion radical ($O_2^{\cdot -}$) is formed by the process of reduction of molecular oxygen mediated by NADPH oxidase (NOX), dual oxidase (DUOX), and xanthine oxidase (XO) or nonenzymatically by redox-reactive compounds such as the semi-ubiquinone compound of the mitochondrial electron transport chain. Reaction 2: superoxide anion radical is dismutated by the superoxide dismutase (SOD) to hydrogen peroxide (H_2O_2). Reaction 3: hydrogen peroxide is scavenged by the enzyme glutathione (GSH) peroxidase, catalase, and thioredoxin (Trx) peroxidase. Reaction 4: some transition metals (e.g., Fe^{2+}, Cu^+ and others) can break down hydrogen peroxide to the hydroxyl radical (·OH) (Fenton reaction). Reaction 5: the hydroxyl radical can abstract an electron from polyunsaturated fatty acid to give rise to a carbon-centered lipid radical (L·). Reaction 6: the lipid radical can interact with molecular oxygen to give a lipid peroxyl radical (LOO·). Reaction 7: the lipid peroxyl radical is reduced by the reduced form of vitamin E (T-OH) resulting in the formation of a lipid hydroperoxide. Reaction 8: lipid hydroperoxides can react with Fe^{2+} to form lipid alkoxyl radicals (LO·). Reaction 9: the nitric oxide (NO·) is formed by the enzyme nitric oxide synthases (NOS) from L-arginine and molecular oxygen. Reaction 7: the nitric oxide can react with superoxide anion radical to produce the peroxynitrite ($ONOO^-$)

et al. 2000; Gopalakrishna and Jaken 2000), implicating ROS as important signaling molecules in a variety of cellular events.

Oxidative stress plays a critical role in various pathophysiological processes, including cancer, acute and chronic neurodegenerative disorders (e.g., Alzheimer's and Parkinson's diseases), diabetes mellitus, atherosclerosis, ischemia–reperfusion injury, and autoimmune disease, but is also involved in normal cellular functions such as apoptosis, cell proliferation, and gene expression (Chandra et al. 2000; Kietzmann et al. 2000; Langley and Ratan 2004; Chiu and Dawes 2012). In ischemia–reperfusion injury and degenerative diseases, necrotic and apoptotic processes are controlled by oxidative stress (Coyle and Puttfarcken 1993; Giroux and Scatton 1996). Tissue damage and inflammation produce an array of substances, including ROS, that excite or sensitize nociceptors, eliciting localized pain. It also contributes significantly to the expression of a variety of different inflammatory cytokines, adhesion molecules, and enzymes by activating redox-sensitive transcription factors such as nuclear factor-κB (NF-κB) (Dröge 2000). Regulation of intracellular Ca^{2+} concentration ($[Ca^{2+}]_i$) is critical to most of these processes, but the identity of the oxidative stress-activated Ca^{2+}-permeable channels involved in mediating calcium influx remained a mystery until the recent identification of a family of mammalian homologs of transient receptor potential (TRP) channels.

1.1 Activation Mechanisms for TRPM2 by H_2O_2

The second member of the TRP melastatin subfamily, TRPM2, is a Ca^{2+}-permeable nonselective cation channel that is expressed in the brain, pancreatic β-cells, and immunocytes including monocytes/macrophages and neutrophils (Wehage et al. 2002; Perraud et al. 2003; Clapham et al. 2013). TRPM2 possesses a NudT9-H domain in its C-terminus, which delivers only a low level ADP-ribose pyrophosphatase activity, despite its significant homology with NudT9 ADP-ribose pyrophosphatase. Intracellular ADP-ribose can activate TRPM2 by binding to this C-terminal NudT9-H domain (Perraud et al. 2001), and we have demonstrated that TRPM2 is also activated by ROS such as H_2O_2 (Hara et al. 2002). The first proposed mechanism underlying TRPM2 channel activation by H_2O_2 was production of nicotinamide adenine dinucleotide (NAD^+) and its binding to the C-terminal NudT9-H domain (Hara et al. 2002). However, it is more believed that this H_2O_2-mediated channel opening is transduced by the production of ADP-ribose, a metabolite of NAD^+ (Kolisek et al. 2005; Perraud et al. 2005).

It has been suggested that ADP-ribose is produced from mitochondria in oxidative stress-induced TRPM2 activity (Perraud et al. 2005). Cytosolic or mitochondrial overexpression of ADP-ribose pyrophosphatase, which degrades ADP-ribose, suppresses H_2O_2-induced Ca^{2+} responses, illustrating the role of ADP-ribose as the link between H_2O_2 and TRPM2 activation. However, ADP-ribose is produced not only by mitochondria but also in the nucleus in response to various noxious factors including oxidative stress and involving a pathway that includes poly(ADP-ribose) polymerase-1 (PARP-1)/poly(ADP-ribose) glycohydrolase (PARG) as components (Tanuma et al. 1985; de Murcia and Menissier de Murcia 1994; Oliver et al. 1999; Virag and Szabo 2002). Evidence for a functional role for PARP-1 is derived from experiments on three structurally distinct PARP inhibitors that suppress oxidative stress-dependent activation of TRPM2 (Fonfria et al. 2004). Electrophysiological studies have shown that PARP inhibitors fail to interfere with the activation of TRPM2 by ADP-ribose, indicating that the PARP inhibitors must act upstream from the interaction between TRPM2 and ADP-ribose. Additionally, it has also been reported that the PARP-deficient DT40 cell line, which expresses TRPM2, exhibits impaired Ca^{2+} responses to oxidative stress (Buelow et al. 2008).

1.2 Physiological Roles of ROS-Activated TRPM2

1.2.1 Cell Death

Since the first report on the involvement of TRPM2-evoked $[Ca^{2+}]_i$ elevation in cell death (Hara et al. 2002), substantial evidence has accrued supporting the proposal that TRPM2 is an important factor in cell death induced by oxidative stress (Zhang et al. 2003, 2006; Fonfria et al. 2005). TRPM2-expressing HEK cells are susceptible to cell death induced by exposure to H_2O_2 through the elevation of $[Ca^{2+}]_i$ (Hara et al. 2002; Zhang et al. 2003). This has been consistently demonstrated in cells that endogenously express TRPM2, including neuronal cells, monocytes, and insulinoma cells. H_2O_2-induced cell death is attenuated by reducing TRPM2 expression using specific antisense oligonucleotides, RNA interference, and pharmacological inhibitors (Hara et al. 2002; Fonfria et al. 2005; Zhang et al. 2006). TRPM2-specific antisense also almost abolishes tissue necrosis factor α (TNFα)-evoked $[Ca^{2+}]_i$ oscillations in RIN-5F cells and significantly suppresses TNF-α-induced death in RIN-5F cells (Hara et al. 2002). These results suggest an important involvement of TRPM2 in cell death mediated by TNFα-activated Ca^{2+} channels.

TRPM2 channels have been implicated in neuronal damage induced by oxidants, TNFα, and amyloid β-peptide (Aβ, a major component of senile plaques) and are thus important in the pathogenesis of Alzheimer's disease (AD) (Hara et al. 2002; Halliwell 2006). Kaneko et al. (2006) have demonstrated that TRPM2 expressed in rat cortical neurons is critically involved in H_2O_2-induced Ca^{2+} influx that causes neuronal cell death. Accumulating evidence suggests that a failure of Ca^{2+} homeostasis plays a critical role in the neuropathology of AD. The direct neurotoxicity of Aβ and enhanced vulnerability of cells to excitotoxicity are both attenuated when cells are incubated in Ca^{2+}-free medium, suggesting the involvement of Ca^{2+} influx in this process (Mattson et al. 1993). Several paradigmatic mechanisms for Aβ neurotoxicity involve the production of ROS such as H_2O_2 and NO, as well as excitotoxicity due to intracellular Ca^{2+} accumulation (Marchesi 2011). Aβ is known to stimulate microglia and astrocytes (Goodwin et al. 1997; Rossi and Bianchini 1996; Wallace et al. 1997; Akama et al. 1998), and the activation of microglia releases cytokines such as interleukin-1α (IL-1α), IL-1β, and TNFα (Yamabe et al. 1994; Walker et al. 1995; Dickson et al. 1988). These lines of circumstantial evidence are consistent with the idea that TNFα released from Aβ-activated microglia triggers neuronal cell death via TRPM2 in AD. Fonfria et al. (2005) have also suggested that activation of TRPM2, functionally expressed in primary cultures of rat striatum, contributes to Aβ- and oxidative stress-induced striatal cell death. Thus, TRPM2 activity likely contributes to neuronal cell death in pathophysiological circumstances where ROS are abundant. Little is known about the intracellular mechanisms that regulate oxidative stress-induced cell death via TRPM2, although we recently revealed that the EF-hand motif-containing protein EFHC1, mutation of which causes juvenile myoclonic epilepsy (JME) via mechanisms including neuronal apoptosis, associates with TRPM2 and positively regulates its channel activity and H_2O_2-induced cell death (Katano et al. 2012).

1.2.2 Immunological Function of TRPM2

A major physiological role of TRPM2 is to regulate the signal cascades leading to chemokine production in monocytes/macrophages. In monocytes, we showed that H_2O_2 induces TRPM2 activation alongside a Ca^{2+}-independent activation of extracellular signal-regulated kinase (Erk) as an initial response phase. Subsequently, activated TRPM2 mediates H_2O_2-activated Ca^{2+} influx, which drives the autophosphorylation of Ca^{2+}-sensitive proline-rich tyrosine kinase 2 (Pyk2), therein amplifying Erk activation in a Ras-dependent manner. This amplified Erk activity upregulates the transcription of the CXCL8/CXCL2 gene by inducing nuclear translocation of RelA, an NF-κB subunit in human monocytes (Yamamoto et al. 2008). ROS, including H_2O_2, are generated and released from various cell types at sites of inflammation. Using a *Trpm2* knockout (KO) mouse, we have revealed that TRPM2-mediated chemokine production in monocytes and macrophages is an important mechanism in the progressive severity of dextran sulfate sodium-induced ulcerative colitis (Yamamoto et al. 2008). Recently, Wehrhahn et al. (2010) have shown that lipopolysaccharide (LPS) and TNF-α both significantly upregulate TRPM2 expression and channel activity in human monocytes, resulting in the production of cytokines including IL-6, IL-8, IL-10, and TNF-α. A more recent study has shown that *Trpm2* KO mice are extremely susceptible to infection with the bacterial pathogen *Listeria monocytogenes* (Knowles et al. 2011). TRPM2 channels seem to have manifold contributions to pathogenesis in immunity and inflammation (Knowles et al. 2013).

1.2.3 Modulation of Cardiac Ischemia–Reperfusion Injury

Myocardial ischemia–reperfusion (I/R) injury is characterized by acute inflammation associated with the augmentation of oxidative stress. In particular, neutrophils accumulating in the heart play a central role in the early phase of I/R injury (Romson et al. 1983; Vinten-Johansen 2004). I/R of the heart also causes myocardial Ca^{2+} overload, leading to contractile dysfunction and cell death (Carrozza et al. 1992; Dong et al. 2006). Thus, oxidative stress, neutrophils, and Ca^{2+} homoeostasis are key factors in the development of myocardial I/R injury. We have recently revealed that myocardial infarction following I/R, but not ischemia alone, is reduced in *Trpm2* KO mice compared to wild-type (WT) mice and cardiac contractile function is also improved in *Trpm2* KO mice (Hiroi et al. 2013). TRPM2 is highly expressed in polymorphonuclear leucocytes (PMNs) rather than in the heart, and the number of neutrophils in the reperfused area following ischemia is reduced in *Trpm2* KO mice. When WT or *Trpm2* KO PMNs were administered to the *Trpm2* KO heart ex vivo through the perfusate or in vivo by intravenous injection, WT PMNs precipitated an enlargement of the infarct size, whereas *Trpm2* KO-derived PMNs did not. Thus, we have shown that the accumulation of TRPM2-activated neutrophils in the reperfused area is likely to play a crucial role in myocardial I/R injury. Yang et al. (2006) have demonstrated the involvement of TRPM2 in oxidative stress-induced cardiomyocyte death in cultured ventricular myocytes. In ex vivo experiments, the administration of *Trpm2* KO PMNs to WT hearts induces greater increases in infarct size than those in *Trpm2* KO hearts, suggesting that TRPM2 expressed in the heart partly contributes to myocardial infarction.

In contrast to our findings, an independent *Trpm2* KO mouse model has been used to suggest that TRPM2 channels protect the heart from I/R injury by decreasing ROS generation and enhancing ROS scavenging (Miller et al. 2013). The reasons for the discrepancy in these results are not obvious. Future studies using tissue-specific *Trpm2* KO mice would provide an unequivocal answer to this contentious issue.

1.3 ROS Sensing by Other TRP Channels

The cellular signals initiated by ROS involve the posttranslational modification of specific amino acid residues on signaling proteins, notably the sulfhydryl groups on cysteine (Cys) residues (Satoh and Lipton 2007). Certain members of the TRPC and TRPV subfamilies, including TRPC5 and TRPV1, are activated by ROS through modification of Cys free sulfhydryl groups (Yoshida et al. 2006; Chuang and Lin 2009). TRPC5 is also activated by the reducing agent dithiothreitol (DTT) and by extracellular reduced thioredoxin, both of which cleave a disulfide bridge in the predicted extracellular loop adjacent to the ion-selectivity filter of TRPC5 (Xu et al. 2008). TRPC3 shows insensitivity or, if at all, only marginal sensitivity to H_2O_2 but is sensitive to *tert*-butyl hydroperoxide at relatively low concentrations, although the mechanism for this sensitivity is unclear (Balzer et al. 1999). More recently, TRPA1 channel activation has been shown to occur by H_2O_2 through modification of N-terminal Cys residues (Takahashi et al. 2008a; Sawada et al. 2008; Bessac et al. 2008; Nilius et al. 2012).

2 Sensing of RNS by TRP Channels

Among the gaseous signaling molecules, nitric oxide (NO) is the most extensively studied. Its biological significance and the systems by which it is generated were first revealed in the 1980s (Furchgott and Zawadzki 1980; Palmer et al. 1988; Sakuma et al. 1988), and it is now known to regulate a variety of biological events, including vascular relaxation and neurotransmission. Conversely, excessive generation of NO and NO-derived reactive nitrogen species (RNS) has been implicated in a number of pathological conditions (Reiter 2006). Cyclic GMP (cGMP) is the canonical mediator of NO signaling. However, the importance of a cGMP-independent signaling pathway involving protein S-nitrosylation is becoming increasingly recognized (Jaffrey et al. 2001; Hess et al. 2005). S-nitrosylation of Cys is readily reversible with high spatial and temporal specificity. The NADH-dependent oxidoreductase, S-nitrosoglutathione reductase, specifically catalyzes the denitrosylation of S-nitrosoglutathione, by which protein S-nitrosylation is regulated in the cellular equilibrium between S-nitrosylated proteins and S-nitrosoglutathione. Thioredoxin also mediates direct denitrosylation of multiple

S-nitrosylated proteins. Thus, the temporal and spatial regulation of S-nitrosylation and denitrosylation confers specificity to NO-based cellular signaling (Benhar et al. 2009).

2.1 Regulation of TRP Channels by NO Via Cys S-Nitrosylation

Some TRP channels are potently regulated by Cys modifications, including Cys S-nitrosylation by NO in heterologous systems and bovine aortic endothelial cells (Yoshida et al. 2006). Labeling and functional assays performed with Cys mutants have shown that Cys553 and nearby Cys558 on the N-terminal side of the putative pore-forming region between the fifth and sixth transmembrane domains (S5 and S6) are essential for mouse TRPC5 activation in response to the NO donor S-nitroso-N-acetyl-DL-penicillamine (SNAP). In *Drosophila melanogaster* Shaker voltage-gated K^+ channels, the activation gate formed by S6 residues near the intracellular entrance of the pore cavity has been identified (del Camino and Yellen 2001). Given the longer S5–S6 linkers in TRPC5, the TRPC5 S5–S6 linker with modified Cys553 and Cys558 may be invaginated toward the cytoplasm to reach the S6 activation gate. The corresponding Cys sites in TRPC1, TRPC4, TRPV1, TRPV3, and TRPV4 are potential targets of nitrosylation leading to channel activation in heterologous expression systems. Although the differences in maximal $[Ca^{2+}]_i$ responses [Δ ratio (340/380)] to SNAP in control cells and cells heterologously expressing either TRPC1 or TRPC4β (a splice isoform of TRPC4) were not statistically significant, a larger fraction (7–9 %) of the TRP-expressing cells showed a Δ ratio (340/380) >0.5 when compared to control cells (2–5 %). Heterologous co-expression of TRPC4β and TRPC5 gave SNAP-induced responses comparable to those in cells expressing TRPC5 alone, whereas co-expression of TRPC1 and TRPC5 gave slightly suppressed (albeit still robust) responses. Co-immunoprecipitation of TRPC5 with TRPC1 or TRPC4β suggests that heteromultimeric TRPC5/TRPC1 and TRPC5/TRPC4β channels also have sensitivity to NO (Yoshida et al. 2006). The thermosensor TRP channels TRPV1, TRPV3, and TRPV4 also show SNAP-induced activation, as predicted from conserved Cys residues in the corresponding regions of these homologs. Indeed, the substitution of two conserved Cys residues in TRPV1 leads to significantly suppressed responses to SNAP and Cys S-nitrosylation (Yoshida et al. 2006). Notably, the sensitivity of TRPV1 to H^+ and heat is enhanced by SNAP but abolished by the mutations, despite normal surface expression and intact control H^+ and heat responses (Yoshida et al. 2006). Thus, channel activation regulated by nitrosylation is conserved among a number of TRP channels that belong to different subfamilies.

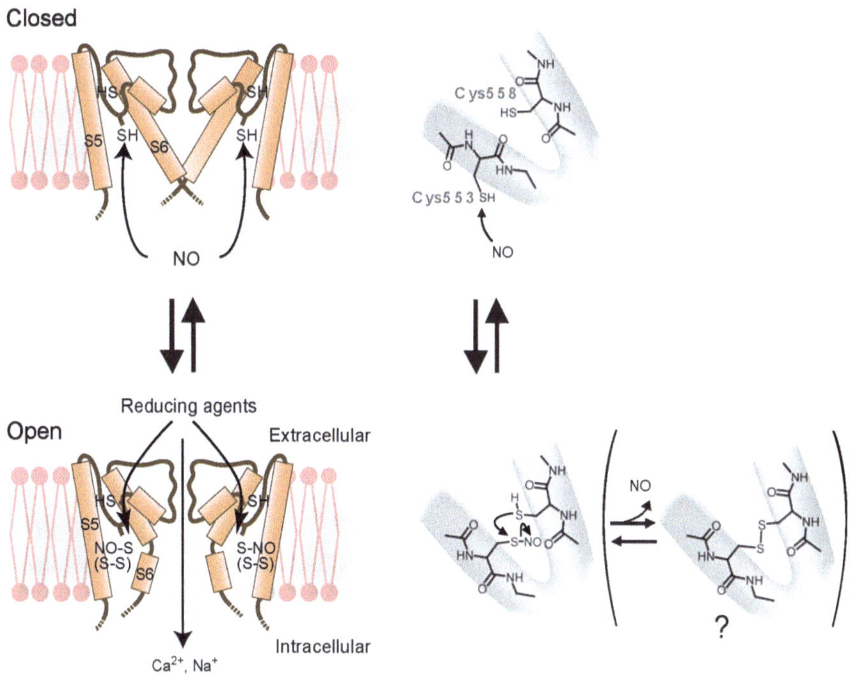

Fig. 2 Model for activation ofTRPC5 by NO. Possible protein conformation changes and chemical reactions during activation of TRPC5 by NO. NO modify the free sulfhydryl group of Cys553 accessible from the cytoplasmic side to open the activation gate. The modified Cys553 can be further attacked nucleophilically by the free sulfhydryl group of Cys558 to form an intramolecular disulfide bond, which may stabilize the channel in activation states

2.2 NO Sensing by TRPC5 Channel

SNAP-activated TRPC5 channels are not entirely inactivated by ascorbate, which reduces S-nitrosothiols, but not disulfides, to thiols. However, DTT, which reduces both S-nitrosothiols and disulfides to thiols, fully suppressed SNAP-activated TRPC5 channel activity, suggesting that both nitrosylation and disulfide bond formation are involved in SNAP-induced TRPC5 activation. S-nitrosylation is abolished by mutation of Cys553 in TRPC5, but not by mutation of Cys558 (Jaffrey et al. 2001). As proposed for the acid–base catalysis of hemoglobin nitrosylation in proteins with high NO sensitivity, basic and acidic amino acids surrounding S-nitrosylated Cys may enhance the nucleophilicity of the sulfhydryl (SH) group and therefore the S-nitrosylation of this group of proteins (Hess et al. 2005). Charged residues flanking Cys553 and Cys558 in TRPC5 may confer a susceptibility to being modified by NO. It is possible that the TRPC5 channel is opened via S-nitrosylation of Cys553, with a subsequent nucleophilic attack on nitrosylated Cys553 by the free SH group of Cys558 forming a disulfide bond that stabilizes this open state (Fig. 2). However, the NO sensitivity of TRPC5 channels has been disputed by several groups (Xu et al. 2008; Wong et al. 2010) and may be influenced by culture conditions, drug administration protocols, cell density during measurements, levels of antioxidants, or other experimental conditions that may affect the modification state of TRPC5 proteins.

NO is produced by nitric oxide synthase (NOS), of which two constitutive isoforms (neuronal NOS (nNOS) and endothelial NOS (eNOS)) and one inducible isoform (iNOS) have been cloned. NO produced by NOS is involved in diverse physiological processes including vascular smooth muscle relaxation. Increases in $[Ca^{2+}]_i$ in endothelial cells induce NO production via eNOS, which triggers vascular smooth muscle relaxation. However, the molecular entities and/or mechanisms that mediate the increased $[Ca^{2+}]_i$ required for eNOS activation remain elusive. Vasodilators such as ATP, substance P, and acetylcholine can increase endothelial cell $[Ca^{2+}]_i$ and activate TRP channels, suggesting that TRP channels may be involved in these $[Ca^{2+}]_i$ increases. NO-evoked $[Ca^{2+}]_i$ increases were observed in bovine aortic endothelial cells (BAECs), and NO-induced increases in $[Ca^{2+}]_i$ were suppressed by TRPC5-specific siRNA and expression of a TRPC5 dominant-negative mutant, indicating that NO-evoked $[Ca^{2+}]_i$ increases in BAECs are attributable to Ca^{2+} influx through endogenous TRPC5 channels. ATP, a vasodilator, evoked $[Ca^{2+}]_i$ increases and NO production in BAECs, which were suppressed by TRPC5-specific siRNA, suggesting that TRPC5 is also involved in these signaling processes. Additionally, these ATP-evoked $[Ca^{2+}]_i$ increases and NO production were suppressed by the NOS inhibitor, N^G-nitro-L-arginine, and by an eNOS-specific siRNA, implicating eNOS-derived NO in the ATP-evoked $[Ca^{2+}]_i$ influx via TRPC5. We have since confirmed the S-nitrosylation of endogenous TRPC5 after ATP stimulation (Yoshida et al. 2006). These results demonstrate the activation of native TRPC5 channels by nitrosylation via eNOS following ATP receptor stimulation in endothelial cells. Subsequent Ca^{2+} influx through nitrosylated TRPC5 channels may then mediate positive feedback regulation of Ca^{2+}-dependent NO production.

2.3 NO Sensing by Other Channels

Recent reports have shown that SNAP and another NO donor, (6)-(E)-Ethyl-2-[(E)-hydroxyimino]-5-nitro-3-hexeneamide (NOR3), activate human TRPA1 in heterologous systems and mouse TRPA1 in dissociated sensory neurons (Sawada et al. 2008; Takahashi et al. 2008a; Miyamoto et al. 2009). Functional characterization of site-directed Cys mutants of TRPA1 has demonstrated that Cys421, Cys641, and Cys665 are responsible for human TRPA1 activation by NO (Takahashi et al. 2008a). Cys421, Cys641, and Cys665 are located, respectively, in the eleventh and seventeenth ankyrin repeat (AnkR) domains and the N-terminal cytoplasmic region between the seventeenth AnkR domain and S1 (Gaudet 2008) and are reversibly bound by methylglyoxal, a reactive metabolite accumulating in diabetes and uremia apparently leading to subsequent formation of disulfide bonds and TRPA1 activation (Eberhardt et al. 2012).

Cys modifications are also central to the ability of several regulatory factors to mediate the activation of TRPV1 (Yoshida et al. 2006). The alignment of amino acid sequences surrounding Cys553 and Cys558 of TRPC5 with counterpart sequences reveals Cys residues conserved on the N-terminal side of the putative pore-forming region, which is located between the fifth and sixth transmembrane domains in TRPV1. Indeed, TRPV1 channels are activated by NO itself, and a TRPV1 mutant with substitutions at these conserved Cys residues exhibits a significantly blunted response to NO. NO also enhances the sensitivity of TRPV1 to H^+ and heat, suggesting that nitrosylation-induced Ca^{2+} entry through TRPV1 is involved in heat and pain sensation (Yoshida et al. 2006).

Injection of (±)-(E)-4-ethyl-2[(E)-hydroxyimino]-S-nitro-3-hexenamide (NOR3) into the mouse hind paw after phospholipase C (PLC) and protein kinase A (PKA) pathway activation (which sensitizes nociceptors, including TRPV1 and TRPA1) causes nociceptive behavior (Miyamoto et al. 2009). Interestingly, a decrease in nociception is observed in mice lacking both TRPV1 and TRPA1 but not in animals where only one of these genes is deleted (Miyamoto et al. 2009), despite in vitro results showing that the NO donor can activate both ion channels. Note that due to solubility issues with NO donors, limitations exist in using NO donors at relatively high concentrations for behavioral experiments ($<$10-fold the in vitro EC_{50}). For capsaicin, injections of capsaicin at $>$1,000$\times$ the in vitro EC_{50} are typically used to observe acute pain behavior. It is therefore not surprising that significant acute nocifensive behavior was not observed following administration of NO donors. Another potential explanation for the lack of phenotype in individual KO mice is due to functional compensation for each other. Indeed, TRPV1 and TRPA1 have overlapping expression in a subset of DRG neurons (Story et al. 2003) and could thereby share all or part of their intracellular signaling pathway in vivo. Recently, icilin, an agonist of TRPA1 and TRPM8, was found to trigger shaking and hyperthermia, which require NO production and NMDA receptor activation (Ding et al. 2008; Werkheiser et al. 2009). In this context, it would be interesting to explore the mechanism underlying concerted regulation of Ca^{2+} and NO signaling by TRP channels, NMDA receptors, and nNOS in neurons.

2.4 Regulation of TRP Channels Via the NO/cGMP/PKG Pathway

A principal consequence of activating the NO/cGMP/PKG cascade in blood vessels is vasorelaxation, which is mediated by phosphorylation of proteins that regulate intracellular Ca^{2+} levels and the sensitivity of the contractile machinery. Several distinct mechanisms have been proposed for the reduction in $[Ca^{2+}]_i$ caused by PKG (Lincoln et al. 2001; Feil et al. 2003). Increased activity of the BK_{Ca} channels following PKG activation has been reported to induce membrane hyperpolarization, thereby decreasing the rate of Ca^{2+} entry into vascular smooth muscle cells (VSMCs) through voltage-dependent Ca^{2+} channels (VDCCs). Phosphorylation of the inositol 1,4,5-trisphosphate receptor (IP_3R)-associated protein, IRAG (IP_3R-associated PKG substrate), is thought to inhibit agonist-induced Ca^{2+} release from internal stores. Phosphorylation of phospholamban by PKG has also been proposed to increase the activity of sarcoplasmic Ca^{2+}-ATPase, facilitating the active transport of Ca^{2+} into internal stores, and thereby decreasing $[Ca^{2+}]_i$. In many different types of blood vessel, activation of PKG has been found to inhibit vasoconstrictor-induced Ca^{2+} influx through plasmalemmal pathways other than those employing VDCCs (Karaki et al. 1997). However, little information exists regarding the molecular components of such Ca^{2+} influx pathways, and, accordingly, it remains unclear how phosphorylation via PKG activation reduces the rate of Ca^{2+} influx.

TRPC6 is a Ca^{2+}-permeable channel regulated negatively by PKG-mediated phosphorylation in the NO/cGMP signaling pathway in heterologous systems and in A7r5 rat vascular myocytes (Takahashi et al. 2008b). Macroscopic and single-channel current recordings using patch-clamp techniques have demonstrated that SNAP-induced inhibition of receptor-activated TRPC6 currents is abolished by pharmacological blockade of cGMP/PKG signaling with 1H-[1,2,4]oxadiazolo [4,3-a]quinoxalin-1-one (ODQ), 2,3,9,10,11,12-hexahydro-10R-methoxy-2,9-dimethyl-1-oxo-9S,12R-epoxy-1H-diindolo[1, 2,3-fg:3′,2′,1′-kl]pyrrolo[3,4-i][1,6] benzodiazocine-10-carboxylic acid, methyl ester (KT5823), or a membrane-permeable PKG inhibitory peptide, DT3. It is also ablated by site-directed alanine mutation of a PKG phosphorylation site [threonine (Thr) 69] within the N-terminal cytoplasmic region of TRPC6. The critical involvement of Thr69 in PKG phosphorylation is confirmed by ^{32}P-incorporation assays using WT and alanine-substitution mutant TRPC6 proteins. Similar negative regulation of TRPC6-like currents by the NO/cGMP/PKG pathway has been observed in A7r5 VSMCs. Indeed, vasopressin-evoked membrane depolarization in these cells, which is expected secondarily to activate VDCCs, is significantly slowed and attenuated after application of SNAP. The TRPC6 protein is abundantly expressed in various types of VSMCs and has been shown to be a constituent of vasoconstrictor-activated cation channels, which increase Ca^{2+} entry into VSMCs via direct Ca^{2+} permeation or by secondary activation of a VDCC and/or Na^{+}–Ca^{2+} exchanger (Inoue et al. 2006; Dietrich et al. 2007; Poburko et al. 2007). Thus, it is highly possible that, in a direct or indirect manner (i.e., via changes in membrane potential or an increase in intracellular Na^{+} concentration), PKG-mediated mechanisms may work as a universal negative feedback to regulate neurohormonal Ca^{2+} mobilization across the VSMC membrane. This mechanism may be physiologically important in vascular tissues where NO is constantly released from vascular endothelial cells or nitrergic nerves.

3 Sensing of RCS by TRP Channels

Another form of posttranslational modifications is the addition of reactive carbonyl species (RCS) to proteins, a process generically termed "protein carbonylation." The most reactive and common form of these carbonyl groups are aldehydes. α,β-unsaturated aldehydes such as 15-deoxy-$\Delta^{12,14}$-prostaglandin J_2 (15d-PGJ_2) and 4-hydroxy-(2*E*)-nonenal (4-HNE) are reactive aldehydes generated from polyunsaturated fatty acid oxidation. Because of the presence of electron-withdrawing functional groups, the double bond of these compounds serves as a site for Michael addition with the sulfur atom of Cys. Through the modification of key Cys sulfhydryls, these electrophiles can mediate various biological actions including functional regulation of the IκB kinase (IKK) β-subunit, the nuclear factor (NF)-κB p65 subunit (Rossi et al. 2000; Straus et al. 2000), kelch-like ECH-associated protein (Keap1; Eggler et al. 2005), thioredoxin (Shibata et al. 2003), and peroxisome proliferator-activated receptor (PPAR)γ (Shiraki et al. 2005). Thus, highly conserved redox reactions of Cys sulfhydryls can be elementary molecular processes in the same way as phosphorylation reactions of threonine, serine, and tyrosine residues are in cell signaling.

Using recombinant expression systems and isolated sensory neurons from TRPA1 KO mice, 15d-PGJ_2 has been shown to activate TRPA1, with a number of N-terminal Cys residues being candidates for the mediation of this effect (Takahashi et al. 2008a; Andersson et al. 2008; Maher et al. 2008; Taylor-Clark et al. 2008b). TRPA1 was originally identified as a sensor of pungent natural compounds, as well as for low temperatures, receptor stimulation, and the presence of cannabinoids (Jordt et al. 2004; Bandell et al. 2004; Bautista et al. 2006). TRPA1-activating pungent compounds include cinnamaldehyde, allyl isothiocyanate and α,β-unsaturated aldehydes (from plants such as mustard, onion, cinnamon, and wasabi), and allicin, the pungent compound from garlic. These compounds are potentially susceptible to nucleophilic attack by a Cys residue sulfhydryl group. More recent investigations of various noxious compounds led to the conclusion that electrophilic compounds that covalently modify Cys residues through mechanisms such as Michael addition are indeed potent activators of TRPA1 channels (Hinman et al. 2006; Macpherson et al. 2007). Other α,β-unsaturated aldehyde derivatives (e.g., the endogenous autacoids 4-oxononenal and 4-HNE) and nitrosylated oleic acid also display activating effects on TRPA1 channels (Trevisani et al. 2007; Taylor-Clark et al. 2008a, 2009). Interestingly, TRPV1 also shows sensitivity to pungent compounds from onion and garlic (e.g., allicin), with the covalent modification of a single Cys residue located in the N-terminal region thought to be pivotal in this phenomenon (Salazar et al. 2008).

4 Sensing of O_2 by TRP Channels

Of the gaseous molecules, O_2 is the most well known to physicians, scientists, and laymen alike as an essential physical requirement. O_2 functions primarily as a terminal acceptor of electrons on mitochondrial electron transport. Most of the O_2 consumed in this process is reduced to generate water through the actions of cytochrome oxidase. The remainder is used to generate compounds that exert potent biological actions, including prostaglandins, ROS, and gaseous molecules such as NO and CO (Suematsu et al. 2003). In the last decade, O_2 itself has been increasingly recognized to mediate many physiological and pathophysiological processes including proliferation of stem cells, ischemia injury, and tumor progression (Csete

2006; Swartz et al. 2007). Thus, O_2 is required not only for cellular respiration but also as a signaling molecule and an essential substrate in the formation of other signaling molecules. However, O_2 also exerts toxicity causing aging, respiratory disorders, and eventually death in a high O_2 (hyperoxic) environment. Because of the ambivalent physiological nature of O_2, aerobic life-forms must adapt to hyperoxic and hypoxic (low O_2) environments by sensing O_2 availability and transmitting this information to adaptive effector systems.

Cellular responses to changes in O_2 availability can be acute or chronic (López-Barneo et al. 2001). Acute responses rely mainly on O_2-regulated ion channels, which mediate adaptive changes in cell excitability, contractility, and secretory activity (Gonzalez et al. 1994; Neubauer and Sunderram 2004; Weir et al. 2005). Chronic responses depend on the modulation of transcription factors such as hypoxia-inducible factor (HIF), which determines the expression of numerous genes encoding growth factors, enzymes, and transporters (Semenza and Wang 1992; Schofield and Ratcliffe 2004; Webb et al. 2009). O_2-regulated ion channels and transcription factors are part of a widely operating signaling system that helps to provide an appropriate amount of O_2 to the tissues and to protect the cells against toxicity damage due to excess or deficient O_2.

In mammals, catecholaminergic neurons in the carotid bodies (located near the carotid artery bifurcations) and brainstem rapidly detect changes in partial O_2 pressure (PO_2) in arterial blood (Gonzalez et al. 1994; Neubauer and Sunderram 2004). It is known that BK_{Ca}, TASK, and K_V K^+ channels are involved in the mechanism of arterial O_2 sensing in the carotid bodies (Gonzalez et al. 1994; Weir et al. 2005). Hypoxia inhibits K^+ channels by several mechanisms, including CO production by heme oxygenase and intracellular ATP depletion that depolarizes glomus cells. This inhibition leads to activation of VDCCs, exocytosis, and the excitation of carotid sinus nerves. In contrast, hyperoxia attenuates depolarization and inhibits exocytosis (Gonzalez et al. 1994; Weir et al. 2005). The chemosensory inputs of the carotid sinus nerve are carried toward the medullary centers that regulate the ventilatory pattern. The local O_2 tension is also rapidly detected by other tissues, including vagal and sensory neurons in the airway and lungs; cerebral neurons; chromaffin cells of the fetal adrenal medulla; and smooth muscle cells in the pulmonary resistance arteries, fetoplacental arteries, systemic arteries, and ductus arteriosus (Lee and Milhorn 1975; Waypa and Schumacker 2010). However, detection of hypoxia by these tissues remains to be fully characterized. Recently, a class of TRP channels has been demonstrated to act as cell-borne sensor of changes in O_2 availability (Aarts et al. 2003; Weissmann et al. 2006; Takahashi et al. 2011; Numata et al. 2013).

4.1 Anoxia-Sensing Mediated by TRPM7 Channels in the Brain

Excitotoxicity during brain ischemia triggers neuronal death and neurological disability, yet these are not prevented by antiexcitotoxic therapy (AET) in humans. The failure of AET in the face of a clear role for excitotoxicity in acute neurological disorders is paradoxical (Birmingham 2002; Ikonomidou and Turski 2002). In addressing this problem, Aarts et al. (2003) have revealed that, in heterologous systems and cortical neurons isolated from rats, the TRPM7 channel (termed a "chanzyme" because it possess a channel and α-kinase domain (Runnels et al. 2001; Nadler et al. 2001)) is activated by oxygen–glucose deprivation (OGD) through the production of ROS and RNS, permitting Ca^{2+} uptake that further stimulates ROS and TRPM7 activation. Suppressing TRPM7 expression in rat cortical neurons prevents anoxic neuronal death even in the absence of AET, indicating that TRPM7 is an essential mediator of anoxic death. It is possible that patients enrolled in failed trials studying the use of AET for stroke or traumatic brain injury are selected to have severe injuries (Morris et al. 1999) or that these disorders in humans, by their nature, induce severe ischemia. Therefore, future treatment of such disorders may also require the inhibition of TRPM7. Indeed, it has been shown that suppression of hippocampal TRPM7 by intrahippocampal injections of viral vectors bearing shRNA specific for TRPM7 renders neurons resistant to ischemic death after brain ischemia and preserves neuronal morphology and function in rats (Sun et al. 2009). TRPM7 suppression also prevents ischemia-induced deficits in long-term potentiation and preserves performance in fear-associated and spatial–navigational memory tasks. Thus, regional suppression of TRPM7 is feasible and well tolerated and inhibits delayed neuronal death in vivo in an animal model.

4.2 Hypoxia-Sensing by TRPC6 Channels in Pulmonary Smooth Muscle Cells

Regional alveolar hypoxia causes local vasoconstriction in the lungs, shifting blood flow from hypoxic to normoxic areas, thereby maintaining gas exchange. This mechanism is known as hypoxic pulmonary vasoconstriction (HPV) (Jeffery and Wanstall 2001; Weissmann et al. 2001; Schermuly et al. 2005). Disturbances in HPV can cause life-threatening hypoxemia, whereas chronic hypoxia triggers vascular remodeling in the lungs and pulmonary hypertension (Sartori et al. 1999). In studying signaling cascades of this vitally important mechanism, Weissmann et al. have shown that cation influx and currents induced in smooth muscle cells by severe hypoxia (1 % O_2) are largely absent in precapillary pulmonary arteries of *Trpc6* KO mice, although recombinant TRPC6 cannot be activated by hypoxia (Weissmann et al. 2006). Hypoxia-induced TRPC6 activation in smooth muscle cells is mediated by DAG accumulation, probably as a result of phospholipase activation. TRPC6 appears to be a key regulator of acute HPV, because this regulatory mechanism is absent in *Trpc6* KO mice, whereas the pulmonary vasoconstrictor response to the thromboxane mimetic, U46619, is unchanged. Accordingly, induction of regional hypoventilation results in severe arterial hypoxemia in *Trpc6* KO mice, but not in WT mice. Notably, chronic hypoxia-induced pulmonary hypertension is independent of TRPC6 activity. Thus, TRPC6 plays a unique and indispensable role in acute HPV. Manipulation of TRPC6 function may thus offer a therapeutic strategy for the control of pulmonary hemodynamics and gas exchange.

Besides the acute HPV, mutations in the *Trpc6* gene are linked with focal segmental glomerulosclerosis (FSGS), a proteinuric disease characterized by dysregulated function of renal glomerular epithelial cells (podocytes) (Reiser et al. 2005; Winn et al. 2005; Kanda et al. 2011). Some studies have indicated that hypoxia affects glomerular cells and plays an important role in the initiation and progression of various renal diseases (Brukamp et al. 2007; Neusser et al. 2010). Kambham et al. (2001) have shown that the combination of focal segmental glomerulosclerosis and glomerulomegaly in obesity-related glomerulopathy resembles the secondary forms of focal segmental glomerulosclerosis arising in conditions of chronic hypoxia. More recently, it has been reported that TRPC6 is upregulated by chronic hypoxia in human mesangial cells, and the upregulation of TRP6 is involved in the Ca^{2+} signaling and actin assembly (Liao et al. 2012). Thus, TRPC6 activation by hypoxia is highly interesting in regard to the initiation and the progression of renal diseases.

4.3 O_2-Sensing by the TRPA1 Channel in Vagal and Sensory Neurons

Sensory and vagal afferent neurons, which project nerve endings throughout the body, are thought to detect hypoxia in organs such as the airway, lungs, and heart after ischemia and other conditions of low O_2 supply (Howe et al. 1981; De Sanctis et al. 1991; Longhurst et al. 2001; Gruss et al. 2006). However, the characteristics and mechanisms by which sensory and vagal neurons detect hypoxia have yet to be fully defined (Longhurst et al. 2001).

Systematic evaluation of TRP channels using reactive disulfides with different redox potentials has revealed that TRPA1 can sense O_2 (Takahashi et al. 2011). O_2 sensing by TRPA1 is based upon proline (Pro) hydroxylation by Pro hydroxylases (PHDs) and direct oxidation of Cys residues. In normoxia, PHDs hydroxylate a conserved proline residue, Pro394, within the tenth AnkR domain of TRPA1, inhibiting its activity. In hypoxia, the decrease in O_2 concentration diminishes PHD activity, activating TRPA1 by relieving it from the inhibitory action of Pro hydroxylation. This recovery of TRPA1 activity can also be achieved by insertion of fresh, unmodified TRPA1 proteins into the plasma membrane or by dehydroxylation of modified proteins, although the mechanism for this is as yet uncharacterized. In hyperoxia, O_2 activates TPRA1 by oxidizing Cys633, Cys856, or both. Cys633 is located within the seventeenth AnkR domain while Cys856 is found in the intracellular linker region between S4 and S5. TRPA1 can assume at least two oxidized forms during hyperoxia: a relatively unstable form (state 1) readily reversed by glutathione and a relatively stable form (state 2). SH groups on the key Cys residues (Cys633 and Cys856) may be modified to sulfenic acid (S-OH) in state 1 and form disulfide bonds (S–S) in state 2. This oxidation mechanism activates TRPA1 by overriding the inhibition of this channel by Pro hydroxylation. In mice, disruption of the *Trpa1* gene abolishes cationic currents induced by hyperoxia and mild hypoxia (15 % O_2) in vagal and sensory neurons and thereby impedes the enhancement of in vivo vagal discharges induced by these stimuli. These results suggest the existence of a novel O_2-sensing mechanism mediated by TRPA1.

The vagal nerve conveys sensory information regarding organ status to the central nervous system, in addition to providing output to these same organs. Enhanced discharges in vagal afferents induce respiratory, cardiac, and vascular responses (Meller and Gebhart 1992; Longhurst et al. 2001; Kubin et al. 2006). Chemicals encountered in the airway are detected by airway vagal C fibers (Kubin et al. 2006). Recently, TRPA1 has been shown to sense environmental irritants, thus initiating defensive reflexes such as coughing and respiratory depression in the C fibers (Bessac and Jordt 2008; Bessac et al. 2008; Nassenstein et al. 2008). Interestingly, TRPA1 activation shows an inverted bell-shaped O_2-dependency curve with its lowest point at a PO_2 of 137 mmHg, slightly below atmospheric PO_2 (159 mmHg) (Takahashi et al. 2011). Considering that tracheal PO_2 (149 mmHg) is comparable to atmospheric PO_2 (Cottrell 1988), it is highly possible that TRPA1 expressed in the trachea is slightly but significantly activated to act as a frontline defense against mild hyperoxia in the atmosphere.

5 Sensing of Other Gaseous Molecules by TRP Channels

The origin of gaseous molecules is shown in Fig. 3. Hydrogen sulfide (H_2S) is increasingly recognized as a gaseous modulator of various biological functions including nociception (Szabó 2007; Gadalla and Snyder 2010). NaHS activates capsaicin-sensitive sensory neurons in isolated rat urinary bladder (Patacchini et al. 2005), and although the mechanisms underlying the action of H_2S are yet to be clarified, its pharmacological profile hints at the involvement of TRPA1. Indeed, Streng et al. (2008) have shown that NaHS causes activation of human and mouse TRPA1 in heterologous expression systems, providing further evidence implicating TRPA1 as the molecular target for H_2S in the bladder. In this context, it is interesting to note that bladder inflammation can be triggered by TRPA1 activation (Cox 1979; McMahon and Abel 1987) and that potential pathogens, such as *Escherichia coli*, can produce H_2S (Berglin and Carlsson 1985). Recently, Miyamoto et al. (2011) have shown that NaHS-evoked increases in $[Ca^{2+}]_i$ are inhibited by the removal of external Ca^{2+} and by the TRPA1-specific inhibitor, HC-030031, suggesting that H_2S stimulates sensory neurons via activation of TRPA1. Furthermore, the hyperalgesia/allodynia induced in mice by intraplantar administration of a low concentration (100 pmol) of NaHS is significantly suppressed by pre-administration of the TRPA1-specific antagonist, AP18, and by silencing TRPA1 channels in sensory neurons (Okubo et al. 2012). Thus, H_2S-induced mechanical hyperalgesia and allodynia require activation of TRPA1 channels.

In humans, gaseous CO_2 produces a pungent sensation, as noted by the Scottish philosopher Alexander Bain more than 100 years ago (Cain and Murphy 1980). A variety of sensory structures and receptors mediate the responses to CO_2 in different

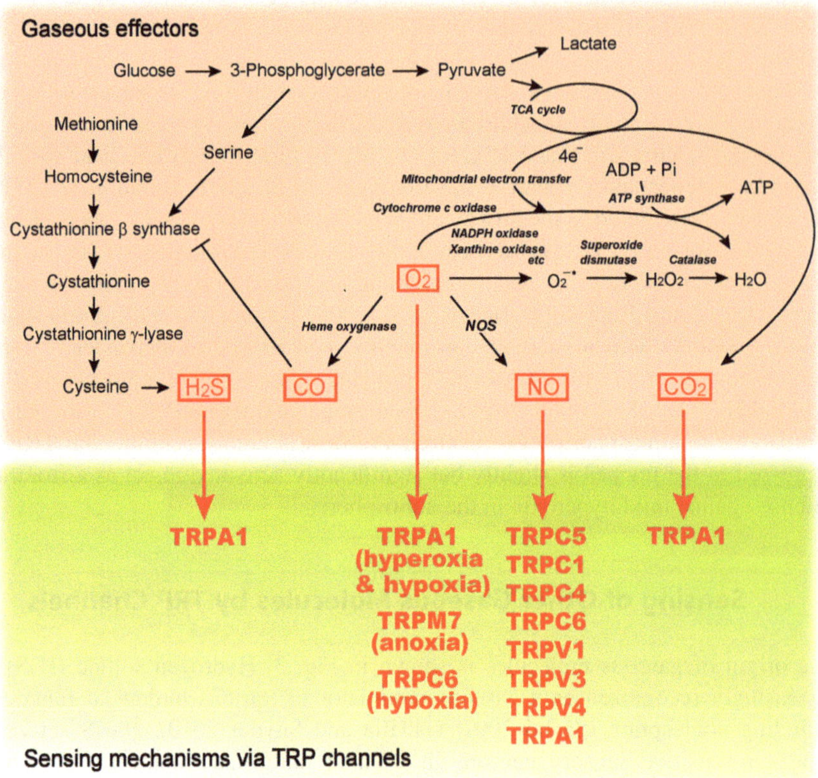

Fig. 3 A sensing mechanism for gaseous molecules linked to metabolic pathways. TRP channels mediate sensing mechanism for H_2S, O_2 (anoxia, hypoxia, or hyperoxia), NO, and CO_2

organisms (Luo et al. 2009). For example, in flies, gaseous CO_2 is detected by gustatory receptors on the antenna, whereas dissolved CO_2 is detected on the proboscis: CO_2 is either aversive or attractive depending on the sensory structure activated (Suh et al. 2004; Fischler et al. 2007; Jones et al. 2007; Kwon et al. 2007). In mice, ingested CO_2 is sensed by taste receptors in the mouth (Chandrashekar et al. 2009), and blood CO_2 is detected by K^+ channels in the brainstem (Trapp et al. 2008) and by acid-sensing ion channel (ASIC)-1a in the amygdala (Ziemann et al. 2009). Atmospheric levels of CO_2 are detected by guanylyl cyclase D, which is expressed in a subset of olfactory sensory neurons in mice (Luo et al. 2009). However, this latter system is not conserved in humans and most other primates (Young et al. 2007). It is widely believed that the noxious sensation of CO_2 is due to activation of the trigeminal nerve fibers that innervate the nasal and oral cavities (Silver and Moulton 1982; Steen et al. 1992). Recently, Wang et al. (2010) have shown that CO_2 specifically activates a subpopulation of trigeminal neurons expressing a functional *Trpa1* gene. CO_2-induced activation of TRPA1 is

Table 1 Activation of TRP homologs by ROS, RNS, RCS, and gasotransmitters

TRPs	Agonists (environment factors)	References
C3	*tert*-Butyl hydroperoxide	Balzer et al. (1999)
C5	H_2O_2	Yoshida et al. (2006)
	NO	Yoshida et al. (2006)
C6	Hypoxia	Weissmann et al. (2006)
M2	H_2O_2	Hara et al. (2002), Kolisek et al. (2005), Perraud et al. (2005)
M7	Anoxia	Aarts et al. (2003)
V1	H_2O_2	Yoshida et al. (2006), Chuang and Lin (2009)
	NO	Yoshida et al. (2006), Miyamoto et al. (2009)
	Allicin	Salazar et al. (2008)
V3	NO	Yoshida et al. (2006)
V4	H_2O_2	Yoshida et al. (2006)
	NO	Yoshida et al. (2006)
A1	H_2O_2	Takahashi et al. (2008a), Sawada et al. (2008), Bessac et al. (2008)
	NO	Sawada et al. (2008), Takahashi et al. (2008a), Miyamoto et al. (2009)
	15d-PGJ_2	Takahashi et al. (2008a), Andersson et al. (2008), Maher et al. (2008), Taylor-Clark et al. (2008b)
	4-Oxononenal	Trevisani et al. (2007), Taylor-Clark et al. (2008a, 2009)
	Hypoxia	Takahashi et al. (2011)
	O_2	Takahashi et al. (2011)
	H_2S	Streng et al. (2008), Miyamoto et al. (2011), Okubo et al. (2012)

downstream of intracellular acidification, consistent with our observation that TRPA1 is activated by H^+ (Takahashi et al. 2008a). Thus, TRPA1 makes an important contribution to nociceptive responses to CO_2.

Conclusions

TRP channels can respond to multiple activation triggers and therefore serve as multimodal signal detectors. As mentioned above, TRP homologs have been demonstrated to act as sensors of ROS, RNS, RCS, and various gaseous messenger molecules (Table 1). Some TRP channels including TRPM2, TRPC5, TRPV1, and TRPA1 can sense ROS-induced redox changes. The TRPM2 channel, which is activated by ROS via production of ADP-ribose as a second messenger, mediates not only cell death but also immune and inflammatory responses. Using *Trpm2* KO mice developed in our laboratory, it has been demonstrated that TRPM2 channel activation aggravates various oxidative stress-related inflammatory diseases such as DSS-induced colitis and I/R injury. However, these observations are occasionally in conflict with results from other groups using independently developed *Trpm2* KO mice. A specific inhibitor of the TRPM2 channel is still not available, but its development would be a useful addition to our armory of tools for the further analysis of the implication between TRPM2 activation and inflammatory diseases. Furthermore, the physiological and/or pathological roles of the other ROS-sensitive TRP channels, which are activated by Cys modification, remain to be elucidated.

An important aspect in the multimodality of the activation sensitivity of TRP channels is its role in signal integration and amplification. When a TRP channel that is part of a specific signaling cascade is activated by downstream or upstream constituents (small molecule messengers/enzymes/scaffolding proteins) of the cascade, in addition to the immediate upstream activation trigger, the cascade is equipped with positive feedback or feed-forward loops. This mechanism, which is capable of ensuring the fidelity of cellular responses and minimizing the variation in their magnitude, may contribute to the synchronization of responses in neighboring cells that comprise functional domains in tissues. For example, vasodilator receptor stimulation in vascular endothelial cells causes Ca^{2+} influx via NO-activated TRPC5 channels that can amplify the production of NO by eNOS, resulting in the enhancement of NO production in nearby endothelial cells and NO-dependent relaxation of smooth muscle cells. The intercellular amplification of NO production should eventually lead to vasodilation synchronized at a whole-tissue level in the local vasculature.

The accumulating evidence summarized here strongly suggests that TRPV1 and TRPA1 are sensors that transduce gaseous signals into electrical signals in sensory and vagal neurons. The chemosensory inputs to these neurons are propagated toward the central nervous system to induce pain sensation or to change ventilatory patterns. However, full understanding of the roles of Ca^{2+} influx via TRPV1 and TRPA1 in controlling Ca^{2+} signaling pathways in neurons remains elusive. Considering that the PHD–HIF pathway is central to chronic hypoxia responses that increase red blood cell mass and stimulate new blood vessel growth (Semenza and Wang 1992; Webb et al. 2009), it would be interesting to examine the effect of Ca^{2+} influx through O_2-sensitive TRPA1 channels on the functional regulation of PHD and HIF. Studies of TRP channels have been extended dramatically from the simple functional description of single molecules to the holistic analysis and integration of the molecular systems controlled by TRP channels. Our understanding of the activation of TRP channels is now stepping forward from functional identification of single molecules to analysis and integration of molecular systems.

References

Aarts M, Iihara K, Wei WL, Xiong ZG, Arundine M, Cerwinski W, MacDonald JF, Tymianski M (2003) A key role for TRPM7 channels in anoxic neuronal death. Cell 115:863–877

Akama KT, Albanese C, Pestell RG, Van Eldik LJ (1998) Amyloid β-peptide stimulates nitric oxide production in astrocytes through an NFκB-dependent mechanism. Proc Natl Acad Sci U S A 95:5795–5800

Andersson DA, Gentry C, Moss S, Bevan S (2008) Transient receptor potential A1 is a sensory receptor for multiple products of oxidative stress. J Neurosci 28:2485–2494

Balzer M, Lintschinger B, Groschner K (1999) Evidence for a role of Trp proteins in the oxidative stress-induced membrane conductances of porcine aortic endothelial cells. Cardiovasc Res 42:543–549

Bandell M, Story GM, Hwang SW, Viswanath V, Eid SR, Petrus MJ, Earley TJ, Patapoutian A (2004) Noxious cold ion channel TRPA1 is activated by pungent compounds and bradykinin. Neuron 41:849–857

Bautista DM, Jordt SE, Nikai T, Tsuruda PR, Read AJ, Poblete J, Yamoah EN, Basbaum AI, Julius D (2006) TRPA1 mediates the inflammatory actions of environmental irritants and proalgesic agents. Cell 124:1269–1282

Benhar M, Forrester MT, Stamler JS (2009) Protein denitrosylation: enzymatic mechanisms and cellular functions. Nat Rev Mol Cell Biol 10:721–732

Berglin EH, Carlsson J (1985) Potentiation by sulfide of hydrogen peroxide-induced killing of *Escherichia coli*. Infect Immun 49:538–543

Bessac BF, Jordt SE (2008) Breathtaking TRP channels: TRPA1 and TRPV1 in airway chemosensation and reflex control. Physiology 23:360–370

Bessac BF, Sivula M, von Hehn CA, Escalera J, Cohn L, Jordt SE (2008) TRPA1 is a major oxidant sensor in murine airway sensory neurons. J Clin Invest 118:1899–1910

Birmingham K (2002) Future of neuroprotective drugs in doubt. Nat Med 8:5

Brukamp K, Jim B, Moeller MJ, Haase VH (2007) Hypoxia and podocyte-specific Vhlh deletion confer risk of glomerular disease. Am J Physiol Renal Physiol 293:F1397–F1407

Buelow B, Song Y, Scharenberg AM (2008) The poly(ADP-ribose) polymerase PARP-1 is required for oxidative stress-induced TRPM2 activation in lymphocytes. J Biol Chem 283:24571–24583

Cain WS, Murphy CL (1980) Interaction between chemoreceptive modalities of odour and irritation. Nature 284:255–257

Carrozza JP Jr, Bentivegna LA, Williams CP, Kuntz RE, Grossman W, Morgan JP (1992) Decreased myofilament responsiveness in myocardial stunning follows transient calcium overload during ischemia and reperfusion. Circ Res 71:1334–1340

Chandra J, Samali A, Orrenius S (2000) Triggering and modulation of apoptosis by oxidative stress. Free Radic Biol Med 29:323–333

Chandrashekar J, Yarmolinsky D, von Buchholtz L, Oka Y, Sly W, Ryba NJ, Zuker CS (2009) The taste of carbonation. Science 326:443–445

Chiu J, Dawes IW (2012) Redox control of cell proliferation. Trends Cell Biol 22:592–601

Chuang HH, Lin S (2009) Oxidative challenges sensitize the capsaicin receptor by covalent cysteine modification. Proc Natl Acad Sci U S A 106:20097–20102

Clapham DE, Nilius B, Owsianik G (2013) Transient receptor potential channels. Last modified on 27/03/2013. IUPHAR database (IUPHAR-DB), http://www.iuphar-db.org/DATABASE/FamilyMenuForward?familyId=78. Accessed on 16 May 2013

Cottrell JJ (1988) Altitude exposures during aircraft flight. Flying higher. Chest 93:81–84

Cox PJ (1979) Cyclophosphamide cystitis–identification of acrolein as the causative agent. Biochem Pharmacol 28:2045–2049

Coyle JT, Puttfarcken P (1993) Oxidative stress, glutamate, and neurodegenerative disorders. Science 262:689–695

Csete M (2006) Oxygen in the cultivation of stem cells. Ann N Y Acad Sci 1049:1–8

de Murcia G, Menissier de Murcia J (1994) Poly(ADP-ribose) polymerase: a molecular nick-sensor. Trends Biochem Sci 19:172–176

De Sanctis GT, Green FH, Remmers JE (1991) Ventilatory responses to hypoxia and hypercapnia in awake rats pretreated with capsaicin. J Appl Physiol 70:1168–1174

del Camino D, Yellen G (2001) Tight steric closure at the intracellular activation gate of a voltage-gated K^+ channel. Neuron 32:649–656

Dickson DW, Farlo J, Davies P, Crystal H, Fuld P, Yen SH (1988) Alzheimer's disease. A double-labeling immunohistochemical study of senile plaques. Am J Pathol 132:86–101

Dietrich A, Kalwa H, Fuchs B, Grimminger F, Weissmann N, Gudermann T (2007) In vivo TRPC functions in the cardiopulmonary vasculature. Cell Calcium 42:233–244

Ding Z, Gomez T, Werkheiser JL, Cowan A, Rawls SM (2008) Icilin induces a hyperthermia in rats that is dependent on nitric oxide production and NMDA receptor activation. Eur J Pharmacol 578:201–208

Dong Z, Saikumar P, Weinberg JM, Venkatachalam MA (2006) Calcium in cell injury and death. Annu Rev Pathol 1:405–434

Dröge W (2000) Free radicals in the physiological control of cell function. Physiol Rev 82:47–95

Eberhardt MJ, Filipovic MR, Leffler A, de la Roche J, Kistner K, Fischer MJ, Fleming T, Zimmermann K, Ivanovic-Burmazovic I, Nawroth PP, Bierhaus A, Reeh PW, Sauer SK (2012) Methylglyoxal activates nociceptors through transient receptor potential channel A1 (TRPA1): a possible mechanism of metabolic neuropathies. J Biol Chem 287:28291–28306

Eggler AL, Liu G, Pezzuto JM, van Breemen RB, Mesecar AD (2005) Modifying specific cysteines of the electrophile-sensing human Keap1 protein is insufficient to disrupt binding to the Nrf2 domain Neh2. Proc Natl Acad Sci U S A 102:10070–10075

Feil R, Lohmann SM, de Jonge H, Walter U, Hofmann F (2003) Cyclic GMP-dependent protein kinases and the cardiovascular system: insights from genetically modified mice. Circ Res 93:907–916

Fischler W, Kong P, Marella S, Scott K (2007) The detection of carbonation by the Drosophila gustatory system. Nature 448:1054–1057

Fonfria E, Marshall IC, Benham CD, Boyfield I, Brown JD, Hill K, Hughes JP, Skaper SD, McNulty S (2004) TRPM2 channel opening in response to oxidative stress is dependent on activation of poly(ADP-ribose) polymerase. Br J Pharmacol 143:186–192

Fonfria E, Marshall IC, Boyfield I, Skaper SD, Hughes JP, Owen DE, Zhang W, Miller BA, Benham CD, McNulty S (2005) Amyloid beta-peptide(1-42) and hydrogen peroxide-induced toxicity are mediated by TRPM2 in rat primary striatal cultures. J Neurochem 95:715–23

Furchgott RF, Zawadzki JV (1980) The obligatory role of endothelial cells in the relaxation of arterial smooth muscle by acetylcholine. Nature 288:373–376

Gadalla MM, Snyder SH (2010) Hydrogen sulfide as a gasotransmitter. J Neurochem 113:14–26

Gaudet R (2008) A primer on ankyrin repeat function in TRP channels and beyond. Mol BioSyst 4:372–379

Giroux C, Scatton B (1996) Ischemic stroke: treatment on the horizon. Eur Neurol 36:61–64

Gonzalez C, Almaraz L, Obeso A, Rigual R (1994) Carotid body chemoreceptors: from natural stimuli to sensory discharges. Physiol Rev 74:829–898

Goodwin JL, Kehrli ME Jr, Uemura E (1997) Integrin Mac-1 and β-amyloid in microglial release of nitric oxide. Brain Res 768:279–286

Gopalakrishna R, Jaken S (2000) Protein kinase C signaling and oxidative stress. Free Radic Biol Med 28:1349–1361

Gruss M, Ettorre G, Stehr AJ, Henrich M, Hempelmann G, Scholz A (2006) Moderate hypoxia influences excitability and blocks dendrotoxin sensitive K^+ currents in rat primary sensory neurones. Mol Pain 2:12

Halliwell B (2006) Oxidative stress and neurodegeneration: where are we now? J Neurochem 97:1634–1658

Hara Y, Wakamori M, Ishii M, Maeno E, Nishida M, Yoshida T, Yamada H, Shimizu S, Mori E, Kudoh J, Shimizu N, Kurose H, Okada Y, Imoto K, Mori Y (2002) LTRPC2 Ca^{2+}-permeable channel activated by changes in redox status confers susceptibility to cell death. Mol Cell 9:163–173

Hess DT, Matsumoto A, Kim SO, Marshall HE, Stamler JS (2005) Protein S-nitrosylation: purview and parameters. Nat Rev Mol Cell Biol 6:150–166

Hinman A, Chuang HH, Bautista DM, Julius D (2006) TRP channel activation by reversible covalent modification. Proc Natl Acad Sci U S A 103:19564–19568

Hiroi T, Wajima T, Negoro T, Ishii M, Nakano Y, Kiuchi Y, Mori Y, Shimizu S (2013) Neutrophil TRPM2 channels are implicated in the exacerbation of myocardial ischaemia/reperfusion injury. Cardiovasc Res 97:271–81

Howe A, Pack RJ, Wise JC (1981) Arterial chemoreceptor-like activity in the abdominal vagus of the rat. J Physiol 320:309–318

Ikonomidou C, Turski L (2002) Why did NMDA receptor antagonists fail clinical trials for stroke and traumatic brain injury? Lancet Neurol 1:383–386

Inoue R, Jensen LJ, Shi J, Morita H, Nishida M, Honda A, Ito Y (2006) Transient receptor potential channels in cardiovascular function and disease. Circ Res 99:119–131

Jaffrey SR, Erdjument-Bromage H, Ferris CD, Tempst P, Snyder SH (2001) Protein S-nitrosylation: a physiological signal for neuronal nitric oxide. Nat Cell Biol 3:193–197

Jeffery TK, Wanstall JC (2001) Pulmonary vascular remodeling: a target for therapeutic intervention in pulmonary hypertension. Pharmacol Ther 92:1–20

Jones WD, Cayirlioglu P, Kadow IG, Vosshall LB (2007) Two chemosensory receptors together mediate carbon dioxide detection in Drosophila. Nature 445:86–90

Jordt SE, Bautista DM, Chuang HH, McKemy DD, Zygmunt PM, Höqestätt ED, Meng ID, Julius D (2004) Mustard oils and cannabinoids excite sensory nerve fibres through the TRP channel ANKTM1. Nature 427:260–265

Kambham N, Markowitz GS, Valeri AM, Lin J, D'Agati VD (2001) Obesity-related glomerulopathy: an emerging epidemic. Kidney Int 59:1498–1509

Kanda S, Harita Y, Shibagaki Y, Sekine T, Igarashi T, Inoue T, Hattori S (2011) Tyrosine phosphorylation-dependent activation of TRPC6 regulated by PLC-γ1 and nephrin: effect of mutations associated with focal segmental glomerulosclerosis. Mol Biol Cell 22:1824–1835

Kaneko S, Kawakami S, Hara Y, Wakamori M, Itoh E, Minami T, Takada Y, Kume T, Katsuki H, Mori Y, Akaike A (2006) A critical role of TRPM2 in neuronal cell death by hydrogen peroxide. J Pharmacol Sci 101:66–76

Karaki H, Ozaki H, Hori M, Mitsui-Saito M, Amano K, Harada K, Miyamoto S, Nakazawa H, Won KJ, Sato K (1997) Calcium movements, distribution, and functions in smooth muscle. Pharmacol Rev 49:157–230

Katano M, Numata T, Aguan K, Hara Y, Kiyonaka S, Yamamoto S, Miki T, Sawamura S, Suzuki T, Yamakawa K, Mori Y (2012) The juvenile myoclonic epilepsy-related protein EFHC1 interacts with the redox-sensitive TRPM2 channel linked to cell death. Cell Calcium 51:179–185

Kietzmann T, Fandrey J, Acker H (2000) Oxygen radicals as messengers in oxygen-dependent gene expression. New Physiol Sci 15:202–208

Knowles H, Heizer JW, Li Y, Chapman K, Ogden CA, Andreasen K, Shapland E, Kucera G, Mogan J, Humann J, Lenz LL, Morrison AD, Perraud AL (2011) Transient receptor potential melastatin 2 (TRPM2) ion channel is required for innate immunity against listeria monocytogenes. Proc Natl Acad Sci U S A 108:11578–83

Knowles H, Li Y, Perraud AL (2013) The TRPM2 ion channel, an oxidative stress and metabolic sensor regulating innate immunity and inflammation. Immunol Res 55:241–248

Kolisek M, Beck A, Fleig A, Penner R (2005) Cyclic ADP-ribose and hydrogen peroxide synergize with ADP-ribose in the activation of TRPM2 channels. Mol Cell 18:61–69

Kubin L, Alheid GF, Zuperku EJ, McCrimmon DR (2006) Central pathways of pulmonary and lower airway vagal afferents. J Appl Physiol 101:618–627

Kulkarni AC, Kuppusamy P, Parinandi N (2007) Oxygen, the lead actor in the pathophysiologic drama: enactment of the trinity of normoxia, hypoxia, and hyperoxia in disease and therapy. Antioxid Redox Signal 9:1717–1730

Kwon JY, Dahanukar A, Weiss LA, Carlson JR (2007) The molecular basis of CO_2 reception in Drosophila. Proc Natl Acad Sci U S A 104:3574–3578

Langley B, Ratan RR (2004) Oxidative stress-induced death in the nervous system: cell cycle dependent or independent? J Neurosci Res 77:621–629

Lee LY, Milhorn HT Jr (1975) Central ventilatory responses to O_2 and CO_2 at three levels of carotid chemoreceptor stimulation. Respir Physiol 25:319–333

Liao C, Yang H, Zhang R, Sun H, Zhao B, Gao C, Zhu F, Jiao J (2012) The upregulation of TRPC6 contributes to Ca^{2+} signaling and actin assembly in human mesangial cells after chronic hypoxia. Biochem Biophys Res Commun 421:750–756

Lincoln TM, Dey N, Sellak H (2001) Invited review: cGMP-dependent protein kinase signaling mechanisms in smooth muscle: from the regulation of tone to gene expression. J Appl Physiol 91:1421–1430

Longhurst JC, Tjen-A-Looi SC, Fu LW (2001) Cardiac sympathetic afferent activation provoked by myocardial ischemia and reperfusion. Mechanisms and reflexes. Ann N Y Acad Sci 940:74–95

López-Barneo J, Pardal R, Ortega-Sáenz P (2001) Cellular mechanism of oxygen sensing. Annu Rev Physiol 63:259–287

Luo M, Sun L, Hu J (2009) Neural detection of gases–carbon dioxide, oxygen–in vertebrates and in vertebrates. Curr Opin Neurobiol 19:354–361

Macpherson LJ, Dubin AE, Evans MJ, Marr F, Schultz PG, Cravatt BF, Patapoutian A (2007) Noxious compounds activate TRPA1 ion channels through covalent modification of cysteines. Nature 445:541–545

Maher M, Ao H, Banke T, Nasser N, Wu NT, Breitenbucher JG, Chaplan SR, Wickenden AD (2008) Activation of TRPA1 by farnesyl thiosalicylic acid. Mol Pharmacol 73:1225–1234

Marchesi VT (2011) Alzheimer's dementia begins as a disease of small blood vessels, damaged by oxidative-induced inflammation and dysregulated amyloid metabolism: implications for early detection and therapy. FASEB J 25:5–13

Mattson MP, Barger SW, Cheng B, Lieberburg I, Smith-Swintosky VL, Rydel RE (1993) β-Amyloid precursor protein metabolites and loss of neuronal Ca^{2+} homeostasis in Alzheimer's disease. Trends Neurosci 16:409–414

McMahon SB, Abel C (1987) A model for the study of visceral pain states: chronic inflammation of the chronic decerebrate rat urinary bladder by irritant chemicals. Pain 28:109–127

Meller ST, Gebhart GF (1992) A critical review of the afferent pathways and the potential chemical mediators involved in cardiac pain. Neuroscience 48:501–524

Miller BA, Wang J, Hirschler-Laszkiewicz I, Gao E, Song J, Zhang XQ, Koch WJ, Madesh M, Mallilankaraman K, Gu T, Chen SJ, Keefer K, Conrad K, Feldman AM, Cheung JY (2013) The second member of transient receptor potential-melastatin channel family protects hearts from ischemia-reperfusion injury. Am J Physiol Heart Circ Physiol 304:H1010–H1022

Miyamoto T, Dubin AE, Petrus MJ, Patapoutian A (2009) TRPV1 and TRPA1 mediate peripheral nitric oxide-induced nociception in mice. PLoS One 4:e7596. doi:10.1371/journal.pone.0007596

Miyamoto R, Otsuguro K, Ito S (2011) Time- and concentration-dependent activation of TRPA1 by hydrogen sulfide in rat DRG neurons. Neurosci Lett 499:137–142

Morris GF, Bullock R, Marshall SB, Marmarou A, Maas A, Marshall LF (1999) Failure of the competitive N-methyl-D-aspartate antagonist Selfotel (CGS 19755) in the treatment of severe head injury: results of two phase III clinical trials. The Selfotel investigators. J Neurosurg 91:737–743

Nadler MJ, Hermosura MC, Inabe K, Perraud AL, Zhu Q, Stokes AJ, Kurosaki T, Kinet JP, Penner R, Scharenberg AM, Fleig A (2001) LTRPC7 is a Mg · ATP-regulated divalent cation channel required for cell viability. Nature 411:590–595

Nassenstein C, Kwong K, Taylor-Clark T, Kollarik M, Macglashan DM, Braun A, Undem BJ (2008) Expression and function of the ion channel TRPA1 in vagal afferent nerves innervating mouse lungs. J Physiol 586:1595–1604

Neubauer JA, Sunderram J (2004) Oxygen-sensing neurons in the central nervous system. J Appl Physiol 96:367–374

Neusser MA, Lindenmeyer MT, Moll AG, Segerer S, Edenhofer I, Sen K, Stiehl DP, Kretzler M, Gröne HJ, Schlöndorff D, Cohen CD (2010) Human nephrosclerosis triggers a hypoxia-related glomerulopathy. Am J Pathol 176:594–607

Nilius B, Appendino G, Owsianik G (2012) The transient receptor potential channel TRPA1: from gene to pathophysiology. Pflugers Arch Eur J Physiol 464:425–458

Numata T, Ogawa N, Takahashi N, Mori Y (2013) TRP channels as sensors of oxygen availability. Pflugers Arch Eur J Physiol 465:1075–85. doi:10.1007/s00424-013-1237-9

Okubo K, Matsumura M, Kawaishi Y, Aoki Y, Matsunami M, Okawa Y, Sekiguchi F, Kawabata A (2012) Hydrogen sulfide-induced mechanical hyperalgesia and allodynia require activation of both Ca_V 3.2 and TRPA1 channels in mice. Br J Pharmacol 166:1738–1743

Oliver FJ, Menissier-de Murcia J, Nacci C, Decker P, Andriantsitohaina R, Muller S, de la Rubia G, Stoclet JC, de Murcia G (1999) Resistance to endotoxic shock as a consequence of defective NF-κB activation in poly (ADP-ribose) polymerase-1 deficient mice. EMBO J 18:4446–4454

Orrenius S, Zhivotovsky B, Nicotera P (2003) Regulation of cell death: the calcium-apoptosis link. Nat Rev Mol Cell Biol 4:552–565

Palmer RM, Ashton DS, Moncada S (1988) Vascular endothelial cells synthesize nitric oxide from L-arginine. Nature 333:664–666

Patacchini R, Santicioli P, Giuliani S, Maggi CA (2005) Pharmacological investigation of hydrogen sulfide (H_2S) contractile activity in rat detrusor muscle. Eur J Pharmacol 509:171–177

Perraud AL, Fleig A, Dunn CA, Bagley LA, Launay P, Schmitz C, Stokes AJ, Zhu Q, Bessman MJ, Penner R, Kinet JP, Scharenberg AM (2001) ADP-ribose gating of the calcium-permeable LTRPC2 channel revealed by Nudix motif homology. Nature 411:595–599

Perraud AL, Schmitz C, Scharenberg AM (2003) TRPM2 Ca^{2+} permeable cation channels: from gene to biological function. Cell Calcium 33:519–531

Perraud AL, Takanishi CL, Shen B, Kang S, Smith MK, Schmitz C, Knowles HM, Ferraris D, Li W, Zhang J, Stoddard BL, Scharenberg AM (2005) Accumulation of free ADP-ribose from mitochondria mediates oxidative stress-induced gating of TRPM2 cation channels. J Biol Chem 280:6138–6148

Poburko D, Liao CH, Lemos VS, Lin E, Maruyama Y, Cole WC, van Breemen C (2007) Transient receptor potential channel 6-mediated, localized cytosolic [Na^+] transients drive Na^+/Ca^{2+} exchanger-mediated Ca^{2+} entry in purinergically stimulated aorta smooth muscle cells. Circ Res 101:1030–1038

Reiser J, Polu KR, Möller CC, Kenlan P, Altintas MM, Wei C, Faul C, Herbert S, Villegas I, Avila-Casado C, McGee M, Sugimoto H, Brown D, Kalluri R, Mundel P, Smith PL, Clapham DE, Pollak MR (2005) TRPC6 is a glomerular slit diaphragm-associated channel required for normal renal function. Nat Genet 37:739–744

Reiter TA (2006) NO · chemistry: a diversity of targets in the cell. Redox Rep 11:194–206

Romson JL, Hook BG, Kunkel SL, Abrams GD, Schork MA, Lucchesi BR (1983) Reduction of the extent of ischemic myocardial injury by neutrophil depletion in the dog. Circulation 67:1016–1023

Rossi F, Bianchini E (1996) Synergistic induction of nitric oxide by β-amyloid and cytokines in astrocytes. Biochem Biophys Res Commun 225:474–478

Rossi A, Kapahi P, Natoli G, Takahashi T, Chen Y, Karin M, Santoro MG (2000) Anti-inflammatory cyclopentenone prostaglandins are direct inhibitors of IkappaB kinase. Nature 403:103–108

Runnels LW, Yue L, Clapham DE (2001) TRP-PLIK, a bifunctional protein with kinase and ion channel activities. Science 291:1043–1047

Sakuma I, Stuehr DJ, Gross SS, Nathan C, Levi R (1988) Identification of arginine as a precursor of endothelium-derived relaxing factor. Proc Natl Acad Sci U S A 85:8664–8667

Salazar H, Llorente I, Jara-Oseguera A, García-Villegas R, Munari M, Gordon SE, Islas LD, Rosenbaum T (2008) A single N-terminal cysteine in TRPV1 determines activation by pungent compounds from onion and garlic. Nat Neurosci 11:255–261

Sartori C, Allemann Y, Trueb L, Delabays A, Nicod P, Scherrer U (1999) Augmented vasoreactivity in adult life associated with perinatal vascular insult. Lancet 353:2205–2207
Satoh T, Lipton SA (2007) Redox regulation of neuronal survival mediated by electrophilic compounds. Trends Neurosci 30:37–45
Sawada Y, Hosokawa H, Matsumura K, Kobayashi S (2008) Activation of transient receptor potential ankyrin 1 by hydrogen peroxide. Eur J Neurosci 27:1131–1142
Schermuly RT, Dony E, Ghofrani HA, Pullamsetti S, Savai R, Roth M, Sydykov A, Lai YJ, Weissmann N, Seeger W, Grimminger F (2005) Reversal of experimental pulmonary hypertension by PDGF inhibition. J Clin Invest 115:2811–2821
Schofield CJ, Ratcliffe PJ (2004) Oxygen sensing by HIF hydroxylases. Nat Rev Mol Cell Biol 5:343–354
Semenza GL, Wang GL (1992) A nuclear factor induced by hypoxia via de novo protein synthesis binds to the human erythropoietin gene enhancer at a site required for transcriptional activation. Mol Cell Biol 12:5447–5454
Shibata T, Yamada T, Ishii T, Kumazawa S, Nakamura H, Masutani H, Yodoi J, Uchida K (2003) Thioredoxin as a molecular target of cyclopentenone prostaglandins. J Biol Chem 278:26046–26054
Shiraki T, Kamiya N, Shiki S, Kodama TS, Kakizuka A, Jingami H (2005) Alpha, beta-unsaturated ketone is a core moiety of natural ligands for covalent binding to peroxisome proliferator-activated receptor gamma. J Biol Chem 280:14145–14153
Silver WL, Moulton DG (1982) Chemosensitivity of rat nasal trigeminal receptors. Physiol Behav 28:927–931
Steen KH, Reeh PW, Anton F, Handwerker HO (1992) Protons selectively induce lasting excitation and sensitization to mechanical stimulation of nociceptors in rat skin, in vitro. J Neurosci 12:86–95
Story GM, Peier AM, Reeve AJ, Eid SR, Mosbacher J, Hricik TR, Earley TJ, Hergarden AC, Andersson DA, Hwang SW, McIntyre P, Jegla T, Bevan S, Patapoutian A (2003) ANKTM1, a TRP-like channel expressed in nociceptive neurons, is activated by cold temperatures. Cell 112:819–829
Straus DS, Pascual G, Li M, Welch JS, Ricote M, Hsiang CH, Sengchanthalangsy LL, Ghosh G, Glass CK (2000) 15-deoxy-delta 12,14-prostaglandin J_2 inhibits multiple steps in the NF-κB signaling pathway. Proc Natl Acad Sci U S A 97:4844–4849
Streng T, Axelsson HE, Hedlund P, Andersson DA, Jordt SE, Bevan S, Andersson KE, Högestätt ED, Zygmunt PM (2008) Distribution and function of the hydrogen sulfide-sensitive TRPA1 ion channel in rat urinary bladder. Eur Urol 53:391–400
Suematsu M, Suganuma K, Kashiwagi S (2003) Mechanistic probing of gaseous signal transduction in microcirculation. Antioxid Redox Signal 5:485–492
Suh GS, Wong AM, Hergarden AC, Wang JW, Simon AF, Benzer S, Axel R, Anderson DJ (2004) A single population of olfactory sensory neurons mediates an innate avoidance behaviour in Drosophila. Nature 431:854–859
Sun HS, Jackson MF, Martin LJ, Jansen K, Teves L, Cui H, Kiyonaka S, Mori Y, Jones M, Forder JP, Golde TE, Orser BA, Macdonald JF, Tymianski M (2009) Suppression of hippocampal TRPM7 protein prevents delayed neuronal death in brain ischemia. Nat Neurosci 12:1300–1307
Swartz HM, Khan N, Khramtsov VV (2007) Use of electron paramagnetic resonance spectroscopy to evaluate the redox state in vivo. Antioxid Redox Signal 9:1757–1771
Szabó C (2007) Hydrogen sulphide and its therapeutic potential. Nat Rev Drug Discov 6:917–935
Takahashi N, Mizuno Y, Kozai D, Yamamoto S, Kiyonaka S, Shibata T, Uchida K, Mori Y (2008a) Molecular characterization of TRPA1 channel activation by cysteine-reactive inflammatory mediators. Channels (Austin) 2:287–298
Takahashi S, Lin H, Geshi N, Mori Y, Kawarabayashi Y, Takami N, Mori MX, Honda A, Inoue R (2008b) Nitric oxide-cGMP-protein kinase G pathway negatively regulates vascular transient receptor potential channel TRPC6. J Physiol 586:4209–4223

Takahashi N, Kuwaki T, Kiyonaka S, Numata T, Kozai D, Mizuno Y, Yamamoto S, Naito S, Knevels E, Carmeliet P, Oga T, Kaneko S, Suga S, Nokami T, Yoshida J, Mori Y (2011) TRPA1 underlies a sensing mechanism for O_2. Nat Chem Biol 7:701–711

Tanuma S, Yagi T, Johnson GS (1985) Endogenous ADP ribosylation of high mobility group proteins 1 and 2 and histone H1 following DNA damage in intact cells. Arch Biochem Biophys 237:38–42

Taylor-Clark TE, McAlexander MA, Nassenstein C, Sheardown SA, Wilson S, Thornton J, Carr MJ, Undem BJ (2008a) Relative contributions of TRPA1 and TRPV1 channels in the activation of vagal bronchopulmonary C-fibres by the endogenous autacoid 4-oxononenal. J Physiol 586:3447–3459

Taylor-Clark TE, Undem BJ, Macglashan DW Jr, Ghatta S, Carr MJ, McAlexander MA (2008b) Prostaglandin-induced activation of nociceptive neurons via direct interaction with transient receptor potential A1 (TRPA1). Mol Pharmacol 73:274–281

Taylor-Clark TE, Ghatta S, Bettner W, Undem BJ (2009) Nitrooleic acid, an endogenous product of nitrative stress, activates nociceptive sensory nerves via the direct activation of TRPA1. Mol Pharmacol 75:820–829

Trapp S, Aller MI, Wisden W, Gourine AV (2008) A role for TASK-1 (KCNK3) channels in the chemosensory control of breathing. J Neurosci 28:8844–8850

Trevisani M, Siemens J, Materazzi S, Bautista DM, Nassini R, Campi B, Imamachi N, Andrè E, Patacchini R, Cottrell GS, Gatti R, Basbaum AI, Bunnett NW, Julius D, Geppetti P (2007) 4-Hydroxynonenal, an endogenous aldehyde, causes pain and neurogenic inflammation through activation of the irritant receptor TRPA1. Proc Natl Acad Sci U S A 104:13519–13524

Vinten-Johansen J (2004) Involvement of neutrophils in the pathogenesis of lethal myocardial reperfusion injury. Cardiovasc Res 61:481–497

Virag L, Szabo C (2002) The therapeutic potential of poly(ADP-ribose) polymerase inhibitors. Pharmacol Rev 54:375–429

Walker DG, Kim SU, McGeer PL (1995) Complement and cytokine gene expression in cultured microglial derived from postmortem human brains. J Neurosci Res 40:478–493

Wallace MN, Geddes JG, Farquhar DA, Masson MR (1997) Nitric oxide synthase in reactive astrocytes adjacent to β-amyloid plaques. Exp Neurol 144:266–272

Wang YY, Chang RB, Liman ER (2010) TRPA1 is a component of the nociceptive response to CO_2. J Neurosci 30:12958–12963

Waypa GB, Schumacker PT (2010) Hypoxia-induced changes in pulmonary and systemic vascular resistance: where is the O_2 sensor? Respir Physiol Neurobiol 174:201–211

Webb JD, Coleman ML, Pugh CW (2009) Hypoxia, hypoxia-inducible factors (HIF). HIF hydroxylases and oxygen sensing. Cell Mol Life Sci 66:3539–3554

Wehage E, Eisfeld J, Heiner I, Jüngling E, Zitt C, Lückhoff A (2002) Activation of the cation channel long transient receptor potential channel 2 (LTRPC2) by hydrogen peroxide. A splice variant reveals a mode of activation independent of ADP-ribose. J Biol Chem 277:23150–23156

Wehrhahn J, Kraft R, Harteneck C, Hauschildt S (2010) Transient receptor potential melastatin 2 is required for lipopolysaccharide-induced cytokine production in human monocytes. J Immunol 184:2386–93

Weir EK, López-Barneo J, Buckler KJ, Archer SL (2005) Acute oxygen-sensing mechanisms. N Engl J Med 353:2042–2055

Weissmann N, Winterhalder S, Nollen M, Voswinckel R, Quanz K, Ghofrani HA, Schermuly RT, Seeger W, Grimminger F (2001) NO and reactive oxygen species are involved in biphasic hypoxic vasoconstriction of isolated rabbit lungs. Am J Physiol Lung Cell 280:L638–L645

Weissmann N, Dietrich A, Fuchs B, Kalwa H, Ay M, Dumitrascu R, Olschewski A, Storch U, Mederosy Schnitzler M, Ghofrani HA, Schermuly RT, Pinkenburg O, Seeger W, Grimminger F, Gudermann T (2006) Classical transient receptor potential channel 6 (TRPC6) is essential for hypoxic pulmonary vasoconstriction and alveolar gas exchange. Proc Natl Acad Sci U S A 103:19093–19098

Werkheiser J, Cowan A, Gomez T, Henry C, Parekh S, Chau S, Baron DA, Rawls SM (2009) Icilin-induced wet-dog shakes in rats are dependent on NMDA receptor activation and nitric oxide production. Pharmacol Biochem Behav 92:543–548

Winn MP, Conlon PJ, Lynn KL, Farrington MK, Creazzo T, Hawkins AF, Daskalakis N, Kwan SY, Ebersviller S, Burchette JL, Pericak-Vance MA, Howell DN, Vance JM, Rosenberg PB (2005) A mutation in the TRPC6 cation channel causes familial focal segmental glomerulosclerosis. Science 308:1801–1804

Wong CO, Sukumar P, Beech DJ, Yao X (2010) Nitric oxide lacks direct effect on TRPC5 channels but suppresses endogenous TRPC5-containing channels in endothelial cells. Pflugers Arch 460:121–130

Xu SZ, Sukumar P, Zeng F, Li J, Jairaman A, English A, Naylor J, Ciurtin C, Majeed Y, Milligan CJ, Bahnasi YM, Al-Shawaf E, Porter KE, Jiang LH, Emery P, Sivaprasadarao A, Beech DJ (2008) TRPC channel activation by extracellular thioredoxin. Nature 451:69–72

Yamabe T, Dhir G, Cowan EP, Wolf AL, Bergey GK, Krumholz A, Barry E, Hoffman PM, Dhib-Jalbut S (1994) Cytokine-gene expression in measles-infected adult human glial cells. J Neuroimmunol 49:171–179

Yamamoto S, Shimizu S, Kiyonaka S, Takahashi N, Wajima T, Hara Y, Negoro T, Hiroi T, Kiuchi Y, Okada T, Kaneko S, Lange I, Fleig A, Penner R, Nishi M, Takeshima H, Mori Y (2008) TRPM2-mediated Ca^{2+} influx induces chemokine production in monocytes that aggravates inflammatory neutrophil infiltration. Nat Med 14:738–747

Yang KT, Chang WL, Yang PC, Chien CL, Lai MS, Su MJ, Wu ML (2006) Activation of the transient receptor potential M2 channel and poly(ADP-ribose) polymerase is involved in oxidative stress-induced cardiomyocyte death. Cell Death Differ 13:1815–1826

Yoshida T, Inoue R, Morii T, Takahashi N, Yamamoto S, Hara Y, Tominaga M, Shimizu S, Sato Y, Mori Y (2006) Nitric oxide activates TRP channels by cysteine S-nitrosylation. Nat Chem Biol 2:596–607

Young JM, Waters H, Dong C, Fülle HJ, Liman ER (2007) Degeneration of the olfactory guanylyl cyclase D gene during primate evolution. PLoS One 2:e884

Zhang W, Chu X, Tong Q, Cheung JY, Conrad K, Masker K, Miller BA (2003) A novel TRPM2 isoform inhibits calcium influx and susceptibility to cell death. J Biol Chem 278:16222–16229

Zhang W, Hirschler-Laszkiewicz I, Tong Q, Conrad K, Sun SC, Penn L, Barber DL, Stahl R, Carey DJ, Cheung JY, Miller BA (2006) TRPM2 is an ion channel that modulates hematopoietic cell death through activation of caspases and PARP cleavage. Am J Physiol Cell Physiol 290:C1146–C1159

Ziemann AE, Allen JE, Dahdaleh NS, Drebot II, Coryell MW, Wunsch AM, Lynch CM, Faraci FM, Howard MA 3rd, Welsh MJ, Wemmie JA (2009) The amygdala is a chemosensor that detects carbon dioxide and acidosis to elicit fear behavior. Cell 139:1012–1021

Photosensitive TRPs

Roger C. Hardie

Contents

R.C. Hardie (✉)
Department Physiology Development and Neuroscience, Cambridge University, Cambridge, UK
e-mail: rch14@cam.ac.uk

B. Nilius and V. Flockerzi (eds.), *Mammalian Transient Receptor Potential (TRP) Cation Channels*, Handbook of Experimental Pharmacology 223, DOI 10.1007/978-3-319-05161-1_4,

Abstract

The *Drosophila* "transient receptor potential" channel is the prototypical TRP channel, belonging to and defining the TRPC subfamily. Together with a second TRPC channel, *trp-like* (TRPL), TRP mediates the transducer current in the fly's photoreceptors. TRP and TRPL are also implicated in olfaction and Malpighian tubule function. In photoreceptors, TRP and TRPL are localised in the ~30,000 packed microvilli that form the photosensitive "rhabdomere"—a light-guiding rod, housing rhodopsin and the rest of the phototransduction machinery. TRP (but not TRPL) is assembled into multimolecular signalling complexes by a PDZ-domain scaffolding protein (INAD). TRPL (but not TRP) undergoes light-regulated translocation between cell body and rhabdomere. TRP and TRPL are also found in photoreceptor synapses where they may play a role in synaptic transmission. Like other TRPC channels, TRP and TRPL are activated by a G protein-coupled phospholipase C (PLCβ4) cascade. Although still debated, recent evidence indicates the channels can be activated by a combination of PIP_2 depletion and protons released by the PLC reaction. PIP_2 depletion may act mechanically as membrane area is reduced by cleavage of PIP_2's bulky inositol headgroup. TRP, which dominates the light-sensitive current, is Ca^{2+} selective ($P_{Ca}:P_{Cs}$ >50:1), whilst TRPL has a modest Ca^{2+} permeability ($P_{Ca}:P_{Cs}$ ~5:1). Ca^{2+} influx via the channels has profound positive and negative feedback roles, required for the rapid response kinetics, with Ca^{2+} rapidly facilitating TRP (but not TRPL) and also inhibiting both channels. In *trp* mutants, stimulation by light results in rapid depletion of microvillar PIP_2 due to lack of Ca^{2+} influx required to inhibit PLC. This accounts for the "transient receptor potential" phenotype that gives the family its name and, over a period of days, leads to light-dependent retinal degeneration. Gain-of-function *trp* mutants with uncontrolled Ca^{2+} influx also undergo retinal degeneration due to Ca^{2+} cytotoxicity. In vertebrate retina, mice knockout studies suggest that TRPC6 and TRPC7 mediate a PLCβ4-activated transducer current in intrinsically photosensitive retinal ganglion cells, expressing melanopsin. TRPA1 has been implicated as a "photo-sensing" TRP channel in human melanocytes and light-sensitive neurons in the body wall of *Drosophila*.

Keywords

Phototransduction • Photoreceptors • Drosophila • Phospholipase C

1 Introduction

The history of the transient receptor potential (TRP) ion channel family (Hardie 2011; Minke 2010; Montell 2011) dates back to the discovery of a spontaneously occurring mutant in a laboratory culture (Cosens and Manning 1969). Although the photoreceptors could generate a response, this decayed to baseline over a few seconds of continuous illumination rendering the flies effectively blind. Cosens and Manning called their mutant "A-type"; fortunately, Baruch Minke and colleagues, who studied this mutant extensively (Barash et al. 1988; Minke 1977, 1982; Minke et al. 1975), provided the more descriptive name, "transient receptor potential" or *trp*, which ultimately christened the entire gene family. When the *trp* gene was cloned, its sequence indicated a transmembrane (TM) protein with 8 TM helices (later revised to 6 plus a pore loop), with topology reminiscent of known receptor/transporter/channel proteins (Montell and Rubin 1989). However, there were no homologies to proteins in available databases and its function remained unknown. Intracellular recordings from *trp* mutant photoreceptors had suggested that the light-sensitive channels were normal (Minke 1982), apparently ruling out the possibility that the gene encoded the light-sensitive channel; instead, based on the ability of La^{3+} to mimic the *trp* phenotype, Minke and Selinger (1991) proposed it might represent a Ca^{2+} transporter required for refilling Ca^{2+} stores. Then, using whole-cell voltage-clamp recordings to determine ionic selectivity, Hardie and Minke (1992) showed that the primary defect in the *trp* mutant was a drastic reduction in the Ca^{2+} permeability of the light-sensitive channels themselves (Fig. 1). It was concluded that the normal light-sensitive current comprised two distinct conductances: one is highly Ca^{2+} selective and encoded by the *trp* gene and second is a channel responsible for the residual light-sensitive current in *trp* mutants. We now know the latter is encoded by a homologous gene: *trp-like* or *trpl* (Niemeyer et al. 1996; Phillips et al. 1992; Reuss et al. 1997). *Drosophila* TRP thus became the prototypical member of the TRP ion channel family and defines the "canonical" TRPC family, the first of which (TRPC1) was cloned by homology a few years later (Wes et al. 1995; Zhu et al. 1995).

This chapter primarily concerns the prototypical TRP channel and its homologue TRPL, which together mediate the light-sensitive current in *Drosophila* retinal photoreceptors (Niemeyer et al. 1996; Reuss et al. 1997). Related TRP channels that may mediate light-sensitive currents in atypical photoreceptor cells in other systems are also discussed at the end.

2 Gene and Protein Structure

The *trp* gene is located on the distal end of the third of *Drosophila's* four chromosomes (cytogenetic map: 99C6-7). A single transcript with 15 exons encodes a 1,275 amino-acid protein with 6 transmembrane domains (S1–6) and a putative pore loop between S5 and S6 (Fig. 2). The *trpl* gene is on the second arm of chromosome 2 (46B2) and encodes a shorter peptide with 900 amino acids. Overall *trpl* shares ~40 % identity to *trp*; most of the divergence is within the C termini, and within the S1–6 the homology is much closer (~70 %). Within TM helices S3–S6, as first recognised by Phillips et al. (1992), there is also ~40 % identity with voltage-gated Ca^{2+} channels, although the positively charged voltage-sensing residues in S4 are replaced by polar residues. TRPs belong to the overarching superfamily of voltage and cyclic nucleotide-gated (CNG) ion channels, and like K channels and

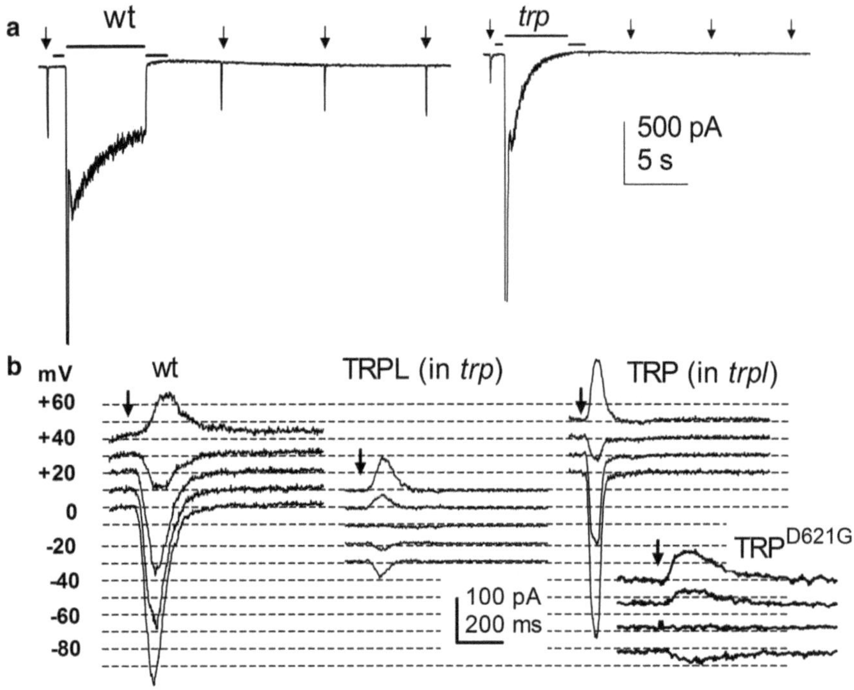

Fig. 1 *The transient receptor potential phenotype*. (**a**) In response to a prolonged stimulus (5 s step containing ~10^5 effective photons), wild-type photoreceptors show a rapid peak-plateau transition, relaxing to a steady state that is maintained indefinitely. Responses to brief test flashes (*arrows*) recover rapidly after the adapting light is turned off. In a *trp* mutant the response decays to baseline and sensitivity to test flashes is abolished, recovering only slowly in the dark (not shown). Shown here under whole-cell voltage-clamp conditions, this is the original *transient receptor potential* phenotype, which gives the TRP-channel family its name. The response decay is due to the failure of Ca^{2+}-dependent inhibition of PLC, resulting in depletion of the entire pool of PIP_2 in the microvilli (see Sect. 6). (**b**) Reversal potential (E_{rev}) measurements. Responses to light flashes (*arrows*) in photoreceptors clamped at different holding potentials (in 10 mV steps) under bi-ionic conditions (130 mM Cs_i, 10 mM Ca_o). A large shift in E_{rev} from ~35 mV in wild type to 0 mV in *trp* (expressing only TRPL channels) indicates a ~10-fold reduction in relative Ca^{2+} permeability. In *trpl* mutants (expressing only TRP channels), E_{rev} is shifted to +42 mV. The wild-type response is a weighted response of both channels. A point mutation in the TRP channel (TRP^{D621G}) virtually eliminates Ca^{2+} permeation, shifting E_{rev} to −70 mV (Reprinted with modification from Reuss et al. (1997) with permission from Elsevier)

CNG channels, each gene encodes a subunit of a putative tetrameric ion channel. Although there is some evidence that TRP and TRPL may form heteromultimers when heterologously expressed (Xu et al. 1997), in the photoreceptors they are believed to form homo-tetramers in vivo (Katz et al. 2013; Reuss et al. 1997). Like other TRPC channels the N-termini of TRP and TRPL contain a coiled-coil region and 4 ankyrin repeats of unknown function, whilst the C-terminus includes the highly conserved "EWKFAR" motif of the "TRP box" found in several TRP subfamilies and one (TRP) or two (TRPL) calmodulin-binding sites. TRP has a

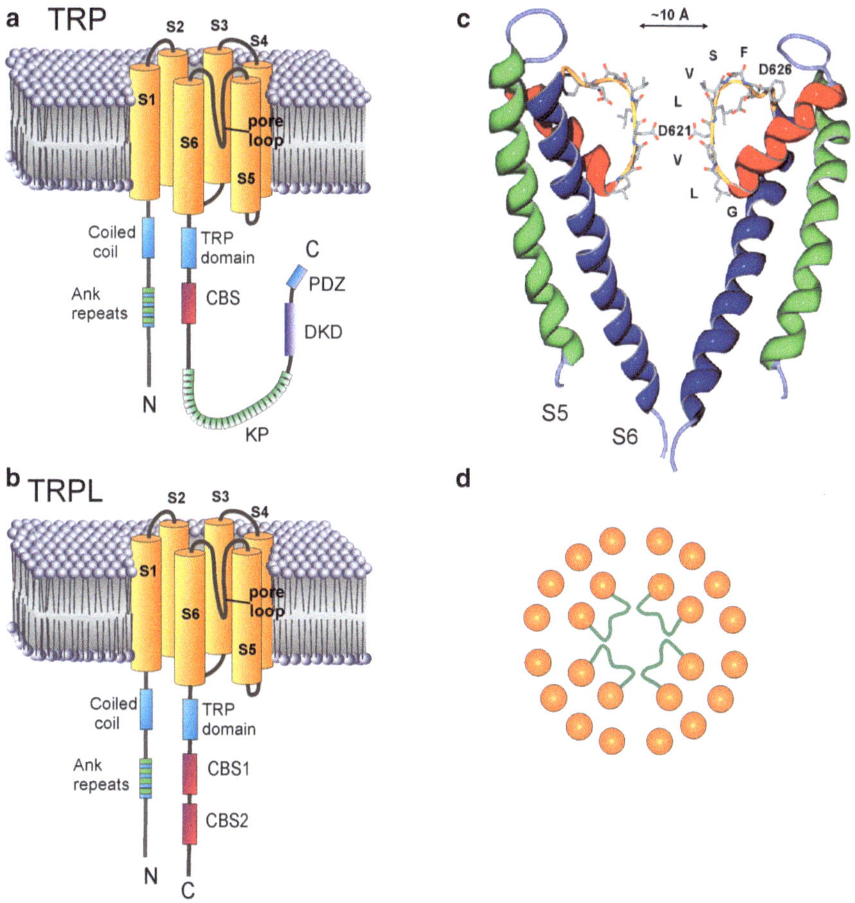

Fig. 2 *TRP and TRPL channels*. (**a**) dTRP is a channel subunit with 6 transmembrane helices (S1–S6), with a pore loop between S5 and S6. The N-terminus contains a coiled-coil region and 4 ankyrin repeats. The "TRP domain" with the "EWKFAR" motif is a highly conserved region found in all TRPC channels. The C-terminus also contains a CaM binding site (CBS), a proline-rich region with 29 KP repeats, multiple repeats of a hydrophilic 8–9 peptide sequence DKDKKP (A/G)D and a PDZ-domain-binding motif required for binding to the INAD protein. (**b**) TRPL has similar N-terminal and transmembrane regions, but the C-terminal is distinct. The only recognised domains are two calmodulin-binding sites (CBS1 and CBS2). (**c**) Molecular model of the pore region of the TRP channel (2 subunits shown) based on the crystal structure of KcsA. An acidic residue (aspartate 621) in the pore loop is responsible for the high Ca^{2+} selectivity. (**d**) Like voltage-gated K^+ channels and CNG channels, the TRP channel is believed to be a tetrameric assembly (probably homomeric) of 4 such subunits (**a** and **b** reprinted with modification from Hardie and Postma (2008) with permission from Elsevier; (**c**) Reprinted from Liu et al. (2007) with permission from the Journal of Neuroscience)

conspicuously long C-terminal domain containing a proline-rich region, 8–9 peptide repeats of unknown function (Montell and Rubin 1989) and a PDZ-binding motif that binds to the scaffolding protein INAD (Peng et al. 2008; Shieh and Zhu

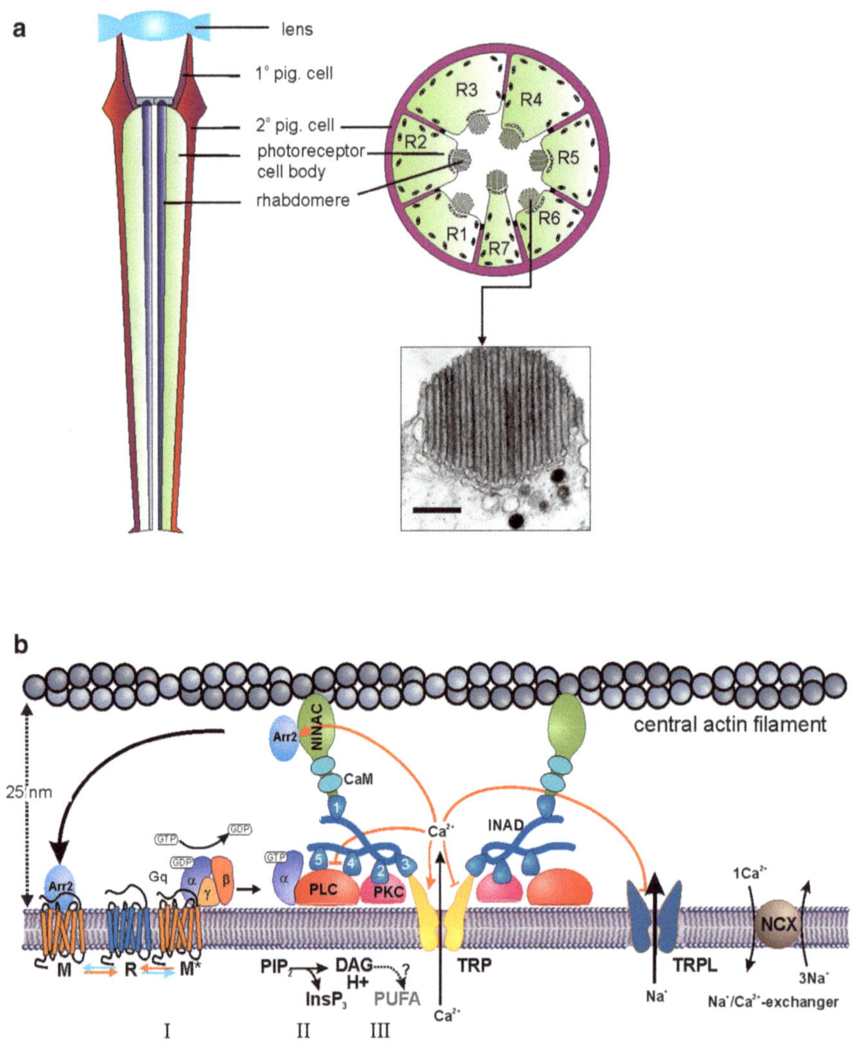

Fig. 3 Drosophila retina and phototransduction cascade. (**a**) *Left*: Diagram of ommatidium showing two (out of a total of eight) photoreceptors with their microvillar rhabdomeres. Light is focussed on the rhabdomere tips by the overlying lens (~16 μm in diameter). Each ommatidium is optically screened from its neighbours by pigment cells. The cross section (*right*) shows the geometrically precise array of rhabdomeres in the six R1–6 cells and a central photoreceptor R7 (a UV receptor). The electron micrograph shows a single rhabdomere (scale bar 500 nm) with tightly packed microvilli, each of which contain multiple copies of all major components of the transduction cascade. (**b**) Components of the transduction cascade shown approximately to scale in a schematic section of a "half" microvillus with its central actin filament. Photoisomerisation of rhodopsin (R) to metarhodopsin (M) activates G_q via GTP–GDP exchange (Step I), releasing the $G_q\alpha$ subunit; $G_q\alpha$ activates phospholipase C (PLC), generating $InsP_3$ and DAG and a proton from PIP_2 (Step II). Two classes of light-sensitive channels (TRP and TRPL) are activated, putatively by a combination of PIP_2 depletion and acidification (Step III). DAG is also a potential precursor for polyunsaturated fatty acids (PUFAs). Exogenously applied PUFAs activate channels, but the lipase necessary for their generation in situ has not been demonstrated. Ca^{2+} influx, primarily through the TRP channels, feeds back at multiple targets, e.g., via PKC and CaM, thereby

1996). The *Drosophila* genome includes a third TRPC channel—TRPγ (Xu et al. 2000). However, unlike TRP and TRPL, TRPγ is not particularly eye-enriched (Jors et al. 2006) and it seems doubtful whether it plays a role in phototransduction.

3 Expression

Both TRP and TRPL are highly enriched in adult photoreceptors and were once considered photoreceptor-specific. However, genetic evidence also implicates them in olfaction (Stortkuhl et al. 1999) and Malphigian tubule function (Macpherson et al. 2005), whilst microarray data suggests limited expression in the brain (Flyatlas.org). TRPL, but not TRP, is also expressed in the larval eye, or Bolwig's organ (Rosenbaum et al. 2011). In adult photoreceptors, both channel proteins are predominantly localised in the phototransduction compartment, a rod-like structure (rhabdomere) consisting of ~30,000 tightly packed microvilli (Fig. 3). Each microvillus probably contains about 20–30 TRP and 2–3 TRPL channels, representing channel densities of ~100/μm^2 and ~10/μm^2, respectively. TRPL, but not TRP, translocates reversibly from rhabdomere to cell body in response to illumination (Bahner et al. 2002), and a recent study also reported TRP and TRPL in the photoreceptor synaptic terminal where the channels may play a role in regulating transmitter release (Astorga et al. 2012).

During pupal development, which lasts ~100 h, TRP protein is first detected by immunocytochemistry at ~45 h (Satoh et al. 2005). However, functional TRP-channel activity is only detectable from ~82 h, whilst functional TRPL channels are first detected even later at ~90 h (Hardie et al. 1993). Little is known about the secretory pathway and targeting; however, TRP trafficking requires both the Rab GTPase Rab11 (Satoh et al. 2005) and a novel chaperone protein (XPORT) both of which are required for successful targeting of both TRP and rhodopsin to the rhabdomeric membrane (Rosenbaum et al. 2011).

Despite many attempts, it has proved very difficult to express TRP heterologously. In the few apparently successful attempts (Gillo et al. 1996; Vaca et al. 1994; Xu et al. 1997), the reported biophysical properties did not closely match those of the native *trp*-dependent current. Any doubts this might raise over the identity of TRP as the light-sensitive channel were dispelled with the identification of a unique aspartate residue within the pore loop as the major determinant of Ca^{2+} permeation (Liu et al. 2007). Targeted mutagenesis of this residue resulted in

Fig. 3 (continued) regulating, inter alia, the channels, PLC and M-Arr2 (arrestin). Ca^{2+} is extruded by the Na^+/Ca^{2+} exchanger (NCX). Several components of the cascade, including TRP, protein kinase C (PKC) and PLC, are assembled into signalling complexes by the scaffolding protein, INAD, which may be linked to the central F-actin filament via the NINAC class III myosin. INAD contains 5 PDZ domains (1–5) (Reprinted with modification from Yau and Hardie (2009) with permission of Elsevier)

systematic alteration of pore properties in vivo demonstrating that TRP indeed forms a pore-forming subunit of the *Drosophila* light-sensitive channel (see Sect. 5.1).

By contrast TRPL has been successfully expressed in many expression systems (Gillo et al. 1996; Hambrecht et al. 2000; Hardie et al. 1997; Harteneck et al. 1995; Hu et al. 1994; Lan et al. 1996). Its biophysical properties are essentially indistinguishable from those of the native TRPL-dependent current isolated in *trp* mutants (Chyb et al. 1999; Hardie et al. 1997), with the exception that in many expression systems it is constitutively active (for recent discussion, see Lev et al. 2012).

4 Interacting Proteins

4.1 Scaffolding Protein INAD

Along with other key components of the cascade, TRP channels are organised into multimolecular signalling complexes by the "*inactivation no afterpotential D*" (INAD) protein (Fig. 3), a scaffolding protein with five PDZ domains (Chevesich et al. 1997; Shieh and Niemeyer 1995; Shieh and Zhu 1996; Tsunoda et al. 1997). PDZ domains typically bind to a terminal three amino-acid motif on the target protein; however, the interaction of TRP with INAD appears to involve up to 14 amino acids in the C-terminal (Peng et al. 2008). In vitro evidence suggests the interaction may be "bidentate", with PDZ3 forming a classical association with the terminal 3 amino acids (GWL) of TRP and the upstream residues interacting with the PDZ4/5 supramodule (Liu et al. 2011). As well as TRP, INAD binds stoichiometrically (1:1) with PLC and PKC (Huber et al. 1996a; Li and Montell 2000; Tsunoda et al. 1997). INAD has also been reported to interact with NINAC, a class III myosin that may link the complex to the actin cytoskeleton (Hicks et al. 1996; Wes et al. 1999): calmodulin; retinophilin (Mecklenburg et al. 2010) and the immunophilin FKBP59 (Goel et al. 2001; Huber 2001; Montell 1999). Although TRPL has also been reported to bind INAD (Xu et al. 1998), this has been questioned, and a quantitative study suggests that at most only a small fraction of TRPL protein could associate with INAD (Paulsen et al. 2000). INAD can form homomultimers in vitro (Xu et al.1998), and together with other potential linkages (e.g., via TRP subunits which form tetramers), it is possible that extended complexes are formed.

Molecular scaffolds like the INAD complex promote signalling specificity and might facilitate rapid kinetics by minimising diffusional delays. In apparent support of this, null *inaD* mutants have reduced sensitivity and slow responses (Scott and Zuker 1998); however, these phenotypes could also simply reflect the greatly reduced levels of microvillar PLC found in *inaD* mutants (Tsunoda et al. 2001). Whether or not the integrity of the scaffolding complex is critical for rapid excitation, INAD is required to maintain high concentrations of TRP, PLC and PKC in the rhabdomere in the correct stoichiometry (Cook et al. 2000; Huber et al. 1996a). In fact, TRP and INAD are mutually required for the long-term stability of the scaffolding complex; thus TRP protein disappears from the rhabdomere in *inaD* mutants over a period of days, whilst INAD protein is reciprocally destabilised in *trp* mutants (Li and Montell 2000; Tsunoda et al. 1997).

Structural studies of the fourth and fifth PDZ domains in INAD show that PDZ4/5 can exist in two redox states due to the reversible light and PKC-dependent formation of a Cys-Cys disulfide bridge which disrupts the binding site—possibly for TRP or PLC (Liu et al. 2011; Mishra et al. 2007). Mutants in which PDZ4/5 is locked in one or the other state show specific defects in response to inactivation and/or activation. This raises the possibility that rather than acting as just a passive scaffold, INAD may regulate signal transduction on a millisecond timescale through dynamic switching between conformational states (Liu et al. 2011; Mishra et al. 2007).

4.2 INAF

The *inaF* mutant has a response decay phenotype very similar to *trp* (Li et al. 1999). When first cloned the *inaF* gene was considered novel without vertebrate homologues; however, a second transcript was later identified with a transmembrane motif that is conserved from flies to humans, but of unknown function (Cheng and Nash 2007). Using an HA-tagged *inaF* transgene, INAF was found to localise to the rhabdomeres and to co-immunoprecipitate with TRP (Cheng and Nash 2007). In null *inaF* mutants TRP protein levels are reduced to ~10 %, but the residual protein still localises to the rhabdomeres (Li et al. 1999). Despite this residual protein, the *inaF* phenotype is as severe as in null *trp* mutants, perhaps implicating INAF as an auxiliary subunit required for TRP-channel function, (Li et al. 1999).

4.3 Calmodulin

Both TRP and TRPL are calmodulin (CaM)-binding proteins, and both TRP and TRPL are potently regulated by Ca^{2+} (see below); however, surprisingly little is known of CaM's role in this regulation.

TRPL has two CaM-binding sites (CBS1 and CBS2) in its C-terminus, but their roles are still unclear. In vitro studies give somewhat conflicting results: Warr and Kelly (1996) reported that CBS1 is conventional, binding CaM in a Ca^{2+}-dependent fashion, whilst CBS2 binds the Ca^{2+}-free form of CaM, with dissociation occurring at high (>10 μM) Ca^{2+} concentrations. However, Trost et al. (1999) reported that both sites bind CaM in the presence of Ca^{2+}, but with slightly different Ca^{2+} dependencies (50 % binding at 100 nM and 3.3 μM Ca^{2+}, respectively, for CBS1 and CBS2). TRPL channels are strongly inhibited by Ca^{2+} with an IC_{50} of 1 μM or lower (Obukhov et al. 1998; Parnas et al. 2007; Reuss et al. 1997); however, CaM's role, if any, in this regulation is unclear. Scott et al. (1997) reported that Ca^{2+}-dependent inhibition of TRPL channel activity in vivo was impaired in *trpl* mutants lacking either CBS1 or CBS2; however, our lab was unable to reproduce this result using the same transgenic flies (Hardie R.C. unpublished). When expressed in Sf9 cells, the inhibitory action of Ca^{2+} on TRPL was unaffected by the CaM inhibitor calmidazolium (Obukhov et al. 1998). There are, however, reports of facilitation of TRPL channels by Ca^{2+}-CaM in heterologous expression systems (Lan et al. 1998; Trost et al. 1999).

The TRP protein has one CBS in the C-terminus, and peptide fragments from this region bind CaM in a Ca^{2+}-dependent manner in vitro (Chevesich et al. 1997). TRP channels are both positively (EC_{50} 300 nM) and negatively ($IC_{50} \sim 1$ μM) regulated by Ca^{2+} (Chu et al. 2013a; Gu et al. 2005), but again, whether CaM is involved in this regulation is not known.

4.4 Phosphorylation

Both TRP and TRPL proteins have multiple phosphorylation sites. TRP is a target for PKC, which is also a component of the INAD scaffolding complex (Huber et al. 1996b, 1998; Liu et al. 2000). Popescu et al. (2006) identified Ser982 on the TRP C-terminus as an in vitro target for PKC and found a mild slow-deactivation phenotype (albeit rather less severe than in a PKC null mutant) in Ser to Ala mutants of this site. More recently, 21 phosphorylation sites on the native TRP protein were identified by mass spectrometry, mainly located in the C-terminus (Voolstra et al. 2010). Several of these, including Ser982, were phosphorylated in a light-dependent manner, but surprisingly none were found to be compromised in a PKC null mutant, indicating either that they were not PKC targets or that they can be redundantly phosphorylated by additional kinases. Although most sites were phosphorylated in response to light, one in particular (Ser936) was dephosphorylated in response to illumination (Voolstra et al. 2010). Apart from the mild phenotype in TRP^{S982A} mutants, no function has been attributed to any of these sites.

In vitro studies of recombinant TRPL peptide fragments found that CaM binding to CBS1 was prevented by PKA-dependent phosphorylation, whilst phosphorylation by PKA was inhibited by prior PKC-dependent phosphorylation (Warr and Kelly 1996). Again no function has been attributed to this regulation. Recently, mass spectrometry identified eight C-terminal, light-dependent phosphorylation sites in TRPL. Mutation of these sites did not affect the electrophysiological response to light, but led to reduced TRPL content and partial mislocalisation out of the rhabdomere (Cerny et al. 2013).

4.5 Miscellaneous Interacting Proteins

At least three further proteins have been reported to co-immunoprecipitate with TRP and/or TRPL, although a direct interaction has not been demonstrated.

1. FKBP59, the *Drosophila* homologue of the human FK506-binding protein, was identified as an INAD binding partner in a yeast-2-hybrid screen and also found to co-immunoprecipitate with TRPL (Goel et al. 2001). The function of this

Table 1 Relative permeabilities of TRP and TRPL channels determined under bi-ionic conditions from reversal potential measurements of the light-induced currents

	$P_{Ca}:P_{Cs}$	$P_{Mg}:P_{Cs}$	$P_{Ba}:P_{Cs}$	$P_{Mn}:P_{Cs}$	$P_{Na}:P_{Cs}$	$P_{Li}:P_{Cs}$
TRP	56.9	15.8	56	17.7	1.3	0.89
TRPL	7.3	3.3	7.7	3.6	0.84	0.8

Data from Reuss et al. (1997) and Liu et al. (2007)

interaction has not been studied in vivo, but when co-expressed in Sf9 cells, FKBP59 induced a graded inhibition of TRPL activity in fura-2 Ca^{2+} influx assays.

2. Moesin, which is the sole member of the ezrin-radixin-moesin (ERM) family in *Drosophila*, co-immunopreciptates with both TRP and TRPL (Chorna-Ornan et al. 2005). ERM proteins promote actin-membrane interactions (Fehon et al. 2010) and are required for the development and maintenance of the microvilli (Karagiosis and Ready 2004). Upon illumination and PIP_2 depletion, moesin becomes dephosphorylated, dissociates from TRP and/or TRPL and migrates reversibly from the base of the rhabdomere to cytosolic regions (Chorna-Ornan et al. 2005; Sengupta et al. 2013).
3. TRP (but not TRPL) interacts with XPORT, which is a resident ER protein in the secretory pathway and which was recently identified as chaperone for both TRP and rhodopsin (Rosenbaum et al. 2011).

5 Biophysical Description of Channel Function, Permeation and Gating

5.1 Ionic Selectivity

The biophysical properties of the native TRP and TRPL conductances in vivo have been characterised in null *trp* and *trpl* mutants to isolate the respective currents (e.g., Reuss et al. 1997) (Fig. 1). Both channels permeate a broad range of monovalent and divalent ions including Ca^{2+}, Mg^{2+}, Ba^{2+}, Sr^{2+} and Mn^{2+} (Table 1). TRP and TRPL differ most notably in their selectivity for divalent ions: TRP channels are highly selective for Ca^{2+} ($P_{Ca}:P_{Cs}$ >50:1), and of the entire TRP superfamily, only the truly Ca^{2+}-selective TRPV5 and TRPV6 have a higher selectivity. TRPL has a Ca^{2+} selectivity ($P_{Ca}:P_{Cs}$ ~ 4:1) more typical of most TRP channels, including the broader TRPC subfamily (Hardie and Minke 1992; Liu et al. 2007; Reuss et al. 1997).

The high Ca^{2+} selectivity of TRP has been attributed to a single negatively charged aspartate residue (Asp^{621}) located in the pore loop. Neutralising Asp^{621} (by mutation to glycine or asparagine) eliminates Ca^{2+} permeation as well as the block by Mg^{2+} and ruthenium red leaving a monovalent cation channel (Liu et al. 2007). Increasing the side-chain length but maintaining the positive charge (Asp > Glu) barely affected Ca^{2+} selectivity but decreased the effective pore size, suggesting the residue also resides at the narrowest part of the channel. By analogy with other Ca^{2+}-selective channels, it seems likely that four such Asp^{621} residues (one from each subunit) coordinate a Ca^{2+}-binding site in the mouth of the pore (Figs. 1 and 3). A conserved charged residue (Glu) in the equivalent position was also found to be responsible for Ca^{2+} selectivity in mammalian TRPC3 (Poteser et al. 2011).

5.2 Single-Channel Properties

Under physiological conditions noise analysis of native currents in photoreceptors suggests single-channel conductances of ~8 pS for TRP and 35 pS for TRPL in the presence of divalent cations (1.5 mM Ca_o and 4 mM Mg_o) or ~30 pS (TRP) and 68 pS (TRPL) under divalent-free conditions (Hardie et al. 1997; Henderson et al. 2000; Reuss et al. 1997). In excised patch single-channel recordings from rhabdomeric membrane, values of 57 pS (TRP) and 49 pS (TRPL) have been reported under divalent-free conditions (Delgado and Bacigalupo 2009). When expressed in S2 cells, TRPL had a single-channel conductance of 110 pS in the presence of 4 mM Mg and 0 Ca (Parnas et al. 2007).

Based on permeability to a series of organic cations, the pore diameter of TRP was estimated at ~8 Å (Liu et al. 2007). Pore diameter has not been systematically measured for TRPL, although it was reported to be essentially impermeant to tetraethylamine and N-methyl-D-glucamine (Parnas et al. 2009a).

Open times of both channels are brief: TRPL openings are characterised by two time constants (~0.1 ms and 0.5–2 ms) both in single-channel recordings in S2 cells and in noise analysis from spontaneously active TRPL channels in *trp* mutants (Hardie and Minke 1994; Hardie et al. 1997; Parnas et al. 2007). Based on power spectra of native channels, TRP-channel kinetics are even faster, and only a single time constant of ~0.5 ms can be resolved (Raghu et al. 2000b). In either case the very brief channel openings mean that physiological response to illumination should not be limited by channel kinetics.

5.3 Voltage Dependence and Divalent Ion Open-Channel Block

Under physiological conditions, in the presence of divalent cations, TRPL channels show pronounced outward rectification, whilst TRP channels display complex dual inward and outward rectifying characteristics. In both cases voltage-dependent divalent ion open-channel block appears to be responsible.

5.3.1 TRP Channels

For TRP channels, the dual inward/outward rectification is abolished and the effective single-channel conductance increased 10-fold under divalent-free conditions (Hardie and Mojet 1995). Using Mg^{2+} as the blocking ion, IC_{50} (under Ca^{2+} free conditions) was reported as ~280 μM. The dual rectification was interpreted as voltage-dependent relief of flickering open-channel block (OCB) by divalent ions with both depolarisation and hyperpolarisation sufficient to drive the blocking ions from the pore. Under physiological conditions the voltage-dependent block is relatively weak around resting potential (−70 mV) but intensifies as the cell depolarises. This means that the effective single-channel conductance should decrease as the cell depolarises in response to light, potentially representing an economical mechanism for light adaptation (Hardie and Mojet 1995).

5.3.2 TRPL channels

The *I–V* relation of the TRPL channel is strongly outwardly rectifying in physiological solutions but becomes essentially linear under divalent-free conditions. The outward rectification under physiological conditions is thus interpreted as voltage-dependent relief of OCB (Parnas et al. 2007). In many cases of OCB, the kinetics of block and unblock are too fast to be resolved resulting in a reduction in effective single-channel conductance (flickering block) as appears to be the case for the TRP channel. However, in a detailed single-channel analysis of heterologously expressed TRPL channels, removal of divalent ions or depolarisation only increased open probability and open time without affecting single-channel conductance (Parnas et al. 2007). The divalent ion OCB described by Parnas et al. (2007) had an IC_{50} of 163 nM for Ca_o at −70 mV and ~2 mM for Mg^{2+}.

5.4 Pharmacology

Like many Ca^{2+} channels and several mammalian TRP channels, TRP, but not TRPL, channels are completely blocked by low μM levels of La^{3+} and ruthenium red (Liu et al. 2007). The highest affinity blocker reported for TRPL is cinnamyl-dihydroxy-cyanocinnamate, with an IC_{50} of 1 μM (Chyb et al. 1999). A variety of compounds, which are activators of some of the thermo-TRPs, inhibit TRPL channels at much higher concentrations, including carvacrol (IC_{50} 357 μM), thymol (1 mM), eugenol (3 mM), cinnamaldehyde (2 mM) and menthol (1.8 mM) (Parnas et al. 2009b). We recently found that a range of amphiphilic weak bases were all reversible inhibitors of both TRP and TRPL activity. Although the potency varied widely (IC_{50} for TRPL: trifluoperazine 5 μM, chlorpromazine 25 μM, imipramine 100 μM, procaine 5 mM) after correction for partitioning and pK_a values (in an attempt to predict concentration in the lipid membrane), the IC_{50} values converged near a common value of ~5 mM, suggesting their action could be attributed to the effect of the weak bases on physico-chemical properties of the lipid bilayer (Hardie and Franze 2012).

6 Physiological Functions

The primary function of the *Drosophila* TRP and TRPL channels is clear: TRP is the major light-sensitive channel in the photoreceptors of the retina. It accounts for up to 95 % of the light-sensitive current, the remainder carried by its homologue TRPL (Reuss et al. 1997).

6.1 Mechanism of Activation

Phototransduction in *Drosophila* has long been recognised as a G protein-coupled PLC-based cascade (Bloomquist et al. 1988; Pak 1995). All the essential elements of the cascade including the channels are localised within ~30,000 microvilli together forming the light-guiding rhabdomere (Fig. 3). As in other photoreceptors the cascade is initiated by absorption of a photon by a chromophore (3-OH *11-cis* retinaldehyde) covalently attached to rhodopsin (R). Photoisomerisation of *11-cis* retinal to the *all-trans* configuration converts R to the active metarhodopsin state (M), which catalytically activates a heterotrimeric Gq protein releasing the active Gqα subunit (Scott et al. 1995). A handful, perhaps 5–10, Gqα are released by each activated M within the same microvillus during a brief but variable latency period (15–100 ms) and diffuse in the plane of the membrane before binding to and activating a PLC molecule (Hardie et al. 2002). Each of the 5–10 PLC molecules thus activated then hydrolyses PIP_2 at a rate of ~1,000 molecules per second yielding $InsP_3$, DAG and a proton. The final electrical response in response to absorption of a single photon is a quantum bump, ~10 pA in amplitude, probably representing the opening of most of the 20 or so TRP channels in a single microvillus. Under voltage-clamp conditions, the output of the cell as a whole represents the summation of the quantum bumps independently generated over the 30,000 microvilli (see Hardie 2012 for recent detailed review).

Which consequences of PLC activity are responsible for channel activation remains debated. Most evidence suggests that $InsP_3$-induced Ca^{2+} release plays no role (Hardie 1995; Hardie and Raghu 1998) with phototransduction and adaptation appearing completely normal in $InsP_3$ receptor mutants (Acharya et al. 1997; Raghu et al. 2000a).

6.1.1 Diacylglycerol and Polyunsaturated Fatty Acids

TRP and TRPL channels are constitutively active in mutants of diacylglycerol kinase (DGK), encoded by the *rdgA* gene (Raghu et al. 2000b). Because DGK metabolises DAG to phosphatidic acid (PA), this is consistent with the notion that the constitutive activity in *rdgA* mutants is due to the build-up of DAG by basal PLC activity (Raghu et al. 2000b). Genetic interactions between *rdgA* and *norpA* (encoding PLC) or *Gαq* (Gqα subunit) also support an excitatory role for DAG, namely, in hypomorphic *Gαq* and *norpA* mutants, responses to light are greatly reduced due to diminished PLC activity, but when combined with the *rdgA* mutation (*rdgA,norpA* or *rdgA;Gαq* double mutants), the residual responses are greatly facilitated (Hardie et al. 2002).

Since DAG activates several mammalian TRPC channels (Hofmann et al. 1999), DAG seems a plausible candidate for the excitatory second messenger; however, there are problems with this suggestion (Katz and Minke 2009; Raghu 2006; Raghu and Hardie 2009). First, conversion of DAG to PA by DGK is the first step in PIP_2 resynthesis, and hence phenotypes reported in *rdgA* might also reflect reduced PIP_2 levels (Garcia-Murillas et al. 2006; Raghu and Hardie 2009). In addition, PA potently facilitates PI(4)P 5-kinase (Cockcroft 2009; Jenkins et al. 1994), so the final stage of PIP_2 resynthesis may also be compromised in *rdgA* mutants. Second, DGK appears to immunolocalise predominantly to the endoplasmic reticulum and not the microvilli where other components of transduction are localised (Masai et al. 1997). Third, attempts to activate TRP and TRPL channels with exogenous DAG have, in general, been unsuccessful (but see Delgado and Bacigalupo 2009).

By contrast polyunsaturated fatty acids (PUFAs), such as arachidonic and linolenic acid (which, in principle, might be released from DAG by a DAG lipase), robustly activate native TRP and TRPL channels as well as heterologously expressed TRPL channels (Chyb et al. 1999; Estacion et al. 2001; Parnas et al. 2009a) with an EC_{50} of ~10 μM. Apparent genetic evidence for PUFAs as the endogenous excitatory messenger came from mutants of a DAG lipase gene, *inaE*, which have greatly reduced sensitivity to light (Leung et al. 2008). However, the *inaE* DAG lipase is of the *sn*-1 class, which releases monoacylglycerols rather than PUFAs from DAG; whilst the INAE protein immunolocalises to the cell body with, at most, isolated traces in the rhabdomeres (Leung et al. 2008). There is also no evidence that light results in generation of PUFAs in the photoreceptors. As discussed below it is possible that PUFAs may be acting as surrogate agonists, mimicking the effects of PIP_2 hydrolysis on the physico-chemical properties of the lipid bilayer.

6.1.2 PIP_2 Depletion, Protons, and Bilayer Mechanics

In addition to generating DAG and $InsP_3$, hydrolysis of PIP_2 by PLC has two further actions: (1) it depletes PIP_2 and (2) it releases a proton. A recent study indicated that these two neglected consequences of PLC activity can act combinatorially to activate the channels (Huang et al. 2010). Thus, following depletion of PIP_2 in the photoreceptors, both TRP and TRPL channels could be rapidly and reversibly activated by protonophores, whilst heterologously expressed TRPL channels in S2 cells could be activated by acidification of the cytosolic surface of inside-out patches (Fig. 4). This indicates that a combination of PIP_2 depletion and acidification is sufficient to activate the light-sensitive channels (Huang et al. 2010).

These results appear to suggest that the channels are activated by protons and that PIP_2 binding to the channel inhibits this, either allosterically or by masking a protonatable site. Arguing against this however, PIP_2 applied to the cytosolic surface of inside-out patches actually activated TRPL channels expressed in S2 cells (Huang et al. 2010)—although the opposite result was reported in Sf9 cells (Estacion et al. 2001). An alternative possibility is that by removing its bulky and

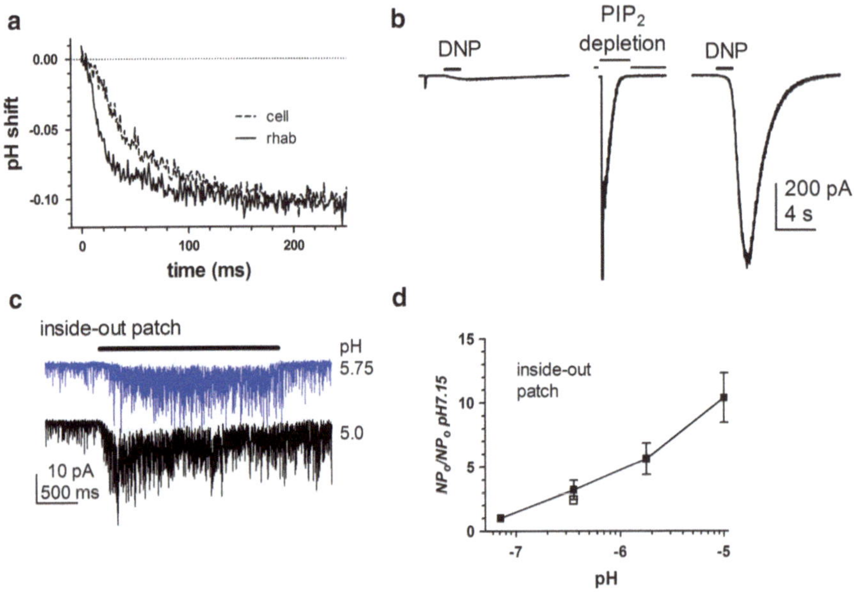

Fig. 4 *Activation by protons and PIP_2 depletion.* (**a**) Upon illumination (at time zero), there is a rapid acidification of the photoreceptor (measured with the fluorescent pH indicator, HPTS, loaded via whole-cell recording electrode: the signal is significantly faster when imaged from the rhabdomere than from the whole cell). (**b**) Under control conditions (*left*) brief application of the protonophore 2,4 dinitrophenol (DNP, 100 μM) activates no channels (whole-cell recording) in a *trp* mutant. However, after PIP_2 depletion induced by a 5 s light pulse, the same dose of DNP rapidly activates the light-sensitive channels. (**c**) Patch-clamp recording from an excised inside-out patch containing TRPL channels expressed in *Drosophila* S2 cells. Acidification of the bath (cytosolic surface of patch) rapidly and reversibly activated the channels. (**d**) pH dependence of channel activation (Reprinted with modification from Huang et al. (2010) with permission from Elsevier)

charged headgroup, hydrolysis of PIP_2 results in a change in the mechanical properties of the lipid bilayer, as has been suggested, albeit controversially, for the PLC-mediated activation of heterologously expressed TRPC6 (Spassova et al. 2006). A similar idea was also championed by Parnas et al. (2009a) who suggested that lipid activators such as PUFAs might exert their effect by modulating channel interactions with the lipid bilayer rather than on the channel per se. In support of this they found that heterologously expressed TRPL channels were influenced by other manipulations expected to alter bilayer properties including facilitation by osmotic stretch and suppression by GsMTx-4 toxin, a mechanosensitive ion channel blocker which acts on the channel/lipid boundary (Suchyna et al. 2000). They also found that the divalent ion-dependent outward rectification of TRPL channels was linearised by PUFA application and proposed that activation of channels by such lipid-mediated interaction was achieved by removal of open-channel block (Parnas et al. 2009a).

In support of a "photomechanical" mechanism, photoreceptors were recently shown to contract in response to light (Hardie and Franze 2012). The contractions, measured using atomic force microscopy on isolated retina, were faster than the electrical response (latency <5 ms); were strictly dependent upon PLC activity; and in the absence of Ca^{2+} influx (which normally inhibits PLC) were saturated at intensities equivalent to ~1 absorbed photon per microvillus. The contractions seem most parsimoniously explained by the decrease in membrane area due to hydrolysis of the large headgroup of PIP_2 from the inner leaflet, thereby increasing membrane tension and resulting in constriction of the microvilli. Furthermore, light responses, whether mediated by TRP or TRPL channels, were reversibly increased or suppressed by hypo- and hyper-osmotic solutions, respectively, indicating that the native channels were sensitive to membrane tension (Fig. 5). To ask whether light-induced hydrolysis of PIP_2 generated sufficient mechanical force to activate ion channels, a known mechano-sensitive channel (gramicidin) was incorporated into the membranes of *trpl;trp* double mutant photoreceptors lacking all native light-sensitive channels. Remarkably, such photoreceptors now responded to light by increased activity of the ectopic gramicidin channels with similar intensity dependence to the measured contractions (Hardie and Franze 2012).

Given the dual requirement for PIP_2 depletion and protonophores, a working hypothesis would be that the altered physical membrane environment following PIP_2 hydrolysis is energetically favourable for a conformational state of the channel that can be directly activated by protonation.

6.2 Ca^{2+} Influx

As well as mediating the cell's electrical response to light, TRP and to some extent TRPL channels are responsible for massive Ca^{2+} influx during the light response. Measurements of fractional Ca^{2+} currents indicate that ~30 % of the light-sensitive current is carried by Ca^{2+} ions, broadly in line with GHK predictions (Chu et al. 2013b). As a result, the concentration of Ca^{2+} reached transiently within each microvillus during a quantum bump probably exceeds 1 mM, saturating even the lowest affinity Ca^{2+} indicators (Oberwinkler and Stavenga 2000; Postma et al. 1999). During maintained illumination in a light-adapted state, this relaxes to a global steady-state concentration of around 5–10 μM due to the action of a potent Na^+/Ca^{2+} exchanger.

The Ca^{2+} influx, particularly via the dominant TRP channels, plays essential positive and negative feedback roles in regulating the gain and kinetics of the light response including most aspects of light adaptation (Gu et al. 2005; Hardie 1991; Henderson et al. 2000; Reuss et al. 1997). The importance of Ca^{2+} influx is shown by removing extracellular Ca^{2+} from the bath, in which case both rising and falling phases of the response become at least tenfold slower, and light adaptation is abolished. The positive feedback is primarily, if not exclusively, reflected in facilitation of TRP-channel activity, whilst both TRP and TRPL channels are inhibited by Ca^{2+} (Reuss et al. 1997). Although Ca^{2+} influx influences many

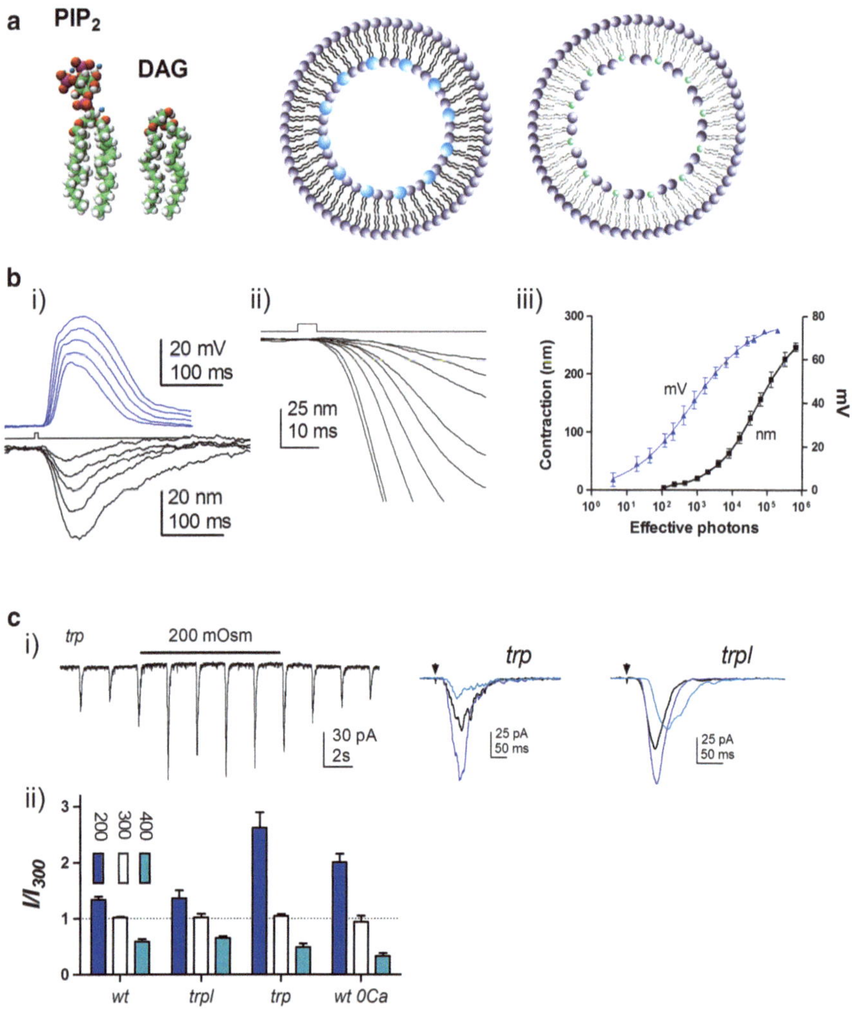

Fig. 5 *PIP_2 depletion and photomechanical response*. (**a**) The bulky, charged headgroup of PIP_2 is cleaved by PLC leaving DAG, with its much smaller footprint, in the membrane. Cartoon cross-sections of a microvillus indicate how removal of the inositol headgroup (*blue*) from the inner leaflet would decrease crowding, thereby increasing membrane tension, with the result that the microvillar diameter may contract. (**b**) (1) Contractions of the photoreceptors measured by AFM (lower traces) to moderate intensity flashes (200–800 effective photons); upper traces show electrical responses recorded to identical stimuli. (2) Family of contractions measured to 5 ms flashes of increasing intensity: latency to brightest flashes is less than 5 ms (cf ~ 6 ms for electrical response). (3) Response intensity function for contractions (nm) and electrical response (mV) mediated by TRP channels recorded under current clamp conditions. (**c**) (i) *Left*: Whole-cell voltage-clamp light responses in a *trp* mutant to a series of brief flashes are rapidly and reversibly facilitated by hypo-osmotic solution (200 mOsm). *Right*: Representative superimposed flash responses mediated by both TRPL channels (in *trp* mutant) and TRP channels (in *trpl* mutant) are suppressed by hyper-osmotic (400 mOsm) and facilitated by hypo-osmotic (200 mOsm) solutions. Bar graph (ii) shows normalised responses for wild type (wt), *trp* and *trpl,* as well as wt in Ca^{2+}-free solution to exclude indirect effect by change in Ca^{2+} (Reprinted with modification from Hardie and Franze (2012) with permission from Science)

molecular targets, both positive and negative feedback effects may be mediated directly on the channels themselves (Gu et al. 2005; Hardie 1995). Using the transmembrane Na^+ gradient to control cytosolic Ca^{2+} via the Na^+/Ca^{2+} exchanger equilibrium, the IC_{50} for both TRP and TRPL channel inhibition was estimated at ~ 1μM (Gu et al. 2005), whilst TRP, but not TRPL, channel activity, induced by exogenous agonist (PUFAs), was facilitated by Ca^{2+} with an EC_{50} of ~ 300 nM (Chu et al. 2013a).

Other Ca^{2+}-dependent targets modulated by influx via the TRP and TRPL channels include Ca^{2+}-CaM-dependent acceleration of arrestin binding to activated metarhodopsin (Liu et al. 2008) and Ca^{2+}-dependent inhibition of PLC, which appears to be mediated via PKC—possibly indirectly via phosphorylation of INAD (Gu et al. 2005). In addition, Ca^{2+} at low (submicromolar) concentrations can facilitate PLC (Hardie 2005; Katz and Minke 2012; Running Deer et al. 1995), although it is unclear whether this is significant under physiological conditions. There are also at least two Ca^{2+}-dependent steps in the visual pigment cycle: CaMKII-dependent phosphorylation of Arr2 and Ca-CaM-dependent dephosphorylation of rhodopsin by rhodopsin phosphatase (*rdgC* gene) (Byk et al. 1993; Steele et al. 1992; Vinos et al. 1997). Although neither appear to contribute directly to the electrophysiological response kinetics, both are required for normal pigment recycling and long-term cell survival (see Sect. 8).

7 Channel Mutant Phenotypes

7.1 *trp* Phenotype

The explanation for the classical transient receptor potential phenotype, namely, the decay of the response to baseline during continuous illumination (Fig. 1), has been controversial. Originally it was interpreted as the depletion of $InsP_3$-sensitive Ca^{2+} stores due to reduced Ca^{2+} influx (Cook and Minke 1999; Minke and Selinger 1991) and later as Ca^{2+}-dependent inhibition of the TRPL channels mediating the residual response (Scott et al. 1997). However, subsequent studies failed to support a role of intracellular Ca^{2+} stores in phototransduction (Acharya et al. 1997; Hardie 1996; Raghu et al. 2000a; Ranganathan et al. 1994), whilst the *trp* phenotype was found to be accentuated in the absence of Ca^{2+} influx (Cook and Minke 1999; Hardie et al. 2001). What then causes the phenotype that effectively christened the TRP ion channel family? An answer came using a genetically targeted PIP_2-sensitive ion channel (Kir2.1) to monitor endogenous PIP_2 levels in vivo. This showed that the decay of the response in *trp* mutants was strictly paralleled by the rapid depletion of PIP_2, whilst the recovery of sensitivity in the dark followed the time course of PIP_2 resynthesis (Hardie et al. 2001, 2004). The underlying cause for PIP_2 depletion and response decay appears to be the failure of Ca^{2+} and PKC-dependent inhibition of PLC, which is normally dependent upon Ca^{2+} influx via the TRP channels (Gu et al. 2005; Hardie et al. 2001). Without this inhibition, PLC activity is sufficient to hydrolyse all PIP_2 in the rhabdomere within as little as a second of relatively modest illumination (Hardie et al. 2001, 2004).

The absence of the major Ca^{2+} influx channel, and the resultant PIP_2 depletion, has many secondary phenotypic consequences including lack of light adaptation, lack of positive feedback, reduced single-photon responses (quantum bumps) as well as light-dependent retinal degeneration (see below).

7.2 *trpl* Phenotype

The TRP channel dominates the wild-type light response, and in whole-cell voltage-clamp recordings there is little obvious phenotype in *trpl* null mutants beyond predictable changes in permeation properties and pharmacology (Niemeyer et al. 1996; Reuss et al. 1997). However, studies under more physiological conditions, using either the electroretinogram (ERG) or the intracellular voltage recordings, revealed further subtle phenotypes. These include a reduced plateau potential, high-frequency (~100 Hz) oscillations superimposed on the response and an impaired ability to light adapt to very dim background lights (Bahner et al. 2002; Leung et al. 2000). With the possible exception of the reduced ability to light adapt, it remains unclear how these phenotypes relate to the known properties of the two channels. The oscillations in the *trpl* mutant ERG are probably synaptic in origin and may reflect a role of TRPL channels in synaptic transmission (Astorga et al. 2012).

8 Role in Hereditary and Acquired Diseases

TRP channels are the primary route for Ca^{2+} influx in the photoreceptor and play a major role in Ca^{2+} homeostasis, balanced by an NCX Na^{+}/Ca^{2+} exchanger which is the major mechanism for Ca^{2+} extrusion (Wang et al. 2005b). A variety of retinal pathologies have been attributed both to *reduced* Ca^{2+} influx in mutants where TRP-channel activity is reduced or eliminated and to *excessive* Ca^{2++} influx under conditions where channels are constitutively activated or fail to inactivate on cessation of the light stimulus.

Mutants in upstream components of the cascade, e.g., *Gαq* or PLC, result in failure to activate channels and typically undergo light-dependent retinal degeneration over a course of one to two weeks. This is believed to result from failure of one or more Ca^{2+}-dependent steps in the visual pigment cycle resulting in the build-up of potentially toxic complexes consisting of arrestin (Arr2) bound to phosphorylated M (MPP-Arr2). Thus, the pigment cycle normally involves photoisomerisation of R to M, whose activity is terminated by Ca^{2+}-dependent binding to Arr2 (Dolph et al. 1993; Liu et al. 2008). At the same time M becomes phosphorylated by Rh kinase. Although Rh kinase is probably not Ca^{2+}-dependent per se, phosphorylated rhodopsin is dephosphorylated by a Ca^{2+}-CaM-dependent phosphatase encoded by the *rdgC* gene so that rhodopsin becomes hyperphosphorylated in the absence of Ca^{2+} (Lee and Montell 2001; Steele et al. 1992; Vinos et al. 1997). Arr2 itself is also rapidly phosphorylated by CaMKII (Matsumoto et al. 1994). This is important for two reasons: first, *un*phosphorylated Arr2-MPP is a target for clathrin-mediated endocytosis which is believed to trigger apoptosis (Kiselev et al. 2000) and, second, Arr2 can only dissociate from R once Arr2 has been phosphorylated (Alloway and Dolph 1999). Red light can photoreisomerise M back to R at any point during the cycle and thus also prevents build-up of toxic MPP-Arr2.

trp mutants themselves also undergo retinal degeneration. There is evidence suggesting that defects in the pigment cycle contribute to the degeneration (Lee et al. 2013; Wang et al. 2005a); however, recently evidence was found for an additional mechanism, namely, the profound PIP_2 depletion that occurs when Ca^{2+} influx is compromised (Sengupta et al. 2013). MPP-Arr2-mediated degeneration can be simply prevented by rearing flies under red light (which reconverts M to R); this rescues degeneration in *norpA* and *rdgC,* but not in *trp*. Instead degeneration under red light was rescued by genetic elimination of PLC (in a *norpA*;*trp* double mutant), whilst the intensity dependence of degeneration quantitatively matched that of PIP_2 depletion. PIP_2 depletion and degeneration were also associated with depolymerisation of the actin cytoskeleton of the microvilli, possibly triggered by dephosphorylation of moesin (Sengupta et al. 2013).

Excessive Ca^{2+} influx is widely implicated in excito-toxicity, and, not surprisingly, overactivity of the photoreceptor TRP channels also triggers retinal degeneration. Again this can be due to defects in upstream signalling or in the TRP channel itself. For example, light-dependent retinal degeneration is found in *arr2* mutants, attributable to persistent channel activity due to failure to inactivate M (Dolph et al. 1993). Particularly severe degeneration occurs in DAG kinase (*rdgA*) mutants, which are characterised by constitutive TRP and TRPL channel activity (Raghu et al. 2000b). However, the most direct and striking evidence implicating the cytotoxic effect of Ca^{2+} influx via TRP channels comes from the dominant Trp^{365} mutant, which has a point mutation (Phe550Ile) in S5 that results in a constitutively active channel and massive retinal degeneration even in the dark (Hong et al. 2002; Yoon et al. 2000). Interestingly this aromatic residue is widely conserved amongst the TRP family, and the equivalent mutation in the yeast TRPY1 channel is also a gain-of-function mutation with increased constitutive activity (Su et al. 2007).

9 Other "Photosensitive" TRP Channels

9.1 ipRGCs

Vertebrate rods and cones are the so-called ciliary photoreceptors which diverged from the microvillar photoreceptors characteristic of *Drosophila* and many other invertebrates at least 550 million years ago. It had long been assumed that rods and cones, which signal using cGMP-gated channels, were the only photoreceptors in vertebrate eyes. However, recently it has become recognised that there are additional classes of light-sensing cell in most if not all vertebrate retinae (Berson 2003; Do and Yau 2010; Fu et al. 2005; Hankins et al. 2008). The best known are the so-called intrinsically photosensitive retinal ganglion cells (ipRGCs) that express a visual pigment, melanopsin, with closer homology to invertebrate opsins than to rod or cone opsins (Provencio et al. 2000). Along with a variety of other molecular similarities, this suggests that ipRGCs share a common evolutionary origin with the microvillar photoreceptors of *Drosophila* and most invertebrates (Arendt 2003).

Unlike rods and cones, which hyperpolarise in response to illumination, ipRGCs depolarise slowly, generating action potentials that adapt at best sluggishly thus giving an output that reflects the intensity of ambient illumination (Berson 2003; Do and Yau 2010; Fu et al. 2005; Hankins et al. 2008). This seems suited to their major functions of controlling the pupillary light reflex and entraining circadian rhythms via projections to the olivary pretectal nucleus and suprachiasmatic nucleus. Until recently there were only hints as to the underlying transduction cascade. For example, when expressed in *Xenopus* oocytes, melanopsin is capable of activating a Gq-based PLC cascade and a mammalian TRPC channel (Panda et al. 2005). In native ipRGCs the light-activated current can be blocked by PLC inhibitors (Graham et al. 2008) and inhibitors (albeit rather unspecific) of TRPC channels (Warren et al. 2006). Most convincingly however, it was recently demonstrated that the light-activated current in native ipRGCs was abolished in mouse knockouts lacking PLCβ4 as well as double knockouts lacking both TRPC6 and TRPC7. Interestingly, PLCβ4 is more closely related to PLC in *Drosophila* photoreceptors (*norpA*) than it is to other vertebrate PLC isoforms (Ferreira and Pak 1994). Along with the apparent involvement of TRPC channels, there are thus close parallels with phototransduction in *Drosophila*. There is, as yet, little or no evidence for the downstream pathway linking PLC activation to TRPC channel activation in the ipRGCs. In heterologous expression systems TRPC6 and TRPC7 can be activated by DAG (Hofmann et al. 1999; Okada et al. 1999). Putative native TRPC6/7 channels, in other mammalian cells, such as the vascular smooth muscle, are also reported to be activated by DAG, possibly in synergy with $InsP_3$ and PIP_2 depletion (Albert and Large 2003, 2006; Ju et al. 2010).

9.2 TRPA1

Finally, recent studies in both vertebrate and invertebrates have suggested roles for TRPA1 channels in some atypical "dermal photoreceptors".

9.2.1 Melanocytes

Melanin synthesis in human epidermal melanocytes (HEM) is stimulated by absorption of UV light. Short-wavelength (UVB) light may stimulate melanin synthesis downstream of DNA damage itself (Lin and Fisher 2007), but the mechanism underlying response to longer-wavelength UVA light is poorly understood. In a recent study UVA light was shown to activate a current in HEM cells, which was inhibited by TRPA1 antagonists (camphor or HC-030031) and almost eliminated by TRPA1 miRNA (Bellono et al. 2013). Both miRNA and TRPA1 antagonists also inhibited early melanin synthesis. Based on pharmacology, activation appeared to be via a G protein and PLC-based signalling pathway. It also required pre-incubation of cells with retinaldehyde suggesting the involvement of an opsin-based pigment (Bellono and Oancea 2013; Wicks et al. 2011).

9.2.2 Larval Photo-Avoidance Behaviour

Drosophila larvae have a primitive eye (Bolwig's organ), which mediates larval phototaxis. However, even after these have been genetically deleted, larvae were shown to avoid bright UV light. This photo-avoidance behaviour is mediated by a class of sensory neuron (class IV neurons) with dendrites that tile essentially the entire body surface and was abolished in mutants lacking TRPA1 channels (Xiang et al. 2010). The response also requires an orphan GPCR (Gr28b), which is annotated as a gustatory receptor. Whether this represents a previously unrecognised visual pigment or whether it might respond to a chemical generated via UV illumination is unclear, as is the transduction pathway linking the receptor to TRPA1 activation. Interestingly, a *Caenorhabditis elegans* homologue of Gr28b—*lite1*—was also found to be required for UV photo-avoidance behaviour in *C. elegans*; however, in this case the transduction cascade involves guanylate cyclase culminating in activation of cGMP-gated channels (Liu et al. 2010).

Concluding Remarks

The unique experimental advantages of the *Drosophila* retina for both genetic and functional analysis not only led to the discovery of the TRP ion channel family but have also provided key insight into many aspects of TRP-channel function. From its discovery it was clear that *Drosophila* TRP was activated downstream of PLC (Hardie and Minke 1992). This is now regarded as a common feature for all TRPC channels, and in fact many, if not most, other TRP channels can be regulated in one way or another by PLC. Nevertheless, exactly which consequence of PLC activity is responsible for activation is often far from clear. TRPC2, 3, 6 and 7 are reportedly activated by DAG (Hofmann et al. 1999; Lucas et al. 2003; Okada et al. 1999); but whether this action is direct remains debated (Lemonnier et al. 2008), and no DAG binding domain has been identified. Roles for $InsP_3$ and PIP_2 have also been reported (Ju et al. 2010). The mechanism of activation of TRPC4, 5 and 1 is yet more mysterious, though complex roles for PIP_2 have been proposed (Otsuguro et al. 2008; Trebak et al. 2009). Even though it is debatable whether PUFAs are physiologically relevant agonists in the photoreceptors, the discovery that TRP and TRPL could be activated by PUFAs was the first indication that TRPC channels were regulated by lipid messengers (Chyb et al. 1999; Hardie 2003). This is again believed to be a common feature of not only all TRPCs but also many other TRPs (Beech 2012; Hardie 2007).

Whether the latest working hypothesis for physiological activation of TRP and TRPL, a combination of membrane tension and protons (Hardie and Franze

2012), is more generally applicable remains to be seen. However, mechanical gating has been repeatedly suggested, albeit often controversially, for many members of the TRP family (Barritt and Rychkov 2005; Christensen and Corey 2007; Liedtke 2007; Lin and Corey 2005; Maroto et al. 2005; Mederos y Schnitzler et al. 2008; Patel et al. 2010; Quick et al. 2012; Sharif-Naeini et al. 2008; Spassova et al. 2006; Su et al. 2007; Yin and Kuebler 2010). Protons are also known modulators of several TRP channels, and although most reports concern extracellular sites—as in TRPV1 and TRPC5 (Ryu et al. 2007; Semtner et al. 2007)—intracellular protons have been implicated as inhibitors of gating of TRPM2 (Du et al. 2009) and activation of TRPA1 (Wang et al. 2010). More generally, like many, if not all, TRP channels, *Drosophila* TRP and TRPL are polymodally regulated, with multiple signals including lipids, protons, Ca^{2+} and membrane tension all playing potentially important roles.

Although the physiological roles of many vertebrate TRP channels are still often only poorly understood, along with the electrical signal (depolarisation), Ca^{2+} influx is probably the single most important physiological consequence of TRP-channel activation. Again, for the light-sensitive *Drosophila* TRP channel, this has been clear from its discovery, with Ca^{2+} influx having profound roles acting on multiple targets, with actions ranging from the millisecond timescale of the light response to long-term cell survival or degeneration. Nevertheless, as with all members of the TRP family, many questions remain to be answered, not least, a final resolution to the question of how the channels are activated. The mechanism of dual (positive and negative) Ca^{2+}-dependent regulation is also surprisingly poorly understood, as is the role of extensive phosphorylation of both proteins.

References

Acharya JK, Jalink K, Hardy RW, Hartenstein V, Zuker CS (1997) $InsP_3$ receptor is essential for growth and differentiation but not for vision in *Drosophila*. Neuron 18:881–887

Albert AP, Large WA (2003) Synergism between inositol phosphates and diacylglycerol on native TRPC6-like channels in rabbit portal vein myocytes. J Physiol 552:789–795

Albert AP, Large WA (2006) Signal transduction pathways and gating mechanisms of native TRP-like cation channels in vascular myocytes. J Physiol 570:45–51

Alloway PG, Dolph PJ (1999) A role for the light-dependent phosphorylation of visual arrestin. Proc Natl Acad Sci U S A 96:6072–6077

Arendt D (2003) Evolution of eyes and photoreceptor cell types. Int J Dev Biol 47:563–571

Astorga G, Hartel S, Sanhueza M, Bacigalupo J (2012) TRP, TRPL and cacophony channels mediate Ca(2+) influx and exocytosis in photoreceptors axons in Drosophila. PLoS One 7: e44182

Bahner M, Frechter S, Da Silva N, Minke B, Paulsen R, Huber A (2002) Light-regulated subcellular translocation of *Drosophila* TRPL channels induces long-term adaptation and modifies the light-induced current. Neuron 34:83–93

Barash S, Suss E, Stavenga DG, Rubinstein CT, Selinger Z, Minke B (1988) Light reduces the excitation efficiency in the nss mutant of the sheep blowfly Lucilia. J Gen Physiol 92:307–330

Barritt G, Rychkov G (2005) TRPs as mechanosensitive channels. Nat Cell Biol 7:105–107

Beech DJ (2012) Integration of transient receptor potential canonical channels with lipids. Acta physiologica (Oxford, England) 204:227–237
Bellono NW, Oancea E (2013) UV light phototransduction depolarizes human melanocytes. Channels (Austin, Tex 7)
Bellono NW, Kammel LG, Zimmerman AL, Oancea E (2013) UV light phototransduction activates transient receptor potential A1 ion channels in human melanocytes. Proc Natl Acad Sci U S A 110:2383–2388
Berson DM (2003) Strange vision: ganglion cells as circadian photoreceptors. Trends Neurosci 26:314–320
Bloomquist BT, Shortridge RD, Schneuwly S, Pedrew M, Montell C, Steller H, Rubin G, Pak WL (1988) Isolation of putative phospholipase C gene of Drosophila, *norpA* and its role in phototransduction. Cell 54:723–733
Byk T, BarYaacov M, Doza YN, Minke B, Selinger Z (1993) Regulatory arrestin cycle secures the fidelity and maintenance of the fly photoreceptor cell. Proc Natl Acad Sci U S A 90:1907–1911
Cerny AC, Oberacker T, Pfannstiel J, Weigold S, Will C, Huber A (2013) Mutation of light-dependent phosphorylation sites of the drosophila transient receptor potential-like (TRPL) ion channel affects its subcellular localization and stability. J Biol Chem 288:15600–15613
Cheng Y, Nash HA (2007) *Drosophila* TRP channels require a protein with a distinctive motif encoded by the inaF locus. Proc Natl Acad Sci U S A 104:17730–17734
Chevesich J, Kreuz AJ, Montell C (1997) Requirement for the PDZ domain protein, INAD, for localization of the TRP store-operated channel to a signalling complex. Neuron 18:95–105
Chorna-Ornan I, Tzarfaty V, Ankri-Eliahoo G, Joel-Almagor T, Meyer NE, Huber A, Payre F, Minke B (2005) Light-regulated interaction of Dmoesin with TRP and TRPL channels is required for maintenance of photoreceptors. J Cell Biol 171:143–152
Christensen AP, Corey DP (2007) TRP channels in mechanosensation: direct or indirect activation? Nat Rev Neurosci 8:510–521
Chu B, Liu CH, Sengupta S, Gupta A, Raghu P, Hardie RC (2013a) Common mechanisms regulating dark noise and quantum bump amplification in Drosophila photoreceptors. J Neurophysiol 109:2044–2055
Chu B, Postma M, Hardie RC (2013b) Fractional Ca(2+) currents through TRP and TRPL channels in Drosophila photoreceptors. Biophys J 104:1905–1916
Chyb S, Raghu P, Hardie RC (1999) Polyunsaturated fatty acids activate the *Drosophila* light-sensitive channels TRP and TRPL. Nature 397:255–259
Cockcroft S (2009) Phosphatidic acid regulation of phosphatidylinositol 4-phosphate 5-kinases. Biochim Biophys Acta 1791:905–912
Cook B, Minke B (1999) TRP and calcium stores in Drosophila phototransduction. Cell Calcium 25:161–171
Cook B, BarYaacov M, BenAmi HC, Goldstein RE, Paroush Z, Selinger Z, Minke B (2000) Phospholipase C and termination of G-protein-mediated signalling in vivo. Nat Cell Biol 1465–7392(2):296–301
Cosens DJ, Manning A (1969) Abnormal electroretinogram from a *Drosophila* mutant. Nature 224:285–287
Delgado R, Bacigalupo J (2009) Unitary recordings of TRP and TRPL channels from isolated Drosophila retinal photoreceptor rhabdomeres: activation by light and lipids. J Neurophysiol 101:2372–2379
Do MT, Yau KW (2010) Intrinsically photosensitive retinal ganglion cells. Physiol Rev 90:1547–1581
Dolph PJ, Ranganathan R, Colley NJ, Hardy RW, Socolich M, Zuker CS (1993) Arrestin function in inactivation of G protein-coupled receptor rhodopsin in vivo. Science 260:1910–1916
Du J, Xie J, Yue L (2009) Modulation of TRPM2 by acidic pH and the underlying mechanisms for pH sensitivity. J Gen Physiol 134:471–488

Estacion M, Sinkins WG, Schilling WP (2001) Regulation of *Drosophila* transient receptor potential-like (TRPL) channels by phospholipase C-dependent mechanisms. J Physiol 530:1–19

Fehon RG, McClatchey AI, Bretscher A (2010) Organizing the cell cortex: the role of ERM proteins. Nat Rev Mol Cell Biol 11:276–287

Ferreira PA, Pak WL (1994) Bovine phospholipase C highly homologous to the NorpA protein of *Drosophila* is expressed specifically in cones. J Biol Chem 269:3129–3131

Fu Y, Liao HW, Do MT, Yau KW (2005) Non-image-forming ocular photoreception in vertebrates. Curr Opin Neurobiol 15:415–422

Garcia-Murillas I, Pettitt T, Macdonald E, Okkenhaug H, Georgiev P, Trivedi D, Hassan B, Wakelam M, Raghu P (2006) *lazaro* encodes a lipid phosphate phosphohydrolase that regulates phosphatidylinositol turnover during Drosophila phototransduction. Neuron 49:533–546

Gillo B, Chorna I, Cohen H, Cook B, Manistersky I, Chorev M, Arnon A, Pollock JA, Selinger Z, Minke B (1996) Coexpression of *Drosophila* TRP and TRP-like proteins in *Xenopus* oocytes reconstitutes capacitative Ca^{2+} entry. Proc Natl Acad Sci U S A 93:14146–14151

Goel M, Garcia R, Estacion M, Schilling WP (2001) Regulation of *Drosophila* TRPL channels by immunophilin FKBP59. J Biol Chem 276:38762–38773

Graham DM, Wong KY, Shapiro P, Frederick C, Pattabiraman K, Berson DM (2008) Melanopsin ganglion cells use a membrane-associated rhabdomeric phototransduction cascade. J Neurophysiol 99:2522–2532

Gu Y, Oberwinkler J, Postma M, Hardie RC (2005) Mechanisms of light adaptation in *Drosophila* photoreceptors. Curr Biol 15:1228–1234

Hambrecht J, Zimmer S, Flockerzi V, Cavalie A (2000) Single-channel currents through transient-receptor-potential- like (TRPL) channels. Pflugers Archiv Eur J Physiol 440:418–426

Hankins MW, Peirson SN, Foster RG (2008) Melanopsin: an exciting photopigment. Trends Neurosci 31:27–36

Hardie RC (1991) Whole-cell recordings of the light-induced current in *Drosophila* photoreceptors: evidence for feedback by calcium permeating the light sensitive channels. Proc R Soc Lond B 245:203–210

Hardie RC (1995) Photolysis of caged Ca^{2+} facilitates and inactivates but does not directly excite light-sensitive channels in *Drosophila* photoreceptors. J Neurosci 15:889–902

Hardie RC (1996) INDO-1 measurements of absolute resting and light-induced Ca^{2+} concentration in *Drosophila* photoreceptors. J Neurosci 16:2924–2933

Hardie RC (2003) Regulation of TRP channels via lipid second messengers. Annu Rev Physiol 65:735–759

Hardie RC (2005) Inhibition of phospholipase C activity in Drosophila photoreceptors by 1,2-bis (2-aminophenoxy)ethane N, N, N', N'-tetraacetic acid (BAPTA) and di-bromo BAPTA. Cell Calcium 38:547–556

Hardie RC (2007) TRP channels and lipids: from Drosophila to mammalian physiology. J Physiol 578:9–25

Hardie RC (2011) A brief history of trp: commentary and personal perspective. Pflugers Arch 461:493–498

Hardie RC (2012) Phototransduction mechanisms in *Drosophila* microvillar photoreceptors. WIREs Membr Transp Signal 1:2162–2187. doi:10.1002/wmts.20

Hardie RC, Franze K (2012) Photomechanical responses in Drosophila photoreceptors. Science 338:260–263

Hardie RC, Minke B (1992) The *trp* gene is essential for a light-activated Ca^{2+} channel in *Drosophila* photoreceptors. Neuron 8:643–651

Hardie RC, Minke B (1994) Spontaneous activation of light-sensitive channels in *Drosophila* photoreceptors. J Gen Physiol 103:389–407

Hardie RC, Mojet MH (1995) Magnesium-dependent block of the light-activated and *trp*-dependent conductance in *Drosophila* photoreceptors. J Neurophysiol 74:2590–2599

Hardie RC, Postma M (2008) Phototransduction in microvillar photoreceptors of Drosophila and other invertebrates. In: Albright TD, Masland, R (eds) The senses—a comprehensive reference (eds. Basbaum K, Shepherd, Westheimer), vision vol 1. Academic Press, Oxford, pp 77–130

Hardie RC, Raghu P (1998) Activation of heterologously expressed *Drosophila* TRPL channels: Ca^{2+} is not required and $InsP_3$ is not sufficient. Cell Calcium 24:153–163

Hardie RC, Peretz A, Pollock JA, Minke B (1993) Ca^{2+} limits the development of the light response in *Drosophila* photoreceptors. Proc Biol Sci 252:223–229

Hardie RC, Reuss H, Lansdell SJ, Millar NS (1997) Functional equivalence of native light-sensitive channels in the *Drosophila* trp^{301} mutant and TRPL cation channels expressed in a stably transfected *Drosophila* cell line. Cell Calcium 21:431–440

Hardie RC, Raghu P, Moore S, Juusola M, Baines RA, Sweeney ST (2001) Calcium influx via TRP channels is required to maintain PIP_2 levels in *Drosophila* photoreceptors. Neuron 30:149–159

Hardie RC, Martin F, Cochrane GW, Juusola M, Georgiev P, Raghu P (2002) Molecular basis of amplification in *Drosophila* phototransduction. Roles for G protein, phospholipase C, and diacylglycerol kinase. Neuron 36:689–701

Hardie RC, Gu Y, Martin F, Sweeney ST, Raghu P (2004) In vivo light-induced and basal phospholipase C activity in *Drosophila* photoreceptors measured with genetically targeted phosphatidylinositol 4,5-bisphosphate-sensitive ion channels (Kir2.1). J Biol Chem 279:47773–47782

Harteneck C, Obukhov AG, Zobel A, Kalkbrenner F, Schultz G (1995) The *Drosophila* cation channel Trpl expressed in insect Sf9 cells is stimulated by agonists of G-protein-coupled receptors. FEBS Lett 358:297–300

Henderson SR, Reuss H, Hardie RC (2000) Single photon responses in *Drosophila* photoreceptors and their regulation by Ca^{2+}. J Physiol London 524:179–194

Hicks JL, Liu X, Williams DS (1996) Role of the NinaC proteins in photoreceptor cell structure: ultrastructure of *ninaC* deletion mutants and binding to actin filaments. Cell Motil Cytoskeleton 35:367–379

Hofmann T, Obukhov AG, Schaefer M, Harteneck C, Gudermann T, Schultz G (1999) Direct activation of human TRPC6 and TRPC3 channels by diacylglycerol. Nature 397:259–263

Hong YS, Park S, Geng C, Baek K, Bowman JD, Yoon J, Pak WL (2002) Single amino acid change in the fifth transmembrane segment of the TRP Ca2+ channel causes massive degeneration of photoreceptors. J Biol Chem 277:33884–33889

Hu Y, Vaca L, Zhu X, Birnbaumer L, Kunze DL, Schilling WP (1994) Appearance of a novel Ca^{2+} influx pathway in Sf9 insect cells following expression of the transient receptor potential-like (*trpl*) protein of *Drosophila*. Biochem Biophys Res Commun 201:1050–1056

Huang J, Liu CH, Hughes SA, Postma M, Schwiening CJ, Hardie RC (2010) Activation of TRP channels by protons and phosphoinositide depletion in *Drosophila* photoreceptors. Curr Biol 20:189–197

Huber A (2001) Scaffolding proteins organize multimolecular protein complexes for sensory signal transduction. Eur J Neurosci 14:769–776

Huber A, Sander P, Gobert A, Bahner M, Hermann R, Paulsen R (1996a) The transient receptor potential protein (Trp), a putative store-operated Ca^{2+} channel essential for phosphoinositide-mediated photoreception, forms a signaling complex with NorpA, InaC and InaD. EMBO J 15:7036–7045

Huber A, Sander P, Paulsen R (1996b) Phosphorylation of the InaD gene product, a photoreceptor membrane protein required for recovery of visual excitation. J Biol Chem 271:11710–11717

Huber A, Sander P, Bahner M, Paulsen R (1998) The TRP Ca^{2+} channel assembled in a signaling complex by the PDZ domain protein INAD is phosphorylated through the interaction with protein kinase C (ePKC). FEBS Lett 425:317–322

Jenkins GH, Fisette PL, Anderson RA (1994) Type I phosphatidylinositol 4-phosphate 5-kinase isoforms are specifically stimulated by phosphatidic acid. J Biol Chem 269:11547–11554

Jors S, Kazanski V, Foik A, Krautwurst D, Harteneck C (2006) Receptor-induced activation of Drosophila TRP{gamma} by polyunsaturated fatty acids. J Biol Chem 281:29693–29702

Ju M, Shi J, Saleh SN, Albert AP, Large WA (2010) Ins(1,4,5)P3 interacts with PIP2 to regulate activation of TRPC6/C7 channels by diacylglycerol in native vascular myocytes. J Physiol 588:1419–1433

Karagiosis SA, Ready DF (2004) Moesin contributes an essential structural role in Drosophila photoreceptor morphogenesis. Development 131:725–732

Katz B, Minke B (2009) Drosophila photoreceptors and signaling mechanisms. Front Cell Neurosci 3:2

Katz B, Minke B (2012) Phospholipase C-mediated suppression of dark noise enables single-photon detection in Drosophila photoreceptors. J Neurosci 32:2722–2733

Katz B, Oberacker T, Richter D, Tzadok H, Peters M, Minke B, Huber A (2013) The Drosophila TRP and TRPL are assembled as homomultimeric channels in vivo. J Cell Sci 126(14):3121–3133

Kiselev A, Socolich M, Vinos J, Hardy RW, Zuker CS, Ranganathan R (2000) A molecular pathway for light-dependent photoreceptor apoptosis in Drosophila. Neuron 28:139–152

Lan L, Bawden MJ, Auld AM, Barritt GJ (1996) Expression of Drosophila Trpl cRNA In *Xenopus-laevis* oocytes leads to the appearance of a Ca^{2+} channel activated by Ca^{2+} and Calmodulin, and by guanosine 5'[gamma-thio]triphosphate. Biochem J 316:793–803

Lan L, Brereton H, Barritt GJ (1998) The role of calmodulin-binding sites in the regulation of the Drosophila TRPL cation channel expressed in Xenopus laevis oocytes by Ca2+, inositol 1,4,5-trisphosphate and GTP-binding proteins. Biochem J 330:1149–1158

Lee SJ, Montell C (2001) Regulation of the rhodopsin protein phosphatase, RDGC, through interaction with calmodulin. Neuron 32:1097–1106

Lee J, Song M, Hong S (2013) Negative regulation of the novel norpA(P24) suppressor, diehard4, in the endo-lysosomal trafficking underlies photoreceptor cell degeneration. PLoS Genet 9: e1003559

Lemonnier L, Trebak M, Putney JW Jr (2008) Complex regulation of the TRPC3, 6 and 7 channel subfamily by diacylglycerol and phosphatidylinositol-4,5-bisphosphate. Cell Calcium 43:506–514

Leung HT, Geng C, Pak WL (2000) Phenotypes of trpl mutants and interactions between the transient receptor potential (TRP) and TRP-like channels in Drosophila. J Neurosci 20:6797–6803

Leung HT, Tseng-Crank J, Kim E, Mahapatra C, Shino S, Zhou Y, An L, Doerge RW, Pak WL (2008) DAG lipase activity is necessary for TRP channel regulation in *Drosophila* photoreceptors. Neuron 58:884–896

Lev S, Katz B, Minke B (2012) The activity of the TRP-like channel depends on its expression system. Channels (Austin Tex) 6:86–93

Li HS, Montell C (2000) TRP and the PDZ protein, INAD, form the core complex required for retention of the signalplex in *Drosophila* photoreceptor cells. J Cell Biol 150:1411–1422

Li CJ, Geng CX, Leung HT, Hong YS, Strong LLR, Schneuwly S, Pak WL (1999) INAF, a protein required for transient receptor potential Ca^{2+} channel function. Proc Natl Acad Sci U S A 96:13474–13479

Liedtke W (2007) TRPV channels' role in osmotransduction and mechanotransduction. Handb Exp Pharmacol 179:473–487

Lin SY, Corey DP (2005) TRP channels in mechanosensation. Curr Opin Neurobiol 15:350–357

Lin JY, Fisher DE (2007) Melanocyte biology and skin pigmentation. Nature 445:843–850

Liu M, Parker LL, Wadzinski BE, Shieh BH (2000) Reversible phosphorylation of the signal transduction complex in *Drosophila* photoreceptors. J Biol Chem 275:12194–12199

Liu CH, Wang T, Postma M, Obukhov AG, Montell C, Hardie RC (2007) In vivo identification and manipulation of the Ca^{2+} selectivity filter in the *Drosophila* transient receptor potential channel. J Neurosci 27:604–615

Liu CH, Satoh AK, Postma M, Huang J, Ready DF, Hardie RC (2008) Ca^{2+}-dependent metarhodopsin inactivation mediated by Calmodulin and NINAC myosin III. Neuron 59:778–789

Liu J, Ward A, Gao J, Dong Y, Nishio N, Inada H, Kang L, Yu Y, Ma D, Xu T, Mori I, Xie Z, Xu XZ (2010) C. elegans phototransduction requires a G protein-dependent cGMP pathway and a taste receptor homolog. Nat Neurosci 13:715–722

Liu W, Wen W, Wei Z, Yu J, Ye F, Liu CH, Hardie RC, Zhang M (2011) The INAD scaffold is a dynamic, redox-regulated modulator of signaling in the Drosophila eye. Cell 145:1088–1101

Lucas P, Ukhanov K, Leinders-Zufall T, Zufall F (2003) A diacylglycerol-gated cation channel in vomeronasal neuron dendrites is impaired in TRPC2 mutant mice: mechanism of pheromone transduction. Neuron 40:551–561

Macpherson MR, Pollock VP, Kean L, Southall TD, Giannakou ME, Broderick KE, Dow JA, Hardie RC, Davies SA (2005) Transient receptor potential-like (TRPL) channels are essential for calcium signalling and fluid transport in a Drosophila epithelium. Genetics 169:1541–1552

Maroto R, Raso A, Wood TG, Kurosky A, Martinac B, Hamill OP (2005) TRPC1 forms the stretch-activated cation channel in vertebrate cells. Nat Cell Biol 7:179–185

Masai I, Suzuki E, Yoon CS, Kohyama A, Hotta Y (1997) Immunolocalization of Drosophila eye-specific diacylgylcerol kinase, rdgA, which is essential for the maintenance of the photoreceptor. J Neurobiol 32:695–706

Matsumoto H, Kurien BT, Takagi Y, Kahn ES, Kinumi T, Komori N, Yamada T, Hayashi F, Isono K, Pak WL et al (1994) Phosrestin I undergoes the earliest light-induced phosphorylation by a calcium/calmodulin-dependent protein kinase in Drosophila photoreceptors. Neuron 12:997–1010

Mecklenburg KL, Takemori N, Komori N, Chu B, Hardie RC, Matsumoto H, O'Tousa JE (2010) Retinophilin is a light-regulated phosphoprotein required to suppress photoreceptor dark noise in *Drosophila*. J Neurosci 30:1238–1249

Mederos y Schnitzler M, Storch U, Meibers S, Nurwakagari P, Breit A, Essin K, Gollasch M, Gudermann T (2008) Gq-coupled receptors as mechanosensors mediating myogenic vasoconstriction. EMBO J 27:3092–3103

Minke B (1977) Drosophila mutant with a transducer defect. Biophys Struct Mech 3:59–64

Minke B (1982) Light-induced reduction in excitation efficiency in the trp mutant of Drosophila. J Gen Physiol 79:361–385

Minke B (2010) The history of the Drosophila TRP channel: the birth of a new channel superfamily. J Neurogenet 24:216–233

Minke B, Selinger Z (1991) Inositol lipid pathway in fly photoreceptors: excitation, calcium mobilization and retinal degeneration. Prog Retinal Res 11:99–124

Minke B, Wu C, Pak WL (1975) Induction of photoreceptor voltage noise in the dark in *Drosophila* mutant. Nature 258:84–87

Mishra P, Socolich M, Wall MA, Graves J, Wang Z, Ranganathan R (2007) Dynamic Scaffolding in a G Protein-Coupled Signaling System. Cell 131:80–92

Montell C (1999) Visual transduction in *Drosophila*. Annu Rev Cell Dev Biol 15:231–268

Montell C (2011) The history of TRP channels, a commentary and reflection. Pflugers Arch 461:499–506

Montell C, Rubin GM (1989) Molecular characterization of Drosophila *trp* locus, a putative integral membrane protein required for phototransduction. Neuron 2:1313–1323

Niemeyer BA, Suzuki E, Scott K, Jalink K, Zuker CS (1996) The *Drosophila* light-activated conductance is composed of the two channels TRP and TRPL. Cell 85:651–659

Oberwinkler J, Stavenga DG (2000) Calcium transients in the rhabdomeres of dark- and light-adapted fly photoreceptor cells. J Neurosci 20:1701–1709

Obukhov AG, Schultz G, Luckhoff A (1998) Regulation of heterologously expressed transient receptor potential-like channels by calcium ions. Neuroscience 85:487–495

Okada T, Inoue R, Yamazaki K, Maeda A, Kurosaki T, Yamakuni T, Tanaka I, Shimizu S, Ikenaka K, Imoto K, Mori Y (1999) Molecular and functional characterization of a novel mouse transient receptor potential protein homologue TRP7 - Ca2 + - permeable cation channel that is constitutively activated and enhanced by stimulation of G protein-coupled receptor. J Biol Chem 274:27359–27370

Otsuguro K, Tang J, Tang Y, Xiao R, Freichel M, Tsvilovskyy V, Ito S, Flockerzi V, Zhu MX, Zholos AV (2008) Isoform-specific inhibition of TRPC4 channel by phosphatidylinositol 4,5-bisphosphate. J Biol Chem 283:10026–10036

Pak WL (1995) Drosophila in vision research: The Friedenwald lecture. Invest Ophthalmol Vis Sci 36:2340–2357

Panda S, Nayak SK, Campo B, Walker JR, Hogenesch JB, Jegla T (2005) Illumination of the melanopsin signaling pathway. Science 307:600–604

Parnas M, Katz B, Minke B (2007) Open channel block by Ca^{2+} underlies the voltage dependence of *Drosophila* TRPL channel. J Gen Physiol 129:17–28

Parnas M, Katz B, Lev S, Tzarfaty V, Dadon D, Gordon-Shaag A, Metzner H, Yaka R, Minke B (2009a) Membrane lipid modulations remove divalent open channel block from TRP-Like and NMDA channels. J Neurosci 29:2371–2383

Parnas M, Peters M, Dadon D, Lev S, Vertkin I, Slutsky I, Minke B (2009b) Carvacrol is a novel inhibitor of Drosophila TRPL and mammalian TRPM7 channels. Cell Calcium 45:300–309

Patel A, Sharif-Naeini R, Folgering JR, Bichet D, Duprat F, Honore E (2010) Canonical TRP channels and mechanotransduction: from physiology to disease states. Pflugers Arch 460:571–581

Paulsen R, Bahner M, Huber A (2000) The PDZ assembled "transducisome" of microvillar photoreceptors: the TRP/TRPL problem. Pflugers Arch 439:R181–R183

Peng L, Popescu DC, Wang N, Shieh BH (2008) Anchoring TRP to the INAD macromolecular complex requires the last 14 residues in its carboxyl terminus. J Neurochem 104:1526–1535

Phillips AM, Bull A, Kelly LE (1992) Identification of a Drosophila gene encoding a calmodulin-binding protein with homology to the trp phototransduction gene. Neuron 8:631–642

Popescu DC, Ham A-JL, Shieh B-H (2006) Scaffolding protein INAD regulates deactivation of vision by promoting phosphorylation of transient receptor potential by rye protein kinase C in *Drosophila*. J Neurosci 26:8570–8577

Postma M, Oberwinkler J, Stavenga DG (1999) Does Ca^{2+} reach millimolar concentrations after single photon absorption in *Drosophila* photoreceptor microvilli? Biophys J 77:1811–1823

Poteser M, Schleifer H, Lichtenegger M, Schernthaner M, Stockner T, Kappe CO, Glasnov TN, Romanin C, Groschner K (2011) PKC-dependent coupling of calcium permeation through transient receptor potential canonical 3 (TRPC3) to calcineurin signaling in HL-1 myocytes. Proc Natl Acad Sci U S A 108:10556–10561

Provencio I, Rodriguez IR, Jiang G, Hayes WP, Moreira EF, Rollag MD (2000) A novel human opsin in the inner retina. J Neurosci 20:600–605

Quick K, Zhao J, Eijkelkamp N, Linley JE, Rugiero F, Cox JJ, Raouf R, Gringhuis M, Sexton JE, Abramowitz J, Taylor R, Forge A, Ashmore J, Kirkwood N, Kros CJ, Richardson GP, Freichel M, Flockerzi V, Birnbaumer L, Wood JN (2012) TRPC3 and TRPC6 are essential for normal mechanotransduction in subsets of sensory neurons and cochlear hair cells. Open Biol 2:120068

Raghu P (2006) Regulation of *Drosophila* TRPC channels by protein and lipid interactions. Semin Cell Dev Biol 17:646–653

Raghu P, Hardie RC (2009) Regulation of Drosophila TRPC channels by lipid messengers. Cell Calcium 45:566–573

Raghu P, Colley NJ, Webel R, James T, Hasan G, Danin M, Selinger Z, Hardie RC (2000a) Normal phototransduction in *Drosophila* photoreceptors lacking an $InsP_3$ receptor gene. Mol Cell Neurosci 15:429–445

Raghu P, Usher K, Jonas S, Chyb S, Polyanovsky A, Hardie RC (2000b) Constitutive activity of the light-sensitive channels TRP and TRPL in the *Drosophila* diacylglycerol kinase mutant, *rdgA*. Neuron 26:169–179

Ranganathan R, Bacskai BJ, Tsien RY, Zuker CS (1994) Cytosolic calcium transients: spatial localization and role in Drosophila photoreceptor cell function. Neuron 13:837–848

Reuss H, Mojet MH, Chyb S, Hardie RC (1997) In vivo analysis of the *Drosophila* light-sensitive channels, TRP and TRPL. Neuron 19:1249–1259

Rosenbaum EE, Brehm KS, Vasiljevic E, Liu CH, Hardie RC, Colley NJ (2011) XPORT-dependent transport of TRP and rhodopsin. Neuron 72:602–615

Running Deer JL, Hurley JB, Yarfitz SL (1995) G protein control of *Drosophila* photoreceptor phospholipase C. J Biol Chem 270:12623–12628

Ryu S, Liu B, Yao J, Fu Q, Qin F (2007) Uncoupling proton activation of vanilloid receptor TRPV1. J Neurosci 27:12797–12807

Satoh AK, O'Tousa JE, Ozaki K, Ready DF (2005) Rab11 mediates post-Golgi trafficking of rhodopsin to the photosensitive apical membrane of Drosophila photoreceptors. Development 132:1487–1497

Scott K, Zuker CS (1998) Assembly of the *Drosophila* phototransduction cascade into a signalling complex shapes elementary responses. Nature 395:805–808

Scott K, Becker A, Sun Y, Hardy R, Zuker C (1995) G_q a protein function in vivo: genetic dissection of its role in photoreceptor cell physiology. Neuron 15:919–927

Scott K, Sun YM, Beckingham K, Zuker CS (1997) Calmodulin regulation of *Drosophila* light-activated channels and receptor function mediates termination of the light response in vivo. Cell 91:375–383

Semtner M, Schaefer M, Pinkenburg O, Plant TD (2007) Potentiation of TRPC5 by Protons. J Biol Chem 282:33868–33878

Sengupta S, Barber TR, Xia H, Ready DF, Hardie RC (2013) Depletion of PtdIns(4,5)P2 underlies retinal degeneration in Drosophila trp mutants. J Cell Sci 126:1247–1259

Sharif-Naeini R, Dedman A, Folgering JH, Duprat F, Patel A, Nilius B, Honore E (2008) TRP channels and mechanosensory transduction: insights into the arterial myogenic response. Pflugers Arch 456:529–540

Shieh BH, Niemeyer B (1995) A novel protein encoded by the InaD gene regulates recovery of visual transduction in Drosophila. Neuron 14:201–210

Shieh BH, Zhu MY (1996) Regulation of the TRP Ca^{2+} channel by INAD in Drosophila photoreceptors. Neuron 16:991–998

Spassova MA, Hewavitharana T, Xu W, Soboloff J, Gill DL (2006) A common mechanism underlies stretch activation and receptor activation of TRPC6 channels. Proc Natl Acad Sci U S A 103:16586–16591

Steele FR, Washburn T, Rieger R, O'Tousa JE (1992) Drosophila retinal degeneration C (rdgC) encodes a novel serine/threonine protein phosphatase. Cell 69:669–676

Stortkuhl KF, Hovemann BT, Carlson JR (1999) Olfactory adaptation depends on the trp Ca2+ channel in Drosophila. J Neurosci 19:4839–4846

Su Z, Zhou X, Haynes WJ, Loukin SH, Anishkin A, Saimi Y, Kung C (2007) Yeast gain-of-function mutations reveal structure-function relationships conserved among different subfamilies of transient receptor potential channels. Proc Natl Acad Sci U S A 104:19607–19612

Suchyna TM, Johnson JH, Hamer K, Leykam JF, Gage DA, Clemo HF, Baumgarten CM, Sachs F (2000) Identification of a peptide toxin from Grammostola spatulata spider venom that blocks cation-selective stretch-activated channels. J Gen Physiol 115:583–598

Trebak M, Lemonnier L, Dehaven WI, Wedel BJ, Bird GS, Putney JW Jr (2009) Complex functions of phosphatidylinositol 4,5-bisphosphate in regulation of TRPC5 cation channels. Pflugers Arch 457:757–769

Trost C, Marquart A, Zimmer S, Philipp S, Cavalie A, Flockerzi V (1999) Ca2+-dependent interaction of the trpl cation channel and calmodulin. FEBS Lett 451:257–263

Tsunoda S, Sierralta J, Sun YM, Bodner R, Suzuki E, Becker A, Socolich M, Zuker CS (1997) A multivalent PDZ-domain protein assembles signalling complexes in a G-protein-coupled cascade. Nature 388:243–249

Tsunoda S, Sun Y, Suzuki E, Zuker C (2001) Independent anchoring and assembly mechanisms of INAD signaling complexes in *Drosophila* photoreceptors. J Neurosci 21:150–158

Vaca L, Sinkins WG, Hu Y, Kunze DL, Schilling WP (1994) Activation of recombinant trp by thapsigargin in Sf9 insect cells. Am J Physiol Cell Physiol 267:C1501–C1505

Vinos J, Jalink K, Hardy RW, Britt SG, Zuker CS (1997) A G protein-coupled receptor phosphatase required for rhodopsin function. Science 277:687–690

Voolstra O, Beck K, Oberegelsbacher C, Pfannstiel J, Huber A (2010) Light-dependent phosphorylation of the Drosophila transient receptor potential (TRP) ion channel. J Biol Chem 285:14275–14284

Wang T, Jiao Y, Montell C (2005a) Dissecting independent channel and scaffolding roles of the *Drosophila* transient receptor potential channel. J Cell Biol 171:685–694

Wang T, Xu H, Oberwinkler J, Gu Y, Hardie RC, Montell C (2005b) Light activation, adaptation, and cell survival Functions of the Na^{+}/Ca^{2+} exchanger CalX. Neuron 45:367–378

Wang YY, Chang RB, Liman ER (2010) TRPA1 Is a Component of the Nociceptive Response to CO2. J Neurosci 30:12958–12963

Warr CG, Kelly LE (1996) Identification and characterization of two distinct calmodulin-binding sites in the Trpl ion-channel protein of *Drosophila melanogaster*. Biochem J 314:497–503

Warren EJ, Allen CN, Brown RL, Robinson DW (2006) The light-activated signaling pathway in SCN-projecting rat retinal ganglion cells. Eur J Neurosci 23:2477–2487

Wes PD, Chevesich J, Jeromin A, Rosenberg C, Stetten G, Montell C (1995) TRPC1, a human homolog of a Drosophila store-operated channel. Proc Natl Acad Sci U S A 92:9652–9656

Wes PD, Xu XZ, Li HS, Chien F, Doberstein SK, Montell C (1999) Termination of phototransduction requires binding of the NINAC myosin III and the PDZ protein INAD. Nat Neurosci 2:447–453

Wicks NL, Chan JW, Najera JA, Ciriello JM, Oancea E (2011) UVA phototransduction drives early melanin synthesis in human melanocytes. Curr Biol 21:1906–1911

Xiang Y, Yuan Q, Vogt N, Looger LL, Jan LY, Jan YN (2010) Light-avoidance-mediating photoreceptors tile the Drosophila larval body wall. Nature 468:921–926

Xu XZS, Li HS, Guggino WB, Montell C (1997) Coassembly of TRP and TRPL produces a distinct store-operated conductance. Cell 89:1155–1164

Xu XZS, Choudhury A, Li XL, Montell C (1998) Coordination of an array of signaling proteins through homo- and heteromeric interactions between PDZ domains and target proteins. J Cell Biol 142:545–555

Xu XZS, Chien F, Butler A, Salkoff L, Montell C (2000) TRP gamma, a Drosophila TRP-related subunit, forms a regulated cation channel with TRPL. Neuron 26:647–657

Yau KW, Hardie RC (2009) Phototransduction motifs and variations. Cell 139:246–264

Yin J, Kuebler WM (2010) Mechanotransduction by TRP channels: general concepts and specific role in the vasculature. Cell Biochem Biophys 56:1–18

Yoon J, Ben-Ami HC, Hong YS, Park S, Strong LL, Bowman J, Geng C, Baek K, Minke B, Pak WL (2000) Novel mechanism of massive photoreceptor degeneration caused by mutations in the trp gene of Drosophila. J Neurosci 20:649–659

Zhu X, Chu PB, Peyton M, Birnbaumer L (1995) Molecular cloning of a widely expressed human homologue for the Drosophila trp gene. FEBS Lett 373:193–198

TRPs in Taste and Chemesthesis

Stephen D. Roper

Contents

Abstract

TRP channels are expressed in taste buds, nerve fibers, and keratinocytes in the oronasal cavity. These channels play integral roles in transducing chemical stimuli, giving rise to sensations of taste, irritation, warmth, coolness, and pungency. Specifically, TRPM5 acts downstream of taste receptors in the taste transduction pathway. TRPM5 channels convert taste-evoked intracellular Ca^{2+} release into membrane depolarization to trigger taste transmitter secretion. PKD2L1 is expressed in acid-sensitive (sour) taste bud cells but is unlikely to be the transducer for sour taste. TRPV1 is a receptor for pungent chemical stimuli

S.D. Roper (✉)
Department of Physiology and Biophysics, Miller School of Medicine, University of Miami, 1600 NW 10th Ave., Miami, FL 33136, USA
e-mail: sroper@med.miami.edu

B. Nilius and V. Flockerzi (eds.), *Mammalian Transient Receptor Potential (TRP) Cation Channels*, Handbook of Experimental Pharmacology 223, DOI 10.1007/978-3-319-05161-1_5,

such as capsaicin and for several irritants (chemesthesis). It is controversial whether TRPV1 is present in the taste buds and plays a direct role in taste. Instead, TRPV1 is expressed in non-gustatory sensory afferent fibers and in keratinocytes of the oronasal cavity. In many sensory fibers and epithelial cells lining the oronasal cavity, TRPA1 is also co-expressed with TRPV1. As with TRPV1, TRPA1 transduces a wide variety of irritants and, in combination with TRPV1, assures that there is a broad response to noxious chemical stimuli. Other TRP channels, including TRPM8, TRPV3, and TRPV4, play less prominent roles in chemesthesis and no known role in taste, *per se*. The pungency of foods and beverages is likely highly influenced by the temperature at which they are consumed, their acidity, and, for beverages, their carbonation. All these factors modulate the activity of TRP channels in taste buds and in the oronasal mucosa.

Keywords
TRPM5 • TRPV1 • TRPA1 • Gustation • Chemesthesis • Sour • Salty • Irritants • Spices • Carbonated sodas • Oronasal cavity • Trigeminal

1 Introduction

Consistent with their role in sensory mechanisms elsewhere in the body, TRP channels are key players in the chemical senses. This chapter reviews findings over the past decades up to the present about TRP channels in two chemical senses—taste and chemesthesis. The other major chemical sense, olfaction, although closely linked with chemesthesis (Cain and Murphy 1980; Frasnelli et al. 2007), is discussed elsewhere, in the chapter "TRPs in Olfaction." Taste comprises sensations of sweet, sour, salty, bitter, umami, and possibly fat, as generated in the peripheral end organs, the taste buds. Chemesthesis refers to the more general chemical sensitivity of the skin and mucous membranes, perceived as irritation, pungency, cooling, warmth, or heat. The following pages catalog and discuss the different TRP channels that may be involved in gustation and chemesthesis. Of the entire superfamily of TRP channels, only a limited number have known functions in taste and chemesthesis. A review of the literature reveals that certain of these TRP channels figure more prominently in taste and others in chemesthesis. Simply enumerating references to a given TRP channel in abstracts of these publications shows that the channel that is mentioned most often in taste literature is TRPM5, and in chemesthesis is TRPV1 (Fig. 1). This might indicate that these particular channels are the main players in those senses, or at least the ones most thoroughly investigated to date.

After briefly introducing the cellular organization of taste buds and the anatomy of the chemesthetic sense (next), I will present and describe the expression pattern and functional properties of TRPM5, TRPV1, and other TRP channels involved in

Fig. 1 *"Word clouds" showing the prominence of literature references to TRP channels in taste (top) and chemesthesis (bottom).* Abstracts of publications describing TRP channels in studies on taste and chemesthesis were searched. The size of the font is proportional to the number of times a given TRP channel is mentioned in these abstracts (see http://www.wordle.net/). This representation gives an approximation of how well-studied are these TRP channels in taste and chemesthesis. For taste, TRPM5 > TRPV1 ≫ PKD2L1 ~ TRPA1 > PKD1L3 > TRPMb > TRPV4 ≫ TRPV3. For chemesthesis, TRPV1 ≫ TRPA1 > TRPM8 ≫ TRPM5 > TRPV4 > TRPV2 > TRPV3

taste and chemesthesis in the order of how often they are cited in the literature, as shown in Fig. 1.

2 What Types of Cells Are Found in Taste Buds?

Taste buds represent a community of cells that express different chemosensory transduction proteins and carry out distinct functions. Taste cells are classified into three main categories, originally based on their histology and ultrastructure, but more recently distinguished by their expression patterns of key proteins involved in taste transduction (Chaudhari and Roper 2010). These categories are type I, type II, and type III taste bud cells. TRP channels are differentially expressed among these taste cell types (discussed below).

Type I cells are the most numerous cells in taste buds, comprising ½ to ⅔ of the total population of 70–85 taste cells in rodents (Ma et al. 2007). These cells enwrap adjacent taste cells with lamellar processes. These lamellar processes are endowed with a powerful ecto-ATPase (Bartel et al. 2006). This surface-bound enzyme rapidly breaks down and limits the diffusion and lifetime of ATP, a transmitter released by type II cells during taste excitation (see next). Type I cells also are believed to participate in the spatial buffering of K^+ ions released into the interstitial spaces deep within a taste bud when taste cells are stimulated and they generate bursts of action potentials (Dvoryanchikov et al. 2009). This combination of functions suggests that type I cells represent a type of glial or supporting cell within taste buds (Chaudhari and Roper 2010). Type I cells have also been implicated in salt taste (Vandenbeuch et al. 2008). This function remains to be confirmed with additional studies.

Type II taste cells make up about 20–30 % of the taste bud[1]. These taste cells possess G protein-coupled receptors (GPCRs) for sweet, bitter, and umami taste compounds. Taste GPCRs for sweet and umami compounds consist of heterodimers of T1R1, T1R2, and T1R3. Bitter taste receptors (T2Rs) form a family of about two dozen (depending on species) GPCRs that differ markedly from the T1Rs. Type II cells have also been called Receptor taste cells because they express these taste GPCRs for sweet, umami, or bitter. Some bitter-sensitive Receptor cells also respond to aversively high salt (NaCl, KCl) concentrations (Oka et al. 2013). Receptor (type II) taste cells express TRPM5 channels which are prominently involved in taste transduction, as will be shown below. Upon taste stimulation, these cells secrete ATP via membrane channels (Huang et al. 2007; Romanov et al. 2007; Taruno et al. 2013). ATP is a neurotransmitter that excites gustatory sensory afferents (Bo et al. 1999; Finger et al. 2005). ATP, acting as a paracrine transmitter, also excites adjacent type III taste cells (Huang et al. 2007).

Type III cells represent a small population of cells within taste buds, ~15 %.[1] These cells respond directly to acid (sour) taste stimuli (Huang et al. 2006, 2008b; Chang et al. 2010) Type III cells may also respond to high concentrations of NaCl and KCl (Oka et al. 2013). In addition, type III cells also respond indirectly to sweet, bitter, and umami tastes via the aforementioned purinergic paracrine pathway from Receptor (type II) cells (Tomchik et al. 2007). Type III cells are the only taste cells that have ultrastructural specializations associated with synapses. Correspondingly, these cells selectively express proteins associated with synapses, such as SNAP25, and they store and release the neurotransmitters serotonin and GABA. Because of these characteristics, type III cells have been named Presynaptic cells. Stimulation of Presynaptic (type III) cells releases serotonin and GABA into the interstitial space of the taste bud. Serotonin, and perhaps also GABA, mediates an inhibitory feedback onto Receptor (type II) taste bud cells during taste excitation (Chaudhari and Roper 2010). Presynaptic (type III) cells also express the TRP channel PKD2L1, whose function remains debatable, as will be discussed below.

These cells and their connections are reviewed in Chaudhari and Roper (2010) and Roper (2013), and diagrammed in Fig. 2.

3 What Is the Anatomical Basis for Chemesthesis?

The sense of irritation, pungency, warmth, or cooling elicited by a wide variety of chemical compounds acting on the skin and mucus membranes, coined the "common chemical sense" (Parker 1912), is now termed chemesthesis (Green et al. 1990; Green 2012). This multimodal sense is related to but distinct from taste, olfaction, pain, temperature, and itch (Laska et al. 1997; Green 2000). Examples of

[1] The proportions of types I, II, and III cells stated here are based on taste tissues from mice and rats, immunostained for cell markers that characterize these types (Ma et al. 2007). These proportions vary markedly depending on the location of the taste buds within the oral cavity.

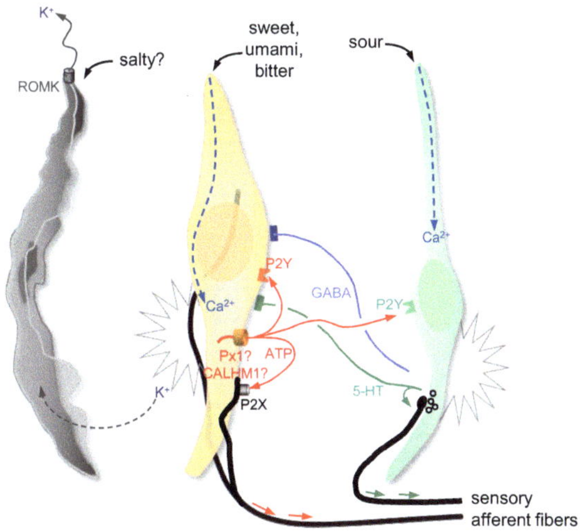

Type I glial-like cell	
neurotransmitter clearance	
GLAST	glutamate reuptake
NTPDase2	ecto-ATPase
NET	norepinephrine uptake
ion redistribution and transport	
ROMK	K^+ homeostasis
other	
OXTR	oxytocin signaling?

Type II Receptor cell	
taste transduction	
T1Rs, T2Rs	taste GPCRs
mGluRs	taste GPCRs
Gα-gus, Gγ13	G protein subunits
PLCβ2	synthesis of IP_3
TRPM5	depolarizing cation current
excitation & transmitter release	
$Na_v1.7$, $Na_v1.3$	action potential generation
Px1	ATP release channel
CALHM1	"

Type III Presynaptic cell	
surface glycoproteins, ion channels	
NCAM	neuronal adhesion
PKD channels	sour taste?
neurotransmitters, synthesis	
AADC	biogenic amine synthesis
GAD67	GABA synthesis
5-HT	neurotransmitter
Chromogranin	vesicle packaging
excitation, transmitter release	
$Na_v1.2$	action potential generation
$Ca_v2.1$, Cav1.2	voltage-gated Ca^{2+} current
SNAP25	SNARE protein, exocytosis

Fig. 2 *The three major classes of taste cells*. This classification incorporates ultrastructural features, patterns of gene expression, and the functions of each of types I, II (Receptor), and III (Presynaptic) taste cells. Type I cells (*grey*) degrade or absorb neurotransmitters. They also may clear extracellular K^+ that accumulates after action potentials (shown as starburst symbols) in Receptor (*yellow*) and Presynaptic (*green*) cells. K^+ may be extruded through an apical K channel such as ROMK. Salty taste may be transduced by some type I cells, but this remains uncertain. Sweet, bitter, and umami taste compounds activate Receptor cells, inducing them to release ATP through pannexin1 (Panx1) (Huang et al. 2007; Romanov et al. 2007) or CALHM1 (Taruno et al. 2013) channels. The extracellular ATP excites ATP receptors (P2X, P2Y) on sensory nerve fibers and on taste cells. Presynaptic cells, in turn, release serotonin (5-HT) and GABA, which inhibit Receptor cells. Sour stimuli (and carbonation, not depicted) directly activate presynaptic cells. Only presynaptic cells form ultrastructurally identifiable synapses with nerves. Starburst symbols denote action potentials. Tables below the cells list some of the proteins that are expressed in a cell type—selective manner. Modified from Chaudhari and Roper (2010)

chemesthesis include the irritating effects of concentrated saline solutions (e.g., seawater) on the nasal mucosa, or painful sensations evoked by a high concentration of CO_2. Other examples are the irritation of wasabi or hot chili peppers (capsaicin), the astringency of tannins in tea, and the unpleasant sensation from methyl salicylate.

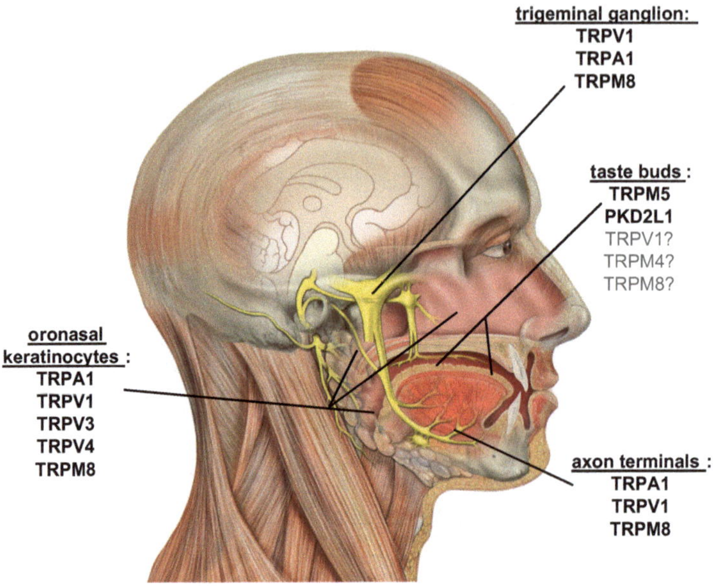

Fig. 3 *The oronasal cavity and its innervation, showing TRP channels involved in taste and chemesthesis*. TRP channels are expressed in sensory ganglion neurons, their axon terminals, and in epithelial keratinocytes throughout the oronasal cavity. The TRP channels illustrated here have known or implied functions in gustation and chemesthesis. Other TRP channels are expressed in these sites, but their functions in taste and chemesthesis are not well understood. Drawing courtesy of Patrick Lynch, Yale University

Chemesthesis also refers to nonirritating sensations such as the cooling of menthol, used in mouthwash, toothpaste, decongestants, and other products. Chemesthesis also includes pleasant sensations, for instance, those produced by aromatic spices such as oregano and mint. Given this broad range of sensations, it is not surprising that the receptor mechanisms subserving chemesthesis are equally diverse and are present in a wide array of sensory structures, including nociceptors, other free nerve endings, and keratinocytes.

Although formally, chemesthesis represents the general chemical sensitivity of the skin and mucus membranes throughout the body, this term commonly refers to sensations in the oronasal cavities (Fig. 3) and eyes (specifically, the cornea). Oronasal chemesthetic signals are believed to be conducted by somatosensory fibers in the trigeminal (V), glossopharyngeal (IX), and vagus (X) nerves. Chemesthetic stimuli are transduced by unmyelinated terminals of fibers traveling in these nerves, by isolated chemosensory cells innervated by afferent axons travelling in these nerves, and possibly by epithelial keratinocytes, as discussed below.

4 TRP Channels in Taste Buds

4.1 TRPM5 Channels Are Downstream Effectors for G Protein-Coupled Taste Receptors

The first TRP channel identified unambiguously in taste tissue was TRPM5. Consequently, there is an abundant literature on the expression and function of this channel in taste cells. Antibodies to TRPM5 and transgenic mice expressing GFP on the TRPM5 promoter have proven to be key markers for certain taste bud cells and have been invaluable for investigating taste bud structure and function.

4.1.1 TRPM5 Expression in Taste Buds

Margolskee and his colleagues (Perez et al. 2002) used single-cell PCR to search for RNAs differentially distributed among taste cells expressing or lacking gustducin, a taste-specific G_α protein associated with sweet and bitter transduction. Their search identified TRPM5 in a subset of taste bud cells. Perez et al. (2002) and, shortly after, Zhang et al. (2003) confirmed that this TRP channel is selectively expressed in cells that also possess transduction machinery for sweet and bitter tastes. Transgenic mice expressing GFP on the TRPM5 promoter and/or anti-TRPM5 immunostaining verified the association of TRPM5 with Receptor (type II) taste cells (Clapp et al. 2006; Bezencon et al. 2007; Kaske et al. 2007). Consequently, TRPM5 immunostaining or GFP expression driven by TRPM5 has become a reliable and robust marker for Receptor (type II) taste cells.

4.1.2 TRPM5 Function in Taste Buds

Based on its cellular location, Perez et al. (2002) presumed that TRPM5 participated in transduction of sweet and bitter tastes. Heterologous expression of their transcripts in CHO cells and *Xenopus* oocytes led the researchers to conclude that this TRP channel mediated Ca^{2+} influx during taste transduction. They surmised that the immediate events following gustatory activation of taste GPCRs was an IP3-mediated depletion of intracellular Ca^{2+} stores, and that this depletion triggered TRPM5 to open. Shortly following that publication, Montell and his laboratory (Hofmann et al. 2003), Liu and Liman (2003), and Prawitt et al. (2003) clarified that TRPM5 was a monovalent cation channel that was impermeable to Ca^{2+}. These researchers and Zhang et al. (2007) also reported that this channel was triggered open by a rise in, not a depletion of, intracellular Ca^{2+} consequent to taste stimulation. This is now accepted as how TRPM5 participates in taste transduction (Liman 2007). Interestingly, TRPM5 is one of only two TRP channels (the other being TRPM4) that do not permeate Ca^{2+}. They are selectively permeable to monovalent cations. Because Na^+ and K^+ ions permeate TRPM5 channels, this channel is believed to generate depolarizing receptor potentials in Receptor (type II) cells. The consensus chemotransduction pathway for taste GPCRs is outlined in Fig. 4.

Huang and Roper (2010) demonstrated the importance of TRPM5 for taste transmitter secretion, the final step in the above transduction pathway. They showed that during taste-evoked responses, the depolarization generated by TRPM5 acts in

Fig. 4 *Canonical transduction pathway for sweet, bitter, and umami taste stimuli*. These taste compounds activate G protein-coupled taste receptors that are expressed in taste Receptor (type II) cells. Taste receptor activation leads to IP3-mediated intracellular Ca^{2+} release, TRPM5 opening, and ultimately, transmitter secretion from Receptor cells. Reviewed in Chaudhari and Roper (2010)

sweet, bitter, or umami taste agonists
↓
bind to taste GPCRs (T1Rs, T2Rs)
↓
activates PLCβ2
↓
synthesizes IP3
↓
activates IP3R3 receptors
↓
releases intracellular Ca^{2+}
↓
opens TRPM5
↓
causes Na^{+} influx
↓
depolarizes cell
↓
releases transmitter

concert with Ca^{2+} released from intracellular stores to elicit non-vesicular ATP secretion, presumably through pannexin 1 and/or CAHLM1 channels (Huang et al. 2007; Romanov et al. 2007; Huang and Roper 2010; Taruno et al. 2013).

4.1.3 Genetic Ablation of Trpm5: Knockout Studies in Taste

Initial reports of genetically modified mice lacking functional TRPM5 protein showed the mice lacked normal taste responses to sweet, bitter, or umami compounds (Zhang et al. 2003). This finding cemented a role for TRPM5 in taste transduction. Later studies that used a different *Trpm5* knockout mouse strain reported that taste responses were significantly reduced, but not entirely absent (Damak et al. 2006; Oliveira-Maia et al. 2009). Those studies underlined the importance of TRPM5 in taste but also revealed taste transduction mechanisms for sweet, bitter, and umami that are independent of TRPM5. Genetically engineered mice lacking TRPM5 also have a substantially reduced response to aversively high concentrations of sodium and potassium salts (Oka et al. 2013). Specifically how TRPM5 channels participate in aversive salt taste transduction is not presently known. Lastly, Liu et al. (2011) showed that knockout mice lacking TRPM5 had reduced taste responses to linoleic acid, indicating that this TRP channel is involved in the chemotransduction pathway for fatty taste in rodents. The receptors for fatty taste are currently being hotly pursued. Whether fatty is a basic taste is currently actively debated.[2]

[2] It remains arguable whether fatty acids or fats in general stimulate a primary taste quality in humans, as opposed to the sensations of olfaction and texture that fats elicit (Tucker and Mattes 2012).

4.1.4 Pharmacological Block of TRPM5 Channels in Taste Buds

In addition to genetic knockout experiments, researchers have used pharmacological agents to block TRPM5 channel activity and assay how this affects taste. Talavera et al. (2008) showed that quinine, a pharmacological antagonist of TRPM5, reduced sweet-evoked gustatory nerve responses in mice, consistent with the role in taste transduction outlined above. To confirm that TRPM5 was the proximate target for quinine, these researchers showed that quinine had no effect in *Trpm5* knockout mice.[3] These findings may be related to the ability of the bitterness of quinine to reduce sweet, a taste quality transduced by TRPM5 (Lawless 1979; Keast and Breslin 2003; Frank et al. 2005). Sweet/bitter mixture suppression due to central cognitive effects is equally plausible (Kroeze and Bartoshuk 1985). Curiously, Talavera et al. (2008) found that quinine did not significantly inhibit taste nerve responses evoked by umami stimuli, even though umami, like sweet and bitter taste transduction, is believed to utilize TRPM5 channels. A likely explanation is that, like sweet, umami taste is transduced via multiple parallel, TRPM5-dependent and TRPM5-independent pathways (Talavera et al. 2008; Tokita and Boughter 2012; Kusuhara et al. 2013).

Other studies have tested alternative TRPM5 blockers. Palmer et al. (2010) demonstrated that triphenylphosphine oxide (TPPO), a selective antagonist for TRPM5, blocked currents in excised patches of mouse taste cells. Taste-evoked responses were not investigated in that study. Topical application of high concentrations of TPPO (~100× IC50) onto the rat tongue during taste stimulation altered salt responses and reduced gustatory nerve responses to the bitter tastant quinine (Ren et al. 2013). Findings showing responses to aversively strong salt stimulation are reduced in *Trpm5* knockout mice (Oka et al. 2013) are consistent with these effects of TPPO. However, using quinine as a bitter taste stimulus is complicated given that Talavera et al. (2008) had shown that quinine itself is a TRPM5 antagonist. It is fair to say that the detailed actions of TPPO on taste responses have not been well characterized.

Finally, another approach to blocking TRPM5 has been to eliminate inward current through this channel using ion substitutions. As previously cited, Huang and Roper (2010) eliminated TRPM5 function in isolated mouse taste Receptor (type II) cells by replacing Na^+ in the bathing medium with $NMDG^+$, a cation that does not permeate TRPM5 channels. Exchanging Na^+ with $NMDG^+$ blocked taste-evoked transmitter (ATP) release from the cells, consistent with the proposed role for TRPM5 in taste transduction.

4.1.5 Factors That Modulate TRPM5 Channel Activity in Taste Buds

Acidic solutions potently block TRPM5 when this channel is expressed in HEK cells. The effect of acid has an IC50 at pH 6.2 with nearly complete block at pH 5.4

[3] Talavera et al. (2008) used the same strain of *Trpm5* knockout mice that had been shown to exhibit reduced, but reliable, TRPM5-independent sweet responses (Damak et al. 2006). Quinine did not alter these TRPM5-independent responses.

(Liu et al. 2005). Intracellular acidification does not affect TRPM5 channels. The pronounced effect of low pH on TRPM5 is curious insofar as one might predict acidic solutions might reduce tastes that are transduced by TRPM5 channels, namely sweet, bitter, and umami. However, combining acid (sour) taste with sweet or bitter results in variable and complex outcomes, including no interaction, suppression, or enhancement, depending on concentrations of the two stimuli (Schiffman et al. 2000; Green et al. 2010; see Keast and Breslin 2003). A possible explanation for these disparate effects of acid stimulation in taste mixtures is that TRPM5 channels in situ may have different properties than when expressed in HEK cells, such as forming channel multimers that are pH insensitive or that TRPM5 channels in situ are situated on the basolateral surface of taste cells and protected from fluctuations of pH in the oral cavity, as suggested by Liu et al. (2005). Alternatively, there may be overriding central neural mechanisms (Savant and McDaniel 2004). Furthermore, acidic compounds will stimulate Presynaptic (type III) taste cells, thereby activating inhibitory paracrine pathways (serotonergic, GABAergic) in taste buds (Fig. 2), complicating any direct actions of low pH on TRPM5 channels in Receptor (type II) cells.

Temperature also modulates TRPM5 activity. Warming HEK cells expressing TRPM5 from 15° to 35 °C activates this channel (Talavera et al. 2005). This means that temperatures to which taste buds are routinely exposed when one consumes cold or hot food and beverages can affect TRPM5. Talavera et al. (2005) showed that this heat sensitivity of TRPM5 indeed confers a temperature modulation of certain taste responses. Specifically, warming taste stimuli to 35 °C enhanced responses to sweet compounds in chorda tympani nerve recordings from mice. Sweet taste responses in *Trpm5* knockout mice were unaltered by temperature changes. This temperature sensitivity of TRPM5 is believed to explain how warming or cooling the tongue gives rise to gustatory nerve responses and taste sensations even in the absence of chemical stimuli (Sato et al. 1975; Cruz and Green 2000). Warming the tongue would be expected to activate TRPM5 directly, bypassing taste receptors and IP3-mediated Ca^{2+} release and eliciting transmitter release. This simple mechanism cannot completely explain thermal taste, however, because cooling the tongue, which would have the opposite effect and suppress TRPM5 channels, elicits sour or bitter tastes (Cruz and Green 2000). Furthermore, warming solutions of taste stimuli selectively enhanced responses to sweet compounds, yet TRPM5 also transduces bitter and umami compounds. Why sweet taste is specifically enhanced was not explained. Lastly, presenting a taste stimulus at different temperatures onto the tongue elicits different responses in gustatory neurons in the brain stem (Wilson and Lemon 2013). This temperature modulation of taste-evoked responses was attributed in part to the thermal sensitivity of TRPM5.

Other regulators of TRPM5 include phosphatidylinositol-4,5-bisphosphate (PIP2) (Liu and Liman 2003) and arachidonic acid (Oike et al. 2006). How or whether these intracellular factors figure into taste modulation via alterations in TRPM5 channel activity during taste transduction remains undetermined.

4.1.6 TRPM4 in Taste Buds

There is only very indirect evidence that another member of the melastatin-related branch of TRP channels, TRPM4, is present in taste buds. TRPM4 channels, like TRPM5, do not pass Ca^{2+}, which sets these channels apart from other TRP channels. Neither the transcripts for TRPM4 nor the protein itself has been detected in taste buds to date. Yet, using patch clamp recordings on taste bud cells from mice lacking TRPM5 channels, researchers have observed Ca^{2+}-activated currents with properties suggestive of TRPM4 (Zhang et al. 2007). A role for this current in taste cells, or even whether it is generated by TRPM4, is not known.

4.1.7 Conclusions Regarding TRPM4 and TRPM5 Channels in Taste Buds[4]

There is strong, unambiguous, and compelling evidence that exists for TRPM5 channels in a subset of taste bud cells, namely in Receptor (type II) taste cells. This Ca^{2+}-activated monovalent cation channel functions as a downstream effector to link taste GPCR activation and release of intracellular Ca^{2+} with membrane depolarization during taste stimulation. TRPM5 is found in many chemoreceptor

[4] TRPM5 channels are also found in chemical-sensing cells that express "taste" GPCRs but that are located outside the taste buds. For instance, solitary chemosensory cells in the upper air tract express taste receptors and TRPM5 channels (Kaske et al. 2007; Lin et al. 2008). These cells are discussed in greater detail later in this chapter. Also, isolated receptor cells in the lower respiratory tract that are innervated by vagal sensory fibers ("brush cells") express bitter taste receptors and other components of the taste receptor transduction pathway, including TRPM5 (Kaske et al. 2007; Tizzano et al. 2010; Krasteva et al. 2011). In the intestinal tract, nutrient-sensing cells express taste GPCRs as well as TRPM5 channels (Wu et al. 2002; Fonfria et al. 2006; Bezencon et al. 2007; Jang et al. 2007; Kidd et al. 2008; Kokrashvili et al. 2009; Young et al. 2009; Janssen et al. 2011). These gut chemoreceptor cells may employ TRPM5 in a similar transduction pathway as do taste cells. Also, TRPM5 and taste receptors are expressed in the pancreas (Taniguchi 2004; Fonfria et al. 2006; Reimann et al. 2008; Nakagawa et al. 2009; Colsoul et al. 2010). TRPM5 channels are required for normal glucose-stimulated insulin secretion from the pancreas (Uchida and Tominaga 2011); *Trpm5* knockout mice have impaired glucose tolerance, and pancreatic islets from these mice show defective glucose-induced insulin release (Colsoul et al. 2010). Nakagawa et al. (2009) showed that the insulin-secreting pancreatic β cell line, MIN6, that had previously been shown to express TRPM5 channels (Prawitt et al. 2003) expresses sweet taste receptors. They reported that artificial sweeteners and glucose promoted insulin secretion from these cells. These data reinforce the notion that there may be a transduction pathway in pancreatic β cells resembling that in taste Receptor (Type II) cells. [Curiously, without referring to the earlier study showing TRPM5 expression in MIN6 cells (Prawitt et al. 2003), Nakagawa et al. (2009) reported that TRPM5 was *not* present in their MIN6 cells.] Finally, taste receptors and TRPM5 are co-expressed in spermatids (Iwatsuki et al. 2010; Li and Zhou 2012; Meyer et al. 2012; Mosinger et al. 2013). The function of this chemoreceptor transduction pathway remains to be elucidated, though it appears to be involved in spermatid differentiation and maturation (Mosinger et al. 2013).

As a generality, one might argue that many chemosensory cells throughout the body that express "taste" GPCRs also express TRPM5 and likely mobilize intracellular Ca^{2+} in a manner similar to the canonical taste transduction pathway (Fig. 4). Perhaps the nomenclature for "taste" receptor genes, "TASRs," should be reconsidered and renamed to apply more broadly to chemical sensors situated far distant from the end organs of taste in the oral cavity.

cells located outside of taste buds that also have "taste" GPCRs, suggesting a common linkage between TRPM5 and taste receptors. Evidence for expression of the related channel, TRPM4, in taste buds is considerably weaker.

4.2 Do Taste Bud Cells Express TRPV1 Channels?

The next most intensely studied TRP channel after TRPM5 in taste is TRPV1, the capsaicin receptor. Two sensations related to gustation are believed to involve TRPV1—pungency and salty taste. Pungent sensations elicited by capsaicin, the active ingredient of hot chili peppers, were investigated long before its cognate receptor TRPV1 (formerly, VR1) was cloned (Szolcsanyi 1977; Rozin et al. 1981; Lawless et al. 1985). The consensus is that this spicy ingredient mainly affects somatosensory afferents (nociceptors, see chemesthesis, below). However, there may be some ill-undefined interactions between capsaicin and taste bud cells. More recently, salt taste has been postulated to involve TRPV1 or a TRPV1-like channel (Lyall et al. 2004). Despite these associations between TRPV1 and taste, the notion that this channel is present and functional in taste buds is arguable, as reviewed below.

4.2.1 TRPV1 Expression in Taste Buds

It is debatable whether TRPV1 is expressed in taste bud cells. The findings are equivocal. TRPV1 transcripts have been identified using RT-PCR on RNA extracted from rat taste buds (Liu and Simon 2001; Lyall et al. 2004; Moon et al. 2010). It is impossible in these experiments, however, to rule out contamination from somatosensory nerve terminals (e.g., trigeminal) that might contribute TRPV1 RNA (Whitehead et al. 1999; Tohda et al. 2001) or contamination from epithelial cells that also express TRPV1 (Marincsak et al. 2009). Highlighting this problem, Ishida et al. (2002) and Kido et al. (2003) independently observed immunostaining for TRPV1 protein (then named VR1) in nerve fibers surrounding and penetrating into rat taste buds. Both laboratories specifically noted the absence of TRPV1-immunopositive taste bud cells. Those findings were contradicted by another group who claimed to find TRPV1 immunostaining in rat taste bud cells that were also immunopositive for taste receptors (Gu et al. 2009; Moon et al. 2010). These latter authors offered no explanation for the discrepancy between their findings and those of previous researchers. It must be noted that because their tissue source was rat lingual samples, Gu et al. (2009) and Moon et al. (2010) were unable to conduct the gold standard control—immunostaining tissue from *Trpv1*-knockout animals. Failing this control and given the nonselectivity of some anti-TRPV1 antibodies (Everaerts et al. 2009), the claim of TRPV1-immunopositive taste cells should be viewed with some reservation. The contradictory findings remain unresolved.

If gustatory sensory afferents, *per se*, express TRPV1, as claimed in some of the above publications, one might expect to find evidence for TRPV1 expression in the parent neurons—geniculate and petrosal ganglion cells. Given that geniculate ganglion neurons predominantly (but not exclusively) innervate taste buds, that ganglion is well suited to investigate TRPV1 expression in gustatory afferents. A comprehensive study of cranial ganglia from rats failed to reveal expression of TRPV1 in geniculate ganglion neurons, though expression was abundant in the neighboring trigeminal ganglion (Matsumoto et al. 2001). In contrast, Katsura et al. (2006) reported TRPV1 immunostaining and in situ hybridization for TRPV1 in roughly 20 % of rat geniculate ganglion cells, a finding that might be consistent with the aforementioned TRPV1 expression in fibers innervating taste buds. Yet, a portion of geniculate ganglion neurons innervates the skin of the outer ear via the posterior auricular nerve. This population might include TRPV1-expressing somatosensory nociceptors.[5] Katsura et al. (2006) did not refer to or explain their discrepancy with the earlier findings of Matsumoto et al. (2001). As explained above, rat tissue does not lend itself to immunostaining controls using knockout tissue. Furthermore, it was not clear in Katsura et al. (2006) that control probes (sense vs. antisense) were included in their in situ hybridization. In short, the absence of critical controls for in situ hybridization and immunostaining substantially weakens findings claiming expression of TRPV1 in the geniculate ganglion.

In an interesting alternative approach to exploring expression of TRPV1 in sensory fibers that innervate taste buds and specifically in chorda tympani nerve fibers, Hiura et al. (1990) injected a limited number of mice ($n = 3$) with capsaicin shortly after birth. This procedure destroys neurons that express TRPV1. Using electron microscopy, they counted the number of axons remaining in the chorda tympani nerves. Injecting capsaicin caused no loss of chorda tympani axons relative to uninjected controls, leading Hiura et al. to conclude that geniculate ganglion neurons and their axons in the chorda tympani do not have capsaicin receptors (later identified as TRPV1).[6] In earlier studies using this technique on neonatal rats, Nagy et al. (1982) had shown that peptidergic (Substance P) fibers surrounding the taste buds were destroyed by capsaicin treatment, entirely consistent with the destruction of somatosensory, capsaicin-sensitive fibers (see below). That is, neurons that express TRPV1, such as those in the trigeminal ganglion, are destroyed by neonatal capsaicin injection. Neurons lacking TRPV1 are refractory to this procedure. The above findings stress the absence of TRPV1 in sensory neurons and their axons that innervate taste buds.

---

[5] However, using patch-clamp recordings, Nakamura and Bradley (2011) reported that geniculate ganglion neurons with axons in the posterior auricular nerve specifically were insensitive to capsaicin, a TRPV1 agonist. This finding argues against a population of geniculate ganglion nociceptive neurons dedicated to the posterior auricular nerve.

[6] Nakamura and Bradley (2011) also noted that geniculate ganglion neurons with axons in the chorda tympani nerve were insensitive to capsaicin, complementing the findings of Hiura et al. (1990). The only capsaicin-responsive geniculate ganglion neurons were those innervating the soft palate and having axons in the greater petrosal nerve (Nakamura and Bradley 2011).

4.2.2 TRPV1 Function in Taste Buds

If there is no definitive evidence for the expression of TRPV1 RNA or protein in taste buds, is there instead functional evidence for this channel in the end organs of taste? Some researchers report that in isolated rat taste receptor cells, 50–100 μM capsaicin stimulates Ca^{2+} influx and inhibits K^+ conductance, though not necessarily via TRPV1 (Park et al. 2003; Costa et al. 2005).[7] Additionally, Liu and Simon (2001) found no functional evidence for TRPV1 in isolated rat taste bud cells. Liu and Simon (2001) based their conclusion on the inability of an established TRPV1 antagonist, capsazepine, to block proton-activated taste cell responses. Those findings were supported by their later study where they speculated that capsaicin elicits effects independent of TRPV1, perhaps by partitioning into the plasma membrane and influencing ion channels and GRPCs nonselectively (Costa et al. 2005).

Capsaicin (10 μM) applied to the tongue elicits small, almost negligible responses in gustatory nerves innervating taste buds (Dahl et al. 1997), a finding that does not strongly argue for the existence of capsaicin receptors in taste bud cells or taste fibers. Capsaicin solutions, however, are able to *modulate* taste nerve responses to other compounds, particularly NaCl (Wang et al. 1995; Osada et al. 1997). This modulation was interpreted as being produced indirectly, via stimulation of non-gustatory, capsaicin-sensitive somatosensory nerve terminals in the lingual epithelium (e.g., trigeminal) that released peptides such as Substance P or CGRP onto taste cells, that is, by axon reflex (Finger 1986; Holzer 1988; Wang et al. 1995).

Other researchers extended the above findings and reached a different conclusion. Namely, Lyall et al. (2004) applied the TRPV1 agonists capsaicin and resiniferatoxin topically onto the tongue and also found these agents enhanced gustatory nerve (chorda tympani) responses to NaCl stimulation. This enhancement was specifically in amiloride-insensitive NaCl responses, that component of salt taste that appears to transduce aversive concentrations of Na^+ and K^+ salts (Oka et al. 2013). Capsaicin and resiniferatoxin were ineffective in enhancing responses in *Trpv1* knockout mice. Based on these findings and their RT-PCR data (see above), Lyall et al. (2004) concluded that TRPV1 or a TRPV1-like channel is expressed in taste bud cells and transduces amiloride-insensitive salt taste. However, these conclusions have been challenged by behavioral studies of salt taste in *Trpv1* knockout mice (see below).

Lastly, investigators have conducted physiological tests to search for evidence of TRPV1 in the sensory ganglion cells that innervate taste buds—geniculate ganglion neurons. Nakamura and Bradley (2011) reported that a subset of rat geniculate ganglion neurons, namely, 27 % of the small population of neurons that innervate the soft palate responded to capsaicin (10 μM) with an inward current under patch clamp recordings. Geniculate ganglion neurons that innervated taste buds on the anterior tongue did not respond to capsaicin. Geniculate ganglion cells that responded to capsaicin had features associated with nociceptive neurons, such as large tetrodotoxin-resistance Na^+ currents and immunopositivity for isolectin B_4

[7] These are reasonable concentrations of capsaicin. Five to 100 μM capsaicin, when applied to the tongue, elicits a mild to moderate burning sensation (Simons et al. 2002).

(IB4)-FITC. It was not possible to identify whether geniculate ganglion cells that responded to capsaicin innervated taste buds *per se*. In short, the findings of Katsura et al. (2006) and Nakamura and Bradley (2011), taken together, could be interpreted as suggesting that about ¼ of the geniculate ganglion neurons have characteristics of nociceptors and that these neurons innervate the soft palate, not the tongue. Because both those studies were conducted on rats, it was not possible to validate the observations in animals lacking TRPV1 channels (i.e., knockout, next).

4.2.3 Genetic Ablation of Trpv1, Knockout Studies in Taste

Mice normally avoid drinking solutions containing high concentrations of capsaicin (100 μM), presumably due to the irritating effect generated by stimulating somatosensory TRPV1 channels (chemesthesis, see below). Not surprisingly, *Trpv1* knockout mice consume solutions of 100 μM capsaicin as readily as they do water (Costa et al. 2005). Yet, despite the absence of TRPV1 channels and the absence of capsaicin-evoked avoidance behavior, capsaicin was nonetheless able to alter taste responses in knockout mice. Specifically, 100 μM capsaicin reduced the strong taste preference for sucrose solutions in *Trpv1* knockout mice (Costa et al. 2005). The significance of these findings is that capsaicin acts on targets other than TRPV1. Reports of how capsaicin acts on taste buds and gustatory fibers should be interpreted with caution.

Trpv1 knockout mice have been studied intensively vis-à-vis the hypothesis that this vanilloid receptor transduces salt (NaCl) taste (above). Two independent studies showed that there are no changes, or only ambiguous ones, in salt taste behavior in *Trpv1* knockout mice (Ruiz et al. 2006; Treesukosol et al. 2007). Further investigations substantiated these findings and indicated that wild-type and *Trpv1* knockout mice did not differ in their perceived saltiness of moderate concentrations of NaCl solutions (Smith et al. 2012). Only when NaCl concentrations became sufficiently high to elicit irritation, possibly through nociceptors or osmoreceptors, were differences between wild-type and *Trpv1* knockout mice evident. Put simply, there is no evidence for alterations in salt taste in *Trpv1* knockout mice. Once again, the findings do not vouch for the presence of TRPV1 in taste buds or sensory fibers innervating taste buds.

4.2.4 Conclusions Regarding TRPV1 Channels in Taste[8]

In sum, neither the molecular expression of nor physiological evidence for TRPV1 in sensory gustatory afferent neurons or taste cells is particularly compelling.

[8] Parenthetically, other gustatory-related sensations attributed to TRPV1 channels include metallic taste (Riera et al. 2009) and aversive off-tastes of artificial sweeteners (Riera et al. 2008). These conclusions were reached by investigations using heterologous expression systems and taste behavioral assays. The conclusions from the behavioral studies were verified using *Trpv1* knockout mice. How or whether those findings specifically implicate TRPV1 in taste bud cells or gustatory afferents remains to be examined in greater detail. TRPV1 also appears to play some role in the consumption of ethanol, though whether this involves taste per se is unlikely. Specifically, *Trpv1* knockout mice have a higher preference for and consumption of ethanol solutions. Moreover, *Trpv1* knockout mice have greater tolerance for the inebriating action of ethanol (Blednov and Harris 2009).

RT-PCR evidence for TRPV1 expression in taste buds may stem from contaminating fragments of somatosensory nerve fibers. Immunohistochemical findings lack critical controls using *Trpv1* knockout tissue. However, the lack of evidence for TRPV1 expression in taste buds or gustatory afferent neurons does not, of course, prove that this TRP channel is not present. Until and unless more complete experiments are conducted on knockout mice lacking TRPV1, or single-cell RT-PCR on RNA taken from isolated taste bud cells (DeFazio et al. 2006), a role of this channel in taste transduction remains uncertain, at best.

4.3 Are PKD2L1 Channels the Long-Sought Sour Transducer?

4.3.1 PKD2L1 Expression in Taste Buds

LopezJimenez et al. (2006) were the first to identify PKD2L1 ("polycystic kidney disease 2-Like 1," aka TRPP2) channels co-expressed with the transmembrane protein PKD1L3 ("polycystic kidney disease 1-Like 3") in a subpopulation of mouse circumvallate taste bud cells. These cells were distinct from those that transduce sweet, bitter, and umami compounds. Consequently, LopezJimenez et al. (2006) surmised that PKD2L1 together with PKD1L3 was involved in acid (sour) or salty taste. In situ hybridization and immunostaining showed that PKD2L1 is found in taste buds throughout the tongue and palate. By contrast, PKD1L3 expression is limited to taste buds in the posterior tongue (circumvallate and foliate taste buds) (Huang et al. 2006; Ishimaru et al. 2006). Kataoka et al. (2008) showed that PKD2L1 immunostaining specifically marked Presynaptic (type III) taste cells.

4.3.2 PKD2L1/PKD1L3 Function in Taste Buds

Shortly after its discovery in taste cells, the combination of PKD2L1 and PKD1L3 was expressed in HEK293T cells and shown to form a cation channel activated by acid (sour-tasting) solutions (Ishimaru et al. 2006). Moreover, mice in which Presynaptic (type III) taste cells (and specifically, cells expressing PKD2L1) were selectively ablated lacked responses to acid taste stimuli (Huang et al. 2006). A study on a very limited sample of human subjects ($n = 2$) who exhibited the rare taste disturbance sour ageusia reported that these individuals lacked expression of PKD2L1 and PKD1L3, among other genes (Huque et al. 2009). Interestingly, expressing PKD2L1 in drosophila chemoreceptor cells that normally sense sugars and elicit feeding showed a slight but statistically significant increase in the feeding preference to citric acid (Adachi et al. 2012). These findings all strongly implicated the combination of PKD2L1 and PKD1L3 in acid taste and have led many to conclude that acid taste transduction has finally been explained at the molecular level.

Yet, in the enthusiasm for embracing the view that PKD2L1/PKD1L3 explains acid (sour) taste, important inconsistencies had been overlooked. Namely, heterologously expressed PKD2L1/PKD1L3 has a high threshold for acid activation (e.g., $EC_{50} = \text{pH } 2.9$ with acetic acid) compared with that for sour-sensing taste bud cells, and the expressed channels respond mainly to the *removal* of acid stimulation (offset), not its onset (Ishimaru et al. 2006; Inada et al. 2008; Ishii et al. 2009; Kawaguchi et al. 2010). Further, taste buds in the anterior tongue (fungiform taste buds) and palate, which respond quite well to sour stimuli, lack the PKD1L3 transmembrane protein (Huang et al. 2006; Ishimaru et al. 2006). Finally, many researchers had not taken into account the longstanding awareness that sour taste is not simply activated by acid solutions (i.e., by extracellular protons) (Harvey 1920). Indeed, a proximate stimulus for sourness is *intracellular* acidification (Lyall et al. 2001; Richter et al. 2003; Huang et al. 2008b) produced when fully protonated (i.e., uncharged) acid molecules cross the plasma membrane and dissociate in the cytosol (see (Roper 2007)).[9] In sum, certain details of PKD2L1/PKD1L3 activation do not align well with sour taste activation in taste buds.

4.3.3 Genetic Ablation of Pkd2l1 and Pkd1l3, Knockout Studies in Taste

A final argument challenging the notion that PKD2L1/PKD1L3 channels transduce acid (sour) taste is the finding that genetic knockout of PKD2L1 and/or PKD1L3 in mice reduces acid taste responses only somewhat (Horio et al. 2011) or not at all (Nelson et al. 2010). There is no doubt that these PKD channels are expressed in acid-sensitive taste cells and that ablating taste cells that express PKD2L1 eliminates acid taste (Huang et al. 2006). However, PKD2L1/PKD1L3 channels themselves may play other roles in cellular function apart from transducing acid taste.

4.3.4 Conclusions Regarding PKD2L1 in Sour Taste

In sum, TRP channels may participate in some as-yet unexplained manner to acid taste or to the physiology of acid-sensing taste bud cells, but PKD2L1/PKD1L3 channels *per se* do not appear either to be necessary or sufficient for transducing this taste quality. Mechanisms other than those involving PKD2L1 channels must explain acid taste transduction (Roper 2007; Chang et al. 2010).

---

[9] Organic ("weak") acids such as acetic acid exist as a mixture of protonated and dissociated acid molecules at levels of pH that are sharply sour tasting. Mineral ("strong") acids such as HCl that are fully dissociated in aqueous solution do not readily cross the plasma membrane (i.e., the plasma membrane is tolerably impermeable to protons). Mineral acids are not as sour tasting as organic acids at equivalent pH values. Protons can only gain access to the cytosol via H^+-permeable ion channels and transporters, neither of which are features of the PKD2L1/PKD1L3 dimer. Parenthetically, one such H^+-permeable channel has been identified in Presynaptic (type III) taste bud cells. This channel may contribute to the sour taste of mineral acids (Chang et al. 2010).

4.4 Other, Lesser-Studied TRP Channels in Taste Buds

4.4.1 TRPV4 in Taste Buds

TRPV4 immunostaining was recently identified in zebrafish taste buds (Amato et al. 2012). In other tissues, this TRP channel is associated with sensing cell volume, warm temperatures, and mechanical perturbations (Watanabe et al. 2002; Suzuki et al. 2003; Vriens et al. 2004; Becker et al. 2005; McKemy 2007). The function of TRPV4 channels in zebrafish taste buds was not investigated, but these authors speculated that TRPV4 functioned as a chemoreceptor (Amato et al. 2012). In a separate study, Bradley (2000) argued that certain mammalian taste buds, namely, those in the pharynx, are tonicity detectors. This could provide a possible role for TRPV4 in taste if this TRP channel should be found to be expressed in mammalian pharyngeal taste bud cells. Further, there has been a longstanding comparison between mechano- and osmosensing cutaneous Merkel cells and Merkel-like cells present in taste buds, at least in teleosts and amphibia (Toyoshima 1989; Whitear 1989). In rodents, cutaneous Merkel cells and their innervation in rodents express TRPV4 (Suzuki et al. 2003; Boulais et al. 2009) providing further speculation for a possible mechano- or osmosensing role for TRPV4 in taste buds. Yet, to date, there are no reports of TRPV4 expression in mammalian taste buds. All the aforementioned implied functions for this channel in taste buds are completely speculative.

4.4.2 TRPM8 in Taste Buds

TRPM8 channels are discussed below with reference to chemesthesis. There is one report describing TRPM8 expression in taste buds in the soft palate and pharynx of rats (Sato et al. 2013). Because TRPM8 channels are associated with sensing cool temperatures, these researchers surmised that palatal and pharyngeal taste buds are stimulated during swallowing cold food and drinks.

5 TRP Channels in Chemesthesis

Chemesthesis is often associated with afferent terminals of sensory ganglion neurons innervating the oronasal cavities. A comprehensive quantitative analysis of TRP channels expressed in the neurons from one of these sensory ganglia—mouse trigeminal ganglion neurons—revealed 17 of the possible 28 TRP genes (Vandewauw et al. 2013). A subset of these TRP channels—namely TRPM2,4,7,8; TRPA1; TRPML1; PKD2L1; and TRPV1,2, 4—were present at levels significantly higher than the others, perhaps suggesting these channels have especially important sensory functions. Certain of these highly expressed TRP channels transduce thermosensations and, when activated by plant-derived ligands, stimulate a sense of warmth or cooling (Fig. 5). A selection of these TRP channels is discussed below (also see reviews by Gerhold and Bautista 2009; Nilius and Appendino 2011; Viana 2011; Nilius and Appendino 2013).

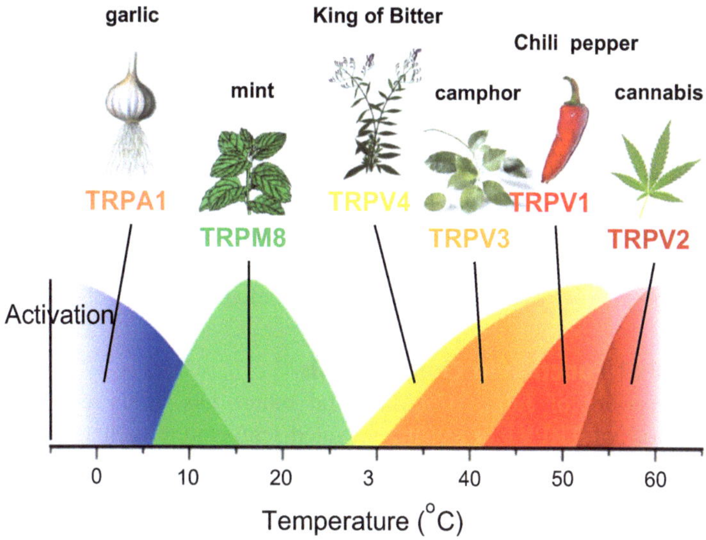

Fig. 5 *Certain TRP channels involved in chemesthesis are also thermoreceptors.* Mammalian TRP ion channels detect a broad range of temperatures. Their in vitro properties predict the thermal zones each channel potentially mediates. A role for TRPV3 and TRPV4 in thermal detection, however, is debatable (Huang et al. 2011). Many of these channels are also activated by plant derivatives, some of which provide distinct sensations of temperature. Examples of these plants are shown at top of figure. Modified from McKemy (2007)

5.1 TRPV1 Channels Play a Major Role in Chemesthesis

By far, TRPV1 is the predominant TRP channel mentioned in the scientific literature of chemesthesis (Fig. 1) and thus will be presented first. A common experience when cooking with hot chili spices is reflex coughing when kitchen vapors are inhaled. The primary irritant in chili peppers is capsaicin, the prototypic TRPV1 agonist. Capsaicin is a well-established chemesthetic irritant of the nasal cavities (Collier and Fuller 1984) as well as a pungent spice for the palate (Nilius and Appendino 2013). TRPV1, however, is involved in responses to a much broader palette of chemical irritants than just capsaicin, as discussed next.

5.1.1 TRPV1 Expression in the Oronasal Cavities

The capsaicin receptor TRPV1 is robustly expressed in somatosensory neurons and their peripheral nerve fibers (e.g., trigeminal) of the oral and nasal cavities (Caterina et al. 1997; Ishida et al. 2002; Kido et al. 2003; O'Hanlon et al. 2007; Yilmaz et al. 2007; Vandewauw et al. 2013). This widespread expression reflects a major role for this TRP channel in chemesthesis, nociception, thermoreception, and the pungency of hot spices (see reviews by Gerhold and Bautista 2009 and Nilius and Appendino 2013). A significant number (between 20 % and 56 %) of trigeminal ganglion neurons are immunopositive for TRPV1 channels (Ichikawa and Sugimoto 2001; Dussor et al. 2003; Bae et al. 2004; Tanimoto et al. 2005). Yet, as previously cited (Everaerts et al. 2009), caution must be taken interpreting positive results with TRPV1 antibodies, especially in tissues where *Trpv1* knockout animals are not available.

TRPV1 expression has also been documented by immunostaining and RT-PCR in keratinocytes and dermal mast cells, including those in the oronasal mucosa (Kido et al. 2003; Stander et al. 2004; Seki et al. 2006; Kawashima et al. 2012; Kun et al. 2012). This suggests a much broader presence of this TRP channel than just in sensory neurons. As mentioned, however, great caution should be taken interpreting immunostaining with anti-TRPV1 antibodies and RT-PCR may detect TRPV1 expression from axon fragments included in epithelial samples.

5.1.2 TRPV1 Function in Chemesthesis

The seminal work of Julius, Caterina, and others has firmly established the polymodal role of TRPV1 in sensing elevated temperature, generating the pungency of chili peppers (capsaicin) and black pepper (piperine), and eliciting respiratory reflexes (e.g., coughing) (Caterina et al. 1997, 1999; Tominaga et al. 1998; Jordt et al. 2000; Trevisani et al. 2004; McNamara et al. 2005; Adcock 2009).[10] TRPV1 is activated by many chemical compounds in addition to capsaicin and piperine, including acids, ammonia, allicin, mustard oil, cannabinoids such as anandamide, and numerous other compounds (Tominaga et al. 1998; Silver et al. 2006; Veronesi and Oortgiesen 2006; Ohta et al. 2007; Dhaka et al. 2009).

Not all acidic irritants stimulate TRPV1 effectively. Volatilized citric acid but not acetic acid elicits TRPV1-mediated respiratory reflexes (Symanowicz et al. 2004; Trevisani et al. 2004). However, acetic acid excites TRPV1 channels in the somatosensory axons that innervate the oral cavity (Arai et al. 2010). Differences between acetic and citric acids may be explained by the relative inability of citric acid, compared with acetic acid, to cross cell membranes and effect intracellular acidosis (Ishii et al. 2012). *Extracellular* protons (e.g., from citric acid) activate TRPV1 (Jordt et al. 2000), but intracellular acidification (e.g., from acetic acid) has the reverse action—it inhibits this channel (Chung et al. 2011).[11] *Intracellular* protons instead activate another TRP channel present in trigeminal axons, TRPA1 (Wang et al. 2010, 2011) (see below). Thus, whether

---

[10] An interesting aside is that TRPV1 in birds lacks the molecular binding domain for capsaicin. Consequently, birds are indifferent to the irritation of capsaicin. This allows birds to consume and disperse seeds of plants such as chili peppers that otherwise repel animals (Jordt and Julius 2002).

[11] This is a simplification. By definition, at equi-pH, citric and acetic acid solutions contribute the same $[H^+]_o$. However, at a given pH, fully protonated acetic acid molecules will more readily cross the cell membrane and contribute to intracellular acidification more effectively than will partially protonated citric acid. For example, given their dissociation constants for the fully protonated, neutral moieties (presumably the molecule that is most membrane-permeant), a 10 mM solution of citric acid ($pK_{a1} = 3.13$) at pH = 3.13 will contribute 0.75 mM H^+ and 4.9 mM triprotonated (neutral, membrane-permeant) acid molecules (H_3Citrate). The remaining 5.1 mM consists of H_2Citrate$^-$, H_1Citrate^{2-}, and Citrate^{3-}. By contrast, the same concentration of acetic acid (10 mM, $pK_a = 4.75$) at this same pH will also contribute 0.75 mM H^+ but a twofold higher concentration (9.8 mM) of the uncharged, membrane-permeant molecule, Hacetate.

TRPV1 channels are stimulated in a given situation may depend on the balance of extracellular tissue pH versus intracellular, cytosolic acidification.

Sensory afferent axons expressing TRPV1 channels not only transmit chemesthetic information orthodromically to the CNS, but they also stimulate surrounding (peripheral) tissues via antidromic axon reflex and release of glutamate and peptides from their terminals (Holzer 1988; Jeftinija et al. 1991; Omote et al. 1998). Peptides secreted by axon reflexes include Substance P and calcitonin gene-related peptide (CGRP). This local effector role of sensory nerve endings underlies inflammatory responses in the skin, such as vasodilation when nociceptors are stimulated. Indeed, Finger (1986) and Wang et al. (1995) speculated that stimulating the tongue with capsaicin releases peptides from lingual nerve fibers that terminate near or within taste buds and modulate gustatory responses. Solitary chemosensory receptor cells in the airway mucosa, discussed below, are innervated by peptidergic trigeminal fibers expressing TRPV1 (Tizzano et al. 2010). Chemesthetic stimulation of these TRPV1-rich trigeminal fibers might similarly be expected to elicit Substance P, CGRP, and glutamate release onto the chemosensory cells via axon reflex, possibly affecting the sensitivity of the solitary chemosensory cells.

A function for TRPV1 channels in keratinocytes and other nonneural tissues in the oronasal cavities is not well studied. As briefly discussed below, there are indications that TRPV1 expressed in nonneural tissues may be involved in inflammatory responses (Fernandes et al. 2012), similar to what has been found for TRPV3 in thermal hyperalgesia (Huang et al. 2008a).

5.1.3 Genetic Ablation of Trpv1, Knockout Studies in Chemesthesis

The original studies on *Trpv1* knockout mice demonstrated the central role of this TRP channel in pain and thermoreception (Caterina et al. 2000; Davis et al. 2000). Decades before *Trpv1* was cloned, capsaicin-sensitive neurons were ablated in rodents by neonatal injections of the capsinoid. Among other sensory phenomena, this pharmacological ablation showed that capsaicin-sensitive fibers were responsible for certain chemesthetic responses. For instance, rats lacking capsaicin-sensitive afferents did not respond to cigarette smoke as do intact rats (Lundberg and Saria 1983). Modern genetic ablation shows that *Trpv1* knockout mice lack responses to certain irritants but maintain normal avoidance to other chemesthetic compounds. For instance, in an isolated mouse tracheal preparation, acid-stimulated secretion of CGRP, a bronchostimulator and index of tracheal irritation, was significantly reduced in *Trpv1* knockout compared with wild-type mice (Kichko and Reeh 2009). Further, Saunders et al. (2013) demonstrated that wild-type C57Bl mice avoid cotton swabs saturated with capsaicin or cyclohexanone, but *Trpv1* knockout mice have a somewhat neutral response to the swabs. Yet, in that same study, *Trpv1* knockout mice had mixed responses or no deficiencies in reacting aversively to benzaldehyde, eugenol, amyl acetate, nicotine, toluene, and acetic acid. Similarly, mice lacking TRPV1 channels maintained normal protective respiratory reflexes against irritation by acetic acid and other volatilized irritants (Symanowicz et al. 2004). In other words, somatosensory afferents express chemesthetic receptors other than TRPV1, as will be discussed below. As Saunders et al. (2013) state, polymodal afferent fibers express multiple TRP channels besides TRPV1 to provide redundant mechanisms for detecting irritants. Eliminating TRPV1 only reduces, but does not eliminate, the capacity to respond to irritant stimuli.

How knocking out TRPV1 expression in *nonneural* tissues impacts chemesthesis has not been investigated substantially. Okada et al. (2011) reported that TRPV1 channels in basal epithelial cells in the cornea participate in inflammation and scarring after chemical irritation (alkali treatment). Corneal inflammation was reduced and healing enhanced in *Trpv1* knockout mice. A solid connection between TRPV1 and neurogenic inflammation, of course, is well established (Cortright and Szallasi 2009).

5.1.4 Modulation of TRPV1 Channels

There is a rich literature on influences that modulate TRPV1 activity, including such varied factors as bradykinin, membrane lipids, lysophosphatidic acid, heat, and acids (Sugiura et al. 2002; Tominaga and Tominaga 2005; Vay et al. 2010; Morales-Lazaro et al. 2013). Of interest for the present chapter, the chemesthetic irritant nicotine sensitizes TRPV1 channels to capsaicin, but curiously not to heat (Liu et al. 2004). This sensitization of TRPV1 channels by nicotine does not appear to affect cough reflexes. That is, the presence of nicotine did not alter capsaicin-evoked coughing (Hansson et al. 1994). Parenthetically, nicotine itself is not believed to stimulate TRPV1 channels (Silver et al. 2006).

Equally important for its role in chemesthesis, protons stimulate TRPV1 and enhance capsaicin activation of this channel (Jordt et al. 2000). This may have implications for how certain beverages and foods interact, discussed below. Briefly, acidic soft drinks such as colas might be expected to potentiate TRPV1-mediated pungency of hot spicy foods. This has not been systematically investigated, although there are numerous anecdotal references to this phenomenon.

Finally, as mentioned above, heat directly activates TRPV1. However, when tested on TRPV1-expressing HEK cells, increasing the temperature (22° to 50 °C) actually decreased the potency for capsaicin (Sprague et al. 2001). It is surprising, then, that human subjects perceive heated solutions of capsaicin (34° to 60 °C) as more pungent than cooled solutions (2° to 18 °C) (Sizer and Harris 1985; Green 1986). Yet, another report recorded little effect of temperature upon capsaicin-evoked oral irritation (Prescott et al. 1993), perhaps explained by methodological differences among the different studies. An altered pungency with increased temperature would have obvious implications for serving food spiced with chili peppers. Interactions between capsaicin and temperature on TRPV1, however, are complicated by their mutual desensitization of the channel.

5.1.5 Conclusions Regarding TRPV1 in Chemesthesis

There is no question that TRPV1 channels in somatosensory afferent fibers underlie responses to many oronasal chemesthetic stimuli, especially volatilized acids, capsaicin, and a variety of other irritants. Yet, TRPV1 immunolocalization to

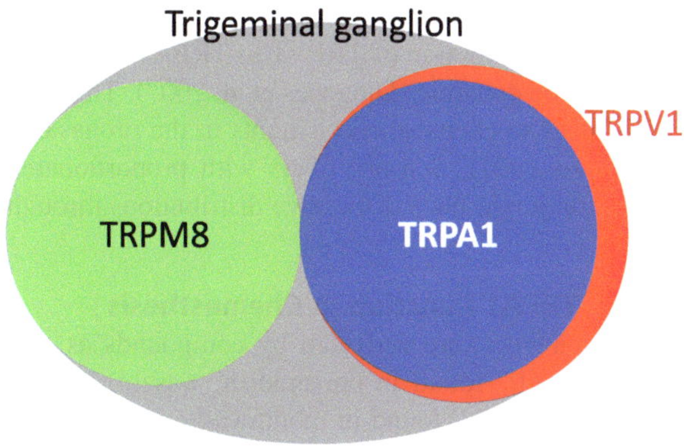

Fig. 6 *Distribution of rat trigeminal ganglion neurons that express TRPM8, TRPA1, and TRPV1*. There is extensive co-expression of TRPA1 and TRPV1 in many of these sensory ganglion cells. A separate population of neurons expresses TRPM8. This Venn diagram was created from data published in Kobayashi et al. (2005)

specific cells and tissues is clouded by uncertainties regarding anti-TRPV1 antibodies. Knockout studies show that receptors other than TRPV1 in somatosensory fibers are also activated by many chemical irritants. Redundant mechanisms for sensing oronasal chemesthetic stimuli provide a safety net to insure potentially harmful chemicals are detected. TRPV1 expression in nonneural cells such as keratinocytes may contribute to chemesthetic responses in some as-yet poorly explained manner, possibly involving inflammatory cellular responses. The ability of acidic foods and beverages to enhance TRPV1-mediated responses in the oronasal cavity is a distinct, but as-yet largely unexplored possibility.

5.2 TRPA1 Channels are Co-Expressed with TRPV1

The next most-studied TRP channel in chemesthesis is TRPA1. TRPA1 channels have been implicated in thermo- (cold), mechano-, and chemosensory reception, as well as in other physiological processes (Nilius et al. 2012). TRPA1 is co-expressed with TRPV1 in many somatosensory cells and elsewhere in the oronasal cavities. The two TRP channels TRPA1 and TRPV1 cooperate to assure that an individual responds to chemesthetic irritants.

5.2.1 TRPA1 Expression in the Oronasal Cavities

As with TRPV1 channels, TRPA1 is prominently expressed in trigeminal neurons and fibers innervating nasal and oral epithelium (Kobayashi et al. 2005; Peyrot des Gachons et al. 2011; Vandewauw et al. 2013). TRPA1 channels are also found in nonneural tissues, including keratinocytes (Anand et al. 2008; Atoyan et al. 2009; Nassini et al. 2012). TRPA1 is often co-expressed with TRPV1 (Story et al. 2003; Kobayashi et al. 2005; Akopian 2011; Vandewauw et al. 2013). For instance, in one study on rat trigeminal ganglion cells, 44 % of the neurons expressed TRPV1 and ~83 % of these cells also expressed TRPA1 (Kobayashi et al. 2005). All cells expressing TRPA1 co-expressed TRPV1. These relationships are illustrated in Fig. 6. Although, to my knowledge, the ratio of TRPV1 to TRPA1 expression from cell to cell in sensory ganglia has not been quantified, it seems unlikely that this is uniformly constant across all neurons. Instead, it seems more plausible that the relative expression of TRPV1 to TRPA1 may vary from neuron to neuron and may even be dynamic (Diogenes et al. 2007). This variation could impart different sensitivities to chemesthetic irritants in the oronasal cavities (Rentmeister-Bryant and Green 1997). Sensory fibers with proportionately different TRPA1/TRPV1 expression might have distinctive distributions throughout the mucosa.

5.2.2 TRPA1 Function in Chemesthesis

TRPA1 channels are activated by compounds as diverse as allyl isothiocyanate ("mustard oil," found in horseradish, wasabi), allicin (garlic), methyl salicylate ("wintergreen oil," found in mouthwashes), cinnamaldehyde (main constituent of cinnamon oil, present in foods, chewing gums, and toothpastes) (Bandell et al. 2004; Jordt et al. 2004; Bautista et al. 2005; Macpherson et al. 2005; Gerhold and Bautista 2009; Nilius and Appendino 2013), and even cannabinoids (Akopian et al. 2008) and ozone (Taylor-Clark and Undem 2010). The inflammatory pain evoked by lipopolysaccharide (LPS) is attributed to activation of TRPA1 channels (Andersson et al. 2012). Interestingly, LPS appears to stimulate TRPA1 not by binding to a particular molecular domain within the channel protein, but instead by inserting into the plasma membrane surrounding the channels and changing the membrane rigidity (Talavera et al. 2013). That is, LPS activates TRPA1 channels via mechanical perturbation rather than chemostimulation. Menthol, long recognized as an agonist of TRPM8 (cool sensing) channels, exerts concentration-dependent actions on TRPA1. Low concentrations of menthol activate TRPA1; high concentrations inhibit this channel (Karashima et al. 2007). Menthol may be influencing TRPA1 activity not only by ligand binding, but, as with LPS, by altering plasma membrane fluidity (Kashiwayanagi et al. 1990).[12]

Interestingly, TRPA1 channels in somatosensory fibers innervating the throat are stimulated by oleocanthal, a pungent compound present in high-quality extra-virgin olive oil. Excitation of these fibers underlies a reflexive cough elicited by consuming high-quality olive oil (Peyrot des Gachons et al. 2011).

High concentrations of CO_2 in the oral cavity and nasal passages elicit a stinging or pungent sensation. This is now understood to be transduced by diffusion of CO_2 across cell membranes of nociceptors, causing cytosolic acidification and activation of TRPA1 channels (Wang et al. 2010). Similarly, heterologously expressed TRPA1 is strongly activated by organic (weak) acids. This is explained by the same mechanism as for CO_2, namely, diffusion of the uncharged, fully protonated acid across the cell membrane, followed by intracellular acidification (Wang et al. 2010).

Stimulating TRPA1 channels in peripheral sensory fibers, as does TRPV1 activation, triggers release of Substance P and CGRP via axon reflex (Trevisani et al. 2007). These peptides initiate inflammatory responses in the surrounding

[12] Parenthetically, altering the physical properties of the plasma membrane is how menthol might be modulating TRPM8 channel activity (Morenilla-Palao et al. 2009).

tissues. One might speculate that activating TRPA1 channels in keratinocytes might trigger ATP release and stimulation of nearby somatosensory nerve endings (Dixon et al. 1999; Rong et al. 2000; Koizumi et al. 2004; Lee and Caterina 2005; Lumpkin and Caterina 2007), similar to how stimulating TRPV1 in uroepithelial keratinocytes causes them to secrete ATP and excite nociceptors (Birder et al. 2002).

TRPA1 is also associated with sensing noxious cold temperatures (Story et al. 2003), but this is more related to its role in nociception and is not further discussed here.

Finally, as mentioned above, TRPA1 and TRPV1 are co-expressed in many cells. The two TRP channels appear to form a complex at the plasma membrane and influence each other's function (Staruschenko et al. 2010; Akopian 2011). For instance, responses of neurons to mustard oil differ depending on the presence of TRPV1 (Salas et al. 2009; Staruschenko et al. 2010), in part, because mustard oil is an agonist for both these TRP channels. Similar modulation of TRPA1 responses to other chemesthetic ligands is likely. Indeed, regional differences in the sensitivity of the oral and nasal cavities to common culinary spices such as wasabi, piperine, and capsaicin may reflect the different distribution pattern of cranial nerves innervating these regions (Rentmeister-Bryant and Green 1997) as well as the presence of nerve fibers co-expressing variable proportions of TRPV1 and TRPA1.

5.2.3 Genetic Ablation of Trpa1, Knockout Studies in Chemesthesis

Mutant mice lacking TRPA1 channels have been used to validate many of the functional studies cited above. For instance, an early study using *Trpa1* knockout mice showed that TRPA1$^{-/-}$ mice consume water containing the irritant mustard oil (allyl isothiocyanate), whereas wild-type mice strongly avoid this solution (Kwan et al. 2006). In more recent studies, researchers showed that trigeminal neurons isolated from *Trpa1* knockout mice do not respond to cinnamaldehyde and only weakly to propionic acid (a weak acid that acidifies the cytosol), whereas neurons isolated from wild-type mice exhibit robust responses (Wang et al. 2011). These data support the argument that intracellular acidification stimulates TRPA1 channels. Buffer solutions saturated with ozone (10 μM) stimulate neurons isolated from vagal ganglion of wild-type mice, but do not do so for vagal ganglion neurons from *Trpa1* knockout mice. This confirms that TRPA1 is a target for ozone stimulation of vagal airway fibers (Taylor-Clark and Undem 2010). Styrene and naphthalene elicit respiratory reflexes (reduce breathing) mediated by nasal trigeminal nerves in wild-type mice but not in *Trpa1* knockout mice (Lanosa et al. 2010).

5.2.4 Conclusions Regarding TRPA1 in Chemesthesis

In summary, TRPA1 channels in somatosensory nerve fibers (and possibly in mucosal epithelial cells, too) are major targets for diverse chemesthetic stimuli. The co-expression of TRPV1 with TRPA1 in many of these same fibers and cells may modify responses to TRPA1 ligands and contribute to polymodal chemesthetic responses.

5.3 TRPM8 Channels Mediate a Cooling Sensation

After TRPV1 and TRPA1, TRPM8 is the next most commonly referenced TRP channel in publications regarding chemesthesis. TRPM8 was initially discovered in prostate epithelial cells and seems to play a poorly understood role in regulating cell growth in that tissue (Tsavaler et al. 2001; Malkia et al. 2011). Shortly after its discovery in the prostate, this channel was identified in cool-sensing thermoreceptive sensory neurons. Pertinent to the present review, TRPM8 was also found to respond to chemesthetic stimuli, specifically the cooling sensation of menthol (e.g., peppermint oil) and the synthetic compound icilin (McKemy et al. 2002; Peier et al. 2002a).

5.3.1 TRPM8 Expression in the Oronasal Cavities

TRPM8 is prominently expressed in trigeminal ganglion cells (Ta et al. 2010; Vandewauw et al. 2013). When explored with single-cell RT-PCR or immunostaining, only a subset of trigeminal neurons express TRPM8 (Nealen et al. 2003; Kobayashi et al. 2005; Kim et al. 2011) (see Fig. 6). TRPM8 immunostaining has been reported in nerve fibers in the nasal cavity (specifically, the olfactory epithelium, (Nakashimo et al. 2010), cornea (Bautista et al. 2007), and oral epithelium (Abe et al. 2005; Sato et al. 2013). Takashima et al. (2007) and Dhaka et al. (2008) genetically engineered mice to express EGFP in axons having TRPM8 channels and demonstrated that fibers with this channel innervate a variety of tissues, including dental pulp and lingual epithelium. Both Takashima et al. (2007) and Dhaka et al. (2008) state that in the tongue, TRPM8-containing fibers surround but do not penetrate into taste buds. This finding is consistent with the immunohistochemical findings of Abe et al. (2005). Contrasting this, Sato et al. (2013) reported TRPM8-immunopositive cells in structures that appear to be taste buds in the soft palate and pharynx from rats. Sato et al. (2013) concluded that taste buds in these regions differ from lingual taste buds.

5.3.2 TRPM8 Function and Genetic Ablation of Trpm8 in Chemesthesis

Naturally-occurring chemesthetic ligands that activate TRPM8, in addition to menthol from mint plants, include extracts from essential oils and aromatic spices used in cooking, mouthwashes, and fragrances, such as geraniol (from rose oil or citronella oil) and eucalyptol (from eucalyptus) (Vetter and Lewis 2011). Knocking out TRPM8 expression in mice results in a marked decrease in menthol-evoked responses recorded from sensory ganglion cells, including trigeminal neurons, isolated from these mice. Moreover, the mice themselves are significantly less sensitive to innocuous cool temperatures or topical application of the cool irritant acetone (Bautista et al. 2007; Colburn et al. 2007; Dhaka et al. 2007). Lastly, mice in which neurons that express TRPM8 were ablated did not respond to innocuous or noxious cold temperatures (Knowlton et al. 2013). These results indicate that although TRPM8 channels *per se* may not transduce temperatures <10 °C, they are co-expressed in sensory neurons with other cold receptors that do respond to these low temperatures.

5.3.3 Conclusions Regarding TRPM8 in Chemesthesis

TRPM8 is expressed in polymodal sensory neurons that respond to cool temperatures and noxious cold, though this channel itself is not responsible for transducing temperatures below ~10 °C. Chemesthetic stimuli that elicit cooling sensations, notably menthol, activate this channel in somatosensory neurons.

5.4 TRPM5 Channels also Play a Role in Chemesthesis

TRPM5 is only moderately expressed in trigeminal neurons and not nearly to the same extent as TRPV1, TRPA1, or TRPM8 and other TRP channels (Vandewauw et al. 2013). A prominent role for TRPM5 in chemesthesis is more likely due to the presence of these TRP channels in cells other than somatosensory neurons. Specifically, respiratory and laryngeal epithelium has a population of isolated, chemical-sensing receptor cells that respond to irritants. These cells exist as individual sensors, not aggregated into clusters such as taste buds. This cell population has been termed the diffuse chemosensory system (Sbarbati and Osculati 2005). Receptor cells in the diffuse chemosensory system, termed *solitary chemosensory cells*, express G protein-coupled taste receptors (specifically, T2R bitter receptors) and downstream effectors, including TRPM5 (Kaske et al. 2007; Lin et al. 2008). Solitary chemosensory cells respond to noxious chemicals and bacterial secretions, release transmitters that excite trigeminal nerve fibers, and elicit protective respiratory reflexes (Finger et al. 2003; Kaske et al. 2007; Tizzano et al. 2010; Barham et al. 2013), presumably by employing TRPM5 channels in a similar manner as with chemotransduction in taste cells. In rodents, solitary chemosensory cells also control access to the pheromone-sensing vomeronasal organ in the nasal cavity.[13] TRPM5 is required for this function (Ogura et al. 2010). Stimulation of solitary chemosensory cells can release transmitters such as acetylcholine onto sensory afferents (Krasteva et al. 2011) and elicit respiratory responses (e.g., sneezing, reduction of respiration) and epithelial inflammation. Conversely, as mentioned above, axon reflex release of glutamate and peptides (substance P, CGRP) from these same afferent terminals may modulate responses of solitary chemosensory cells.

5.4.1 Conclusions Regarding TRPM5 in Chemesthesis

The best-documented role for TRPM5 in chemesthesis is as a downstream transducer channel for "taste" GPCRs expressed in solitary chemoreceptor cells of the airway mucosa and involved in sensing chemicals released by bacteria. Other

[13] Humans do not possess a comparable anatomical structure and the existence of human pheromones is debatable.

TRPM5 functions include control of access to pheromone-sensing cells, a function also mediated by solitary chemoreceptor cells.

5.5 Less-studied TRP Channels in Chemesthesis: TRPV3, TRPV4, and TRPV2

There is a scant literature on the presence and function of TRPV2, TRPV3, and TRPV4 channels in the oronasal cavities. These TRP channels have been associated with sensing temperatures >~25° (Watanabe et al. 2002; Chung et al. 2003; Vriens et al. 2004; McKemy 2007). However, more recent experiments show that *Trpv3*/*Trpv4* double-knockout mice display no marked difference from wild-type mice regarding a preference for thermal gradients or response to heat-evoked nociception (Huang et al. 2011). Additionally, TRPV4 appears to play a role in osmosensing, mechanoreception, and stimulating pain associated with ultraviolet radiation (sunburn) (Suzuki et al. 2003; Vriens et al. 2004; Becker et al. 2005; Moore et al. 2013).

5.5.1 TRPV3,V4 Expression in the Oronasal Cavities

TRPV3 is expressed only at very low levels in trigeminal neurons from mice (Vandewauw et al. 2013), though this channel is present at significantly higher levels in trigeminal neurons from monkey (Xu et al. 2002). TRPV4 is significantly expressed in mouse trigeminal ganglion cells (Vandewauw et al. 2013). TRPV3 and TRPV4 presumably are also present in the nerve terminals of somatosensory fibers that innervate the oral and nasal epithelium. However, Nakashimo et al. (2010) failed to detect TRPV4 immunostaining in nerve fibers of the olfactory epithelium in mice.

Keratinocytes, including in the oral and nasal epithelium, express TRPV3 and TRPV4 channels (Peier et al. 2002b; Xu et al. 2002, 2006; Chung et al. 2003; Bandell et al. 2007; Nakatsuka and Iwai 2009; Kida et al. 2012). Most functional studies (see next) indicate a function for TRPV3 and TRPV4 in keratinocytes.

5.5.2 TRPV3,V4 function in Chemesthesis

TRPV3 in keratinocytes is believed to contribute to the sensations of warmth when consuming foods spiced with ingredients such as oregano, savory, clove, or thyme, representing chemostimulation by carvacrol, eugenol, and thymol (Xu et al. 2006). TRPV3 may also mediate irritation or thermosensations in the nose and throat elicited by inhaling camphor (e.g., Vicks VapoRub); camphor is a TRPV3 agonist (Moqrich et al. 2005) and icilin is an antagonist (Sherkheli et al. 2012). TRPV3 is even believed to transduce the mystical exhilaration induced by certain incenses, though this effect was attributed to TRPV3 channels in the CNS, not in oronasal keratinocytes (Moussaieff et al. 2008). How stimulation of TRPV3 channels in keratinocytes is transmitted to sensory afferent fibers is not well understood, but may involve the release of neuroactive compounds such as nitric oxide and ATP, as with TRPV4 (next) (Lumpkin and Caterina 2007; Mihara et al. 2011).

Few ligands from natural products are known for TRPV4. One is bisandrographolide A extracted from the Asian herb, "King of Bitter" (*Andrographis paniculata*) (Smith et al. 2006). This plant is used in traditional medicine in India and China. It is unlikely that its bitter taste is related to TRPV4 activity insofar as this channel is not known to be expressed in taste bud cells. As with TRPV3, in keratinocytes, TRPV4 channels may participate in thermal sensations (Watanabe et al. 2002; Chung et al. 2003; McKemy 2007). However, TRPV4 also plays a significant role in regulating cell–cell junctions and maintaining the transepidermal barrier. Genetically engineered mice lacking TRPV4 have defective skin barrier function (Sokabe et al. 2010), and activating TRPV4 (e.g., with warm temperatures, 33°) promotes this barrier and its recovery after damage (Kida et al. 2012). The barrier function of TRPV4 in keratinocytes may be related to its ability to regulate cell volume (Strotmann et al. 2000; Becker et al. 2005). By upholding the keratinocyte barrier, TRPV4 may regulate access of topically applied chemical stimuli to underlying somatosensory free nerve terminals. Yet, assigning any one particular role for TRPV4 in chemesthesis is complicated by the polymodal nature of the stimuli that activate this channel and the heterogeneity of cellular processes in which it is involved (Vriens et al. 2004).

5.5.3 Conclusions Regarding TRPV3,V4 in Chemesthesis

TRPV3 and TRPV4 are abundantly expressed in trigeminal neurons, at least in monkeys. However, there is more documentation for a function for these TRP channels in keratinocytes, where they may participate in thermal sensations and, for TRPV4, possibly controlling transepidermal barrier of the oronasal mucosa to chemical compounds. There is little direct information concerning chemesthetic responses transduced by TRPV3 or TRPV4.

5.5.4 TRPV2 and Chemesthesis: Expression, Function, and Genetic Ablation

TRPV2 is a channel that is associated with sensing high temperatures (McKemy 2007). This TRP channel is abundantly expressed in the mouse trigeminal ganglion (Vandewauw et al. 2013), and TRPV2-immunopositive fibers are present in oral and nasal epithelium (Nakashimo et al. 2010; Sasaki et al. 2013). Dendritic and Langerhans cells within the epithelium also are reported to be TRPV2 immunopositive (Shimohira et al. 2009; Sasaki et al. 2013). Yet even given this pattern of expression in fibers and oronasal epithelium, a role for TRPV2 in chemesthesis is not readily apparent. TRPV2 is insensitive to capsaicin and related hot spicy compounds (Caterina et al. 1999). Few natural compounds are known that activate TRPV2. Interestingly, TRPV2 agonists include the non-psychotropic compounds found in *Cannabis sativa*—cannabidiol, Δ9-tetrahydrocannabinol, and cannabinol (Qin et al. 2008).

Instead, TRPV2 channels have been associated with noxious stimulation, especially hot burning pain and mechanical nociception (McKemy 2007). Heterologous expression of rodent TRPV2 showed channel activation at temperatures >52° (Neeper et al. 2007). Yet, *Trpv2* knockout mice have normal responses to heat or mechanical stimulation (Park et al. 2011) and heterologous expression of human TRPV2 did not confer heat sensitivity to HEK cells (Neeper et al. 2007). These findings challenge the notion that this channel senses heat and mechanical nociception.

Concluding Remarks: TRP Channels and Implications for Consuming Food and Beverages

The quantitative study by Vandewauw et al. (2013) indicates that a number of TRP channels not mentioned above are expressed in relatively high levels in the mouse trigeminal ganglion, specifically TRPM2, TRPM4, TRPM7, TRPML1, and PKD2L1. Although there is little information about the function of these channels in chemesthesis, their presence in this ganglion suggests that they may play such a role. Additionally, as a generality, many TRP channels are co-expressed in cells and interact to modify each other's properties (Akopian 2011; Fernandes et al. 2012; see the TRIP Database: http://trpchannel.org/, Shin et al. 2011). Thus, the presence of multiple TRP channels in somatosensory ganglion cells with varying proportions of co-expression may confer a rich variety of responses to a wide spectrum of chemical stimuli in the oronasal cavities.

Additionally, many of the TRP channels discussed in this review on taste and chemesthesis are regulated by pH. Classic examples include TRPV1 channels that are directly activated by extracellular protons (Tominaga et al. 1998) and the binary complex of PKD2L1/PKD1L3 which is stimulated by the withdrawal of protons (Ishimaru et al. 2006; Inada et al. 2008; Ishii et al. 2009; Kawaguchi et al. 2010). By contrast, extracellular acidification rapidly blocks TRPM5 and slowly enhances inactivation (Liu et al. 2005). Protons in the extracellular solution activate human TRPA1 but block rodent TRPA1 (de la Roche et al. 2013). These are but a few examples of how protons act on the extracellular domains of TRP channels to regulate their opening.

Conversely, intracellular protons can also profoundly modulate TRP channels. For instance, cytosolic acidification opens rodent TRPA1 channels (Wang et al. 2010) but inhibits TRPV1 channels (Chung et al. 2011).

Interestingly, soft drinks such as sodas are often quite acidic and they are carbonated. Carbonated drinks release CO_2 that readily passes across cell membranes and acidifies the cytosol. This lowers the intracellular pH of the superficial oral epithelium and possibly somatosensory fibers coursing through this epithelium as well (Carstens et al. 2002). Intracellular acidification activates TRPA1 channels, at least in the rodent (Wang et al. 2010), and would be expected to enhance TRPA1 activity in response to agonists. Furthermore, consuming foods acidified with organic ("weak") acids such as acetic acid (vinegar) would similarly be expected to acidify both the extra- and intracellular milieu of cells (Richter et al. 2003; Roper 2007) and might be expected to modulate TRP channel responses (Fig. 7).

The acidity of many popular soft drinks, especially colas, is achieved by a mineral ("strong") acid, phosphoric acid ($pK_1 = 2.148$). Phosphoric acid is

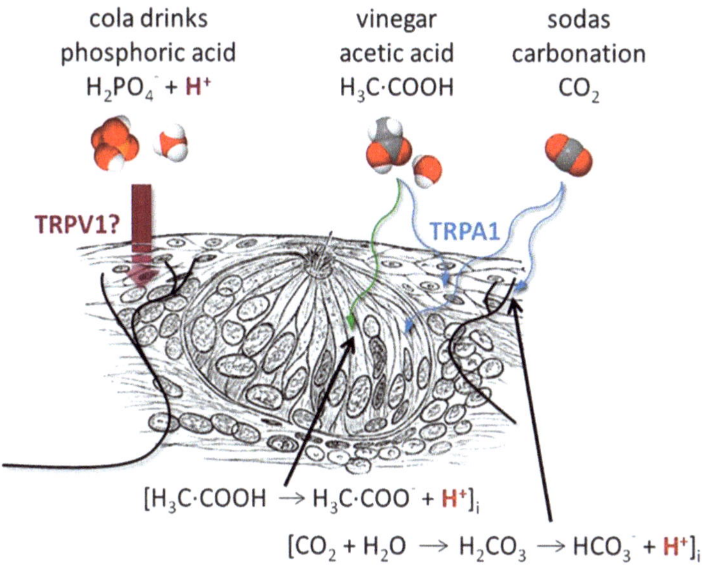

Fig. 7 *Acidic food and drink are likely to influence TRP channel activity in the oronasal epithelium*. Mineral (strong) acids, for example, phosphoric acid in cola drinks, might be expected to penetrate into and acidify the interstitial fluid spaces in the mucosal epithelium (increase $[H^+]_o$). This would enhance TRPV1 channel activity in cells and nerve fibers there. Alternatively, weak acids such as acetic acid or CO_2 from carbonated drinks will diffuse across cell membranes, dissociate inside the cells, and acidify the cytosol (increase $[H^+]_i$). This would be expected to occur inside nerve terminals, keratinocytes, and taste bud cells, as shown here. Intracellular acidification enhances TRPA1 channel activity and suppresses TRPV1 channels. Drawing of taste bud from Parker (1912)

largely dissociated at pH levels found in colas and thus will not readily pass across cell membranes. Instead, protons largely remain in the extracellular spaces. For example, Coca Cola and Pepsi Cola are acidified with phosphoric acid to pH 2.5 (Jain et al. 2007). Thus, assuming no buffering, ~70 % of the phosphoric acid is dissociated and $[H^+]_o$ may reach ~3 mM in the oral cavity.[14] These drinks will deposit extracellular protons in the epithelium and around somatosensory fibers in the oral cavity, depending on how well the acid penetrates into the oral epithelium and how much buffering takes place in the mucosa. It is likely that the interstitial spaces may become quite acidic, at least in the superficial epithelium. Such extracellular acidification of the tissue spaces would be expected to modulate the activity of TRP channels on cells and fibers there. As an example, one would anticipate that the low extracellular pH would enhance TRPV1 activity, as it does in experiments on isolated sensory neurons

[14] As an aside, this low pH contributes little sourness to colas because protons do not readily cross the plasma membrane and stimulate sour taste. To put this pH into perspective, vinegar is ~700 mM acetic acid, pH 2.3 to 2.6 (see discussion in Roper 2007).

(Petersen and LaMotte 1993; Martenson et al. 1994; Liu and Simon 2000) (Fig. 7).

Consequently, by combining carbonation with acidic pH, soda drinks would be expected to have a strong influence on foods containing TRP agonists. Indeed, the oral tingling and irritation of carbonated drinks has been documented (Carstens et al. 2002; Hewson et al. 2009). Further, interactions between TRPV1 and the artificial sweeteners used in many sodas has also been investigated (Riera et al. 2008). Yet, any specific interaction such as enhancement of TRPV1- and TRPA1-mediated pungency by drinking colas has not been systematically studied despite the abundance of anecdotes to this effect on the Internet. This includes a Facebook page dedicated to the burn of sodas and capsaicin (https://www.facebook.com/pages/Eating-spicy-food-then-accidentally-reaching-for-some-soda-to-cool-down/115325535176212). Interactions between carbonation, acidity, and pungency might be generalized beyond those elicited by soft drinks. It is likely that the rich pallet of flavors and spices during a meal is constantly in flux and amplified by interactions between food chemicals, food temperature, and the TRP channels expressed in taste buds, somatosensory fibers, and oral keratinocytes.

Acknowledgments I would like to thank Drs. Sidney A Simon (Duke University), Thomas Finger (University of Colorado School of Medicine), and Emily R Liman (University of Southern California) for their detailed reading of and perceptive remarks on this manuscript during its writing and editing.

References

Abe J, Hosokawa H, Okazawa M, Kandachi M, Sawada Y, Yamanaka K, Matsumura K, Kobayashi S (2005) TRPM8 protein localization in trigeminal ganglion and taste papillae. Brain Res Mol Brain Res 136:91–98

Adachi R, Sasaki Y, Morita H, Komai M, Shirakawa H, Goto T, Furuyama A, Isono K (2012) Behavioral analysis of Drosophila transformants expressing human taste receptor genes in the gustatory receptor neurons. J Neurogenet 26:198–205

Adcock JJ (2009) TRPV1 receptors in sensitisation of cough and pain reflexes. Pulm Pharmacol Ther 22:65–70

Akopian AN (2011) Regulation of nociceptive transmission at the periphery via TRPA1-TRPV1 interactions. Curr Pharm Biotechnol 12:89–94

Akopian AN, Ruparel NB, Patwardhan A, Hargreaves KM (2008) Cannabinoids desensitize capsaicin and mustard oil responses in sensory neurons via TRPA1 activation. J Neurosci 28:1064–1075

Amato V, Vina E, Calavia MG, Guerrera MC, Laura R, Navarro M, De Carlos F, Cobo J, Germana A, Vega JA (2012) TRPV4 in the sensory organs of adult zebrafish. Microsc Res Techniq 75:89–96

Anand U, Otto WR, Facer P, Zebda N, Selmer I, Gunthorpe MJ, Chessell IP, Sinisi M, Birch R, Anand P (2008) TRPA1 receptor localisation in the human peripheral nervous system and functional studies in cultured human and rat sensory neurons. Neurosci Lett 438:221–227

Andersson DA, Gentry C, Bevan S (2012) TRPA1 has a key role in the somatic pro-nociceptive actions of hydrogen sulfide. PLoS One 7:e46917

Arai T, Ohkuri T, Yasumatsu K, Kaga T, Ninomiya Y (2010) The role of transient receptor potential vanilloid-1 on neural responses to acids by the chorda tympani, glossopharyngeal and superior laryngeal nerves in mice. Neuroscience 165:1476–1489
Atoyan R, Shander D, Botchkareva NV (2009) Non-neuronal expression of transient receptor potential type A1 (TRPA1) in human skin. J Invest Dermatol 129:2312–2315
Bae YC, Oh JM, Hwang SJ, Shigenaga Y, Valtschanoff JG (2004) Expression of vanilloid receptor TRPV1 in the rat trigeminal sensory nuclei. J Comp Neurol 478:62–71
Bandell M, Story GM, Hwang SW, Viswanath V, Eid SR, Petrus MJ, Earley TJ, Patapoutian A (2004) Noxious cold ion channel TRPA1 is activated by pungent compounds and bradykinin. Neuron 41:849–857
Bandell M, Macpherson LJ, Patapoutian A (2007) From chills to chilis: mechanisms for thermosensation and chemesthesis via thermoTRPs. Curr Opin Neurobiol 17:490–497
Barham HP, Cooper SE, Anderson CB, Tizzano M, Kingdom TT, Finger TE, Kinnamon SC, Ramakrishnan VR (2013) Solitary chemosensory cells and bitter taste receptor signaling in human sinonasal mucosa. Int Forum Allergy Rhinol 3:450–457
Bartel DL, Sullivan SL, Lavoie EG, Sevigny J, Finger TE (2006) Nucleoside triphosphate diphosphohydrolase-2 is the ecto-ATPase of type I cells in taste buds. J Comp Neurol 497: 1–12
Bautista DM, Movahed P, Hinman A, Axelsson HE, Sterner O, Hogestatt ED, Julius D, Jordt SE, Zygmunt PM (2005) Pungent products from garlic activate the sensory ion channel TRPA1. Proc Natl Acad Sci USA 102:12248–12252
Bautista DM, Siemens J, Glazer JM, Tsuruda PR, Basbaum AI, Stucky CL, Jordt SE, Julius D (2007) The menthol receptor TRPM8 is the principal detector of environmental cold. Nature 448:204–208
Becker D, Blase C, Bereiter-Hahn J, Jendrach M (2005) TRPV4 exhibits a functional role in cell-volume regulation. J Cell Sci 118:2435–2440
Bezencon C, le CJ, Damak S (2007) Taste-signaling proteins are coexpressed in solitary intestinal epithelial cells. Chem Senses 32:41–49
Birder LA, Nakamura Y, Kiss S, Nealen ML, Barrick S, Kanai AJ, Wang E, Ruiz G, De Groat WC, Apodaca G, Watkins S, Caterina MJ (2002) Altered urinary bladder function in mice lacking the vanilloid receptor TRPV1. Nat Neurosci 5:856–860
Blednov YA, Harris RA (2009) Deletion of vanilloid receptor (TRPV1) in mice alters behavioral effects of ethanol. Neuropharmacology 56:814–820
Bo X, Alavi A, Xiang Z, Oglesby I, Ford A, Burnstock G (1999) Localization of ATP-gated P2X2 and P2X3 receptor immunoreactive nerves in rat taste buds. Neuroreport 10(5):1107–1111
Boulais N, Pennec JP, Lebonvallet N, Pereira U, Rougier N, Dorange G, Chesne C, Misery L (2009) Rat Merkel cells are mechanoreceptors and osmoreceptors. PLoS One 4:e7759
Bradley RM (2000) Sensory receptors of the larynx. Am J Med 108:47S–50S
Cain WS, Murphy CL (1980) Interaction between chemoreceptive modalities of odour and irritation. Nature 284:255–257
Carstens E, Iodi Carstens M, Dessirier J-M, O'Mahony M, Simons CT, Sudo M, Sudo S (2002) It hurts so good: oral irritation by spices and carbonated drinks and the underlying neural mechanisms. Food Qual Pref 13:431–443
Caterina MJ, Schumacher MA, Tominaga M, Rosen TA, Levine JD, Julius D (1997) The capsaicin receptor: a heat-activated ion channel in the pain pathway. Nature 389:816–824
Caterina MJ, Rosen TA, Tominaga M, Brake AJ, Julius D (1999) A capsaicin-receptor homologue with a high threshold for noxious heat. Nature 398:436–441
Caterina MJ, Leffler A, Malmberg AB, Martin WJ, Trafton J, Petersen-Zeitz KR, Koltzenburg M, Basbaum AI, Julius D (2000) Impaired nociception and pain sensation in mice lacking the capsaicin receptor. Science 288:306–313
Chang RB, Waters H, Liman ER (2010) A proton current drives action potentials in genetically identified sour taste cells. Proc Natl Acad Sci USA 107:22320–22325
Chaudhari N, Roper SD (2010) The cell biology of taste. J Cell Biol 190:285–296

Chung MK, Lee H, Caterina MJ (2003) Warm temperatures activate TRPV4 in mouse 308 keratinocytes. J Biol Chem 278:32037–32046
Chung S, Kim YH, Koh JY, Nam TS, Ahn DS (2011) Intracellular acidification evoked by moderate extracellular acidosis attenuates transient receptor potential V1 (TRPV1) channel activity in rat dorsal root ganglion neurons. Exp Physiol 96:1270–1281
Clapp TR, Medler KF, Damak S, Margolskee RF, Kinnamon SC (2006) Mouse taste cells with G protein-coupled taste receptors lack voltage-gated calcium channels and SNAP-25. BMC Biol 4:7
Colburn RW, Lubin ML, Stone DJ Jr, Wang Y, Lawrence D, D'Andrea MR, Brandt MR, Liu Y, Flores CM, Qin N (2007) Attenuated cold sensitivity in TRPM8 null mice. Neuron 54:379–386
Collier JG, Fuller RW (1984) Capsaicin inhalation in man and the effects of sodium cromoglycate. Br J Pharmacol 81:113–117
Colsoul B, Schraenen A, Lemaire K, Quintens R, Van Lommel L, Segal A, Owsianik G, Talavera K, Voets T, Margolskee RF, Kokrashvili Z, Gilon P, Nilius B, Schuit FC, Vennekens R (2010) Loss of high-frequency glucose-induced Ca^{2+} oscillations in pancreatic islets correlates with impaired glucose tolerance in Trpm5-/- mice. Proc Natl Acad Sci USA 107: 5208–5213
Cortright DN, Szallasi A (2009) TRP channels and pain. Curr Pharm Des 15:1736–1749
Costa RM, Liu L, Nicolelis MA, Simon SA (2005) Gustatory effects of capsaicin that are independent of TRPV1 receptors. Chem Senses 30:i198–200
Cruz A, Green BG (2000) Thermal stimulation of taste. Nature 403:889–892
Dahl M, Erickson RP, Simon SA (1997) Neural responses to bitter compounds in rats. Brain Res 756:22–34
Damak S, Rong M, Yasumatsu K, Kokrashvili Z, Perez CA, Shigemura N, Yoshida R, Mosinger B, Glendinning JI, Ninomiya Y, Margolskee RF (2006) Trpm5 null mice respond to bitter, sweet, and umami compounds. Chem Senses 31:253–264
Davis JB, Gray J, Gunthorpe MJ, Hatcher JP, Davey PT, Overend P, Harries MH, Latcham J, Clapham C, Atkinson K, Hughes SA, Rance K, Grau E, Harper AJ, Pugh PL, Rogers DC, Bingham S, Randall A, Sheardown SA (2000) Vanilloid receptor-1 is essential for inflammatory thermal hyperalgesia. Nature 405:183–187
de la Roche J, Eberhardt MJ, Klinger AB, Stanslowksy N, Wegner F, Koppert W, Reeh PW, Lampert A, Fischer MJ, Leffler A (2013) The molecular basis for species-specific activation of human TRPA1 by protons involves poorly conserved residues within transmembrane domains 5 and 6. J Biol Chem 288:20280–20292
DeFazio RA, Dvoryanchikov G, Maruyama Y, Kim JW, Pereira E, Roper SD, Chaudhari N (2006) Separate populations of receptor cells and presynaptic cells in mouse taste buds. J Neurosci 26:3971–3980
Dhaka A, Murray AN, Mathur J, Earley TJ, Petrus MJ, Patapoutian A (2007) TRPM8 is required for cold sensation in mice. Neuron 54:371–378
Dhaka A, Earley TJ, Watson J, Patapoutian A (2008) Visualizing cold spots: TRPM8-expressing sensory neurons and their projections. J Neurosci 28:566–575
Dhaka A, Uzzell V, Dubin AE, Mathur J, Petrus M, Bandell M, Patapoutian A (2009) TRPV1 is activated by both acidic and basic pH. J Neurosci 29:153–158
Diogenes A, Akopian AN, Hargreaves KM (2007) NGF up-regulates TRPA1: implications for orofacial pain. J Dent Res 86:550–555
Dixon CJ, Bowler WB, Littlewood-Evans A, Dillon JP, Bilbe G, Sharpe GR, Gallagher JA (1999) Regulation of epidermal homeostasis through P2Y2 receptors. Br J Pharmacol 127:1680–1686
Dussor GO, Leong AS, Gracia NB, Kilo S, Price TJ, Hargreaves KM, Flores CM (2003) Potentiation of evoked calcitonin gene-related peptide release from oral mucosa: a potential basis for the pro-inflammatory effects of nicotine. Eur J Neurosci 18:2515–2526
Dvoryanchikov G, Sinclair MS, Perea-Martinez I, Wang T, Chaudhari N (2009) Inward rectifier channel, ROMK, is localized to the apical tips of glial-like cells in mouse taste buds. J Comp Neurol 517:1–14

Everaerts W, Sepulveda MR, Gevaert T, Roskams T, Nilius B, De Ridder D (2009) Where is TRPV1 expressed in the bladder, do we see the real channel? Naunyn Schmiedebergs Arch Pharmacol 379:421–425
Fernandes ES, Fernandes MA, Keeble JE (2012) The functions of TRPA1 and TRPV1: moving away from sensory nerves. Br J Pharmacol 166:510–521
Finger TE (1986) Peptide immunohistochemistry demonstrates multiple classes of perigemmal nerve fibers in the circumvallate papilla of the rat. Chem Senses 11:135–144
Finger TE, Bottger B, Hansen A, Anderson KT, Alimohammadi H, Silver WL (2003) Solitary chemoreceptor cells in the nasal cavity serve as sentinels of respiration. Proc Natl Acad Sci USA 100:8981–8986
Finger TE, Danilova V, Barrows J, Bartel DL, Vigers AJ, Stone L, Hellekant G, Kinnamon SC (2005) ATP signaling is crucial for communication from taste buds to gustatory nerves. Science 310:1495–1499
Fonfria E, Murdock PR, Cusdin FS, Benham CD, Kelsell RE, McNulty S (2006) Tissue distribution profiles of the human TRPM cation channel family. J Recept Signal Transduct Res 26:159–178
Frank ME, Formaker BK, Hettinger TP (2005) Peripheral gustatory processing of sweet stimuli by golden hamsters. Brain Res Bull 66:70–84
Frasnelli J, Schuster B, Hummel T (2007) Interactions between olfaction and the trigeminal system: what can be learned from olfactory loss. Cereb Cortex 17:2268–2275
Gerhold KA, Bautista DM (2009) Molecular and cellular mechanisms of trigeminal chemosensation. Ann N Y Acad Sci 1170:184–189
Green BG (1986) Sensory interactions between capsaicin and temperature in the oral cavity. Chem Senses 11:371–382
Green BG (2000) Measurement of sensory irritation of the skin. Am J Contact Dermat 11:170–180
Green BG (2012) Chemesthesis and the chemical senses as components of a "chemofensor complex". Chem Senses 37:201–206
Green BG, Macon JR, Kare MR (1990) Chemical senses volume 2: irritation. Marcel Dekker, New York, NY
Green BG, Lim J, Osterhoff F, Blacher K, Nachtigal D (2010) Taste mixture interactions: suppression, additivity, and the predominance of sweetness. Physiol Behav 101:731–737
Gu XF, Lee JH, Yoo SB, Moon YW, Jahng JW (2009) Intra-oral pre-treatment with capsaicin increases consumption of sweet solutions in rats. Nutr Neurosci 12:149–154
Hansson L, Choudry NB, Karlsson JA, Fuller RW (1994) Inhaled nicotine in humans: effect on the respiratory and cardiovascular systems. J Appl Physiol 76:2420–2427
Harvey RB (1920) The relation between the total acidity, the concentration of the hydrogen ion, and the taste of acid solutions. J Am Chem Soc 42:712–714
Hewson L, Hollowood T, Chandra S, Hort J (2009) Gustatory, olfactory and trigeminal interactions in a model carbonated beverage. Chem Percept 2:94–107
Hiura A, Ishizuka H, Sakamoto Y (1990) Electron microscopic study of the effect of capsaicin on the mouse chorda tympani nerves. Arch Oral Biol 35:913–916
Hofmann T, Chubanov V, Gudermann T, Montell C (2003) TRPM5 is a voltage-modulated and Ca(2+)-activated monovalent selective cation channel. Curr Biol 13:1153–1158
Holzer P (1988) Local effector functions of capsaicin-sensitive sensory nerve endings: involvement of tachykinins, calcitonin gene-related peptide and other neuropeptides. Neuroscience 24:739–768
Horio N, Yoshida R, Yasumatsu K, Yanagawa Y, Ishimaru Y, Matsunami H, Ninomiya Y (2011) Sour taste responses in mice lacking PKD channels. PLoS One 6:e20007
Huang YA, Roper SD (2010) Intracellular Ca^{2+} and TRPM5-mediated membrane depolarization produce ATP secretion from taste receptor cells. J Physiol 588:2343–2350
Huang AL, Chen X, Hoon MA, Chandrashekar J, Guo W, Trankner D, Ryba NJ, Zuker CS (2006) The cells and logic for mammalian sour taste detection. Nature 442:934–938

Huang YJ, Maruyama Y, Dvoryanchikov G, Pereira E, Chaudhari N, Roper SD (2007) The role of pannexin 1 hemichannels in ATP release and cell-cell communication in mouse taste buds. Proc Natl Acad Sci USA 104:6436–6441
Huang SM, Lee H, Chung MK, Park U, Yu YY, Bradshaw HB, Coulombe PA, Walker JM, Caterina MJ (2008a) Overexpressed transient receptor potential vanilloid 3 ion channels in skin keratinocytes modulate pain sensitivity via prostaglandin E2. J Neurosci 28:13727–13737
Huang YA, Maruyama Y, Stimac R, Roper SD (2008b) Presynaptic (Type III) cells in mouse taste buds sense sour (acid) taste. J Physiol 586:2903–2912
Huang SM, Li X, Yu Y, Wang J, Caterina MJ (2011) TRPV3 and TRPV4 ion channels are not major contributors to mouse heat sensation. Mol Pain 7:37
Huque T, Cowart BJ, Dankulich-Nagrudny L, Pribitkin EA, Bayley DL, Spielman AI, Feldman RS, Mackler SA, Brand JG (2009) Sour ageusia in two individuals implicates ion channels of the ASIC and PKD families in human sour taste perception at the anterior tongue. PLoS One 4:e7347
Ichikawa H, Sugimoto T (2001) VR1-immunoreactive primary sensory neurons in the rat trigeminal ganglion. Brain Res 890:184–188
Inada H, Kawabata F, Ishimaru Y, Fushiki T, Matsunami H, Tominaga M (2008) Off-response property of an acid-activated cation channel complex PKD1L3-PKD2L1. EMBO Rep 9: 690–697
Ishida Y, Ugawa S, Ueda T, Murakami S, Shimada S (2002) Vanilloid receptor subtype-1 (VR1) is specifically localized to taste papillae. Brain Res Mol Brain Res 107:17–22
Ishii S, Misaka T, Kishi M, Kaga T, Ishimaru Y, Abe K (2009) Acetic acid activates PKD1L3–PKD2L1 channel–a candidate sour taste receptor. Biochem Biophys Res Commun 385: 346–350
Ishii S, Kishi M, Yamagami K, Okada S, Abe K, Misaka T (2012) The use of mammalian cultured cells loaded with a fluorescent dye shows specific membrane penetration of undissociated acetic acid. Biosci Biotechnol Biochem 76:523–529
Ishimaru Y, Inada H, Kubota M, Zhuang H, Tominaga M, Matsunami H (2006) Transient receptor potential family members PKD1L3 and PKD2L1 form a candidate sour taste receptor. Proc Natl Acad Sci USA 103:12569–12574
Iwatsuki K, Nomura M, Shibata A, Ichikawa R, Enciso PL, Wang L, Takayanagi R, Torii K, Uneyama H (2010) Generation and characterization of T1R2-LacZ knock-in mouse. Biochem Biophys Res Commun 402:495–499
Jain P, Nihill P, Sobkowski J, Agustin MZ (2007) Commercial soft drinks: pH and in vitro dissolution of enamel. Gen Dent 55:150–154
Jang HJ, Kokrashvili Z, Theodorakis MJ, Carlson OD, Kim BJ, Zhou J, Kim HH, Xu X, Chan SL, Juhaszova M, Bernier M, Mosinger B, Margolskee RF, Egan JM (2007) Gut-expressed gustducin and taste receptors regulate secretion of glucagon-like peptide-1. Proc Natl Acad Sci USA 104:15069–15074
Janssen S, Laermans J, Verhulst PJ, Thijs T, Tack J, Depoortere I (2011) Bitter taste receptors and alpha-gustducin regulate the secretion of ghrelin with functional effects on food intake and gastric emptying. Proc Natl Acad Sci USA 108:2094–2099
Jeftinija S, Jeftinija K, Liu F, Skilling SR, Smullin DH, Larson AA (1991) Excitatory amino acids are released from rat primary afferent neurons in vitro. Neurosci Lett 125:191–194
Jordt SE, Julius D (2002) Molecular basis for species-specific sensitivity to "hot" chili peppers. Cell 108:421–430
Jordt SE, Tominaga M, Julius D (2000) Acid potentiation of the capsaicin receptor determined by a key extracellular site. Proc Natl Acad Sci USA 97:8134–8139
Jordt SE, Bautista DM, Chuang HH, McKemy DD, Zygmunt PM, Hogestatt ED, Meng ID, Julius D (2004) Mustard oils and cannabinoids excite sensory nerve fibres through the TRP channel ANKTM1. Nature 427:260–265
Karashima Y, Damann N, Prenen J, Talavera K, Segal A, Voets T, Nilius B (2007) Bimodal action of menthol on the transient receptor potential channel TRPA1. J Neurosci 27:9874–9884

Kashiwayanagi M, Suenaga A, Enomoto S, Kurihara K (1990) Membrane fluidity changes of liposomes in response to various odorants. Complexity of membrane composition and variety of adsorption sites for odorants. Biophys J 58:887–895

Kaske S, Krasteva G, Konig P, Kummer W, Hofmann T, Gudermann T, Chubanov V (2007) TRPM5, a taste-signaling transient receptor potential ion-channel, is a ubiquitous signaling component in chemosensory cells. BMC Neurosci 8:49

Kataoka S, Yang R, Ishimaru Y, Matsunami H, Sevigny J, Kinnamon JC, Finger TE (2008) The candidate sour taste receptor, PKD2L1, is expressed by type III taste cells in the mouse. Chem Senses 33:243–254

Katsura H, Tsuzuki K, Noguchi K, Sakagami M (2006) Differential expression of capsaicin-, menthol-, and mustard oil-sensitive receptors in naive rat geniculate ganglion neurons. Chem Senses 31:681–688

Kawaguchi H, Yamanaka A, Uchida K, Shibasaki K, Sokabe T, Maruyama Y, Yanagawa Y, Murakami S, Tominaga M (2010) Activation of polycystic kidney disease-2-like 1 (PKD2L1)-PKD1L3 complex by acid in mouse taste cells. J Biol Chem 285:17277–17281

Kawashima M, Imura K, Sato I (2012) Topographical organization of TRPV1-immunoreactive epithelium and CGRP-immunoreactive nerve terminals in rodent tongue. Eur J Histochem 56: e21

Keast RSJ, Breslin PAS (2003) An overview of binary taste–taste interactions. Food Qual Prefer 14:111–124

Kichko TI, Reeh PW (2009) TRPV1 controls acid- and heat-induced calcitonin gene-related peptide release and sensitization by bradykinin in the isolated mouse trachea. Eur J Neurosci 29:1896–1904

Kida N, Sokabe T, Kashio M, Haruna K, Mizuno Y, Suga Y, Nishikawa K, Kanamaru A, Hongo M, Oba A, Tominaga M (2012) Importance of transient receptor potential vanilloid 4 (TRPV4) in epidermal barrier function in human skin keratinocytes. Pflugers Arch 463: 715–725

Kidd M, Modlin IM, Gustafsson BI, Drozdov I, Hauso O, Pfragner R (2008) Luminal regulation of normal and neoplastic human EC cell serotonin release is mediated by bile salts, amines, tastants, and olfactants. Am J Physiol Gastrointest Liver Physiol 295:G260–272

Kido MA, Muroya H, Yamaza T, Terada Y, Tanaka T (2003) Vanilloid receptor expression in the rat tongue and palate. J Dent Res 82:393–397

Kim HY, Chung G, Jo HJ, Kim YS, Bae YC, Jung SJ, Kim JS, Oh SB (2011) Characterization of dental nociceptive neurons. J Dent Res 90:771–776

Knowlton WM, Palkar R, Lippoldt EK, McCoy DD, Baluch F, Chen J, McKemy DD (2013) A sensory-labeled line for cold: TRPM8-expressing sensory neurons define the cellular basis for cold, cold pain, and cooling-mediated analgesia. J Neurosci 33:2837–2848

Kobayashi K, Fukuoka T, Obata K, Yamanaka H, Dai Y, Tokunaga A, Noguchi K (2005) Distinct expression of TRPM8, TRPA1, and TRPV1 mRNAs in rat primary afferent neurons with adelta/c-fibers and colocalization with trk receptors. J Comp Neurol 493:596–606

Koizumi S, Fujishita K, Inoue K, Shigemoto-Mogami Y, Tsuda M, Inoue K (2004) Ca^{2+} waves in keratinocytes are transmitted to sensory neurons: the involvement of extracellular ATP and P2Y2 receptor activation. Biochem J 380:329–338

Kokrashvili Z, Rodriguez D, Yevshayeva V, Zhou H, Margolskee RF, Mosinger B (2009) Release of endogenous opioids from duodenal enteroendocrine cells requires Trpm5. Gastroenterology 137:598–606

Krasteva G, Canning BJ, Hartmann P, Veres TZ, Papadakis T, Muhlfeld C, Schliecker K, Tallini YN, Braun A, Hackstein H, Baal N, Weihe E, Schutz B, Kotlikoff M, Ibanez-Tallon I, Kummer W (2011) Cholinergic chemosensory cells in the trachea regulate breathing. Proc Natl Acad Sci USA 108:9478–9483

Kroeze JH, Bartoshuk LM (1985) Bitterness suppression as revealed by split-tongue taste stimulation in humans. Physiol Behav 35:779–783

Kun J, Helyes Z, Perkecz A, Ban A, Polgar B, Szolcsanyi J, Pinter E (2012) Effect of surgical and chemical sensory denervation on non-neural expression of the transient receptor potential vanilloid 1 (TRPV1) receptors in the rat. J Mol Neurosci 48:795–803
Kusuhara Y, Yoshida R, Ohkuri T, Yasumatsu K, Voigt A, Hubner S, Maeda K, Boehm U, Meyerhof W, Ninomiya Y (2013) Taste responses in mice lacking taste receptor subunit T1R1. J Physiol 591:1967–1985
Kwan KY, Allchorne AJ, Vollrath MA, Christensen AP, Zhang DS, Woolf CJ, Corey DP (2006) TRPA1 contributes to cold, mechanical, and chemical nociception but is not essential for hair-cell transduction. Neuron 50:277–289
Lanosa MJ, Willis DN, Jordt S, Morris JB (2010) Role of metabolic activation and the TRPA1 receptor in the sensory irritation response to styrene and naphthalene. Toxicol Sci 115:589–595
Laska M, Distel H, Hudson R (1997) Trigeminal perception of odorant quality in congenitally Anosmic subjects. Chem Senses 22:447–456
Lawless HT (1979) Evidence for neural inhibition in bittersweet taste mixtures. J Comp Physiol Psychol 93:538–547
Lawless H, Rozin P, Shenker J (1985) Effects of oral capsaicin on gustatory, olfactory and irritant sensations and flavor identification in humans who regularly or rarely consume chili pepper. Chemi Senses 10:579–589
Lee H, Caterina MJ (2005) TRPV channels as thermosensory receptors in epithelial cells. Pflugers Arch 451:160–167
Li F, Zhou M (2012) Depletion of bitter taste transduction leads to massive spermatid loss in transgenic mice. Mol Hum Reprod 18:289–297
Liman ER (2007) TRPM5 and taste transduction. Handb Exp Pharmacol 179:287–298
Lin W, Ogura T, Margolskee RF, Finger TE, Restrepo D (2008) TRPM5-expressing solitary chemosensory cells respond to odorous irritants. J Neurophysiol 99:1451–1460
Liu D, Liman ER (2003) Intracellular Ca^{2+} and the phospholipid PIP2 regulate the taste transduction ion channel TRPM5. Proc Natl Acad Sci USA 100:15160–15165
Liu L, Simon SA (2000) Capsaicin, acid and heat-evoked currents in rat trigeminal ganglion neurons: relationship to functional VR1 receptors. Physiol Behav 69:363–378
Liu L, Simon SA (2001) Acidic stimuli activates two distinct pathways in taste receptor cells from rat fungiform papillae. Brain Res 923:58–70
Liu L, Zhu W, Zhang ZS, Yang T, Grant A, Oxford G, Simon SA (2004) Nicotine inhibits voltage-dependent sodium channels and sensitizes vanilloid receptors. J Neurophysiol 91:1482–1491
Liu D, Zhang Z, Liman ER (2005) Extracellular acid block and acid-enhanced inactivation of the Ca^{2+}-activated cation channel TRPM5 involve residues in the S3-S4 and S5-S6 extracellular domains. J Biol Chem 280:20691–20699
Liu P, Shah BP, Croasdell S, Gilbertson TA (2011) Transient receptor potential channel type M5 is essential for fat taste. J Neurosci 31:8634–8642
LopezJimenez ND, Cavenagh MM, Sainz E, Cruz-Ithier MA, Battey JF, Sullivan SL (2006) Two members of the TRPP family of ion channels, Pkd1l3 and Pkd2l1, are co-expressed in a subset of taste receptor cells. J Neurochem 98:68–77
Lumpkin EA, Caterina MJ (2007) Mechanisms of sensory transduction in the skin. Nature 445:858–865
Lundberg JM, Saria A (1983) Capsaicin-induced desensitization of airway mucosa to cigarette smoke, mechanical and chemical irritants. Nature 302:251–253
Lyall V, Alam RI, Phan DQ, Ereso GL, Phan TH, Malik SA, Montrose MH, Chu S, Heck GL, Feldman GM, DeSimone JA (2001) Decrease in rat taste receptor cell intracellular pH is the proximate stimulus in sour taste transduction. Am J Physiol Cell Physiol 281:C1005–C1013
Lyall V, Heck GL, Vinnikova AK, Ghosh S, Phan TH, Alam RI, Russell OF, Malik SA, Bigbee JW, DeSimone JA (2004) The mammalian amiloride-insensitive non-specific salt taste receptor is a vanilloid receptor-1 variant. J Physiol 558:147–159
Ma H, Yang R, Thomas SM, Kinnamon JC (2007) Qualitative and quantitative differences between taste buds of the rat and mouse. BMC Neurosci 8:5

Macpherson LJ, Geierstanger BH, Viswanath V, Bandell M, Eid SR, Hwang S, Patapoutian A (2005) The pungency of garlic: activation of TRPA1 and TRPV1 in response to allicin. Curr Biol 15:929–934

Malkia A, Morenilla-Palao C, Viana F (2011) The emerging pharmacology of TRPM8 channels: hidden therapeutic potential underneath a cold surface. Curr Pharm Biotechnol 12:54–67

Marincsak R, Toth BI, Czifra G, Marton I, Redl P, Tar I, Toth L, Kovacs L, Biro T (2009) Increased expression of TRPV1 in squamous cell carcinoma of the human tongue. Oral Dis 15:328–335

Martenson ME, Ingram SL, Baumann TK (1994) Potentiation of rabbit trigeminal responses to capsaicin in a low pH environment. Brain Res 651:143–147

Matsumoto I, Emori Y, Ninomiya Y, Abe K (2001) A comparative study of three cranial sensory ganglia projecting into the oral cavity: in situ hybridization analyses of neurotrophin receptors and thermosensitive cation channels. Brain Res Mol Brain Res 93:105–112

McKemy DD (2007) Temperature sensing across species. Pflugers Arch 454:777–791

McKemy DD, Neuhausser WM, Julius D (2002) Identification of a cold receptor reveals a general role for TRP channels in thermosensation. Nature 416:52–58

McNamara FN, Randall A, Gunthorpe MJ (2005) Effects of piperine, the pungent component of black pepper, at the human vanilloid receptor (TRPV1). Br J Pharmacol 144:781–790

Meyer D, Voigt A, Widmayer P, Borth H, Huebner S, Breit A, Marschall S, de Angelis MH, Boehm U, Meyerhof W, Gudermann T, Boekhoff I (2012) Expression of Tas1 taste receptors in mammalian spermatozoa: functional role of Tas1r1 in regulating basal Ca(2)(+) and cAMP concentrations in spermatozoa. PLoS One 7:e32354

Mihara H, Boudaka A, Sugiyama T, Moriyama Y, Tominaga M (2011) Transient receptor potential vanilloid 4 (TRPV4)-dependent calcium influx and ATP release in mouse oesophageal keratinocytes. J Physiol 589:3471–3482

Moon YW, Lee JH, Yoo SB, Jahng JW (2010) Capsaicin receptors are colocalized with sweet/bitter receptors in the taste sensing cells of circumvallate papillae. Genes Nutr 5:251–255

Moore C, Cevikbas F, Pasolli HA, Chen Y, Kong W, Kempkes C, Parekh P, Lee SH, Kontchou NA, Yeh I, Jokerst NM, Fuchs E, Steinhoff M, Liedtke WB (2013) UVB radiation generates sunburn pain and affects skin by activating epidermal TRPV4 ion channels and triggering endothelin-1 signaling. Proc Natl Acad Sci USA 110:E3225–E3234

Moqrich A, Hwang SW, Earley TJ, Petrus MJ, Murray AN, Spencer KS, Andahazy M, Story GM, Patapoutian A (2005) Impaired thermosensation in mice lacking TRPV3, a heat and camphor sensor in the skin. Science 307:1468–1472

Morales-Lazaro SL, Simon SA, Rosenbaum T (2013) The role of endogenous molecules in modulating pain through TRPV1. J Physiol 591:3109–3121

Morenilla-Palao C, Pertusa M, Meseguer V, Cabedo H, Viana F (2009) Lipid raft segregation modulates TRPM8 channel activity. J Biol Chem 284:9215–9224

Mosinger B, Redding KM, Parker MR, Yevshayeva V, Yee KK, Dyomina K, Li Y, Margolskee RF (2013) Genetic loss or pharmacological blockade of testes-expressed taste genes causes male sterility. Proc Natl Acad Sci USA 110:12319–12324

Moussaieff A, Rimmerman N, Bregman T, Straiker A, Felder CC, Shoham S, Kashman Y, Huang SM, Lee H, Shohami E, Mackie K, Caterina MJ, Walker JM, Fride E, Mechoulam R (2008) Incensole acetate, an incense component, elicits psychoactivity by activating TRPV3 channels in the brain. FASEB J 22:3024–3034

Nagy JI, Goedert M, Hunt SP, Bond A (1982) The nature of the substance P-containing nerve fibres in taste papillae of the rat tongue. Neuroscience 7:3137–3151

Nakagawa Y, Nagasawa M, Yamada S, Hara A, Mogami H, Nikolaev VO, Lohse MJ, Shigemura N, Ninomiya Y, Kojima I (2009) Sweet taste receptor expressed in pancreatic beta-cells activates the calcium and cyclic AMP signaling systems and stimulates insulin secretion. PLoS One 4:e5106

Nakamura S, Bradley RM (2011) Characteristics of sodium currents in rat geniculate ganglion neurons. J Neurophysiol 106:2982–2991

Nakashimo Y, Takumida M, Fukuiri T, Anniko M, Hirakawa K (2010) Expression of transient receptor potential channel vanilloid (TRPV) 1–4, melastin (TRPM) 5 and 8, and ankyrin (TRPA1) in the normal and methimazole-treated mouse olfactory epithelium. Acta Otolaryngol 130:1278–1286
Nakatsuka M, Iwai Y (2009) Expression of TRPV4 in the stimulated rat oral mucous membrane–nociceptive mechanisms of lingual conical papillae. Okajimas Folia Anat Jpn 86:45–54
Nassini R, Pedretti P, Moretto N, Fusi C, Carnini C, Facchinetti F, Viscomi AR, Pisano AR, Stokesberry S, Brunmark C, Svitacheva N, McGarvey L, Patacchini R, Damholt AB, Geppetti P, Materazzi S (2012) Transient receptor potential ankyrin 1 channel localized to non-neuronal airway cells promotes non-neurogenic inflammation. PLoS One 7:e42454
Nealen ML, Gold MS, Thut PD, Caterina MJ (2003) TRPM8 mRNA is expressed in a subset of cold-responsive trigeminal neurons from rat. J Neurophysiol 90:515–520
Neeper MP, Liu Y, Hutchinson TL, Wang Y, Flores CM, Qin N (2007) Activation properties of heterologously expressed mammalian TRPV2: evidence for species dependence. J Biol Chem 282:15894–15902
Nelson TM, Lopezjimenez ND, Tessarollo L, Inoue M, Bachmanov AA, Sullivan SL (2010) Taste function in mice with a targeted mutation of the pkd113 gene. Chem Senses 35:565–577
Nilius B, Appendino G (2011) Tasty and healthy TR(i)Ps. The human quest for culinary pungency. EMBO Rep 12:1094–1101
Nilius B, Appendino G (2013) Spices: the savory and beneficial science of pungency. Rev Physiol Biochem Pharmacol 164:1–76
Nilius B, Appendino G, Owsianik G (2012) The transient receptor potential channel TRPA1: from gene to pathophysiology. Pflugers Arch 464:425–458
Ogura T, Krosnowski K, Zhang L, Bekkerman M, Lin W (2010) Chemoreception regulates chemical access to mouse vomeronasal organ: role of solitary chemosensory cells. PLoS One 5:e11924
O'Hanlon S, Facer P, Simpson KD, Sandhu G, Saleh HA, Anand P (2007) Neuronal markers in allergic rhinitis: expression and correlation with sensory testing. Laryngoscope 117: 1519–1527
Ohta T, Imagawa T, Ito S (2007) Novel agonistic action of mustard oil on recombinant and endogenous porcine transient receptor potential V1 (pTRPV1) channels. Biochem Pharmacol 73:1646–1656
Oike H, Wakamori M, Mori Y, Nakanishi H, Taguchi R, Misaka T, Matsumoto I, Abe K (2006) Arachidonic acid can function as a signaling modulator by activating the TRPM5 cation channel in taste receptor cells. Biochim Biophys Acta 1761:1078–1084
Oka Y, Butnaru M, von Buchholtz L, Ryba NJ, Zuker CS (2013) High salt recruits aversive taste pathways. Nature 494:472–475
Okada Y, Reinach PS, Shirai K, Kitano A, Kao WW, Flanders KC, Miyajima M, Liu H, Zhang J, Saika S (2011) TRPV1 involvement in inflammatory tissue fibrosis in mice. Am J Pathol 178:2654–2664
Oliveira-Maia AJ, Stapleton-Kotloski JR, Lyall V, Phan TH, Mummalaneni S, Melone P, Desimone JA, Nicolelis MA, Simon SA (2009) Nicotine activates TRPM5-dependent and independent taste pathways. Proc Natl Acad Sci USA 106:1596–1601
Omote K, Kawamata T, Kawamata M, Namiki A (1998) Formalin-induced release of excitatory amino acids in the skin of the rat hindpaw. Brain Res 787:161–164
Osada K, Komai M, Bryant BP, Suzuki H, Goto A, Tsunoda K, Kimura S, Furukawa Y (1997) Capsaicin modifies responses of rat chorda tympani nerve fibers to NaCl. Chem Senses 22:249–255
Palmer RK, Atwal K, Bakaj I, Carlucci-Derbyshire S, Buber MT, Cerne R, Cortes RY, Devantier HR, Jorgensen V, Pawlyk A, Lee SP, Sprous DG, Zhang Z, Bryant R (2010) Triphenylphosphine oxide is a potent and selective inhibitor of the transient receptor potential melastatin-5 ion channel. Assay Drug Dev Technol 8:703–713

Park K, Brown PD, Kim YB, Kim JS (2003) Capsaicin modulates K + currents from dissociated rat taste receptor cells. Brain Res 962:135–143

Park U, Vastani N, Guan Y, Raja SN, Koltzenburg M, Caterina MJ (2011) TRP vanilloid 2 knock-out mice are susceptible to perinatal lethality but display normal thermal and mechanical nociception. J Neurosci 31:11425–11436

Parker GH (1912) The relations of smell, taste, and the common chemical sense in vertebrates. J Acad Natl Sci Phila 15:221–234

Peier AM, Moqrich A, Hergarden AC, Reeve AJ, Andersson DA, Story GM, Earley TJ, Dragoni I, McIntyre P, Bevan S, Patapoutian A (2002a) A TRP channel that senses cold stimuli and menthol. Cell 108:705–715

Peier AM, Reeve AJ, Andersson DA, Moqrich A, Earley TJ, Hergarden AC, Story GM, Colley S, Hogenesch JB, McIntyre P, Bevan S, Patapoutian A (2002b) A heat-sensitive TRP channel expressed in keratinocytes. Science 296:2046–2049

Perez CA, Huang L, Rong M, Kozak JA, Preuss AK, Zhang H, Max M, Margolskee RF (2002) A transient receptor potential channel expressed in taste receptor cells. Nat Neurosci 5: 1169–1176

Petersen M, LaMotte RH (1993) Effect of protons on the inward current evoked by capsaicin in isolated dorsal root ganglion cells. Pain 54:37–42

Peyrot des Gachons C, Uchida K, Bryant B, Shima A, Sperry JB, Dankulich-Nagrudny L, Tominaga M, Smith AB, Beauchamp GK, Breslin PA (2011) Unusual pungency from extra-virgin olive oil is attributable to restricted spatial expression of the receptor of oleocanthal. J Neurosci 31:999–1009

Prawitt D, Monteilh-Zoller MK, Brixel L, Spangenberg C, Zabel B, Fleig A, Penner R (2003) TRPM5 is a transient Ca^{2+}-activated cation channel responding to rapid changes in $[Ca^{2+}]_i$. Proc Natl Acad Sci USA 100:15166–15171

Prescott J, Allen S, Stephens L (1993) Interactions between oral chemical irritation, taste and temperature. Chem Senses 18:389–404

Qin N, Neeper MP, Liu Y, Hutchinson TL, Lubin ML, Flores CM (2008) TRPV2 is activated by cannabidiol and mediates CGRP release in cultured rat dorsal root ganglion neurons. J Neurosci 28:6231–6238

Reimann F, Habib AM, Tolhurst G, Parker HE, Rogers GJ, Gribble FM (2008) Glucose sensing in L cells: a primary cell study. Cell Metab 8:532–539

Ren Z, Rhyu MR, Phan TH, Mummalaneni S, Murthy KS, Grider JR, Desimone JA, Lyall V (2013) TRPM5-dependent amiloride- and benzamil-insensitive NaCl chorda tympani taste nerve response. Am J Physiol Gastrointest Liver Physiol 305:G106–117

Rentmeister-Bryant H, Green BG (1997) Perceived irritation during ingestion of capsaicin or piperine: comparison of trigeminal and non-trigeminal areas. Chem Senses 22:257–266

Richter TA, Caicedo A, Roper SD (2003) Sour taste stimuli evoke Ca^{2+} and pH responses in mouse taste cells. J Physiol 547:475–483

Riera CE, Vogel H, Simon SA, Damak S, le Coutre J (2008) The capsaicin receptor participates in artificial sweetener aversion. Biochem Biophys Res Commun 376:653–657

Riera CE, Vogel H, Simon SA, Damak S, le Coutre J (2009) Sensory attributes of complex tasting divalent salts are mediated by TRPM5 and TRPV1 channels. J Neurosci 29:2654–2662

Romanov RA, Rogachevskaja OA, Bystrova MF, Jiang P, Margolskee RF, Kolesnikov SS (2007) Afferent neurotransmission mediated by hemichannels in mammalian taste cells. EMBO J 26: 657–667

Rong W, Burnstock G, Spyer KM (2000) P2X purinoceptor-mediated excitation of trigeminal lingual nerve terminals in an in vitro intra-arterially perfused rat tongue preparation. J Physiol 524(Pt 3):891–902

Roper SD (2007) Signal transduction and information processing in mammalian taste buds. Pflugers Arch 454:759–776

Roper SD (2013) Taste buds as peripheral chemosensory processors. Semin Cell Dev Biol 2013 (24):71–79

Rozin P, Mark M, Schiller D (1981) The role of desensitization to capsaicin in chili pepper ingestion and preference. Chem Senses 6:23–31
Ruiz C, Gutknecht S, Delay E, Kinnamon S (2006) Detection of NaCl and KCl in TRPV1 Knockout Mice. Chem Senses 31:813–820
Salas MM, Hargreaves KM, Akopian AN (2009) TRPA1-mediated responses in trigeminal sensory neurons: interaction between TRPA1 and TRPV1. Eur J Neurosci 29:1568–1578
Sasaki R, Sato T, Yajima T, Kano M, Suzuki T, Ichikawa H (2013) The distribution of TRPV1 and TRPV2 in the rat pharynx. Cell Mol Neurobiol 33(5):707–714
Sato M, Ogawa H, Yamashita S (1975) Response properties of macaque monkey chorda tympani fibers. J Gen Physiol 66:781–810
Sato T, Fujita M, Kano M, Hosokawa H, Kondo T, Suzuki T, Kasahara E, Shoji N, Sasano T, Ichikawa H (2013) The distribution of transient receptor potential melastatin-8 in the rat soft palate, epiglottis, and pharynx. Cell Mol Neurobiol 33:161–165
Saunders CJ, Winston YL, Patel TD, Muday JA, Silver WL (2013) Dissecting the role of TRPV1 in detecting multiple trigeminal irritants in three behavioral assays for sensory irritation. F1000Res 2:74
Savant L, McDaniel M (2004) Suppression of sourness: A comparative study involving mixtures of organic acids and sugars. Percept Psychophys 66:642–650
Sbarbati A, Osculati F (2005) The taste cell-related diffuse chemosensory system. Prog Neurobiol 75:295–307
Schiffman SS, Sattely-Miller EA, Graham BG, Bennett JL, Booth BJ, Desai N, Bishay I (2000) Effect of temperature, pH, and ions on sweet taste. Physiol Behav 68:469–481
Seki N, Shirasaki H, Kikuchi M, Sakamoto T, Watanabe N, Himi T (2006) Expression and localization of TRPV1 in human nasal mucosa. Rhinology 44:128–134
Sherkheli MA, Gisselmann G, Hatt H (2012) Supercooling agent icilin blocks a warmth-sensing ion channel TRPV3. Scientific World J 2012:982725
Shimohira D, Kido MA, Danjo A, Takao T, Wang B, Zhang JQ, Yamaza T, Masuko S, Goto M, Tanaka T (2009) TRPV2 expression in rat oral mucosa. Histochem Cell Biol 132:423–433
Shin Y-C, Shin S-Y, So I, Kwon D, Jeon J-H (2011) TRIP Database: a manually curated database of protein–protein interactions for mammalian TRP channels. Nucleic Acids Res 39: D356–D361
Silver WL, Clapp TR, Stone LM, Kinnamon SC (2006) TRPV1 receptors and nasal trigeminal chemesthesis. Chem Senses 31:807–812
Simons CT, O'Mahony M, Carstens E (2002) Taste suppression following lingual capsaicin pre-treatment in humans. Chem Senses 27:353–365
Sizer F, Harris N (1985) The influence of common food additives and temperature on threshold perception of capsaicin. Chem Senses 10:279–286
Smith PL, Maloney KN, Pothen RG, Clardy J, Clapham DE (2006) Bisandrographolide from Andrographis paniculata activates TRPV4 channels. J Biol Chem 281:29897–29904
Smith KR, Treesukosol Y, Paedae AB, Contreras RJ, Spector AC (2012) Contribution of the TRPV1 channel to salt taste quality in mice as assessed by conditioned taste aversion generalization and chorda tympani nerve responses. Am J Physiol Regul Integr Comp Physiol 303:R1195–1205
Sokabe T, Fukumi-Tominaga T, Yonemura S, Mizuno A, Tominaga M (2010) The TRPV4 channel contributes to intercellular junction formation in keratinocytes. J Biol Chem 285: 18749–18758
Sprague J, Harrison C, Rowbotham DJ, Smart D, Lambert DG (2001) Temperature-dependent activation of recombinant rat vanilloid VR1 receptors expressed in HEK293 cells by capsaicin and anandamide. Eur J Pharmacol 423:121–125
Stander S, Moormann C, Schumacher M, Buddenkotte J, Artuc M, Shpacovitch V, Brzoska T, Lippert U, Henz BM, Luger TA, Metze D, Steinhoff M (2004) Expression of vanilloid receptor subtype 1 in cutaneous sensory nerve fibers, mast cells, and epithelial cells of appendage structures. Exp Dermatol 13:129–139

Staruschenko A, Jeske NA, Akopian AN (2010) Contribution of TRPV1-TRPA1 interaction to the single channel properties of the TRPA1 channel. J Biol Chem 285:15167–15177
Story GM, Peier AM, Reeve AJ, Eid SR, Mosbacher J, Hricik TR, Earley TJ, Hergarden AC, Andersson DA, Hwang SW, McIntyre P, Jegla T, Bevan S, Patapoutian A (2003) ANKTM1, a TRP-like channel expressed in nociceptive neurons, is activated by cold temperatures. Cell 112:819–829
Strotmann R, Harteneck C, Nunnenmacher K, Schultz G, Plant TD (2000) OTRPC4, a nonselective cation channel that confers sensitivity to extracellular osmolarity. Nat Cell Biol 2:695–702
Sugiura T, Tominaga M, Katsuya H, Mizumura K (2002) Bradykinin lowers the threshold temperature for heat activation of vanilloid receptor 1. J Neurophysiol 88:544–548
Suzuki M, Watanabe Y, Oyama Y, Mizuno A, Kusano E, Hirao A, Ookawara S (2003) Localization of mechanosensitive channel TRPV4 in mouse skin. Neurosci Lett 353:189–192
Symanowicz PT, Gianutsos G, Morris JB (2004) Lack of role for the vanilloid receptor in response to several inspired irritant air pollutants in the C57Bl/6J mouse. Neurosci Lett 362:150–153
Szolcsanyi J (1977) A pharmacological approach to elucidation of the role of different nerve fibres and receptor endings in mediation of pain. J Physiol (Paris) 73:251–259
Ta LE, Bieber AJ, Carlton SM, Loprinzi CL, Low PA, Windebank AJ (2010) Transient Receptor Potential Vanilloid 1 is essential for cisplatin-induced heat hyperalgesia in mice. Mol Pain 6:15
Takashima Y, Daniels RL, Knowlton W, Teng J, Liman ER, McKemy DD (2007) Diversity in the neural circuitry of cold sensing revealed by genetic axonal labeling of transient receptor potential melastatin 8 neurons. J Neurosci 27:14147–14157
Talavera K, Alpizar YA, Startek J (2013) Interaction of lipopolysaccharides with plasma membranes as possible trigger of TRP channel activation. In: Paper presented at ECRO 2013, Leuven, Belgium, 26–29 Aug 2013
Talavera K, Yasumatsu K, Voets T, Droogmans G, Shigemura N, Ninomiya Y, Margolskee RF, Nilius B (2005) Heat activation of TRPM5 underlies thermal sensitivity of sweet taste. Nature 438:1022–1025
Talavera K, Yasumatsu K, Yoshida R, Margolskee RF, Voets T, Ninomiya Y, Nilius B (2008) The taste transduction channel TRPM5 is a locus for bitter-sweet taste interactions. FASEB J 22:1343–1355
Taniguchi K (2004) Expression of the sweet receptor protein, T1R3, in the human liver and pancreas. J Vet Med Sci 66:1311–1314
Tanimoto T, Takeda M, Nasu M, Kadoi J, Matsumoto S (2005) Immunohistochemical co-expression of carbonic anhydrase II with Kv1.4 and TRPV1 in rat small-diameter trigeminal ganglion neurons. Brain Res 1044:262–265
Taruno A, Vingtdeux V, Ohmoto M, Ma Z, Dvoryanchikov G, Li A, Adrien L, Zhao H, Leung S, Abernethy M, Koppel J, Davies P, Civan MM, Chaudhari N, Matsumoto I, Hellekant G, Tordoff MG, Marambaud P, Foskett JK (2013) CALHM1 ion channel mediates purinergic neurotransmission of sweet, bitter and umami tastes. Nature 495:223–226
Taylor-Clark TE, Undem BJ (2010) Ozone activates airway nerves via the selective stimulation of TRPA1 ion channels. J Physiol 588:423–433
Tizzano M, Gulbransen BD, Vandenbeuch A, Clapp TR, Herman JP, Sibhatu HM, Churchill ME, Silver WL, Kinnamon SC, Finger TE (2010) Nasal chemosensory cells use bitter taste signaling to detect irritants and bacterial signals. Proc Natl Acad Sci USA 107:3210–3215
Tohda C, Sasaki M, Konemura T, Sasamura T, Itoh M, Kuraishi Y (2001) Axonal transport of VR1 capsaicin receptor mRNA in primary afferents and its participation in inflammation-induced increase in capsaicin sensitivity. J Neurochem 76:1628–1635
Tokita K, Boughter JD Jr (2012) Sweet-bitter and umami-bitter taste interactions in single parabrachial neurons in C57BL/6J mice. J Neurophysiol 108:2179–2190
Tomchik SM, Berg S, Kim JW, Chaudhari N, Roper SD (2007) Breadth of tuning and taste coding in mammalian taste buds. J Neurosci 27:10840–10848
Tominaga M, Tominaga T (2005) Structure and function of TRPV1. Pflugers Arch 451:143–150

Tominaga M, Caterina MJ, Malmberg AB, Rosen TA, Gilbert H, Skinner K, Raumann BE, Basbaum AI, Julius D (1998) The cloned capsaicin receptor integrates multiple pain-producing stimuli. Neuron 21:531–543

Toyoshima K (1989) Chemoreceptive and mechanoreceptive paraneurons in the tongue. Arch Histol Cytol 52(Suppl):383–388

Treesukosol Y, Lyall V, Heck GL, DeSimone JA, Spector AC (2007) A psychophysical and electrophysiological analysis of salt taste in Trpv1 null mice. Am J Physiol Regul Integr Comp Physiol 292:R1799–1809

Trevisani M, Milan A, Gatti R, Zanasi A, Harrison S, Fontana G, Morice AH, Geppetti P (2004) Antitussive activity of iodo-resiniferatoxin in guinea pigs. Thorax 59:769–772

Trevisani M, Siemens J, Materazzi S, Bautista DM, Nassini R, Campi B, Imamachi N, Andre E, Patacchini R, Cottrell GS, Gatti R, Basbaum AI, Bunnett NW, Julius D, Geppetti P (2007) 4-Hydroxynonenal, an endogenous aldehyde, causes pain and neurogenic inflammation through activation of the irritant receptor TRPA1. Proc Natl Acad Sci USA 104:13519–13524

Tsavaler L, Shapero MH, Morkowski S, Laus R (2001) Trp-p8, a novel prostate-specific gene, is up-regulated in prostate cancer and other malignancies and shares high homology with transient receptor potential calcium channel proteins. Cancer Res 61:3760–3769

Tucker RM, Mattes RD (2012) Are free fatty acids effective taste stimuli in humans? J Food Sci 77:S148–151

Uchida K, Tominaga M (2011) The role of thermosensitive TRP (transient receptor potential) channels in insulin secretion. Endocr J 58:1021–1028

Vandenbeuch A, Clapp TR, Kinnamon SC (2008) Amiloride-sensitive channels in type I fungiform taste cells in mouse. BMC Neurosci 9:1

Vandewauw I, Owsianik G, Voets T (2013) Systematic and quantitative mRNA expression analysis of TRP channel genes at the single trigeminal and dorsal root ganglion level in mouse. BMC Neurosci 14:21

Vay L, Gu C, McNaughton PA (2010) Current perspectives on the modulation of thermo-TRP channels: new advances and therapeutic implications. Expert Rev Clin Pharmacol 3:687–704

Veronesi B, Oortgiesen M (2006) The TRPV1 receptor: target of toxicants and therapeutics. Toxicol Sci 89:1–3

Vetter I, Lewis RJ (2011) Natural product ligands of TRP channels. Adv Exp Med Biol 704:41–85

Viana F (2011) Chemosensory properties of the trigeminal system. ACS Chem Neurosci 2:38–50

Vriens J, Watanabe H, Janssens A, Droogmans G, Voets T, Nilius B (2004) Cell swelling, heat, and chemical agonists use distinct pathways for the activation of the cation channel TRPV4. Proc Natl Acad Sci USA 101:396–401

Wang Y, Erickson RP, Simon SA (1995) Modulation of rat chorda tympani nerve activity by lingual nerve stimulation. J Neurophysiol 73:1468–1483

Wang YY, Chang RB, Liman ER (2010) TRPA1 is a component of the nociceptive response to CO2. J Neurosci 30:12958–12963

Wang YY, Chang RB, Allgood SD, Silver WL, Liman ER (2011) A TRPA1-dependent mechanism for the pungent sensation of weak acids. J Gen Physiol 137:493–505

Watanabe H, Vriens J, Suh SH, Benham CD, Droogmans G, Nilius B (2002) Heat-evoked activation of TRPV4 channels in a HEK293 cell expression system and in native mouse aorta endothelial cells. J Biol Chem 277:47044–47051

Whitear M (1989) Merkel cells in lower vertebrates. Arch Histol Cytol 52(Suppl):415–422

Whitehead MC, Ganchrow JR, Ganchrow D, Yao B (1999) Organization of geniculate and trigeminal ganglion cells innervating single fungiform taste papillae: a study with tetramethylrhodamine dextran amine labeling. Neuroscience 93:931–941

Wilson DM, Lemon CH (2013) Modulation of central gustatory coding by temperature. J Neurophysiol 110:1117–1129

Wu SV, Rozengurt N, Yang M, Young SH, Sinnett-Smith J, Rozengurt E (2002) Expression of bitter taste receptors of the T2R family in the gastrointestinal tract and enteroendocrine STC-1 cells. Proc Natl Acad Sci USA 99:2392–2397

Xu H, Ramsey IS, Kotecha SA, Moran MM, Chong JA, Lawson D, Ge P, Lilly J, Silos-Santiago I, Xie Y, DiStefano PS, Curtis R, Clapham DE (2002) TRPV3 is a calcium-permeable temperature-sensitive cation channel. Nature 418:181–186

Xu H, Delling M, Jun JC, Clapham DE (2006) Oregano, thyme and clove-derived flavors and skin sensitizers activate specific TRP channels. Nat Neurosci 9:628–635

Yilmaz Z, Renton T, Yiangou Y, Zakrzewska J, Chessell IP, Bountra C, Anand P (2007) Burning mouth syndrome as a trigeminal small fibre neuropathy: Increased heat and capsaicin receptor TRPV1 in nerve fibres correlates with pain score. J Clin Neurosci 14:864–871

Young RL, Sutherland K, Pezos N, Brierley SM, Horowitz M, Rayner CK, Blackshaw LA (2009) Expression of taste molecules in the upper gastrointestinal tract in humans with and without type 2 diabetes. Gut 58:337–346

Zhang Y, Hoon MA, Chandrashekar J, Mueller KL, Cook B, Wu D, Zuker CS, Ryba NJ (2003) Coding of sweet, bitter, and umami tastes: different receptor cells sharing similar signaling pathways. Cell 112:293–301

Zhang Z, Zhao Z, Margolskee R, Liman E (2007) The transduction channel TRPM5 is gated by intracellular calcium in taste cells. J Neurosci 27:5777–5786

TRPs and Pain

Jane E. Sexton, Jeffrey Vernon, and John N. Wood

Contents

Abstract

Pain usually occurs as a result of tissue damage and has a role in healing and protection. However, in certain conditions it has no functional purpose and can become chronic and debilitating. A demand for more effective treatments to deal with this highly prevalent problem requires a better understanding of the underlying mechanisms. TRP channels are associated with numerous sensory functions across a wide range of species. Investigation into the expression patterns, electrophysiological properties and the effects of channel deletion in transgenic animal models have produced a great deal of evidence linking these channels to transduction of noxious stimuli as well as signalling within the pain system.

Keywords

TRP channels • Pain • Nociception • Inflammation • Neuropathy • Analgesia

---

J.E. Sexton (✉) • J. Vernon • J.N. Wood
Molecular Nociception Group, Wolfson Institute for Biomedical Research, University College London, London WC1E 6BT, UK
e-mail: jane.sexton.09@ucl.ac.uk

B. Nilius and V. Flockerzi (eds.), *Mammalian Transient Receptor Potential (TRP) Cation Channels*, Handbook of Experimental Pharmacology 223, DOI 10.1007/978-3-319-05161-1_6,

1 Pain and Nociception

Pain is a complex phenomenon which has been described by the International Association for the Study of Pain as 'an unpleasant sensory and emotional experience associated with actual or potential tissue damage, or described in terms of such damage' (http://www.iasp-pain.org). The perception of pain is usually the result of tissue damage caused by a noxious stimulus and as such has a protective function as well as being important for allowing healing of damaged tissue. Specialised damage-sensing neurons that innervate the skin, muscle and viscera are activated by noxious stimuli, and this initial transduction of pain-producing stimuli is known as nociception. Noxious stimuli can be mechanical, thermal or chemical, and they activate nociceptors, a type of sensory afferent neuron whose cell bodies lie in the dorsal and trigeminal root ganglia. The two main categories of nociceptor are Aδ and C fibres which are myelinated and unmyelinated, respectively. As a result of their myelination state and their larger diameter, Aδ fibres have a greater conduction velocity than C fibres and are responsible for the so-called first pain, a pinprick sensation which precedes the burning sensation, and the 'second pain' mediated by small-diameter unmyelinated C fibres. C fibres can be subdivided into peptidergic and non-peptidergic sets according to their expression profiles; peptidergic C fibres express the TrkA receptor and respond to nerve growth factor (NGF), while non-peptidergic nociceptors bind the lectin IB4 and are sensitive to glial-derived neurotrophic factor (GDNF) acting through the c-Ret receptor. Each subtype of sensory neurons expresses a different subset of transient receptor potential (TRP) channels which are linked to a variety of functions (Table 1).

Although pain usually arises from a stimulus, it can occur without noxious input or outlast the initial insult and become chronic. A large-scale survey estimated that approximately 19 % of people in Europe suffer from chronic pain (http://www.britishpainsociety.org/Pain%20in%20Europ%20survey%20report.pdf, British Pain Society 2003). The comorbidities associated with chronic pain are numerous and debilitating, including depression, anxiety and insomnia. It was also reported that 1 in 5 sufferers in Europe have lost, or had to leave, their job as a result of their pain. Chronic pain can occur as a result of either inflammatory or neuropathic processes.

Inflammation associated with sensitisation (reduced pain thresholds) is characterised by redness, swelling and tenderness. Following tissue damage peripheral leucocytes, Schwann cells and endothelia release proinflammatory cytokines, such as tumour necrosis factor-α (TNF-α) and interleukin-1 (IL-1), which leads to an increase in NGF production by macrophages. In turn, NGF forms part of a positive feedback loop whereby it binds to receptors such as TrkA on the peripheral terminals of nociceptors. As a consequence, tyrosine kinases phosphorylate TRPV1 which then is trafficked into the plasma membrane (Zhang et al. 2005a, b). NGF also activates mast cells which release mediators such as prostaglandins, bradykinin, ATP and serotonin (5-HT), adding to the inflammatory milieu. This cycle of events via its effect on TRPV1 function leads to nociceptor sensitisation as well as increased excitability at central terminals resulting in symptoms such as thermal and mechanical hyperalgesia. When these proinflammatory effects persist, chronic

Table 1 Summary of TRP channel expression in sensory neurons and the potential roles in pain

TRP	Peripheral sensory neuron expression	Possible role in pain
TRPV1	Peptidergic small- and medium-diameter sensory neurons (Caterina et al. 1997)	Inflammatory hyperalgesia
TRPV2	Peptidergic small-diameter and medium-diameter fibres (Caterina et al. 1999)	Appears to be upregulated in DRG in certain models of inflammatory pain, though functional significance unclear
TRPV3	Peptidergic small- and medium-diameter sensory neurons (in a subset of TRPV1+ neurons) (Smith et al. 2002); keratinocytes (Peier et al. 2002a, b)	Acute noxious heat
TRPV4	Small- and medium-diameter sensory neurons (Facer et al. 2007)	Inflammatory mechanical and thermal hyperalgesia
TRPA1	Peptidergic small- and medium-diameter sensory neurons (co-expressed with TRPV1) (Story et al. 2003)	Acute noxious cold mechanosensation
TRPM8	Small-diameter, capsaicin-insensitive neurons (higher expression in TG than DRG) (McKemy et al. 2002)	Innocuous cold, noxious cold activated by temperatures 8 °C–28 °C.
TRPC1	Peripheral sensory neurons (Elg et al. 2007). Co-expressed with TRPV4 and TRPC6 (Alessandri-Haber et al. 2009)	Mechanical hyperalgesia
TRPC6	Most peripheral sensory neurons (Quick et al. 2012) co-expressed with TRPV4 and TRPC1 in peripheral sensory neurons (Alessandri-Haber et al. 2009)	Mechanical hyperalgesia

pain can occur and is seen, for example, in conditions such as osteoarthritis where ongoing pain and increased sensitivity in the affected joints are highly detrimental to normal life.

Neuropathic pain is the result of damage, disease or dysfunction within the nervous system. When damage occurs, several key events follow which can lead to spontaneous pain, hyperalgesia (enhanced sensitivity to a noxious stimulus) and allodynia (pain from a normally non-noxious stimulus). For example, there is a loss of trophic support; usually, NGF and other growth factors such as GDNF are taken up by peripheral neurons and retrogradely transported to the cell somata. When peripheral neurons are damaged, the portion of the cell distal to the injury begins to degenerate, and consequently, the peripheral receptors can no longer respond appropriately to growth factor inputs. Further, the contents of the cell, including growth factors and neurotransmitters, are released into the surrounding area. As a result, not only is there aberrant activity and ectopic transduction in the damaged neuron itself, but there is also an increase in excitability of surrounding neurons. Peripheral nociceptors themselves also undergo an increase in excitability and response to a wider range of stimulus intensities. The barrage of peripheral activity which occurs following damage to the nervous system also leads to sensitisation centrally. It is important to note, however, that while animal models of neuropathic pain (Fig. 1) have elucidated many details of underlying mechanisms of pain

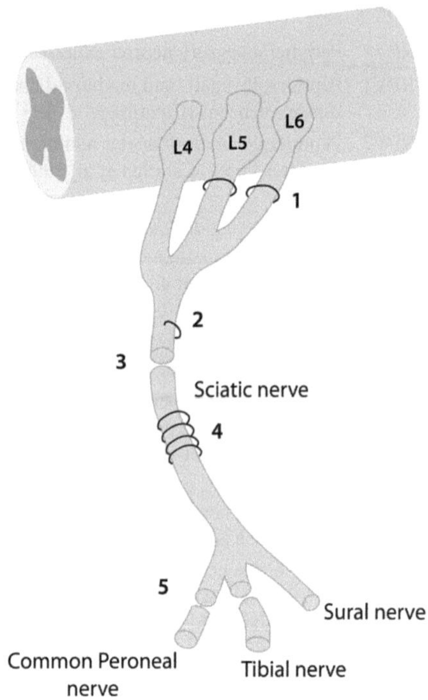

Fig. 1 Models of neuropathic pain. (1) *SNL* spinal nerve ligation; tight ligation of L5 and L6 spinal nerves. (2) *PSNL* partial sciatic nerve ligation; tight ligation of between one third and one half of the sciatic nerve. (3) *SNT* sciatic nerve transection; complete transection of sciatic nerve. (4) *CCI* chronic constriction injury; 3–4 loose ligatures around nerve. (5) *SNI* spared nerve injury; tibial and common peroneal branches of the sciatic nerve are transected distal to its trifurcation

consequent to nerve damage, they have also highlighted that each model can undergo different molecular and cellular changes, albeit with an apparently similar nociceptive phenotype (Koltzenburg and Scadding 2001).

Existing analgesics, centred around opioids and nonsteroidal anti-inflammatory drugs (NSAIDs), are limited in their efficacy and are frequently associated with undesirable side effects and can induce dependence. The development of new, more effective analgesics therefore requires new potential targets and a better understanding of the underlying mechanisms of pain.

2 Heat

A number of TRP channels have been linked to noxious heat and cold transduction. When TRPV1 was first identified, it was named vanilloid receptor subtype 1 (VR1) because of its response to capsaicin, a component of capsicum peppers which is the active ingredient in spicy foods (Caterina et al. 1997). TRPV1 has repeatedly been shown to be activated by noxious thermal stimuli (temperatures >43 °C) and low pH (Caterina et al. 1997, 2000). Interestingly, however, while mice lacking the TRPV1 receptor lose sensitivity to capsaicin as well as low pH, they exhibit almost no loss of sensitivity to acute noxious heat (Davis et al. 2000; Caterina et al. 2000; Woodbury et al. 2004). In contrast, in models of inflammatory pain induced by carrageenan or complete Freund's adjuvant (CFA), thermal hyperalgesia is lost (Caterina et al. 2000; Davis et al. 2000). Consistent with a deficit in inflammation-induced thermal hyperalgesia in TRPV1 null mice, tissue damage leads to elevated TRPV1 expression in sensory neurons of the dorsal root ganglia (DRG) and in lamina I and II of the spinal cord (Amaya et al. 2003; Luo et al. 2004). Following these findings, Mishra et al. (2011) developed a conditional mouse model and found that selective ablation by diphtheria toxin of TRPV1-expressing sensory neurons results in loss of both sensitivity to acute noxious heat and thermal hyperalgesia. Thus the population of TRPV1 expressing neurons is involved in heat sensing, and TRPV1 is principally involved in inflammatory pain.

As TRPV1 has no role in sensing acute noxious heat (Davis et al. 2000), other members of the TRPV family were considered potential candidates for this function. TRPV2, though insensitive to capsaicin, confers sensitivity to high temperatures >52 °C in Xenopus oocytes (Caterina et al. 1999), although the relevance of this finding to mammalian cells has not been demonstrated. TRPV3 and TRPV4 are activated at 33–39 °C and 25–34 °C, respectively, and are expressed in sensory neurons (TRPV3 is co-expressed with TRPV1) and keratinocytes (Liedtke et al. 2000; Smith et al. 2002; Xu et al. 2002; Peier et al. 2002b; Chung et al. 2004). TRPV4 is sensitive to capsaicin, but TRPV3 is not (Smith et al. 2002), and instead it responds the phenylpropanoid eugenol (which is also detected by TRPV1 and TRPV3).

TRPV2 has been studied less intensively than TRPV1, and TRPV2 knockout mice have been available only since 2011 (Park et al. 2011). In these animals (which have reduced viability and survival into adulthood) no hyperalgesia develops after inflammation or spinal nerve ligation. Further evidence against an acute heat-sensing role for TRPV2 is provided by the fact that 82 % of heat-sensitive neurons in TRPV1 null mice do not express TRPV2 (Woodbury et al. 2004). Moreover, in TRPV1 neuron-depleted mice, TRPV2 expression was still found in spite of the complete loss of heat sensitivity (Mishra et al. 2011).

TRPV3 shows the unusual property of hysteresis. Current is progressively activated at temperatures above 28 °C, but if the preparation is cooled even slightly during the experiment, the current deactivates sharply. Moreover, the channel itself undergoes sensitisation after a train of stimuli, whereby the current is evoked more reliably after the temperature is stepped repeatedly from 21 °C to 45 °C (Liu et al. 2011; Peier et al. 2002b). TRPV3 also supports a secondary current, which develops after channel sensitisation, with increased amplitude, loss of rectification and altered permeability to cations (Chung et al. 2005). These observations plausibly account for some features of peripheral sensitisation; however since the expression of TRPV3 is higher in keratinocytes than in nervous tissue, some of the behavioural effects might be only indirectly neuronal. TRPV3 null mice show deficits in detection of noxious heat but not a complete loss of sensitivity, and mice develop thermal hyperalgesia induced by bradykinin and CFA in the same way as wild-type controls (Moqrich et al. 2005).

TRPV4 is the mammalian homologue of the *C. elegans* osmosensor *osm-9*. Osmolar and pH changes occur as a result of inflammation, motivating a study

(Alessandri-Haber et al. 2003) on single fibres of rat sensory nerve in hypotonic solution. This treatment activated a proportion of the fibres, an effect that was enhanced by application of the inflammatory mediator prostaglandin but was reduced by antisense to TRPV4. Knockout mice showed a reduced sensitivity to protons and altered thermal preference, whereas response to noxious heat was unimpaired, and thresholds of mechanical sensation were increased (Liedtke and Friedman 2003). The channel is sensitised by an agonist of protease-activated receptor 2/PAR2, which elicits mechanical hyperalgesia in mice but not in TRPV4 knockout animals. In rat spinal cord, PAR2 agonist stimulated release from the dorsal horn of the nociceptive neuropeptide, substance P, an effect presumed to depend on the expression of TRPV4, though this was not shown directly (Grant et al. 2007). In expression systems, TRPV4 is activated by heat, and the extent of activation is increased in hypotonic conditions (Güler et al. 2002) of the kind that might prevail during inflammation.

Since TRPV3 and TRPV4 have overlapping temperature response profiles, Huang et al. (2011) developed TRPV3/TRPV4 double knockout mice. They found no significant differences in the acute thermal responses or in various thermal preference paradigms between wild-type (WT) and TRPV3/4 null mice. Similarly, minimal differences were seen in null mice treated with a TRPV1 antagonist in both acute and inflammatory models of thermal pain suggesting this channel does not mask a role of TRPV3 and TRPV4 (Huang et al. 2011).

Inflammatory mediators (including bradykinin, ATP, cytokines) are released from endothelium and cells of the immune system in the neighbourhood of the nociceptor, and via receptors on the neuron surface, these factors can engage second messengers that act on TRP channels (Amadesi et al. 2006). Since TRPV1 null mice have deficits in thermal hyperalgesia, the role of TRPV1 in inflammatory pain states has been widely investigated. Pro-inflammatory mediators sensitise TRPV1, leading to induction and maintenance of thermal hyperalgesia, via protein kinase A (PKA) and PKA-mediated phosphorylation of the C-terminal. PKA accomplishes this in a process mediated by PKA-anchoring proteins (AKAP) leading to enhanced gating and increased TRPV1 translocation to the membrane; for example, inhibition of the AKAP79/150 protein prevents sensitisation of TRPV1 by bradykinin and PGE2 (Zhang et al. 2008). Protein kinase C (PKC) is able to reduce the activation threshold via phosphorylation of TRPV1 leading to enhanced gating and potentiation of channel activity (Premkumar and Ahern 2000; Vellani et al. 2001) and is upregulated in DRG during inflammation (Zhou et al. 2003).

PGE2 and PGI2, released following tissue damage, sensitise TRPV1 via PKC- and PKA-dependent mechanisms and are able to reduce the thermal threshold of TRPV1 to ~35 °C (Moriyama et al. 2005). By a similar mechanism of sensitisation, injection of the chemokine CCL3 in mice-induced thermal hyperalgesia (Zhang et al. 2005a, b), while treating cultured trigeminal ganglia (TG) neurons with 5-HT, also potentiates TRPV1 activity (Loyd et al. 2011). Bradykinin-mediated thermal hyperalgesia can be inhibited by administration of the TRPV1 antagonist, capsazepine, as well as by blocking activity of phospholipase C (PLC), PKC and PKA (Ferreira et al. 2004). Interestingly, TRPA1 null mice, and neurons in culture, also show attenuated sensitivity to bradykinin in a similar way to TRPV1 null TG cultures and animal models (Bautista et al. 2006). TRPV1 and TRPV4 are both sensitised by ATP, a known mediator of inflammation. ATP reduces the sensitivity of TRPV3 whereas it has no effects on TRPV2 (Phelps et al. 2010). TRPV3 activation in keratinocytes, by agonists or heat, leads to release of PGE2 and appears to cause increased sensitivity to acute noxious heat and thermal hyperalgesia in a TRPV1-independent manner (Huang et al. 2008). Recruitment of 'silent' afferents which previously did not respond is a feature of sensitisation evoked by inflammation; there appears to be a single report on the role of TRP channels in this process, which concluded, on the basis of work in null mutant mice, that TRPV1 is not involved (Koerber et al. 2010). NGF can increase TRPV1 expression by upregulating its transcription and also via PKC-mediated phosphorylation of the channel (Ji et al. 2002); PI3K can also potentiate its activity and induce thermal hyperalgesia (Zhuang et al. 2004), and this requires tyrosine phosphorylation by Src family kinases (Zhang et al. 2005a, b). When retrogradely transported to the DRG, NGF activates the MAP kinase p38 which increases translation and transport of TRPV1 to the periphery where it contributes to maintenance of thermal hyperalgesia in inflammatory and neuropathic pain states (Ji et al. 2002).

TRPV1 expression is downregulated in damaged neurons following partial or total spinal nerve ligation (SNL) and in sciatic nerve transection (SNT), models of neuropathic pain (Hudson et al. 2001; Fukuoka et al. 2002). It is also upregulated in surrounding, undamaged neurons, for example, in L4 neurons following SNL of L5 (Fukuoka et al. 2002). This upregulation correlates with the thermal hypersensitivity profile observed after SNL (Fukuoka et al. 2002). Thermal nocifensive behaviour is reduced by block of NGF following chronic constriction injury (CCI) (Wilson-Gerwing et al. 2005), an effect which might depend on reduced levels of TRPV1. TRPV1 expression is also altered in human painful neuropathies; a loss of TRPV1-positive neurons in peripheral and sural nerves and in the skin was observed in diabetic and motor neuropathies (Lauria et al. 2006; Facer et al. 2007). An increase in TRPV1 and TRPV3 expression is seen in intact nerves after injury while TRPV4 appears to be unaltered (Facer et al. 2007).

3 Cold

Recently cell ablation studies have highlighted TRPM8-positive neurons as the key cell types involved in noxious cold perception in mice (Pogorzala et al. 2013). The threshold for noxious cold, as distinct from innocuous cool, is considered to occur at temperatures <15 °C. TRPM8 is sensitive to menthol (a cooling compound) and cold, with a threshold of ~25 °C, and is activated by temperatures encompassing the innocuous cool and noxious cold range (McKemy et al. 2002; Peier et al. 2002a). TRPM8 is expressed in a subset of capsaicin-insensitive, small-diameter neurons in both dorsal root and trigeminal ganglia, though at higher levels in the latter (7.4 % vs. 14.8 %), consistent with greater cold sensitivity in structures of the face and head (McKemy et al. 2002). Since TRPV1 expression is often considered a marker for nociceptors, this would suggest that TRPM8 is not involved in detection of noxious stimuli. Indeed, the use of TRPM8 null mice has firmly established a role for the channel in detection of innocuous cold though its role in detection of noxious stimuli is less conclusive. Knowlton et al. (2013) suggested that the cold-plate technique of measuring cold hypersensitivity, used in many of these investigations, was highly variable across earlier studies. Two groups found there was no difference in the nocifensive responses of TRPM8 null mice when compared to WT in a noxious cold-plate test (Bautista et al. 2007; Dhaka et al. 2007). However these same two groups showed a significant decrease in the response of these mice to application of the noxious cooling chemical acetone on the hind paw. Colburn et al. (2007) found a deficit in cold sensitivity in both of these assays, but this deficit was not reversed upon administration of TRPM8 antisense oligonucleotides.

It has emerged more recently that TRPM8 is upregulated in experimental bowel inflammation and that the TRPM8 agonist icilin, which attenuates chemically induced colitis in normal mice, presumably because the channel mediates a local cooling, brings no relief in the knockout animals (Ramachandran et al. 2013). In other circumstances, however, TRPM8 agonists lead to increases of core temperature, an effect that suggests coupling of central channels to the homeostatic mechanisms, whereas a TRPM8 blocker induces hypothermia (Ma et al. 2012; Knowlton et al. 2011). Once again, this example illustrates the apparently paradoxical effect of channels on somatosensation, when they are activated in different tissues by different experimental paradigms. Because TRPV1 is expressed in TRPM8-positive neurons during development but subsequently downregulated, conditional knockout of TRPV1 in sensory neurons also caused ablation of TRPM8. These animals showed significant deficits in responses to acute noxious cold (Mishra et al. 2011).

TRPA1 was the first candidate cold-sensing TRP channel. This channel is expressed in a subset of capsaicin-sensitive, calcitonin gene-related peptide (CGRP) positive neurons and has an activation threshold of 17 °C (Story et al. 2003) which is close to the cold temperature considered to be painful. Interestingly these neurons do not appear to co-express TRPM8 (Story et al. 2003).

TRPA1 is also activated by chemicals such as mustard oil (allyl isothiocyanate) which, when applied to the skin, elicits a burning or pricking sensation and causes aversive behaviour which is lost in TRPA1 null mice (Jordt et al. 2004; Bandell et al. 2004; Bautista et al. 2006; Kwan et al. 2006). Jordt et al. (2004) found that the majority (96 %) of cultured trigeminal neurons from rat which were sensitive to the mustard oil component allyl isothiocyanate (AITC) were insensitive to cold and that the remaining 4 % were sensitive to menthol, suggesting a role for TRPM8 in their cold sensitivity (Jordt et al. 2004). Also, when human embryonic kidney (HEK293) cells were transfected with TRPA1, although a response to mustard oil (AITC) was clearly observed, no responses to cold temperature of 5 °C were seen. Furthermore, it was shown that 5 % of the cold-sensitive neurons in culture were insensitive to both menthol and AITC, something corroborated by Babes et al. (2004), indicating another mechanism for noxious cold sensitivity, independent of TRPA1 and TRPM8.

Kwan et al. (2006) reported that mice lacking the TRPA1 channel showed decreased responsiveness to noxious cold temperatures on a cold plate and reduced sensitivity to acetone application when compared with WT mice. Using the same behavioural assays, Bautista et al. (2006) observed no such difference. Furthermore, Bautista et al. (2006) cultured WT and TRPA1 null trigeminal neurons and found there was no difference in the magnitude of current in response to application of a noxious cold stimulus. It has been proposed that this discrepancy may be the result of different techniques and experimental setup (Kwan et al. found a greater deficit in the response of female mice while Bautista et al. only used males). Also, when considering the conflicting evidence provided by these models, the design of the knockout model must be considered. For example, some knockout constructs involve deletion of an entire gene while others may merely result in insertion of a cassette or premature stop codon. The latter may lead to generation of a truncated form of the protein. As a result, two knockout models of the same gene may vary substantially in their expression of that gene which may potentially impact upon the phenotype exhibited by each model. Indeed, TRP channels are known to heteromultimerise, and since the TRPA1 'null' mice used by both aforementioned groups are believed to express a truncated form of the channel, it has been suggested that this truncated form may exert an effect on other implicated channels, thus affecting the phenotype of the null mice (Foulkes and Wood 2007).

The difficulty in clarifying the role of TRPA1 in noxious cold detection led to investigations of its underlying properties. The release of Ca^{2+} from intracellular stores activates TRPA1 by a PLC-dependent mechanism, while the presence of extracellular Ca^{2+} is able to augment the response of TRPA1 to agonists such as AITC (Jordt et al. 2004). Zurborg et al. (2007) investigated this further and found that an EF-hand domain within the channel subunit is required for intracellular Ca^{2+}-mediated activation of TRPA1. They demonstrated that cells expressing TRPA1 showed responses to cold; however, in EF-hand domain mutants, responses to cold were not abolished but rather were reduced to levels of control cells which did not express TRPA1. They reasoned, therefore, that TRPA1 is activated by an increase in intracellular Ca^{2+} seen upon cooling rather than being directly activated by cold thus providing a potential explanation for the inconsistencies seen in vitro and in vivo. Subsequently, however, Karashima et al. (2009) used heterologous expression studies to argue that TRPA1 could be activated by cold in the absence of both intra- and extracellular Ca^{2+}.

Cold hyperalgesia is a symptom of diseases such as rheumatoid arthritis (Jahanshaki et al. 1989), and cold allodynia is a common feature of many neuropathic pain states including those caused by traumatic nerve injury and post-herpetic neuralgia (Jørum et al. 2003). Since TRPM8 and TRPA1 have both been implicated in sensitivity to innocuous and noxious cold, their roles in these phenomena have been investigated. Caspani et al. (2009) reported nociceptive behaviour in response to menthol, a TRPM8 agonist, in mice following CCI surgery, suggesting a normally cool stimulus mediating a noxious effect via

TRPM8 in a neuropathic pain model. Similarly, Colburn et al. (2007) found that acetone application resulted in a reduced response in TRPM8 null mice after CCI surgery and CFA injection when compared to WT; an even greater allodynic phenotype was seen after CCI in mice with ablation of TRPM8 expressing sensory neurons (Knowlton et al. 2013). In spite of this, reports of changes in TRPM8 expression after CCI surgery remain contentious (Proudfoot et al. 2006; Caspani et al. 2009), and no changes in expression following CFA-induced inflammation were seen in one report (Obata et al. 2005). No upregulation of TRPM8 mRNA or protein was seen following SNL, and administration of antisense oligonucleotides to TRPM8 had no effect on SNL-induced cold hyperalgesia, although antisense oligonucleotide block of TRPA1 was able to attenuate this behaviour (Obata et al. 2005; Katsura et al. 2006; Stucky et al. 2009). Indeed, though TRPA1 expression is decreased in rats and mice after CCI (Caspani et al. 2009) and in injured L5 nerves in SNL, it is upregulated in intact L4 nerves and DRG in this model (Obata et al. 2005; Katsura et al. 2006). Cold allodynia resulting from spared nerve injury (SNI) surgery and CFA injection in rats was diminished by administration of TRPA1 antagonists (Stucky et al. 2009; del Camino et al. 2010). TRPA1 null mice exhibited reduced nocifensive responses to formalin, a chemical inducer of inflammation (Macpherson et al. 2007; McNamara et al. 2007; Stucky et al. 2009) and TG cultures from null mice show attenuated responses to bradykinin (Bautista et al. 2006).

Using HEK293 and cultured neurons as well as behavioural models, del Camino et al. (2010) found that even noxious cold temperatures were, alone, unable to evoke significant channel activity but that in the presence of an agonist, such as AITC, even low levels of innocuous cool evoked large currents and nocifensive responses. Moreover, the responses to noxious cold plate were comparable in WT and TRPA1 null mice (del Camino et al. 2010) suggesting that TRPA1 may play a role in mediating cold sensitivity only in pathological conditions, such as those mimicked in animal models of inflammatory and neuropathic pain.

4 Mechanical

The role of TRP channels in mechanosensation has been summarised in a recent review (Eijkelkamp et al. 2013). Very interestingly, TRPV1 global and conditional knockout mice show normal responses to mechanical stimulation even in models of inflammatory pain. A CCI model of neuropathic pain in rats, however, showed ipsilateral upregulation of TRPV1 protein in lamina I and II of the spinal cord and mechanical allodynia behaviour which was attenuated by a specific TRPV1 antagonist, BCTC (Kanai et al. 2005). TRPV3 null mice show no deficits in mechanical responses nor any differences in mechanical hyperalgesia induced by CFA or bradykinin compared to WTs (Moqrich et al. 2005). Though it does not respond to stretching of the membrane in vitro, and thus may not be gated by mechanical stimulation (Strotmann et al. 2000), TRPV4 does appear to play a role in mechanical hyperalgesia in inflammatory and neuropathic pain. In a variety of rat models of neuropathic pain, including diabetic and peripheral neuropathies, the reduction in mechanical threshold was reversed by intrathecal administration of TRPV4 antisense oligonucleotides while TRPV4 null mice did not exhibit mechanical hypersensitivity to the same extent as WTs (Alessandri-Haber et al. 2008). Interestingly, however, the expression of TRPV4 does not appear to be upregulated in these rat models of neuropathic pain. TRPV4 is frequently co-expressed with TRPC1 and TRPC6 in DRG, and it has been proposed that the channels may act in concert to mediate mechanical hyperalgesia in sensitised nociceptive neurons. Induction of mechanical hyperalgesia in inflammatory and neuropathic pain models was reversed by administration of antisense oligonucleotides to TRPC6 and, in certain models, TRPC1 (Alessandri-Haber et al. 2009). In contrast with this it has recently been shown that TRPC3 and TRPC6 double knockout mice do not have deficits in sensitivity to noxious mechanical pressure (Quick et al. 2012).

Drosophila larvae lacking expression of the TRPA1 homologue, *painless*, are insensitive to noxious mechanical stimuli (Tracey et al. 2003); similar to its role in noxious cold, however, the role of TRPA1 in transduction of noxious mechanical stimuli is in debate. Kwan et al. (2006) found a deficit in response to repeated application of 'high force' innocuous and noxious mechanical stimulation in TRPA1 null mice though Bautista et al. (2006) focus on their finding that there is no variation in the mechanical thresholds of these mice. A loss of slowly and intermediate adapting currents in small-diameter, non-peptidergic fibres was seen in DRG neurons taken from TRPA1 null mice (Vilceanu and Stucky 2010; Brierley et al. 2011); it had previously been suggested that slowly adapting currents from such neurons are associated with noxious mechanosensation (Drew et al. 2007). Mustard oil-induced mechanical hyperalgesia was inhibited by TRPA1 antisense oligonucleotides and TRPA1 antagonists (Perin-Martins et al. 2013), while CFA- and SNL-induced mechanical hyperalgesia is attenuated by TRPA1 antagonists (Petrus et al. 2007; Eid et al. 2008) but not by antisense oligonucleotides (Obata et al. 2005). On the basis of transfection of HEK293 cells with TRPA1, however, it seemed that the channel alone is not sufficient to confer mechanical sensitivity (Vilceanu and Stucky 2010); hence, it is possible that TRPA1 does not contribute to acute noxious mechanosensitivity but rather to the maintenance of mechanical hyperalgesia (Petrus et al. 2007). Alternatively, Brierley et al. (2011) suggest that TRPA1 confers mechanical sensitivity to a specific set of small-diameter fibres and that there are other DRG neurons which do not require the channel to confer mechanosensitivity.

A point mutation (N855S) in the S4 transmembrane segment of TRPA1 causes an autosomal dominant heritable pain condition known as familial episodic pain syndrome (FEPS) (Kremeyer et al. 2010). This mutation leads to greater inward currents following channel activation at resting neuronal membrane potentials and manifests as crippling upper body pain which begins in infancy and consists of attacks with a duration of ~1.5 h. Interestingly, the attacks are described as involving a sensation of mechanical pain which is initiated by a number of factors including cold.

5 Central Pain Pathways

The architecture of the central pain-processing pathway is generally agreed (Ossipov et al. 2010; Perl 2011), although refinements of the wiring have been proposed to account for the range of pain sensations evoked by different stimuli (Craig 2003). The primary afferent nociceptor terminates centrally on relay neurons and interneurons of the dorsal horn in the spinal cord, mostly in laminae I and II. Most of these terminals are glutamatergic, but some are peptidergic. Relay neurons within the cord project to spinothalamic neurons, which course through the brainstem and midbrain, and synapse onto different nuclei of the thalamus (Fig. 2). Lateral branches of the ascending pathway also terminate within brainstem structures, the periaqueductal grey (PAG) and the rostroventral medulla (RVM). Thalamic neurons are wired to cortical regions mediating sensory and emotive aspects of pain, respectively, the somatosensory cortex and anterior cingulate cortex. A spinoparabrachial pathway travels from the spinal cord to limbic structures. Descending pathways (Fig. 2) originating in the cortex and amygdala, and modulated by outputs of the PAG, medial RVM and locus ceruleus, affect the activity of the dorsal horn neurons via neurotransmitters including GABA and serotonin, which can antagonise or facilitate the sensation of pain, in some cases leading to the generation of efferent signals along the nociceptor into the periphery.

TRP channels of the A, C, V and M subfamilies are detectable in the spinal cord and brain. Topical and systemic application of capsaicin dominates the study of central TRP channels. TRPV1 may occur postsynaptically in lamina II of the dorsal horn, especially in lumbar segments, in glia as well as in neurons, and co-localises with substance P (Spicarová and Palecek 2008). TRPV1 activity in the cord causes release of substance P, ultimately antagonising peripheral inputs by activating interneurons (Ferrini et al. 2007). The inflammatory mediator bradykinin is released from endothelia and glia following injury (Hausmann 2003) and mediates pain via sensitisation of TRPA1 co-localised with B2R receptors on DRG neurons (Wang et al. 2008), but there is no report of a similar co-expression in central neurons.

TRPV1 shows robust in vitro response to inflammatory mediators, but the special characteristics of the central nervous system (CNS) immune response—generally weaker, involving a different repertoire of cell types and excluding many serum proteins by reason of the blood–spinal cord and blood–brain barriers—imply that the signalling pathways will differ from those in the periphery (Hausmann 2003). Nonetheless, microglial activation has been correlated with peripheral pain states including those caused by peripheral nerve trauma and bone cancer pain (Watkins et al. 2001a, b; Xu et al. 2006), and TRPV1 activity leads to microglial cell death (Kim et al. 2006), suggesting a coupling of central pain and inflammatory pathways. In the same way, spinal cord trauma leads to release of factors, some of which are not found after peripheral nerve damage, to mediate oxidative stress, inflammatory, energetic, apoptotic and lipid metabolic change (Kuner 2010; Yip and Malaspina 2012). It is not clear which if any of these factors subserve nociceptive functions by acting on TRP channels or in which direction the effect occurs. TRPV1 may stand upstream of some inflammatory responses, as suggested by the

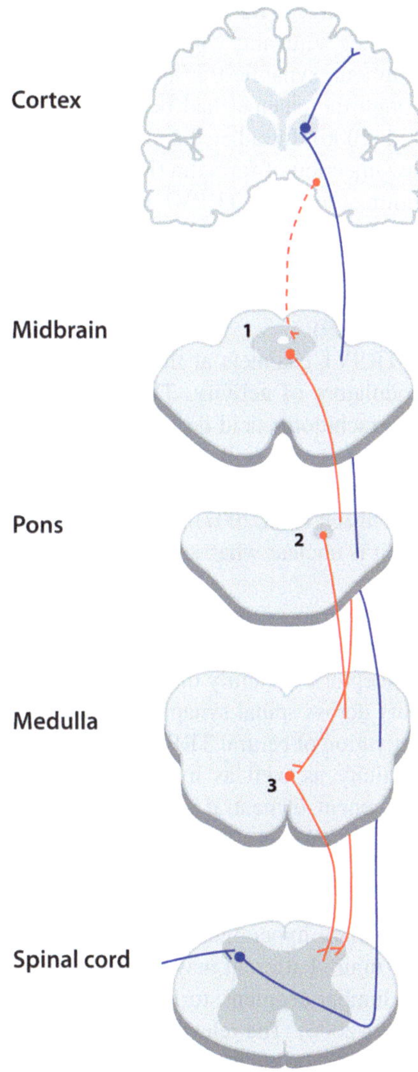

Fig. 2 Central pain processing. Ascending tracts (*blue*) and their peripheral inputs. Descending tracts (*red*) and modulatory pathways. (1) PAG; (2) Locus ceruleus; (3) RVM

finding that systemic capsaicin upregulated B1R expression on spinal cord microglia (Talbot et al. 2012). The secondary injury to the blood capillaries that follows spinal cord trauma is preceded by upregulation of the TRPM4 channel on the vessels (Gerzanich et al. 2009). Capillary fragmentation does not necessarily contribute to pain sensation, though it impairs neurological function. This channel overexpressed in HEK293 cells shows a graduated response to temperatures between 15 °C and 35 °C (Talavera et al. 2005) but must respond to other, unidentified outputs from injury in vivo to mediate the reported effects on haemorrhage. TRPV1 is implicated in the central response to a plantar injection of Freund's adjuvant; within substantia gelatinosa of spinal cord isolated after CFA treatment, synaptic transmission is inhibited by a TRPV1-selective antagonist, as though the channel had become activated by the inflammatory stimulus (Lappin et al. 2006). Intrathecal application of TRV1 antagonists alleviates the response to paw injection of formalin (Kanai et al. 2006), supporting the same conclusion. Oral administration of TRPV1 antagonists with different degrees of CNS permeance shows that central TRP channels contribute to pain responses (Cui et al. 2006). Intrathecal application of the TRPV1 agonist RTX produces analgesia, attributable to selective ablation of TRPV1-expressing central nerve terminals (Jeffry et al. 2009).

TRPV1 channels at the central terminals of nociceptors are exposed to CNS modulators of activity. The endogenous ligands of TRPV1 include 12-HPETE (an arachidonic acid derivative), AEA (anandamide) and NADA (*N*-arachidonoyl dopamine) (O'Neill et al. 2012). The endocannabinoid anandamide promotes calcium entry through TRPV1, via a mechanism that differs from that of capsaicin (Fischbach et al. 2007). The affinity of anandamide for the CB1 receptor is higher, and it is unclear what sequence of events would lead to effects on TRPV1 function in situ.

Neurotransmitters and peptides (e.g. substance P) with modulatory effect on the circuitry of relay neurons and interneurons are released by the central terminals of nociceptors. A family of lipid mediators, the resolvins, lowers transmission probability across spinal synapses via effects on TRPV1 and TRPA1 (Park et al. 2011). Activation of central TRPV1 channels by capsaicin is correlated with altered spinal plasticity as well as hyperalgesic behaviour. This altered plasticity is a likely component of central sensitisation, whereby innocuous inputs are interpreted as noxious (Willis 2009). Sensitisation of spinal—and perhaps supraspinal—responses is proposed to account for a sensation of pain that persists beyond the initial stimulus (Park et al. 1995: Sang et al. 1996). This pain might be ectopic (secondary hyperalgesia) or evoked in response to innocuous stimuli (allodynia). The injured afferent may additionally re-route, projecting to a synapse on a different lumbar segment, to facilitate input from undamaged afferents (Campbell and Meyer 2006). The relevant biological mechanism leading to facilitation is likely to be nociceptor burst activity, repeatedly stimulating the central synapse, following peripheral injury (Campbell and Meyer 2006); this proposed peripheral origin does not exclude the action of mediators within the cord sensitising TRP channels. Following synaptic plasticity, upregulation of cyclooxygenase (COX)-2 and other neuronal signalling pathways leads to ongoing pain beyond the initial stimulus (Rivat et al. 2010). Distinct features of central sensitisation include secondary hyperalgesia and mechanical allodynia, whereby the abnormally enhanced response occurs in undamaged tissue. The TRPV1 agonist gingerol can relieve secondary hyperalgesia after central application in rat model of spinal nerve injury (Gauthier et al. 2012).

The reactive metabolite methylglyoxal (MG) occurs in all cell types as a by-product of glycolysis and lipid peroxidation and might be an endogenous ligand of the TRPA1 channel. MG is normally detoxified by a glyoxalase pathway, but deficiency of this route is suggested by the high MG plasma levels in diabetes mellitus patients. The effect of MG on depressing the compound action potential of the sciatic nerve preparation is abolished in TRPA1 null mice (Eberhardt et al. 2012). While diabetes patients show reduced conduction velocity in the peroneal nerve (Hyllienmark et al. 2013), and patients with diabetic neuropathy can experience an increase in sensitivity to mechanical and thermal stimuli, the relevance of the MG observations to pathology is currently unclear.

Pain-associated plastic changes are best documented in spinothalamic tract neurons of the spinal cord (Willis 2002), but thalamic and cortical plasticity also is believed to contribute to sensitisation (Fu et al. 2008). Although potentiation of spinal transmission can be demonstrated, this is not conclusively a cellular analogue of pain sensation. Nonetheless, the data of Hjornevik et al. (2010) suggest that high-frequency stimulation of the spine produces both LTP and alterations in opioid receptor activity in higher brain areas.

Supraspinal involvement of TRP channels in pain are harder to demonstrate. Nonetheless, a coherent body of work with channel agonists and antagonists, applied centrally or peripherally, suggests a role for TRPV1 in descending modulation. TRPV1 mRNA occurs on cell bodies and synapses in brainstem structures involved in pain, namely the PAG and the RVM. Binding sites for the TRPV1 radioligand [3H]RTX occur in these locations as well as in the cortex, thalamus, hypothalamus, cerebellar cortex, locus ceruleus and spinal cord (Roberts et al. 2004) and the somatosensory cortex, anterior cingulate and amygdala (Steenland et al. 2006). Radioligand affinity for its target varies between structures (Szabo et al. 2002), which implies that accessory factors may be required for receptor availability. LTP in the amygdala is reported to be TRPV1 dependent (Zschenderlein et al. 2011). On the other hand, a recent publication on a TRPV1 reporter mouse claims expression restricted to the hippocampus, PAG, hypothalamus and midbrain only (Cavanaugh et al. 2011).

Accumulating data also indicate a role for TRPC channels in synaptic plasticity. LTP elicited at hippocampal interneuron synapses can be blocked by a nonspecific TRP channel inhibitor, and accompanying evidence on the calcium dependence of these channels implicates the TRPC family (Topolnik et al. 2006). RNAi experiments in hippocampal slice culture have shown that TRPC3 is required for dendritic spine formation in the presence of brain-derived neurotrophic factor (BDNF) (Amaral and Pozzo-Miller 2007). The influx of sodium and calcium ions through TRPC3 is proposed to stand upstream of NMDA receptor activation, leading to enhanced synaptic transmission in the form of LTP (Minichiello 2009). In hippocampal neuron culture, TRPC5 and TRPC6 activity mediates the phosphorylation of Akt by BDNF; this may have consequences for synaptic plasticity, since both Akt phosphorylation and LTP induction protocols can lead to the transient insertion of calcium-permeable AMPA receptors in the neuronal membrane (Fortin et al. 2012). Finally, the TRPC5 knockout mouse, which resists pilocarpine induction of seizures, is defective in LTP induction (Phelan et al. 2013), suggesting that plasticity and regulated excitability might be coupled outputs of TRP-mediated calcium homeostasis.

TRPM, the melastatin family of TRP channels, currently includes eight members. TRPM3 is highly expressed in the kidney, brain, spinal cord and testis and in some experimental preparations works as an osmosensor (Grimm et al. 2003). This function would be consistent with a role in kidney physiology. TPRM3 in beta cells is moreover activated by pregnenolone sulphate (Wagner et al. 2008), a cholesterol derivative synthesised in the adrenal and the brain, and standing upstream of the glucocorticoids, mineralocorticoids and sex steroids. Pregnenolone sulphate (PS) was proposed as a pain mediator in the same year that TRPM3 was cloned, on quite independent grounds related to expression of enzymes in the spinal cord (Patte-Mensah et al. 2003). The association of the channel and the steroid motivated a study that showed heat activation of TRPM3 expressed in HEK cells and the potentiation of this effect by low doses of PS. In the tail immersion test, TRPM3 -/- mice tolerated temperatures up to 57 °C to a greater extent than wild-type mice. A different behavioural response to PS was not reported by this study (Vriens et al. 2011), and so the relevance of the steroid as an endogenous ligand for TRPM3-mediated nociception remains to be demonstrated.

The PAG is connected monosynaptically to the RVM, in which the activity of a population of pain modulatory OFF cells is correlated with analgesia. Injection of TRPV1 agonists into the PAG enhances the excitability of the OFF cells and suppresses thermoception in the plantar test (Maione et al. 2006). This is discrepant with the effects of capsaicin in the periphery. Consistent with this report, infusions of a TRPV1 antagonist into the PAG enhance the response of rats to thermal stimulation of the paw (Starowicz et al. 2007). The activation of TRP channels in the PAG plausibly modulates pain responses, since RVM neurons project to the dorsal horn, synapsing with the primary nociceptive afferents as well as with the relay and interneurons. The antinociceptive projections are GABAergic and glycinergic, but it is unclear that they originate from the OFF cells (Ossipov et al. 2010). The effect of capsaicin on excitation in the ventrolateral PAG depends on glutamate receptors (Palazzo et al. 2002). McGaraughty et al. (2003) recorded neuronal responses to infused capsaicin simultaneously with tail-flick latencies at 52 °C. They reported that the OFF neurons in dorsal PAG are active during analgesia only after some two hours of ON neuronal activity and pain-avoiding behaviour and suggested that pain reduction occurred only after desensitisation of TRPV1 by capsaicin.

Capsaicin microinjected into the ACC increases the activity of selected neurons, while repressing others (Steenland et al. 2006). The correlation of this TRPV1-mediated neuronal activity with behaviourally relevant stimuli is unknown, but since ACC is connected to the PAG, a descending effect on nociception is presumed.

Systemic injection of capsaicin increases excitability within the locus ceruleus, a structure that responds to noxious stimuli (Hajós et al. 1987), is reciprocally connected with the RVM and sends a noradrenergic descending projection to the dorsal horn (Ossipov et al. 2010). The noradrenergic output can be suspended after peripheral nerve injury (Rahman et al. 2008), and this presumably facilitates pain sensation, but the involvement of TRP channels in this process is not known.

6 Analgesics

The roles of TRP channels in mediating sensitivity to noxious stimuli make them attractive targets for analgesics. Most drug development studies have focussed on TRPV1 and TRPA1, which are clearly linked to aspects of peripheral nociceptor function. For example, Honore et al. (2000) found repeated dosing of two distinct TRPV1 antagonists was able to abolish spontaneous pain and thermal and mechanical behaviour responses in animal models of inflammatory and neuropathic pain. Capsaicin causes sensitisation of TRPV1; however, application of high-dose patches of ~8 % to the skin can be used to cause desensitisation and is used as a treatment for patients suffering with neuropathic pain. A review of six studies involving 2,073 participants with post-herpetic neuralgia or HIV neuropathy found these high-concentration capsaicin patches were effective at inducing pain relief (Derry et al. 2013). Pain ratings of sufferers of chronic lower back pain, a common and notoriously difficult to treat condition, are reduced by application of a capsicum plaster (Frerick et al. 2003). The minimal side effects resulting from this topically applied TRPV1-mediated analgesic provide further support for the potential role of TRP channels in the development of improved analgesics. Low doses of icilin and menthol, both TRPM8 agonists, are able to reduce mechanical and thermal hypersensitivity caused by peripheral nerve injury (Proudfoot et al. 2006). Antagonism of TRPV1 and TRPA1 during the acute phase of pancreatitis reduced pain behaviour and inflammation as well as preventing progression of the pathological changes seen in chronic pancreatitis though these treatments were unable to reverse pain behaviours in models of established chronic pancreatitis (Schwartz et al. 2013). TRPA1 has been shown to play a role in hyperalgesia, and the antagonist HC-030031 attenuates inflammatory- and neuropathy-induced mechanical hypersensitivity (Eid et al. 2008).

In spite of the evidence that TRP channels are involved in sensing noxious stimuli, their potential as analgesic drug targets has not been fulfilled in a substantial way. The widespread expression of TRP channels, together with their ability to heteromultimerise increases the likelihood that such drugs may also induce a number of off-target effects.

Acknowledgments We thank the Wellcome Trust and the Medical Research Council for generous support.

References

Alessandri-Haber N, Yeh JJ, Boyd AE, Parada CA, Chen X, Reichling DB, Levine JD (2003) Hypotonicity induces TRPV4-mediated nociception in rat. Neuron 39:497–511

Alessandri-Haber N, Dina OA, Joseph EK, Reichling DB, Levine JD (2008) Interaction of transient receptor potential vanilloid 4, integrin, and SRC tyrosine kinase in mechanical hyperalgesia. J Neurosci 28:1046–1057

Alessandri-Haber N, Dina OA, Chen X, Levine JD (2009) TRPC1 and TRPC6 channels cooperate with TRPV4 to mediate mechanical hyperalgesia and nociceptor sensitization. J Neurosci 29:6217–6228

Amadesi S, Cottrell GS, Divino L, Chapman K, Grady EF, Bautista F, Karanjia R, Barajas-Lopez C, Vanner S, Vergnolle N, Bunnett NW (2006) Protease-activated receptor 2 sensitizes TRPV1 by protein kinase Cepsilonand A-dependent mechanisms in rats and mice. J Physiol 575:555–571

Amaral MD, Pozzo-Miller L (2007) TRPC3 channels are necessary for brain-derived neurotrophic factor to activate a nonselective cationic current and to induce dendritic spine formation. J Neurosci 27:5179–5189

Amaya F, Oh-hashi K, Naruse Y, Iijima N, Ueda M, Shimosato G, Tominaga M, Tanaka Y, Tanaka M (2003) Local inflammation increases vanilloid receptor 1 expression within distinct subgroups of DRG neurons. Brain Res 963:190–196

Babes A, Zorzon D, Reid G (2004) Two populations of cold-sensitive neurons in rat dorsal root ganglia and their modulation by nerve growth factor. Eur J Neurosci 20:2276–2282

Bandell M, Story GM, Hwang SW, Viswanath V, Eid SR, Petrus MJ, Earley TJ, Patapoutian A (2004) Noxious cold ion channel TRPA1 is activated by pungent compounds and bradykinin. Neuron 41:849–857

Bautista DM, Jordt SE, Nikai T, Tsuruda PR, Read AJ, Poblete J, Yamoah EN, Basbaum AI, Julius D (2006) TRPA1 mediates the inflammatory actions of environmental irritants and proalgesic agents. Cell 124:1269–1282

Bautista DM, Siemens J, Glazer JM, Tsuruda PR, Basbaum AI, Stucky CL, Jordt SE, Julius D (2007) The menthol receptor TRPM8 is the principal detector of environmental cold. Nature 448:204–208

Brierley SM, Castro J, Harrington AM, Hughes PA, Page AJ, Rychkov GY, Blackshaw LA (2011) TRPA1 contributes to specific mechanically activated currents and sensory neuron mechanical hypersensitivity. J Physiol 589:3575–3593

British Pain Society (2003) Pain in Europe Survey, Janet Fricker for Mundipharma International Ltd. http://www.britishpainsociety.org/Pain%20in%20Europ%20survey%20report.pdf

Campbell JN, Meyer RA (2006) Mechanisms of neuropathic pain. Neuron 52:77–92

Caspani O, Zurborg S, Labuz D, Heppenstall PA (2009) The contribution of TRPM8 and TRPA1 channels to cold allodynia and neuropathic pain. PLoS One 4:e7383

Caterina MJ, Schumacher MA, Tominaga M, Rosen TA, Levine JD, Julius D (1997) The capsaicin receptor: a heat-activated ion channel in the pain pathway. Nature 389:816–824

Caterina MJ, Rosen TA, Tominaga M, Brake AJ, Julius D (1999) A capsaicin-receptor homologue with a high threshold for noxious heat. Nature 398:436–441

Caterina MJ, Leffler A, Malmberg AB, Martin WJ, Trafton J, Petersen-Zeitz KR, Koltzenburg M, Basbaum AI, Julius D (2000) Impaired nociception and pain sensation in mice lacking the capsaicin receptor. Science 288:306–13

Cavanaugh DJ, Chesler AT, Jackson AC, Sigal YM, Yamanaka H, Grant R, O'Donnell D, Nicoll RA, Shah NM, Julius D, Basbaum AI (2011) Trpv1 reporter mice reveal highly restricted brain distribution and functional expression in arteriolar smooth muscle cells. J Neurosci 31:5067–5077

Chung MK, Lee H, Mizuno A, Suzuki M, Caterina MJ (2004) TRPV3 and TRPV4 mediate warmth-evoked currents in primary mouse keratinocytes. J Biol Chem 279:21569–21575

Chung MK, Güler AD, Caterina MJ (2005) Biphasic currents evoked by chemical or thermal activation of the heat-gated ion channel, TRPV3. J Biol Chem 280:15928–15941

Colburn RW, Lubin ML, Stone DJ, Wang Y, Lawrence D, D'Andrea MR, Brandt MR, Liu Y, Flores CM, Qin N (2007) Attenuated cold sensitivity in TRPM8 null mice. Neuron 54:379–386

Craig AD (2003) Pain mechanisms: labeled lines versus convergence in central processing. Annu Rev Neurosci 26:1–30

Cui M, Honore P, Zhong C, Gauvin D, Mikusa J, Hernandez G, Chandran P, Gomtsyan A, Brown B, Bayburt EK, Marsh K, Bianchi B, McDonald H, Niforatos W, Neelands TR, Moreland RB, Decker MW, Lee CH, Sullivan JP, Faltynek CR (2006) TRPV1 receptors in the CNS play a key role in broad-spectrum analgesia of TRPV1 antagonists. J Neurosci 26:9385–9393

Davis JB, Gray J, Gunthorpe MJ, Hatcher JP, Davey PT, Overend P, Harries MH, Latcham J, Clapham C, Atkinson K, Hughes SA, Rance K, Grau E, Harper AJ, Pugh PL, Rogers DC, Bingham S, Randall A, Sheardown SA (2000) Vanilloid receptor-1 is essential for inflammatory thermal hyperalgesia. Nature 405:183–187

del Camino D, Murphy S, Heiry M, Barrett LB, Earley TJ, Cook CA, Petrus MJ, Zhao M, D'Amours M, Deering N, Brenner GJ, Costigan M, Hayward NJ, Chong JA, Fanger CM, Woolf CJ, Patapoutian A, Moran MM (2010) TRPA1 contributes to cold hypersensitivity. J Neurosci 30:15165–15174

Derry S, Sven-Rice A, Cole P, Tan T, Moore RA (2013) Topical capsaicin (high concentration) for chronic neuropathic pain in adults. Cochrane Database Syst Rev 2, CD007393

Dhaka A, Murray AN, Mathur J, Earley TJ, Petrus MJ, Patapoutian A (2007) TRPM8 is required for cold sensation in mice. Neuron 54:371–378

Drew LJ, Rugiero F, Cesare P, Gale JE, Abrahamsen B, Bowden S, Heinzmann S, Robinson M, Brust A, Colless B, Lewis RJ, Wood JN (2007) High-threshold mechanosensitive ion channels blocked by a novel conopeptide mediate pressure-evoked pain. PLoS One 2:e515

Eberhardt MJ, Filipovic MR, Leffler A, de la Roche J, Kistner K, Fischer MJ, Fleming T, Zimmermann K, Ivanovic-Burmazovic I, Nawroth PP, Bierhaus A, Reeh PW, Sauer SK (2012) Methylglyoxal activates nociceptors through transient receptor potential channel A1 (TRPA1): a possible mechanism of metabolic neuropathies. J Biol Chem 287:28291–28306

Eid SR, Crown ED, Moore EL, Liang HA, Choong KC, Dima S, Henze DA, Kane SA, Urban MO (2008) HC-030031, a TRPA1 selective antagonist, attenuates inflammatory- and neuropathy-induced mechanical hypersensitivity. Mol Pain 4:48

Eijkelkamp N, Quick K, Wood JN (2013) Transient receptor potential channels and mechanosensation. AnnuRev Neurosci 36:519–546

Elg S, Marmigere F, Mattsson JP, Ernfors P (2007) Cellular subtype distribution and developmental regulation of TRPC channel members in mouse dorsal root ganglion. J Comp Neurol 503 (1):35–46

Facer P, Casula MA, Smith GD, Benham CD, Chessell IP, Bountra C, Sinisi M, Birch R, Anand P (2007) Differential expression of the capsaicin receptor TRPV1 and related novel receptors TRPV3, TRPV4 and TRPM8 in normal human tissues and changes in traumatic and diabetic neuropathy. BMC Neurol 7:11

Ferreira J, da Silva GL, Calixto JB (2004) Contribution of vanilloid receptors to the overt nociception induced by B2 kinin receptor activation in mice. Br J Pharmacol 141:787–794

Ferrini F, Salio C, Vergnano AM, Merighi A (2007) Vanilloid receptor-1 (TRPV1)-dependent activation of inhibitory neurotransmission in spinal substantia gelatinosa neurons of mouse. Pain 129:195–209

Fischbach T, Greffrath W, Nawrath H, Treede RD (2007) Effects of anandamide and noxious heat on intracellular calcium concentration in nociceptive drg neurons of rats. J Neurophysiol 98:929–938

Fortin DA, Srivastava T, Dwarakanath D, Pierre P, Nygaard S, Derkach VA, Soderling TR (2012) Brain-derived neurotrophic factor activation of CaM-kinase kinase via transient receptor potential canonical channels induces the translation and synaptic incorporation of GluA1-containing calcium-permeable AMPA receptors. J Neurosci 32:8127–8137

Foulkes T, Wood JN (2007) Mechanisms of cold pain. Channels (Austin) 1:154–160

Frerick H, Keitel W, Kuhn U, Schmidt S, Bredehorst A, Kuhlmann M (2003) Topical treatment of chronic low back pain with a capsicum plaster. Pain 106:59–64

Fu Y, Han J, Ishola T, Scerbo M, Adwanikar H, Ramsey C, Neugebauer V (2008) PKA and ERK, but not PKC, in the amygdala contribute to pain-related synaptic plasticity and behavior. Mol Pain 4:26
Fukuoka T, Tokunaga A, Tachibana T, Dai Y, Yamanaka H, Noguchi K (2002) VR1, but not P2X (3), increases in the spared L4 DRG in rats with L5 spinal nerve ligation. Pain 99:111–120
Gauthier ML, Beaudry F, Vachon P (2013) Intrathecal [6]-Gingerol administration alleviates peripherally induced neuropathic pain in male Sprague-Dawley Rats. Phytother Res 27 (8):1251–1254
Gerzanich V, Woo SK, Vennekens R, Tsymbalyuk O, Ivanova S, Ivanov A, Geng Z, Chen Z, Nilius B, Flockerzi V, Freichel M, Simard JM (2009) De novo expression of Trpm4 initiates secondary hemorrhage in spinal cord injury. Nat Med 15:185–191
Grant AD, Cottrell GS, Amadesi S, Trevisani M, Nicoletti P, Materazzi S, Altier C, Cenac N, Zamponi GW, Bautista-Cruz F, Lopez CB, Joseph EK, Levine JD, Liedtke W, Vanner S, Vergnolle N, Geppetti P, Bunnett NW (2007) Protease-activated receptor 2 sensitizes the transient receptor potential vanilloid 4 ion channel to cause mechanical hyperalgesia in mice. J Physiol 578:715–733
Grimm C, Kraft R, Sauerbruch S, Schultz G, Harteneck C (2003) Molecular and functional characterization of the melastatin-related cation channel TRPM3. J Biol Chem 278 (24):21493–21501
Güler AD, Lee H, Iida T, Shimizu I, Tominaga M, Caterina M (2002) Heat-evoked activation of the ion channel, TRPV4. J Neurosci 22:6408–6414
Hajós M, Jancsó G, Engberg G (1987) Capsaicin-induced excitation of locus coeruleus neurons. Acta Physiol Scand 129:415–420
Hausmann ON (2003) Post-traumatic inflammation following spinal cord injury. Spinal Cord 41:369–378
Hjornevik T, Schoultz BW, Marton J, Gjerstad J, Drzezga A, Henriksen G, Willoch F (2010) Spinal long-term potentiation is associated with reduced opioid neurotransmission in the rat brain. Clin Physiol Funct Imaging 30:285–293
Honore P, Chandran P, Hernandez G, Gauvin DM, Mikusa JP, Zhong C, Joshi SK, Ghilardi JR, Sevcik MA, Fryer RM, Segreti JA, Gomtsyan A, Lee C-H, Kort ME, Reilly RM, Surowy CS, Kym PR, Mantyh PW, Sullivan JP, Jarvis MF, Faltynek CR (2000) Repeated dosing of ABT-102, a potent and selective TRPV1 antagonist, enhances TRPV1-mediated analgesic activity in rodents, but attenuates antagonist-induced hyperthermia. Pain 99:111–120
Huang SM, Lee H, Chung MK, Park U, Yu YY, Bradshaw HB, Coulombe PA, Walker JM, Caterina MJ (2008) Overexpressed transient receptor potential vanilloid 3 ion channels in skin keratinocytes modulate pain sensitivity via prostaglandin E2. J Neurosci 28:13727–13737
Huang SM, Li X, Yu Y, Wang J, Caterina MJ (2011) TRPV3 and TRPV4 ion channels are not major contributors to mouse heat sensation. Mol Pain 7:37
Hudson LJ, Bevan S, Wotherspoon G, Gentry C, Fox A, Winter J (2001) VR1 protein expression increases in undamaged DRG neurons after partial nerve injury. Eur J Neurosci 13:2105–2114
Hyllienmark L, Alstrand N, Jonsson B, Ludvigsson J, Cooray G, Wahlberg-Topp J (2013) Early electrophysiological abnormalities and clinical neuropathy: a prospective study in patients with type 1 diabetes. Diabetes Care doi:10.2337/dc12-2226
Jahanshaki M, Pitt P, Williams I (1989) Pain avoidance in rheumatoid arthritis. J Psychosomat Res 33:579–589
Jeffry JA, Yu SQ, Sikand P, Parihar A, Evans MS, Premkumar LS (2009) Selective targeting of TRPV1 expressing sensory nerve terminals in the spinal cord for long lasting analgesia. PLoS One 4:e7021
Ji RR, Samad TA, Jin SX, Schmoll R, Woolf CJ (2002) p38 MAPK activation by NGF in primary sensory neurons after inflammation increases TRPV1 levels and maintains heat hyperalgesia. Neuron 36:57–68

Jordt SE, Bautista DM, Chuang HH, McKemy DD, Zygmunt PM, Högestätt ED, Meng ID, Julius D (2004) Mustard oils and cannabinoids excite sensory nerve fibres through the TRP channel ANKTM1. Nature 427:260–265

Jørum E, Warncke T, Stubhaug A (2003) Cold allodynia and hyperalgesia in neuropathic pain: the effect of N methyl-D-aspartate (NMDA) receptor antagonist ketamine-a double-blind, cross-over comparison with alfentanil and placebo. Pain 101:229–235

Kanai Y, Nakazato E, Fujiuchi A, Hara T, Imai A (2005) Involvement of an increased spinal TRPV1 sensitization through its up-regulation in mechanical allodynia of CCI rats. Neuropharmacology 49:977–984

Kanai Y, Hara T, Imai A (2006) Participation of the spinal TRPV1 receptors in formalin-evoked pain transduction: a study using a selective TRPV1 antagonist, iodo-resiniferatoxin. J Pharm Pharmacol 58:489–493

Karashima Y, Talavera K, Everaerts W, Janssens A, Kwan KY, Vennekens R, Nilius B, Voets T (2009) TRPA1 acts as a cold sensor in vitro and in vivo. Proc Natl Acad Sci U S A 106:1273–1278

Katsura H, Obata K, Mizushima T, Yamanaka H, Kobayashi K, Dai Y, Fukuoka T, Tokunaga A, Sakagami M, Noguchi K (2006) Antisense knock down of TRPA1, but not TRPM8, alleviates cold hyperalgesia after spinal nerve ligation in rats. Exp Neurol 200:112–123

Kim SR, Kim SU, Oh U, Jin BK (2006) Transient receptor potential vanilloid subtype 1 mediates microglial cell death in vivo and in vitro via Ca2+-mediated mitochondrial damage and cytochrome c release. J Immunol 177:4322–4329

Knowlton WM, Daniels RL, Palkar R, McCoy DD, McKemy DD (2011) Pharmacological blockade of TRPM8 ion channels alters cold and cold pain responses in mice. PLoS One 6: e25894

Knowlton WM, Palkar R, Lippoldt EK, McCoy DD, Baluch F, Chen J, McKemy DD (2013) A sensory-labeled line for cold: TRPM8-expressing sensory neurons define the cellular basis for cold, cold pain, and cooling mediated analgesia. J Neurosci 33:2837–2848

Koerber HR, McIlwrath SL, Lawson JJ, Malin SA, Anderson CE, Jankowski MP, Davis BM (2010) Cutaneous C-polymodal fibers lacking TRPV1 are sensitized to heat following inflammation, but fail to drive heat hyperalgesia in the absence of TPV1 containing C-heat fibers. Mol Pain 6:58

Koltzenburg M, Scadding J (2001) Neuropathic pain. Curr Opin Neurol 14:641–647

Kremeyer B, Lopera F, Cox JJ, Momin A, Rugiero F, Marsh S, Woods CG, Jones NG, Paterson KJ, Fricker FR, Villegas A, Acosta N, Pineda-Trujillo NG, Ramírez JD, Zea J, Burley MW, Bedoya G, Bennett DL, Wood JN, Ruiz-Linares A (2010) A gain-of-function mutation in TRPA1 causes familial episodic pain syndrome. Neuron 66:671–680

Kuner R (2010) Central mechanisms of pathological pain. Nat Med 16:1258–1266

Kwan KY, Allchorne AJ, Vollrath MA, Christensen AP, Zhang DS, Woolf CJ, Corey DP (2006) TRPA1 contributes to cold, mechanical, and chemical nociception but is not essential for hair-cell transduction. Neuron 50:277–289

Lappin SC, Randall AD, Gunthorpe MJ, Morisset V (2006) TRPV1 antagonist, SB-366791, inhibits glutamatergic synaptic transmission in rat spinal dorsal horn following peripheral inflammation. Eur J Pharmacol 540:73–81

Lauria G, Morbin M, Lombardi R, Capobianco R, Camozzi F, Pareyson D, Manconi M, Geppetti P (2006) Expression of capsaicin receptor immunoreactivity in human peripheral nervous system and in painful neuropathies. J Peripher Nerv Syst 11:262–271

Liedtke W, Friedman JM (2003) Abnormal osmotic regulation in trpv4-/- mice. Proc Natl Acad Sci U S A 100:13698–13703

Liedtke W, Choe Y, Martí-Renom MA, Bell AM, Denis CS, Sali A, Hudspeth AJ, Friedman JM, Heller S (2000) Vanilloid receptor-related osmotically activated channel (VR-OAC), a candidate vertebrate osmoreceptor. Cell 103:525–535

Liu B, Yao J, Zhu MX, Qin F (2011) Hysteresis of gating underlines sensitization of TRPV3 channels. J Gen Physiol 138:509–520

Loyd DR, Weiss G, Henry MA, Hargreaves KM (2011) Serotonin increases the functional activity of capsaicinsensitive rat trigeminal nociceptors via peripheral serotonin receptors. Pain 152:2267–2276
Luo H, Cheng J, Han JS, Wan Y (2004) Change of vanilloid receptor 1 expression in dorsal root ganglion and spinal dorsal horn during inflammatory nociception induced by complete Freund's adjuvant in rats. Neuroreport 15:655–658
Ma S, Yu H, Zhao Z, Luo Z, Chen J, Ni Y, Jin R, Ma L, Wang P, Zhu Z, Li L, Zhong J, Liu D, Nilius B (2012) Activation of the cold-sensing TRPM8 channel triggers UCP1-dependent thermogenesis and prevents obesity. J Mol Cell Biol 4:88–96
Macpherson LJ, Xiao B, Kwan KY, Petrus MJ, Dubin AE, Hwang S, Cravatt B, Corey DP, Patapoutian A (2007) An ion channel essential for sensing chemical damage. J Neurosci 27:11412–11415
Maione S, Bisogno T, de Novellis V, Palazzo E, Cristino L, Valenti M, Petrosino S, Guglielmotti V, Rossi F, Di Marzo V (2006) Elevation of endocannabinoid levels in the ventrolateral periaqueductal grey through inhibition of fatty acid amide hydrolase affects descending nociceptive pathways via both cannabinoid receptor type 1 and transient receptor potential vanilloid type-1 receptors. J Pharmacol Exp Ther 316:969–982
McGaraughty S, Chu KL, Bitner RS, Martino B, El Kouhen R, Han P, Nikkel AL, Burgard EC, Faltynek CR, Jarvis MF (2003) Capsaicin infused into the PAG affects rat tail flick responses to noxious heat and alters neuronal firing in the RVM. J Neurophysiol 90:2702–2710
McKemy DD, Neuhausser WM, Julius D (2002) Identification of a cold receptor reveals a general role for TRP channels in thermosensation. Nature 416:52–58
McNamara CR, Mandel-Brehm J, Bautista DM, Siemens J, Deranian KL, Zhao M, Hayward NJ, Chong JA, Julius D, Moran MM, Fanger CM (2007) TRPA1 mediates formalin-induced pain. Proc Natl Acad Sci U S A 104:13525–30
Minichiello L (2009) TrkB signalling pathways in LTP and learning. Nat Rev Neurosci 10:850–860
Mishra SK, Tisel SM, Orestes P, Bhangoo SK, Hoon MA (2011) TRPV1-lineage neurons are required for thermal sensation. EMBO J 30:582–593
Moqrich A, Hwang SW, Earley TJ, Petrus MJ, Murray AN, Spencer KS, Andahazy M, Story GM, Patapoutian A (2005) Impaired thermosensation in mice lacking TRPV3, a heat and camphor sensor in the skin. Science 307:1468–1472
Moriyama T, Higashi T, Togashi K, Iida T, Segi E, Sugimoto Y, Tominaga T, Narumiya S, Tominaga M (2005) Sensitization of TRPV1 by EP1 and IP reveals peripheral nociceptive mechanism of prostaglandins. Mol Pain 1:3
O'Neill J, Brock C, Olesen AE, Andresen T, Nilsson M, Dickenson AH (2012) Unravelling the mystery of capsaicin: a tool to understand and treat pain. Pharmacol Rev 64:939–971
Obata K, Katsura H, Mizushima T, Yamanaka H, Kobayashi K, Dai Y, Fukuoka T, Tokunaga A, Tominaga M, Noguchi K (2005) TRPA1 induced in sensory neurons contributes to cold hyperalgesia after inflammation and nerve injury. J Clin Invest 115:2393–2401
Ossipov MH, Dussor GO, Porreca F (2010) Central modulation of pain. J Clin Invest 120:3779–3787
Palazzo E, de Novellis V, Marabese I, Cuomo D, Rossi F, Berrino L, Maione S (2002) Interaction between vanilloid and glutamate receptors in the central modulation of nociception. Eur J Pharmacol 439:69–75
Park KM, Max MB, Robinovitz E, Gracely RH, Bennett GJ (1995) Effects of intravenous ketamine, alfentanil, or placebo on pain, pinprick hyperalgesia, and allodynia produced by intradermal capsaicin in human subjects. Pain 63:163–172
Park CK, Xu ZZ, Liu T, Lü N, Serhan CN, Ji RR (2011) Resolvin D2 is a potent endogenous inhibitor for transient receptor potential subtype V1/A1, inflammatory pain, and spinal cord synaptic plasticity in mice: distinct roles of resolvin D1, D2, and E1. J Neurosci 31:18433–18438

Patte-Mensah C, Kappes V, Freund-Mercier MJ, Tsutsui K, Mensah-Nyagan AG (2003) Cellular distribution and bioactivity of the key steroidogenic enzyme, cytochrome P450side chain cleavage, in sensory neural pathways. J Neurochem 86:1233–1246

Peier AM, Moqrich A, Hergarden AC, Reeve AJ, Andersson DA, Story GM, Earley TJ, Dragoni I, McIntyre P, Bevan S, Patapoutian A (2002a) A TRP channel that senses cold stimuli and menthol. Cell 108:705–715

Peier AM, Reeve AJ, Andersson DA, Moqrich A, Earley TJ, Hergarden AC, Story GM, Colley S, Hogenesch JB, McIntyre P, Bevan S, Patapoutian A (2002b) A heat-sensitive TRP channel expressed in keratinocytes. Science 296:2046–2049

Perin-Martins A, Teixeira JM, Tambeli CH, Parada CA, Fischer L (2013) Mechanisms underlying transient receptor potential ankyrin 1 (TRPA1)-mediated hyperalgesia and edema. J Peripher Nerv Syst 18:62–74

Perl ER (2011) Pain mechanisms: a commentary on concepts and issues. Prog Neurobiol 94:20–38

Petrus M, Peier AM, Bandell M, Hwang SW, Huynh T, Olney N, Jegla T, Patapoutian A (2007) A role of TRPA1 in mechanical hyperalgesia is revealed by pharmacological inhibition. Mol Pain 3:40

Phelan KD, Shwe UT, Abramowitz J, Wu H, Rhee SW, Howell MD, Gottschall PE, Freichel M, Flockerzi V, Birnbaumer L, Zheng F (2013) Canonical transient receptor channel 5 (TRPC5) and TRPC1/4 contribute to seizure and excitotoxicity by distinct cellular mechanisms. Mol Pharmacol 83:429–438

Phelps CB, Wang RR, Choo SS, Gaudet R (2010) Differential regulation of TRPV1, TRPV3, and TRPV4 sensitivity through a conserved binding site on the ankyrin repeat domain. J Biol Chem 285:731–740

Pogorzala LA, Mishra SK, Hoon MA (2013) The cellular code for mammalian thermosensation. J Neurosci 33(13):5533–5541

Premkumar LS, Ahern GP (2000) Induction of vanilloid receptor channel activity by protein kinase C. Nature 408:985–990

Proudfoot CJ, Garry EM, Cottrell DF, Rosie R, Anderson H, Robertson DC, Fleetwood-Walker SM, Mitchell R (2006) Analgesia mediated by the TRPM8 cold receptor in chronic neuropathic pain. Curr Biol 16:1591–1605

Quick K, Zhao J, Eijkelkamp N, Linley JE, Rugiero F, Cox JJ, Raouf R, Gringhuis M, Sexton JE, Abramowitz J, Taylor R, Forge A, Ashmore J, Kirkwood N, Kros CJ, Richardson GP, Freichel M, Flockerzi V, Birnbaumer L, Wood JN (2012) TRPC3 and TRPC6 are essential for normal mechanotransduction in subsets of sensory neurons and cochlear hair cells. Open Biol 2:120068

Rahman W, D'Mello R, Dickenson AH (2008) Peripheral nerve injury-induced changes in spinal alpha(2)-adrenoceptor-mediated modulation of mechanically evoked dorsal horn neuronal responses. J Pain 9:350–359

Ramachandran R, Hyun E, Zhao L, Lapointe TK, Chapman K, Hirota CL, Ghosh S, McKemy DD, Vergnolle N, Beck PL, Altier C, Hollenberg MD (2013) TRPM8 activation attenuates inflammatory responses in mouse models of colitis. Proc Natl Acad Sci U S A 110:7476–7481

Rivat C, Becker C, Blugeot A, Zeau B, Mauborgne A, Pohl M, Benoliel JJ (2010) Chronic stress induces transient spinal neuroinflammation, triggering sensory hypersensitivity and long-lasting anxiety-induced hyperalgesia. Pain 150:358–368

Roberts JC, Davis JB, Benham CD (2004) [3H]Resiniferatoxin autoradiography in the CNS of wild-type and TRPV1 null mice defines TRPV1 (VR-1) protein distribution. Brain Res 995:176–183

Sang CN, Gracely RH, Max MB, Bennett GJ (1996) Capsaicin-evoked mechanical allodynia and hyperalgesia cross nerve territories. Evidence for a central mechanism. Anesthesiology 85:491–496

Schwartz ES, La JH, Scheff NN, Davis BM, Albers KM, Gebhart GF (2013) TRPV1 and TRPA1 antagonists prevent the transition of acute to chronic inflammation and pain in chronic pancreatitis. J Neurosci 33:5603–5611

Smith GD, Gunthorpe MJ, Kelsell RE, Hayes PD, Reilly P, Facer P, Wright JE, Jerman JC, Walhin JP, Ooi L, Egerton J, Charles KJ, Smart D, Randall AD, Anand P, Davis JB (2002) TRPV3 is a temperature-sensitive vanilloid receptor-like protein. Nature 418:186–190
Spicarová D, Palecek J (2008) The role of spinal cord vanilloid (TRPV1) receptors in pain modulation. Physiol Res 57(Suppl 3):S69–S77
Starowicz K, Maione S, Cristino L, Palazzo E, Marabese I, Rossi F, de Novellis V, Di Marzo V (2007) Tonic endovanilloid facilitation of glutamate release in brainstem descending antinociceptive pathways. J Neurosci 27:13739–13749
Steenland HW, Ko SW, Wu LJ, Zhuo M (2006) Hot receptors in the brain. Mol Pain 2:34
Story GM, Peier AM, Reeve AJ, Eid SR, Mosbacher J, Hricik TR, Earley TJ, Hergarden AC, Andersson DA, Hwang SW, McIntyre P, Jegla T, Bevan S, Patapoutian A (2003) ANKTM1, a TRP-like channel expressed in nociceptive neurons, is activated by cold temperatures. Cell 112:819–829
Strotmann R, Harteneck C, Nunnenmacher K, Schultz G, Plant TD (2000) OTRPC4, a nonselective cation channel that confers sensitivity to extracellular osmolarity. Nat Cell Biol 2:695–702
Stucky CL, Dubin AE, Jeske NA, Malin SA, McKemy DD, Story GM (2009) Roles of transient receptor potential channels in pain. Brain Res Rev 60:2–23
Szabo T, Biro T, Gonzalez AF, Palkovits M, Blumberg PM (2002) Pharmacological characterization of vanilloid receptor located in the brain. Brain Res Mol Brain Res 98:51–57
Talavera K, Yasumatsu K, Voets T, Droogmans G, Shigemura N, Ninomiya Y, Margolskee RF, Nilius B (2005) Heat activation of TRPM5 underlies thermal sensitivity of sweet taste. Nature 438:1022–1025
Talbot S, Dias JP, Lahjouji K, Bogo MR, Campos MM, Gaudreau P, Couture R (2012) Activation of TRPV1 by capsaicin induces functional kinin B(1) receptor in rat spinal cord microglia. J Neuroinflammation 9:16
Topolnik L, Azzi M, Morin F, Kougioumoutzakis A, Lacaille JC (2006) mGluR1/5 subtype-specific calcium signalling and induction of long-term potentiation in rat hippocampal oriens/alveus interneurones. J Physiol 575(Pt 1):115–131
Tracey WD, Wilson RI, Laurent G, Benzer S (2003) painless, a Drosophila gene essential for nociception. Cell 113:261–273
Vellani V, Mapplebeck S, Moriondo A, Davis JB, McNaughton PA (2001) Protein kinase C activation potentiates gating of the vanilloid receptor VR1 by capsaicin, protons, heat and anandamide. J Physiol 534:813–825
Vilceanu D, Stucky CL (2010) TRPA1 mediates mechanical currents in the plasma membrane of mouse sensory neurons. PLoS One 5:e12177
Vriens J, Owsianik G, Hofmann T, Philipp SE, Stab J, Chen X, Benoit M, Xue F, Janssens A, Kerselaers S, Oberwinkler J, Vennekens R, Gudermann T, Nilius B, Voets T (2011) TRPM3 is a nociceptor channel involved in the detection of noxious heat. Neuron 70:482–494
Wagner TF, Loch S, Lambert S, Straub I, Mannebach S, Mathar I, Düfer M, Lis A, Flockerzi V, Philipp SE, Oberwinkler J (2008) Transient receptor potential M3 channels are ionotropic steroid receptors in pancreatic beta cells. Nat Cell Biol Dec 10(12):1421–1430
Wang S, Dai Y, Fukuoka T, Yamanaka H, Kobayashi K, Obata K, Cui X, Tominaga M, Noguchi K (2008) Phospholipase C and protein kinase A mediate bradykinin sensitization of TRPA1: a molecular mechanism of inflammatory pain. Brain 131:1241–1251
Watkins LR, Milligan ED, Maier SF (2001a) Glial activation: a driving force for pathological pain. Trends Neurosci 24:450–455
Watkins LR, Milligan ED, Maier SF (2001b) Spinal cord glia: new players in pain. Pain 93:201–205
Willis WD (2002) Long-term potentiation in spinothalamic neurons. Brain Res Brain Res Rev 40:202–214
Willis WD (2009) The role of TRPV1 receptors in pain evoked by noxious thermal and chemical stimuli. Exp Brain Res 196:5–11

Wilson-Gerwing TD, Dmyterko MV, Zochodne DW, Johnston JM, Verge VM (2005) Neurotrophin-3 suppresses thermal hyperalgesia associated with neuropathic pain and attenuates transient receptor potential vanilloid receptor-1 expression in adult sensory neurons. J Neurosci 25:758–67

Woodbury CJ, Zwick M, Wang S, Lawson JJ, Caterina MJ, Koltzenburg M, Albers KM, Koerber HR, Davis BM (2004) Nociceptors lacking TRPV1 and TRPV2 have normal heat responses. J Neurosci 24:6410–6415

Xu H, Ramsey IS, Kotecha SA, Moran MM, Chong JA, Lawson D, Ge P, Lilly J, Silos-Santiago I, Xie Y, DiStefano PS, Curtis R, Clapham DE (2002) TRPV3 is a calcium-permeable temperature-sensitive cation channel. Nature 418:181–186

Xu JT, Xin WJ, Zang Y, Wu CY, Liu XG (2006) The role of tumor necrosis factor-alpha in the neuropathic pain induced by Lumbar 5 ventral root transection in rat. Pain 123:306–321

Yip PK, Malaspina A (2012) Spinal cord trauma and the molecular point of no return. Mol Neurodegener 7:6

Zhang N, Inan S, Cowan A, Sun R, Wang JM, Rogers TJ, Caterina M, Oppenheim JJ (2005a) A proinflammatory chemokine, CCL3, sensitizes the heat- and capsaicin-gated ion channel TRPV1. Proc Natl Acad Sci U S A 102:4536–4541

Zhang X, Huang J, McNaughton PA (2005b) NGF rapidly increases membrane expression of. TRPV1 heat-gated ion channels EMBO J 24(24):4211–4223

Zhang X, Li L, McNaughton PA (2008) Proinflammatory mediators modulate the heat-activated ion channel TRPV1 via the scaffolding protein AKAP79/150. Neuron 59:450–461

Zhou Y, Li GD, Zhao ZQ (2003) State-dependent phosphorylation of epsilon-isozyme of protein kinase C in adult rat dorsal root ganglia after inflammation and nerve injury. J Neurochem 85:571–580

Zhuang ZY, Xu H, Clapham DE, Ji RR (2004) Phosphatidylinositol 3-kinase activates ERK in primary sensory neurons and mediates inflammatory heat hyperalgesia through TRPV1 sensitization. J Neurosci 24:8300–8309

Zschenderlein C, Gebhardt C, von Bohlen Und Halbach O, Kulisch C, Albrecht D (2011) Capsaicin-induced changes in LTP in the lateral amygdala are mediated by TRPV1. PLoS One 6:e16116

Zurborg S, Yurgionas B, Jira JA, Caspani O, Heppenstall PA (2007) Direct activation of the ion channel TRPA1 by Ca2+. Nat Neurosci 10:277–279

TRPs in Hearing

Damiano Zanini and Martin C. Göpfert

Contents

Abstract

Hearing is a particularly sensitive form of mechanosensation that relies on dedicated ion channels transducing sound-induced vibrations that hardly exceed Brownian motion. Attempts to molecularly identify these auditory transduction channels have put the focus on TRPs in ears. In *Drosophila*, hearing has been shown to involve TRPA, TRPC, TRPN, and TRPV subfamily members, with candidate auditory transduction channels including NOMPC (=TRPN1) and the TRPVs Nan and Iav. In vertebrates, TRPs are unlikely to form auditory transduction channels, yet most TRPs are expressed in inner ear tissues, and mutations in TRPN1, TRPVA1, TRPML3, TRPV4, and TRPC3/TRPC6 have

D. Zanini • M.C. Göpfert (✉)
Department of Cellular Neurobiology, University of Göttingen, Julia-Lermontowa-Weg 3, 37077 Göttingen, Germany

DFG Priority Program 1608 and Collaborative Research Center 889
e-mail: mgoepfe@gwdg.de

B. Nilius and V. Flockerzi (eds.), *Mammalian Transient Receptor Potential (TRP) Cation Channels*, Handbook of Experimental Pharmacology 223, DOI 10.1007/978-3-319-05161-1_7,

been implicated in inner ear function. Starting with a brief introduction of fly and vertebrate auditory anatomies and transduction mechanisms, this review summarizes our current understanding of the auditory roles of TRPs.

Keywords
Drosophila • Hair cell • Cochlea • Mechanotransduction • Vestibular • Force-gating • Osmosensing • Varitint-Waddler mouse

1 Introduction

Hearing impairment is the most prevalent sensorineural disorder, afflicting two out of 1,000 newborns (Hilgert et al. 2009). This ratio increases progressively with age, reaching 3.5 per 1,000 during adolescence (Morton and Nance 2006). Worldwide, 275 million people are estimated to suffer from hearing disorders (Cuajungco et al. 2007), and 35 % of the adult Americans display balance impairments (Agrawal et al. 2009) that also signal deficits in inner ear function. Apart from environmental factors, various genetic factors can cause hearing and balance disorders: about half of the cases of profound early-onset deafness in children seem to arise from genetic defects (Marazita et al. 1993), and up to 1 % of the genes comprised by the human genome are thought to be required for hearing (Friedman and Griffith 2003).

Over the past decade, the genetic dissection of hearing and balance has progressed rapidly, and molecular key constituents of inner ear sensory cells and their transduction machineries have been defined (e.g., Petit and Richardson 2009; Gillespie and Müller 2009; Kazmierczak and Müller 2012). Ion channels that are involved in, e.g., ion homeostasis (Zdebik et al. 2009) and synaptic transmission (Moser et al. 2013) in the inner ear, have been identified, and roles of TRPs in hearing and balance have been described (e.g., Cuajungco et al. 2007). The study of auditory channelopathies has gained particular momentum by the fact that the mysterious ion channels that mediate auditory transduction still remain unidentified (Kazmierczak and Müller 2012). Evidence that TRPs may form auditory transduction channels has come from *Drosophila*, yet whether, and if so, which TRPs mediate transduction in the fly's hearing organ is still a matter of debate (Effertz et al. 2012; Lehnert et al. 2013). In vertebrates, TRPs are unlikely to serve as the auditory transduction channels (Kazmierczak and Müller 2012), and although nearly all TRPs have been shown to be expressed in inner ear tissues, their requirements for hearing are largely undefined (e.g., Cuajungco et al. 2007). Before discussing what is known about TRPs in hearing, we will briefly introduce the ears of *Drosophila* and mammals and summarize how they transduce mechanical stimuli.

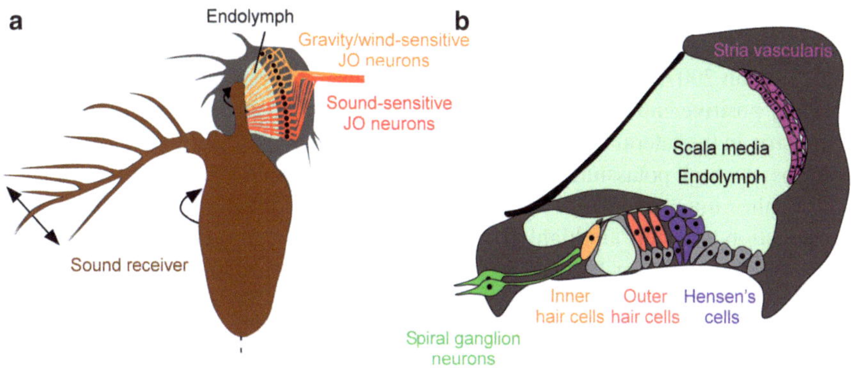

Fig. 1 *Auditory anatomies.* (**a**) Sketch of the antennal ear of *Drosophila*, depicting the sound receiver and Johnston's organ neurons. In the presence of sound, the receiver twists back and forth (*arrows*) about its longitudinal axis, thereby mechanically activating the neurons. About half of the ca. 500 neurons comprised by Johnston's organ preferentially respond to antennal vibrations and serve hearing. The other half of the neurons preferentially responds to sustained antennal deflections and serves the detection of gravity and wind. (**b**) Sketch of the organ of Corti, the sensory organ in the mammalian cochlea, depicting spiral ganglion neurons, hair cells, and supporting Hensen's cells, and the stria vascularis that generates the endolymph in the scala media and the endocochlear potential. Two hair cell types can be distinguished: outer hair cells whose electromotility mainly drives mechanical amplification and inner hair cells that primarily serve sensory roles and are densely innervated by afferent spiral ganglion neurons

2 Auditory Anatomies and Sound Transduction

Hearing is a multistep process that commences with the conversion of sound into structural vibrations ("sound reception") that are coupled to—and elicit ionic currents ("auditory transduction") in—auditory sensory cells and finally trigger action potentials ("auditory encoding"). In *Drosophila*, sound reception is mediated by the antenna's distal part (Fig. 1), which analogous to our eardrum vibrates in response to sound (Göpfert and Robert 2003), and is also deflected by gravity and wind (Kamikouchi et al. 2009; Yorozu et al. 2009) (Fig. 1a). Inside the ear, vibrations of the sound receiver are coupled to sensory cells, i.e., hair cells in vertebrates and Johnston's organ (JO) neurons in the fly (Fig. 1a, b).

Unlike JO neurons, hair cells use actin-based stereocilia instead of true, microtubule-based (kino)cilia for mechanotransduction, yet both cell types seem to be evolutionarily related (Fritzsch et al. 2007) and serve dual, transducing and actuating roles: transduction in hair cells and JO neurons seems to be mediated by nonselective cation channels that are directly activated mechanically by the pull of elastic elements that are known as gating springs (Corey and Hudspeth 1979; Ohmori 1985; Howard and Hudspeth 1988; Albert et al. 2007; Kernan 2007; Lu et al. 2009). Single-channel gating energies hardly exceed thermal noise

(Nadrowski et al. 2008), and the sensory organelles that bear the channels, i.e., JO neuron cilia and hair cell stereocilia, are bathed in an endolymph that is potassium rich (Kernan 2007). In the cochlea, this high potassium concentration, along with a strongly positive endocochlear potential, is generated by the epithelium of the stria vascularis in the lateral wall of the scala media (Zdebik et al. 2009) (Fig. 1b). In JO neurons, the high potassium concentration seems generated by supporting cells that enwrap the cilia of the sensory cells (Roy et al. 2013) (Fig. 1a). Vertebrate auditory transduction channels are highly permeable to K^+, Ca^{2+}, and fluorescent dyes like FM1-43, and they are blocked by, e.g., amiloride and aminoglycosides (Corey 2006). The pore is estimated to vary between 1.3 and 1.5 nm (Farris et al. 2004), and the conductance is large, ranging between 100 and some 300 pS (Crawford et al. 1991; Geleoc et al. 1997; Ricci et al. 2003) and varying considerably from cell to cell (Ricci et al. 2003).

In addition to transducing mechanical vibrations, hair cells and JO neurons are motile and actively assist the vibrations on a cycle-by-cycle basis, thereby boosting the mechanical input of the ear (Robles and Ruggero 2001; Göpfert and Robert 2003; Hudspeth 2008; Nadrowski et al. 2008). This positive feedback amplification actively modulates the ear's mechanics in an intensity- and frequency-dependent manner, introducing a compressive nonlinearity that enhances the mechanical sensitivity and frequency-selectivity of the ear (Hudspeth 2008; Nadrowski et al. 2011). By-products of this nonlinear amplification are distortion product otoacoustic emissions (DPOAEs), which are widely used for hearing screens (Janssen et al. 2006). Excessive amplification, in turn, can give rise to spontaneous otoacoustic emissions that arise from self-sustained feedback oscillations in the ear (Hudspeth 2008). For hair cells, two forms of motility have been described: a prestin-based electromotility, which is known from mammalian outer hair cells (Zheng et al. 2000) and seems also present in birds (Beurg et al. 2013), and an active motility of the stereociliar hair bundle that arises from the interplay between the transduction channels and associated adaptation motors (Hudspeth 2008). The latter transducer-based mechanism also explains the motility of JO neurons in the fly (Nadrowski et al. 2008).

Whereas JO neurons bear axons and generate action potentials, sound encoding in vertebrates takes place downstream of hair cells in spiral ganglion neurons that transmit auditory information to the auditory brainstem (Fig. 1b). Apart from DPOAEs, auditory brainstem responses (ABRs) that can be recorded noninvasively via external electrodes are widely used to assess mammalian hearing (Sininger 1993).

3 TRPs in *Drosophila* Hearing

The antennal ear of *Drosophila* has proved a valuable system for studying auditory TRP channel functions. Nine of the fly's 16 TRP channels are reportedly expressed in JO (Fig. 2), including the TRPC subfamily members TRP and TRPL (Senthilan et al. 2012); the TRPV subfamily members Nanchung (Nan) (Kim et al. 2003a, b)

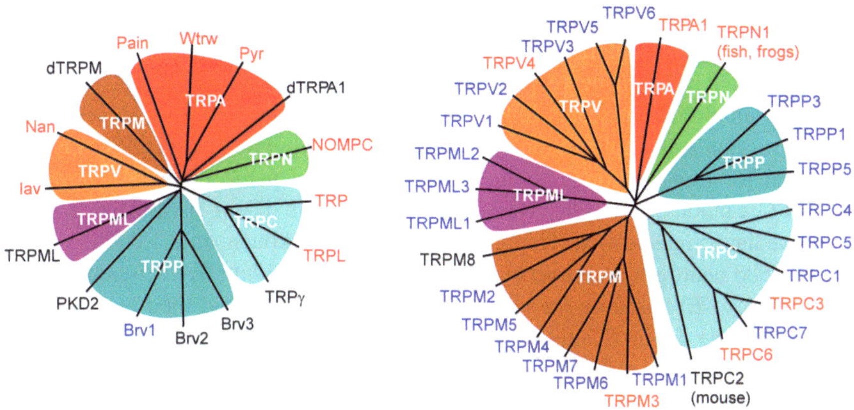

Fig. 2 *TRPs in hearing*. Phylogenetic trees of fly and human TRP channels indicating members that are reportedly expressed in ears (*blue*) and members that have been implicated auditory function (*red*). The tree of human TRPs includes mouse TRPC2 and TRPN to depict their phylogenetic relation. The fly tree was adopted from Gallio et al. (2011) and the human tree from Nilius and Owsianik (2011)

and inactive (Iav) (Gong et al. 2004); the TRPN1 channel No Mechanoreceptor Potential C (NOMPC) (Eberl et al. 2000; Kamikouchi et al. 2009); the TRPP subfamily member Brivido1 (Brv1) (Gallio et al. 2011); and the TRPA subfamily members Water Witch (Wtrw) (Senthilan et al. 2012), Pyrexia (Pyx), and Painless (Pain) (Sun et al. 2009). At least six of these TRPs—Nan, Iav, NOMPC, Pain, and Brivido1—are expressed in JO neurons (Kim et al. 2003a, b; Gong et al. 2004; Kamikouchi et al. 2009; Gallio et al. 2011; Sun et al. 2009), and disruption of at least eight TRPs (Trp, Trpl, Nan, Iav, NOMPC, Wtrw, Pyx, and Pain) reportedly affects JO function (Eberl et al. 2001; Kim et al. 2003a, b; Gong et al. 2004; Sun et al. 2009; Senthilan et al. 2012). Particular attention has been paid to the NOMPC TRPN1 channel and the TRPVs Nan and Iav, which are considered candidates for the fly's auditory transduction channels. In the following, we will focus on these latter TRPs and discuss their roles in hearing.

3.1 NOMPC

NOMPC localizes to the tips of the mechanosensory cilia of virtually all the ca. 500 sensory neurons of the fly's JO (Lee et al. 2010; Cheng et al. 2010; Liang et al. 2011). NOMPC is a *bona fide* mechanotransduction channel that can be directly activated mechanically and can confer mechanosensitivity to cells (Yan et al. 2013; Gong et al. 2013). The 29N-terminal ankyrin residues of NOMPC form a one coiled spring, raising the possibility that NOMPC forms both a mechanotransduction channel and a gating spring (Howard and Bechstedt 2004; Sotomayor et al. 2005; Lee et al. 2006; Liang et al. 2013; Zanini and Göpfert 2013).

Loss of NOMPC affects JO neuron function, as witnessed by a loss of mechanical amplification (Göpfert et al. 2006) and a reduced amplitude and sensitivity of sound-evoked afferent nerve potentials (Eberl et al. 2000; Effertz et al. 2011). Because amplification seems linked to transduction (Nadrowski et al. 2008), the loss of mechanical amplification in *nompC* mutants suggests that NOMPC might be required for transduction. Judging from the residual nerve potentials, however, NOMPC cannot be the only transduction channel in JO or it modulates rather than mediates transduction (Göpfert et al. 2006; Kamikouchi et al. 2009).

The 500 neurons comprised by JO can be subdivided into five different classes (labeled A to E) based on their positions inside the organ and their axonal target regions in the brain (Kamikouchi et al. 2006). Measurements of intracellular calcium signals and targeted cell ablations revealed that whereas the neurons of classes A and B preferentially respond to sound-induced antennal vibrations and mediate hearing, those of classes C and E preferentially respond to maintained antennal deflections as imposed by gravity and wind (Kamikouchi et al. 2009; Yorozu et al. 2009) (Fig. 1a). Ablating only the former auditory JO neurons was shown to copy the auditory phenotypes of *nompC* mutants, including the loss of mechanical amplification and the reduced amplitude and sensitivity of the nerve response (Effertz et al. 2011). When only the gravity-/wind-sensitive neurons were ablated, however, amplification and auditory sensitivity remained unchanged. Measurements of compound calcium responses confirmed that NOMPC is dispensable for the mechanosensory function of the gravity-/wind-sensitive—but not the sound-sensitive—JO neurons (Effertz et al. 2012). Judging from promoter-fusion constructs (Liu et al. 2007; Kamikouchi et al. 2009), it seems that auditory and gravity-/wind-sensitive JO neurons may express different NOMPC isoforms, possibly explaining the coexistence of NOMPC-dependent and NOMPC-independent mechanosensory neurons in JO.

Measurements of mechanical correlates of channel gating revealed that JO neurons harbor at least two different types of mechanically gated ion channels (Effertz et al. 2012): the less sensitive channel type is NOMPC-independent and mainly affiliated with the gravity-/wind-sensitive JO neurons, whereas the more sensitive channel type primarily resides in auditory JO neurons and requires NOMPC for its mechanical gating and integrity. To further test the requirement of NOMPC for auditory transduction, Lehnert et al. (2013) recorded from the giant fiber neuron—a second-order interneuron that is electrically coupled to an unknown number of auditory JO neurons that belong to classes A and B (Kamikouchi et al. 2009; Lehnert et al. 2013). Consistent with previous findings, loss of NOMPC was found to abolish sensitive sound responses, yet notwithstanding the reported requirement of NOMPC for the function of auditory JO neurons, some insensitive sound responses remained (Lehnert et al. 2013). One possible explanation is that NOMPC mediates transduction in some but not all of the fly's ca. 250 auditory JO neurons: class B neurons, for example, seem dispensable for sensitive hearing (Kamikouchi et al. 2009), indicating that their function might be independent of NOMPC. An alternative explanation is that the mechanical gating of the sensitive NOMPC-dependent channel type observed by Effertz et al. (2011)

primarily promotes mechanical amplification in auditory JO neurons (Lehnert et al. 2013): Ca^{2+} entering trough NOMPC, for example, might modulate motor proteins. Such NOMPC-dependent mechanical preamplification could facilitate the mechanical gating of downstream ion channels that mediate transduction by contributing most of the transduction currents in JO (Lehnert et al. 2013). Clearly, more work is needed to distinguish between these alternatives (Zanini and Göpfert 2013), including fine-scale analyses of the NOMPC-dependence of single JO neuron subpopulations down to single cells. Also the TRP dependence of transduction in the gravity-/wind-sensitive class C and D neurons remains to be tested: judging from mutant defects in gravitaxis, but not auditory behaviors, it seems that these neurons specifically require the TRPA subfamily member Pain (Sun et al. 2009).

3.2 Nan and Iav

Like NOMPC, the fly's two TRPV subfamily members Nan and Iav are also expressed in virtually all 500 JO neurons, yet they localize to the proximal region of the cilium (Kim et al. 2003a, b; Gong et al. 2004), downstream of NOMPC. Disrupting the *iav* gene abolishes Nan protein and vice versa, indicating that these two TRPVs form a heteromeric Nan-Iav channel in the neurons of JO (Gong et al. 2004). When expressed heterologously, each Nan and Iav can be activated by hypotonicity, indicating that they are mechanosensitive (Kim et al. 2003a, b; Gong et al. 2004). Whether mechanical stimuli activate Nan and Iav directly or via second messengers has not been determined yet.

Null mutations in either *nan* or *iav* completely eliminate sound-evoked nerve responses, raising the possibility that *Drosophila* auditory transduction is mediated by TRPVs (Kim et al. 2003a, b; Gong et al. 2004). In the absence of Nan-Iav, however, mechanical feedback amplification by JO neurons is not lost but increases, as witnessed by self-sustained feedback oscillations and an excessive amplification gain (Göpfert et al. 2006). The mere persistence of mechanical amplification points to an intact sound transduction, simply because this nonlinear and frequency-specific cycle-by-cycle amplification requires information about sound intensity, frequency, and phase. Residual sound-evoked calcium signals in JO neurons indeed persist if Nan-Iav is lost (Kamikouchi et al. 2009), but recordings from a downstream neuron suggest that transduction currents might be lost (Lehnert et al. 2013).

3.3 Models of TRP Channel Function in Fly Auditory Transduction

To explain the auditory phenotypes observed in *nompC*, *nan*, and *iav* mutants, two different models have been proposed (Fig. 3):

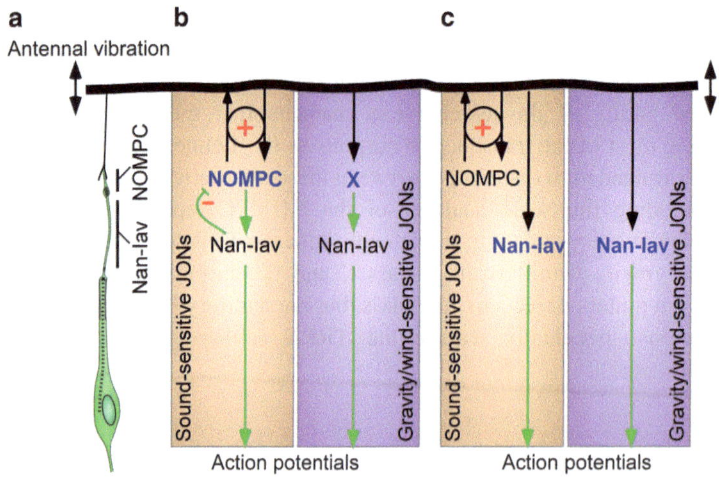

Fig. 3 *Models of TRP function in Drosophila auditory transduction.* (**a**) Sketch of a JO neuron that is mechanically coupled to the antennal receiver. The respective subcellular localizations of NOMPC and Nan-Iav are depicted. (**b**) "NOMPC transducer model": in sound-sensitive JO neurons, antennal vibrations are coupled to the NOMPC channel, which transduces the vibrations into cellular signals and, by interacting with adaptation motors, provides mechanical signal amplification (+). Nan-Iav serves signal propagation/amplification downstream of transduction and negatively regulates mechanical amplification via NOMPC. Gravity-/wind-sensitive JO neurons, which do not provide amplification, are endowed with a second type of transduction channel (X) and also use Nan-Iav for downstream signal propagation/amplification. (**c**) "Nan-Iav transduction model": antennal vibrations are coupled to NOMPC, which contributes to mechanical amplification. The amplified vibrations are then mechanically coupled to Nan-Iav, which transduces them into cellular signals in both the sound- and the gravity-/wind-sensitive neurons of JO

The "NOMPC transducer model" (Göpfert et al. 2006; Effertz et al. 2011, 2012) posits that NOMPC mediates sensitive sound transduction and mechanical amplification in certain auditory neurons of JO. The second, less sensitive transduction channel type occurs primarily in the gravity-/wind-sensitive neurons, which contribute the residual nerve responses that persist in *nompC* nulls. Nan-Iav is required for downstream electrical signal amplification/propagation, explaining the loss of sound-evoked nerve potentials in *nan* and *iav* nulls. In addition, Nan-Iav is assumed to negatively regulate mechanical amplification in auditory JO neurons via NOMPC, explaining why amplification becomes excessive when Nan-Iav is disrupted but is abolished when Nan-Iav and NOMPC are both lost (Göpfert et al. 2006).

The "Nan-Iav transducer model" (Lehnert et al. 2013), by contrast, posits that mechanical stimuli acting on JO neurons directly activate—and are transduced by—Nan-Iav. NOMPC primarily serves mechanical amplification in auditory JO neurons, thereby facilitating transduction by enhancing the mechanical stimuli that are acting on Nan-Iav.

In line with the latter "Nan-Iav transduction" model, one would expect that Nan-Iav is directly activated by mechanical stimuli. And in line with the "NOMPC transducer model," one would predict that Nan-Iav is activated by voltage; judging from response latencies, ions flowing through NOMPC must activate Nan-Iav within microseconds, which seems too fast for diffusible messengers (Lehnert et al. 2013). Defining the activation mechanisms of the Nan-Iav channel thus will be crucial for testing the models, which both provide useful concepts to further dissect the auditory roles of NOMPC and Nan-Iav, not to mention all the other TRPs that occur in the fly's ear.

4 TRPs in Vertebrate Hearing

In vertebrates, nearly all TRPs seem to be expressed inside the ear (Fig. 2): using qPCR analysis, Gabashvili et al. (2007) found 21 TRPs in the inner ears of diverse vertebrates, ranging from zebrafish to humans. Asai et al. (2010) analyzed the expression of 33 TRPs including splice variants in the developing mouse cochlea and found all except TRPC7 and TRPM8. So far, only six of these TRPs have been implicated in hearing, i.e., TRPN1 (Sidi et al. 2003), TRPA1 (Corey et al. 2004), TRPV4 (Liedtke et al. 2000), TRPML3 (Di Palma et al. 2002), TRPC3 (Wong et al. 2013), and TRPC6 (Quick et al. 2012). An indirect link seems to exist between TRPV5 and TRPV6 and deafness and vestibular dysfunction in Pendred syndrome caused by mutations in the $Cl^-/HCO3^-$ exchanger pendrin (= SLC26A4) (Everett et al. 1997): in pendrin knockout mice, acidification of the endolymph due to impaired $HCO3^-$ secretion reportedly leads to a pathological elevation of the endolymphatic Ca^{2+} concentration by inhibiting Ca^{2+} reabsorption via TRPV5 and TRPV6 (Nakaya et al. 2007; Wangemann et al. 2007). As far as the mysterious hair cell transduction channels are concerned, it now seems unlikely that they are TRPs: a tetraspan superfamily member, ThMS (tetraspan membrane protein of hair cell stereocilia) was recently identified as a regulatory subunit of mouse auditory transduction channels (Xiong et al. 2012). The pore-forming subunit of these channels still need to be identified, with strong candidates including transmembrane channel-like proteins (Kawashima et al. 2011; Kim and Fettiplace 2013) and ion channels of the piezo family (Coste et al. 2010, 2012).

4.1 TRPN1

TRPN1 is expressed in hair cells of certain fish and frogs but absent from reptilian, avian, and mammalian genomes (Cuajungco et al. 2007). In the zebrafish ear, Morpholino-mediated knockdown of TRPN1 reportedly abolishes sound-evoked electrical potentials (Sidi et al. 2003), yet further confirmation of this observation seems needed. In frog hair cells, TRPN1 localizes to the kinocilium but not to the stereociliar hair bundle that harbors the transduction channels (Shin et al. 2005). Recent evidence suggests that the kinocilia of zebrafish hair cells serve mechanosensory roles during development (Kindt et al. 2012), raising the possibility that TRPN1 mediates some form of kinociliar mechanosensitivity. Such function of TRPN1 would seem consistent with those in *Caenorhabditis elegans* and *Drosophila*, where TRPN1 acts as a mechanotransduction channel, localizes to cilia, and seems to interact with microtubules (Kang et al. 2010; Cheng et al. 2010; Yan et al. 2013; Liang et al. 2013). The imaging approach Beurg et al. (2009) used to narrow down the transduction sites in hair cell stereocilia might help to test the roles of TRPN1 in kinociliar mechanosensitivity.

4.2 TRPA1

TRPA1 is broadly expressed in inner ear, including hair cells as well as Hensen's cells and other supporting cells (Corey et al. 2004; Kwan et al. 2006). Judging from qPCR, TRPA1 transcripts are particularly abundant at the cochlear apex, where expression levels seem higher than those of all the other TRPs (Asai et al. 2010). Notwithstanding this strong and broad expression, the auditory relevance of TRPA1 remains unclear: using short interfering RNA (siRNA) in mammals and Morpholino antisense oligonucleotides in zebrafish, Corey et al. (2004) provided in vitro evidence that TRPA1 might be a subunit of the hair cell transduction channels. TRPA1 knockout mice (Bautista et al. 2006; Kwan et al. 2006) and zebrafish TRPA1 mutants (Prober et al. 2008), however, turned out to have normal hair cell function and to be neither hearing nor balance impaired. Recent evidence suggests that TRPA1 forms functional channels in outer hair cells and that these channels retain their function when mechanotransduction is disturbed (Stepanyan et al. 2011). This apparent functionality of TRPA1, along with the lack of mutant auditory phenotypes, suggests that functionally redundant proteins might compensate for the loss of TRPA1 inside the ear: because TRPA1 and TRPV1 reportedly can interact and regulate each other (Akoipan et al. 2007; Salas et al. 2009), this compensation was speculated to involve an upregulation of TRPV1 (Asai et al. 2010).

4.3 TRPV4

In the mouse inner ear, TRPV4 is expressed in cochlear and vestibular hair cells, the utricular macula, spiral ganglion neurons, and in the cochlear stria vascularis that generates the ionic milieu of the scala media and the endocochlear potential (Liedtke et al. 2000; Takumida et al. 2005). Also, hair cells of zebrafish were recently reported to express TRPV4 (Amato et al. 2012). Like the *Drosophila* TRPV channels, TRPV4 is activated by hypotonicity, implicating it in cellular osmoregulation (Liedtke et al. 2000; Strotmann et al. 2000). Osmotic stimuli reportedly activate TRPV4 indirectly via a second-messenger cascade (Watanabe et al. 2003; Vriens et al. 2004), which contrasts with the direct mechanical activation of the hair cell transduction channels.

Judging from ABRs and DPOAEs, homozygous TRPV4 knockout mice hear normally when two (Tabuchi et al. 2005) to three (Cuajungco et al. 2007) months old. At the age of 6 months, however, hearing is impaired (Tabuchi et al. 2005): this late-onset hearing loss, which seems unlikely to reflect deficits in transduction, reportedly manifests itself as a drop in auditory sensitivity and an increased susceptibility to acoustic trauma (Tabuchi et al. 2005). Intriguingly, TRPV3 maps into the locus DFNA25 (Greene et al. 2001), one of the autosomal dominant deafness (DFNA) loci that typically associates with delayed-onset and progressive hearing loss (Van Camp et al. 1997). DFNA25 has been linked to mutations in the vesicular glutamate transporter VGLUT3 (Ruel et al. 2008), yet it seems that hearing impairments can also associate with mutations in TRPV4: by screening patients with various neuropathies, Zimoń et al. (2010) identified a family with sensorineural hearing loss that carries an autosomal dominant missense mutations in TRPV4 (Zimoń et al. 2010).

4.4 TRPML3

In the mouse inner ear, TRPML3 is expressed in hair cells, the stria vascularis, and the epithelial lining of the vestibular endolymph spaces (Di Palma et al. 2002; Nagata et al. 2008; van Aken et al. 2008; Castiglioni et al. 2011). Within hair cells, TRPML localizes to the basal regions of the mechanosensory stereocilia (van Aken et al. 2008).

Mutations in TRPML cause hearing loss and vestibular dysfunction in varitint-waddler (Va) mice (Deol 1954; Di Palma et al. 2002; Kim et al. 2003a, b). Two independent missense mutations in TRPML have been identified in VA mice: *Va*, in which an alanine residue at amino acid position 419 in the predicted pore region in the fifth transmembrane domain is substituted with proline (A419P); and *Va*^{J}, in which a isoleucine at amino acid position 362 in the second extracellular loop region is substituted with threonine (I362T) (Di Palma et al. 2002). The *Va* mutation is a gain-of-function mutation (Grimm et al. 2007; Kim et al. 2007; Xu et al. 2007), and the *Va*^{J} mutation, which arose in the *Va* background, seems to weaken the phenotypes caused by *Va* (Di Palma et al. 2002; van Aken et al. 2008).

Deafness in the *Va* and *Va*^{J} mutants associates with disorganized stereociliar hair bundles of inner and outer cochlear hair cells and a strong reduction of the endocochlear potential (Cable and Steel 1998; Di Palma et al. 2002). Heterologous expression revealed that the mutations A419P and I362T + A419P cause cell death by constituently activating TRPML3 (Grimm et al. 2007; Kim et al. 2007, 2008; Xu et al. 2007; Nagata et al. 2008; van Aken et al. 2008). This constituent TRPML activation presumably explains the ultimate degeneration of hair cells, the organ of Corti, and the stria vascularis observed in *Va* and *Va*^{J} mice (Deol 1954; Cable and Steel 1998): similar leak currents have been documented for hair cells of homozygous *Va*/*Va*^{J} mutants, and in both the homozygous and heterozygous *Va*^{J}/+mutants, hair cell transduction currents are reduced (van Aken et al. 2008).

Unlike VA mice, TRPML knockout mice display normal inner ear anatomies and auditory and vestibular functions, indicating that TRPML3 itself is dispensable for the function of the ear (Jörs et al. 2010; Castiglioni et al. 2011); whereas hearing and balance defects in Va mice seem to result from a constitutive and cytotoxic activation of TRPML3 (Jörs et al. 2010), the auditory relevance of native TRPML3 is thus unclear: one possibility is that TRPML3 is a nonessential subunit of a heteromeric channel, with putative partners including, e.g., TRPML2 (Jörs et al. 2010) and TRPV5 (Guo et al. 2013).

4.5 TRPC3 and TRPC6

TRPC3 and TRPC6 are both expressed in cochlear hair cells and spiral ganglion neurons (Raybould et al. 2007; Phan et al. 2010; Quick et al. 2012). TRPC3 transcripts are enriched in the cochlea's high-frequency region, whereas TRPC6 transcripts seem equally distributed along the cochlear length (Asai et al. 2010). TRPC3 can form homomeric channels, but also assembles into heteromeric channels with other TRPCs including TRPC6 (Goel et al. 2002; Gudermann et al. 2004).

No auditory phenotypes have hitherto been reported for single TRPC6 knockouts, whereas TRPC3 knockout mice hear better than normal, being about twice as sensitive to sounds at intermediate frequencies between ca. 8 and 16 kHz (Wong et al. 2013). Measurements of DPOAEs revealed that this auditory hypersensitivity—or hyperacusis—associates with excess mechanical amplification, indicating that TRPC3 might negatively regulate outer hair cell electromotility. This regulation may involve Ca^{2+} entering through TRPC3: upon the depletion of cytosolic Ca^{2+} in isolated hair cells, Ca^{2+} reentry is significantly diminished in the absence of TRPC3 (Wong et al. 2013).

Whereas the loss of TRPC3 facilitates hearing, hearing worsens if TRPC3 and TRPC6 are both lost (Quick et al. 2012): in TRPC3/TRPC6 double knockout mice, the sensitivity to high-frequency sounds is strongly diminished, dropping by as much as ca. 30 dB. This high-frequency hearing loss is not accompanied by apparent alterations in hair cell numbers and morphologies, but hair cell transduction currents are impaired: outer hair cells from the cochlear high-frequency region display current amplitudes that are about 75 % smaller than normal, indicating that transduction in the respective hair cells is facilitated by the concerted function of TRPC3/TRPC6. Intriguingly, a mechanical activation of TRPC3 that is augmented by TRPC6 has been observed in a certain cell line that is derived from dorsal root ganglion neurons (Quick et al. 2012). Whether such mechanical activation also occurs in hair cells—and whether TRPC3 and TRPC6 localize to hair cell stereocilia—remains to be seen.

Conclusion

Despite considerable progress over the past years, much remains to be learned about the roles of TRPs in ears. Work on *Drosophila* has shown that auditory transduction and amplification can depend on TRPs, yet whether, and if so, which TRPs mediate transduction in the fly's hearing organ remains to be seen. Work on vertebrates, in turn, revealed that nearly all TRPs are expressed in ears, yet the auditory relevance of most of them is still unclear: linking auditory phenotypes to TRPs is often difficult, presumably because of functional redundancy. The use of double knockouts thus seems promising, as was exemplified by work on TRPC3/TRPC6. Another problem in assigning auditory phenotypes to TRPs might be the genetic and functional diversity of the mechanosensory cells within the ears: the fly's JO neurons harbor at least two mechanically gated channel types, and it seems conceivable that also several such channel types occur in vertebrate inner ears. Even though TRPs seem rather unlikely to form transduction channels in vertebrate hair cells, the abundance of TRPs in ears will continue to make their study an enticing endeavor, with much room for discovery.

References

Agrawal Y, Carey JP, Della Santina CC, Schubert MC, Minor LB (2009) Disorders of balance and vestibular function in US adults: data from the National Health and Nutrition Examination Survey, 2001–2004. Arch Intern Med 169:938–944

Akoipan AN, Ruparel NB, Jeske NA, Hargraeves KM (2007) Transient receptor potential TRPA1 channel desensitization in sensory neurons is agonist dependent and regulated by TRPV1-directed internalization. J Physiol 583:175–193

Albert JT, Nadrowski B, Göpfert MC (2007) Mechanical signatures of transducer gating in the *Drosophila* ear. Curr Biol 17:1000–1006

Amato V, Viña E, Calavia MG, Guerrera MC, Laurà R, Navarro M, De Carlos F, Cobo J, Germanà A, Vega JA (2012) TRPV4 in the sensory organs of adult zebrafish. Microsc Res Tech 75:89–96

Asai Y, Holt JR, Géléoc GS (2010) A quantitative analysis of the spatiotemporal pattern of transient receptor potential gene expression in the developing mouse cochlea. J Assoc Res Otolaryngol 11:27–37

Bautista DM, Jordt SE, Nikai T, Tsuruda PR, Read AJ, Poblete J, Yamoah EN, Basbaum AI, Julius D (2006) TRPA1 mediates the inflammatory actions of environmental irritants and proalgesic agents. Cell 124:1269–1282

Beurg M, Fettiplace R, Nam JH, Ricci AJ (2009) Localization of inner hair cell mechanotransducer channels using high-speed calcium imaging. Nat Neurosci 12:553–558

Beurg M, Tan X, Fettiplace R (2013) A prestin motor in chicken auditory hair cells: active force generation in a nonmammalian species. Neuron 79(1):69–81

Cable J, Steel KP (1998) Combined cochleo-saccular and neuroepithelial abnormalities in the Varitint-waddler-J (*Va*^{J}) mouse. Hear Res 123:125–136

Castiglioni AJ, Remis NN, Flores EN, García-Añoveros J (2011) Expression and vesicular localization of mouse Trpml3 in stria vascularis, hair cells, and vomeronasal and olfactory receptor neurons. J Comp Neurol 519:1095–1114

Cheng LE, Song W, Looger LL, Jan LY, Jan YN (2010) The role of the TRP channel NompC in *Drosophila* larval and adult locomotion. Neuron 67:373–380

Coste B, Mathur J, Schmidt M, Earley TJ, Ranade S, Petrus MJ, Dubin AE, Patapoutian A (2010) Piezo1 and Piezo2 are essential components of distinct mechanically activated cation channels. Science 330:55–60

Coste B, Xiao B, Santos JS, Syeda R, Grandl J, Spencer KS, Kim SE, Schmidt M, Mathur J, Dubin AE, Montal M, Patapoutian A (2012) Piezo proteins are pore-forming subunits of mechanically activated channels. Nature 483:176–181

Corey DP, Hudspeth AJ (1979) Ionic basis of the receptor potential in a vertebrate hair cell. Nature 281:675–677

Corey DP, Garcia-Anoveros J, Holt JR, Kwan KY, Lin SY, Vollrath MA, Amalfitano A, Cheung EL, Derfler BH, Duggan A, Geleoc GS, Gray PA, Hoffman MP, Rehm HL, Tamasauskas D, Zhang DS (2004) TRPA1 is a candidate for the mechanosensitive transduction channel of vertebrate hair cells. Nature 432:723–730

Corey DP (2006) What is the hair cell transduction channel? J Physiol 576:23–28

Crawford AC, Evans MG, Fettiplace R (1991) The actions of calcium on the mechano-electrical transducer current of turtle hair cells. J Physiol (London) 434:369–398

Cuajungco MP, Grimm C, Heller S (2007) TRP channels as candidates for hearing and balance abnormalities in vertebrates. Biochim Biophys Acta 1772:1022–1027

Di Palma F, Belyantseva IA, Kim HJ, Vogt TF, Kachar B, Noben-Trauth K (2002) Mutations in Mcoln3 associated with deafness and pigmentation defects in varitint–waddler (Va) mice. Proc Natl Acad Sci U S A 99:14994–14999

Deol MS (1954) The anomalies of the labyrinth of the mutants varitint-waddler, shaker-2 and jerker in the mouse. J Genet 52:562–588

Eberl DF, Hardy RW, Kernan MJ (2000) Genetically similar transduction mechanisms for touch and hearing in *Drosophila*. J Neurosci 20:5981–5988

Effertz T, Wiek R, Göpfert MC (2011) NompC TRP channel is essential for *Drosophila* sound receptor function. Curr Biol 21:592–597

Effertz T, Nadrowski B, Piepenbrock D, Albert JT, Göpfert MC (2012) Direct gating and mechanical integrity of Drosophila auditory transducers require TRPN1. Nat Neurosci 15:1198–2000

Everett LA, Glaser B, Beck JC, Idol JR, Buchs A, Heyman M, Adawi F, Hazani E, Nassir E, Baxevanis AD, Sheffield VC, Green ED (1997) Pendred syndrome is caused by mutations in a putative sulphate transporter gene (PDS). Nat Genet 17:411–422

Farris HE, Leblanc CL, Goswami J, Ricci AJ (2004) Probing the pore of the auditory hair cell mechanotransducer channel in turtle. J Physiol 558:769–792

Friedman TB, Griffith AJ (2003) Human nonsyndromic sensorineural deafness. Annu Rev Genomics Hum Genet 4:341–402

Fritzsch B, Beisel KW, Pauley S, Soukup G (2007) Molecular evolution of the vertebrate mechanosensory cell and ear. Int J Dev Biol 51:663–678

Beisel KW, Pauley S, Soukup G (2007) Molecular evolution of the vertebrate mechanosensory cell and ear. Int J Dev Biol 51:663–678

Gabashvili IS, Sokolowski BH, Morton CC, Giersch AB (2007) Ion channel gene expression in the inner ear. J Assoc Res Otolaryngol 8:305–328

Gallio M, Ofstad TA, Macpherson LJ, Wang JW, Zuker CS (2011) The coding of temperature in the *Drosophila* brain. Cell 144:614–624

Geleoc GS, Lennan GW, Richardson GP, Kros CJ (1997) A quantitative comparison of mechanoelectrical transduction in vestibular and auditory hair cells of neonatal mice. Proc R Soc Lond B 264:611–621

Gillespie PG, Müller U (2009) Mechanotransduction by hair cells: models, molecules, and mechanisms. Cell 139:33–44

Goel M, Sinkins WG, Schilling WP (2002) Selective association of TRPC channel subunits in rat brain synaptosomes. J Biol Chem 277:48303–48310

Gong J, Wang Q, Wang Z (2013) NOMPC is likely a key component of Drosophila mechanotransduction channels. Eur J Neurosci 38(1):2057–64

Gong Z, Son W, Chung YD, Kim J, Shin DW, McClung CA, Lee Y, Lee HW, Chang DJ, Kaang BK, Cho H, Oh U, Hirsh J, Kernan MJ, Kim C (2004) Two interdependent TRPV channel subunits, Inactive and Nanchung, mediate hearing in *Drosophila*. J Neurosci 24:9059–9066

Göpfert MC, Robert D (2003) Motion generation by *Drosophila* mechanosensory neurons. Proc Natl Acad Sci U S A 100:5514–5519

Göpfert MC, Albert JT, Nadrowski B, Kamikouchi A (2006) Specification of auditory sensitivity by *Drosophila* TRP channels. Nat Neurosci 9:999–1000

Greene CC, McMillan PM, Barker SE, Kurnool P, Lomax MI, Burmeister M, Lesperance MM (2001) DFNA25, a novel locus for dominant nonsyndromic hereditary hearing impairment, maps to 12q21-24. Am J Hum Genet 68:254–260

Grimm C, Cuajungco MP, Van Aken AFJ, Schnee M, Jörs S, Kros CJ, Ricci AJ, Heller S (2007) A helix-breaking mutation in TRPML3 leads to constitutive activity underlying deafness in the varitint-waddler mouse. Proc Natl Acad Sci U S A 104:19583–19588

Gudermann T, Hofmann T, Mederos y Schnitzler M, Dietrich A (2004) Activation, subunit composition and physiological relevance of DAG-sensitive TRPC proteins. Novartis Found Symp 258:103–118

Guo Z, Grimm C, Becker L, Ricci AJ, Heller S (2013) A novel ion channel formed by interaction of TRPML3 with TRPV5. PLoS One 8:e58174

Howard J, Hudspeth AJ (1988) Compliance of the hair bundle associated with gating of mechanoelectrical transduction channels in the bullfrog's saccular hair cell. Neuron 1:189–199

Howard J, Bechstedt S (2004) Hypothesis: a helix of ankyrin repeats of the NOMPC-TRP ion channel is the gating spring of mechanoreceptors. Curr Biol 14:R224–R226

Hilgert N, Smith RJ, Van Camp G (2009) Forty six genes causing nonsyndromic hearing impairment: which ones should be analysed in DNA diagnostics? Mutat Res 681:189–196

Hudspeth AJ (2008) Making an effort to listen: mechanical amplification in the ear. Neuron 59:530–545

Janssen T, Niedermeyer HP, Arnold W (2006) Diagnostics of the cochlear amplifier by means of distortion product otoacoustic emissions. ORL J Otorhinolaryngol Relat Spec 68:334–339

Jörs S, Grimm C, Becker L, Heller S (2010) Genetic inactivation of Trpml3 does not lead to hearing and vestibular impairment in mice. PLoS One 5:e14317

Kamikouchi A, Shimada T, Ito K (2006) Comprehensive classification of the auditory sensory projections in the brain of the fruit fly *Drosophila melanogaster*. J Comp Neurol 499:317–356

Kamikouchi A, Inagaki HK, Effertz T, Hendrich O, Fiala A, Göpfert MC, Ito K (2009) The neural basis of *Drosophila* gravity-sensing and hearing. Nature 458:165–171

Kawashima Y, Géléoc GS, Kurima K, Labay V, Lelli A, Asai Y, Makishima T, Wu DK, Della Santina CC, Holt JR, Griffith AJ (2011) Mechanotransduction in mouse inner ear hair cells requires transmembrane channel-like genes. J Clin Invest 121:4796–4809

Kang L, Gao J, Schafer WR, Xie Z, Xu XZ (2010) C. elegans TRP family protein TRP-4 is a pore-forming subunit of a native mechanotransduction channel. Neuron 67:381–391

Kazmierczak P, Müller U (2012) Sensing sound: molecules that orchestrate mechanotransduction by hair cells. Trends Neurosci 35:220–229

Kernan MJ (2007) Mechanotransduction and auditory transduction in *Drosophila*. Pflugers Arch 454:703–720

Kim HJ, Jackson T, Noben-Trauth K (2003a) Genetic analyses of the mouse deafness mutations varitint-waddler (Va) and jerker (Espnje). J Assoc Res Otolaryngol 4:83–90

Kim HJ, Li Q, Tjon-Kon-Sang S, So I, Kiselyov K, Muallem S (2007) Gain-of-function mutation in TRPML3 causes the mouse varitint-waddler phenotype. J Biol Chem 282:36138–36142

Kim HJ, Li Q, Tjon-Kon-Sang S, So I, Kiselyov K, Soyombo AA, Muallem S (2008) A novel mode of TRPML3 regulation by extracytosolic pH absent in the varitint-waddler phenotype. EMBO J 27:1197–1205

Kim J, Chung YD, Park DY, Choi S, Shin DW, Soh H, Lee HW, Son W, Yim J, Park CS, Kernan MJ, Kim C (2003b) A TRPV family ion channel required for hearing in *Drosophila*. Nature 424:81–84

Kim KX, Fettiplace R (2013) Developmental changes in the cochlear hair cell mechanotransducer channel and their regulation by transmembrane channel-like proteins. J Gen Physiol 141:141–148

Kindt KS, Finch G, Nicolson T (2012) Kinocilia mediate mechanosensitivity in developing zebrafish hair cells. Dev Cell 23:329–341

Kwan KY, Allchorne AJ, Vollrath MA, Christensen AP, Zhang DS, Woolf CJ, Corey DP (2006) TRPA1 contributes to cold, mechanical, and chemical nociception but is not essential for hair-cell transduction. Neuron 50:277–289

Lee G, Abdi K, Jiang Y, Michaely P, Bennett V, Marszalek PE (2006) Nanospring behaviour of ankyrin repeats. Nature 440:246–249

Lee J, Moon S, Cha Y, Chung YD (2010) *Drosophila* TRPN(=NOMPC) channel localizes to the distal end of mechanosensory cilia. PLoS One 5:e11012

Lehnert BP, Baker AE, Gaudry Q, Chiang AS, Wilson RI (2013) Distinct roles of TRP channels in auditory transduction and amplification in *Drosophila*. Neuron 77:115–128

Liang X, Madrid J, Saleh HS, Howard J (2011) NOMPC, a member of the TRP channel family, localizes to the tubular body and distal cilium of *Drosophila* campaniform and chordotonal receptor cells. Cytoskeleton 68:1–7

Liang X, Madrid J, Gärtner R, Verbavatz JM, Schiklenk C, Wilsch-Bräuninger M, Bogdanova A, Stenger F, Voigt A, Howard J (2013) A NOMPC-dependent membrane-microtubule connector is a candidate for the gating spring in fly mechanoreceptors. Curr Biol 23:755–763

Liedtke W, Choe Y, Marti-Renom M, Beli AM, Denis CS, Sali A, Hudspeth AJ, Friedman JM, Heller S (2000) Vanilloid receptor-related osmotically activated channel (VR-OAC) a candidate vertebrate osmoreceptor. Cell 105:525–535

Liu L, Li Y, Wang R, Yin C, Dong Q, Hing H, Kim C, Welsh MJ (2007) *Drosophila* hygrosensation requires the TRP channels water witch and nanchung. Nature 450:294–298

Lu Q, Senthilan PR, Effertz T, Nadrowski B, Göpfert MC (2009) Using *Drosophila* for studying fundamental processes in hearing. Integr Comp Biol 49:674–680

Marazita ML, Ploughman LM, Rawlings B, Remington E, Arnos KS, Nance WE (1993) Genetic epidemiological studies of early-onset deafness in the U.S. school-age population. Am J Med Genet 46:486–491

Morton CC, Nance W (2006) Newborn hearing screening—a silent revolution. N Engl J Med 354:2151–2164

Moser T, Predoehl F, Starr A (2013) Review of hair cell synapse defects in sensorineural hearing impairment. Otol Neurotol 34(6):995–1004

Nadrowski B, Albert JT, Göpfert MC (2008) Transducer-based force generation explains active process in *Drosophila* hearing. Curr Biol 18:1365–1372

Nadrowski B, Effertz T, Senthilan PR, Göpfert MC (2011) Antennal hearing in insects – new findings, new questions. Hear Res 273:7–13

Nagata K, Zheng L, Madathany T, Castiglioni AJ, Bartles JR, Garcia-Anoveros J (2008) The varitint-waddler (Va) deafness mutation in TRPML3 generates constitutive, inward rectifying currents and causes cell degeneration. Proc Natl Acad Sci U S A 105:353–358

Nakaya K, Harbidge DG, Wangemann P, Schultz BD, Green ED, Wall SM, Marcus DC (2007) Lack of pendrin HCO3- transport elevates vestibular endolymphatic [Ca^{2+}] by inhibition of acid-sensitive TRPV5 and TRPV6 channels. Am J Physiol Renal Physiol 292:F1314–F1321

Nilius B, Owsianik G (2011) The transient receptor potential family of ion channels. Genome Biol 12:218

Ohmori H (1985) Mechanoelectrical transduction currents in isolated vestibular hair cells of the chick. J Physiol 359:189–217

Petit C, Richardson GP (2009) Linking genes underlying deafness to hair-bundle development and function. Nat Neurosci 12:703–710

Phan PA, Tadros SF, Kim Y, Birnbaumer L, Housley GD (2010) Developmental regulation of TRPC3 ion channel expression in the mouse cochlea. Histochem Cell Biol 133:437–448

Prober DA, Zimmerman S, Myers BR, McDermott BM Jr, Kim SH, Caron S, Rihel J, Solnica-Krezel L, Julius D, Hudspeth AJ, Schier AF (2008) Zebrafish TRPA1 channels are required for chemosensation but not for thermosensation or mechanosensory hair cell function. J Neurosci 28:10102–10110

Quick K, Zhao J, Eijkelkamp N, Linley JE, Rugiero F, Cox JJ, Raouf R, Gringhuis M, Sexton JE, Abramowitz J, Taylor R, Forge A, Ashmore J, Kirkwood N, Kros CJ, Richardson GP, Freichel M, Flockerzi V, Birnbaumer L, Wood JN (2012) TRPC3 and TRPC6 are essential for normal mechanotransduction in subsets of sensory neurons and cochlear hair cells. Open Biol 2:120068

Raybould NP, Jagger DJ, Kanjhan R, Greenwood D, Laslo P, Hoya N, Soeller C, Cannell MB, Housley GD (2007) TRPC-like conductance mediates restoration of intracellular Ca^{2+} in cochlear outer hair cells in the guinea pig and rat. J Physiol 579:101–113

Ricci AJ, Crawford AC, Fettiplace R (2003) Tonotopic variation in the conductance of the hair cell mechanotransducer channel. Neuron 40:983–990

Robles L, Ruggero MA (2001) Mechanics of the mammalian cochlea. Physiol Rev 81:1305–1352

Roy M, Sivan-Loukianova E, Eberl DF (2013) Cell-type-specific roles of Na+/K + ATPase subunits in *Drosophila* auditory mechanosensation. Proc Natl Acad Sci U S A 110:181–186

Ruel J, Emery S, Nouvian R, Bersot T, Amilhon B, Van Rybroek JM, Rebillard G, Lenoir M, Eybalin M, Delprat B, Sivakumaran TA, Giros B, El Mestikawy S, Moser T, Smith RJ, Lesperance MM, Puel JL (2008) Impairment of SLC17A8 encoding vesicular glutamate transporter-3, VGLUT3, underlies nonsyndromic deafness DFNA25 and inner hair cell dysfunction in null mice. Am J Hum Genet 83:278–292

Salas MM, Hargraeves KM, Akopian AN (2009) TRPA1-mediated responses in trigeminal sensory neurons: interaction between TRPA1 and TRPV1. Eur J Neurosci 29:1568–1578

Senthilan PR, Piepenbrock D, Ovezmyradov G, Nadrowski B, Bechstedt S, Pauls S, Winkler M, Möbius W, Howard J, Göpfert MC (2012) *Drosophila* auditory organ genes and genetic hearing defects. Cell 150:1042–1054

Shin JB, Adams D, Paukert M, Siba M, Sidi S, Levin M, Gillespie PG, Gründer S (2005) Xenopus TRPN1 (NOMPC) localizes to microtubule-based cilia in epithelial cells, including inner-ear hair cells. Proc Natl Acad Sci U S A 102:12572–12577

Sidi S, Friedrich RW, Nicolson T (2003) NompC TRP channel required for vertebrate sensory hair cell mechanotransduction. Science 301:96–99

Sininger YS (1993) Auditory brain stem response for objective measures of hearing. Ear Hear 14:23–30

Sotomayor M, Corey DP, Schulten K (2005) In search of the hair-cell gating spring elastic properties of ankyrin and cadherin repeats. Structure 13:669–682

Stepanyan RS, Indzhykulian AA, Vélez-Ortega AC, Boger ET, Steyger PS, Friedman TB, Frolenkov GI (2011) TRPA1-mediated accumulation of aminoglycosides in mouse cochlear outer hair cells. J Assoc Res Otolaryngol 12:729–740

Strotmann R, Harteneck C, Nunnenmacher K, Schultz G, Plant TD (2000) OTRPC4, a nonselective cation channel that confers sensitivity to extracellular osmolarity. Nat Cell Biol 2:695–702

Sun Y, Liu L, Ben-Shahar Y, Jacobs JS, Eberl DF, Welsh MJ (2009) TRPA channels distinguish gravity sensing from hearing in Johnston's organ. Proc Natl Acad Sci U S A 106:13606–13611

Tabuchi K, Suzuki M, Mizuno A, Hara A (2005) Hearing impairment in TRPV4 knockout mice. Neurosci Lett 382:304–308

Takumida M, Kubo N, Ohtani M, Suzuka Y, Anniko M (2005) Transient receptor potential channels in the inner ear: presence of transient receptor potential channel subfamily 1 and 4 in the guinea pig inner ear. Acta Otolaryngol 125:929–934

van Aken AF, Atiba-Davies M, Marcotti W, Goodyear RJ, Bryant JE, Richardson GP, Noben-Trauth K, Kros CJ (2008) TRPML3 mutations cause impaired mechano-electrical transduction and depolarization by an inward-rectifier cation current in auditory hair cells of varitint-waddler mice. J Physiol 586:5403–5418

Van Camp G, Willems PJ, Smith RJ (1997) Nonsyndromic hearing impairment: unparalleled heterogeneity. Am J Hum Genet 60:758–764

Vriens J, Watanabe H, Janssens A, Droogmans G, Voets T, Nilius B (2004) Cell swelling, heat, and chemical agonists use distinct pathways for the activation of the cation channel TRPV4. Proc Natl Acad Sci U S A 101:396–401

Wangemann P, Nakaya K, Wu T, Maganti RJ, Itza EM, Sanneman JD, Harbidge DG, Billings S, Marcus DC (2007) Loss of cochlear HCO3- secretion causes deafness via endolymphatic acidification and inhibition of Ca2+ reabsorption in a Pendred syndrome mouse model. Am J Physiol Renal Physiol 292:F1345–F1353

Watanabe H, Vriens J, Prenen J, Droogmans G, Voets T, Nilius B (2003) Anandamide and arachidonic acid use epoxyeicosatrienoic acids to activate TRPV4 channels. Nature 424:434–438

Wong AC, Birnbaumer L, Housley GD (2013) Canonical transient receptor potential channel subtype 3-mediated hair cell Ca(2+) entry regulates sound transduction and auditory neurotransmission. Eur J Neurosci 37:1478–1486

Xiong W, Grillet N, Elledge HM, Wagner TF, Zhao B, Johnson KR, Kazmierczak P, Müller U (2012) TMHS is an integral component of the mechanotransduction machinery of cochlear hair cells. Cell 151:1283–1295

Xu H, Delling M, Li L, Dong X, Clapham DE (2007) Activating mutation in a mucolipin transient receptor potential channel leads to melanocyte loss in varitint waddler mice. Proc Natl Acad Sci U S A 104:18321–18326

Yan Z, Zhang W, He Y, Gorczyca D, Xiang Y, Cheng LE, Meltzer S, Jan LY, Jan YN (2013) *Drosophila* NOMPC is a mechanotransduction channel subunit for gentle-touch sensation. Nature 493:221–225

Yorozu S, Wong A, Fischer BJ, Dankert H, Kernan MJ, Kamikouchi A, Ito K, Anderson DJ (2009) Distinct sensory representations of wind and near-field sound in the *Drosophila* brain. Nature 458:201–205

Zanini D, Göpfert MC (2013) Mechanosensation: tethered ion channels. Curr Biol 23:R349–R351

Zdebik AA, Wangemann P, Jentsch TJ (2009) Potassium ion movement in the inner ear: insights from genetic disease and mouse models. Physiology (Bethesda) 24:307–316

Zheng J, Shen W, He DZ, Long KB, Madison LD, Dallos P (2000) Prestin is the motor protein of cochlear outer hair cells. Nature 405:149–155

Zimoń M, Baets J, Auer-Grumbach M, Berciano J, Garcia A, Lopez-Laso E, Merlini L, Hilton-Jones D, McEntagart M, Crosby AH, Barisic N, Boltshauser E, Shaw CE, Landouré G, Ludlow CL, Gaudet R, Houlden H, Reilly MM, Fischbeck KH, Sumner CJ, Timmerman V, Jordanova A, Jonghe PD (2010) Dominant mutations in the cation channel gene transient receptor potential vanilloid 4 cause an unusual spectrum of neuropathies. Brain 133:1798–1809

TRPs in Olfaction

Frank Zufall

Contents

Abstract

The mammalian olfactory system has become an excellent model system to understand the function of transient receptor potential (TRP) channels within their native cellular and circuit environment. The discovery that the canonical TRP channel TRPC2 is highly expressed in sensory neurons of the vomeronasal organ (VNO) has led to major advances in our understanding of the cellular and molecular processes underlying signal transduction of pheromones and other molecular cues that play an essential role in the control of instinctive decisions and innate social behaviors. TRPC2 knockout mice provide a striking example that the loss of function of a single gene can cause severe alterations in a variety of social interactions including the display of aggression, social dominance, and sexual behaviors. There is mounting evidence that TRPC2 is not the only TRP

F. Zufall (✉)
Department of Physiology and Center for Integrative Physiology and Molecular Medicine, University of Saarland School of Medicine, 66424 Homburg, Germany
e-mail: frank.zufall@uks.eu

B. Nilius and V. Flockerzi (eds.), *Mammalian Transient Receptor Potential (TRP) Cation Channels*, Handbook of Experimental Pharmacology 223, DOI 10.1007/978-3-319-05161-1_8,

channel expressed in cells of the olfactory system but that other TRP channel subtypes such as TRPC1, TRPC4, TRPC6, TRPM4, and TRPM5 could also play important functional roles in mammalian olfaction. Here, I review such findings and discuss future areas for investigation.

Keywords

Olfactory • Sensory • Pheromone • Long-lasting neuronal activation • I_{can} • Ca^{2+}-activated cation channel • Rostral migratory stream • Olfactory ensheathing cell • TRPM8 • TRPA1 • Modulation

1 Introduction

The mammalian olfactory system has become an excellent model system to understand the function of TRP channels within their native cellular and circuit environment. The discovery that a canonical TRP channel, TRPC2, is highly expressed in sensory neurons of the vomeronasal organ (VNO) has led to major advances in our understanding of the cellular and molecular processes underlying signal transduction of pheromones and other molecular cues that play an essential role in the control of instinctive decisions and innate social behaviors. TRPC2 knockout mice provided a striking example that the loss of function of a single gene can cause severe alterations in a variety of social interactions in mice including the display of aggression, social dominance, and sexual behaviors. Vomeronasal sensory neurons (VSNs) also serve as a valuable model to understand activation and Ca^{2+}-regulation mechanisms of TRP channels within small neuronal compartments such as dendrites and microvilli. More recently, there has been mounting evidence that TRPC2 is not the only TRP channel expressed in cells of the olfactory system and that TRP channel function could be widespread for mammalian olfaction. In this chapter, I review the available evidence for the expression of TRP channels in mammalian olfaction and the current understanding of the functional significance of these findings. As such, these studies are part of a larger set of investigations aimed at understanding general TRP channel function in sensory systems (for reviews see Damann et al. 2008; Montell 2011; Eijkelkamp et al. 2013). In order to appreciate recent developments in this field, it is necessary to gain some understanding on the cellular and molecular organization of the mammalian sense of smell.

2 Peripheral Olfactory System

The peripheral olfactory system in most mammals comprises millions of chemosensory neurons that, in rodents, are organized into four distinct sensory structures. These include (1) a sensory epithelium in the main olfactory system and its first processing station in the main olfactory bulb (MOB); (2) a sensory epithelium in the VNO and a first processing station in the accessory olfactory bulb (AOB); (3) a sensory epithelium in the Grueneberg ganglion and its first processing station in the necklace region of the olfactory bulb, located at the border between MOB and AOB; and (4) a sensory epithelium in the septal organ (of Masera) with a first processing station in specific regions of the MOB. Several comprehensive reviews have covered this basic organization of the mammalian olfactory system (Mombaerts 2004; Munger et al. 2009; Tirindelli et al. 2009). Sensory neuron populations located in each of these four structures are further subdivided into a variety of independent subsystems, each one characterized by distinct families of chemoreceptors, signal transduction pathways, axonal projection patterns in the olfactory bulb, and roles in odor-guided behaviors. With respect to TRP channel function, the vomeronasal or accessory olfactory system of mice has received specific attention because of the finding that TRPC2 plays a critical role in pheromonal signaling of VSNs and the regulation of social and sexual behaviors (for reviews see Dulac and Torello 2003; Liman and Zufall 2004; Zufall et al. 2005; Brennan and Zufall 2006; Liman and Dulac 2007; Yildirim and Birnbaumer 2007; Kiselyov et al. 2010; Chamero et al. 2012). I will therefore begin by discussing the function of TRPC2 in the VNO. Outside the olfactory system, TRPC2 has been investigated in rodent sperm function (Jungnickel et al. 2001; Sutton et al. 2004; Stamboulian et al. 2005), erythroblasts (Chu et al. 2002, 2004), red blood cells (Hirschler-Laszkiewicz et al. 2012), and rat thyroid cells (Lof et al. 2012; Sukumaran et al. 2012, 2013; Viitanen et al. 2013).

2.1 Vomeronasal Organ

Mouse VSNs express members of two large families of G protein-coupled receptor genes. VSNs with cell bodies in the apical layer of the vomeronasal epithelium (VNE) express receptor genes of the *Vmn1r* repertoire, coexpress the G protein Gαi2, and project their axons to the anterior AOB; VSNs with cell bodies in the basal layer of the VNE express receptor genes of the unrelated *Vmn2r* repertoire, coexpress Gαo, and project their axons to the posterior AOB (for a review see Chamero et al. 2012). The basal VNE layer is further subdivided in apical and basal parts, based on expression of a family of nine nonclassical class I major histocompatibility complex (MHC) genes termed *H2-Mv* genes (Ishii et al. 2003; Loconto et al. 2003). All of these VSNs, as far as we know, express TRPC2 (Liman et al. 1999). Aside from these neurons, additional subsets of VSNs have been identified that express members of the formyl peptide receptor (FPR) family of G protein-coupled receptors (Riviere et al. 2009; Liberles et al. 2009). It is currently unknown whether FPR-positive VSNs also express TRPC2 or whether they employ other types of cation channels.

2.1.1 TRPC2

2.1.1.1 Gene

For the initial identification and genomic organization of *Trpc2*, the reader is referred to two previous reviews (Liman and Dulac 2007; Yildirim and Birnbaumer 2007). *Trpc2* is unique among other TRPCs in that functional open reading frames are no longer present in the genomes of Old World monkeys and humans (Liman and Innan 2003; Yildirim et al. 2003). *Trpc2* was first identified as a pseudogene by searching an expressed sequence tag database (Wes et al. 1995), and its presence in other genomes was later confirmed through sequencing of partial clones from mice and cows (Zhu et al. 1996; Wissenbach et al. 1998). Full-length sequences of rat (Liman et al. 1999) and mouse (Vannier et al. 1999) *Trpc2* were subsequently reported. Of the mouse primary transcript, four variants that apparently result from alternative splicing have been described: *Trpc2a* and *Trpc2b* (Vannier et al. 1999) and *Trpc2α* and *Trpc2β* (Hofmann et al. 2000). With few exceptions (Chu et al. 2005), little progress has been made during the past decade in understanding the functional significance of *Trpc2* splice variants (Liman and Dulac 2007; Yildirim and Birnbaumer 2007). Recent work reported the identification of two distinct genes at the vertebrate *Trpc2* locus, bona fide *Trpc2*, and a gene that has been named *XNDR* (Frankenberg et al. 2011).

2.1.1.2 Expression

A breakthrough came with the demonstration that *Trpc2* is highly expressed in VNE of rats, where it represents as much as 1/10,000 of the mRNA (Liman et al. 1999; Liman and Dulac 2007). Furthermore, immunolabeling demonstrated intense staining in VSN microvilli (Liman et al. 1999), specialized structures for the detection of chemosignals. The localization of TRPC2 to VSN microvilli was later confirmed at the ultrastructural level (Menco et al. 2001). This localization suggested a role for TRPC2 in pheromone-induced signaling in VSNs (Liman et al. 1999), which was tested and indeed demonstrated when $Trpc2^{-/-}$ mice were generated (Leypold et al. 2002; Stowers et al. 2002; Lucas et al. 2003). Low levels of TRPC2 have been detected in the main olfactory epithelium (Liman et al. 1999, and see below), testes, heart, brain, liver, spleen, and erythrocytes (Wissenbach et al. 1998; Hofmann et al. 2000; Liman and Dulac 2007).

2.1.1.3 The Channel Protein and Structural Aspects

Kyte–Doolittle analysis of the amino acid sequence of TRPC2 identifies seven hydrophobic regions with the potential to form six transmembrane segments (Yildirim and Birnbaumer 2007). However, because of difficulties in expressing TRPC2 in heterologous cells, very little is still known about TRPC2 protein and structural aspects of the channel. One study indicated that TRPC2 is largely trapped in the endoplasmic reticulum when expressed in HEK 293 cells and does not seem to interact with other members of the TRPC family (Hofmann et al. 2002; but see Chu et al. 2004). As a result, it is not yet clear whether native TRPC2 channels form homomeric or heteromeric complexes, and no structure–activity studies of TRPC2 have been performed.

2.1.1.4 Interacting Proteins

TRPC2 is coexpressed with type III inositol 1,4,5-trisphosphate receptor (IP_3R) in the microvillous layer of the VNE and both molecules associate in a protein–protein interaction process (Brann et al. 2002). Type II IP_3R immunoreactivity is present in the supporting cell zone but not in the sensory cells (Brann et al. 2002). TRPC2 also forms protein–protein interactions with Homer family scaffolding proteins, particularly Homer 1b/c, and with the chaperone receptor transporting protein RTP1 (Mast et al. 2010). It has been proposed that TRPC2 activation is physically linked to type III IP_3R via Homer (Mast et al. 2010), but no functional measurements in native VSNs have tested this model thus far. Similarly, in CHO-S cells there is evidence that TRPC2, IP_3R, phospholipase C gamma, and erythropoietin interact to form a signaling complex (Tong et al. 2004). More recently TRPC2 has also been shown to bind the stromal interaction molecule STIM1, indicating that TRPC2 could be linked to store-operated Ca^{2+} release (Worley et al. 2007). In native VSNs, however, it must be kept in mind that TRPC2 is mostly expressed in the sensory microvilli of VSNs, a neuronal compartment that lacks Ca^{2+} store organelles such as endoplasmic reticulum and mitochondria.

The long cytoplasmic N-terminus of TRPC2 contains ankyrin (Liman et al. 1999; Vannier et al. 1999), calmodulin (CaM) (Yildirim et al. 2003), and enkurin binding domains (Sutton et al. 2004). TRPC2 also has a CaM-binding site on its C-terminus (Yildirim et al. 2003). Strong inhibition by Ca^{2+}-calmodulin of native TRPC2 channels and of pheromone-evoked currents in VSNs has been demonstrated (see below, Spehr et al. 2009).

2.1.1.5 Biophysical Description of Channel Function, Permeation, and Gating

Because most investigators have failed to express TRPC2 in heterologous cells (see Liman and Dulac 2007), the biophysical properties of TRPC2 still remain unclear. Early results using COS-M6 cells concluded that TRPC2 is a store depletion-activated capacitative Ca^{2+} entry channel that can be readily activated not only after stimulation with an agonist but also by store depletion in the absence of an agonist using thapsigargin (Vannier et al. 1999). However, studies from our group searching for pheromone-sensitive ion channels at dendritic tips of native VSNs came to a very different conclusion (Lucas et al. 2003; Zufall et al. 2005; Spehr et al. 2009). This work identified a 42 pS cation channel that is permeable for Na^+, Cs^+, and Ca^{2+} but not for *N*-methyl-D-glucamine (Lucas et al. 2003). In excised inside-out membrane patches, this channel can be activated by application of several analogs of diacylglycerol but not by IP_3 or Ca^{2+} (Lucas et al. 2003) and, once activated by diacylglycerol, application of Ca^{2+}/CaM causes channel closure (Spehr et al. 2009). This conductance requires TRPC2 as it is nearly absent from TRPC2-deficient VSNs (Lucas et al. 2003). Furthermore, the biophysical and pharmacological properties of the diacylglycerol-activated currents are closely similar to the currents stimulated by pheromonal ligands, leading us to conclude that TRPC2 is required in VSNs for a pheromone-sensitive cation channel that is activated via receptor-operated G protein coupling of phospholipase C and subsequent formation of diacylglycerol (Lucas et al. 2003; Zufall et al. 2005). An essential role of the G protein Gαo in this cascade was recently demonstrated for VSNs of the basal layer (Chamero et al. 2011).

2.1.1.6 Physiological Function in Native VSNs and Lessons from Knockouts

In the absence of a reliable expression system for TRPC2, the generation of TRPC2-deficient mice paved the way for subsequent investigations in native VSNs and whole animals and became crucial for gaining insight into the role of TRPC2 in mouse vomeronasal function. Two independently generated mouse lines with a global *Trpc2* deficiency provided evidence that TRPC2 plays a central role in mediating pheromone-evoked excitation of VSNs (Leypold et al. 2002; Stowers et al. 2002) and enabled initial investigations into the nature of the signal transduction mechanisms of VSNs (see above, Lucas et al. 2003; Zufall et al. 2005). The mutant mice are both viable and fertile, thus ruling out an essential role of TRPC2 in sperm function (Leypold et al. 2002; Stowers et al. 2002). These mice afforded the opportunity to examine the role of the VNO in the generation of innate sexual and social behaviors. TRPC2 mutant males and nursing females are docile and fail to initiate aggressive attacks on intruder males (Leypold et al. 2002; Stowers et al. 2002). Male–female sexual behavior appears normal, but TRPC2 mutants males also vigorously mount other males (Leypold et al. 2002; Stowers et al. 2002). Since these original descriptions, numerous additional behavioral studies have been performed with these mice, thus gaining further insight into the behavioral consequences of the *Trpc2* deletion (Chamero et al. 2007; Kimchi et al. 2007; Hasen and Gammie 2009, 2011; Haga et al. 2010; Papes et al. 2010; Ferrero et al. 2013). Together, these results clearly demonstrate that loss-of-function mutation of a single gene, *Trpc2*, has severe and complex consequences for social behavior in mice. An important next step will be to generate conditional *Trpc2* mutations that will allow for cell-specific or time-dependent deletion of this channel.

Concerning the interpretation of behavioral data in *Trpc2*$^{-/-}$ mice, it is still unclear whether all TRPC2-deficient VSNs are completely silent when stimulated by chemical ligands or whether other mechanisms can contribute to electrical signaling under these circumstances (Kelliher et al. 2006; Zhang et al. 2010). Further evidence that global *Trpc2*$^{-/-}$ mice may retain partial VNO function has come from a study indicating a requirement for Ca^{2+}-activated chloride channels in the activation of mouse VSNs (Kim et al. 2011). In the absence of TRPC2, the activation of the chloride conductance, perhaps via Ca^{2+} release from intracellular stores, is sufficient to trigger action potential activity in VSNs (Kim et al. 2011). Furthermore, the Ca^{2+}-activated potassium channel SK3 and the G protein-activated potassium channel GIRK seem to be part of an independent pathway for VNO activation (Kim et al. 2012). It will be necessary to gain a more detailed understanding of the role of Ca^{2+} signaling, Ca^{2+} stores, and IP_3R function in VSNs in order to resolve these remaining questions.

2.1.1.7 Role in Hereditary and Acquired Disease

The finding that the *Trpc2* gene is a pseudogene in humans (Wes et al. 1995; Liman and Innan 2003; Yildirim et al. 2003) excludes a potential role in hereditary and acquired human disease, making TRPC2 a unique case among all other TRPCs.

2.2 Main Olfactory Epithelium

The rodent MOE comprises four major cell types: sensory neurons, supporting cells, microvillous cells, and basal progenitor cells. Among the sensory neurons, classical or canonical olfactory sensory neurons (OSNs) represent the vast majority of sensory neurons in the MOE. Each canonical OSN expresses a member of the odor receptor (OR) gene family and a cAMP signaling cascade that consists of the G protein subunit $G\alpha_{olf}$, type III adenylyl cyclase (ACIII), and the cyclic nucleotide-gated channel subunits CNGA2, CNGA4, and CNGB1b, which form a cAMP-gated cation channel (Munger et al. 2009). Besides canonical OSNs, several other sensory neuron populations exist in MOE. These differ from canonical OSNs with respect to the receptors and signal transduction elements they express, the chemosignals they detect, and the axonal projections and target regions in the olfactory bulb. Therefore, these additional subsets of sensory neurons represent distinct subsystems of the MOE (Munger et al. 2009). They include GC-D OSNs that express the particulate guanylyl cyclase type D and the cGMP-specific CNG channel subunit CNGA3 (Zufall and Munger 2010) and TAAR OSNs that express a member of the trace amine-associated receptor family (Liberles and Buck 2006; Pacifico et al. 2012). In addition to these recently characterized subpopulations, the MOE comprises cell types that are characterized by the expression of specific TRP channel isoforms, indicating further functional diversity in the main olfactory system. These cell types will be discussed in the following chapters.

2.2.1 TRPM5

With respect to TRP channels in the MOE, current interest has focused on a large group of ciliated OSNs that express the Ca^{2+}-activated, monovalent-selective cation channel TRPM5 (Lin et al. 2007; Oshimoto et al. 2013). Using a transgenic mouse line in which a TRPM5-GFP construct is expressed under the control of *Trpm5* promoter sequences (Clapp et al. 2006), it was shown that GFP-positive neurons are enriched in the ventrolateral MOE and project axons to the ventral, lateral, and medial MOB (Lin et al. 2007). Surprisingly, such TRPM5+ OSNs express components of both phosphatidylinositide- and cAMP-mediated signaling cascades, including PLCβ2, the G protein subunit Gγ13, and the CNGA2 channel subunit (Lin et al. 2007). They also express olfactory marker protein (Lin et al. 2007).

The role that each cascade plays in cellular responses to chemostimulation remains unclear. TRPM5+ OSNs may respond to pheromones or other semiochemicals. Mouse urine, the putative pheromone 2,5-dimethylpyrazine, and the social cue (methylthio)methanethiol (MTMT) each elicit increased c-fos expression in periglomerular cells associated with MOB glomeruli receiving TRPM5+ neuron innervation (Lin et al. 2007). Population responses (measured as an electroolfactogram) to the semiochemical 2,5-dimethylpyrazine, but not the environmental odor lilial, are reduced in *Trpm5* null mice in the presence, but not absence, of an adenylyl cyclase inhibitor (Lin et al. 2007). Thus, both pathways may be required for the transduction of certain stimuli by TRPM5+ OSNs, but the precise role of TRPM5 and the TRPM5+ OSNs in olfaction or pheromone detection eagerly awaits functional characterization of individual TRPM5+ neurons.

Besides TRPM5+ OSNs, there is now evidence for the existence of several types of TRPM5-expressing microvillous cells in MOE (Lin et al. 2007, 2008; Hansen and Finger 2008). None of these cells have an axon that projects to the olfactory bulb. The function of these cells is not yet understood. Thus, TRPM5+ microvillous cells are clearly distinct from ciliated TRPM5+ OSNs.

2.2.2 TRPC6

As indicated in the previous paragraph, several populations of microvillous cells, with properties and functions that are not yet clear, have been described in the rodent olfactory epithelium. On the basis of immunohistochemical staining, one of these populations comprising about 5 % of all olfactory cells seems to express TRPC6 in their microvilli (Elsaesser et al. 2005). These cells also express phospholipase C β2, type III IP_3R, and ecto-5′-nucleotidase CD73 (Pfister et al. 2012), and they are negative for the expression of TRPM5 (Hansen and Finger 2008). However, TRPC6-deficient mice have not been used thus far to evaluate the specificity of TRPC6 antibody labeling nor have they been examined to investigate the biophysical properties and function of the TRPC6 channel in these cells. A potential involvement of these cells in olfactory neurogenesis and homeostasis, possibly driven by extracellular ATP and adenosine (Pfister et al. 2012), would make such experiments timely.

2.2.3 TRPC2

Besides TRPM5- and TRPC6-expressing cells in MOE, there is now also at least some preliminary evidence for the presence of a small subpopulation of MOE cells that are characterized by the expression of TRPC2 (Mombaerts 2008). Generation of a mouse strain with a targeted mutation in the *Trpc2* locus resulting in expression of axonal markers revealed these cells and showed projections of axons of TRPC2-expressing neurons from the MOE to specific glomeruli of the olfactory bulb (Mombaerts 2008). It is not yet clear whether these cells represent displaced microvillous VSNs or whether they are ciliated OSNs. The chemosensory role of these neurons, if any, remains to be investigated. Thus, it appears that TRPC2-expressing cells have some function outside the VNO and comprise yet another independent subsystem in the MOE (Firestein 2009).

2.3 Grueneberg Ganglion

The Grueneberg ganglion, located at the rostral tip of the nasal cavity close to the opening of the naris, consists of a cluster of sensory cells that express olfactory marker protein and project their axons to regions of the necklace system of glomeruli in the olfactory bulb (Munger et al. 2009). At least some Grueneberg ganglion neurons (GGNs) are chemodetectors and seem to sense alarm pheromones (Brechbuhl et al. 2013), but GGNs also function as finely tuned cold sensors (Mamasuew et al. 2008; Schmid et al. 2010). Although it is now well established that GGNs express several elements of a cGMP second messenger pathway, including the membrane-bound guanylyl cyclase GC-G, the cGMP-specific CNG channel CNGA3, and the cGMP-stimulated phosphodiesterase PDE2 (see references in Schmid et al. 2010), the fact that these cells produce graded intracellular Ca^{2+} increases evoked by cooling has raised interest in a potential role of TRP channels in this function. However, in TRPA1-deficient mice, there was no difference in cold-evoked Ca^{2+} signal compared to controls, and chemicals that normally activate another temperature-sensitive TRP channel, TRPM8, were unable to evoke a Ca^{2+} transient in these cells (Schmid et al. 2010). TRPC5-deficient mice also showed normal cold-evoked Ca^{2+} responses in GGNs (Schmidt et al. unpublished observation). Thus, the ion channels underlying cold detection in these neurons are still to be defined.

3 Olfactory Ensheathing Cells

The axons of OSNs form the olfactory nerve (also known as cranial nerve I) which transmits neural activity from the periphery of the olfactory system to the first processing center in the brain, the MOB. OSN axons are unusual in that they can regenerate and reinnervate the MOB following injury, a process that occurs throughout life. These axons are ensheathed by a type of glial cell, the olfactory ensheathing cells (OECs), that are thought to provide an environment permissive for axon growth. There is evidence that Ca^{2+} signaling in OECs may play an important role in the ability of these cells to promote neurite regrowth (Hayat et al. 2003). Therefore, the mechanisms underlying Ca^{2+} entry into OECs are of considerable interest. Few studies have examined whether TRP channel function underlies Ca^{2+} signaling in OECS. Davies et al. (2004) used high extracellular Ca^{2+} (20 mM) to stimulate enhanced Ca^{2+} entry into OECs and characterized the pharmacological properties of this normally constitutively active Ca^{2+} entry pathway. On the basis of its pharmacological profile, with an order of inhibitor potency of $Lu^{3+} > La^{3+} =$ econazole $= Gd^{3+} >$ SKF96365 $> Cd^{2+}$, the authors conclude that an as yet unidentified member of the TRPC family underlies this Ca^{2+} entry (Davies et al. 2004). OECs also exhibit cooling-evoked Ca^{2+} signals and these signals persist in TRPM8- and TRPA1-deficient mice (Stavermann et al. 2012). No systematic studies on TRP channel expression in OECs have been reported thus far.

4 Olfactory Bulb

4.1 Main Olfactory Bulb

The MOB receives sensory input from all canonical OSNs as well as from those OSNs that express TRPM5 channels or trace amine-associated receptors (Munger et al. 2009). Within a spherical structure of neuropil called a glomerulus, OSN axons converge onto several classes of mitral/tufted cells, the principal output neurons of MOB (Shepherd et al. 2004). The MOB network also includes several classes of GABAergic interneurons (such as periglomerular cells, short axon cells, and granule cells) and at least one class of glutamatergic local interneuron, the external tufted cells (Wachowiak and Shipley 2006).

Few systematic studies on TRP channel expression in MOB neurons are available. It is generally thought that TRPC4 is the most abundant TRPC channel expressed in MOB (Lein et al. 2007; Zechel et al. 2007; Stroh et al. 2012). Recent work analyzed TRP channel mRNA expression in identified external tufted, mitral, and granule cells (Dong et al. 2012). Consistent with some of the physiological studies discussed below, these authors identified 12 TRP channel mRNAs in mitral cells, 9 TRP channel mRNAs in external tufted cells, and 7 TRPC channel mRNAs in granule cells (Dong et al. 2012). This work provides a rich source of potential candidate channels for understanding TRP channel function in MOB neurons.

Perhaps the best evidence for a physiological role of TRP channels in MOB neurons comes from a recent study investigating the function of the most abundant type of synaptic connection in vertebrate MOB, the reciprocal dendrodendritic synapse between mitral/tufted and granule cells (Stroh et al. 2012). These synapses produce a long-lasting depolarization upon suprathreshold activation that is absent from granule cells deficient for the NMDA receptor subunit NR1. Both the long-lasting depolarization and the accompanying intracellular Ca^{2+} increase are also absent in granule cells from double knockout mice deficient for the TRPC channel subtypes 1 and 4 (Stroh et al. 2012). Taken together, these results provide clear evidence that both TRPC1 and TRPC4 contribute to slow, NR1-dependent synaptic transmission in MOB, including the Ca^{2+} dynamics required for asynchronous transmitter release from these interneurons (Stroh et al. 2012).

Several other studies suggest that TRP channels may mediate G protein-coupled receptor responses in multiple classes of MOB neurons and play important roles in the processing of olfactory signals, yet in none of these cases, TRP channel gene deletion has been employed to determine which TRP channels underlie such responses. First, group I metabotropic glutamate receptor (mGluR1) activation enhances Ca^{2+}-dependent nonselective cation currents (I_{CAN}) and rhythmic bursting in external tufted cells (Dong et al. 2009). Such slow rhythmic oscillations may play a key role in respiratory-driven activity in the MOB (Dong et al. 2009). Second, serotonin (5-HT) has been shown to modulate external tufted cells, and this modulation is also thought to be due to a nonselective inward current mediated by an as yet unknown TRP channel (Liu et al. 2012). Third, there is evidence for a role of I_{CAN} in two other types of MOB interneurons. For example, Blanes cells,

GABAergic interneurons located in the granule cell layer, generate afterdepolarizations and persistent firing following brief depolarization. The pharmacological properties of these effects are consistent with a role for I_{CAN} (Pressler and Strowbridge 2006). Furthermore, in granules cells, muscarinic receptor activation produces an afterdepolarization with properties also consistent with I_{CAN} (Pressler et al. 2007).

Hence, centrifugal modulation has multiple effects on the MOB network, and there is good evidence that TRP channels could play important roles in these functions. Systematic gene deletion studies will be required to determine precisely which TRP channels underlie these effects.

4.2 Accessory Olfactory Bulb

The AOB, located caudally to the MOB, receives sensory input from all VSNs. As in the MOB, sensory neuron axons converge onto mitral cells, the principal output neurons of AOB, within glomerular neuropil structures and the AOB network involves multiple types of interneurons (Zufall et al. 2008).

Much less is known about TRP channel expression and function in AOB neurons compared to those of MOB. Based on the Allen Brain Atlas (http://mouse.brain-map.org), nine TRPC and TRPM candidates are expressed in the AOB mitral cells alone (TRPC1, TRPC3, TRPC5, TRPC6, TRPC7, TRPM2, TRPM4, TRPM5, and TRPM7) (Shpak et al. 2012). A single study has addressed the physiological role of TRP channels in AOB (Shpak et al. 2012). This work described sustained firing lasting up to several minutes in responses to brief electrical stimulation in AOB mitral cells recorded in mouse brain slices. This sustained excitation was found to be mediated by a nonselective cationic current. Combination of single-cell gene expression analysis, pharmacology, and immunocytochemistry led the authors to conclude that this current was at least partially mediated by TRPM4 channels activated by Ca^{2+} influx (Shpak et al. 2012). The authors speculate that this current could serve to modulate AOB neural activity in a context-dependent manner.

5 Rostral Migratory Stream

MOB and AOB neurons are continuously replenished during adulthood through lifelong neurogenesis. An important site of such neurogenesis is the subventricular zone where neuronal precursor cells (neuroblasts) are generated and then migrate over long distances through the rostral migratory stream to become interneurons in MOB and AOB (Whitman and Greer 2009). Ca^{2+} signaling in neuroblasts is likely to regulate cell migration. While Ca^{2+} signaling mechanisms in neuroblasts are still incompletely understood, a recent study concluded that a TRP channel, possibly TRPC1, provides an important Ca^{2+} source for modulating neuroblast migration (Turner and Sontheimer 2013). The Ca^{2+}-activated K^+ channel KCa3.1 serves as a downstream target for TRP-mediated Ca^{2+} entry in these cells. Both the use of TRP channel blockers and blockade of KCa3.1 caused a strong decrease in cell migration in these neuroblasts (Turner and Sontheimer 2013).

6 Higher Olfactory Centers

MOB mitral/tufted cells project to numerous downstream structures in the central nervous system, including the anterior olfactory nucleus, olfactory tubercle, piriform cortex, amygdala, and entorhinal cortex. With the exception of a few studies (von Bohlen Und Halbach et al. 2005; Meis et al. 2007), there are no systematic investigations examining TRP channel expression and function in these brain structures.

AOB output is transmitted to the bed nucleus of the stria terminalis, the bed nucleus of the accessory olfactory tract, the posteroventral medial amygdala and the more rostral and posterior-dorsal subdivisions of the medial amygdaloid nucleus, and the posteromedial cortical amygdala (Brennan and Zufall 2006). Connections exist from the amygdala to hypothalamic nuclei for the regulation of endocrine status. The evidence for TRP channel expression and function in neuroendocrine centers of the brain will be summarized in chapter XX (Leinders-Zufall and Boehm, TRPs in endocrinology).

Concluding Remarks

In conclusion, the mammalian olfactory system has been a rewarding neural system to investigate TRP channel function in native neurons and at the whole animal level. Although these investigations were initially largely focused on the role of TRPC2 in the accessory olfactory system, this review has summarized a number of recent developments indicating that TRP channels other than TRPC2 are likely to play important roles at multiple places in the olfactory system. These include primary sensory neurons, neurons of the olfactory bulb and higher olfactory centers, glial cells that contribute to axonal regeneration, and even some stem cell populations that are required for adult neurogenesis and network homeostasis. A major future challenge is to develop conditional gene targeting approaches in which subsets of cell populations expressing specific TRP channel subtypes can be visualized and specific channel subunits be deleted in a cell-specific or time-dependent manner in order to determine the precise functions of these channels within their native cellular environment.

Acknowledgments Work in the author's laboratory is supported by Deutsche Forschungsgemeinschaft grants Sonderforschungsbereich 894, Schwerpunktprogramm 1392, and Graduiertenkolleg 1362 and by the National Institutes of Health (DC005633).

Conflict of Interest The author declares that he has no conflict of interest.

References

Brann JH, Dennis JC, Morrison EE, Fadool DA (2002) Type-specific inositol 1,4,5-trisphosphate receptor localization in the vomeronasal organ and its interaction with a transient receptor potential channel, TRPC2. J Neurochem 83:1452–1460. doi:10.1186/1471-2202-11-61

Brechbuhl J, Moine F, Klaey M et al (2013) Mouse alarm pheromone shares structural similarity with predator scents. Proc Natl Acad Sci U S A 110:4762–4767. doi:10.1073/pnas.1214249110

Brennan PA, Zufall F (2006) Pheromonal communication in vertebrates. Nature 444:308–315. doi:10.1038/nature05404

Chamero P, Marton TF, Logan DW et al (2007) Identification of protein pheromones that promote aggressive behaviour. Nature 450:899–902

Chamero P, Katsoulidou V, Hendrix P et al (2011) G protein Gαo is essential for vomeronasal function and aggressive behavior in mice. Proc Natl Acad Sci U S A 108:12898–12903. doi:10.1073/pnas.1107770108

Chamero P, Leinders-Zufall T, Zufall F (2012) From genes to social communication: molecular sensing by the vomeronasal organ. Trends Neurosci 35:597–606. doi:10.1016/j.tins.2012.04.011

Chu X, Cheung JY, Barber DL et al (2002) Erythropoietin modulates calcium influx through TRPC2. J Biol Chem 277:34375–34382

Chu X, Tong Q, Cheung JY et al (2004) Interaction of TRPC2 and TRPC6 in erythropoietin modulation of calcium influx. J Biol Chem 279:10514–10522. doi:10.1074/jbc.M308478200

Chu X, Tong Q, Wozney J et al (2005) Identification of an N-terminal TRPC2 splice variant which inhibits calcium influx. Cell Calcium 37:173–182

Clapp TR, Medler KF, Damak S, Margolskee RF, Kinnamon SC (2006) Mouse taste cells with G protein-coupled taste receptors lack voltage-gated calcium channels and SNAP-25. BMC Biol 4:7

Damann N, Voets T, Nilius B (2008) TRPs in our senses. Curr Biol 18:R880–889. doi:10.1016/j.cub.2008.07.063

Davies R, Hayat S, Wigley CB, Robbins J (2004) The calcium influx pathway in rat olfactory ensheathing cells shows TRPC channel pharmacology. Brain Res 1023:154–156. doi:10.1016/j.brainres.2004.07.032

Dong HW, Hayar A, Callaway J, Yang XH, Nai Q, Ennis M (2009) Group I mGluR activation enhances Ca^{2+}-dependent nonselective cation currents and rhythmic bursting in main olfactory bulb external tufted cells. J Neurosci 29:11943–11953. doi:10.1523/JNEUROSCI.0206-09.2009

Dong HW, Davis JC, Ding S, Nai Q, Zhou FM, Ennis M (2012) Expression of transient receptor potential (TRP) channel mRNAs in the mouse olfactory bulb. Neurosci Lett 524:49–54. doi:10.1016/j.neulet.2012.07.013

Dulac C, Torello AT (2003) Molecular detection of pheromone signals in mammals: from genes to behaviour. Nat Rev Neurosci 4:551–562

Eijkelkamp N, Quick K, Wood JN (2013) Transient receptor potential channels and mechanosensation. Annu Rev Neurosci 36:519–546. doi:10.1146/annurev-neuro-062012-170412

Elsaesser R, Montani G, Tirindelli R, Paysan J (2005) Phosphatidyl-inositide signalling proteins in a novel class of sensory cells in the mammalian olfactory epithelium. Eur J Neurosci 21:2692–2700

Ferrero DM, Moeller LM, Osakada T et al (2013) A juvenile mouse pheromone inhibits sexual behaviour through the vomeronasal system. Nature 502:368–371. doi:10.1038/nature12579

Firestein SJ (2009) Symposium overview: Noncanonical transduction pathways in olfaction: new views on olfactory signaling. Ann NY Acad Sci 1170:161–163. doi:10.1111/j.1749-6632.2009.04882.x

Frankenberg S, Schneider NY, Fletcher TP, Shaw G, Renfree MB (2011) Identification of two distinct genes at the vertebrate TRPC2 locus and their characterisation in a marsupial and a monotreme. BMC Mol Biol 12:39. doi:10.1186/1471-2199-12-39

Haga S, Hattori T, Sato T et al (2010) The male mouse pheromone ESP1 enhances female sexual receptive behaviour through a specific vomeronasal receptor. Nature 466:118–122

Hansen A, Finger TE (2008) Is TRPM5 a reliable marker for chemosensory cells? Multiple types of microvillous cells in the main olfactory epithelium of mice. BMC Neurosci 9:115

Hasen NS, Gammie SC (2009) Trpc2 gene impacts on maternal aggression, accessory olfactory bulb anatomy and brain activity. Genes Brain Behav 8:639–649

Hasen NS, Gammie SC (2011) Trpc2-deficient lactating mice exhibit altered brain and behavioral responses to bedding stimuli. Behav Brain Res 217:347–353

Hayat S, Thomas A, Afshar F, Sonigra R, Wigley CB (2003) Manipulation of olfactory ensheathing cell signaling mechanisms: effects on their support for neurite regrowth from adult CNS neurons in coculture. Glia 44:232–241. doi:10.1002/glia.10299

Hirschler-Laszkiewicz I, Zhang W, Keefer K et al (2012) Trpc2 depletion protects red blood cells from oxidative stress-induced hemolysis. Exp Hematol 40:71–83. doi:10.1016/j.exphem.2011.09.006

Hofmann T, Schaefer M, Schultz G, Gudermann T (2000) Cloning, expression and subcellular localization of two novel splice variants of mouse transient receptor potential channel 2. Biochem J 351:115–122

Hofmann T, Schaefer M, Schultz G, Gudermann T (2002) Subunit composition of mammalian transient receptor potential channels in living cells. Proc Natl Acad Sci U S A 99:7461–7466. doi:10.1073/pnas.102596199

Ishii T, Hirota J, Mombaerts P (2003) Combinatorial coexpression of neural and immune multigene families in mouse vomeronasal sensory neurons. Curr Biol 13:394–400

Jungnickel MK, Marrero H, Birnbaumer L, Lemos JR, Florman HM (2001) Trp2 regulates entry of Ca^{2+} into mouse sperm triggered by egg ZP3. Nat Cell Biol 3:499–502. doi:10.1038.3507.5.0

Kelliher KR, Spehr M, Li XH, Zufall F, Leinders-Zufall T (2006) Pheromonal recognition memory induced by TRPC2-independent vomeronasal sensing. Eur J Neurosci 23:3385–3390. doi:10.1111/j.1460-9568.2006.04866.x

Kim S, Ma L, Yu CR (2011) Requirement of calcium-activated chloride channels in the activation of mouse vomeronasal neurons. Nat Commun 2:365. doi:10.1038/ncomms1368

Kim S, Ma L, Jensen KL, Kim MM, Bond CT, Adelman JP, Yu CR (2012) Paradoxical contribution of SK3 and GIRK channels to the activation of mouse vomeronasal organ. Nat Neurosci 15:1236–1244. doi:10.1038/nn.3173

Kimchi T, Xu J, Dulac C (2007) A functional circuit underlying male sexual behaviour in the female mouse brain. Nature 448:1009–1014

Kiselyov K, van Rossum DB, Patterson RL (2010) TRPC channels in pheromone sensing. Vitam Horm 83:197–213. doi:10.1016/S0083-6729(10)83008-0

Lein ES, Hawrylycz MJ, Ao N et al (2007) Genome-wide atlas of gene expression in the adult mouse brain. Nature 445:168–176

Leypold BG, Yu CR, Leinders-Zufall T, Kim MM, Zufall F, Axel R (2002) Altered sexual and social behaviors in trp2 mutant mice. Proc Natl Acad Sci U S A 99:6376–6381. doi:10.1073/pnas.082127599

Liberles SD, Buck LB (2006) A second class of chemosensory receptors in the olfactory epithelium. Nature 442:645–650

Liberles SD, Horowitz LF, Kuang D et al (2009) Formyl peptide receptors are candidate chemosensory receptors in the vomeronasal organ. Proc Natl Acad Sci U S A 106:9842–9847

Liman ER, Dulac C (2007) TRPC2 and the molecular biology of pheromone detection in mammals. In: Liedtke WB, Heller S (eds) TRP ion channel function in sensory transduction and cellular signaling cascades. CRC Press, Boca Raton (FL)

Liman ER, Innan H (2003) Relaxed selective pressure on an essential component of pheromone transduction in primate evolution. Proc Natl Acad Sci U S A 100:3328–3332

Liman ER, Zufall F (2004) Transduction channels in the vomeronasal organ. In: Frings S, Bradley J (eds) Transduction channels in sensory cells. Wiley-VHC, Weinheim, Germany, pp 135–152
Liman ER, Corey DP, Dulac C (1999) TRP2: a candidate transduction channel for mammalian pheromone sensory signaling. Proc Natl Acad Sci U S A 96:5791–5796
Lin W, Margolskee R, Donnert G, Hell SW, Restrepo D (2007) Olfactory neurons expressing transient receptor potential channel M5 (TRPM5) are involved in sensing semiochemicals. Proc Natl Acad Sci U S A 104:2471–2476
Lin W, Ogura T, Margolskee RF, Finger TE, Restrepo D (2008) TRPM5-expressing solitary chemosensory cells respond to odorous irritants. J Neurophysiol 99:1451–1460
Liu S, Aungst JL, Puche AC, Shipley MT (2012) Serotonin modulates the population activity profile of olfactory bulb external tufted cells. J Neurophysiol 107:473–483. doi:10.1152/jn.00741.2011
Loconto J, Papes F, Chang E et al (2003) Functional expression of murine V2R pheromone receptors involves selective association with the M10 and M1 families of MHC class Ib molecules. Cell 112:607–618
Lof C, Sukumaran P, Viitanen T et al (2012) Communication between the calcium and cAMP pathways regulate the expression of the TSH receptor: TRPC2 in the center of action. Mol Endocrinol 26:2046–2057. doi:10.1210/me.2012-1171
Lucas P, Ukhanov K, Leinders-Zufall T, Zufall F (2003) A diacylglycerol-gated cation channel in vomeronasal neuron dendrites is impaired in TRPC2 mutant mice: mechanism of pheromone transduction. Neuron 40:551–561
Mamasuew K, Breer H, Fleischer J (2008) Grueneberg ganglion neurons respond to cool ambient temperatures. Eur J Neurosci 28:1775–1785. doi:10.1111/j.1460-9568.2008.06465.x
Mast TG, Brann JH, Fadool DA (2010) The TRPC2 channel forms protein-protein interactions with Homer and RTP in the rat vomeronasal organ. BMC Neurosci 11:61. doi:10.1186/1471-2202-11-61
Meis S, Munsch T, Sosulina L, Pape HC (2007) Postsynaptic mechanisms underlying responsiveness of amygdaloid neurons to cholecystokinin are mediated by a transient receptor potential-like current. Mol Cell Neurosci 35:356–367
Menco BP, Carr VM, Ezeh PI, Liman ER, Yankova MP (2001) Ultrastructural localization of G-proteins and the channel protein TRP2 to microvilli of rat vomeronasal receptor cells. J Comp Neurol 438:468–489
Mombaerts P (2004) Genes and ligands for odorant, vomeronasal and taste receptors. Nat Rev Neurosci 5:263–278
Mombaerts P (2008) TRPC2: expression outside the mouse VNO. Int Symp Olfaction Taste (ISOT) XV abstr 125:39
Montell C (2011) The history of TRP channels, a commentary and reflection. Pflugers Arch 461:499–506. doi:10.1007/s00424-010-0920-3
Munger SD, Leinders-Zufall T, Zufall F (2009) Subsystem organization of the mammalian sense of smell. Annu Rev Physiol 71:115–140. doi:10.1146/annurev.physiol.70.113006.100608
Oshimoto A, Wakabayashi Y, Garske A et al (2013) Potential role of transient receptor potential channel M5 in sensing putative pheromones in mouse olfactory sensory neurons. PLoS One 8: e61990. doi:10.1371/journal.pone.0061990
Pacifico R, Dewan A, Cawley D, Guo C, Bozza T (2012) An olfactory subsystem that mediates high-sensitivity detection of volatile amines. Cell Rep 2:76–88. doi:10.1016/j.celrep.2012.06.006
Papes F, Logan DW, Stowers L (2010) The vomeronasal organ mediates interspecies defensive behaviors through detection of protein pheromone homologs. Cell 141:692–703
Pfister S, Dietrich MG, Sidler C, Fritschy JM, Knuesel I, Elsaesser R (2012) Characterization and turnover of CD73/IP_3R3-positive microvillar cells in the adult mouse olfactory epithelium. Chem Senses 37:859–868. doi:10.1093/chemse/bjs069
Pressler RT, Strowbridge BW (2006) Blanes cells mediate persistent feedforward inhibition onto granule cells in the olfactory bulb. Neuron 49:889–904

Pressler RT, Inoue T, Strowbridge BW (2007) Muscarinic receptor activation modulates granule cell excitability and potentiates inhibition onto mitral cells in the rat olfactory bulb. J Neurosci 27:10969–10981

Riviere S, Challet L, Fluegge D, Spehr M, Rodriguez I (2009) Formyl peptide receptor-like proteins are a novel family of vomeronasal chemosensors. Nature 459:574–577

Schmid A, Pyrski M, Biel M, Leinders-Zufall T, Zufall F (2010) Grueneberg ganglion neurons are finely tuned cold sensors. J Neurosci 30:7563–7568. doi:10.1523/JNEUROSCI.0608-10.2010

Shepherd GM, Chen WR, Greer CA (2004) Olfactory Bulb. In: Shepherd GM (ed) The synaptic organization of the brain, 5th edn. Oxford University Press, New York, NY, pp 165–216

Shpak G, Zylbertal A, Yarom Y, Wagner S (2012) Calcium-activated sustained firing responses distinguish accessory from main olfactory bulb mitral cells. J Neurosci 32:6251–6262. doi:10.1523/JNEUROSCI.4397-11.2012

Spehr J, Hagendorf S, Weiss J, Spehr M, Leinders-Zufall T, Zufall F (2009) Ca^{2+}-calmodulin feedback mediates sensory adaptation and inhibits pheromone-sensitive ion channels in the vomeronasal organ. J Neurosci 29:2125–2135

Stamboulian S, Moutin MJ, Treves S et al (2005) Junctate, an inositol 1,4,5-triphosphate receptor associated protein, is present in rodent sperm and binds TRPC2 and TRPC5 but not TRPC1 channels. Dev Biol 286:326–337

Stavermann M, Buddrus K, St John JA, Ekberg JA, Nilius B, Deitmer JW, Lohr C (2012) Temperature-dependent calcium-induced calcium release via $InsP_3$ receptors in mouse olfactory ensheathing glial cells. Cell Calcium 52:113–123

Stowers L, Holy TE, Meister M, Dulac C, Koentges G (2002) Loss of sex discrimination and male-male aggression in mice deficient for TRP2. Science 295:1493–1500

Stroh O, Freichel M, Kretz O, Birnbaumer L, Hartmann J, Egger V (2012) NMDA receptor-dependent synaptic activation of TRPC channels in olfactory bulb granule cells. J Neurosci 32:5737–5746

Sukumaran P, Lof C, Kemppainen K et al (2012) Canonical transient receptor potential channel 2 (TRPC2) as a major regulator of calcium homeostasis in rat thyroid FRTL-5 cells: importance of protein kinase C delta (PKCdelta) and stromal interaction molecule 2 (STIM2). J Biol Chem 287:44345–44360

Sukumaran P, Lof C, Pulli I, Kemppainen K, Viitanen T, Tornquist K (2013) Significance of the transient receptor potential canonical 2 (TRPC2) channel in the regulation of rat thyroid FRTL-5 cell proliferation, migration, adhesion and invasion. Mol Cell Endocrinol 374:10–21. doi:10.1016/j.mce.2013.03.026

Sutton KA, Jungnickel MK, Wang Y, Cullen K, Lambert S, Florman HM (2004) Enkurin is a novel calmodulin and TRPC channel binding protein in sperm. Dev Biol 274:426–435

Tirindelli R, Dibattista M, Pifferi S, Menini A (2009) From pheromones to behavior. Physiol Rev 89:921–956

Tong Q, Chu X, Cheung JY et al (2004) Erythropoietin-modulated calcium influx through TRPC2 is mediated by phospholipase Cgamma and IP_3R. Am J Physiol Cell Physiol 287:C1667–1678

Turner KL, Sontheimer H (2013) KCa3.1 modulates neuroblast migration along the rostral migratory stream (RMS) in vivo. Cereb Cortex. Advance Access published May 15, 2013. doi:10.1093/cercor/bht090

Vannier B, Peyton M, Boulay G et al (1999) Mouse trp2, the homologue of the human trpc2 pseudogene, encodes mTrp2, a store depletion-activated capacitative Ca^{2+} entry channel. Proc Natl Acad Sci U S A 96:2060–2064

Viitanen TM, Sukumaran P, Lof C, Tornquist K (2013) Functional coupling of TRPC2 cation channels and the calcium-activated anion channels in rat thyroid cells: implications for iodide homeostasis. J Cell Physiol 228:814–823. doi:10.1002/jcp.24230

von Bohlen Und Halbach O, Hinz U, Unsicker K, Egorov AV (2005) Distribution of TRPC1 and TRPC5 in medial temporal lobe structures of mice. Cell Tissue Res 322:201–206. doi:10.1007/s00441-005-0004-4

Wachowiak M, Shipley MT (2006) Coding and synaptic processing of sensory information in the glomerular layer of the olfactory bulb. Semin Cell Dev Biol 17:411–423

Wes PD, Chevesich J, Jeromin A, Rosenberg C, Stetten G, Montell C (1995) TRPC1, a human homolog of a Drosophila store-operated channel. Proc Natl Acad Sci U S A 92:9652–9656

Whitman MC, Greer CA (2009) Adult neurogenesis and the olfactory system. Prog Neurobiol 89:162–175

Wissenbach U, Schroth G, Philipp S, Flockerzi V (1998) Structure and mRNA expression of a bovine trp homologue related to mammalian trp2 transcripts. FEBS Lett 429:61–66

Worley PF, Zeng W, Huang GN et al (2007) TRPC channels as STIM1-regulated store-operated channels. Cell Calcium 42:205–211

Yildirim E, Birnbaumer L (2007) TRPC2: molecular biology and functional importance. In: Nilius B, Flockerzi V (eds) Handbook of Experimental Pharmacology. Springer, Berlin, pp 53–75. doi:10.1007/978-3-540-34891-7_3

Yildirim E, Dietrich A, Birnbaumer L (2003) The mouse C-type transient receptor potential 2 (TRPC2) channel: alternative splicing and calmodulin binding to its N terminus. Proc Natl Acad Sci U S A 100:2220–2225

Zechel S, Werner S, von Bohlen Und Halbach O (2007) Distribution of TRPC4 in developing and adult murine brain. Cell Tissue Res 328:651–656. doi:10.1007/s00441-007-0388-4

Zhang P, Yang C, Delay RJ (2010) Odors activate dual pathways, a TRPC2 and a AA-dependent pathway, in mouse vomeronasal neurons. Am J Physiol Cell Physiol 298:C1253–1264. doi:10.1152/ajpcell.00271.2009

Zhu X, Jiang M, Peyton M, Boulay G, Hurst R, Stefani E, Birnbaumer L (1996) trp, a novel mammalian gene family essential for agonist-activated capacitative Ca^{2+} entry. Cell 85:661–671

Zufall F, Munger SD (2010) Receptor guanylyl cyclases in mammalian olfactory function. Mol Cell Biochem 334:191–197. doi:10.1007/s11010-009-0325-9

Zufall F, Ukhanov K, Lucas P, Liman ER, Leinders-Zufall T (2005) Neurobiology of TRPC2: from gene to behavior. Pflugers Arch 451:61–71. doi:10.1007/s00424-005-1432-4

Zufall F, Leinders-Zufall T, Puche A (2008) Accessory olfactory system. In: Squire LR (ed) The senses: a comprehensive reference. Academic, Oxford, pp 783–814

Part VI

General Topics

Evolutionarily Conserved, Multitasking TRP Channels: Lessons from Worms and Flies

Kartik Venkatachalam, Junjie Luo, and Craig Montell

Contents

Abstract

The Transient Receptor Potential (TRP) channel family is comprised of a large group of cation-permeable channels, which display an extraordinary diversity of roles in sensory signaling. TRPs allow animals to detect chemicals, mechanical force, light, and changes in temperature. Consequently, these channels control a

K. Venkatachalam (✉)
Department of Integrative Biology and Pharmacology, University of Texas School of Medicine, Houston, TX 77030, USA
e-mail: kartik.venkatachalam@uth.tmc.edu

J. Luo
Department of Biological Chemistry, The Johns Hopkins University School of Medicine, Baltimore, MD 21205, USA

Department of Molecular, Cellular, and Developmental Biology, Neuroscience Research Institute, University of California at Santa Barbara, Santa Barbara, CA 93106, USA

C. Montell (✉)
Department of Molecular, Cellular, and Developmental Biology, Neuroscience Research Institute, University of California at Santa Barbara, Santa Barbara, CA 93106, USA
e-mail: craig.montell@lifesci.ucsb.edu

B. Nilius and V. Flockerzi (eds.), *Mammalian Transient Receptor Potential (TRP) Cation Channels*, Handbook of Experimental Pharmacology 223, DOI 10.1007/978-3-319-05161-1_9,

plethora of animal behaviors. Moreover, their functions are not limited to the classical senses, as they are cellular sensors, which are critical for ionic homeostasis and metabolism. Two genetically tractable invertebrate model organisms, *Caenorhabditis elegans* and *Drosophila melanogaster*, have led the way in revealing a wide array of sensory roles and behaviors that depend on TRP channels. Two overriding themes have emerged from these studies. First, TRPs are multitasking proteins, and second, many functions and modes of activation of these channels are evolutionarily conserved, including some that were formerly thought to be unique to invertebrates, such as phototransduction. Thus, worms and flies offer the potential to decipher roles for mammalian TRPs, which would otherwise not be suspected.

Keywords

Drosophila TRP channels • Sensory transduction • Invertebrate TRP channels • Vision • Chemosensation • Thermosensation • Nociception • Mechanosensation • Courtship • Disease models

1 Introduction

TRP channels include six transmembrane segments and are subdivided into seven subfamilies on the basis of amino acid homology (Fig. 1) (Venkatachalam and Montell 2007). The channels that are the most homologous to the original TRP protein (Montell and Rubin 1989), which functions in Drosophila phototransduction, belong to the TRPC subfamily. The TRPCs as well as four other subfamilies (TRPV, TRPM, TRPA, and TRPN) constitute the "group 1 TRPs" (Fig. 1) (Montell 2005). Two other subfamilies (TRPP and TRPML) that are relatively similar to each other in sequence, but are more distantly related group 1 TRPs, are referred to as "group 2 TRPs."

Due to the broad roles of TRPs in sensory signaling, they are critically important in allowing animals to sense a changing environment. As such, it is not surprising that these channels are ancient, evolutionarily conserved proteins that function in a wide range of metazoan organisms. These include invertebrates, such as worms (Kahn-Kirby and Bargmann 2006; Xiao and Xu 2009), arachnids (e.g., ticks), insects (e.g., flies, mosquitoes and bees) (Matsuura et al. 2009; Wolstenholme et al. 2011), and vertebrates such as zebrafish, mice, and humans (Venkatachalam and Montell 2007; Wu et al. 2010). However, the range of TRP channels in some ancient organisms, such as protozoa, is limited to group 2 TRPs (Dong et al. 2010; Wolstenholme et al. 2011).

In invertebrates, the functions of TRP channels have been studied most extensively in two genetically tractable model organisms—the fruit fly, *Drosophila melanogaster* (Fowler and Montell 2013), and the roundworm, *C. elegans* (Xiao and Xu 2011). All seven subfamilies are represented in flies and worms, although

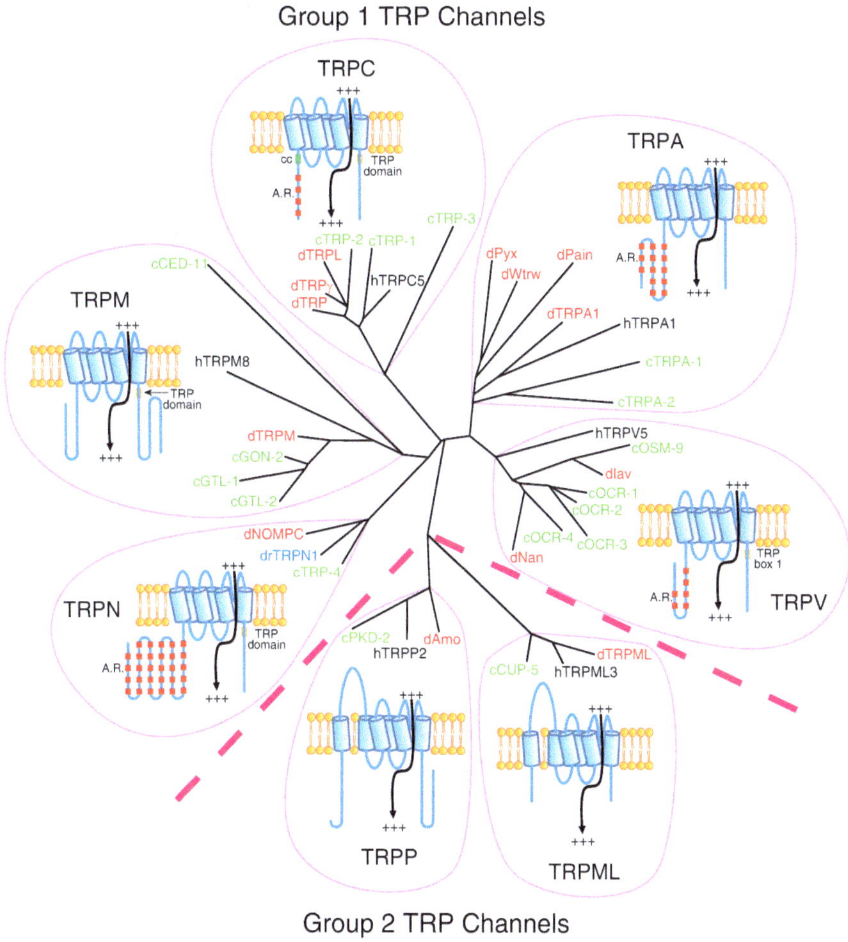

Fig. 1 Phylogenetic tree and cartoons of TRP channels from *C. elegans* (c; *green*), *Drosophila melanogaster* (d; *red*), and *Homo sapiens* (h; *black*). The TRPN channel from *Danio rerio* (dr; zebrafish; *blue*) is also included. To generate the tree, we used the predicted transmembrane segments of each TRP channel (http://www.cbs.dtu.dk/services/TMHMM/) in combination with the following online program: http://www.genome.jp/tools/clustalw/. Indicated on the TRP cartoons are ankyrin repeats (A.R.), coiled coil domains (cc), and the TRP domains, which include TRP boxes 1 and 2 (Montell 2001, 2005). TRPV proteins include only TRP box 1, rather than the full TRP domain. The N- and C-termini are the *left* and *right ends* of the TRP structures. The TRP termini are situated on the cytoplasmic side of the lipid bilayer

these organisms have fewer TRP channels than humans. Nevertheless, the TRPA subfamily is the largest in insects, even though only one representative exists in mice and humans. This notable difference might reflect a particularly important role for TRPA channels in sensing environmental chemicals and changes in temperature, since poikilothermic animals such as insects are particularly sensitive to heat and cold and are subjected to a very complex repertoire of compounds in their surroundings.

Characterization of TRPs in worms and flies underscore the theme that individual TRP channels do not respond to one type of sensory stimuli. Rather, a single TRP channel is capable of sensing a surprisingly broad range of sensory input. In this regard, Drosophila TRPA1 is a particularly notable polymodal sensor, as it functions in the avoidance of noxious volatile and nonvolatile chemicals, intense light, excessively warm temperatures, and small temperature differences in the comfortable range.

2 Sensory Transduction

The peripheral nervous system in Drosophila is composed of four general types of sensory elements. These include (1) external sense organs, such as chemosensory and mechanosensory bristles (sensilla); (2) chordotonal organs, which serve in part as stretch receptors; (3) multidendritic neurons; and (4) photoreceptor cells. *C. elegans* senses the external world through sensillar organs and a variety of isolated sensory neurons.

2.1 Light Sensation

2.1.1 Role of TRPs in Image Formation in Drosophila

In 1969, Cosens and Manning described a Drosophila mutant characterized by a loss of a sustained light response (Cosens and Manning 1969). Using a simple field recording, the electroretinogram (ERG), the flies displayed only a transient response to light. The basis for the *transient receptor potential* (*trp*) phenotype was enigmatic, despite a series of studies over the next two decades (Montell 2011). The first indication as to the function of TRP emerged from the cloning of the *trp* gene, and the observation that the TRP protein had a predicted transmembrane topology similar to the limited number of ion channels and transporters known at the time (Montell and Rubin 1989). A subsequent report demonstrated that loss of *trp* resulted in the rapid decrease the light-activated Ca^{2+} conductance (Hardie and Minke 1992). Together, these findings supported the model that *trp* encodes a Ca^{2+} permeable channel.

In vitro biophysical analyses of TRP, and a related channel identified in 1992 (TRP-Like; TRPL) (Harteneck et al. 1995; Hu et al. 1994; Phillips et al. 1992; Vaca et al. 1994; Xu et al. 1997), indicated that both TRP and TRPL were cation channels. Direct in vivo evidence demonstrating that TRP is a Ca^{2+} permeable channel was obtained by manipulation of the selectivity filter, resulting in a dramatic decrease in the light-induced Ca^{2+} conductance (Liu et al. 2007a).

TRP and TRPL are activated via a signaling pathway that couples light stimulation of rhodopsin with a heterotrimeric G protein (Gq) that engages a phospholipase

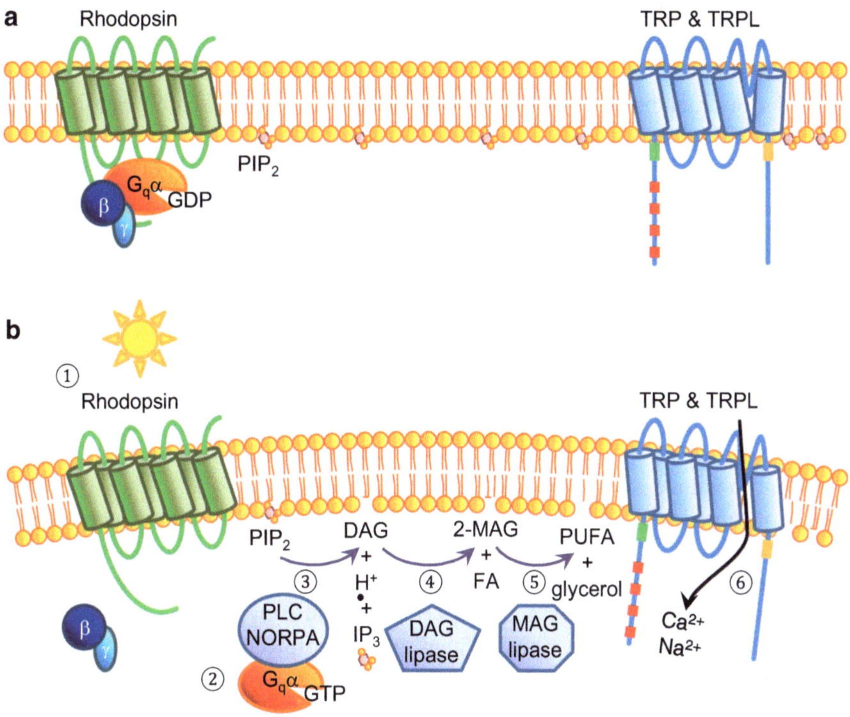

Fig. 2 Model for activation of Drosophila TRP and TRPL. (**a**) The fly phototransduction cascade is not active, and the TRP and TRPL channels are closed in photoreceptor cells maintained in the *dark*. (**b**) The light-activated phototransduction cascade. (1) Light activation of rhodopsin. (2) GTP is exchanged for GDP on the Gα subunit ($G_q\alpha$; also known as Gα49B) and the βγ subunit dissociates from the α subunit. (3) Activation of PLC (phospholipase Cβ) catalyzes the hydrolysis of PIP_2 (phosphoinositide-4,5-bisphosphate (PIP_2) into DAG (diacylglycerol), IP_3 (inositol-1,4,5-trisphosphate), and H^+ (proton). (4) The major products of the DAG lipase (encoded by the *inaE* locus) are 2-MAG (2-monoacylglycerol) and a saturated FA (fatty acid). Not shown are the minor products: 1-MAG and a polyunsaturated FA (PUFA). (5) Activity of an unknown MAG lipase catalyzes the production of a PUFA and glycerol. (6) TRP and TRPL are activated, leading to influx of Ca^{2+} and Na^+. The cleavage of PIP_2 (step 3) is proposed to contribute to activation of TRP and TRPL through a change in curvature of the plasma membrane, thereby creating mechanical force

Cβ (PLCβ) encoded by the *norpA* locus (Bloomquist et al. 1988; Inoue et al. 1985) (Fig. 2). PLCβ catalyzes the hydrolysis of phosphoinositide-4,5-bisphosphate (PIP_2) into inositol-1,4,5-trisphosphate (IP_3), diacylglycerol (DAG), and a proton (H^+) (Huang et al. 2010). The enzymatic activity of NORPA is required for gating of TRP and TRPL since replacement of a single residue that is critical for phospholipase C activity eliminates the light response (Wang et al. 2008).

The activation mechanism of TRP and TRPL has been scrutinized extensively and is still not fully resolved. Nevertheless, it is clear that neither IP_3 nor the IP_3-

receptor is important for channel activation (Acharya et al. 1997; Raghu et al. 2000). According to one model, TRP and TRPL are activated by polyunsaturated fatty acids (PUFAs) such as arachidonic acid and linoleic acid, which are derived from metabolism of DAG (Chyb et al. 1999). Another proposal is that a decline in PIP_2, which is inhibitory, and a decrease in pH, due to the production of H^+, are the two signals necessary for channel activation (Huang et al. 2010).

The question then arises as to why PIP_2 depletion leads to channel activation. A provocative concept is that light stimulation activates TRP and TRPL through mechanical gating (Hardie and Franze 2012) (Fig. 2b). PIP_2 hydrolysis appears to cause minute mechanical contractions of the photoreceptor cell membrane due to removal of the bulky head group of the PIP_2, and these membrane contractions may activate the TRP/TRPL channels (Hardie and Franze 2012). In further support of the concept that TRP and TRPL are mechanically gated, incorporation of a mechanically activated monovalent cation channel (gramicidin) into the membranes of photoreceptor cells dissociated from the *trpl;trp* double mutant is sufficient to induce a light conductance (Hardie and Franze 2012).

The mechanisms of phototransduction in the Drosophila photoreceptors are in stark contrast to those occurring in vertebrate rods and cones. Whereas Drosophila phototransduction involves depolarization of the photoreceptors, light hyperpolarizes vertebrate rods and cones (Fu and Yau 2007). On the other hand, the intrinsically photosensitive retinal ganglion cells (ipRGCs) in the mammalian retina, which participate in circadian entrainment and the pupillary light response (Berson et al. 2002; Provencio et al. 2000; Schmidt et al. 2011), use a phototransduction cascade that employs TRPC6 and TRPC7 and bears a striking resemblance to the Drosophila phototransduction cascade (Xue et al. 2011).

2.1.2 Non-image Light Sensing in Worms and Flies

Drosophila larvae also sense light and do so through two types of phototransduction cascades, both of which depend on TRP channels. However, in contrast to the adult visual system, which functions in image detection, larval phototransduction participates in phototaxis only. The two discrete larval phototransduction cascades take place in separate body parts and are activated by different light intensities. Low to moderate light is received by the Bolwig's organ, which appears to use a cascade that is initiated by rhodopsin and culminates with activation of the TRP and TRPL channels, since these signaling proteins are expressed in this tissue (Petersen and Stowers 2011; Sprecher and Desplan 2008).

Intense bright light is detected by class IV multidendritic sensory neurons that tile the Drosophila larval body wall (Xiang et al. 2010). This light avoidance requires TRPA1, but remarkably, it appears to be independent of rhodopsin, PLC, and the flavin-based light sensor, cryptochrome (Xiang et al. 2010). Rather, the photoresponse of the class IV neurons requires a member of the gustatory receptor family (GR28b) (Xiang et al. 2010), which is the Drosophila homolog of LITE-1—a protein that functions in light detection in *C. elegans* (Liu et al. 2010; Ward et al. 2008). Whether *Gr28b* homozygous mutant flies have a defect in light avoidance behavior has not been reported. Other open questions are whether

GR28b binds a chromophore, the spectral sensitivity of this light receptor, and how activity of GR28b is coupled to TRPA1. One possibility is that GR28b engages a heterotrimeric G protein. However, at least some members of the insect GR family have a topology opposite to classical G protein-coupled receptors (Zhang et al. 2011).

2.2 Chemical Senses

Volatile chemicals are detected through the sense of smell, while nonvolatile chemicals are sensed through contact chemosensation, which includes the sense of taste.

2.2.1 Odor and CO_2 Detection (Non-contact Chemosensation)

The first TRP channel shown to have a role in invertebrate chemosensation is OSM-9. This TRPV protein couples to an odorant receptor, ODR-10, and plays a regulatory role in *C. elegans* olfaction (Colbert et al. 1997). The *osm-9* mutants are defective in chemotaxis towards a subset of olfactory cues that are mediated by the AWA neurons. These mutants also exhibit reduced avoidance to benzaldehyde, which depends on ASH neurons, and display diminished olfactory adaptation to some odorants detected by AWC neurons (Colbert et al. 1997).

Drosophila use two primary organs to sense volatile chemical, the third antennal segment and the maxillary palp (Vosshall and Stocker 2007). The olfactory receptor neurons (ORNs) that extend from these organs terminate in the antennal lobe, which send projection neurons into higher brain centers (Vosshall and Stocker 2007). The two main classes of channels that are critical for Drosophila olfaction are seven-transmembrane ionotropic receptors, referred to "olfactory receptors" (ORs) (Clyne et al. 1999; Gao and Chess 1999; Robertson et al. 2003; Vosshall et al. 1999), and ionotropic receptors (IRs) (Benton et al. 2009), which are distantly related to glutamate receptors. Several Drosophila TRP channels also participate in olfaction, although they are not the primary olfactory detectors. For instance, the classical TRP plays a role in olfactory adaptation (Störtkuhl et al. 1999). In addition, TRPA1 is expressed in ORNs and is required for avoiding the naturally occurring insect repellent, citronellal (Kwon et al. 2010a). Fly TRPA1 is not effectively activated by citronellal directly. Rather, it is activated through a Gq and PLC (NORPA)-dependent signaling cascade. However, the TRPA1 expressed by the mosquito, *Anopheles gambiae*, which is an insect vector that spreads malaria, is directly and potently activated by citronellal (Kwon et al. 2010a). Both TRP and TRPL are co-expressed in CO_2-sensitive olfactory receptor neurons (ORNs) in the antennae and contribute to CO_2 avoidance, possibly through a Gq/PLC signaling cascade (Badsha et al. 2012). Another TRPA channel, Painless (Tracey et al. 2003), functions in projection neurons emanating from the olfactory glomeruli, where it participates in an olfactory circuit that inhibits male-male courtship (Wang et al. 2011).

2.2.2 Taste (Contact Chemosensation)

The sense of taste in flies is mediated through gustatory receptor neurons (GRNs) housed in hairlike projections referred to as sensilla. Gustatory sensilla are distributed on multiple body parts, including the main gustatory organ (labellum), the legs, wing margins, and ovipositor. Taste in flies depends predominately on a large family of gustatory receptors (GRs), which are related to ORs, and at least one IR (Montell 2009; Zhang et al. 2013b). This is in contrast to mammalian taste, which is mediated largely through a TRP channel (Pérez et al. 2002; Zhang et al. 2003). Nevertheless, at least three TRP channels participate in Drosophila taste sensation (TRPA1, Painless, and TRPL) as described below.

TRPA1 is expressed in the main taste organ, the labellum, and is critical for sensing the naturally occurring plant compound, aristolochic acid (Kim et al. 2010). Activation of TRPA1 through this chemical depends on the same Gq and PLC that is required for phototransduction in the compound eye and for activating TRPA1 in ORNs. TRPA1 is also expressed in internal mouthparts where it is employed to detect a subset of aversive chemicals prior to ingestion (Kang et al. 2010a). As with mammalian TRPA1, Drosophila TRPA1 is a receptor for electrophilic chemicals, such as allyl isothiocyanate (AITC) (Bandell et al. 2004; Hinman et al. 2006; Jordt et al. 2004; Kang et al. 2010a; Macpherson et al. 2007). Avoidance of AITC—the pungent ingredient of wasabi—also appears to depend on Painless (Al-Anzi et al. 2009). However, whether Painless is required in GRNs or postsynaptic to these afferent neurons is unclear. In *C. elegans*, the TRPV channel, OSM-9, mediates avoidance behavior towards the bitter chemical quinine (Hilliard et al. 2004).

The TRPL channel functions in diet-induced changes in taste preference through a mechanism that employs reversible changes in GRNs (Zhang et al. 2013c). Animals including flies tend to avoid some unpalatable compounds, if more appealing options are available in the diet. Indeed, flies will reject the bitter tasting compound, camphor. However, after long-term exposure to camphor, the animals reduce their distaste to this food ingredient. TRPL is expressed in GRNs that mediate an avoidance response and is directly activated by camphor. Persistent exposure to TRPL causes ubiquitination and downregulation of TRPL, thereby diminishing camphor rejection (Zhang et al. 2013c). If camphor is removed from the diet for a long time, the concentration of TRPL gradually increases, thereby restoring the original aversion to this food additive. Thus, the reversible changes in TRPL levels provide a mechanism whereby an animal can change their food preferences as a result of long-term alterations in the diet.

2.3 Thermal Sensation

Animals avoid noxious heat and cold for survival and seek out the ideal temperature zone to maximize comfort. Responding to small differences in temperature is especially important for poikilothermic organisms, since their body temperature equilibrates with the environment. Consequently, in invertebrate animals, changes in temperature affect the rate of growth and development, and aging. In addition, since it is warmer during the day than at night, daily temperature fluctuations can set circadian rhythms. Seminal work on mammalian somatosensation established that channels such as TRPV1 are thermal sensors and can be directly activated by changes in temperature (Caterina et al. 1999; Dhaka et al. 2006). We now know that roles for TRPs as thermosensors are evolutionarily conserved, and invertebrate thermoTRPs sense changes in temperature for multiple purposes, beyond the avoidance of noxious heat and cold.

2.3.1 Nociception

2.3.1.1 Responding to Warm and Hot Temperatures

Three channels, all within the TRPA subfamily, contribute to nociceptive responses of Drosophila larvae to excessive heat. Normally, wild-type larvae move along a surface parallel to their body axis. However, if they are exposed to a heat probe with a threshold around 40 °C, they quickly roll in a corkscrew fashion around their body axis. This writhing response depends on Painless expression in multidendritic neurons in the body wall (Tracey et al. 2003). There are three Painless isoforms, one of which includes a long N-terminus with eight ankyrin repeats (Painless^{P103}) and another with a very short N-terminus (Painless^{P60}) (Fig. 3a). Expression of PainlessP^{103} rescues the *pain* mutant phenotype, but not Painless^{P60} (Hwang et al. 2012). Painless^{P103} appears to be a direct thermosensor, as it is activated with a threshold (≥42 °C) (Sokabe et al. 2008) similar to that needed to initiate the rolling response (Tracey et al. 2003).

Another TRPA channel, Pyrexia, is activated directly by warm temperatures (≥40 °C) and enable larvae to manage the thermal stress of relatively high temperatures (Lee et al. 2005). In the absence of Pyrexia, the larvae are paralyzed by exposure to 40 °C for several minutes (Lee et al. 2005).

TRPA1 participates in the response to high temperatures in both larvae and adult flies (Neely et al. 2011). The channel is activated with a threshold of ~25–29 °C (Viswanath et al. 2003) and is used as a sensor in larvae and adults to avoid temperature above this threshold, which are uncomfortable but not acutely dangerous (Hamada et al. 2008; Kwon et al. 2008; Rosenzweig et al. 2005). In adults, TRPA1-mediated temperature sensation in this range (≥25 °C) is detected by sensory neurons in the brain (anterior cell neurons) (Hamada et al. 2008). There are four TRPA1 isoforms (Kwon et al. 2010a; Zhong et al. 2012), two of which include a 37 amino acid domain between the N-terminal 13 ankyrin repeat and the transmembrane segments (Fig. 3b). These two isoforms, TRPA1-A and TRPA1-D, are directly activated by increases in temperature, and the 37 amino acid "TRP ankyrin cap" (TAC) domain appears to be essential for thermal activation (Kang et al. 2012; Zhong et al. 2012) (note that the TRPA1-A and TRPA1-D isoforms are referred to as TRPA1-B and TRPA1-A, respectively, in one study)(Kang et al. 2012). The TRPA1-A isoform is remarkably sensitive to temperature activation, as it has a Q_{10} of ~130 (Kang et al. 2012). TRPA1-C does not appear to be a direct thermosensor. Yet, expression of this isoform also rescues the *trpA1*

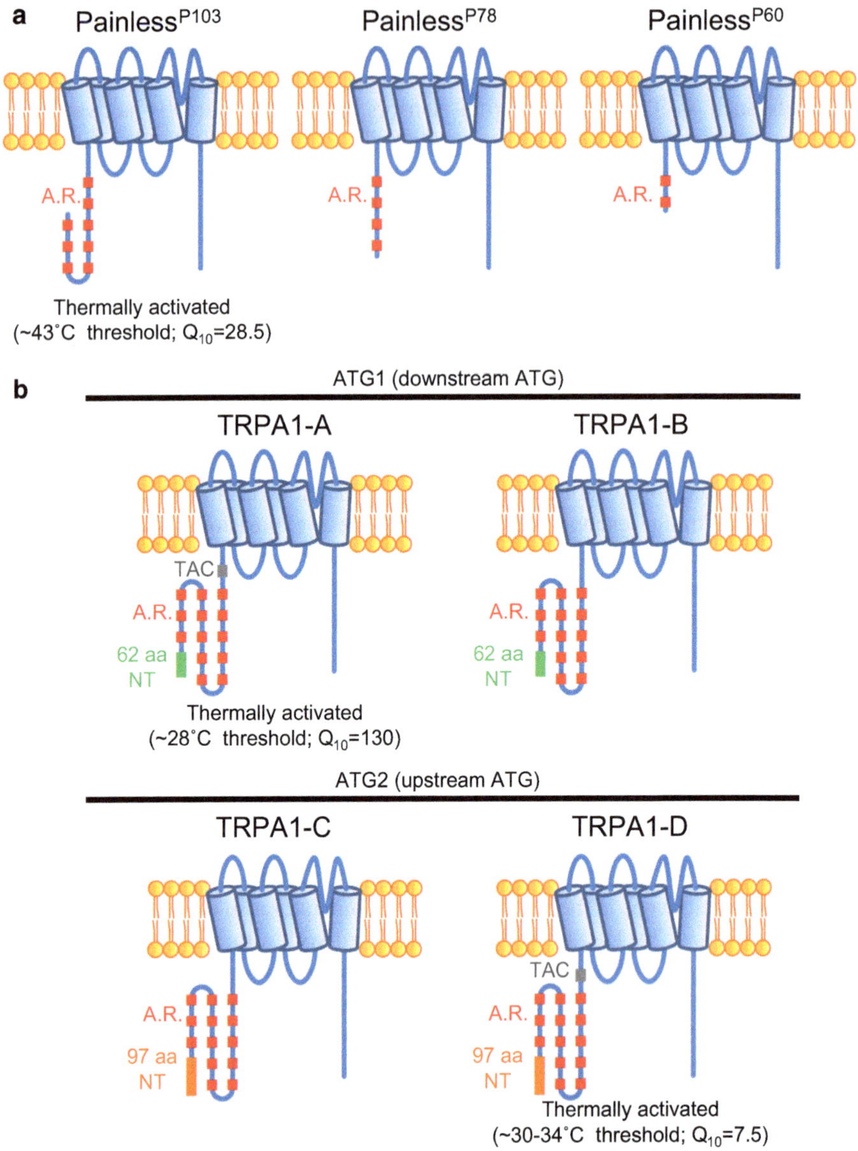

Fig. 3 Protein isoforms of Painless and Drosophila TRPA1. (**a**) Painless. (**b**) TRPA1. A.R. ankyrin repeats, NT N-terminal domain, TAC TRP ankyrin cap domain

nociceptive phenotype (Zhong et al. 2012). These findings indicate that there are both direct and indirect mechanisms for thermal activation of TRPA1 (see below).

The mosquito vector for malaria, *Anopheles gambiae*, expresses a TRPA1 homolog in small coeloconic sensilla in the antenna, which is heat activated (≥25 °C) (Wang et al. 2009). These sensilla house thermosensitive neurons that are activated as the temperature rises above ~25–40 °C. Thus, *A. gambiae* appear to use peripheral neurons in the antenna to detect warm temperatures (Wang et al. 2009).

C. elegans employ two parallel pathways for noxious heat avoidance (≥25 °C). The first uses the neuropeptide receptor, NPR-1, and the second depends on the two TRPV proteins, OSM-9 and OCR-2 (Glauser et al. 2011). Whether OSM-9 and OCR-2 are heat activated is not known.

2.3.1.2 Responding to Excessively Cool Temperatures

Temperatures slightly below 18 °C are avoided by Drosophila larvae and adults. When given a choice between 18 °C and lower temperatures, the animals will migrate towards 18 °C. In larvae, this behavior depends in part on the TRPL channel (Rosenzweig et al. 2008) in chordotonal neurons (Kwon et al. 2010b), but not the TRP channel (Kwon et al. 2010b). The fly genome encodes two TRPV channels, Inactive and Nanchung, and at least one of these channels, Inactive, contributes to the avoidance of cool temperatures (Kwon et al. 2010b). Whether Nanchung also functions in this behavior is unclear due to the severe defect in locomotion, which is also associated with this mutation.

In adult flies, cool sensation is mediated via thermosensory neurons in the antenna (arista and the sacculus) and requires three proteins, referred to as Brivido1–3 (Gallio et al. 2011). These proteins contain either eight (Brv1) or ten predicted transmembrane domains (Brv2 and Brv3) (Gallio et al. 2011) and are most akin to mammalian polycystin1 proteins, which are part of the TRPP channel complex and thought to regulate the activity of the channels. The last six transmembrane segments are related to TRPP channels. Since it is not known whether the individual Brv proteins are cation channels or whether all three proteins are subunits of a cool-sensing channel, these proteins are not currently listed as TRPP proteins.

2.3.2 Temperature Control in the Comfortable Range

Animals including Drosophila are capable of discriminating small temperature differences in the comfortable range (≤1 °C). Although temperatures from 18 to 24 °C are comfortable for Drosophila larvae, 18 °C is preferred, and they will select 18 °C over 19–24 °C. TRPA1 is required for fine temperature discrimination in this range (Kwon et al. 2008). However, activation of TRPA1 is indirect since the comfortable temperatures (18–24 °C) are below the threshold for direct thermal activation of the channel (24–29 °C) (Viswanath et al. 2003). Rather, TRPA1 is activated downstream of phototransduction-like signaling cascade that depends on a rhodopsin (Rh1) (Shen et al. 2011) and the same heterotrimeric G protein and PLC that function in phototransduction (Kwon et al. 2008). Although thermal discrimination in the comfortable range depends on a rhodopsin, the concentration of rhodopsin in the thermosensory neurons is extremely low, precluding efficient photon capture. Consequently, light exposure does not impact on selection of 18° over 19–24 °C. This thermosensory signaling cascade may facilitate amplification of small temperature differences in the comfortable range and allow for thermal adaptation to 19–24 °C, if the larvae are unable to identify an area that is 18 °C within their thermal landscape.

2.3.3 Temperature Control of Circadian Rhythm

Circadian rhythms in most animals are entrained principally by day/night cycles (Allada and Chung 2010; Hardin 2011; Kwon et al. 2011; Peschel and Helfrich-Forster 2011; Reppert and Weaver 2002). In some poikilothermic organisms, such as Drosophila, circadian rhythms can also be set by higher and lower relative temperatures that simulate day/night periods (Glaser and Stanewsky 2007). TRPA1 is one molecular sensor that contributes to the temperature control of circadian rhythms in flies (Lee and Montell 2013). TRPA1 is expressed and functions in a subset of pacemaker neurons in the adult brain, and loss of this channel alters, but does not eliminate entrainment.

Recently, Pyrexia has also been reported to contribute to temperature synchronization of circadian rhythm through expression in peripheral sensory neurons in chordotonal organs (Wolfgang et al. 2013). Pyrexia affected temperature synchronization to relatively low temperatures (16–20 °C) but not at higher temperatures (21–29 °C). Since Pyrexia is activated by temperatures with a threshold near 40 °C (Lee et al. 2005), it may function in temperature control of clock synchronization via a thermosensory signaling cascade.

2.3.4 Temperature Control of Aging

Lower body temperature can increase animal lifespan (Conti 2008), and it was formerly assumed that this was due to a general reduction in metabolism. However, this is not the case, at least in *C. elegans*. Worm TRPA-1 is a cold-activated channel, as is mammalian TRPA1, and functions in the intestine to sense a decrease in temperature (Xiao et al. 2013). Cold activation of TRPA-1 leads to stimulation of a Ca^{2+}-sensitive protein kinase C (PKC-2) and nuclear entry of the transcription factor DAF-16/FOXO, which promotes an increase in longevity. Given that expression of TRPA-1 in the intestine is sufficient for cold-induced longevity, this finding demonstrates that a non-excitable tissue functions in thermosensation and lifespan extension (Xiao et al. 2013).

2.4 Mechanosensation

Mechanosensation contributes to gentle and noxious touch, proprioception, sensing gravity, and hearing, and invertebrate TRP channels are important for each of these functions.

2.4.1 Touch, Proprioception, and Gravity Sensation

In *C. elegans* several TRP channels contribute to the response induced by nose touch. These include TRPA-1, which is activated in vitro by applying negative pressure to cells (Kindt et al. 2007), and both TRPV proteins, OSM-9 and OCR-2 (Colbert et al. 1997; Tobin et al. 2002). Since OSM-9 and OCR-2 are required for sensing volatile and nonvolatile compounds, mechanosensation, as well as osmosensation, they are polymodal sensors (Colbert et al. 1997; Tobin et al. 2002). The diversity of roles for these two proteins may reflect their expression in several types of sensory neurons. The two proteins may be subunits of a single channel since they form a complex and mutually affect the spatial distribution of the other protein (Tobin et al. 2002). However, it is not known if TRPA-1, OSM-9, or OCR-2 is responding directly or indirectly to mechanical stimulation.

Drosophila larvae depend on the TRPA channel, Painless, for responding to noxious touch, and this function is mediated through the same nociceptors that sense noxious heat (Tracey et al. 2003). A role for Painless in mechanosensation extends to the Drosophila heart (Senatore et al. 2010). Painless is expressed in the heart where it senses mechanical stress, which in turn regulates heart activity (Senatore et al. 2010). Painless, as well as a second TRPA channel, Pyrexia, and two TRPV channels, Inactive and Nanchung, are also expressed in the Johnston's organ of the second antennal segment, where they participate in gravity sensing (Sun et al. 2009).

An important advance in dissecting the roles of TRP channels in mechanosensation is the demonstration that the *C. elegans* TRPN protein, TRP-4, is a direct mechanosensor (Kang et al. 2010b). Unlike the other six TRP subfamilies, TRPN proteins are not expressed in mammals. However, TRPN proteins are conserved from worms to zebrafish (Sidi et al. 2003; Walker et al. 2000). One of the intriguing features of TRPN channels is the large number of N-terminal ankyrin repeats (29) in flies and worms (Walker et al. 2000), which are proposed to comprise a gating spring (Howard and Bechstedt 2004). The Drosophila TRPN protein, NOMPC, is necessary for touch transduction in adult flies (Walker et al. 2000) and in two of the four classes of multidendritic neurons (II and III) for light touch transduction in larvae (Tsubouchi et al. 2012; Yan et al. 2013). Fly TRPN is also required for larval locomotion and normal proprioception (Cheng et al. 2010). The *C. elegans* TRP-4 protein plays a role in proprioception (Li et al. 2006) and harsh touch (Kang et al. 2010b).

The quintessential hallmark of a mechanotransducer is very rapid activation kinetics (Christensen and Corey 2007; Sharif-Naeini et al. 2008), and the TRP-4-dependent conductance is activated in vivo in less than 1 ms (Kang et al. 2010b). Thus, worm TRP-4 is the first TRP shown to be a direct mechanosensor in vivo. Fly NOMPC also appears to be a mechanosensor, since ectopic expression of this protein in non-mechanosensitive neurons confers touch sensitivity, and this channel is activated rapidly in vitro by applying negative pressure (latency <2 mS) (Yan et al. 2013). Currently, it is not known if any mammalian TRP channel is a direct mechanosensor.

2.4.2 Hearing

In flies hearing plays critical roles in courtship and mating, in addition to the avoidance of predators (Kamikouchi 2013). The mechanotransduction that is essential for audition occurs in the Johnston's organ, which contains chordotonal sensilla (Kernan 2007). Hearing requires both the detection and amplification of sound-evoked signaling, and at least three TRP channels (NOMPC, Iav and Nan) contribute to hearing in adult (Eberl et al. 2000; Effertz et al. 2011, 2012; Gong et al. 2004; Göpfert et al. 2006; Kamikouchi et al. 2009; Kim et al. 2003; Lehnert et al. 2013) and in larval chordotonal organs (Zhang et al. 2013a). Primary sound sensation and sound amplification appear to be distinct processes. Nan and Iav mediate the initial detection of mechanical vibrations, while NOMPC plays a role in sound amplification (Lehnert et al. 2013).

2.4.3 Sensing Moist and Dry Environments

Hygrosensation refers to the ability of an organism to sense moisture or humidity in the environment. In flies, hygrosensation requires a region of the antenna known as the arista (Sayeed and Benzer 1996). The mechanisms for detecting moist and dry air are distinct because the TRPA channel, Waterwitch (Wtrw), is involved in sensing the transition to moist air, whereas Nan mediates the response to dry air (Liu et al. 2007b). However, neurons expressing both channels project to the region of the fly CNS associated with mechanotransduction (Liu et al. 2007b). Thus, hygrosensation may involve humidity-dependent alterations in mechanical stretch in the hygrosensitive neurons.

Drosophila larvae avoid dry surfaces through nocifensive behavior that depends on class IV multidendritic neurons in the body wall, and the TRPA channel, Painless (Johnson and Carder 2012). The larval cuticle is rich in charged glycoproteins (Silvert et al. 1984) that are likely to have an incredibly high affinity towards dry surfaces, thereby causing the cuticle to adhere to such surfaces. Thus, larval propensity to avoid dry surfaces may arise from noxious mechanical stretch as the larval cuticle adheres to a dry substrate during peristalsis.

2.5 Complex Behaviors

2.5.1 Feeding Behavior and Social Influences

Feeding behavior is a product of specialized interactions between chemosensory and somatosensory modalities and involves sensing of the internal metabolic state followed by state-dependent regulation of food seeking and feeding. In vertebrates, neuropeptide Y (NPY) signaling in the arcuate nucleus of the hypothalamus is intimately linked to feeding and energy expenditure (Gruninger et al. 2007; Stanley et al. 1985; Tomaszuk et al. 1996). The NPY receptors in both *C. elegans* and Drosophila larvae play critical roles in determining the selection of social or solitary feeding (de Bono and Bargmann 1998; Wu et al. 2003). Some worm strains exhibit social feeding, which involves avoidance of noxious chemical stimuli in a process requiring the TRP channels, OCR-2 and OSM-9 (de Bono et al. 2002). Additionally, OCR-2 impacts on the biosynthesis of serotonin (Sokolchik et al. 2005; Zhang et al. 2004)—a neurotransmitter with conserved roles in regulating feeding behavior (Magalhaes et al. 2010; Neckameyer 2010; Sze et al. 2000). OCR-2 also functions in a small subset of peripheral sensory neurons in *C. elegans* larvae that mediate the detection of nutrient availability and neuropeptide release (Lee and Ashrafi 2008).

Social feeding in worms involves aggregation, and this causes a drop in O_2 levels. Worms have developed a mechanism for detecting changes in O_2 levels as a part of their strategy to promote aggregation, and both OSM-9 and OCR-2 are required for sensing O_2 (Rogers et al. 2006). Mammals also use O_2-sensing TRP channels, as mouse TRPA1 can be directly activated by O_2 under hyperoxic conditions, through a mechanism involving cysteine modification (Takahashi et al. 2011). It is proposed that TRPA1 may contribute to detecting O_2 toxicity in the mouse (Takahashi et al. 2011).

In Drosophila larvae, an NPY family member (known as NPF in flies) mediates the developmental switch from food-seeking behavior in younger larvae to food-avoiding behavior in late third instar larvae (Xu et al. 2008). In response to food, late-stage larvae also display social behavior as they aggregate in order to dig cooperatively through hard food (Wu et al. 2003). Both the food avoidance and aggregation behavior are impaired in *painless* mutant late-stage larvae (Xu et al. 2008). NPF exerts its effects by inhibiting the Painless conductance in larval sensory neurons (Xu et al. 2008). The normal role for Painless in food aversion is consistent with the general function of invertebrate TRPA channels and mammalian TRPA1 in mediating avoidance behavior in response to noxious stimuli.

2.5.2 Nicotine-Dependent Behavior

Exposure of worms to nicotine, which is the addictive ingredient in tobacco, causes a variety of behaviors that are reminiscent of those induced in mammals (Dwoskin et al. 1999). These include increased activity to acute exposure, sensitization to intermittent administration of nicotine, and withdrawal upon removal of chronic nicotine (Feng et al. 2006). These behaviors are greatly diminished in worms missing either of two TRPC channels, TRP-1 or TRP-2. Expression of these channels is required in command neurons, which is consistent with their role in locomotion (Feng et al. 2006). Roles for TRP channels in nicotine sensitivity may be conserved throughout animal phylogeny since mouse TRPA1 is activated by nicotine and contributes in vivo to the irritant effects of nicotine (Talavera et al. 2009).

2.5.3 Courtship and Mating

Courtship and mating involves a complex interplay between multiple senses, including hearing, touch, chemosensation, and vision. Drosophila Painless contributes to mating, as the mutant females display enhanced sexual receptivity (Sakai et al. 2009). The mutant females copulate with a shorter latency than do wild-type females, once the males initiate courtship behavior. A role for Painless in sexual receptivity requires expression of the channel in GABAergic and/or cholinergic neurons (Sakai et al. 2009). Thus, it is unclear if Painless modulates female receptivity by acting in inhibitory or excitatory neurons. Painless also functions in courtship in males, as expression of the channel in olfactory projection neurons in the brain suppresses male-male courtship (Wang et al. 2011).

C. elegans display mating behavior, which involves males seeking out the vulva of the hermaphrodite, followed by spicule insertion and sperm transfer (Liu and Sternberg 1995). Male *C. elegans* lacking the TRPP2 homolog, *pkd2*, or its interacting protein, PKD1 (TRPP1), are defective in vulva location (Barr et al. 2001; Barr and Sternberg 1999).

2.6 Sperm and Fertilization

A *C. elegans* TRPC homolog, *trp-3*, functions in sperm and is critical for fertility. Upon sperm activation, TRP-3 translocates from an intracellular compartment to the plasma membrane. TRP-3-dependent Ca^{2+} influx is proposed to promote gamete fusion (Xu and Sternberg 2003). Roles for mammalian TRPs in sperm function, including TRPC2, have been suggested (Darszon et al. 2012; Jungnickel et al. 2001). However, because human TRPC2 is a pseudogene (Wes et al. 1995), it does not have a role in human sperm.

Loss of the Drosophila TRPP2 homolog, Amo, affects sperm function (Gao et al. 2003; Watnick et al. 2003). Normally, the beat frequency of sperm increases after they exit the uterus (Köttgen et al. 2011). The sperm then enters the storage organs, which serves to supply sperm for fertilization over an extended time. Amo is localized to the posterior end of mature sperm and is required for beating hyperactivity and for the backward entry of the sperm into the storage organs (Köttgen et al. 2011).

While the roles of the fly and worm TRPPs are distinct, they both function in ciliated cells. Human TRPP1 and TRPP2, which are disrupted in autosomal polycystic kidney disease (ADPKD), are expressed in monocilia (Gallagher et al. 2010). Thus, TRPPs have evolutionarily conserved roles in ciliated cells.

2.7 Metabolism

2.7.1 Ionic Homeostasis

Mg^{2+} homeostasis is essential for animal survival, as low or high intracellular Mg^{2+} impairs a wide range of essential cellular events. Mg^{2+} is absorbed by the intestines and secreted by the kidneys. In humans, mutations in the Mg^{2+} and Ca^{2+} permeable channel, TRPM6, cause familial hypomagnesemia with secondary hypocalcemia (HSH), which leads to seizures and muscle spasms (Schlingmann et al. 2002; Walder et al. 2002).

Roles for TRPM channels for Mg^{2+} homeostasis are conserved in worms and flies. Two *C. elegans* TRPMs (GON-2 and GTL-1) are required for Mg^{2+} uptake by intestinal cells (Teramoto et al. 2005). Worms missing GON-2 and GTL-1 grow poorly under low Mg^{2+} conditions. GON-2 and GTL-1 may form distinct channels, since mutation of just *gon-2* or *gtl-1* result in different phenotypes, and the two TRPMs have different Mg^{2+} sensitivities. GON-2 is inactive when the Mg^{2+} concentration is <1 mM, while GTL-1 is not inhibited by Mg^{2+} to a significant extent and may be constitutively active. Thus, GON-2 only comes into play when the Mg^{2+} concentration is high, thereby maintaining Mg^{2+} homeostasis. A third worm TRPM (GTL-2) is expressed in the excretory cell and is required for Mg^{2+} excretion (Teramoto et al. 2010).

The sole Drosophila TRPM functions in the fly kidney equivalent, the Malpighian tubules, where it serves to remove Mg^{2+} from the hemolymph (Hofmann et al. 2010). The *trpm*-deficient animals are characterized by hypermagnesemia in the hemolymph. Loss of *trpm* also affects intracellular Zn^{2+} homeostasis (Georgiev et al. 2010). Unlike all other fly TRPs, the TRPM channel is absolutely required for viability. The mutant animals die as prepupae or pupae (Georgiev et al. 2010; Hofmann et al. 2010). Supplementation of the food with high Mg^{2+} exacerbated the phenotype, resulting in larval lethality (Hofmann et al. 2010).

2.7.2 Lysosomal Function and a Model for MLIV

Mutations in the human TRPML1 cause a lysosomal storage disorder, mucolipidosis type IV (MLIV), which is characterized by motor impairments, severe cognitive deficits, and blindness (Bargal et al. 2000; Bassi et al. 2000; Sun et al. 2000). The *C. elegans* TRPML homolog, CUP-5, is localized to lysosomes, and elimination of this protein results in defective lysosomal biogenesis and maternal-effect embryonic lethality (Fares and Greenwald 2001; Hersh et al. 2002; Treusch et al. 2004). Due to nutrient deprivation, there appears to be an upregulation of autophagy in *cup-5* mutant cells (Schaheen et al. 2006). However, the autophagy is ineffective since there is a decrease in lysosomal biogenesis (Schaheen et al. 2006).

In Drosophila, mutation of *trpml* results in severe neurodegeneration (Venkatachalam et al. 2008). The neurons die due to a requirement for TRPML for autophagic removal of damaged mitochondria. TRPML is localized to late endosomes and lysosomes and is essential in a variety of cell types for releasing luminal Ca^{2+} and promoting Ca^{2+}-dependent fusion of late endosomes and lysosomes (Venkatachalam et al. 2008, 2013; Wong et al. 2012). The widespread neuronal cell death is a consequence of a role for TRPML in two cell types. First, *trpml* functions in neurons to promote cell survival. Second, *trpml* acts in phagocytic cells such as glia and macrophages to remove early apoptotic neurons (Venkatachalam et al. 2008). In the absence of TRPML activity, the early apoptotic neurons are not removed effectively. This results in accumulation of late apoptotic neurons, release of cytotoxic agents, and magnification of cell death due to a bystander effect (Venkatachalam et al. 2008).

Concluding Remarks

There are two recurring themes that apply to the worm and fly TRP channels. First, these channels are multitaskers. There are many examples that illustrate this concept. The fly TRPL channel contributes to light detection, cool sensation, taste adaptation, and sensitivity to CO_2 (Badsha et al. 2012; Kwon et al. 2010b; Niemeyer et al. 1996; Phillips et al. 1992; Rosenzweig et al. 2008; Zhang et al. 2013c). Painless serves in the detection of noxious heat, noxioius mechanical stimuli, gravity, wasabi, food avoidance stimuli in late-stage larvae, dry surfaces, and in signals that control female receptivity and male-male courtship (Al-Anzi et al. 2006; Johnson and Carder 2012; Sakai et al. 2009; Sun et al. 2009; Tracey et al. 2003; Wang et al. 2011; Xu et al. 2008). The worm TRPV channel OSM-9 functions in olfaction, contact chemosensation, mechanosensation, thermosensation, O_2 sensation, and social feeding (Colbert et al. 1997; de Bono et al. 2002; Rogers et al. 2006). Drosophila TRPA1 is a particularly notable example of a polymodal sensor as it either directly or indirectly senses suboptimal or excessively warm temperature, noxious olfactory and gustatory cues, and light (Kang et al. 2010a, 2012; Kim et al. 2010; Kwon et al. 2008, 2010a; Lee and Montell 2013; Rosenzweig et al. 2005; Viswanath et al. 2003).

Second, the repertoire of functions of worm, fly, and vertebrate TRP channels is far more similar than initially envisioned. Roles for PLC signaling and TRP channels in phototransduction are not peculiar to flies but are conserved in mammalian ipRGCs (Berson et al. 2002; Provencio et al. 2000; Schmidt et al. 2011). Thermally activated TRPs are also employed in vertebrates and invertebrates. However, some TRP channels in flies and mammals exhibit opposite thermal activities. TRPA1 is heat activated in flies (Viswanath et al. 2003), but cold activated in mice (Bandell et al. 2004; Karashima et al. 2009; Kwan et al. 2006; Story et al. 2003). Nevertheless, in some snakes (Gracheva et al. 2010), lizards, and frogs (Saito et al. 2012), TRPA1 is heat activated. The threshold for heat activation for rattlesnake TRPA1 (~28 °C) is similar to fly TRPA1 (~24–29 °C) (Gracheva et al. 2010; Viswanath et al. 2003), rather than mammalian TRPA1. A similar scenario emerges with thermally activated TRPV channels, which are heat activated in mammals (Caterina 2007). A Drosophila TRPV channel, Inactive, functions in the discrimination of 17.5 °C and slightly cooler temperatures (14–16 °C) (Kwon et al. 2010b). TRPV3 in western clawed frogs is cool activated with a threshold around 16 °C (Saito et al. 2011). The mechanisms for detecting some noxious chemicals are also conserved between several vertebrate and invertebrate TRP channels, such as reactive electrophiles (e.g., allyl isothiocyanate), which activate mammalian and Drosophila TRPA1 (Bandell et al. 2004; Hinman et al. 2006; Jordt et al. 2004; Kang et al. 2010a; Macpherson et al. 2007). Evolutionarily conserved functions for TRP channels are not limited to sensory physiology, as worms, flies, and vertebrates use TRPMs in ionic homeostasis: TRPPs in cilia and TRPMLs in lysosomes.

In view of the many functional similarities between invertebrate and vertebrate TRPs, it is intriguing to speculate that some roles that are currently known specifically in worms or flies will turn out to be conserved in mammals. Among the many examples are gravity sensation, hygrosensation, and fertility.

References

Acharya JK, Jalink K, Hardy RW, Hartenstein V, Zuker CS (1997) $InsP_3$ receptor essential for growth and differentiation but not for vision in Drosophila. Neuron 18:881–887

Al-Anzi B, Tracey WD Jr, Benzer S (2006) Response of *Drosophila* to wasabi is mediated by *painless*, the fly homolog of mammalian TRPA1/ANKTM1. Curr Biol 16:1034–1040

Al-Anzi B, Sapin V, Waters C, Zinn K, Wyman RJ, Benzer S (2009) Obesity-blocking neurons in *Drosophila*. Neuron 63:329–341

Allada R, Chung BY (2010) Circadian organization of behavior and physiology in *Drosophila*. Annu Rev Physiol 72:605–624

Badsha F, Kain P, Prabhakar S, Sundaram S, Padinjat R, Rodrigues V, Hasan G (2012) Mutants in Drosophila TRPC channels reduce olfactory sensitivity to carbon dioxide. PLoS One 7:e49848

Bandell M, Story GM, Hwang SW, Viswanath V, Eid SR, Petrus MJ, Earley TJ, Patapoutian A (2004) Noxious cold ion channel TRPA1 is activated by pungent compounds and bradykinin. Neuron 41:849–857

Bargal R, Avidan N, Ben-Asher E, Olender Z, Zeigler M, Frumkin A, Raas-Rothschild A, Glusman G, Lancet D, Bach G (2000) Identification of the gene causing mucolipidosis type IV. Nat Genet 26:118–123

Barr MM, Sternberg PW (1999) A polycystic kidney-disease gene homologue required for male mating behaviour in *C. elegans*. Nature 401:386–389

Barr MM, DeModena J, Braun D, Nguyen CQ, Hall DH, Sternberg PW (2001) The *Caenorhabditis elegans* autosomal dominant polycystic kidney disease gene homologs *lov-1* and *pkd-2* act in the same pathway. Curr Biol 11:1341–1346

Bassi MT, Manzoni M, Monti E, Pizzo MT, Ballabio A, Borsani G (2000) Cloning of the gene encoding a novel integral membrane protein, mucolipidin-and identification of the two major founder mutations causing mucolipidosis type IV. Am J Hum Genet 67:1110–1120

Benton R, Vannice KS, Gomez-Diaz C, Vosshall LB (2009) Variant ionotropic glutamate receptors as chemosensory receptors in *Drosophila*. Cell 136:149–162

Berson DM, Dunn FA, Takao M (2002) Phototransduction by retinal ganglion cells that set the circadian clock. Science 295:1070–1073

Bloomquist BT, Shortridge RD, Schneuwly S, Perdew M, Montell C, Steller H, Rubin G, Pak WL (1988) Isolation of a putative phospholipase C gene of Drosophila, *norpA*, and its role in phototransduction. Cell 54:723–733

Caterina MJ (2007) Transient receptor potential ion channels as participants in thermosensation and thermoregulation. Am J Physiol Regul Integr Comp Physiol 292:R64–R76

Caterina MJ, Rosen TA, Tominaga M, Brake AJ, Julius D (1999) A capsaicin-receptor homologue with a high threshold for noxious heat. Nature 398:436–441

Cheng LE, Song W, Looger LL, Jan LY, Jan YN (2010) The role of the TRP channel NompC in *Drosophila* larval and adult locomotion. Neuron 67:373–380

Christensen AP, Corey DP (2007) TRP channels in mechanosensation: direct or indirect activation? Nat Rev Neurosci 8:510–521

Chyb S, Raghu P, Hardie RC (1999) Polyunsaturated fatty acids activate the *Drosophila* light-sensitive channels TRP and TRPL. Nature 397:255–259

Clyne PJ, Warr CG, Freeman MR, Lessing D, Kim J, Carlson JR (1999) A novel family of divergent seven-transmembrane proteins: candidate odorant receptors in *Drosophila*. Neuron 22:327–338

Colbert HA, Smith TL, Bargmann CI (1997) OSM-9, a novel protein with structural similarity to channels, is required for olfaction, mechanosensation, and olfactory adaptation in *Caenorhabditis elegans*. J Neurosci 17:8259–8269

Conti B (2008) Considerations on temperature, longevity and aging. Cell Mol Life Sci 65:1626–1630

Cosens DJ, Manning A (1969) Abnormal electroretinogram from a *Drosophila* mutant. Nature 224:285–287

Darszon A, Sanchez-Cardenas C, Orta G, Sanchez-Tusie AA, Beltran C, Lopez-Gonzalez I, Granados-Gonzalez G, Trevino CL (2012) Are TRP channels involved in sperm development and function? Cell Tissue Res 349:749–764

de Bono M, Bargmann CI (1998) Natural variation in a neuropeptide Y receptor homolog modifies social behavior and food response in *C. elegans*. Cell 94:679–689

de Bono M, Tobin DM, Davis MW, Avery L, Bargmann CI (2002) Social feeding in *Caenorhabditis elegans* is induced by neurons that detect aversive stimuli. Nature 419:899–903

Dhaka A, Viswanath V, Patapoutian A (2006) TRP ion channels and temperature sensation. Annu Rev Neurosci 29:135–161

Dong XP, Shen D, Wang X, Dawson T, Li X, Zhang Q, Cheng X, Zhang Y, Weisman LS, Delling M, Xu H (2010) PI(3,5)P_2 controls membrane traffic by direct activation of mucolipin Ca^{2+} release channels in the endolysosome. Nat Commun 1:38. doi:10.1038/ncomms1037

Dwoskin LP, Crooks PA, Teng L, Green TA, Bardo MT (1999) Acute and chronic effects of nornicotine on locomotor activity in rats: altered response to nicotine. Psychopharmacology (Berlin) 145:442–451

Eberl DF, Hardy RW, Kernan MJ (2000) Genetically similar transduction mechanisms for touch and hearing in *Drosophila*. J Neurosci 20:5981–5988

Effertz T, Wiek R, Gopfert MC (2011) NompC TRP channel is essential for *Drosophila* sound receptor function. Curr Biol 21:592–597

Effertz T, Nadrowski B, Piepenbrock D, Albert JT, Gopfert MC (2012) Direct gating and mechanical integrity of *Drosophila* auditory transducers require TRPN1. Nat Neurosci 15:1198–1200

Fares H, Greenwald I (2001) Regulation of endocytosis by CUP-5, the *Caenorhabditis elegans* mucolipin-1 homolog. Nat Genet 28:64–68

Feng Z, Li W, Ward A, Piggott BJ, Larkspur ER, Sternberg PW, Xu XZ (2006) A *C. elegans* model of nicotine-dependent behavior: regulation by TRP-family channels. Cell 127:621–633

Fowler MA, Montell C (2013) *Drosophila* TRP channels and animal behavior. Life Sci 92:394–403

Fu Y, Yau KW (2007) Phototransduction in mouse rods and cones. Pflugers Arch 454:805–819

Gallagher AR, Germino GG, Somlo S (2010) Molecular advances in autosomal dominant polycystic kidney disease. Adv Chronic Kidney Dis 17:118–130

Gallio M, Ofstad TA, Macpherson LJ, Wang JW, Zuker CS (2011) The coding of temperature in the *Drosophila* brain. Cell 144:614–624

Gao Q, Chess A (1999) Identification of candidate *Drosophila* olfactory receptors from genomic DNA sequence. Genomics 60:31–39

Gao Z, Ruden DM, Lu X (2003) PKD2 cation channel is required for directional sperm movement and male fertility. Curr Biol 13:2175–2178

Georgiev P, Okkenhaug H, Drews A, Wright D, Lambert S, Flick M, Carta V, Martel C, Oberwinkler J, Raghu P (2010) TRPM channels mediate zinc homeostasis and cellular growth during *Drosophila* larval development. Cell Metab 12:386–397

Glaser FT, Stanewsky R (2007) Synchronization of the *Drosophila* circadian clock by temperature cycles. Cold Spring Harb Symp Quant Biol 72:233–242

Glauser DA, Chen WC, Agin R, Macinnis BL, Hellman AB, Garrity PA, Tan MW, Goodman MB (2011) Heat avoidance is regulated by transient receptor potential (TRP) channels and a neuropeptide signaling pathway in *Caenorhabditis elegans*. Genetics 188:91–103

Gong Z, Son W, Chung YD, Kim J, Shin DW, McClung CA, Lee Y, Lee HW, Chang DJ, Kaang BK, Cho H, Oh U, Hirsh J, Kernan MJ, Kim C (2004) Two interdependent TRPV channel subunits, Inactive and Nanchung, mediate hearing in *Drosophila*. J Neurosci 24:9059–9066

Göpfert MC, Albert JT, Nadrowski B, Kamikouchi A (2006) Specification of auditory sensitivity by *Drosophila* TRP channels. Nat Neurosci 9:999–1000

Gracheva EO, Ingolia NT, Kelly YM, Cordero-Morales JF, Hollopeter G, Chesler AT, Sanchez EE, Perez JC, Weissman JS, Julius D (2010) Molecular basis of infrared detection by snakes. Nature 464:1006–1011

Gruninger TR, LeBoeuf B, Liu Y, Garcia LR (2007) Molecular signaling involved in regulating feeding and other motivated behaviors. Mol Neurobiol 35:1–20

Hamada FN, Rosenzweig M, Kang K, Pulver SR, Ghezzi A, Jegla TJ, Garrity PA (2008) An internal thermal sensor controlling temperature preference in *Drosophila*. Nature 454:217–220

Hardie RC, Franze K (2012) Photomechanical responses in *Drosophila* photoreceptors. Science 338:260–263

Hardie RC, Minke B (1992) The *trp* gene is essential for a light-activated Ca^{2+} channel in Drosophila photoreceptors. Neuron 8:643–651

Hardin PE (2011) Molecular genetic analysis of circadian timekeeping in *Drosophila*. Adv Genet 74:141–173

Harteneck C, Obukhov AG, Zobel A, Kalkbrenner F, Schultz G (1995) The *Drosophila* cation channel *trpl* expressed in *Sf9* cells is stimulated by agonists of G-protein-coupled receptors. FEBS Lett 358:297–300

Hersh BM, Hartwieg E, Horvitz HR (2002) The *Caenorhabditis elegans* mucolipin-like gene *cup-5* is essential for viability and regulates lysosomes in multiple cell types. Proc Natl Acad Sci U S A 99:4355–4360

Hilliard MA, Bergamasco C, Arbucci S, Plasterk RH, Bazzicalupo P (2004) Worms taste bitter: ASH neurons, QUI-1, GPA-3 and ODR-3 mediate quinine avoidance in *Caenorhabditis elegans*. EMBO J 23:1101–1111

Hinman A, Chuang HH, Bautista DM, Julius D (2006) TRP channel activation by reversible covalent modification. Proc Natl Acad Sci U S A 103:19564–19568

Hofmann T, Chubanov V, Chen X, Dietz AS, Gudermann T, Montell C (2010) *Drosophila* TRPM channel is essential for the control of extracellular magnesium levels. PLoS One 5:e10519

Howard J, Bechstedt S (2004) Hypothesis: a helix of ankyrin repeats of the NOMPC-TRP ion channel is the gating spring of mechanoreceptors. Curr Biol 14:R224–R226

Hu Y, Vaca L, Zhu X, Birnbaumer L, Kunze DL, Schilling WP (1994) Appearance of a novel Ca^{2+} influx pathway in Sf9 insect cells following expression of the transient potential-like (*trpl*) protein of *Drosophila*. Biochem Biophys Res Commun 201:1050–1056

Huang J, Liu CH, Hughes SA, Postma M, Schwiening CJ, Hardie RC (2010) Activation of TRP channels by protons and phosphoinositide depletion in *Drosophila* photoreceptors. Curr Biol 20:189–197

Hwang RY, Stearns NA, Tracey WD (2012) The ankyrin repeat domain of the TRPA protein Painless is important for thermal nociception but not mechanical nociception. PLoS One 7: e30090

Inoue H, Yoshioka T, Hotta Y (1985) A genetic study of inositol trisphosphate involvement in phototransduction using *Drosophila* mutants. Biochem Biophys Res Commun 132:513–519

Johnson WA, Carder JW (2012) *Drosophila* nociceptors mediate larval aversion to dry surface environments utilizing both the Painless TRP channel and the DEG/ENaC subunit, PPK1. PLoS One 7:e32878

Jordt SE, Bautista DM, Chuang HH, McKemy DD, Zygmunt PM, Hogestätt ED, Meng ID, Julius D (2004) Mustard oils and cannabinoids excite sensory nerve fibres through the TRP channel ANKTM1. Nature 427:260–265

Jungnickel MK, Marrero H, Birnbaumer L, Lemos JR, Florman HM (2001) Trp2 regulates entry of Ca^{2+} into mouse sperm triggered by egg ZP3. Nat Cell Biol 3:499–502

Kahn-Kirby AH, Bargmann CI (2006) TRP channels in *C. elegans*. Annu Rev Physiol 68:719–736

Kamikouchi A (2013) Auditory neuroscience in fruit flies. Neurosci Res 76:113–118

Kamikouchi A, Inagaki HK, Effertz T, Hendrich O, Fiala A, Gopfert MC, Ito K (2009) The neural basis of *Drosophila* gravity-sensing and hearing. Nature 458:165–171

Kang K, Pulver SR, Panzano VC, Chang EC, Griffith LC, Theobald DL, Garrity PA (2010a) Analysis of *Drosophila* TRPA1 reveals an ancient origin for human chemical nociception. Nature 464:597–600

Kang L, Gao J, Schafer WR, Xie Z, Xu XZ (2010b) *C. elegans* TRP family protein TRP-4 is a pore-forming subunit of a native mechanotransduction channel. Neuron 67:381–391

Kang K, Panzano VC, Chang EC, Ni L, Dainis AM, Jenkins AM, Regna K, Muskavitch MA, Garrity PA (2012) Modulation of TRPA1 thermal sensitivity enables sensory discrimination in *Drosophila*. Nature 481:76–80

Karashima Y, Talavera K, Everaerts W, Janssens A, Kwan KY, Vennekens R, Nilius B, Voets T (2009) TRPA1 acts as a cold sensor in vitro and in vivo. Proc Natl Acad Sci U S A 106:1273–1278

Kernan MJ (2007) Mechanotransduction and auditory transduction in *Drosophila*. Pflugers Arch 454:703–720

Kim J, Chung YD, Park DY, Choi S, Shin DW, Soh H, Lee HW, Son W, Yim J, Park CS, Kernan MJ, Kim C (2003) A TRPV family ion channel required for hearing in *Drosophila*. Nature 424:81–84

Kim SH, Lee Y, Akitake B, Woodward OM, Guggino WB, Montell C (2010) *Drosophila* TRPA1 channel mediates chemical avoidance in gustatory receptor neurons. Proc Natl Acad Sci U S A 107:8440–8445

Kindt KS, Viswanath V, Macpherson L, Quast K, Hu H, Patapoutian A, Schafer WR (2007) *Caenorhabditis elegans* TRPA-1 functions in mechanosensation. Nat Neurosci 10:568–577

Köttgen M, Hofherr A, Li W, Chu K, Cook S, Montell C, Watnick T (2011) *Drosophila* sperm swim backwards in the female reproductive tract and are activated via TRPP2 ion channels. PLoS One 6:e20031

Kwan KY, Allchorne AJ, Vollrath MA, Christensen AP, Zhang DS, Woolf CJ, Corey DP (2006) TRPA1 contributes to cold, mechanical, and chemical nociception but is not essential for hair-cell transduction. Neuron 50:277–289

Kwon Y, Shim HS, Wang X, Montell C (2008) Control of thermotactic behavior via coupling of a TRP channel to a phospholipase C signaling cascade. Nat Neurosci 11:871–873

Kwon Y, Kim SH, Ronderos DS, Lee Y, Akitake B, Woodward OM, Guggino WB, Smith DP, Montell C (2010a) *Drosophila* TRPA1 channel is required to avoid the naturally occurring insect repellent citronellal. Curr Biol 20:1672–1678

Kwon Y, Shen WL, Shim HS, Montell C (2010b) Fine thermotactic discrimination between the optimal and slightly cooler temperatures via a TRPV channel in chordotonal neurons. J Neurosci 30:10465–10471

Kwon I, Choe HK, Son GH, Kim K (2011) Mammalian molecular clocks. Exp Neurobiol 20:18–28

Lee BH, Ashrafi K (2008) A TRPV channel modulates *C. elegans* neurosecretion, larval starvation survival, and adult lifespan. PLoS Genet 4:e1000213

Lee Y, Montell C (2013) *Drosophila* TRPA1 functions in temperature control of circadian rhythm in pacemaker neurons. J Neurosci 33:6716–6725

Lee Y, Lee Y, Lee J, Bang S, Hyun S, Kang J, Hong ST, Bae E, Kaang BK, Kim J (2005) Pyrexia is a new thermal transient receptor potential channel endowing tolerance to high temperatures in *Drosophila melanogaster*. Nat Genet 37:305–310

Lehnert BP, Baker AE, Gaudry Q, Chiang AS, Wilson RI (2013) Distinct roles of TRP channels in auditory transduction and amplification in *Drosophila*. Neuron 77:115–128

Li W, Feng Z, Sternberg PW, Xu XZ (2006) A *C. elegans* stretch receptor neuron revealed by a mechanosensitive TRP channel homologue. Nature 440:684–687

Liu KS, Sternberg PW (1995) Sensory regulation of male mating behavior in *Caenorhabditis elegans*. Neuron 14:79–89

Liu CH, Wang T, Postma M, Obukhov AG, Montell C, Hardie RC (2007a) *In vivo* identification and manipulation of the Ca^{2+} selectivity filter in the *Drosophila* transient receptor potential channel. J Neurosci 27:604–615

Liu L, Li Y, Wang R, Yin C, Dong Q, Hing H, Kim C, Welsh MJ (2007b) *Drosophila* hygrosensation requires the TRP channels water witch and nanchung. Nature 450:294–298
Liu J, Ward A, Gao J, Dong Y, Nishio N, Inada H, Kang L, Yu Y, Ma D, Xu T, Mori I, Xie Z, Xu XZ (2010) *C. elegans* phototransduction requires a G protein-dependent cGMP pathway and a taste receptor homolog. Nat Neurosci 13:715–722
Macpherson LJ, Dubin AE, Evans MJ, Marr F, Schultz PG, Cravatt BF, Patapoutian A (2007) Noxious compounds activate TRPA1 ion channels through covalent modification of cysteines. Nature 445:541–545
Magalhaes CP, de Freitas MF, Nogueira MI, Campina RC, Takase LF, de Souza SL, de Castro RM (2010) Modulatory role of serotonin on feeding behavior. Nutr Neurosci 13:246–255
Matsuura H, Sokabe T, Kohno K, Tominaga M, Kadowaki T (2009) Evolutionary conservation and changes in insect TRP channels. BMC Evol Biol 9:228
Montell C (2001) Physiology, phylogeny and functions of the TRP superfamily of cation channels. Sci STKE 2001(90):re1
Montell C (2005) The TRP superfamily of cation channels. Sci STKE 2005(272):re3
Montell C (2009) A taste of the *Drosophila* gustatory receptors. Curr Opin Neurobiol 19:345–353
Montell C (2011) The history of TRP channels, a commentary and reflection. Pflugers Arch 461:499–506
Montell C, Rubin GM (1989) Molecular characterization of the Drosophila *trp* locus: a putative integral membrane protein required for phototransduction. Neuron 2:1313–1323
Neckameyer WS (2010) A trophic role for serotonin in the development of a simple feeding circuit. Dev Neurosci 32:217–237
Neely GG, Keene AC, Duchek P, Chang EC, Wang QP, Aksoy YA, Rosenzweig M, Costigan M, Woolf CJ, Garrity PA, Penninger JM (2011) *TrpA1* regulates thermal nociception in *Drosophila*. PLoS One 6:e24343
Niemeyer BA, Suzuki E, Scott K, Jalink K, Zuker CS (1996) The Drosophila light-activated conductance is composed of the two channels TRP and TRPL. Cell 85:651–659
Pérez CA, Huang L, Rong M, Kozak JA, Preuss AK, Zhang H, Max M, Margolskee RF (2002) A transient receptor potential channel expressed in taste receptor cells. Nat Neurosci 5:1169–1176
Peschel N, Helfrich-Förster C (2011) Setting the clock–by nature: circadian rhythm in the fruitfly *Drosophila melanogaster*. FEBS Lett 585:1435–1442
Petersen LK, Stowers RS (2011) A Gateway MultiSite recombination cloning toolkit. PLoS One 6: e24531
Phillips AM, Bull A, Kelly LE (1992) Identification of a Drosophila gene encoding a calmodulin-binding protein with homology to the *trp* phototransduction gene. Neuron 8:631–642
Provencio I, Rodriguez IR, Jiang G, Hayes WP, Moreira EF, Rollag MD (2000) A novel human opsin in the inner retina. J Neurosci 20:600–605
Raghu P, Colley NJ, Webel R, James T, Hasan G, Danin M, Selinger Z, Hardie RC (2000) Normal phototransduction in *Drosophila* photoreceptors lacking an $InsP_3$ receptor gene. Mol Cell Neurosci 15:429–445
Reppert SM, Weaver DR (2002) Coordination of circadian timing in mammals. Nature 418:935–941
Robertson HM, Warr CG, Carlson JR (2003) Molecular evolution of the insect chemoreceptor gene superfamily in *Drosophila melanogaster*. Proc Natl Acad Sci U S A 100(Suppl 2):14537–14542
Rogers C, Persson A, Cheung B, de Bono M (2006) Behavioral motifs and neural pathways coordinating O_2 responses and aggregation in *C. elegans*. Curr Biol 16:649–659
Rosenzweig M, Brenman KM, Taylor TD, Phelps P, Patapoutian A, Garrity PA (2005) The *Drosophila* ortholog of vertebrate TRPA1 regulates thermotaxis. Genes Dev 19:419–424
Rosenzweig M, Kang K, Garrity PA (2008) Distinct TRP channels are required for warm and cool avoidance in *Drosophila melanogaster*. Proc Natl Acad Sci U S A 105:14668–14673

Saito S, Fukuta N, Shingai R, Tominaga M (2011) Evolution of vertebrate transient receptor potential vanilloid 3 channels: opposite temperature sensitivity between mammals and western clawed frogs. PLoS Genet 7:e1002041

Saito S, Nakatsuka K, Takahashi K, Fukuta N, Imagawa T, Ohta T, Tominaga M (2012) Analysis of transient receptor potential ankyrin 1 (TRPA1) in frogs and lizards illuminates both nociceptive heat and chemical sensitivities and coexpression with TRP vanilloid 1 (TRPV1) in ancestral vertebrates. J Biol Chem 287:30743–30754

Sakai T, Kasuya J, Kitamoto T, Aigaki T (2009) The *Drosophila* TRPA channel, Painless, regulates sexual receptivity in virgin females. Genes Brain Behav 8:546–557

Sayeed O, Benzer S (1996) Behavioral genetics of thermosensation and hygrosensation in *Drosophila*. Proc Natl Acad Sci U S A 93:6079–6084

Schaheen L, Dang H, Fares H (2006) Basis of lethality in *C. elegans* lacking CUP-5, the Mucolipidosis Type IV orthologue. Dev Biol 293:382–391

Schlingmann KP, Weber S, Peters M, Niemann Nejsum L, Vitzthum H, Klingel K, Kratz M, Haddad E, Ristoff E, Dinour D, Syrrou M, Nielsen S, Sassen M, Waldegger S, Seyberth HW, Konrad M (2002) Hypomagnesemia with secondary hypocalcemia is caused by mutations in TRPM6, a new member of the TRPM gene family. Nat Genet 31:166–170

Schmidt TM, Chen SK, Hattar S (2011) Intrinsically photosensitive retinal ganglion cells: many subtypes, diverse functions. Trends Neurosci 34:572–580

Senatore S, Rami Reddy V, Semeriva M, Perrin L, Lalevee N (2010) Response to mechanical stress is mediated by the TRPA channel painless in the *Drosophila* heart. PLoS Genet 6: e1001088

Sharif-Naeini R, Dedman A, Folgering JH, Duprat F, Patel A, Nilius B, Honore E (2008) TRP channels and mechanosensory transduction: insights into the arterial myogenic response. Pflugers Arch 456:529–540

Shen WL, Kwon Y, Adegbola AA, Luo J, Chess A, Montell C (2011) Function of rhodopsin in temperature discrimination in *Drosophila*. Science 331:1333–1336

Sidi S, Friedrich RW, Nicolson T (2003) NompC TRP channel required for vertebrate sensory hair cell mechanotransduction. Science 301:96–99

Silvert DJ, Doctor J, Quesada L, Fristrom JW (1984) Pupal and larval cuticle proteins of *Drosophila melanogaster*. Biochemistry 23:5767–5774

Sokabe T, Tsujiuchi S, Kadowaki T, Tominaga M (2008) *Drosophila* Painless is a Ca^{2+}-requiring channel activated by noxious heat. J Neurosci 28:9929–9938

Sokolchik I, Tanabe T, Baldi PF, Sze JY (2005) Polymodal sensory function of the *Caenorhabditis elegans* OCR-2 channel arises from distinct intrinsic determinants within the protein and is selectively conserved in mammalian TRPV proteins. J Neurosci 25:1015–1023

Sprecher SG, Desplan C (2008) Switch of rhodopsin expression in terminally differentiated *Drosophila* sensory neurons. Nature 454:533–537

Stanley BG, Leibowitz SF, Eppel N, St-Pierre S, Hoebel BG (1985) Suppression of norepinephrine-elicited feeding by neurotensin: evidence for behavioral, anatomical and pharmacological specificity. Brain Res 343:297–304

Störtkuhl KF, Hovemann BT, Carlson JR (1999) Olfactory adaptation depends on the Trp Ca^{2+} channel in *Drosophila*. J Neurosci 19:4839–4846

Story GM, Peier AM, Reeve AJ, Eid SR, Mosbacher J, Hricik TR, Earley TJ, Hergarden AC, Andersson DA, Hwang SW, McIntyre P, Jegla T, Bevan S, Patapoutian A (2003) ANKTM1, a TRP-like channel expressed in nociceptive neurons, is activated by cold temperatures. Cell 112:819–829

Sun M, Goldin E, Stahl S, Falardeau JL, Kennedy JC, Acierno JS Jr, Bove C, Kaneski CR, Nagle J, Bromley MC, Colman M, Schiffmann R, Slaugenhaupt SA (2000) Mucolipidosis type IV is caused by mutations in a gene encoding a novel transient receptor potential channel. Hum Mol Genet 9:2471–2478

Sun Y, Liu L, Ben-Shahar Y, Jacobs JS, Eberl DF, Welsh MJ (2009) TRPA channels distinguish gravity sensing from hearing in Johnston's organ. Proc Natl Acad Sci U S A 106:13606–13611

Sze JY, Victor M, Loer C, Shi Y, Ruvkun G (2000) Food and metabolic signalling defects in a *Caenorhabditis elegans* serotonin-synthesis mutant. Nature 403:560–564

Takahashi N, Kuwaki T, Kiyonaka S, Numata T, Kozai D, Mizuno Y, Yamamoto S, Naito S, Knevels E, Carmeliet P, Oga T, Kaneko S, Suga S, Nokami T, Yoshida J, Mori Y (2011) TRPA1 underlies a sensing mechanism for O_2. Nat Chem Biol 7:701–711

Talavera K, Gees M, Karashima Y, Meseguer VM, Vanoirbeek JA, Damann N, Everaerts W, Benoit M, Janssens A, Vennekens R, Viana F, Nemery B, Nilius B, Voets T (2009) Nicotine activates the chemosensory cation channel TRPA1. Nat Neurosci 12:1293–1299

Teramoto T, Lambie EJ, Iwasaki K (2005) Differential regulation of TRPM channels governs electrolyte homeostasis in the *C. elegans* intestine. Cell Metab 1:343–354

Teramoto T, Sternick LA, Kage-Nakadai E, Sajjadi S, Siembida J, Mitani S, Iwasaki K, Lambie EJ (2010) Magnesium excretion in *C. elegans* requires the activity of the GTL-2 TRPM channel. PLoS One 5:e9589

Tobin D, Madsen D, Kahn-Kirby A, Peckol E, Moulder G, Barstead R, Maricq A, Bargmann C (2002) Combinatorial expression of TRPV channel proteins defines their sensory functions and subcellular localization in *C. elegans* neurons. Neuron 35:307–318

Tomaszuk A, Simpson C, Williams G (1996) Neuropeptide Y, the hypothalamus and the regulation of energy homeostasis. Horm Res 46:53–58

Tracey WD, Wilson RI, Laurent G, Benzer S (2003) *painless*, a *Drosophila* gene essential for nociception. Cell 113:261–273

Treusch S, Knuth S, Slaugenhaupt SA, Goldin E, Grant BD, Fares H (2004) *Caenorhabditis elegans* functional orthologue of human protein h-mucolipin-1 is required for lysosome biogenesis. Proc Natl Acad Sci U S A 101:4483–4488

Tsubouchi A, Caldwell JC, Tracey WD (2012) Dendritic filopodia, Ripped Pocket, NOMPC, and NMDARs contribute to the sense of touch in *Drosophila* larvae. Curr Biol 22:2124–2134

Vaca L, Sinkins WG, Hu Y, Kunze DL, Schilling WP (1994) Activation of recombinant *trp* by thapsigargin in Sf9 insect cells. Am J Physiol 266:C1501–C1505

Venkatachalam K, Montell C (2007) TRP channels. Annu Rev Biochem 76:387–417

Venkatachalam K, Long A, Elsaesser R, Nikolaeva D, Broadie K, Montell C (2008) Motor deficit in a *Drosophila* model of mucolipidosis type IV due to defective clearance of apoptotic cells. Cell 135:838–851

Venkatachalam K, Wong CO, Montell C (2013) Feast or famine: role of TRPML in preventing cellular amino acid starvation. Autophagy 9:98–100

Viswanath V, Story GM, Peier AM, Petrus MJ, Lee VM, Hwang SW, Patapoutian A, Jegla T (2003) Opposite thermosensor in fruitfly and mouse. Nature 423:822–823

Vosshall LB, Stocker RF (2007) Molecular architecture of smell and taste in *Drosophila*. Annu Rev Neurosci 30:505–533

Vosshall LB, Amrein H, Morozov PS, Rzhetsky A, Axel R (1999) A spatial map of olfactory receptor expression in the *Drosophila* antenna. Cell 96:725–736

Walder RY, Landau D, Meyer P, Shalev H, Tsolia M, Borochowitz Z, Boettger MB, Beck GE, Englehardt RK, Carmi R, Sheffield VC (2002) Mutation of TRPM6 causes familial hypomagnesemia with secondary hypocalcemia. Nat Genet 31:171–174

Walker RG, Willingham AT, Zuker CS (2000) A *Drosophila* mechanosensory transduction channel. Science 287:2229–2234

Wang T, Wang X, Xie Q, Montell C (2008) The SOCS box protein STOPS Is required for phototransduction through its effects on phospholipase C. Neuron 57:56–68

Wang G, Qiu YT, Lu T, Kwon HW, Pitts RJ, Van Loon JJ, Takken W, Zwiebel LJ (2009) *Anopheles gambiae* TRPA1 is a heat-activated channel expressed in thermosensitive sensilla of female antennae. Eur J Neurosci 30:967–974

Wang K, Guo Y, Wang F, Wang Z (2011) Drosophila TRPA channel painless inhibits male-male courtship behavior through modulating olfactory sensation. PLoS One 6:e25890

Ward A, Liu J, Feng Z, Xu XZ (2008) Light-sensitive neurons and channels mediate phototaxis in *C. elegans*. Nat Neurosci 11:916–922

Watnick TJ, Jin Y, Matunis E, Kernan MJ, Montell C (2003) A flagellar polycystin-2 homolog required for male fertility in Drosophila. Curr Biol 13:2179–2184

Wes PD, Chevesich J, Jeromin A, Rosenberg C, Stetten G, Montell C (1995) TRPC1, a human homolog of a *Drosophila* store-operated channel. Proc Natl Acad Sci U S A 92:9652–9656

Wolfgang W, Simoni A, Gentile C, Stanewsky R (2013) The Pyrexia transient receptor potential channel mediates circadian clock synchronization to low temperature cycles in *Drosophila melanogaster*. Proc Biol Sci 280:20130959

Wolstenholme AJ, Williamson SM, Reaves BJ (2011) TRP channels in parasites. Adv Exp Med Biol 704:359–371

Wong CO, Li R, Montell C, Venkatachalam K (2012) *Drosophila* TRPML is required for TORC1 activation. Curr Biol 22:1616–1621

Wu Q, Wen T, Lee G, Park JH, Cai HN, Shen P (2003) Developmental control of foraging and social behavior by the *Drosophila* neuropeptide Y-like system. Neuron 39:147–161

Wu LJ, Sweet TB, Clapham DE (2010) International union of basic and clinical pharmacology. LXXVI. Current progress in the mammalian TRP ion channel family. Pharmacol Rev 62:381–404

Xiang Y, Yuan Q, Vogt N, Looger LL, Jan LY, Jan YN (2010) Light-avoidance-mediating photoreceptors tile the *Drosophila* larval body wall. Nature 468:921–926

Xiao R, Xu XZ (2009) Function and regulation of TRP family channels in *C. elegans*. Pflugers Arch 458:851–860

Xiao R, Xu XZ (2011) *C. elegans* TRP channels. Adv Exp Med Biol 704:323–339

Xiao R, Zhang B, Dong Y, Gong J, Xu T, Liu J, Xu XZ (2013) A genetic program promotes *C. elegans* longevity at cold temperatures via a thermosensitive TRP channel. Cell 152:806–817

Xu XZ, Sternberg PW (2003) A *C. elegans* sperm TRP protein required for sperm-egg interactions during fertilization. Cell 114:285–297

Xu XZ, Li HS, Guggino WB, Montell C (1997) Coassembly of TRP and TRPL produces a distinct store-operated conductance. Cell 89:1155–1164

Xu J, Sornborger AT, Lee JK, Shen P (2008) *Drosophila* TRPA channel modulates sugar-stimulated neural excitation, avoidance and social response. Nat Neurosci 11:676–682

Xue T, Do MT, Riccio A, Jiang Z, Hsieh J, Wang HC, Merbs SL, Welsbie DS, Yoshioka T, Weissgerber P, Stolz S, Flockerzi V, Freichel M, Simon MI, Clapham DE, Yau KW (2011) Melanopsin signalling in mammalian iris and retina. Nature 479:67–73

Yan Z, Zhang W, He Y, Gorczyca D, Xiang Y, Cheng LE, Meltzer S, Jan LY, Jan YN (2013) *Drosophila* NOMPC is a mechanotransduction channel subunit for gentle-touch sensation. Nature 493:221–225

Zhang Y, Hoon MA, Chandrashekar J, Mueller KL, Cook B, Wu D, Zuker CS, Ryba NJ (2003) Coding of sweet, bitter, and umami tastes: different receptor cells sharing similar signaling pathways. Cell 112:293–301

Zhang S, Sokolchik I, Blanco G, Sze JY (2004) *Caenorhabditis elegans* TRPV ion channel regulates 5HT biosynthesis in chemosensory neurons. Development 131:1629–1638

Zhang HJ, Anderson AR, Trowell SC, Luo AR, Xiang ZH, Xia QY (2011) Topological and functional characterization of an insect gustatory receptor. PLoS One 6:e24111

Zhang W, Yan Z, Jan LY, Jan YN (2013a) Sound response mediated by the TRP channels NOMPC, NANCHUNG, and INACTIVE in chordotonal organs of *Drosophila* larvae. Proc Natl Acad Sci U S A 110:13612–13617

Zhang YV, Ni J, Montell C (2013b) The molecular basis for attractive salt-taste coding in *Drosophila*. Science 340:1334–1338

Zhang YV, Raghuwanshi RP, Shen WL, Montell C (2013c) Food-experience induced taste desensitization modulated by the *Drosophila* TRPL channel. Nat Neurosci 16:1468–1476

Zhong L, Bellemer A, Yan H, Honjo K, Robertson J, Hwang RY, Pitt GS, Tracey WD (2012) Thermosensory and non-thermosensory isoforms of *Drosophila melanogaster* TRPA1 reveal heat sensor domains of a thermoTRP channel. Cell Rep 1:43–55

Structural Biology of TRP Channels

Ute A. Hellmich and Rachelle Gaudet

Contents

Abstract

Membrane proteins remain challenging targets for structural biologists, despite recent technical developments regarding sample preparation and structure determination. We review recent progress towards a structural understanding of TRP channels and the techniques used to that end. We discuss available low-resolution structures from electron microscopy (EM), X-ray crystallography, and nuclear magnetic resonance (NMR) and review the resulting insights into TRP channel function for various subfamily members. The recent high-resolution structure of TRPV1 is discussed in more detail in Chapter 11. We also consider the opportunities and challenges of using the accumulating structural information on TRPs and homologous proteins for deducing full-length structures of different TRP channel subfamilies, such as building

U.A. Hellmich (✉) • R. Gaudet (✉)
Harvard University, Cambridge, MA, USA
e-mail: hellmich@fas.harvard.edu; gaudet@mcb.harvard.edu

B. Nilius and V. Flockerzi (eds.), *Mammalian Transient Receptor Potential (TRP) Cation Channels*, Handbook of Experimental Pharmacology 223, DOI 10.1007/978-3-319-05161-1_10,

homology models. Finally, we close by summarizing the outlook of the "holy grail" of understanding in atomic detail the diverse functions of TRP channels.

Keywords
Ankyrin repeat • Coiled coil • α-Kinase • EF hand • Calmodulin

1 Bottlenecks to TRP Channel Structure Determination

In comparison to soluble proteins, high-resolution structures of membrane proteins are vastly underrepresented, comprising ~1% of structures in the protein database (PDB) versus ~30% of the protein-coding genome. Structure determination of membrane proteins can seem daunting—before the first structures of ion channels were determined not so long ago (Doyle et al. 1998; Dutzler et al. 2002), it had been presumed an almost impossible task (Abbott 2002).

TRP channel architecture is similar to other ion channels (Cao et al. 2013b; Liao et al. 2013): a six-transmembrane helix topology (S1 through S6) with a reentrant loop between S5 and S6 forming the channel pore is a recurring structural motif (Yu and Catterall 2004). These channels tetramerize to a 24-helix functional protein complex (Fig. 1). As observed for other ion channels, TRP channel function is strongly influenced by large intracellular domains (Figs. 1 and 2), and the responsiveness to functional modulators, e.g., regulation by phosphoinositides (Nilius et al. 2008), or inhibition by quaternary ammonium ions (Jara-Oseguera et al. 2008), and venom toxins (Siemens et al. 2006), is conserved across ion channel families.

Structural studies on TRP channels are challenging, partially due to the difficulty of producing sufficient quantities of protein in available (and financially sustainable) host systems. While structure determination of important membrane proteins has been facilitated through the use of bacterial, archeal, or thermophilic homologs, no such homologs have been identified for TRPs. Instead TRP channels are heterologously expressed in yeast, insect, or mammalian cells, all more costly and time-consuming than bacterial systems. The fact that neither cell-free nor bacterial expression systems are currently available for full-length TRP channels [except for TRPV3 and TRPM8 (Kol et al. 2013; Zakharian et al. 2010)] shows the importance of carefully selecting the expression system to fulfill the requirements of eukaryotic protein folding and (where applicable) posttranslational modifications. However, TRPs do have many N- and C-terminal cytoplasmic regions for which bacterial systems have yielded soluble fragments for use in X-ray crystallographic and nuclear magnetic resonance spectroscopy (NMR) approaches (Fig. 2). Recent technological advances in the field of electron microscopy (EM) make smaller sample amounts amenable for medium to high-resolution structure determination (Li et al. 2013), as seen for the 3.4Å TRPV1 structure determined by cryoEM [(Cao et al. 2013b; Liao et al. 2013); see chapter 11 for more detailed description (Hellmich and Gaudet 2014)].

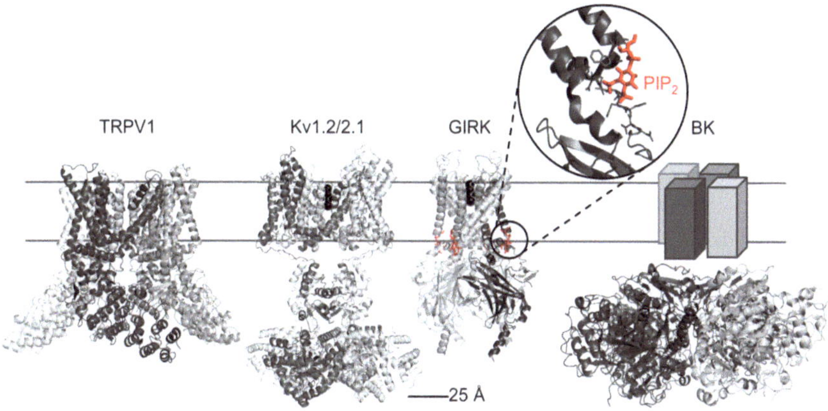

Fig. 1 Structures of representative ion channels. Structures of TRPV1 (pdb: 3j5p) and other ion channels, such as the Kv1.2/2.1 chimera (pdb: 2r9r) or the GIRK channel in complex with PIP_2 (pdb: 3sya), have been determined at high resolution. BK channels (pdb: 3naf), like TRP channels, are only moderately voltage dependent and their function is modulated by large cytoplasmic domains. The TRPV1 construct used to determine the structure has been engineered to remove parts of the N- and C-termini (residues 1–109 and 765–838, respectively) and of the extracellular turret (residues 604–626)

Membrane protein purification is challenging. It requires stripping the lipid bilayer and replacing it by detergents harsh enough to obtain good yields, but mild enough not to compromise protein integrity. Detergent choice is doubly critical for tetrameric TRP channels, as both the 3D structure of each subunit and the correct oligomeric state need to be retained. Finally, the functional integrity of purified proteins has to be verified, which is particularly challenging for many TRPs as their physiological function and naturally occurring ligands are unknown or coarsely characterized. For membrane transporters, substrate binding is often used as a proxy for activity in detergent. With channels, reconstitution into liposomes or planar bilayers is required to detect ion movement across the membrane with electrophysiology or ion flux assays. Functional reconstitution has been described for TRPA1, TRPM8, TRPV1, TRPV2, and TRPV3 (Cao et al. 2013a; Cvetkov et al. 2011; Huynh et al. 2013; Kol et al. 2013; Moiseenkova-Bell et al. 2008; Zakharian et al. 2010).

2 Techniques to Study TRP Channel Structures

Once a source of protein has been identified, the next step is to choose an appropriate structure determination technique. While most available protein structures have been obtained using X-ray crystallography, other powerful techniques exist, such as electron microscopy (EM) or NMR. These can provide information not accessible by crystallographic studies, such as dynamics or, in the case of membrane proteins,

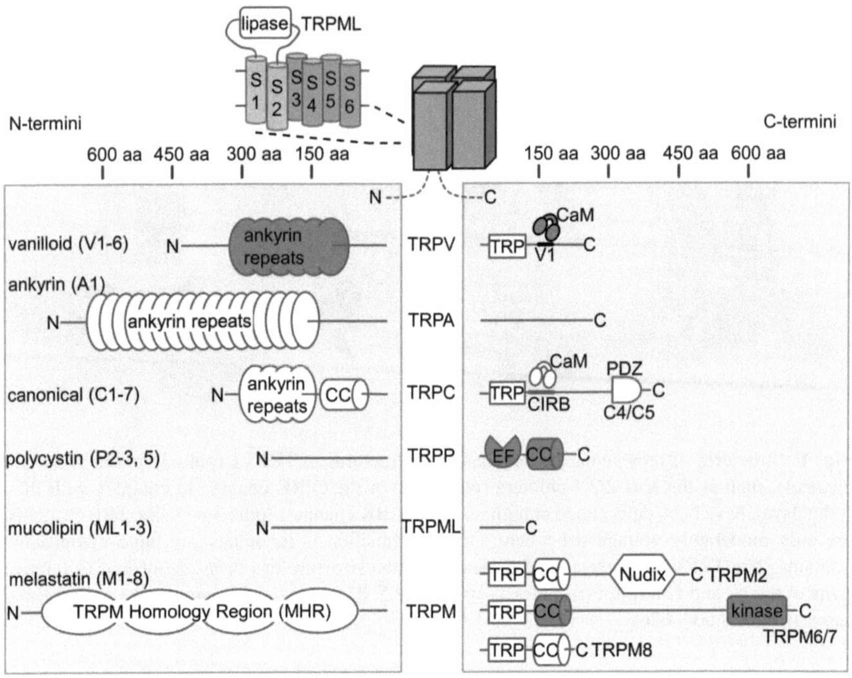

Fig. 2 Schematic overview of TRP channel domain organization. The domain architecture of N- and C-termini, which differs widely across TRP channel subfamilies, is depicted schematically with approximate size scales. TRPML proteins contain a lipase domain between transmembrane helix S1 and S2 (LaPlante et al. 2011). Domains with available high-resolution structural data are *grey*

visualization in lipid bilayers. For TRP channels, current structural information does indeed come from a combination of techniques (Table 1).

Methods to determine 3D structures each have advantages and limitations (see (Minor 2007) for a useful primer). Significant differences include resolution and sample quantity, condition, and preparation. Different levels of resolution are required to answer different questions, and this can guide the choice of method. For example, resolution of ~10–25 Å is adequate to determine approximate molecule shape or observe large domain movements. Detecting secondary structure elements necessitates higher resolution, ~4 Å for β-sheets and ~9 Å for α-helices. Around 3.5 Å, amino acid side chains are discernible, while ~2.5 Å or better enables the observation of bound water molecules or ions. Below we provide brief overviews of EM, X-ray crystallography, and NMR.

2.1 Electron Microscopy

Initially, five low-resolution EM structures—of TRPV1, TRPV4, TRPM2, TRPC3, and TRPA1—were our only experimental glimpse at the overall shapes of full-length TRP channels (Table 1). Recently, TRPV2 joined this collection (Huynh

Table 1 Available structures of TRP channels and TRP channel domains

	Organism	Protein	Fragment	Resolution (Å)	PDB (reference)
EM	Rat	TRPV1	Full-length (cryoEM)	19	Moiseenkova-Bell et al. (2008)
	Rat	TRPV4	Full-length (cryoEM)	35	Shigematsu et al. (2010)
	Mouse	TRPA1	Full-length (negative stain)	16	Cvetkov et al. (2011)
	Human	TRPM2	Full-length (negative stain)	28	Maruyama et al. (2007)
	Mouse	TRPC3	Full-length (cryoEM)	15	Mio et al. (2007a)
	Rat	TRPV1	Near full-length (cryoEM)	3.4	3j5p (Liao et al. 2013)
	Rat	TRPV1	+ DkTx & RTX	3.8	3j5q (Cao et al. 2013b)
	Rat	TRPV1	+ Capsaicin	4.2	3j5r (Cao et al. 2013b)
	rat	TRPV2	Full-length (cryoEM)	13.6	(Huynh et al. 2013)
X-ray crystallography	Rat	TRPV1	ARD + ATP	2.7/3.2	2pnn/2nyj (Lishko et al. 2007)
	Human	TRPV2	ARD	1.7	2f37 (McCleverty et al. 2006)
	Rat	TRPV2	ARD	2.2/1.65/3.1	2eta/2etb/2etc (Jin et al. 2006)
	Mouse	TRPV3	ARD	1.95	4n5q (Shi et al. 2013)
	Human	TRPV4	ARD (+ATP)	2.85 (2.95)	4dx1 (4dx2) (Inada et al. 2012)
	Chicken	TRPV4	ARD	2.3/2.8	3jxi/3jxj (Landoure et al. 2010)
	Mouse	TRPV6	ARD	1.7	2rfa (Phelps et al. 2008)
	Gibberella zeae	TRPGz	Coiled coil	1.25	3vvi (Ihara et al. 2013)
	Rat	TRPM7	Coiled coil	2.01	3e7k (Fujiwara and Minor 2008)
	Human	TRPP2	Coiled coil	1.9/1.9	3hro/3hrn (Yu et al. 2009)
	Human	TRPP3	Coiled coil	2.69	3te3 (Molland et al. 2012)
	Human	TRPP3	Coiled coil	2.8	4gif (Yu et al. 2012)
	Rat	TRPV1	C-terminus + CaM	1.95	3sui (Lau et al. 2012)
	Mouse	TRPM7	α-Kinase (+AMPPNP/ADP·Mg^{2+})	2.8 (2.0/2.4)	1iaj (1ia9/1iah) (Yamaguchi et al. 2001)
NMR	Human	TRPP2	EF hand		2kld/2kle (Schumann et al. 2009)
	Human	TRPP2	EF hand		2kq6 (Petri et al. 2010)
	Human	TRPP2	EF hand		2y4q (unpublished)
	Human	PKD1	PKD domain		1b4r (Bycroft et al. 1999)

et al. 2013). The achievable resolution in EM is generally lower than that of X-ray or NMR structures, although the convergence of ideal technical conditions and sample behavior can now yield atomistic structures (Zhou 2011), including high-resolution cryoEM structures of TRPV1 [(Cao et al. 2013b; Liao et al. 2013); chapter 11 (Hellmich and Gaudet 2014)]. EM has no upper size limits on the macromolecular complex under study—in fact the bigger the molecules, the better the signal—and it does not require crystals, although it does require samples of high purity. The protein is applied as a dilute solution to a carbon film surface. It can then be visualized either by negative staining to yield low-resolution shape information or directly imaged under cryo conditions (cryoEM). Both techniques have been applied to TRP channels.

In cryoEM, the sample is rapidly frozen, i.e., vitrified to avoid ice crystal formation, keeping the protein fully hydrated so that it retains its native conformation. Because the atoms present in biological samples have low electron beam absorption and therefore very low signal to noise, cryoEM requires merging tens of thousands of particle images. Under optimal conditions, especially if particle symmetry can be leveraged into further averaging, cryoEM can yield high-resolution structures.

In negative staining, an electron-rich dye forms a cast around the protein particles as the sample is dried. The particle's outline is visualized with good signal to noise, but the cast inherently limits the resolution. Staining and drying the sample also leads to loss of the protein's hydration shell and may consequentially distort it.

The EM 3D reconstruction process consists of merging tens of thousands of 2D pictures of individual protein particles in different orientation into a single 3D representation. Therefore, a bias in the protein orientation on the grid, which is particularly common in dried negative stained samples, can lead to poor averages and loss of resolution and/or distortions of the 3D volume due to undersampling of some orientations.

2.2 X-ray Crystallography

Most high-resolution structures of TRP channel fragments have been determined by X-ray crystallography (Table 1). We now have an assortment of structures of soluble TRP channel domains that, in combination with the EM structures, begin to delineate larger structural correlations and functional features of various TRP channel subfamilies.

X-ray crystallography requires large quantities of highly pure protein to screen many crystallization conditions as we cannot predict successful conditions. Crystal formation, i.e., the assembly of molecules in an ordered 3D lattice, depends on the chemical and physical properties of the protein and precipitation agents, and the temperature, pH, and other factors such as the presence of ligands and additives (Dumetz et al. 2007; Luft et al. 2011). For membrane proteins, the number of variables increases further, as the nature of the detergent also matters

(Bill et al. 2011). Finally, the protein should be conformationally homogenous: highly flexible regions, such as unstructured N- and C-termini and loops, are often truncated and dozens of variants may need to be tested to find a candidate that crystallizes successfully. This approach can be used to eliminate low complexity sequence regions in TRP channels and increase the chances to obtain high-resolution structures by either EM (Cao et al. 2013b; Liao et al. 2013) or crystallography. Alternatively, ligands or mutations can be used to restrict protein motion. For example, multiple G-protein-coupled receptor crystal structures were obtained after introducing combinations of mutations that significantly increase their melting temperatures (Tate 2012). A general take-home message is that crystallization is facilitated by taking advantage of accumulated biochemical and structural information on a protein to engineer stable conformational states.

2.3 NMR Spectroscopy

Only a few structures of TRP channel fragments have been obtained by NMR (Table 1). However, NMR is advantageous for specific questions, especially regarding conformational dynamics. In addition to high concentration (>100 μM) and large sample volumes, NMR also requires isotope labeling to increase the signal from low-abundance nuclei such as ^{13}C and ^{15}N. Efficient labeling protocols exist for bacteria, yeast, and even insect cells (Saxena et al 2012), but production of isotope-labeled membrane proteins in mammalian cells is generally not economically feasible.

Slow tumbling of large particles in solution degrades NMR signals, imposing size limits on proteins in solution NMR, with few examples of protein structures larger than 35 kDa (see Tugarinov et al. (2005) for one example). This size limitation is particularly acute for membrane proteins because the detergent micelles necessary to maintain a soluble native-like state add significant bulk to the particle size.

Once high-quality NMR spectra with good dispersion of the signals from individual atoms are obtained, determining an NMR structure consists of three steps (Wider 2000): (i) assign each spectral peak to its corresponding ^{13}C, ^{15}N, or ^{1}H atom via through-bond correlations between these peaks, which also yields secondary structure information; (ii) collect through-space correlations linking atom pairs, called the NOEs after the nuclear Overhauser effect allowing nuclei to "see" each other through space to a distance of up to ~5 Å; and (iii) use the resulting constraints to "fold" the protein in silico through iterative molecular dynamics and energy minimization steps, yielding a structure showing atomic details. The NMR assignments also enable further investigations to, for example, probe ligand interactions and protein dynamics. Therefore, NMR spectroscopy has a place in "divide and conquer" approaches to understand details of TRP channel structure, function, and dynamics.

Solid-state NMR, in contrast to solution NMR, does not have inherent size limitations and can thus be applied to microcrystalline samples or membrane proteins reconstituted into lipid bilayers. However, solid-state NMR structure determination is still far from standardized (Hong et al. 2012; Zhao 2012). Membrane proteins are challenging because transmembrane regions contain an overabundance of similar hydrophobic residues in a comparable secondary structure environment, making spectrum analyses difficult. The few heroic examples of membrane protein structures determined with solid-state NMR include the G-protein-coupled receptor CXCR1, kaliotoxin bound to a KcsA-Kv1.3 chimera, and various bacterial retinylidene proteins (Lange et al. 2006; Park et al. 2012; Wang et al. 2013).

3 Structural Information on TRP Channels

In this section, we first discuss the low-resolution electron microscopy (EM) structures of full-length TRP channels and then review the current part list of high-resolution structures of TRP channel fragments, providing additional context about what we know (and do not know) about how these fragments assemble. The high-resolution cryoEM structures of TRPV1 are discussed separately in Chapter 11 (Hellmich and Gaudet 2014).

3.1 Overall Architecture of a TRP Channel: An Electron Microscopy View

TRP proteins share a common topology, with six membrane-spanning segments flanked by the variable cytoplasmic domains (Fig. 2), and in most cases tetramerize to form channels (TRPP channels however form heteromers; see below). The five low-resolution molecular envelopes obtained by single-particle EM reconstructions (Table 1) gave us first glimpses at the possible overall general shape of full-length TRP channels. Low-resolution structures of TRPC3, TRPV1, TRPV2, and TRPV4 have been determined using cryoEM (Huynh et al. 2013; Mio et al. 2007a; Moiseenkova-Bell et al. 2008; Shigematsu et al. 2010), while negative stain EM was used for TRPM2 and TRPA1 (Cvetkov et al. 2011; Maruyama et al. 2007). In most of these cases, functional assays (calcium influx or electrophysiology) were used to confirm that the proteins were active in their heterologous expression system. For the low-resolution structures of TRPV1, TRPV2 and TRPA1, the authors also confirmed that the purified proteins retained their ability to be activated by chemical agonists after reconstitution into liposomes (Cvetkov et al. 2011; Huynh et al. 2013; Moiseenkova-Bell et al. 2008).

Although these EM structures vary in resolution and shape details, they provided some useful frameworks for understanding the overall architecture of TRP channels across subfamilies. A common theme is that they all display some signs of a fourfold symmetry axis; however, their overall dimensions and surface shapes vary significantly. These differences are in part explained by the differences in molecular mass and intracellular domain composition, although some discrepancies are also apparent, as discussed below.

TRPV proteins have a ~400–450-residue N-terminal domain with six ankyrin repeats and a ~150-residue C-terminus that acts as a platform for interactions with other proteins and ligands [Fig. 2; (Lishko et al. 2007; Numazaki et al. 2003; Prescott and Julius 2003)]. TRPV1 and TRPV4 are ~840 and 870 residues, respectively, and their low-resolution EM structures look similar (Moiseenkova-Bell et al. 2008; Shigematsu et al. 2010). TRPV EM structures have a two-domain overall architecture with a larger and a smaller domain connected through four pillars on the four corners of the smaller domain. The overall height of the low-resolution TRPV1 structure is 150 Å. The smaller domain measures ~40 × 60 x 60 Å, while the larger domain, referred to as the "hanging gondola" [a term first coined for the cytoplasmic domains of a voltage-gated potassium channel (Kobertz et al. 2000)], is ~110 × 100 × 100 Å. Such a "hanging gondola" has been observed in other channels including voltage-gated potassium channels (Fig. 1).

Based on superpositions of available crystallographic structures of the N-terminal TRPV1 ankyrin repeat domain (ARD) (Lishko et al. 2007) and the transmembrane domain (TMD) of the Kv1.2 (Long et al. 2005) or MlotiK1 channels (Clayton et al. 2008), the small domain was assigned to the TRPV1 transmembrane region and the larger domain to the cytoplasmic N- and C-termini. The larger domain can readily accommodate four copies of the TRPV1 ARD, with a good fit obtained when the long axis of the ARD is perpendicular to the membrane plane. Interestingly, evidence of fourfold symmetry was observed in both top and bottom views of the TRPV1 and TRPV4 EM structures (Moiseenkova-Bell et al. 2008; Shigematsu et al. 2010), indicating that both the TMD and the cytoplasmic domains adopt a highly symmetric conformation.

The main difference between the low-resolution TRPV1 and TRPV4 EM structures is that in comparison to TRPV1, the "hanging gondola" is shorter in TRPV4. As a result, the ARD crystal structure had to be tilted diagonally relative to the membrane plane to be accommodated in the volume. It is unclear whether these differences represent two conformations of TRPV channels, structural differences between TRPV1 and TRPV4, or are artifacts of the differences in resolution and structure determination methods. The high-resolution TRPV1 structures do show the ARDs much closer to the transmembrane region (Cao et al. 2013b; Liao et al. 2013). The low-resolution TRPV2 structure is more compact (with an overall height of 115 Å) than the low-resolution TRPV1 and TRPV4 structure (Huynh et al. 2013) and closely resembles that of the high-resolution TRPV1 structures.

The observation that all cytoplasmic domains of the TRPV channels are localized very closely to each other in the "hanging gondola" suggested that their N- and C-termini can interact with each other at least in one conformation. The N-terminal TRPV ARDs do not interact in vitro as isolated domains (Phelps et al. 2010), but N- and C-terminal interactions have been observed for TRPV5 and TRPV6 (Chang et al. 2004; Erler et al. 2004), and the N- and C-termini of TRPV1 indeed interact in the high-resolution structures (Cao et al. 2013b; Liao et al. 2013).

A TRPA1 EM structure also shows a two-domain architecture resembling that of TRPV structures with an overall height of 195 Å and fourfold symmetry (Cvetkov et al. 2011). The larger and taller TRPA1 "hanging gondola" measures

~130 × 120 × 120 Å and was accordingly assigned to the larger N-terminal region (~800 residues). TRPA1's smaller domain, designated as the transmembrane region, measures ~70 × 100 × 100 Å and is therefore larger than that of TRPVs. The authors speculate that this could be due to the amphipathic polymer used in place of detergents to stabilize TRPA1 (Cvetkov et al. 2011).

While the low-resolution TRPV1, TRPV2, TRPV4, and TRPA1 have a similar overall architecture, the EM structures of TRPM2 and TRPC3 look markedly different from the aforementioned and each other. The negative stain EM structure TRPM2 (Maruyama et al. 2007) shows an overall bell shape with a few features except a bulb-like extension on its widest end that would correspond to the bell clapper. The calculated molecular mass for this structure is about one-third heavier than expected. The authors attributed this ~200 kDa difference to attached lipids and detergent molecules. However, the protein was purified in dodecylmaltoside, with an isolated micelle size of ~55 kDa, the same detergent used for TRPV4 where no such large discrepancy between expected and observed size was observed (Shigematsu et al. 2010).

To identify the TRPM2 cytoplasmic region, a gold-conjugated antibody fragment against a C-terminal FLAG tag was used and the TMD thus assigned to the apex of the large bell hull structure. However, the "bell clapper" of the TRPM2 structure is tantalizingly similar in size and shape to the small domain assigned to the TMD of the TRPV EM structures.

A cryoEM TRPC3 structure consisting of an intricate mesh of thin rods was reported (Mio et al. 2007a) with a ~200 × 200 × 240 Å volume, extraordinarily large for an expected tetramer mass similar to that of a TRPV1 tetramer. The authors observed smaller particles in high-salt conditions, which they attribute to a tetramer-to-monomer transition (Mio et al. 2007b). Alternatively, the mesh-like structure could be an artifact of trying to contour a volume corresponding to a tetramer when they are instead observing a larger aggregate, with high-salt conditions dissociating the aggregate into isolated tetramers. It is also possible that technical issues with a relatively new automated particle picking algorithm may have led to a distorted reconstruction (Moiseenkova-Bell and Wensel 2009).

In summary, the low-resolution TRPV1, TRPV2, TRPV4, and TRPA1 EM structures are relatively consistent with expectations from our knowledge of distantly related ion channels (Fig. 1). It is noteworthy that only one conformation, with soluble domains forming the hanging gondola, has thus far been observed in low-resolution EM structures. Significant conformational changes in TRPV4 in response to changing levels of phosphoinositides in the cell membrane have been inferred using Förster resonance energy transfer (FRET) between C-termini (Garcia-Elias et al. 2013). In contrast, there was no evidence of conformational changes in TRPV1 in response to capsaicin, heat, or calcium, using a similar experimental strategy (De-la-Rosa et al. 2013). This suggests either that their FRET sensors lack the necessary spatial resolution and/or that there are only small conformational changes upon TRPV1 activation. In agreement with the latter, no large conformational changes were observed when comparing the high-resolution TRPV1 structures in the absence or presence of agonists (Cao et al. 2013b; Liao et al. 2013).

3.2 Current High-Resolution Part List: Structures of TRP Channel Domains

The modularity of ion channels such as TRP channels makes them amenable to a "divide and conquer" approach where high-resolution structures of individual soluble domains are easier or faster to obtain than full-length protein structures (Gaudet 2009). Below, we describe the available "part list" and discuss how the different parts may assemble in the context of full-length tetrameric channels. Because these parts stem from a large number of TRP channels, they allow important insights into the peculiarities and commonalities of various subfamilies that a single structure cannot achieve.

3.2.1 The TRPM7 α-Kinase Domain

TRPM6 and TRPM7 are involved in Mg^{2+} homeostasis in mammals, and their C-terminal α-kinase domain, unique within the TRP superfamily, is implicated in Mg^{2+} sensing [Fig. 2; (Demeuse et al. 2006; Schmitz et al. 2003)]. The crystal structure of the TRPM7 α-kinase domain was the first high-resolution structure of a TRP channel domain and coincidentally the first structure of an α-kinase (Yamaguchi et al. 2001). α-Kinases predominantly target residues in α-helical regions hence their name (Ryazanov et al. 1999). The TRPM7 α-kinase is a Ser/Thr kinase that can be expressed as a soluble protein and retains its activity (Runnels et al. 2001), indicating that it is an independently folding domain.

The TRPM7 α-kinase structure shows a domain-swapped dimer: a short N-terminal α-helix of one kinase subunit reaches into the structural core of the other subunit via an extended peptide region [Fig. 3a; (Yamaguchi et al. 2001)]. As we discuss below, this dimerization mechanism suggests a dimer of dimer arrangement in a tetrameric TRPM7 channel.

The TRPM7 α-kinase structure revealed an unexpected structural similarity to other eukaryotic protein kinases (Yamaguchi et al. 2001), despite the fact that α-kinases do not share all typical sequence motifs of classical kinases (Bates-Withers et al. 2011; Ryazanov et al. 1999). Each individual TRPM7 α-kinase consists of N- and C-terminal lobes connected by a flexible linker region. Similar to classic kinases, the nucleotide-binding pocket is nested in between the two lobes. The TRPM7 α-kinase structure also features a structural zinc-binding site within the C-terminal lobe, which is unique to and conserved in α-kinases.

Within the α-kinase active site, the P-loop (required for nucleotide binding and hydrolysis) and a conserved glycine-rich α-kinase motif are in strikingly similar orientation to the P-loop and activation loop (acting as a structural scaffold for interaction with the substrate) of classical kinases. The glycine-rich α-kinase motif was therefore hypothesized to be the substrate interaction platform (Yamaguchi et al. 2001). Interestingly, point mutations that abrogated TRPM7 activity (Runnels et al. 2001) were not located in the catalytic center of the kinase but rather in this postulated substrate peptide recognition site (Yamaguchi et al. 2001). However, the nature of the substrate(s) cannot be readily inferred from the kinase structure.

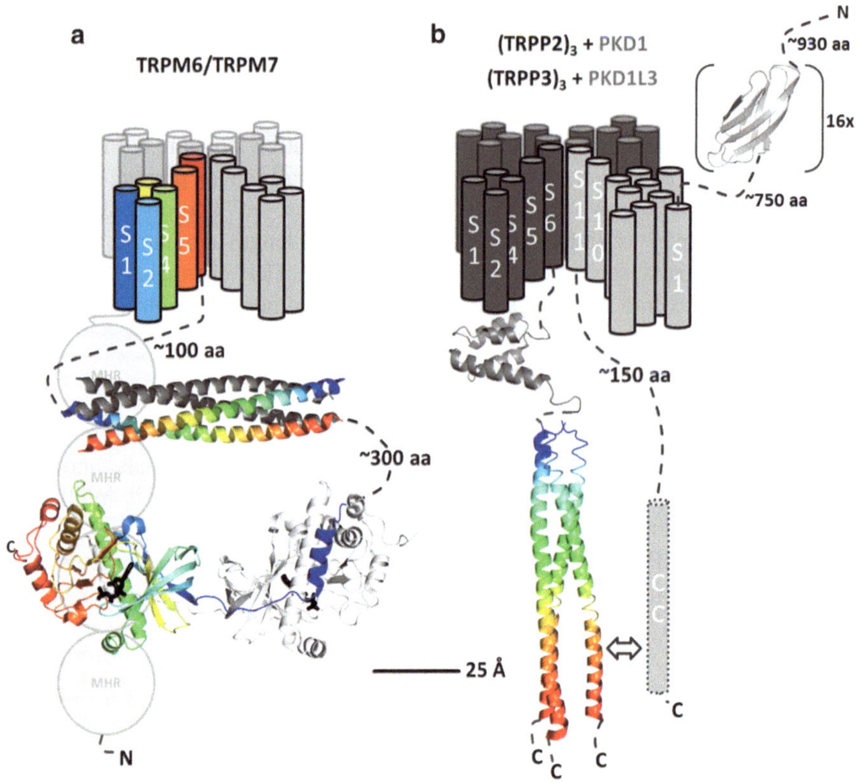

Fig. 3 Overview of spatial organization of TRPM6/M7 channels and TRPP/PKD complexes. (**a**) For TRPM6/M7, transmembrane helices S1–S6 are drawn indicating possible 3D topology of the tetramer in the membrane (based on Kv channel topology). For cytoplasmic domains, the antiparallel coiled coil (pdb: 3e7k) and α-kinase domain (pdb: 1ia9) of TRPM7 are shown; no structure is available for the TRPM homology region (MHR). The coiled coil crystallized as a tetramer, and one antiparallel pair is *colored*. Only one dimer of the α-kinase is shown for clarity. (**b**) For TRPP/PKC complexes, the three TRPP2/P3 subunits are depicted in *dark gray* and the 11-transmembrane helix PKD1/PKD1L3 subunit in *light gray*. One extracellular PKD repeat of PKD1 is depicted (pdb: 1b4r). For TRPP2, the intracellular EF hand (pdb: 2kq6) and the parallel coiled-coil trimer (pdb: 2hrn) are shown

3.2.2 The TRPM7 Coiled Coil

The C-terminal cytoplasmic region of TRPM7 contains a ~50-residue coiled-coil assembly domain that precedes and is separated from the α-kinase domain by ~300 residues (Fig. 3a). This coiled coil is found in all TRPM channels, many of which require it for correct oligomerization and channel activation (Tsuruda et al. 2006). The crystal structure of the isolated TRPM7 coiled coil shows a tetrameric anti-parallel bundle (Fujiwara and Minor 2008).

Before describing the structural details of the TRPM7 coiled coil, it is useful to review the salient features of coiled coils (see (Grigoryan and Keating 2008; Moutevelis and Woolfson 2009) for in-depth reviews). Coiled coils are found in a large number of otherwise functionally and structurally unrelated protein families and consist of two or more α-helices that wind around each other to form a superstructure with helical properties, or supercoil. Coiled coils can be homo- or heteromeric, and the subunits can interact in parallel or antiparallel fashion with regard to the relative orientation of their N- and C-termini. This allows for a broad variety of architectures despite the seemingly simple design of a coiled coil and hence a variety of functions as oligomerization or protein–protein interaction motifs.

Coiled coils have seven-residue or "heptad" repeats, with individual amino acids designated *a* through *g*. Positions *a* and *d* typically feature hydrophobic residues that project into the core of the assembled superstructure. This occurs in a "knob-in-holes" fashion, where an *a* or *d* residue is a "knob" that protrudes into a "hole" formed by side chains of the other helices. This makes for a snug helix–helix interaction surface. In contrast, residues at *e* and *g* positions are usually hydrophilic and on the outside of the coiled coil, forming inter-helical interactions that determine the number and orientation (parallel or antiparallel) of α-helices in the coiled coil. Residues at positions *b*, *c*, and *f* are at the surface, away from inter-helical contacts, and thus have little influence on coiled coil assembly. In other words, positions *a* and *d* provide most of the affinity, whereas positions *e* and *g* provide most of the specificity, although predictive rules for coiled-coil assemblies have yet to be determined (Grigoryan and Keating 2008).

The 77-Å long TRPM7 coiled coil is tetrameric, as expected, but surprisingly is antiparallel, with two strands running in opposite direction to the other two. While it is a classical and well-ordered *a-d* coiled coil with knob-in-hole packing geometry at its center, the termini of the coiled coil show distinct splaying and knob-against-knob packing.

3.2.3 Implications for TRPM Channel Assembly

The available structures of two C-terminal domains of TRPM7 provide some constraints for the assembly of full-length tetrameric TRPM7 channels (Fig. 3a). Because the TRPM7 coiled coil is antiparallel, the fourfold rotational symmetry of the TMD, expected based on homologous voltage-gated potassium channel structures (Long et al. 2005), must be broken in the C-terminal cytoplasmic region. This is in stark contrast to the parallel coiled coil observed in Kv7 channels, which allow for a continuous fourfold rotational symmetry through the entire protein (Howard et al. 2007). However, the TRPM7 TMD is separated by a ~100-residue linker from the coiled coil, which can readily accommodate the divergent symmetry of the TMD and coiled coil.

The antiparallel coiled coil complements the domain-swapped dimer observed for the α-kinase domain (Yamaguchi et al. 2001), suggesting that the four kinase domains in a tetrameric channel assemble as two dimers that may or may not contact each other. It was recently reported that cleavage of the α-kinase domain by caspases leads to increased TRPM7 channel activity in the context of Fas-induced apoptosis (Desai et al. 2012). That the C-terminal kinase domain can influence pore opening implies that the kinase and pore domains communicate, but whether there is direct contact or a long-range network relay remains to be seen. The antiparallel orientation of the TRPM7 coiled coil also raises the question of how this domain is oriented in relation to the rest of the protein (and the membrane plane) and whether its orientation and length remain constant as the TRPM7 channel undergoes activation/deactivation cycles.

Interestingly, sequence alignments of TRPM subfamily members, all of which have a coiled-coil sequence motif at their C-terminus, indicate that their coiled coils vary in length and packing properties, hence clustering subfamily members into two groups (Fujiwara and Minor 2008). This suggests different structural requirements of this domain in different TRPM channels and testable hypotheses of possible hetero-oligomerization partners. Considering that coiled-coil assembly can be influenced by minimal sequence changes (Grigoryan and Keating 2008; Xu and Minor 2009), it is possible that the antiparallel tetramer assembly of the TRPM7 coiled coil is not a conserved feature of all TRPM channels and that some may have other arrangements.

In summary, the TRPM7 coiled-coil and α-kinase domain structures both suggest that at least some TRPM channels do not maintain fourfold rotational symmetry throughout the whole tetramer. This certainly represents a cautionary note in the processing and interpretation of EM structures: fourfold rotational symmetry averaging should not be blindly enforced, but only after clearly observing symmetry in un-averaged preliminary reconstructions.

3.3 Trimeric Coiled Coils and the Assembly of TRPP Channels

Putative coiled coils are also found in the N-terminal cytoplasmic region of TRPC proteins and the C-terminal cytoplasmic region of TRPP proteins (Fig. 2). The TRPP subfamily is somewhat unusual as there is emerging consensus that the functional unit is a heteromeric complex (Fig. 3b). Each complex consists of one PKD1 subunit (polycystic kidney disease 1 or polycystin-1; previously known as TRPP1, although it is not a TRP protein) and three TRPP2 subunits (also referred to as polycystin-2, or polycystic kidney disease 2, PKD2) (Yu et al. 2009). A homologous complex is formed by one polycystic kidney disease 1-like 3 (PKD1L3) subunit and three TRPP3 subunits (also known as polycystic kidney disease 2-like 1 or PKD2L1) (Yu et al. 2012). The PKD1/TRPP2 complex is important for mechanosensation in the kidney [reviewed in (Qamar et al. 2007)], while the acid-sensing PKD1L3/TRPP3 complex is activated by low pH and implicated in sour taste perception (see (Yu et al. 2012) and references therein).

PKD1, PKD1L3, and their homologs have a large extracellular N-terminal region containing up to 16 PKD domains, for which an NMR structure is available (Bycroft et al. 1999), and other subdomains, followed by 11 predicted transmembrane segments, six of which are homologous to the TMD of ion channels (Li et al. 2003). Mutations in the predicted pore region of PKD1L3 affect ion selectivity of the PKD1L3/TRPP3 complex (Yu et al. 2012). Therefore, the

PKD1/TRPP2 and PKD1L3/TRPP3 complexes likely have a tetrameric channel pore domain analogous to homotetrameric TRP channels (Fig. 3b).

The heterotetrameric 1:3 arrangement raises the question of how these channels are assembled, and structures of TRPP coiled coils are providing valuable insights. X-ray structures of the TRPP2 and TRPP3 C-terminal coiled coils have been determined (Molland et al. 2012; Yu et al. 2009; Yu et al. 2012). Consistent with the heteromeric assemblies described above, the coiled coils of TRPP2 and TRPP3 form stable parallel trimers. Intriguingly, the TRPP2 coiled-coil bundle is splayed open at its C-terminus (Yu et al. 2009). Because crystal contacts were observed in this region, the splaying could be a crystal-packing artifact or suggest a physiological mechanism for partial or complete exchange of coiled coil strands as explained below.

PKD1 contains a short cytoplasmic C-terminus with a predicted coiled coil that interacts with the TRPP2 coiled coil in vitro (Qian et al. 1997; Tsiokas et al. 1997). A heteromeric four-helix bundle of a PKD1 and TRPP2 coiled coil has not yet been observed. However, a computational docking study, combined with in vitro cross-linking and mutagenesis, supports a model in which the PKD1 coiled coil replaces one strand of the TRPP2 trimeric coiled coil in the splayed out C-terminal region (Fig. 3b) (Zhu et al. 2011). Furthermore, disruption of these residues prevents assembly of the PKD1/TRPP2 complex in cells (Zhu et al. 2011). Although dynamic exchanges of coiled-coil strands may seem unusual, there is ample precedent for coiled coils as dynamic structures. For example, the long stalk of the dynein motor protein is a two-stranded coiled coil that slides to different helical registries separated by half a heptad, as dynein walks on microtubules (Kon et al. 2009).

PKD1L3, in contrast to PKD1, has a very short C-terminus and no predicted coiled coil. TRPP3 thus interacts with PKD1L3 mainly through interactions in the TMD (Ishimaru et al. 2010). However, disruption of TRPP3 coiled-coil trimerization impairs surface expression of TRPP3/PKD1L3 (Yu et al. 2012). Coiled-coil trimerization is therefore still essential for complex formation, supporting the idea that a TRPP trimer platform is required for a fourth (PKD) subunit to attach itself to form a mature complex.

In summary, the accumulated biochemical and structural data on the TRPP subfamily point to a key role for the coiled coil in the assembly of a functional channel that includes three TRPP subunits and one subunit of a PKD1 family member. The coiled-coil structures also enable studies of the assembly mechanisms of this unusual family of ion channels with important physiological roles including kidney function and sour taste perception.

3.4 Broader Implications of Coiled Coils on TRP Channel Assembly

Recently, the crystal structure of a C-terminal coiled coil of a fungal TRP channel, TRPGz from *Gibberella zeae*, has been determined (Ihara et al. 2013). TRPGz is homologous to the *Saccharomyces cerevisiae* TRPY1, but has a more polymodal activation profile. When compared to vertebrate TRP channels, it shows the highest similarity to members of the TRPC subfamily. TRPGz contains a coiled coil in its C-terminus, the deletion of which disrupts responses to hyperosmotic or temperature shock. In contrast to other TRP channels, deletion of this coiled coil did not lead to oligomerization defects, as TRPGz still formed stable tetramers and retained its responses to voltage. Only the coiled coil shows significant secondary structure within the otherwise unstructured TRPGz C-terminus, as observed with NMR (Ihara et al. 2013). The TRPGz coiled coil was crystallized as a tetramer, with the typical *a* and *d* position hydrophobic knobs. Interestingly, the TRPGz coiled coil does not have exact fourfold rotational symmetry: the four coils come together at a significantly steeper angle than in canonical coiled coils, and the knobs-in-hole packing is only maintained for three out of four helices. In solution, this helical bundle is in equilibrium between monomer, dimer, and tetramer. The precise role of this packing mode is not clear, and it is worth noting that the peptide used for crystallization is very short (20 amino acid residues).

In addition to the TRPM, TRPP, and fungal TRP channels, TRPC subfamily channels also have a potential coiled coil between the ARD and first transmembrane segment (Fig. 2). No high-resolution structures are available for TRPC channels. Biochemical data for TRPC4 suggest that the N-terminal region forms tetramers, although the coiled coil was not directly implicated in this assembly (Lepage et al. 2009). Therefore, it remains to be determined whether the coiled-coil sequence motifs actually form coiled coils, and if so, whether these are heteromeric in assembly and what their stoichiometry is.

As illustrated above, structural studies of isolated coiled coils provide a wealth of information and constraints to better understand TRP channel assembly and function. However, a cautionary note is in order: coiled coils can be inherently dynamic and/or heavily influenced by their environment in both physiologically relevant and artifactual situations. For example, while the full isolated coiled coil of the Kv7.1 channel yielded the expected parallel tetrameric structure (Howard et al. 2007; Wiener et al. 2008), a shorter construct assembled as a non-physiological stable trimer (Xu and Minor 2009). It is therefore possible that, in the context of a full-length channel, oligomerization states of coiled coils other than those observed with short fragments are possible.

3.5 TRP Channel Ankyrin Repeats

The highest number of structures of TRP channel domains to date comes from the ARDs of TRPVs. Ankyrin repeats are one of the most prevalent protein sequence motifs, present in a multitude of unrelated protein families where they are mostly associated with protein–protein interaction functions (Sedgwick and Smerdon 1999). The motif name stems from the protein Ankyrin 1, which contains 24 such repeats (Lux et al. 1990). The basic architecture for this ~33–34 amino acid stretch consists of short tandem antiparallel α-helices (termed inner and outer helix in regard to their location within the ARD concave or “palm” surface; see below), followed by a β-hairpin loop. Individual ankyrin repeats stack against each other with a slight counterclockwise twist from one to the next. This gives them the appearance of a cupped hand: the loops forming the fingers attached to a hand with an open concave palm, and the backside accordingly forms a convex surface (Jacobs and Harrison 1998). The surface residues of the palm and fingers are not particularly conserved, making this region malleable to evolutionary diversification for various interaction partners (Sedgwick and Smerdon 1999).

Ankyrin repeats are found in the cytoplasmic N-terminal region of TRPA, TRPV, TRPC, and TRPN subfamily members (Fig. 2). TRP channel ARDs vary dramatically in size: TRPA1 and TRPN1 have 17 and 29 predicted repeats, respectively, that overall closely match the ankyrin repeat consensus sequence, while TRPC proteins have only four predicted repeats, which, like the six repeats found in TRPV proteins, more loosely follow the consensus sequence motif. So far, ARD structures are only available from the TRPV subfamily.

The ARD structures from TRPV1, TRPV2, TRPV3, TRPV4, and TRPV6 all show six highly conserved repeats (Inada et al. 2012; Jin et al. 2006; Landoure et al. 2010; Lishko et al. 2007; McCleverty et al. 2006; Phelps et al. 2008; Shi et al. 2013). Repeats 5 and 6 have an unusually large angular twist compared to repeats 1–4. Finally, the TRPV ARD finger loops are highly elongated, enhancing the resemblance to a hand (Gaudet 2008a), although it should be noted that finger lengths varies within TRPV channels (Phelps et al. 2008). Variance in finger sequence and length, and of the palm surface residues, may indicate distinct roles for ARDs in TRPV channels, including interactions with different ligands. In TRPV1, finger 3 of the ARD interacts with a β-sheet formed by both the proceeding linker region and the C-terminus of an adjacent subunit (Cao et al. 2013b; Liao et al. 2013).

Recent interesting developments include the structures of the TRPV4 ARD and their implications in understanding the molecular basis of TRPV4-linked genetic diseases. A large number of dominant missense mutations in TRPV4 have emerged that lead to very different channelopathies broadly segregating into two groups: skeletal disorders and degenerative neuropathies (see Nilius and Voets (2013) for a recent review). We currently do not understand the underlying structural implications of the mutants described to date, and even the limited functional data are somewhat contradictory: both loss- or gain-of-function phenotypes have been described for the same mutant (Auer-Grumbach et al. 2010; Landoure et al. 2010). A different approach is to investigate whether neuronal vs. skeletal manifestations of TRPV4-linked diseases may be the consequence of distinct structural or biochemical differences. However, no correlation was observed between disease phenotype and biochemical properties regarding TRPV4 ARD stability and/or ATP binding (Inada et al. 2012).

Mapping known disease-causing TRPV4 mutations (Nilius and Voets 2013) onto the primary and tertiary structure of human TRPV4 allows us to appreciate their relative localization in a new light. While skeletal dysplasia mutations are widely distributed throughout the protein, neuropathy mutations localize primarily, although not exclusively, to the ARD. Furthermore, the skeletal dysplasia mutations located within the ARD map predominantly on its palm surface, whereas the mutants linked

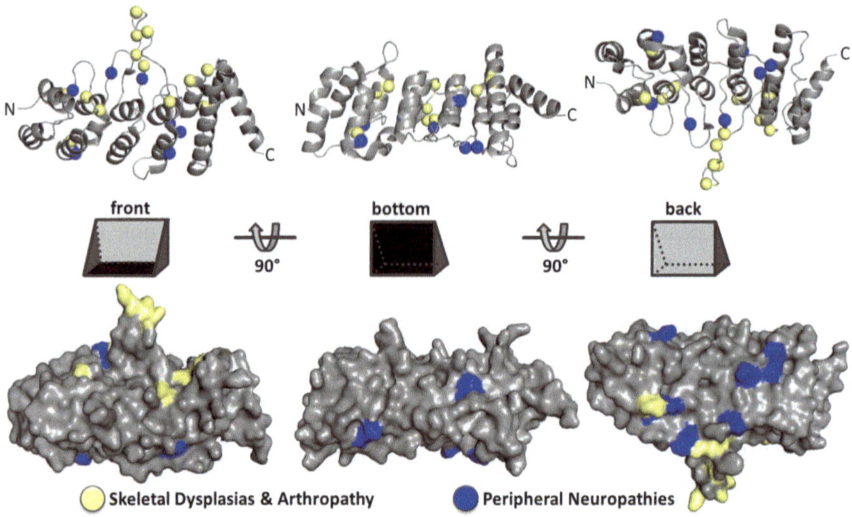

Fig. 4 Crystal structure of the human TRPV4 ARD shown in cartoon and surface representations from three angles (pdb: 4dx1). Locations of disease-causing mutations are mapped onto the structure (*spheres*). Residues described in patients suffering from skeletal dysplasia and arthropathy are predominantly found on the "front" or "palm" surface, and those causing peripheral neuropathies localize preferentially to the "back"

to neuropathies are preferentially located to the surface corresponding to the back of the hand (Fig. 4; Inada et al. 2012; Landoure et al. 2012; Landoure et al. 2010). Interestingly, the back face of the ARD is much more conserved in TRPV4 than it is in TRPV1 (Phelps et al. 2007; Zimon et al. 2010). This suggests that the physical segregation of TRPV4 channelopathy mutations may underlie their different phenotypic outcome, perhaps by differentially affecting intramolecular contacts or intermolecular interactions with cellular partners.

TRPV ARD structures have already been extensively reviewed (Gaudet 2008a, b, 2009), and a few themes are emerging. First, the isolated TRPV ARDs show no evidence of self-oligomerization (Phelps et al. 2008; Phelps et al. 2010). The ARDs instead influence TRPV channel assembly and function through intramolecular contacts with other regions of the TRPV proteins, as illustrated in the TRPV1 structure (Liao et al. 2013). Second, engineered mutations of TRPV ARD surface residues tend to increase the sensitivity and/or basal activity of TRPV channels (Lishko et al. 2007; Phelps et al. 2010), suggesting an overall inhibitory role for this domain on TRPV channel function. Third, there is thus far no evidence of large-scale conformational changes within the ARD itself, even upon ATP ligand binding (Inada et al. 2012), although the finger loops do show inherent flexibility (Inada et al. 2012; Jin et al. 2006). Therefore, how the ARDs influence TRPV channel behavior remains unclear—perhaps a movement of the entire domain relative to the rest of the channel rather than a conformational change within its boundaries is ultimately responsible for its influence on channel function, sensitization, or desensitization. More studies in the context of full-length channels are required to test these hypotheses.

3.6 Structures of TRP Channel Motifs Involved in Calcium Regulation

Many TRP channels are permeable to calcium, and their activation leads to changes in intracellular calcium concentrations. In turn, many of these channels are regulated directly or indirectly by calcium-dependent signaling pathways. Structural biology has begun to yield mechanistic insights into two such calcium regulation mechanisms: (i) calcium-calmodulin (CaM) binding to the C-terminal cytoplasmic region of TRPV1 and (ii) calcium binding to a paired EF hand motif that precedes the coiled coil in the TRPP C-terminus.

EF hands are the most common Ca^{2+}-binding motifs in proteins and typically occur in pairs. The basic architecture is a helix-loop-helix motif with the two helices oriented in a V shape between the thumb and index finger (Fig. 5). Upon calcium binding in the cleft formed by the loop region—by the palm and remaining fingers, using the hand analogy—the thumb moves further away from the index finger, changing the V to an L shape. This exposes otherwise buried residues that can then interact with downstream target sequences, thereby enabling Ca^{2+}-dependent signal propagation (Lewit-Bentley and Rety 2000). The most famous EF hand-containing protein is CaM, a ubiquitous 17 kDa protein with flexibly connected N- and C-terminal lobes comprised of two EF hands that each binds calcium in the μM range (Persechini et al. 1989; Vetter and Leclerc 2003). CaM regulates many ion channels and signaling proteins, including several TRP channels (Zhu 2005). Of note, CaM does not necessarily have to be fully calcified to interact with its targets (Delmas and Brown 2005). This variety in Ca^{2+} occupancy and structural flexibility make CaM a highly versatile regulatory partner.

Ca^{2+} influx through TRPV1 is stimulated by noxious heat, pH, and capsaicin, but prolonged exposure leads to Ca^{2+}-dependent desensitization (Caterina et al. 1997). Understanding the molecular details of desensitization is a major focus of elucidating TRPV1 function. The TRPV1 C-terminus is important for desensitization and thought to mediate both protein and lipid interactions (Cao et al. 2013a; Nieto-Posadas et al. 2012; Numazaki et al. 2003; Prescott and Julius 2003). Recently, the first high-resolution view of such an interaction was obtained through a crystal structure of Ca^{2+}-CaM bound to a TRPV1 C-terminal peptide [Fig.5; (Lau et al. 2012)]. This region is missing from the high-resolution TRPV1 structure (Liao et al. 2013).

A 35-amino acid C-terminal TRPV1 fragment (residues 767–801 in rat) binds CaM in a Ca^{2+}-dependent manner with high affinity ($K_D = 50$ nM), and the fully Ca^{2+}-loaded CaM winds itself around a 10-residue α-helix flanked by extended peptide regions in a 1:1 complex (Fig. 5). The complex is antiparallel: the CaM N-lobe interacts with the TRPV1 peptide's C-terminus and the CaM C-lobe with the peptide's N-terminus. Binding of CaM to the TRPV1 C-terminus follows the

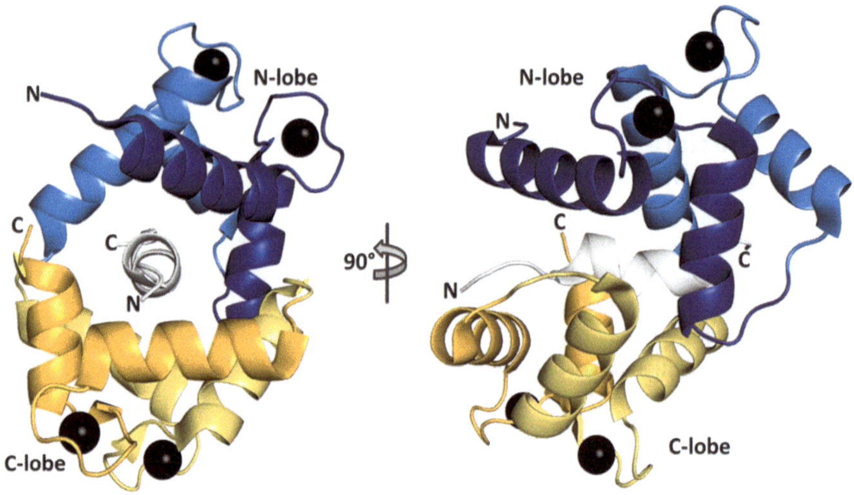

Fig. 5 Crystal structure of the C-terminal fragment of TRPV1 in complex with Ca^{2+}-calmodulin (pdb: 3sui). The fully Ca^{2+}-loaded CaM tightly winds around the TRPV1 peptide, forming an antiparallel complex where the CaM N-lobe interacts with the TRPV1 peptide's C-terminus and the C-lobe with the N-terminus. Each EF hand is in a different shade

classic 1–10 CaM-binding motif organization (Rhoads and Friedberg 1997) where two bulky hydrophobic residues at positions 1 and 10 are the main determinants of the CaM interaction.

Intriguingly, TRPV1 ARD also binds to CaM in a Ca^{2+}-dependent manner (Lishko et al. 2007). It had been suggested that CaM may bridge the ARD and C-terminus and thus lock the channel in a ternary complex (Lishko et al. 2007), similar to what has been observed for SK channels and proposed for Ca_V1 channels (Dick et al. 2008; Schumacher et al. 2001). However, the CaM C-lobe dominates interactions with both the TRPV1 ARD and C-terminus. Accordingly, TRPV1 ARD, Ca^{2+}-CaM, and C-terminus do not form a ternary complex, at least not outside the context of the full-length channel (Lau et al. 2012). While mutation of the Ca^{2+}-CaM binding site on TRPV1 ARD essentially abolished desensitization in response to capsaicin, mutation of the Ca^{2+}-CaM binding site in the TRPV1C-terminus led to slower and reduced TRPV1 desensitization, supporting the idea that Ca^{2+}-CaM binding to the two sites differentially regulates TRPV1 (Lau et al. 2012).

In contrast to CaM-mediated regulation, TRPP channel function is directly regulated by calcium through a C-terminal EF hand motif that precedes the coiled coil [Figs. 2 and 3; (Petri et al. 2010)]. Two groups have independently determined the NMR structure of the TRPP2 EF hand motifs (Petri et al. 2010; Schumann et al. 2009). Schumann and colleagues used a 117-amino acid fragment, which formed a dimer that dissociated upon Ca^{2+} binding. In each subunit they find a pair of helix-loop-helix motifs assigned as a paired EF hand motif that binds two Ca^{2+} with K_Ds in the 50–200 μM range, although only one of these sites was canonical (Schumann et al. 2009).

Petri and colleagues also find a pair of helix-loop-helix motifs, but only observe Ca^{2+} binding to one, and did not observe dimerization of their (40-amino acid shorter) construct [Fig. 3; (Petri et al. 2010)]. However, the structures themselves are different, and Petri and colleagues raise concerns regarding the validity of the Schumann structure, which instead of the typical L-shape shows the two helices almost perfectly in line, thereby distorting the calcium coordination site. If the paired EF hand motif indeed dimerizes, it raises the question of how this assembly fits to the observed trimeric TRPP2 coiled coil (Yu et al. 2009).

In summary, structural data on the interaction of TRP channel regions with Ca^{2+} or Ca^{2+}-dependent regulators are starting to trickle in. While the available structures do not allow a complete view of how TRP channel activity is modulated by these interactions, they provide both tools and constraints for future experiments addressing the molecular mechanisms of Ca^{2+}-mediated TRP channel regulation.

4 Homology Models to Understand TRP Channel Structure and Function

Before a high-resolution 3D structure of a full-length TRP channel was available, considerable efforts were invested by several groups to generate homology models of TRP channels. Valuable lessons can be drawn from these approaches and readily applied to homology models based on the new TRPV1 structures and future TRP channel structures. Structural models can be generally divided into two groups: de novo, or template-free, models which rely on general physical properties of proteins; and template-based, or homology models. Template-based models rely on identifying a template structure through sequence similarity with a protein or protein fragment of known structure.

There have been a lot of developments in both types of modeling. The state of the field is assessed every 2 years through the Critical Assessment of Techniques for Protein Structure Prediction (CASP). The ninth contest, CASP9, in 2011, evaluated over 60,000 template-based models submitted (Mariani et al. 2011). Recent statistical analyses of homology models built using metaservers indicate that sequence identity of >30% between a target sequence and a template provides a very good chance of a model that is accurate enough for things such as site-directed mutagenesis plans (Gront et al. 2012). Sequence identity levels in the 20–30% range yield less accurate models (and the frequency of wrong models increase), whereas identity levels below 20% are generally too low to generate a useable homology model.

Sequence alignments can therefore be used to "thread" the sequence to be modeled onto a known homologous protein structure. One danger for large multidomain proteins such as TRP channels is that a full homology model requires both good templates for each domain to be modeled and untemplated modeling of the connections (either covalent or non-covalent contacts). One must therefore keep in mind that different parts of a homology model warrant different confidence assessments.

Additionally, information from other ion channels should be included in functional and structural assessments. For example, the structure of a GIRK channel bound to PIP_2 [Fig. 1; (Whorton and MacKinnon 2011)] can provide ideas to explore how TRP channels are regulated by phosphoinositides, although its sequence similarity to TRP channels may be too low to be a useful template for homology modeling in the absence of additional constraints.

Considering one of the published TRP channel homology models can serve to illustrate the current challenges. The full-length structure of TRPM8 has been modeled, using Kv2.2/2.1 chimera as a model for the TMD and importin as a model for the N-terminal TRPM homology region (Pedretti et al. 2009), although the 18% sequence identity is below the thresholds described above. The final model contains more than 16% of residues within the disallowed regions of the Ramachandran plot (Pedretti et al. 2009) compared to less than 1% for typical crystal structures, and does not include a C-terminal coiled-coil structure for which there is strong biochemical evidence (Tsuruda et al. 2006). Finally, there is little experimental data on interdomain contacts in TRPM channels to constrain the relative orientation of each domain within the tetrameric assembly.

How can we improve the quality of TRP channel structural models? We need to use the best modeling tools available, but also generate more and better structural templates and accumulate as many other low- or high-resolution constraints such as knowledge of proximities from cross-linking or FRET studies and volumes from EM reconstructions.

Conclusion: The Way Forward in TRP Channel Structural Biology

Although there is continuous progress in TRP channel structural biology, we are still far from the holy grail of seeing the molecular details of a TRP channel in action. In the meantime, many pieces of the structural puzzle have not yet been solved. For example, while we know a lot about the structures of TRPV ARDs, we do not yet have structural insights into the ARDs of TRPC, TRPA, and TRPN channels. Also, a number of TRP domains are unique to their subfamily, such as the N-terminal cytoplasmic TRPM homology region common to all TRPM channels, or the C-terminal Nudix domain of TRPM2. Similarly, structures of TRP channels or their fragments in complex with regulatory partners can provide some insights into the functional consequences of their binding.

It is clear that there are vast differences between TRP channel subfamilies and therefore a need for vast amounts of structural data. While crystal structures will likely be an important component, single-particle electron microscopy techniques are improving and, as recently seen, can attain a resolution that rivals that of X-ray crystallography. NMR can be particularly useful to study questions of dynamics. Finally, techniques for building homology models are also improving, and the availability of scaffolds is rapidly increasing—although it is important to keep in mind that a homology model is simply a model, even when based on a closely related protein, and needs to be thoroughly validated and tested.

Structures of full-length TRP channel will lead to testable hypotheses for biochemical experiments to answer which residues bestow cation selectivity, how the pore is organized, which regions act as sensors or gates, and why certain mutations lead to channelopathies. However, it should be emphasized that a structure never tells the whole story. Ultimately, to completely understand how TRP channels function, we will need structures of at least one representative of each subfamily, as well as experimental evidence of the multiple conformations associated with different functional states to construct a movie of a TRP channel at work.

Acknowledgments We thank members of the Gaudet lab, especially Christina Zimanyi, Sriram Srikant, and Jeffrey MacArthur for feedback and discussions on the manuscript. This work was supported by the National Institutes of Health (Grant R01 GM081340 to R.G.), and U.A.H. is the recipient of an EMBO Long-Term Fellowship.

References

Abbott A (2002) They said it couldn't be done. Nature 418:268–269

Auer-Grumbach M, Olschewski A, Papic L, Kremer H, McEntagart ME, Uhrig S, Fischer C, Frohlich E, Balint Z, Tang B et al (2010) Alterations in the ankyrin domain of TRPV4 cause congenital distal SMA, scapuloperoneal SMA and HMSN2C. Nat Genet 42:160–164

Bates-Withers C, Sah R, Clapham DE (2011) TRPM7, the Mg(2+) inhibited channel and kinase. Adv Exp Med Biol 704:173–183

Bill RM, Henderson PJ, Iwata S, Kunji ER, Michel H, Neutze R, Newstead S, Poolman B, Tate CG, Vogel H (2011) Overcoming barriers to membrane protein structure determination. Nat Biotechnol 29:335–340

Bycroft M, Bateman A, Clarke J, Hamill SJ, Sandford R, Thomas RL, Chothia C (1999) The structure of a PKD domain from polycystin-1: implications for polycystic kidney disease. EMBO J 18:297–305

Cao E, Cordero-Morales JF, Liu B, Qin F, Julius D (2013a) TRPV1 channels are intrinsically heat sensitive and negatively regulated by phosphoinositide lipids. Neuron 77:667–679

Cao E, Liao M, Cheng Y, Julius D (2013b) TRPV1 structures in distinct conformations reveal activation mechanisms. Nature 504:113–118

Caterina MJ, Schumacher MA, Tominaga M, Rosen TA, Levine JD, Julius D (1997) The capsaicin receptor: a heat-activated ion channel in the pain pathway. Nature 389:816–824

Chang Q, Gyftogianni E, van de Graaf SF, Hoefs S, Weidema FA, Bindels RJ, Hoenderop JG (2004) Molecular determinants in TRPV5 channel assembly. J Biol Chem 279:54304–54311

Clayton GM, Altieri S, Heginbotham L, Unger VM, Morais-Cabral JH (2008) Structure of the transmembrane regions of a bacterial cyclic nucleotide-regulated channel. Proc Natl Acad Sci U S A 105:1511–1515

Cvetkov TL, Huynh KW, Cohen MR, Moiseenkova-Bell VY (2011) Molecular architecture and subunit organization of TRPA1 ion channel revealed by electron microscopy. J Biol Chem 286:38168–38176

De-la-Rosa V, Rangel-Yescas GE, Ladron-de-Guevara E, Rosenbaum T, Islas LD (2013) Coarse architecture of the Transient Receptor Potential Vanilloid 1 (TRPV1) ion channel determined by Fluorescence Resonance Energy Transfer (FRET). J Biol Chem 288:29506–29517

Delmas P, Brown DA (2005) Pathways modulating neural KCNQ/M (Kv7) potassium channels. Nat Rev Neurosci 6:850–862

Demeuse P, Penner R, Fleig A (2006) TRPM7 channel is regulated by magnesium nucleotides via its kinase domain. J Gen Physiol 127:421–434

Desai BN, Krapivinsky G, Navarro B, Krapivinsky L, Carter BC, Febvay S, Delling M, Penumaka A, Ramsey IS, Manasian Y et al (2012) Cleavage of TRPM7 releases the kinase domain from the ion channel and regulates its participation in Fas-induced apoptosis. Dev Cell 22:1149–1162

Dick IE, Tadross MR, Liang H, Tay LH, Yang W, Yue DT (2008) A modular switch for spatial Ca2+ selectivity in the calmodulin regulation of CaV channels. Nature 451:830–834

Doyle DA, Morais Cabral J, Pfuetzner RA, Kuo A, Gulbis JM, Cohen SL, Chait BT, MacKinnon R (1998) The structure of the potassium channel: molecular basis of K + conduction and selectivity. Science 280:69–77

Dumetz AC, Snellinger-O'brien AM, Kaler EW, Lenhoff AM (2007) Patterns of protein protein interactions in salt solutions and implications for protein crystallization. Protein Sci 16:1867–1877

Dutzler R, Campbell EB, Cadene M, Chait BT, MacKinnon R (2002) X-ray structure of a ClC chloride channel at 3.0 A reveals the molecular basis of anion selectivity. Nature 415:287–294

Erler I, Hirnet D, Wissenbach U, Flockerzi V, Niemeyer BA (2004) Ca2 + -selective transient receptor potential V channel architecture and function require a specific ankyrin repeat. J Biol Chem 279:34456–34463

Fujiwara Y, Minor DL Jr (2008) X-ray crystal structure of a TRPM assembly domain reveals an antiparallel four-stranded coiled-coil. J Mol Biol 383:854–870

Garcia-Elias A, Mrkonjic S, Pardo-Pastor C, Inada H, Hellmich UA, Rubio-Moscardo F, Plata C, Gaudet R, Vicente R, Valverde MA (2013) Phosphatidylinositol-4,5-biphosphate-dependent rearrangement of TRPV4 cytosolic tails enables channel activation by physiological stimuli. Proc Natl Acad Sci U S A 110:9553–9558

Gaudet R (2008a) A primer on ankyrin repeat function in TRP channels and beyond. Mol Biosyst 4:372–379

Gaudet R (2008b) TRP channels entering the structural era. J Physiol 586:3565–3575

Gaudet R (2009) Divide and conquer: high resolution structural information on TRP channel fragments. J Gen Physiol 133:231–237

Grigoryan G, Keating AE (2008) Structural specificity in coiled-coil interactions. Curr Opin Struct Biol 18:477–483

Gront D, Grabowski M, Zimmerman MD, Raynor J, Tkaczuk KL, Minor W (2012) Assessing the accuracy of template-based structure prediction metaservers by comparison with structural genomics structures. J Struct Funct Genomics 13:213–225

Hellmich UA, Gaudet R (2014) High-resolution views of TRPV1 and their implications for the TRP channel superfamily. In: Nilius B, Flockerzi V (eds) Mammalian Transient Receptor Potential (TRP) cation channels. Springer, Heidelberg

Hong M, Zhang Y, Hu F (2012) Membrane protein structure and dynamics from NMR spectroscopy. Annu Rev Phys Chem 63:1–24

Howard RJ, Clark KA, Holton JM, Minor DL Jr (2007) Structural insight into KCNQ (Kv7) channel assembly and channelopathy. Neuron 53:663–675

Huynh KW, Cohen MR, Chakrapani S, Holdaway HA, Stewart PL, Moiseenkova-Bell VY (2013) Structural Insight into the Assembly of TRPV Channels. Structure 22(2):260–268

Ihara M, Hamamoto S, Miyanoiri Y, Takeda M, Kainosho M, Yabe I, Uozumi N, Yamashita A (2013) Molecular bases of multimodal regulation of a fungal transient receptor potential (TRP) channel. J Biol Chem 288:15303–15317

Inada H, Procko E, Sotomayor M, Gaudet R (2012) Structural and biochemical consequences of disease-causing mutations in the ankyrin repeat domain of the human TRPV4 channel. Biochemistry 51:6195–6206

Ishimaru Y, Katano Y, Yamamoto K, Akiba M, Misaka T, Roberts RW, Asakura T, Matsunami H, Abe K (2010) Interaction between PKD1L3 and PKD2L1 through their transmembrane domains is required for localization of PKD2L1 at taste pores in taste cells of circumvallate and foliate papillae. FASEB J 24:4058–4067

Jacobs MD, Harrison SC (1998) Structure of an IkappaBalpha/NF-kappaB complex. Cell 95:749–758

Jara-Oseguera A, Llorente I, Rosenbaum T, Islas LD (2008) Properties of the inner pore region of TRPV1 channels revealed by block with quaternary ammoniums. J Gen Physiol 132:547–562

Jin X, Touhey J, Gaudet R (2006) Structure of the N-terminal ankyrin repeat domain of the TRPV2 ion channel. J Biol Chem 281:25006–25010

Kobertz WR, Williams C, Miller C (2000) Hanging gondola structure of the T1 domain in a voltage-gated K(+) channel. Biochemistry 39:10347–10352

Kol S, Braun C, Thiel G, Doyle DA, Sundstrom M, Gourdon P, Nissen P (2013) Heterologous expression and purification of an active human TRPV3 ion channel. FEBS J 280(23): 6010–6021

Kon T, Imamula K, Roberts AJ, Ohkura R, Knight PJ, Gibbons IR, Burgess SA, Sutoh K (2009) Helix sliding in the stalk coiled coil of dynein couples ATPase and microtubule binding. Nat Struct Mol Biol 16:325–333

Landoure G, Zdebik AA, Martinez TL, Burnett BG, Stanescu HC, Inada H, Shi Y, Taye AA, Kong L, Munns CH et al (2010) Mutations in TRPV4 cause Charcot-Marie-Tooth disease type 2C. Nat Genet 42:170–174

Landoure G, Sullivan JM, Johnson JO, Munns CH, Shi Y, Diallo O, Gibbs JR, Gaudet R, Ludlow CL, Fischbeck KH et al (2012) Exome sequencing identifies a novel TRPV4 mutation in a CMT2C family. Neurology 79:192–194

Lange A, Giller K, Hornig S, Martin-Eauclaire MF, Pongs O, Becker S, Baldus M (2006) Toxin-induced conformational changes in a potassium channel revealed by solid-state NMR. Nature 440:959–962

LaPlante JM, Falardeau JL, Brown EM, Slaugenhaupt SA, Vassilev PM (2011) The cation channel mucolipin-1 is a bifunctional protein that facilitates membrane remodeling via its serine lipase domain. Exp Cell Res 317:691–705

Lau SY, Procko E, Gaudet R (2012) Distinct properties of Ca2+-calmodulin binding to N- and C-terminal regulatory regions of the TRPV1 channel. J Gen Physiol 140:541–555

Lepage PK, Lussier MP, McDuff FO, Lavigne P, Boulay G (2009) The self-association of two N-terminal interaction domains plays an important role in the tetramerization of TRPC4. Cell Calcium 45:251–259

Lewit-Bentley A, Rety S (2000) EF-hand calcium-binding proteins. Curr Opin Struct Biol 10:637–643

Li A, Tian X, Sung SW, Somlo S (2003) Identification of two novel polycystic kidney disease-1-like genes in human and mouse genomes. Genomics 81:596–608

Li X, Mooney P, Zheng S, Booth CR, Braunfeld MB, Gubbens S, Agard DA, Cheng Y (2013) Electron counting and beam-induced motion correction enable near-atomic-resolution single-particle cryo-EM. Nat Methods 10:584–590

Liao M, Cao E, Julius D, Cheng Y (2013) Structure of the TRPV1 ion channel determined by electron cryo-microscopy. Nature 504:107–112

Lishko PV, Procko E, Jin X, Phelps CB, Gaudet R (2007) The ankyrin repeats of TRPV1 bind multiple ligands and modulate channel sensitivity. Neuron 54:905–918

Long SB, Campbell EB, Mackinnon R (2005) Crystal structure of a mammalian voltage-dependent Shaker family K+channel. Science 309:897–903

Luft JR, Snell EH, Detitta GT (2011) Lessons from high-throughput protein crystallization screening: 10 years of practical experience. Expert Opin Drug Discov 6:465–480

Lux SE, John KM, Bennett V (1990) Analysis of cDNA for human erythrocyte ankyrin indicates a repeated structure with homology to tissue-differentiation and cell-cycle control proteins. Nature 344:36–42

Mariani V, Kiefer F, Schmidt T, Haas J, Schwede T (2011) Assessment of template based protein structure predictions in CASP9. Proteins 79(Suppl 10):37–58

Maruyama Y, Ogura T, Mio K, Kiyonaka S, Kato K, Mori Y, Sato C (2007) Three-dimensional reconstruction using transmission electron microscopy reveals a swollen, bell-shaped structure of transient receptor potential melastatin type 2 cation channel. J Biol Chem 282:36961–36970

McCleverty CJ, Koesema E, Patapoutian A, Lesley SA, Kreusch A (2006) Crystal structure of the human TRPV2 channel ankyrin repeat domain. Protein Sci 15:2201–2206

Minor DL Jr (2007) The neurobiologist's guide to structural biology: a primer on why macromolecular structure matters and how to evaluate structural data. Neuron 54:511–533

Mio K, Ogura T, Kiyonaka S, Hiroaki Y, Tanimura Y, Fujiyoshi Y, Mori Y, Sato C (2007a) The TRPC3 channel has a large internal chamber surrounded by signal sensing antennas. J Mol Biol 367:373–383

Mio K, Ogura T, Kiyonaka S, Mori Y, Sato C (2007b) Subunit dissociation of TRPC3 ion channel under high-salt condition. J Electron Microsc 56:111–117

Moiseenkova-Bell VY, Wensel TG (2009) Hot on the trail of TRP channel structure. J Gen Physiol 133:239–244

Moiseenkova-Bell VY, Stanciu LA, Serysheva II, Tobe BJ, Wensel TG (2008) Structure of TRPV1 channel revealed by electron cryomicroscopy. Proc Natl Acad Sci U S A 105:7451–7455

Molland KL, Paul LN, Yernool DA (2012) Crystal structure and characterization of coiled-coil domain of the transient receptor potential channel PKD2L1. Biochim Biophys Acta 1824:413–421

Moutevelis E, Woolfson DN (2009) A periodic table of coiled-coil protein structures. J Mol Biol 385:726–732

Nieto-Posadas A, Picazo-Juarez G, Llorente I, Jara-Oseguera A, Morales-Lazaro S, Escalante-Alcalde D, Islas LD, Rosenbaum T (2012) Lysophosphatidic acid directly activates TRPV1 through a C-terminal binding site. Nat Chem Biol 8:78–85

Nilius B, Voets T (2013) The puzzle of TRPV4 channelopathies. EMBO Rep 14:152–163

Nilius B, Owsianik G, Voets T (2008) Transient receptor potential channels meet phosphoinositides. EMBO J 27:2809–2816

Numazaki M, Tominaga T, Takeuchi K, Murayama N, Toyooka H, Tominaga M (2003) Structural determinant of TRPV1 desensitization interacts with calmodulin. Proc Natl Acad Sci U S A 100:8002–8006

Park SH, Das BB, Casagrande F, Tian Y, Nothnagel HJ, Chu M, Kiefer H, Maier K, De Angelis AA, Marassi FM et al (2012) Structure of the chemokine receptor CXCR1 in phospholipid bilayers. Nature 491:779–783

Pedretti A, Marconi C, Bettinelli I, Vistoli G (2009) Comparative modeling of the quaternary structure for the human TRPM8 channel and analysis of its binding features. Biochim Biophys Acta 1788:973–982

Persechini A, Moncrief ND, Kretsinger RH (1989) The EF-hand family of calcium-modulated proteins. Trends Neurosci 12:462–467

Petri ET, Celic A, Kennedy SD, Ehrlich BE, Boggon TJ, Hodsdon ME (2010) Structure of the EF-hand domain of polycystin-2 suggests a mechanism for Ca2+-dependent regulation of polycystin-2 channel activity. Proc Natl Acad Sci U S A 107:9176–9181

Phelps CB, Procko E, Lishko PV, Wang RR, Gaudet R (2007) Insights into the roles of conserved and divergent residues in the ankyrin repeats of TRPV ion channels. Channels (Austin) 1:148–151

Phelps CB, Huang RJ, Lishko PV, Wang RR, Gaudet R (2008) Structural analyses of the ankyrin repeat domain of TRPV6 and related TRPV ion channels. Biochemistry 47:2476–2484

Phelps CB, Wang RR, Choo SS, Gaudet R (2010) Differential regulation of TRPV1, TRPV3, and TRPV4 sensitivity through a conserved binding site on the ankyrin repeat domain. J Biol Chem 285:731–740

Prescott ED, Julius D (2003) A modular PIP2 binding site as a determinant of capsaicin receptor sensitivity. Science 300:1284–1288

Qamar S, Vadivelu M, Sandford R (2007) TRP channels and kidney disease: lessons from polycystic kidney disease. Biochem Soc Trans 35:124–128
Qian F, Germino FJ, Cai Y, Zhang X, Somlo S, Germino GG (1997) PKD1 interacts with PKD2 through a probable coiled-coil domain. Nat Genet 16:179–183
Rhoads AR, Friedberg F (1997) Sequence motifs for calmodulin recognition. FASEB J 11: 331–340
Runnels LW, Yue L, Clapham DE (2001) TRP-PLIK, a bifunctional protein with kinase and ion channel activities. Science 291:1043–1047
Ryazanov AG, Pavur KS, Dorovkov MV (1999) Alpha-kinases: a new class of protein kinases with a novel catalytic domain. Curr Biol 9:R43–R45
Saxena K, Dutta A, Klein-Seetharaman J, Schwalbe H (2012) Isotope labeling in insect cells. Methods Mol Biol 831:37–54
Schmitz C, Perraud AL, Johnson CO, Inabe K, Smith MK, Penner R, Kurosaki T, Fleig A, Scharenberg AM (2003) Regulation of vertebrate cellular Mg2+ homeostasis by TRPM7. Cell 114:191–200
Schumacher MA, Rivard AF, Bachinger HP, Adelman JP (2001) Structure of the gating domain of a Ca2+-activated K+channel complexed with Ca2+/calmodulin. Nature 410:1120–1124
Schumann F, Hoffmeister H, Bader R, Schmidt M, Witzgall R, Kalbitzer HR (2009) Ca2+--dependent conformational changes in a C-terminal cytosolic domain of polycystin-2. J Biol Chem 284:24372–24383
Sedgwick SG, Smerdon SJ (1999) The ankyrin repeat: a diversity of interactions on a common structural framework. Trends Biochem Sci 24:311–316
Shi DJ, Ye S, Cao X, Zhang R, Wang K (2013) Crystal structure of the N-terminal ankyrin repeat domain of TRPV3 reveals unique conformation of finger 3 loop critical for channel function. Protein Cell 4:942–950
Shigematsu H, Sokabe T, Danev R, Tominaga M, Nagayama K (2010) A 3.5-nm structure of rat TRPV4 cation channel revealed by Zernike phase-contrast cryoelectron microscopy. J Biol Chem 285:11210–11218
Siemens J, Zhou S, Piskorowski R, Nikai T, Lumpkin EA, Basbaum AI, King D, Julius D (2006) Spider toxins activate the capsaicin receptor to produce inflammatory pain. Nature 444:208–212
Tate CG (2012) A crystal clear solution for determining G-protein-coupled receptor structures. Trends Biochem Sci 37:343–352
Tsiokas L, Kim E, Arnould T, Sukhatme VP, Walz G (1997) Homo- and heterodimeric interactions between the gene products of PKD1 and PKD2. Proc Natl Acad Sci U S A 94:6965–6970
Tsuruda PR, Julius D, Minor DL Jr (2006) Coiled coils direct assembly of a cold-activated TRP channel. Neuron 51:201–212
Tugarinov V, Choy WY, Orekhov VY, Kay LE (2005) Solution NMR-derived global fold of a monomeric 82-kDa enzyme. Proc Natl Acad Sci U S A 102:622–627
Vetter SW, Leclerc E (2003) Novel aspects of calmodulin target recognition and activation. Eur J Biochem 270:404–414
Wang S, Munro RA, Shi L, Kawamura I, Okitsu T, Wada A, Kim SY, Jung KH, Brown LS, Ladizhansky V (2013) Solid-state NMR spectroscopy structure determination of a lipid-embedded heptahelical membrane protein. Nat Methods 10:1007–1012
Whorton MR, MacKinnon R (2011) Crystal structure of the mammalian GIRK2 K+channel and gating regulation by G proteins, PIP2, and sodium. Cell 147:199–208
Wider G (2000) Structure determination of biological macromolecules in solution using nuclear magnetic resonance spectroscopy. Biotechniques 29:1278–1282
Wiener R, Haitin Y, Shamgar L, Fernandez-Alonso MC, Martos A, Chomsky-Hecht O, Rivas G, Attali B, Hirsch JA (2008) The KCNQ1 (Kv7.1) COOH terminus, a multitiered scaffold for subunit assembly and protein interaction. J Biol Chem 283:5815–5830

Xu Q, Minor DL Jr (2009) Crystal structure of a trimeric form of the K(V)7.1 (KCNQ1) A-domain tail coiled-coil reveals structural plasticity and context dependent changes in a putative coiled-coil trimerization motif. Protein Sci 18:2100–2114

Yamaguchi H, Matsushita M, Nairn AC, Kuriyan J (2001) Crystal structure of the atypical protein kinase domain of a TRP channel with phosphotransferase activity. Mol Cell 7:1047–1057

Yu FH and Catterall WA (2004) The VGL-chanome: a protein superfamily specialized for electrical signaling and ionic homeostasis. Sci STKE (253):re15

Yu Y, Ulbrich MH, Li MH, Buraei Z, Chen XZ, Ong AC, Tong L, Isacoff EY, Yang J (2009) Structural and molecular basis of the assembly of the TRPP2/PKD1 complex. Proc Natl Acad Sci U S A 106:11558–11563

Yu Y, Ulbrich MH, Li MH, Dobbins S, Zhang WK, Tong L, Isacoff EY, Yang J (2012) Molecular mechanism of the assembly of an acid-sensing receptor ion channel complex. Nat Commun 3:1252

Zakharian E, Cao C, Rohacs T (2010) Gating of transient receptor potential melastatin 8 (TRPM8) channels activated by cold and chemical agonists in planar lipid bilayers. J Clin Neurosci 30:12526–12534

Zhao X (2012) Protein structure determination by solid-state NMR. Top Curr Chem 326:187–213

Zhou ZH (2011) Atomic resolution cryo electron microscopy of macromolecular complexes. Adv Protein Chem Struct Biol 82:1–35

Zhu MX (2005) Multiple roles of calmodulin and other Ca(2+)-binding proteins in the functional regulation of TRP channels. Pflugers Arch 451:105–115

Zhu J, Yu Y, Ulbrich MH, Li MH, Isacoff EY, Honig B, Yang J (2011) Structural model of the TRPP2/PKD1 C-terminal coiled-coil complex produced by a combined computational and experimental approach. Proc Natl Acad Sci U S A 108:10133–10138

Zimon M, Baets J, Auer-Grumbach M, Berciano J, Garcia A, Lopez-Laso E, Merlini L, Hilton-Jones D, McEntagart M, Crosby AH et al (2010) Dominant mutations in the cation channel gene transient receptor potential vanilloid 4 cause an unusual spectrum of neuropathies. Brain 133:1798–1809

High-Resolution Views of TRPV1 and Their Implications for the TRP Channel Superfamily

Ute A. Hellmich and Rachelle Gaudet

Contents

Abstract

The first high-resolution structures of a near-full-length TRP channel were recently described, structures of the noxious heat receptor TRPV1 in the absence or presence of vanilloid agonists and a spider toxin. Here we briefly review the salient features, including the overall architecture, agonist binding sites, and conformational changes related to channel pore gating. We also discuss some of the structures' implications for the TRP channel family and a few of the many questions still left unanswered.

Keywords

TRPV1 • Cryoelectron microscopy • Resiniferatoxin • Gating • Double-knot toxin • TRP box

The painful sensation evoked by noxious hot temperatures and pungent vanilloid compounds such as capsaicin (found in chili peppers) and resiniferatoxin (RTX; found in *Euphorbia resinifera* cacti-like plants) has been attributed to the activation

U.A. Hellmich (✉) • R. Gaudet (✉)
Harvard University, Cambridge, MA, USA
e-mail: hellmich@fas.harvard.edu; gaudet@mcb.harvard.edu

B. Nilius and V. Flockerzi (eds.), *Mammalian Transient Receptor Potential (TRP) Cation Channels*, Handbook of Experimental Pharmacology 223, DOI 10.1007/978-3-319-05161-1_11,

of the transient receptor potential of subtype vanilloid 1 (TRPV1) ion channel, making TRPV1 an important therapeutic pharmacological target (Szallasi and Sheta 2012). TRPV1 is expressed in peripheral nociceptor neurons, where its activation initiates action potentials leading to an eventual burning pain sensation. TRPV1 is not only activated by natural plant vanilloids but also by endogenous compounds related to arachidonic acid metabolites (Pingle et al. 2007; Starowicz et al. 2007) and by animal venom toxins including the vanillotoxin or double-knot toxin (DkTx) from *Psalmopoeus cambridgei* spiders [(Siemens et al. 2006); this book]. Three recent structures of TRPV1 (Cao et al. 2013b; Liao et al. 2013), one in the absence of ligands, the "apo" state, one in the presence of both DkTx and RTX, and one in the presence of capsaicin, reveal molecular mechanisms for channel gating and provide a 3D scaffold onto which to interpret or reinterpret decades of accumulated physiological, pharmacological, and biochemical data on TRPV1 and other TRP channels. Here we provide a brief introduction to these structures.

1 Three-Dimensional Structure Determination of TRPV1 by Cryoelectron Microscopy

TRPV1 is the namesake member of the vanilloid or TRPV subfamily of TRP channels. The homotetrameric TRPV channels contain a highly variable N-terminal cytoplasmic region that precedes six N-terminal ankyrin repeats, followed by a ~70-residue highly conserved linker region, a transmembrane domain formed by six membrane-spanning segments and a long pore-forming loop between the fifth and sixth segments, and a ~150-residue cytoplasmic C-terminal region.

To determine high-resolution structures of rat TRPV1, some modifications were introduced within its primary sequence, deleting regions predicted to be largely unstructured. Such protein engineering is commonly used in structural biology to improve sample homogeneity. Both the very N- and C-termini were removed, residues 1–109 and 765–838, respectively. In addition, the extracellular turret between the S5 and pore helices (residues 604–626), which includes the only glycosylation site (Rosenbaum et al. 2002), was deleted. Engineered TRPV1 was expressed in HEK293 cells, extracted in dodecylmaltoside detergent, purified, and stabilized by exchanging the detergent with amphipols (Popot et al. 2011), before structure determination. Importantly, this construct retained sensitivity to heat, capsaicin, pH, and DkTx, as assessed in HEK293 cells by electrophysiology (Liao et al. 2013).

The TRPV1 structures were determined using single-particle reconstruction from cryoelectron microscopy (cryoEM) images. CryoEM has always had the theoretical ability to reach atomic resolution but has been plagued by technical challenges that render images much "fuzzier" than their theoretical limit, greatly limiting the resolution of the resulting structures except in rare special cases such as very well-ordered specimens of highly symmetric viral particles (Grigorieff and Harrison 2011). However, the cryoEM technique has seen dramatic technical developments in the last few years (Li et al. 2013), and the TRPV1 structures are some of the first structures demonstrating how these improvements can yield essentially atomic resolution for large molecular complexes with low symmetry for which little-to-no structural information was previously available (Henderson 2013). The technical developments can be summarized in two major advances. The new direct electron detectors have very high sensitivity, low noise, and very fast readout, allowing the accumulation of images as movies (i.e., with a time component) rather than as still images (time averaged). This then enables corrections for the electron beam-induced motion of the sample during image accumulation, generating images that are now exceptionally sharp. The new data format also enables clever processing at different steps in the structure-determination process. At first, averaging all frames of the "movies" yields high signal-to-noise to facilitate accurate semiautomated particle selection. Then eliminating the early frames which have the most beam-induced motion and later frames with too much sample damage produces the most accurate images for classification, averaging, and structure determination. Finally, an important point is that the TRPV1 structure determination was a multistep iterative process with very conservative assessments of data quality, following the strictest standards in the field. Ultimately, fewer than 37 % of the TRPV1 particles from the high-resolution dataset were selected to calculate the high-resolution maps. The published structures therefore likely represent only one of multiple conformational states present in the sample. However, this should not be viewed as "cherry-picking" data; it is instead a judicious choice to use only the most similar and best defined particles to yield the highest resolution map—analogous to "crystallization without crystals" (Frank 2006).

2 Overall Architecture of TRPV1

The TRPV1 structures are tetrameric (Fig. 1), as expected from biochemistry and sequence similarity to the ligand-gated and voltage-gated channel superfamily (Kedei et al. 2001). Notably, the fourfold symmetry was strictly enforced during data processing to obtain the TRPV1 structures, providing fourfold averaging to increase signal intensity. The fact that averaging improved the overall resolution indicates that the four subunits within each TRPV1 tetramer are indeed largely structurally equivalent. However, averaging would also conceal any potential local asymmetries.

The transmembrane domain of each TRPV1 subunit includes six transmembrane helices as well as a shorter pore helix. Similarly to voltage-gated ion channels such as Kv1.2 (Long et al. 2005), the transmembrane helices of each subunit form two distinct "bundles." Viewed from the top, the S1–S4 bundles are located on the periphery of the channel, while the four S5–S6 bundles tetramerize to form the central ion pore (Fig. 1c). Importantly, S4 from one subunit interacts with S5 and S6 of the anticlockwise adjacent subunit; in other words, each peripheral S1–S4 bundle is in close proximity to the S5–S6 bundle of another subunit. Similarly, the S1 helix of one subunit abuts the S5 and pore helices of the anticlockwise neighbor. This

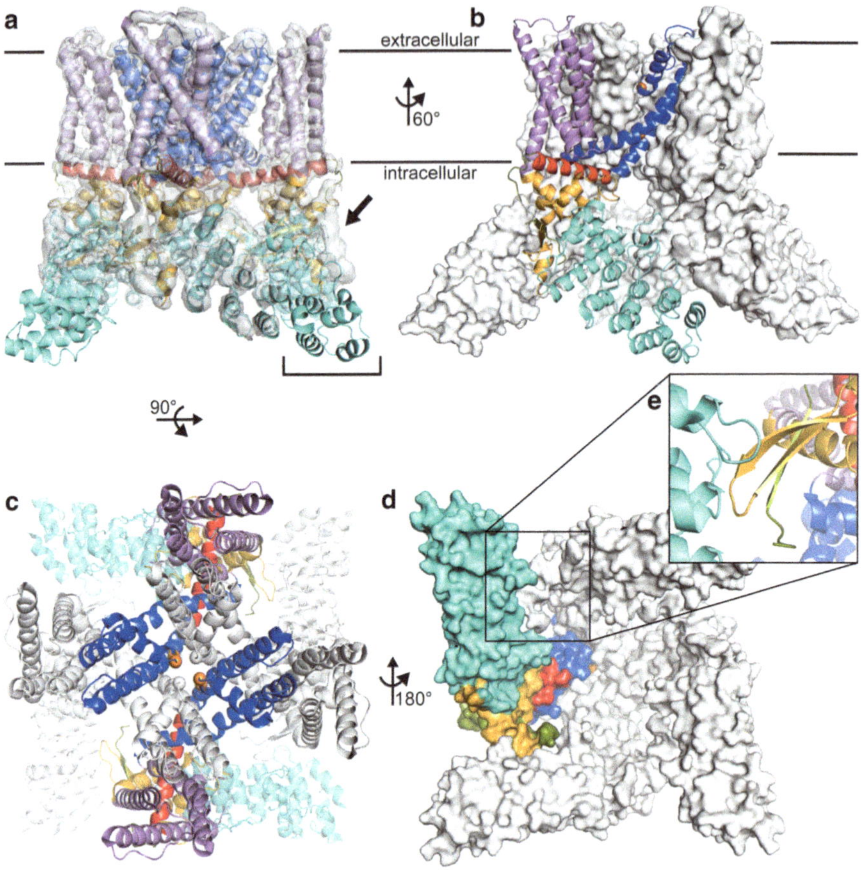

Fig. 1 High-resolution TRPV1 cryoEM structure. (**a**) The cryoEM density (transparent surface) is superimposed on the TRPV1 molecular model (pdb: 3j5p). Note that no density was observed for the first two ankyrin repeats (*bracket*). An *arrow* points to unmodeled cryoEM density near the ARD-S1 linker. (**b**) One subunit is shown as *colored cartoon*, while the other three are shown as *white surface*. The view is rotated 60° on the vertical axis relative to (**a**). Both (**a**) and (**b**) represent "side" views, parallel to the membrane plane. The approximate location of the membrane is indicated by *black bars*. (**c**) View of the TRPV1 tetramer from the extracellular "top" face. Two subunits are *colored* and two are *white*, to highlight the intertwining of the transmembrane domains. (**d**) Intracellular "bottom" face of the TRPV1 tetramer as a surface representation, emphasizing the "bowl" shape of the intracellular domains. One subunit is *colored*. (**e**) Inset shows the relatively small interface between an ARD and the β-sheet of the adjacent subunit. All panels follow the same *color scheme*: ARD, *cyan*; ARD-S1 linker, *yellow*; S1–S4 bundle, *lilac*; S5–S6 pore region, *blue*; TRP box helix, *red*; C-terminus, *green*

same intertwined subunit organization of the transmembrane domain was first observed in the Shaker voltage-gated channel (Long et al. 2005).

As in many tetrameric ion channels, the S5–S6 region of TRPV1, including the intervening pore helix and reentrant loop, forms an "inverted teepee"-shaped pore

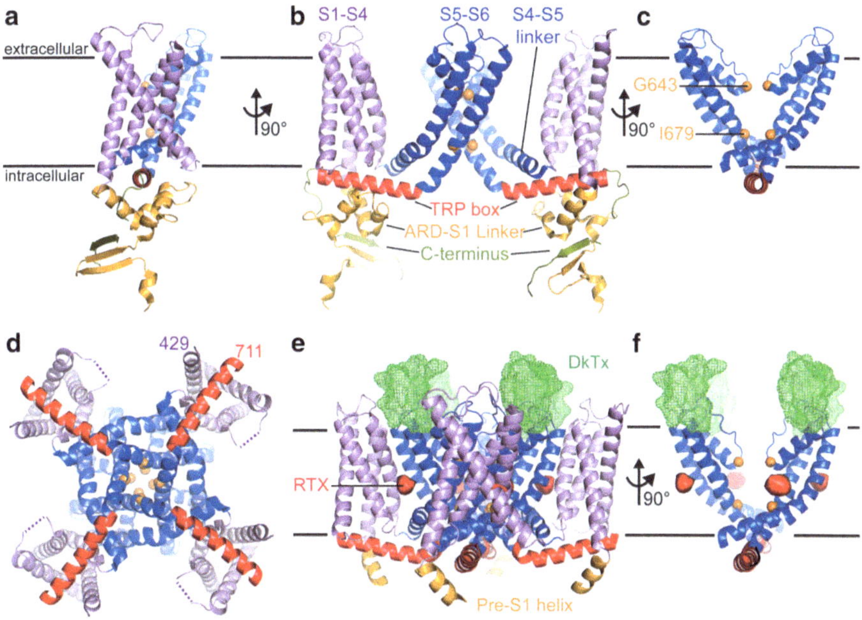

Fig. 2 Details of the TRPV1 structure. (**a**, **b**) The TRP box is straddled by the S1–S4 helix bundle from above and buttressed by the ARD-S1 linker from below, placing it in a perfect position to communicate peripheral signals to the ion channel pore. The S4–S5 linker sits approximately perpendicularly above the TRP box and leads to the S5-pore helix-S6 pore region. The ARD was omitted for clarity. Only one (**a**) or two (**b**) subunits are shown for simplicity, viewed from the membrane plane. (**c**) The arrangement of the pore formed by constrictions at residues G643 at the C-terminus of the pore helix and I679 in helix S6 leading into the TRP box helix. (**d**) Cartoon representation of the transmembrane domain tetramer (residues 429–711) viewed from the intracellular side shows the TRP box helices arranged as spokes on a wheel. Gate residues 643 and 679 are shown as *spheres*. (**e**) The transmembrane region of the RTX and DkTx-bound TRPV1 tetramer. The RTX density, shown as a surface, is nudged in a crevasse formed by the S1–S4 bundle of one subunit and the pore region of the adjacent subunit. Each DkTx knot, shown as a mesh surface, binds the top of the S4 and pore helices from one subunit and the top of S6 from the adjacent subunit. (**f**) The open pore of the RTX and DkTx-bound TRPV1 structure (compare to panel **c**). *Color coding* is the same as in Fig. 1, with RTX in *red* and DkTx in *green*. Panels (**a**)–(**d**) represent the apo structure (pdb: 3j5p) and panels (**e**, **f**) the RTX and DkTx-bound structure (pdb: 3j5q)

(Fig. 2c). In contrast to the tall selectivity filter of potassium channels first observed in KcsA (Doyle et al. 1998), the TRPV1 pore contains a rather short constricted region that likely acts as the selectivity filter (Fig. 2c), consistent with the fact that TRPV1 is a rather nonselective cation channel (Caterina et al. 1997). The TRPV1 pore features two constrictions at glycine 643 and isoleucine 679, which have been assigned as the upper and lower gates, respectively (Fig. 2; more on gating below). In the extracellular (or “upper”) half of the membrane, the short pore helices form a funnel-like shape that tapers from the extracellular face of the channel into the membrane center. At the closest approach of the pore helix C-termini, the four glycine 643 residues point their carbonyl groups towards the center of the pore, forming the upper gate constriction and most likely contributing to cation selectivity. S6 residue tyrosine 671 sits just below glycine 643, consistent with its influence on ion selectivity (Mohapatra et al. 2003). The lower gate is located in the middle of the S6 helix, which lines the pore. The tilt of the S6 helices relative to the plane of the membrane means that they have a single crossing point corresponding to the lower gate, about 15 Å below the upper gate (Fig. 2b). This lower gate is formed by the four hydrophobic isoleucine 679 side chains. Of note, the results of a cysteine-accessibility scan of the TRPV1 S6 helix (Salazar et al. 2009) are in partial but not complete agreement with the cryoEM structure. Overall, the two gates of the TRPV1 pore are placed well within the center of the membrane plane, and the tilted arrangement of the four S5–S6 bundles allows for an aperture-like widening and narrowing of the two gates in response to stimuli.

Helix S6 leads C-terminally into the TRP box, a ~25-residue conserved sequence feature of TRPV, TRPC, and TRPM channels (as well as TRPN channels) implicated in channel gating and regulation (Venkatachalam and Montell 2007; Wu et al. 2010). Residues 690–711 of the TRPV1 TRP box form a long helix parallel to the membrane plane that emanates from the channel center and points towards the periphery, like a spoke on a wheel (Fig. 2d). Interestingly, this means that the N-terminal half of the TRP box helix sits just under the S4–S5 linker, while its C-terminal half lies under the S1–S4 bundle of its own subunit. The S4–S5 linker crosses the TRP box helix approximately perpendicularly, suggesting that movement of the TRP box will influence the S4–S5 linker and thus the pore, and vice versa (Fig. 2a–d). Factors acting on the S4–S5 linker—such as capsaicin or RTX (see below)—will therefore influence the orientation of the TRP box.

The S1–S4 bundle straddles the TRP box helix, with helices S1 and S4 on one side and S2 and S3 on the other (Fig. 2). In voltage-gated channels, the S1–S4 bundle forms the voltage sensor domain, with four highly conserved voltage-sensing arginines on the S4 helix that move across the membrane in response to a voltage stimulus (Catterall 2010). TRP channels are much less voltage-sensitive than voltage-gated channels. Accordingly, few if any basic residues are present in the S4 helix of TRP channels. In TRPV1, the S1–S4 bundle adopts a structure analogous to that of a voltage sensor domain in a depolarized (activated) state (Long et al. 2005), both in the absence or presence of agonists, suggesting that the S1–S4 bundle is essentially static (Cao et al. 2013b). Moreover, its core packs many aromatic residues, further suggesting that it is not a dynamic structure. However, slight rigid-body motions of the entire bundle could readily be transmitted to the channel gates, with the TRP box helix acting as a lever to affect the S4–S5 linker and channel pore.

The TRPV1 intracellular regions, which are both N- and C-terminal to the transmembrane domain in the primary sequence, come together to form an upside-down bowl-shaped structure below the ion channel pore, resembling the "hanging gondola" (Kobertz et al. 2000) described for a large number of ion channels. This bowl opens a large cavity that is accessible to the cytoplasm (Fig. 1d). Although the importance of this intracellular shape is not yet clear,

TRP channels interact with a large number of accessory molecules and proteins which may access their binding sites through this opening.

The N- and C-termini interact with each other through small interfaces (Fig. 1e) and connect to the transmembrane domain mainly through the TRP box helix (Figs. 1 and 2). In fact, the ~70-residue linker region between the N-terminal ankyrin repeat domain (ARD) and the S1 helix, or "ARD-S1 linker," is clearly the hub of these interactions. The ARD-S1 linker, comprised of several short helices and a β-hairpin, wraps the TRP box helix from below, with a short "pre-S1" helix well-positioned for allosteric communications between the intracellular and transmembrane domains (Fig. 2). The ARD-S1-linker effectively extends the ARD, forming a docking site for the exposed face of the sixth ankyrin repeat. In addition a β-sheet formed from the ARD-S1-linker and the C-terminus is in contact with the ARD of an adjacent subunit (Fig. 1e).

The overall structure of the TRPV1 intracellular domains, when viewed from the intracellular side, resembles a right-handed four-bladed pinwheel with each blade corresponding to one ARD (Fig. 1d). Of note, the recently published low-resolution cryoEM structure of TRPV2 shows a left-handed pinwheel-shaped volume (Huynh et al. 2014). However, the structure determination method used for TRPV2, angular reconstitution, arbitrarily assigns handedness, requiring a separate test to determine absolute handedness (Rosenthal and Henderson 2003). Therefore, angular reconstitution can lead to an inadvertent inversion of the resulting 3D volume, particularly for low-resolution symmetric structures where chiral features such as right-handed α-helices are not distinguishable. From our own simple docking, we conclude that the TRPV1 atomic model (Liao et al. 2013) indeed fits very well into an inverted TRPV2 volume, somewhat better than it fits in the published TRPV2 volume. Most importantly, this does suggest that TRPV1 and TRPV2 have very similar structures.

There is extensive accumulated knowledge about the structure of the TRPV ARDs (Gaudet 2008; Hellmich and Gaudet 2014). The TRPV1 ARD consists of six ankyrin repeats, each containing a pair of antiparallel α-helices, the "inner" and "outer" helices, followed by a "finger" loop extending at a ~90° angle relative to the helical axes. The repeats stack side-by-side such that the inner helices and fingers form a concave surface (Lishko et al. 2007). This concave surface is often a site of ligand interactions in other ankyrin repeat-containing proteins (Gaudet 2008). As predicted by a number of biochemical experiments (Gaudet 2008, 2009), the TRPV1 ARDs do not interact directly with each other. Rather, as mentioned above, ARD interactions are mediated by a three-stranded β-sheet that connects the convex face of ankyrin repeat 6 from one subunit at one end and the concave ARD face at finger 3 and ankyrin repeats 3 and 4 of the adjacent subunit at the other end. Two β-strands were assigned to the ARD-S1 linker and the third β-strand to the C-terminus (Figs. 1e and 2a, b; Liao et al. 2013).

However, the density in the β-sheet region is not as well defined as in the transmembrane region, preventing the unequivocal assignment of the sequence register of these strands, and from our own inspection of the publicly available density maps (Lawson et al. 2011), we believe that the current assignment should be considered tentative. Except for this tentative C-terminal β-strand, most of the

53 residues C-terminal to the TRP box could not be unambiguously resolved in the EM density map. Furthermore, the construct used for the structure determination of TRPV1 lacks the most C-terminal 73 residues, including a region that can interact with calmodulin (Lau et al. 2012; Numazaki et al. 2003) and is also important for the regulation of TRPV1 by phosphoinositides (Cao et al. 2013a; Prescott and Julius 2003).

The N-terminal 109 residues were also deleted from the construct and replaced by a maltose-binding protein affinity tag that was then removed by proteolysis before TRPV1 structure determination (Liao et al. 2013). Surprisingly, the EM map shows no density for residues 110–195, corresponding to the first two ankyrin repeats, although the other four repeats are quite well resolved (Fig. 1a). This is unexpected because isolated TRPV channel ankyrin repeats have been very amenable to crystallographic trials and form stable, compact domains without much discernable intrinsic dynamics (Gaudet 2008). It is unclear why no density was observed for the first two ankyrin repeats, and the purification protocol suggests the possibility that they were proteolysed and therefore missing from the EM sample. Interestingly, the low-resolution TRPV2 3D volume does accommodate all six ankyrin repeats, both in the published model (Huynh et al. 2014) and in our own docking into the inverted volume.

The intracellular domains of TRPV1, while well-conserved within the TRPVs, are not common to the other TRP subfamilies. More specifically, while some elements, such as ankyrin repeats, recur in other TRP subfamilies, the TRPV1 sequence regions forming connections between intracellular domains and connecting them to the transmembrane domains (aside from the TRP box) are unique to the TRPV subfamily. Therefore, while the specific arrangement of the TRPV1 intracellular domains can serve as a scaffold to model other TRPVs and interpret biochemical and functional data, it provides only conceptual suggestions of how the other TRP channel intracellular domains are arranged. For example, the ARD-S1 linker region of TRPA1 harbors highly conserved cysteine residues that are covalently modified by electrophilic agonists to activate the channel (Hinman et al. 2006; Kang et al. 2010; Macpherson et al. 2007). The TRPA1 linker has no significant sequence similarity to TRPV1, and its architecture may not follow that of TRPV1 but could similarly form a complex docking structure for the large 17-repeat TRPA1 ARD.

3 Two Gates to the TRPV1 Pore

In addition to the TRPV1 structure in its apo state, two structures in the presence of potent agonists provide insights into ligand gating of the channel (Cao et al. 2013b). A comparison of all three available structures suggests the presence of two gates, an upper and a lower gate (Fig. 2; Cao et al. 2013b; Liao et al. 2013). A structure of TRPV1 bound to both RTX and DkTx, two strong TRPV1 agonists, has the widest pore, which appears to be wide enough to represent a conducting state. Another structure of TRPV1 in the presence of 50 μM capsaicin showed different conformational changes that suggest a shift to a semi-open, possibly corresponding to the previously observed flickering open state (Cao et al. 2013b; Hui et al. 2003).

DkTx is a toxin from *Ornithoctonus huwena* tarantulas that irreversibly activates TRPV1 and leads to enhanced Ca^{2+} influx through the pore, thereby eliciting a pain response (Siemens et al. 2006). This peptide toxin belongs to the family of inhibitor cysteine knot (ICK) toxins and consists of two ICKs with nearly identical sequences connected by a short linker to form a potent bifunctional molecule. Four knots—thought to be from two DkTx molecules—are bound to one TRPV1 tetramer (Fig. 2e, f; Cao et al. 2013b). Each knot is docked at the extracellular membrane interface, interacting with the S4 and pore helices of one subunit and the S6 helix of a second subunit. DkTx is thus perfectly placed to act upon the ion channel pore. Interestingly, DkTx is in close proximity to the deleted extracellular turret (residues 604–626), raising the possibility that it can interact with these residues in wild-type TRPV1. This is consistent with the moderately reduced EC_{50} of DkTx for the deletion construct over wild-type TRPV1 (Liao et al. 2013).

Residues that affect TRPV1 sensitivity to vanilloid compounds such as capsaicin, RTX, and the antagonist capsazepine had been identified primarily in transmembrane segments S3–S4 (e.g., Chou et al. 2004; Gavva et al. 2004; Jordt et al. 2000; Phillips et al. 2004), leading to an expectation that vanilloids would bind to the S1–S4 bundle. EM density consistent with RTX was found in a hydrophobic, membrane-embedded cavity formed by helices S3 and S4 and the S4–S5 linker of one subunit, as well as helices S5 and S6 of the adjacent subunit (Fig. 2e). When viewing TRPV1 from the extracellular side, this vanilloid-binding pocket sits roughly below the DkTx-binding site. The structure determined in the presence of capsaicin showed a smaller EM density feature in the same pocket, consistent with one common binding site for vanilloids (Cao et al. 2013b). This is consistent with previous data indicating that capsaicin approaches its binding site from the cytoplasmic side (Jung et al. 1999). Of note, neither the RTX nor capsaicin density was well defined enough to distinguish the orientation of the ligands in the pocket. Additionally, the authors also observed some density in the same site in the apo structure, possibly corresponding to a lipid or detergent molecule (Cao et al. 2013b), suggesting that this site is in exchange with lipid molecules. TRPV1 is also activated by lipids and lipophilic molecules such as anandamide and diacylglycerols. Consistently, previous studies already indicated a shared binding site for diacylglycerols and capsaicin, and anandamide and RTX, respectively (Woo et al. 2008; Zygmunt et al. 1999).

In both agonist-bound TRPV1 structures, the pore appears open. Capsaicin mainly increased the diameter of the lower gate, while the interplay of DkTx and RTX increased the diameter of both upper and lower gates (Cao et al. 2013b). Because the structural effects on TRPV1 of either DkTx or RTX on their own have not yet been investigated, some assumptions are required to develop a model of allosteric gating by TRPV1 ligands. The effects on the upper gate were assigned to DkTx based on its extracellular position. Similarly, vanilloids were inferred to act mainly through the lower gate. A simplified gating model is illustrated in Fig. 3. In the apo state, the TRPV1 pore is closed at both the upper and lower gates (G643 and

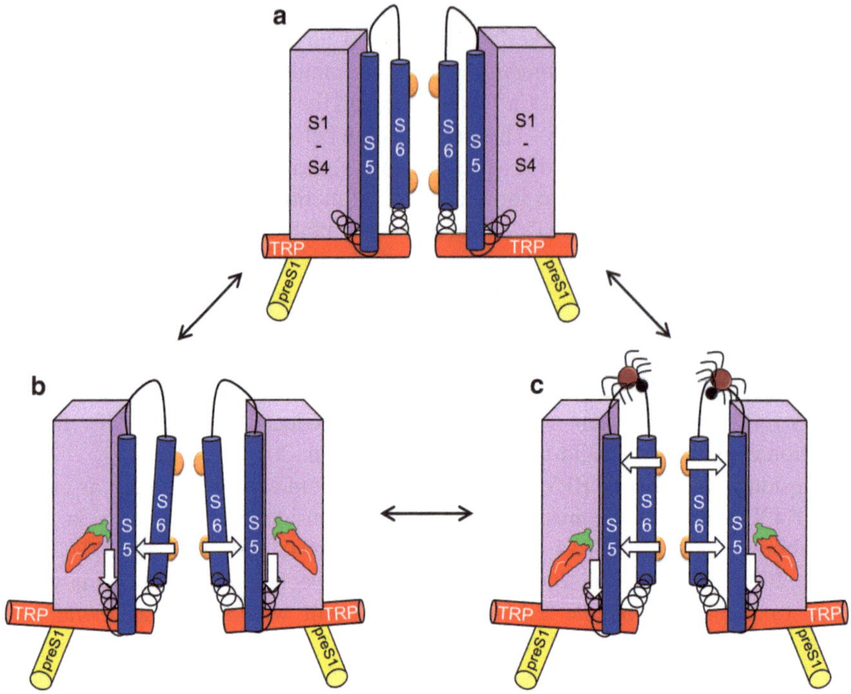

Fig. 3 Model of TRPV1 channel gating by capsaicin, RTX, and DkTx. (**a**) The TRPV1 pore is closed at both the upper and lower gates in the absence of ligands. (**b**, **c**) Upon agonist binding, the pore expands to allow ion flux. Different gates are affected depending on ligand type. (**b**) Capsaicin (*chili pepper*) acts upon the S4–S5 linker that lies on top of the TRP box. The TRP box thus communicates capsaicin binding to the pore and leads to opening of the lower gate in S6. (**c**) Upon binding of both RTX (*chili pepper*) and DkTx (*spider*), both the upper and lower gates open. RTX is presumed to have the same effect as capsaicin, thus opening the lower gate. DkTx binds from the extracellular side and could thus be responsible for the opening of the outer pore. Because it also interacts with the S6 helix, DkTx could also affect the lower gate. *Color coding* follows Figs. 1 and 2

I679, see also Fig. 2c). Vanilloid binding between the S1–S4 and S5–S6 bundles affects the S4–S5 linker, leading to a slight tilt of the TRP box helix and iris-like opening of the connecting S6 helix lining the pore, opening the lower gate. In contrast, DkTx approaches from the extracellular side, leading to opening of the upper gate through binding and slight displacement of the pore helix (Fig. 2c, f). Because DkTx also interacts with helix S6, it may affect the lower gate as well.

The current high-resolution structures have yet to elucidate the physiologically important mechanism of heat sensing by TRPV1. Mutagenesis experiments have indicated the involvement of a number of regions in TRPV1 channel gating by heat, including the pore (Grandl et al. 2010), the extracellular turrets [(Cui et al. 2012; Yang et al. 2010)-results which have been controversial (Yao et al. 2010)], as well as the ankyrin repeats and ARD-S1 linker (Yao et al. 2011). These results suggest that either or both gates could be modulated by the temperature-induced conformational changes.

4 Intracellular Signaling Mechanisms

The intracellular domains of TRP channels are the most variable regions within the TRP protein family. They are also known to be important for channel function—deletions and mutations are often deleterious to channel function, even causing various channelopathies (Nilius and Owsianik 2010; Nilius and Voets 2013). The lack of similarity of these domains and their arrangements to that of other proteins has largely prevented the reliable prediction of their 3D connections and arrangements, although such connections were anticipated.

The structures of TRPV1 do indeed show that the N-terminal region of one subunit is connected to the C-terminal region of the adjacent subunit, with the whole intracellular assembly forming a large inverted-bowl-shaped structure below the channel pore (Fig. 1d). However, the connections between the subunits are relatively small in surface area, leaving open the possibility that they are transient rather than permanent. Also, the density of the intracellular domains is less well defined than the transmembrane domain, suggesting that it is either less symmetric or more flexible, or both. The highly conserved concave surface of the TRPV1 ARD, which has been implicated in ligand interactions (Gaudet 2008; Lishko et al. 2007; Phelps et al. 2007; Phelps et al. 2010), is largely exposed to the inside of the bowl, including several of the residues that cause constitutive activity and/or lack of desensitization. However, some of the concave surface is involved in the intracellular assembly, interacting with the three-stranded β-sheet (Fig. 1d, e). It is possible that the main role of the TRPV1 ARD is to interact with other regions within the channel. Alternatively, there may be physiological states of the channel in which the ARD detaches from the linker to interact with other cellular partners.

The C-terminal region of TRPV1 implicated in PIP_2 regulation (Cao et al. 2013a; Prescott and Julius 2003) and calmodulin binding (Lau et al. 2012; Numazaki et al. 2003) was not included in the construct used to determine these first structures. The only structural element assigned to the C-terminal region is a single β-strand. There is also unassigned density on the outside of the bowl near the ARD-S1 linker (Fig. 1a). Therefore, there are many possible locations for the remaining ~100 C-terminal residues, which could either form a membrane-proximal collar on the outside of the currently available structure or form a plug underneath the intracellular pore opening. Similar speculations can be made for the ~110 missing N-terminal residues, although in contrast to the C-terminus, their removal has little consequence in channel function (Jung et al. 2002; Vlachova et al. 2003).

Conclusion

In summary, the high-resolution TRPV1 structures determined by cryoEM represent a huge leap in our understanding of the TRPV channel architecture.

The TRP box seems perfectly positioned as a nexus to communicate information about stimuli that interact with TRPV1's many domains to the two gates that guard the pore. The TRPV1 intracellular domains form an upside-down bowl structure underneath the pore, with the ARD-S1 linker connecting ARDs to each other and to the TRP box helix. Based on sequence conservation in the TRP channel family, a similar TRP box helix is most likely present at least in the TRPC, TRPM, and TRPN family members. However, the structure of the TRPV1 intracellular domains are likely to only be representative of the TRPV subfamily, and visualizing the intracellular domain architecture of the other subfamilies will require additional structural studies. Finally, it is worth emphasizing that although crystallography will still play a part in discovering atomic-level details crucial for pharmacology, the TRPV1 structures also represent a breakthrough in high-resolution structure determination by single-particle cryoEM, which will likely play a transformative role in future TRP channel structural biology.

Acknowledgments We thank Andres Leschziner, and members of the Gaudet and Leschziner labs for insightful discussions. This work was supported by the National Institutes of Health (Grant R01 GM081340 to R.G.), and U.A.H. is the recipient of an EMBO Long-Term Fellowship.

References

Cao E, Cordero-Morales JF, Liu B, Qin F, Julius D (2013a) TRPV1 channels are intrinsically heat sensitive and negatively regulated by phosphoinositide lipids. Neuron 77:667–679

Cao E, Liao M, Cheng Y, Julius D (2013b) TRPV1 structures in distinct conformations reveal activation mechanisms. Nature 504:113–118

Caterina MJ, Schumacher MA, Tominaga M, Rosen TA, Levine JD, Julius D (1997) The capsaicin receptor: a heat-activated ion channel in the pain pathway. Nature 389:816–824

Catterall WA (2010) Ion channel voltage sensors: structure, function, and pathophysiology. Neuron 67:915–928

Chou MZ, Mtui T, Gao YD, Kohler M, Middleton RE (2004) Resiniferatoxin binds to the capsaicin receptor (TRPV1) near the extracellular side of the S4 transmembrane domain. Biochemistry 43:2501–2511

Cui Y, Yang F, Cao X, Yarov-Yarovoy V, Wang K, Zheng J (2012) Selective disruption of high sensitivity heat activation but not capsaicin activation of TRPV1 channels by pore turret mutations. J Gen Physiol 139:273–283

Doyle DA, Morais Cabral J, Pfuetzner RA, Kuo A, Gulbis JM, Cohen SL, Chait BT, MacKinnon R (1998) The structure of the potassium channel: molecular basis of K+ conduction and selectivity. Science 280:69–77

Frank J (2006) Three-dimensional electron microscopy of macromolecular assemblies. Oxford University Press, Oxford

Gaudet R (2008) A primer on ankyrin repeat function in TRP channels and beyond. Mol Biosyst 4:372–379

Gaudet R (2009) Divide and conquer: high resolution structural information on TRP channel fragments. J Gen Physiol 133:231–237

Gavva NR, Klionsky L, Qu Y, Shi L, Tamir R, Edenson S, Zhang TJ, Viswanadhan VN, Toth A, Pearce LV et al (2004) Molecular determinants of vanilloid sensitivity in TRPV1. J Biol Chem 279:20283–20295

Grandl J, Kim SE, Uzzell V, Bursulaya B, Petrus M, Bandell M, Patapoutian A (2010) Temperature-induced opening of TRPV1 ion channel is stabilized by the pore domain. Nat Neurosci 13:708–714
Grigorieff N, Harrison SC (2011) Near-atomic resolution reconstructions of icosahedral viruses from electron cryo-microscopy. Curr Opin Struct Biol 21:265–273
Hellmich UA, Gaudet R (2014) Structural biology of TRP channels. In: Nilius B, Flockerzi V (eds) Mammalian transient receptor potential (TRP) cation channels. Springer, Heidelberg
Henderson R (2013) Structural biology: ion channel seen by electron microscopy. Nature 504:93–94
Hinman A, Chuang HH, Bautista DM, Julius D (2006) TRP channel activation by reversible covalent modification. Proc Natl Acad Sci U S A 103:19564–19568
Hui K, Liu B, Qin F (2003) Capsaicin activation of the pain receptor, VR1: multiple open states from both partial and full binding. Biophys J 84:2957–2968
Huynh KW, Cohen MR, Chakrapani S, Holdaway HA, Stewart PL, Moiseenkova-Bell VY (2014) Structural insight into the assembly of TRPV channels. Structure 22:260–268
Jordt SE, Tominaga M, Julius D (2000) Acid potentiation of the capsaicin receptor determined by a key extracellular site. Proc Natl Acad Sci U S A 97:8134–8139
Jung J, Hwang SW, Kwak J, Lee SY, Kang CJ, Kim WB, Kim D, Oh U (1999) Capsaicin binds to the intracellular domain of the capsaicin-activated ion channel. J Neurosci 19:529–538
Jung J, Lee SY, Hwang SW, Cho H, Shin J, Kang YS, Kim S, Oh U (2002) Agonist recognition sites in the cytosolic tails of vanilloid receptor 1. J Biol Chem 277:44448–44454
Kang K, Pulver SR, Panzano VC, Chang EC, Griffith LC, Theobald DL, Garrity PA (2010) Analysis of Drosophila TRPA1 reveals an ancient origin for human chemical nociception. Nature 464:597–600
Kędei N, Szabo T, Lile JD, Treanor JJ, Olah Z, Iadarola MJ, Blumberg PM (2001) Analysis of the native quaternary structure of vanilloid receptor 1. J Biol Chem 276:28613–28619
Kobertz WR, Williams C, Miller C (2000) Hanging gondola structure of the T1 domain in a voltage-gated K(+) channel. Biochemistry 39:10347–10352
Lau SY, Procko E, Gaudet R (2012) Distinct properties of Ca2+-calmodulin binding to N- and C-terminal regulatory regions of the TRPV1 channel. J Gen Physiol 140:541–555
Lawson CL, Baker ML, Best C, Bi C, Dougherty M, Feng P, van Ginkel G, Devkota B, Lagerstedt I, Ludtke SJ et al (2011) EMDataBank.org: unified data resource for CryoEM. Nucl Acids Res 39:D456–D464
Li X, Mooney P, Zheng S, Booth CR, Braunfeld MB, Gubbens S, Agard DA, Cheng Y (2013) Electron counting and beam-induced motion correction enable near-atomic-resolution single-particle cryo-EM. Nat Meth 10:584–590
Liao M, Cao E, Julius D, Cheng Y (2013) Structure of the TRPV1 ion channel determined by electron cryo-microscopy. Nature 504:107–112
Lishko PV, Procko E, Jin X, Phelps CB, Gaudet R (2007) The ankyrin repeats of TRPV1 bind multiple ligands and modulate channel sensitivity. Neuron 54:905–918
Long SB, Campbell EB, Mackinnon R (2005) Crystal structure of a mammalian voltage-dependent Shaker family K+ channel. Science 309:897–903
Macpherson LJ, Dubin AE, Evans MJ, Marr F, Schultz PG, Cravatt BF, Patapoutian A (2007) Noxious compounds activate TRPA1 ion channels through covalent modification of cysteines. Nature 445:541–545
Mohapatra DP, Wang SY, Wang GK, Nau C (2003) A tyrosine residue in TM6 of the Vanilloid Receptor TRPV1 involved in desensitization and calcium permeability of capsaicin-activated currents. Mol Cell Neurosci 23:314–324
Nilius B, Owsianik G (2010) Transient receptor potential channelopathies. Pflugers Archiv 460:437–450
Nilius B, Voets T (2013) The puzzle of TRPV4 channelopathies. EMBO Rep 14:152–163

Numazaki M, Tominaga T, Takeuchi K, Murayama N, Toyooka H, Tominaga M (2003) Structural determinant of TRPV1 desensitization interacts with calmodulin. Proc Natl Acad Sci U S A 100:8002–8006

Phelps CB, Procko E, Lishko PV, Wang RR, Gaudet R (2007) Insights into the roles of conserved and divergent residues in the ankyrin repeats of TRPV ion channels. Channels (Austin) 1:148–151

Phelps CB, Wang RR, Choo SS, Gaudet R (2010) Differential regulation of TRPV1, TRPV3, and TRPV4 sensitivity through a conserved binding site on the ankyrin repeat domain. J Biol Chem 285:731–740

Phillips E, Reeve A, Bevan S, McIntyre P (2004) Identification of species-specific determinants of the action of the antagonist capsazepine and the agonist PPAHV on TRPV1. J Biol Chem 279:17165–17172

Pingle SC, Matta JA, Ahern GP (2007) Capsaicin receptor: TRPV1 a promiscuous TRP channel. Handb Exp Pharmacol 179:155–171

Popot JL, Althoff T, Bagnard D, Baneres JL, Bazzacco P, Billon-Denis E, Catoire LJ, Champeil P, Charvolin D, Cocco MJ et al (2011) Amphipols from A to Z. Ann Rev Biophys 40:379–408

Prescott ED, Julius D (2003) A modular PIP2 binding site as a determinant of capsaicin receptor sensitivity. Science 300:1284–1288

Rosenbaum T, Awaya M, Gordon SE (2002) Subunit modification and association in VR1 ion channels. BMC Neurosci 3:4

Rosenthal PB, Henderson R (2003) Optimal determination of particle orientation, absolute hand, and contrast loss in single-particle electron cryomicroscopy. J Mol Biol 333:721–745

Salazar H, Jara-Oseguera A, Hernandez-Garcia E, Llorente I, Arias O II, Soriano-Garcia M, Islas LD, Rosenbaum T (2009) Structural determinants of gating in the TRPV1 channel. Nat Struct Mol Biol 16:704–710

Siemens J, Zhou S, Piskorowski R, Nikai T, Lumpkin EA, Basbaum AI, King D, Julius D (2006) Spider toxins activate the capsaicin receptor to produce inflammatory pain. Nature 444:208–212

Starowicz K, Nigam S, Di Marzo V (2007) Biochemistry and pharmacology of endovanilloids. Pharmacol Ther 114:13–33

Szallasi A, Sheta M (2012) Targeting TRPV1 for pain relief: limits, losers and laurels. Expert Opin Investig Drugs 21:1351–1369

Venkatachalam K, Montell C (2007) TRP channels. Ann Rev Biochem 76:387–417

Vlachova V, Teisinger J, Susankova K, Lyfenko A, Ettrich R, Vyklicky L (2003) Functional role of C-terminal cytoplasmic tail of rat vanilloid receptor 1. J Neurosci 23:1340–1350

Woo DH, Jung SJ, Zhu MH, Park CK, Kim YH, Oh SB, Lee CJ (2008) Direct activation of transient receptor potential vanilloid 1(TRPV1) by diacylglycerol (DAG). Mol Pain 4:42

Wu LJ, Sweet TB, Clapham DE (2010) International union of basic and clinical pharmacology. LXXVI. Current progress in the mammalian TRP ion channel family. Pharmacol Rev 62:381–404

Yang F, Cui Y, Wang K, Zheng J (2010) Thermosensitive TRP channel pore turret is part of the temperature activation pathway. Proc Natl Acad Sci U S A 107:7083–7088

Yao J, Liu B, Qin F (2010) Pore turret of thermal TRP channels is not essential for temperature sensing. Proc Natl Acad Sci U S A 107:E125; author reply E126–E127

Yao J, Liu B, Qin F (2011) Modular thermal sensors in temperature-gated transient receptor potential (TRP) channels. Proc Natl Acad Sci U S A 108:11109–11114

Zygmunt PM, Petersson J, Andersson DA, Chuang H, Sorgard M, Di Marzo V, Julius D, Hogestatt ED (1999) Vanilloid receptors on sensory nerves mediate the vasodilator action of anandamide. Nature 400:452–457

Physiological Functions and Regulation of TRPC Channels

Hwei Ling Ong, Lorena B. de Souza, Kwong Tai Cheng, and Indu S. Ambudkar

Contents

H.L. Ong • L.B. de Souza • K.T. Cheng
Secretory Physiology Section, Molecular Physiology and Therapeutics Branch, National Institute of Dental and Craniofacial Research, National Institutes of Health, Bethesda, MD 20892, USA

I.S. Ambudkar (✉)
Secretory Physiology Section, Molecular Physiology and Therapeutics Branch, National Institute of Dental and Craniofacial Research, National Institutes of Health, Building 10, Room 1N-113, Bethesda, MD 20892, USA
e-mail: indu.ambudkar@nih.gov

B. Nilius and V. Flockerzi (eds.), *Mammalian Transient Receptor Potential (TRP) Cation Channels*, Handbook of Experimental Pharmacology 223, DOI 10.1007/978-3-319-05161-1_12,

Abstract
The TRP-canonical (TRPC) subfamily, which consists of seven members (TRPC1–TRPC7), are Ca^{2+}-permeable cation channels that are activated in response to receptor-mediated PIP_2 hydrolysis via store-dependent and store-independent mechanisms. These channels are involved in a variety of physiological functions in different cell types and tissues. Of these, TRPC6 has been linked to a channelopathy resulting in human disease. Two key players of the store-dependent regulatory pathway, STIM1 and Orai1, interact with some TRPC channels to gate and regulate channel activity. The Ca^{2+} influx mediated by TRPC channels generates distinct intracellular Ca^{2+} signals that regulate downstream signaling events and consequent cell functions. This requires localization of TRPC channels in specific plasma membrane microdomains and precise regulation of channel function which is coordinated by various scaffolding, trafficking, and regulatory proteins.

Keywords
TRPC channels • Ca^{2+} signaling • Protein complex • Trafficking • Regulation • Function

1 Introduction

TRPC channels were first identified as molecular components of the store-operated calcium entry (SOCE) channels (Ambudkar et al. 2007; Parekh and Putney 2005; Venkatachalam and Montell 2007). SOCE is an ubiquitous Ca^{2+} entry mechanism that is activated in response to stimulation of plasma membrane receptors coupled to phosphatidylinositol 4,5-bisphosphate (PIP_2) hydrolysis, inositol 1,4,5-triphosphate (IP_3) generation, and IP_3 receptor (IP_3R)-mediated Ca^{2+} release from the endoplasmic reticulum (ER). The primary trigger for activation of SOCE is depletion of the ER-Ca^{2+} store, while refilling of this store leads to inactivation. The first store-operated Ca^{2+} current to be identified (I_{CRAC}) was the inwardly rectifying and highly Ca^{2+}-selective current that was measured in mast cells and T lymphocytes (Hoth et al. 1993; Hoth and Penner 1992; Parekh and Penner 1997). The channel mediating this current was termed calcium release-activated calcium (CRAC) channel. Later studies revealed currents with varying electrophysiological characteristics in other cell types (Liu et al. 2004; Parekh and Putney 2005). TRPC channels were proposed as possible molecular components of such channels and indeed, several TRPC members have been reported to contribute to SOCE, although data for some TRPCs are not very consistent. TRPC1 was the first mammalian TRPC channel to be cloned (Wes et al. 1995; Zhu et al. 1995), and early studies established that when activated by conditions resulting in store depletion, it is required for the generation of a relatively Ca^{2+}-selective cation current that was termed I_{SOC} (store-operated Ca^{2+} current; Liu et al. 2003) to distinguish it from I_{CRAC}. TRPC1 has been most consistently demonstrated to contribute to SOCE in a variety of cell types (Ambudkar et al. 2007; Beech 2005), although heterologous expression of the channel does not always result in consistent functions.

The critical mechanism that senses the status of ER-[Ca^{2+}] and regulates activation of plasma membrane channels mediating SOCE remained a challenge for more than two decades. This component has now been elucidated, with the discovery of STIM1 as the ER Ca^{2+}-sensor protein involved in regulating the plasma membrane channels. Further, Orai1 has been established as the pore-mediating component of CRAC channels. Of further interest is the finding that activation of TRPC channels following store depletion is dependent not only on STIM1 but also on Orai1 (discussed in detail below). The exact mechanism(s) that regulate TRPC channels in the non-store-operated mode is not yet clearly elucidated, although diacylglycerol (DAG), a product of PIP_2 hydrolysis, has been suggested as an endogenous ligand.

2 Physiological Functions of TRPC Channels

The physiological functions ascribed to TRPCs have been determined in cell cultures and animal models. Some human diseases are also associated with loss or gain of channel function. In cell lines and primary cell cultures, endogenous TRPC channel function has been assessed by decreasing protein expression using shRNA or siRNA (Table 1). TRPC1-mediated Ca^{2+} entry regulates endogenous glioma Cl^- channels to facilitate cell migration by promoting cell shape and volume changes (Cuddapah et al. 2013). The channel is also vital for maintaining permeability of endothelial cell barrier, promoting wound healing following injury to the intestinal epithelial layer and protection against cell cytotoxicity (Bollimuntha et al. 2005b; Paria et al. 2004). Other physiological functions that have been attributed to TRPC1 include cell proliferation and synaptic plasticity (Fiorio Pla et al. 2005; Li et al. 2012a; McGurk et al. 2011). Knocking down endogenous TRPC2 levels or expression of a dominant-negative isoform of TRPC2 in rat thyroid FRTL-5 cells severely impacted cell proliferation and migration, as well as cellular adhesion (Sukumaran et al. 2013). TRPC2 and anoctamin 1 have been proposed to function synergistically to modulate iodide transport in thyroid cells (Viitanen et al. 2013). TRPC3 is involved in proliferation and differentiation of various cell types, such as myoblasts, cardiac fibroblasts, and primary T cells (Harada et al. 2012; Wenning et al. 2011; Woo et al. 2010). In some cells, more than one TRPC channels have been shown to regulate the same physiological event. For example, TRPC3, TRPC4, and TRPC5 facilitate in vitro endothelial tube formation by promoting proliferation of endothelial cells (Antigny et al. 2012). TRPC5 and TRPC6 regulate migration of fibroblasts and kidney podocytes in an antagonistic manner, whereby TRPC5 activates Rac1 to promote motility but TRPC6 activates RhoA to inhibit motility (Tian et al. 2010). TRPC5 also plays an important role in facilitating migration of vascular smooth muscle cells, as well as regulating neurite

Table 1 Cell-based assays used to delineate physiological functions of TRPC1–7 channels

TRPC	Physiological function	Protein manipulation	Cell type	References
1	Cell migration	Knockdown	Human malignant gliomas	Cuddapah et al. (2013)
		Overexpression, knockdown	Intestinal epithelial cells	Bomben et al. (2011), Rao et al. (2006)
	Cell proliferation	Knockdown	Neural stem, hippocampal neural progenitor cells	Fiorio Pla et al. (2005), Li et al. (2012a), McGurk et al. (2011)
	Synaptic plasticity	Knockdown	Neuromuscular junctions	McGurk et al. (2011)
2	Cell proliferation, migration	Knockdown, dominant negative	Rat thyroid cells	Sukumaran et al. (2013)
	Iodide transport	Knockdown	Rat thyroid cells	Viitanen et al. (2013)
3	Cell proliferation	Knockdown	Muscular dysgenic myoblasts, cardiac fibroblasts, endothelial cells, pontine neurons, primary T cells	Antigny et al. (2012), Harada et al. (2012), Li et al. (2005), Wenning et al. (2011), Woo et al. (2010)
	Cell differentiation	Knockdown	Muscular dysgenic myoblasts, cardiac fibroblasts	Harada et al. (2012), Woo et al. (2010)
4	Cell proliferation	Knockdown	Endothelial cells	Antigny et al. (2012)
5	Cell migration	Knockdown	Fibroblasts, kidney podocytes	Tian et al. (2010)
		Antibody block	Vascular smooth muscle cells	Greka et al. (2003), Tian et al. (2010), Xu et al. (2006)
	Cell proliferation	Knockdown	Endothelial cells	Antigny et al. (2012)
	Neurite extension, growth cone morphology	Dominant-negative, interacting protein knockdown	Hippocampal neurons	Greka et al. (2003), Tian et al. (2010), Xu et al. (2006)
6	Cell migration	Knockdown	Fibroblasts, kidney podocytes	Tian et al. (2010)
	Cell proliferation	Knockdown	Pontine neurons, prostate cancer epithelial cells	Li et al. (2005), Thebault et al. (2006)

extension and growth cone morphology of hippocampal neurons (Greka et al. 2003; Tian et al. 2010; Xu et al. 2006). TRPC6 regulates the growth of pontine neurons (in addition to TRPC3) (Li et al. 2005) and prostate cancer epithelial cells (Thebault et al. 2006).

The physiological functions of TRPC channels has been revealed by studies using knockout mouse models (discussed in further detail in Chapters 2 to 8 of volume 1,

"TRPC1", "TRPC2", "TRPC3: A Multifunctional Signaling Molecule", "TRPC4- and TRPC4-Containing Channels"; "TRPC5", "TRPC6: Physiological Function and Pathophysiological Relevance" and "Transient Receptor Potential Canonical 7: A Diacylglycerol-Activated Non-selective Cation Channel"). In addition to knockout mouse models, mice expressing dominant-negative isoforms of TRPC3, TRPC4, and TRPC6 reveal that these channels are involved in the development of cardiac hypertrophy via a calcineurin-NFAT signaling pathway (Wu et al. 2010). Various dominant-negative isoforms of TRPCs have been generated, e.g., pore-dead channel created by mutations in the pore region. Others, such as the TRPC3 N-terminus (amino acids 1–302), also exert dominant-negative effects when overexpressed by disrupting channel assembly since TRPCs interact via their N-terminal regions (Balzer et al. 1999). In the case of TRPC4, an N-terminal fragment that includes the first ankyrin-like repeat has been used (Schindl et al. 2008). In other studies, TRPC function has been revealed by using disease models. A mouse model for Parkinson's disease (PD) shows the vital role of TRPC1 in maintaining calcium homeostasis, promoting neuronal survival to limit neuronal degeneration, and possibly slowing down or preventing the onset of progression of PD. PD-associated symptoms are ameliorated by heterologous expression of TRPC1 in neuronal cells or in vivo intranigral injection of TRPC1-containing adenovirus particles in PD mouse model (Selvaraj et al. 2009, 2012).

A number of other mouse models also reveal important information regarding TRPC channel regulation. Physiological functions attributed to TRPC1 are severely affected in caveolin-deficient (Cav-1$^{-/-}$) and Homer1-deficient (Homer1$^{-/-}$) mice. Knocking out caveolin-1 (Cav-1) results in mislocalization of TRPC1 due to aberrant trafficking, leading to impaired channel function that significantly reduces salivary gland fluid secretion (Pani et al. 2012). Loss of Homer1, a scaffolding protein that mediates TRPC1 interaction with the IP_3R, causes aberrant calcium signaling resulting in skeletal myopathy (Stiber et al. 2008). The *Mecp2* mutant mice are a model system for Rett syndrome, which is caused by loss-of-function mutations in the *Mecp2* gene. These mice display sensory and motor abnormalities due to loss of TRPC3 function in hippocampal neurons, although potential contributions from TRPC6 or TRPC7 have not been ruled out (Li et al. 2012b). Interestingly, in some studies, an increase of TRPC expression and function has been proposed to underlie disease onset and/or progression. Studies with Duchenne muscular dystrophy (*mdx*) mice demonstrate an increase in TRPC1-mediated Ca^{2+} influx induces muscle damage (Gervasio et al. 2008; Williams and Allen 2007). Expression of TRPC1, TRPC5, and TRPC6 is significantly elevated in adrenal medulla of Ossabaw miniature pigs are used to study the metabolic syndrome or pre-diabetes state (Hu et al. 2009). Perturbations in Ca^{2+} signaling and homeostasis have been correlated with increased TRPC3 expression and function in cardiomyocytes obtained from muscle LIM protein knockout mice (model for the myocardial disorder, dilated cardiomyopathy) (Kitajima et al. 2011) and spontaneously hypertensive rats (model for hypertension) (Adebiyi et al. 2012; Bush et al. 2006; Noorani et al. 2011). TRPC7 has been implicated in myocardial apoptosis failure as its expression is upregulated in Dahl salt-sensitive rats with heart failure and has been proposed to be a novel target for treatment of heart failure (Satoh et al. 2007).

The genes encoding TRPC channels have also been linked to various human diseases, such as cardiovascular, pulmonary, and neurological, as well as cancer (Nilius and Owsianik 2010). For example, *trpc5* and *trpc6* loci are linked with infantile hypertrophic pyloric stenosis, a very common condition of stomach obstruction that is characterized by projectile vomiting. Increased *trpc6* promoter activity and TRPC6 expression have been linked to the development of idiopathic pulmonary arterial hypertension, which is caused by excessive proliferation of pulmonary artery smooth muscle cells. Nonetheless, the only TRPC-related channelopathy reported so far is focal and segmental glomerulosclerosis (FSGS), which is linked to a mutation of the *trpc6* gene. These mutations resulted in alterations of residues in the N- and C-termini, leading to significantly elevated TRPC6-mediated calcium signaling that may affect channel interaction with podocyte structural proteins, leading to defects in the filtration barrier. Alternatively, the elevated calcium signaling mediated via TRPC6 may lead to apoptosis, resulting in a defective permeability barrier (Mukerji et al. 2007; Nilius and Owsianik 2010). Nonetheless, some TRPC6 variants linked to FSGS have also been reported to not cause any change in channel activity (Reiser et al. 2005).

3 TRPC Channel Complexes

Much of the initial insights into TRPC protein interactions are based on studies with the *Drosophila* TRP channel which is localized in the *Drosophila* eye and plays a critical role in phototransduction (Venkatachalam and Montell 2007). This TRP channel resides in a multiprotein signalplex with proteins that are important for proper channel assembly, retention, activity, regulation of phototransduction, and downstream signaling. The scaffolding protein INAD forms the core of this complex since it has the ability, via multiple PDZ domains, to bind to numerous signaling proteins and serve as a platform for their interaction with TRP and regulation of channel function (Venkatachalam and Montell 2007). Critical amino acid sequences that are conserved in TRP channel families appear to be involved in these various, but specific, protein–protein interactions. These include the coiled-coiled domain, ankyrin repeat region, calmodulin- and lipid-binding domains, as well as other less well-characterized protein binding domains. Since mammalian TRPC proteins share many of the same structural components as the *Drosophila* TRP channel, it has been hypothesized that the individual TRPC protein is also capable of forming homomeric or heteromeric interactions with other TRPC channels and signaling proteins. It is now well established that a number of key signaling and scaffolding proteins are associated with mammalian TRPC channels (Ambudkar et al. 2006; Ambudkar and Ong 2007; Kiselyov et al. 2007).

3.1 TRPC1

TRPC1 interacts with other TRPCs to form channels with diverse properties, ranging from relatively Ca^{2+}-selective to non-selective (Ca^{2+} vs. Na^{+}) (Cheng et al. 2013). In human submandibular gland (HSG) cells, TRPC1 contributes to a relatively Ca^{2+}-selective cation channel, possibly via a homomeric TRPC channel (Liu et al. 2004). Several TRPCs are endogenously expressed in cells, for example, TRPC1 and TRPC3 in HEK293 cells and neuronal cells (Zhu et al. 1995, 1996). Based on the association of TRPC1 and TRPC3 in heterologous expression systems, it can be suggested that the endogenous channels can also assemble in heteromeric complexes. Indeed, endogenous heteromeric TRPC channels have been described in different cell types: e.g., TRPC1 + TRPC3 in HSY cells (Liu et al. 2005), TRPC1 + TRPC3 + TRPC7 in HEK293 cells (Zagranichnaya et al. 2005), TRPC1 + TRPC4 in mesangial cells (Sours-Brothers et al. 2009) and endothelial cells (Sundivakkam et al. 2012), and TRPC1 + TRPC5 in neuronal cells, vascular endothelial cells, and vascular smooth muscle cells (Goel et al. 2002; Shi et al. 2012; Strubing et al. 2001; Xu et al. 2006). TRPC1 forms a macromolecular complex with TRPC6, SERCA, and IP_3R following passive depletion of the ER-Ca^{2+} stores in human platelets (Redondo et al. 2008).

TRPC1 also interacts with non-TRPC channels, such as Orai1 (Cheng et al. 2008; Lu et al. 2010), TRPV4 (Ma et al. 2010, 2011), and TRPV6 (Schindl et al. 2012). The association with Orai1 is a critical determinant of TRPC1 function (further discussed below). Although it is unclear whether there is a physical interaction between the two channels, studies have clearly established that Orai1 and TRPC1 form distinct STIM1-gated channels in the membrane that are activated following store depletion (Cheng et al. 2008; Lu et al. 2010). In a recent study, a splice variant of TRPC1 has been shown to regulate the activity of Orai1 (Ong et al. 2013). TRPC1 + TRPV4 forms a heteromeric channel involved in SOCE in vascular smooth muscle cells as well as endothelial cells (Ma et al. 2010, 2011). TRPC1 can also interact with and negatively regulate TRPV6 channel activity, without generation of a heteromeric channel, in HEK293 cells (Schindl et al. 2012).

In addition to calcium channels, TRPC1 interacts with a wide range of signaling proteins, as well as scaffolding and trafficking proteins (Table 2). The TRPC1 signaling complex contains key Ca^{2+} signaling proteins that function upstream in the agonist-activated signaling cascade, such as PLC, CaM, $G_{q/11}$, IP_3R, PMCA, SERCA, and STIM1 (Cheng et al. 2008; Heo et al. 2012; Huang et al. 2006; Lockwich et al. 2000; Lu et al. 2010; Ng et al. 2009; Ong et al. 2007; Pani et al. 2009; Redondo et al. 2008; Selvaraj et al. 2012 ; Singh et al. 2002; Sundivakkam et al. 2009; Tang et al. 2001; Yuan et al. 2003). Such findings have led to the proposal that TRPC1 channel complexes are composed of proteins from both ER and plasma membranes and likely represent cellular microdomains where these two membranes are in close proximity to each other. The interaction with STIM1 is critically required for channel activation following store depletion. Additionally, TRPC1 activity is also regulated via its binding to IP_3R as it has the CaM-/IP_3R-binding (CIRB) domain in the C-terminus. Both CaM and IP_3R bind competitively to TRPC1 to modulate channel activity, with IP_3R involved in channel activation and CaM regulating the Ca^{2+}-dependent feedback inhibition (Singh et al. 2002; Tang et al. 2001). It is interesting that the STIM1- and IP_3R-binding domains lie in close proximity in the C-terminus of TRPC1. However, it is yet unclear whether STIM1 and the IP_3R are simultaneously involved in activation

Table 2 Interacting partners for TRPC1–7 channels

Channel	Other channels	Signaling proteins	Scaffolding and trafficking proteins
TRPC1	TRPC1[a] TRPC3 (Xu et al. 1997) TRPC4, TRPC5 (Strubing et al. 2001) TRPC6 (Strubing et al. 2003) TRPC7 (Zagranichnaya et al. 2005) TRPV4 (Ma et al. 2010) TRPV6 (Schindl et al. 2012) Orai1 (Cheng et al. 2008)	IP_3R, CaM, $G_{q/11}$ (Lockwich et al. 2000) PLCγ (Tu et al. 2005) PMCA (Singh et al. 2002) SERCA (Redondo et al. 2008) STIM1 (Huang et al. 2006)	β-tubulin (Bollimuntha et al. 2005a) Cav-1 (Lockwich et al. 2000) Enkurin (Sutton et al. 2004) Homer (Yuan et al. 2003) MxA (Lussier et al. 2005) RhoA (Mehta et al. 2003) SNAP-25, VAMP (Redondo et al. 2004)
TRPC2	TRPC6 (Chu et al. 2004; Tong et al. 2004)	CaM (Tang et al. 2001) epoR, IP_3R, PLCγ (Chu et al. 2004; Tong et al. 2004) STIM1 (Huang et al. 2006)	Enkurin (Sutton et al. 2004) Homer1 (Yuan et al. 2003) RTP1 (Mast et al. 2010)
TRPC3	TRPC1, TRPC4, TRPC5 (Strubing et al. 2003) TRPC6, TRPC7 (Hofmann et al. 2002) Orai1 (Liao et al. 2007)	IP_3R (Kiselyov et al. 1999) CaM (Zhang et al. 2001) CSR (Bandyopadhyay et al. 2012) $G_{q/11,}$ PLCβ (Lockwich et al. 2001) PLCγ (Patterson et al. 2002) PMCA (Kim et al. 2006a) RACK1 (Bandyopadhyay et al. 2008) SERCA (Lockwich et al. 2001)	AP-2, clathrin, dynamin, synaptotagmin (Lockwich et al. 2008) Cav-1, Ezrin (Lockwich et al. 2001) Homer (Kim et al. 2006a) MxA (Lussier et al. 2005) RACK (Bandyopadhyay et al. 2008) SNARES, syntaxin, VAMP2 (Singh et al. 2004)
TRPC4	TRPC1 (Strubing et al. 2001) TRPC3, TRPC6(Strubing et al. 2003) TRPC5 (Hofmann et al. 2002)	Fyn (Odell et al. 2005) IP_3R, CaM (Tang et al. 2001) PLCβ (Tang et al. 2000) Protein 4.1 (Cioffi et al. 2005) SESTD1 (Miehe et al. 2010) STIM1 (Huang et al. 2006)	Cav-1 (Murata et al. 2007) Homer (Yuan et al. 2003) MxA (Lussier et al. 2005) NHERF (Tang et al. 2000) ZOI (Song et al. 2005)
TRPC5	TRPC1 (Strubing et al. 2001) TRPC4 (Hofmann et al. 2002)	IP_3R, CaM (Tang et al. 2001) NCS-1 (Hui et al. 2006) SESTD1 (Miehe et al. 2010) STIM1 (Huang et al. 2006)	AP-2, clathrin, dynamin (Goel et al. 2005) EB50, NHERF (Tang et al. 2000) Enkurin(Sutton et al. 2004) Homer (Yuan et al. 2003) MxA (Lussier et al. 2005) PI(3)K, PIP(5)K, Rac1 (Bezzerides et al. 2004) Stathmin (Greka et al. 2003)
TRPC6	TRPC1, TRPC4, TRPC5 (Strubing et al. 2003) TRPC2 (Chu et al. 2004) TRPC3, TRPC7	IP_3R, CaM, Calcineurin (Tang et al. 2001) FKBP12 (Kim and Saffen 2005)	Clathrin, dynamin (Goel et al. 2005) MxA (Lussier et al. 2005) PI(3)K, PTEN (Monet

(continued)

Table 2 (continued)

Channel	Other channels	Signaling proteins	Scaffolding and trafficking proteins
	(Hofmann et al. 2002) Orai1 (Liao et al. 2007)	Fyn (Hisatsune et al. 2004) $G_{\alpha q/11}$ (Bandyopadhyay et al. 2005) mAChR, PKC (Kim and Saffen 2005) PLCγ (Hirschler-Laszkiewicz et al. 2009) SERCA (Redondo et al. 2008)	et al. 2012) Rab9, Rab11 (Cayouette et al. 2010) RhoA (Tian et al. 2010) Syntaxin (Bandyopadhyay et al. 2005)
TRPC7	TRPC1 (Zagranichnaya et al. 2005) TRPC3 (Hofmann et al. 2002) TRPC5 (Saleh et al. 2008) TRPC6 (Hofmann et al. 2002)	cGMP-dependent protein kinase (Yuasa et al. 2011) IP_3R, CaM (Tang et al. 2001)	MxA (Lussier et al. 2005)

Channel–protein interactions are shown using methods such as immunoprecipitation, yeast two hybrid assays, GST fusion protein pull-down, and microscopy imaging techniques (e.g., immunofluorescence, TIRF, and FRET)

[a]Ambudkar and Singh (unpublished results)

of TRPC1. It is interesting to speculate that the level and type of physiological stimuli may have an impact on the channel regulation. Interaction between TRPC1 and IP_3R has been reported to be mediated by RhoA in endothelial cells (Mehta et al. 2003) and Homer1 in HEK293 cells (Kiselyov et al. 2007). Additionally, RhoA (Mehta et al. 2003) and other proteins such as Cav-1 (Ambudkar et al. 2006; Brazer et al. 2003; Lockwich et al. 2000) and β-tubulin (Bollimuntha et al. 2005a) affect surface expression of TRPC1. TRPC1 interaction with Cav-1 and RhoA is suggested to mediate its localization in lipid raft domains where TRPC1 channels are assembled and activated in response to store depletion (Pani et al. 2008). Further studies will be required to establish the exact contributions of each interacting protein in the regulation of TRPC1.

3.2 TRPC2

While the human *Trpc2* is a pseudogene and does not form a functional channel (Wes et al. 1995; Zhu et al. 1995), TRPC2 in other mammals (e.g., rat, bovine, and mouse) forms functional channels in different cell types and tissues, such as the vomeronasal organ (VNO), testis, spleen, and liver (Liman et al. 1999 ; Vannier et al. 1999; Wissenbach et al. 1998). Few studies have looked at the interactions between TRPC2 and other TRPC channels and various signaling proteins. When heterologously expressed in HEK293 cells, TRPC2 interacts with endogenous

Homer1 and IP_3R (Yuan et al. 2003), but not with other TRPCs (Hofmann et al. 2002). Nonetheless, TRPC2 has been shown to interact with TRPC6 and signaling proteins, erythropoietin receptor, IP_3R, and PLCγ in primary erythroblasts (Chu et al. 2004; Tong et al. 2004). Additionally, TRPC2 forms a signaling complex with the receptor-transporting protein 1 (RTP1), Homer1, and IP_3R in the VNO (Mast et al. 2010). Other signaling proteins that interact with TRPC2 include STIM1 (Huang et al. 2006) and CaM (Tang et al. 2001; Yildirim et al. 2003). TRPC2 has been reported to co-localize with anoctamin 1 in the vomeronasal epithelium (Dibattista et al. 2012), although the interaction between the two proteins has not been confirmed using other techniques such as immunoprecipitation, FRET and TIRF.

3.3 TRPC3

While TRPC3, TRPC6, and TRPC7 share considerable homology in their amino acid sequences, as well as modes of activation, their physiological properties and function are quite distinct (Owsianik et al. 2006; Putney 2005). Depending on the level of expression and its heteromeric interactions with other TRPC channels, TRPC3 can form both store-independent and store-dependent channels in different cell types. As shown in Table 2, TRPC3 interacts with almost every member of the TRPC subfamily, as well as TRPM4 (Park et al. 2008) and Orai1 (Liao et al. 2007; Woodard et al. 2010). A fairly comprehensive list of TRPC3-associated proteins was identified in an earlier proteomic study, including proteins associated with Ca^{2+} entry and signaling, neural growth, vesicle fusion, mitochondria, endocytosis, actin cytoskeleton, and microtubules (Lockwich et al. 2008). As noted above, TRPC1 + TRPC3 and TRPC1 + TRPC3 + TRPC7 contribute to SOCE. TRPC3 has been suggested to act concertedly with TRPC1 to mediate SOCE in H19-7 hippocampal neuronal cells (Wu et al. 2004). Store dependence of TRPC3 might also be mediated by its interactions with Orai1 (Liao et al. 2007) and STIM1, the latter likely dependent on an interaction of TRPC3 with TRPC1 (Yuan et al. 2007). In addition to activation via the G protein/PLC-mediated pathway, heteromeric TRPC3 + TRPC4 channels in porcine aortic endothelial cells are redox activated (Poteser et al. 2006). Further research is required to delineate the molecular interactions involved in regulating TRPC3 channel assembly and function.

As seen with TRPC1, TRPC3 also interacts with a number of key Ca^{2+} signaling proteins involved in receptor-stimulated Ca^{2+} mobilization, such as PIP_2 hydrolysis (PLCβ, $G_{q/11}$), IP_3R, and the calcium-sensing receptor (CSR) (Table 2). SERCA and PMCA pumps also co-immunoprecipitate with TRPC3 (Bandyopadhyay et al. 2005; Kiselyov et al. 2007; Lockwich et al. 2001, 2008). Further, scaffolding proteins such as Homer or RACK1 interact with TRPC3 and modulate its interaction with IP_3R (Bandyopadhyay et al. 2008; Kiselyov et al. 2007). A number of protein interactions are involved in plasma membrane localization of TRPC3 (further discussed below).

3.4 TRPC4

TRPC4 is most closely related to TRPC5, sharing 65 % amino acid identity, but both proteins diverge in the last 220 amino acids. There is general consensus that TRPC4 forms an SOC channel even though it has been shown to form constitutively active or store-independent channels in some studies (Parekh and Putney 2005; Venkatachalam and Montell 2007). Heteromeric interactions have been described between TRPC4 and other TRPCs (Table 2) (Alvarez et al. 2008; Ambudkar et al. 2006; Ambudkar and Ong 2007; Antoniotti et al. 2006; Chen et al. 2009; Cheung et al. 2011; Murata et al. 2007; Phelan et al. 2013; Poteser et al. 2006; Puram et al. 2011; Riccio et al. 2009; Sabourin et al. 2009; Sundivakkam et al. 2012; Woo et al. 2008; Zimmermann et al. 2011). As described above for TRPC1 and TRPC3, TRPC4 heteromultimerizes with TRPC6 and, via its direct interaction with STIM1, forms a TRPC4 + TRPC6 channel that is store-dependent (Yuan et al. 2007). In intestinal smooth muscle cells, TRPC4 and TRPC6 channels are simultaneously activated by muscarinic receptors and contribute independently to the muscarinic receptor-induced cation current. Therefore, TRPC4 and TRPC6 channels couple muscarinic receptors to depolarization of intestinal smooth muscle cells and voltage-activated Ca^{2+} influx and contraction, thereby accelerating small intestinal motility in vivo (Ambudkar 2009; Tsvilovskyy et al. 2009). The interaction between STIM1 and TRPC4 was proposed to be the activation mechanism of the heteromeric TRPC1 + TRPC4 channels in glomerular mesangial cells (Sours-Brothers et al. 2009). Another protein vital for TRPC4 activity is protein 4.1, which functionally links TRPC4 to the actin cytoskeleton and spectrin in endothelial cells (Cioffi et al. 2005). Protein 4.1 and another adaptor protein, SESTD1, have been proposed to stabilize TRPC4 in a macromolecular complex associated with the cytoskeleton. SESTD1 associates with both TRPC4 and TRPC5 via the CIRB domain and functions to couple TRPC channel activity to lipid signaling (Miehe et al. 2010).

Signaling proteins involved in interactions with TRPC4 include the PDZ-domain proteins NHERF and ZO1 via the "VTTRL" sequence in the C-terminus of TRPC4 and PLC (Tang et al. 2000), as well as fyn (Odell et al. 2005). The dynamic interplay between tyrosine kinases, TRPC4 and NHERF, regulates cell surface expression and activation of the channel. TRPC4 also associates with the caveolae where growth factor receptor signaling proteins as well as NHERF-binding proteins, such as ezrin, are localized (Torihashi et al. 2002). It has been suggested that the interaction with NHERF and Z01 provide a scaffold to position the channel in the apical or lateral regions of polarized cells such as endothelial cells.

3.5 TRPC5

Heterologously expressed TRPC5 forms a non-selective channel that can be activated by receptor stimulation but not store depletion in HEK293 (Schaefer et al. 2000), PC12 (Ohta et al. 2004), and murine stomach cells (Lee et al. 2003) or directly by Ca^{2+} in HEK293 cells (Blair et al. 2009; Gross et al. 2009). TRPC5 can potentially form multimeric channels with other TRPCs (Table 2), e.g., TRPC1 + TRPC5 in neurons, vascular endothelial cells, and vascular smooth muscle cells (Goel et al. 2002; Shi et al. 2012; Strubing et al. 2001). Heterologously expressed TRPC5 forms a heteromeric channel with TRPC4 (Schindl et al. 2008). TRPC5 also has the sequence "VTTRL" in its C-terminus that mediates its interaction with the PDZ-binding proteins, NHERF and ezrin/moesin/radixin-binding phosphoprotein 50 (EBP50) (Obukhov and Nowycky 2004; Tang et al. 2000). NHERF mediates TRPC5 association with PLCβ and also regulates surface expression of TRPC5, whereas EBP50 links the channel to the actin cytoskeleton and modulates its activation kinetics following cell stimulation. Two CaM-binding sites located in the C-terminus of TRPC5 are involved in modulating channel activity (Tang et al. 2001). While myosin light chain kinase (MLCK) and PKC have been shown to regulate TRPC5 function, it is not clear whether these kinases exert their effects directly on the channel or indirectly by modulating the status of the actin cytoskeleton. Inhibition of MLCK activity adversely impacts channel activation, whereas PKC regulates channel desensitization following agonist stimulation. Additionally, activation of MLCK by Ca^{2+}/CaM has been proposed to prolong channel activity by enhancing surface expression of TRPC5 (Kim et al. 2006b; Shimizu et al. 2006). Trafficking of TRPC5 to specific sites in the hippocampal neurons is determined by its interaction with the exocyst component protein stathmin-2, SNARE proteins, and other trafficking proteins such as dynamin, clathrin, and MxA (Goel et al. 2005; Greka et al. 2003). Moreover, the neuronal calcium sensor-1 (NCS-1) binds to the C-terminus of TRPC5 (Hui et al. 2006) and is involved in retardation of neurite outgrowth by TRPC5 homomeric channel (Bezzerides et al. 2004).

3.6 TRPC6

TRPC6 has been widely shown to be activated by DAG and not by internal Ca^{2+} store depletion (Dietrich et al. 2005; Putney 2005). Nonetheless, several studies report that activation of TRPC6 by store depletion is mediated by its association with Orai1 (Liao et al. 2007) and TRPC4 (which directly binds STIM1) (Yuan et al. 2007). Heteromeric TRPC6 channels have also been reported in different cell types, such as TRPC3 + TRPC6 in pontine neurons (Li et al. 2005) and prostate cancer epithelial cells (Thebault et al. 2006) and TRPC6 + TRPC7 in A7r5 cells (Maruyama et al. 2006). TRPC6 channel activity is determined via its interactions with different signaling proteins. The tyrosine kinase, fyn, interacts with TRPC6 and modulates channel activity via tyrosine phosphorylation in COS-7 cells (Hisatsune et al. 2004). Stimulation of neuronal PC12D cells with acetylcholine results in formation of a multiprotein complex of TRPC6, M1 mAChR and PKC, and DAG production. While DAG activates TRPC6, DAG-activated PKC phosphorylates the channel to inhibit it (Kim and Saffen 2005).

TRPC6 also undergoes trafficking to the plasma membrane, and several proteins that associate with the channel have a role in this process, such as enkurin (Sutton et al. 2004), actinin, actin, and drebrin (Goel et al. 2005), and endocytic vesicle-associated proteins (Goel et al. 2005; Lussier et al. 2005). TRPC6 also contains the conserved CIRB domain in the C-terminus, and CaM reportedly regulates TRPC6 activation (Tang et al. 2001; Yuan et al. 2003).

3.7 TRPC7

Since the first isolation of TRPC7 by screening the fetal brain and caudate nucleus cDNA libraries (Nagamine et al. 1998), there are relatively few studies that report its properties and function. Both store-dependent and store-independent modes of activation, as well as constitutive activation, have been reported for TRPC7 (Numaga et al. 2007). Multimeric TRPC1 + TRPC3 + TRPC7 channels function as SOC channels, whereas TRPC3 + TRPC7 channels appear to be DAG-activated channels, in HEK293 cells (Zagranichnaya et al. 2005). Additionally, function of TRPC7 has been reported to be modulated by cGMP-dependent protein kinase 1α (Yuasa et al. 2011), CaM, IP_3R, and PIP_2 (Ju et al. 2010; Mery et al. 2001; Tang et al. 2001; Yuan et al. 2003). Little is known about the mechanisms regulating the trafficking and localization of TRPC7, even though it has been shown to interact with IP_3R, CaM, and MxA (Table 2).

4 Regulation of TRPC Channel Function by Intracellular Ca^{2+} Store Depletion

As discussed above, all TRPC channels are activated in response to stimulation of plasma membrane receptors that result in PIP_2 hydrolysis. Some TRPCs are regulated by store depletion induced following stimulation by physiological agonists as well as treatment of cells with passively depleting agents such as thapsigargin and cyclopiazonic acid. Furthermore, in these cases, channel function is blocked by conditions that inhibit SOCE, such as the application of 1 μM Gd^{3+} and 10 μM 2APB. Typically, TRPC1 and TRPC4 have been suggested to be store-operated while TRPCs 3, 5, 6, and 7 have shown to be store-independent. The mechanisms by which store-independent regulation of TRPC channels occurs, presumably via PIP_2 hydrolysis or DAG, are not very well established. Here we will summarize the presently available data on the regulation of TRPC channels by store depletion.

4.1 Role of STIM1

Considerable progress has been made regarding the TRPC channels that contribute to SOCE. In 2005, STIM1 was identified as the ER calcium sensor that regulates

SOCE. STIM1 is diffusely localized in the ER in resting conditions, and upon Ca^{2+} store depletion, it aggregates and translocates to the periphery of the cells where it interacts with both Orai1 and TRPC channels in specialized ER-plasma membrane (PM) junctional domains. In these regions, the ER and plasma membrane come in close proximity to each other (Cheng et al. 2013; Hogan et al. 2010; Liou et al. 2005; Roos et al. 2005). The Orai channel family is comprised of three isoforms (Orais 1, 2, and 3), all of which have four transmembrane domains. Orai1 has been suggested to function as a tetramer (Hogan et al. 2010; Ji et al. 2008; Penna et al. 2008) and more recently as a hexamer based on crystal structure (Hou et al. 2012). The discovery of STIM1 and Orai1 led to the identification of the long sought-after components of the CRAC channel. STIM1 and Orai1 together are sufficient to reconstitute CRAC channel activity, with the C-terminal SOAR domain (aa 344–442) in STIM1 being the region involved in gating Orai1 and generating I_{CRAC} (Hogan et al. 2010; Yuan et al. 2009). Numerous reports have demonstrated that STIM1 also interacts with members of the TRPC channel family and that it is necessary for gating TRPC channels (Cheng et al. 2013; Lee et al. 2010). Furthermore, TRPC heteromers that contain either TRPC1 or TRPC4 can be activated by STIM1. Thus, TRPC3 or TRPC6, likely non-store-operated channels, can appear to be regulated by STIM1 if they are assembled in the channel with TRPC1 or TRPC4 (Huang et al. 2006). The critical role of STIM1 in TRPC regulation [discussed in reviews by Cheng et al. (2013), Worley et al. (2007), and Lee et al. (2010)] is shown by the following data: (1) STIM1 and TRPCs co-immunoprecipitate and this association increases following store depletion; (2) the binding of STIM1 and TRPC1 has been confirmed by GST-fusion protein pull-down assays; (3) TRPC-mediated Ca^{2+} entry in response to store depletion is completely abolished by the knockdown of endogenous STIM1; (4) co-expression of TRPC1 and STIM1 induces an increase in store depletion-induced Ca^{2+} influx as well as I_{SOC}; (5) endogenous TRPC function is suppressed by heterologous expression of dominant-negative STIM1 constructs; and (6) the STIM1D76A mutant, which induces constitutive Orai1 activation, also mediates spontaneous TRPC channel function. In aggregate, all these data provide convincing support that STIM1 regulates TRPC channel activation and function. Structure-function analysis of STIM1 has revealed crucial information regarding STIM1 domains involved in the interaction with TRPCs. The ERM (ezrin/radixin/moesin) domain (aa 251–535) located within the STIM1 cytosolic carboxyl terminus has been shown to bind selectively to some TRPC channels, e.g., TRPC1, TRPC2 and TRPC4, but not TRPC3, TRPC6, and TRPC7. As mentioned above, channels that cannot bind to STIM1 can be regulated by it if they are assembled in a heteromeric channel complex with TRPCs that bind STIM1. Nonetheless, it is notable that several studies with heterologous expression of TRPCs and STIM1 have failed to demonstrate the involvement of these channels in SOCE. It might be important to consider the assembly of TRPC channel complexes in such studies as other components might be essential in the regulation of these channels.

Binding of the ERM domain of STIM1 to TRPC channels is not sufficient for channel activation. A lysine-rich domain (referred to as polybasic tail or K domain) located at the C-terminal end of STIM1 has been established as the region that is involved in gating TRPC channels. Deletion of the STIM1-K domain affected TRPC channel activity but not binding to STIM1. The mechanism underlying the gating of TRPC channels by STIM1 has been revealed in a study demonstrating that the positively charged lysine residues (^{684}KK^{685}) in STIM1 interact electrostatically with negatively charged conserved aspartate residues in TRPC1 (^{639}DD^{640}), which leads to gating of the channel (Zeng et al. 2008). When the negative charges in TRPC1 are neutralized by substituting lysine (K) with alanine (A), channel activation by STIM1 is blocked. Moreover, swapping the charges between TRPC1 and STIM1 induces recovery of channel gating and function, providing conclusive evidence for the gating of TRPC1 by STIM1. Remarkably, the negatively charged sequence in TRPC1 C-terminus is highly conserved among TRPC family members, including TRPC3, TRPC4, TRPC5, and TRPC6 (Zeng et al. 2008). Thus, it was proposed that other TRPCs also have the inherent capacity to be gated by STIM1, although not all TRPCs bind directly to STIM1. Further studies need to be carried out to conclusively establish which TRPCs can bind to and are gated by STIM1, especially with regards to endogenous TRPC channels. Such information is crucial for understanding how store-dependent TRPC channels are assembled and regulated.

4.2 Role of Orai1

A very significant finding reported by several groups of researchers is that TRPC channel activation is not only dependent on STIM1 but also requires Orai1. Conclusive findings show that stimulation of cells results in dynamic assembly of TRPC1, STIM1, and Orai1 in a ternary complex in the ER-PM junctional domains, which is required for the activation of both Orai1 as well as TRPC1 channels. The TRPC1–STIM1–Orai1 complex, associated with SOCE, can be detected in HSG cells (Ong et al. 2007), mouse pulmonary arterial smooth muscle cells (Ng et al. 2009), human parathyroid (Lu et al. 2010), human liver cell (Zhang et al. 2010), and rat kidney fibroblast (Almirza et al. 2012). Assembly of this complex is mediated via STIM1, as knockdown of STIM1 prevents clustering of TRPC1 with Orai1. Knockdown of TRPC1 results in attenuation of function, while knockdown of Orai1 or STIM1 results in complete loss of SOCE. Furthermore, overexpression of pore-deficient, dominant-negative mutants of Orai1 (R91W, E106Q) abrogate Ca^{2+} entry due to TRPC1-STIM1 (Cheng et al. 2008; Ong et al. 2007). The exact mechanism by which Orai1 determines TRPC function has been a matter of much debate. It was suggested that Orai1 can physically interact with the C- and N-termini of both TRPC3 and TRPC6 channels and modulate channel sensitivity to store depletion and STIM1 (Liao et al. 2007). Hence, these investigators proposed that the endogenous SOCE channel pore is contributed by TRPC channels with Orai1 functioning as the regulatory subunit. Alternatively, TRPC channels have been proposed to modify Orai1 function.

The key question of whether TRPC and Orai1 contribute to a single channel pore (formation of a heteromeric TRPC + Orai1 channel) or are two distinct channels which independently contribute to SOCE has been resolved recently for TRPC1 and TRPC5. In a cell line where endogenous TRPC1 contributes to SOCE, TRPC1 and Orai1 form two distinct channels: a relatively Ca^{2+}-selective channel mediating I_{SOC} is composed of STIM1/TRPC1 and a highly Ca^{2+}-selective channel mediating I_{CRAC} is formed by STIM1/Orai1 (Cheng et al. 2011). The smaller conductance of I_{CRAC} is masked by the larger STIM1/TRPC1-mediated current that gets activated under the same conditions. Hence, the I_{SOC} attributed to STIM1/TRPC1 includes a small contribution from STIM1/Orai1 I_{CRAC}. Further, native I_{CRAC} is detected when TRPC1 channel function is suppressed in these cells by expression of the STIM1-KK/EE mutant, which can gate Orai1 but not TRPC1. More importantly, Ca^{2+} entry through STIM1/Orai1 facilitates TRPC1 channel trafficking and triggers TRPC1 insertion into the plasma membrane. Membrane insertion of TRPC is attenuated by removal of extracellular Ca^{2+}, blocking I_{CRAC} with 1 μM Gd^{3+}, knockdown of Orai1 or overexpression of dominant-negative mutant of Orai1 (E106Q) that lacks a functional pore (Cheng et al. 2011). These data define the functional role of Orai1 and provide novel insights into the regulation and activation of TRPC1 in SOCE. Regulated surface insertion of TRPC1 by Orai1 can provide a rapid modulation and amplification of SOCE-facilitated Ca^{2+} signals that could selectively impact regulation of cell function (Cheng et al. 2013). Ca^{2+} entry via Orai1 has also been shown to facilitate TRPC5 activity (Gross et al. 2009). In this case, Ca^{2+} coming into the cell via Orai1 directly activates the TRPC5 channel. Detailed studies have not been done along these lines for other TRPCs to either demonstrate or rule out a secondary effect of Orai1 on channel function. It is important to note that heterologous expression might not yield similar data to that with endogenous channels, especially when several molecular components and regulatory mechanisms concertedly determine channel function.

5 Modulation of TRPC Channels by Membrane Trafficking

Localization of TRPC channels in specific plasma membrane microdomains allows the generation of precise intracellular Ca^{2+} signals that modulate downstream signaling events and consequent cell functions. The amplitude and duration of intracellular Ca^{2+} signals can be varied by regulating Ca^{2+} influx via TRPC channels, which can be enhanced by increasing the number of active channels at the cell surface either by driving channel trafficking to the plasma membrane or by prolonging channel retention at the cell surface. Major modes of regulating Ca^{2+} entry include constitutive and regulated vesicular trafficking mechanisms as well as the rates of protein synthesis and degradation. The constitutive and regulated trafficking processes determine the surface expression of TRPC channels by (1) increasing exocytosis and/or recycling to the plasma membrane or (2) reducing endocytosis and/or increasing channel retention in the plasma membrane.

5.1 TRPC1

Studies of the TRPC1 complex identified several interacting proteins that are involved in vesicle trafficking, membrane fusion, and cytoskeletal and actin rearrangement, such as clathrin, dynamin, Sec1, synapsin-2, Cav-1, and RhoA (Table 2). The TRPC1 signaling complex is localized in distinct cholesterol-rich plasma membrane domains known as lipid rafts. Disruption of lipid rafts with cholesterol-depleting agents like methyl-β-cyclodextrin (MβCD) decreased SOCE in salivary gland cells (Lockwich et al. 2000) and vascular smooth muscle cells (Bergdahl et al. 2003), suggesting lipid raft integrity is a prerequisite for TRPC1 localization and function. Cav-1 is a cholesterol-binding protein found within the caveolae, which are caveolin-containing lipid rafts present in the plasma membrane. Cav-1 plays an important role in the trafficking and function of TRPC1 (Brazer et al. 2003; Kwiatek et al. 2006; Lockwich et al. 2000; Pani et al. 2009, 2012). The present model proposes that Cav-1 functions as a scaffolding protein that facilitates assembly of the TRPC1 signaling complex and acts synergistically with Orai1 and STIM1 to regulate TRPC1 channel activity (Ong and Ambudkar 2012; Pani et al. 2009). In resting cells, TRPC1 is controlled by constitutive trafficking mechanisms. Following trafficking to the cell periphery, TRPC1 associates with Cav-1 but remains inactive and does not get inserted into the plasma membrane. When cells are stimulated by physiological agonists and the ER-Ca^{2+} stores are depleted, STIM1 translocates to the plasma membrane and activates the Orai1 channel. The Orai1-mediated Ca^{2+} influx drives the recruitment of TRPC1 into the plasma membrane. TRPC1 dissociates from Cav-1 and interacts with and is activated by STIM1. Dissociation of TRPC1 from Cav-1 is an essential step in the activation of TRPC1 by STIM1 since C-terminal ^{684}KK^{685} residues of STIM1 responsible for gating TRPC1 also releases the channel from Cav-1 (Pani et al. 2009; Zeng et al. 2008). In addition to Cav-1, Homer1 also interacts with TRPC1 in the C-terminus (aa 644–650), a region that lies just upstream of the STIM1-gating site (aa 639–640). Homer1 forms a dynamic complex with TRPC1 and IP_3R. Following cell stimulation, the TRPC1/Homer1/IP_3R complex disassembles, resulting in channel activation.

Local changes in the cytoskeleton or microtubules also contribute to the trafficking of TRPC1 (Bollimuntha et al. 2005a; Mehta et al. 2003). In retinal epithelial cells, β-tubulin has been shown to interact with TRPC1 and to be required for channel translocation to the plasma membrane (Bollimuntha et al. 2005a). RhoA, a monomeric GTPase protein responsible for actin cytoskeleton dynamics, associates with TRPC1 and IP_3R in endothelial cells following stimulation with thrombin. Assembly of the TRPC1/IP_3R complex, as well as trafficking to the plasma membrane, is dependent of RhoA and actin polymerization since SOCE is attenuated following treatment with C3 transferase protein that inactivates Rho or expression of a Rho dominant mutant (Mehta et al. 2003). Enkurin, a CaM-binding protein, interacts with TRPC1 and TRPC5 in sperm and has been suggested to function as an adaptor protein that tethers signaling proteins to TRPC channels (Sutton et al. 2004). Proteins involved in vesicle docking and fusion have also been reported to interact with TRPC1 and regulate channel activity. Nevertheless, the relevance of these various components in the intracellular trafficking of TRPC1 has yet to be identified.

5.2 TRPC3

As described earlier for TRPC1, the interactions of TRPC3 with several proteins are vital for its proper trafficking and cellular localization. These include PLCγ (van Rossum et al. 2005), Cav-1 (Lockwich et al. 2001), VAMP2 (Singh et al. 2004), RFN24 (Lussier et al. 2008), and Homer1 (Kim et al. 2006a). Surface expression of the TRPC3 channel requires interaction with PLCγ and PIP_2, which anchors the channel in the plasma membrane (van Rossum et al. 2005). Homer1 has been reported to stabilize the interaction between TRPC3 and IP_3R, determining the rate of TRPC3 translocation to and retrieval from the plasma membrane (Kim et al. 2006a; Kiselyov et al. 2007). Both Homer1 and junctate may function synergistically to facilitate the interaction between TRPC3 and IP_3R which leads to channel activation. It is possible that the components involved in TRPC3 trafficking depend on the cell type and the spatial constraints within the cell.

Cell surface expression of TRPC3 is regulated by VAMP2-mediated fusion of mobile intracellular vesicles containing TRPC3 with the plasma membrane. Expression of TRPC3 in the plasma membrane increases following stimulation with carbachol, and this increase is abolished by treatment with tetanus toxin, which inhibits VAMP2 activity (Singh et al. 2004). Likewise, status of the actin cytoskeleton has also been reported to affect TRPC3 localization and function. Conditions that result in enhancement or stabilization of the cortical actin layer, such as treatment with jasplakinolide or calyculin A, promote internalization of TRPC3 signaling complex with a consequent decrease of TRPC3 function (Lockwich et al. 2001). TRPC3-interacting proteins may also influence the trafficking and surface expression of the channel. These include clathrin, dynamin, AP-2, syntaxin, synaptotagmin-1 (Lockwich et al. 2008), MxA (Lussier et al. 2005), and RACK1 (Bandyopadhyay et al. 2008) (Table 2). Additional studies are required to resolve the role of the TRPC3-interacting proteins involved in constitutive and regulated trafficking of the channel.

5.3 TRPC4

Although several studies have reported the association of TRPC4 with scaffolding and trafficking proteins (Table 2), the mechanisms regulating TRPC4 localization in the plasma membrane have not been fully elucidated. A dynamic interplay between tyrosine kinases, TRPC4, and NHERF regulates surface expression and activation of TRPC4 channels (Tang et al. 2000). The protein tyrosine kinase, fyn, phosphorylates TRPC4 following stimulation by the epidermal growth factor (EGF), increasing its interaction with NHERF, as well as its insertion into the plasma membrane (Odell et al. 2005). As mentioned above, TRPC4 forms a heteromeric complex with TRPC1 to mediate SOCE in endothelial cells. Loss of Cav-1 impairs surface expression of both TRPC4 and TRPC1, significantly reduces association of the heteromeric complex with IP_3R, and inhibits agonist-induced Ca^{2+} entry in these cells. Hence, Cav-1 is proposed to function as a scaffold that facilitates the interactions between TRPC4, TRPC1, and IP_3R.

5.4 TRPC5

Proteomic analysis of TRPC5-binding partners revealed the interactions of TRPC5 with proteins involved in vesicle trafficking and scaffolding (Table 2), such as dynamin, clathrin, AP-2 (Goel et al. 2005), and MxA (Lussier et al. 2005). Interaction of TRPC5 with the exocyst component protein, stathmin 2, targets homomeric channels to the growth cone of hippocampal neurons (Greka et al. 2003). In resting neuronal cells, TRPC5 is localized in intracellular vesicles close to the plasma membrane. Following stimulation with growth factors, TRPC5-containing vesicles are rapidly translocated and inserted to the plasma membrane, thereby increasing channel function constitutively. Trafficking of TRPC5 and insertion into the plasma membrane requires phosphatidylinositide 3-kinase (PI(3)K), Rac1, and phosphatidylinositol 4-phosphate 5-kinase (PIP(5)K). Interestingly, Rac1 initiates the insertion of homomeric TRPC5 but not the heteromeric TRPC1 + TRPC5 channels into the plasma membrane. This may be due to homomeric channels being localized in the growth cones to modulate elongation, whereas heteromeric channels are localized in the neurites (Bezzerides et al. 2004). It is also shown that TRPC5 participates in a molecular complex with Rac1 in fibroblasts and kidney podocytes and Ca^{2+} influx mediated by TRPC5 activates Rac1 (Tian et al. 2010). In aggregate, these studies show that components of the TRPC5 signaling complex determine its physiological function by influencing channel trafficking, localization, and activation.

5.5 TRPC6

There is a paucity of information on the proteins that interact with TRPC6 and regulate its trafficking to and localization in the plasma membrane (Table 2). Surface expression of TRPC6 is enhanced following cell stimulation by muscarinic receptor agonists or passive depletion of the ER-Ca^{2+} stores by thapsigargin (Cayouette et al. 2004). The GTPases, Rab9 and Rab11, have been shown to regulate the intracellular trafficking of TRPC6 in HeLa cells (Cayouette et al. 2010). In cells cotransfected with Rab9, TRPC6 shows a diffuse localization through the cell as well as partial colocalization with Rab9 containing vesicles. However, when Rab11 is overexpressed, TRPC6 is predominantly present at the cell periphery. Surface expression of TRPC6, as well as the channel activity, increases following the expression of a dominant-negative mutant of Rab9 (S21N) and Rab11, whereas channel activity decreases when dominant-negative mutant of Rab11 (S25N) is expressed. In aggregate, these data suggest that the intracellular trafficking of TRPC6 is through early endosomes and late endosomes, where the channel interacts with Rab9-containing vesicles and the channel is translocated to the plasma membrane via Rab11-containing vesicles (Cayouette et al. 2010). PI(3)K and PTEN have also been reported to regulate the trafficking and activation of TRPC6 channels. PTEN-dependent inhibition of PI(3)K reduced translocation of TRPC6 to the plasma membrane, as well as TRPC6-mediated Ca^{2+} influx in T6.11 cells. Previous studies have reported the interaction of TRPC6 with other proteins that are involved in vesicle trafficking, such as MxA (Lussier et al. 2005), RhoA (Tian et al. 2010), syntaxin (Bandyopadhyay et al. 2005), clathrin, and dynamin (Goel et al. 2005). MxA (which also interacts with other TRPCs except TRPC2) has been shown to modulate TRPC6-mediated Ca^{2+} entry in response to cell stimulation. The importance of such interactions in modulating surface expression and activity of TRPC6 remain to be fully delineated in future studies.

5.6 TRPC2 and TRPC7

There is relatively less information regarding the protein interactions and trafficking of TRPC2 and TRPC7. Similar to other TRPC channels, TRPC2 interacts with enkurin (Sutton et al. 2004) and Homer (Yuan et al. 2003). It has also been shown that the chaperone receptor-transporting protein 1 (RTP1) regulates the surface expression and channel activity of TRPC2 in HEK 293 cells. In cells cotransfected with RTP1, the surface expression of TRPC2, as well as the channel activity, is increased relative to cells expressing TRPC2 alone (Mast et al. 2010). A previous study demonstrates that TRPC7 interacts with MxA, a member of the dynamin superfamily (Lussier et al. 2005).

Conclusion

In summary, TRPC channels are regulated downstream from receptor-coupled PLC activation. These channels contribute to a wide variety of cellular function. Loss or gain of channel function has resulted in aberrant physiology in human and mouse. The physiological function and regulation of TRPC channels are influenced by their physical and functional interactions with numerous channels and proteins involved in the signaling, scaffolding, and trafficking processes. Further studies are required to delineate the exact steps involved in assembling TRPC channels with their accessory proteins to form functional signaling complexes in discrete ER-PM junctional regions. Understanding the various modes and mechanisms involved in TRPC channel function can provide potentially important targets for treatment of a number of diseases.

References

Adebiyi A, Thomas-Gatewood CM, Leo MD, Kidd MW, Neeb ZP, Jaggar JH (2012) An elevation in physical coupling of type 1 inositol 1,4,5-trisphosphate (IP3) receptors to transient receptor potential 3 (TRPC3) channels constricts mesenteric arteries in genetic hypertension. Hypertension 60:1213–1219

Almirza WH, Peters PH, van Zoelen EJ, Theuvenet AP (2012) Role of Trpc channels, Stim1 and Orai1 in PGF(2alpha)-induced calcium signaling in NRK fibroblasts. Cell Calcium 51:12–21

Alvarez J, Coulombe A, Cazorla O, Ugur M, Rauzier JM, Magyar J, Mathieu EL, Boulay G, Souto R, Bideaux P, Salazar G, Rassendren F, Lacampagne A, Fauconnier J, Vassort G (2008) ATP/UTP activate cation-permeable channels with TRPC3/7 properties in rat cardiomyocytes. Am J Physiol Heart Circ Physiol 295:H21–H28

Ambudkar IS (2009) Unraveling smooth muscle contraction: the TRP link. Gastroenterology 137:1211–1214

Ambudkar IS, Ong HL (2007) Organization and function of TRPC channelosomes. Pflugers Arch 455:187–200

Ambudkar IS, Bandyopadhyay BC, Liu X, Lockwich TP, Paria B, Ong HL (2006) Functional organization of TRPC-Ca^{2+} channels and regulation of calcium microdomains. Cell Calcium 40:495–504

Ambudkar IS, Ong HL, Liu X, Bandyopadhyay BC, Cheng KT (2007) TRPC1: the link between functionally distinct store-operated calcium channels. Cell Calcium 42:213–223

Antigny F, Girardin N, Frieden M (2012) Transient receptor potential canonical channels are required for in vitro endothelial tube formation. J Biol Chem 287:5917–5927

Antoniotti S, Pla AF, Barral S, Scalabrino O, Munaron L, Lovisolo D (2006) Interaction between TRPC channel subunits in endothelial cells. J Recept Signal Transduct Res 26:225–240

Balzer M, Lintschinger B, Groschner K (1999) Evidence for a role of Trp proteins in the oxidative stress-induced membrane conductances of porcine aortic endothelial cells. Cardiovasc Res 42:543–549

Bandyopadhyay BC, Swaim WD, Liu X, Redman RS, Patterson RL, Ambudkar IS (2005) Apical localization of a functional TRPC3/TRPC6-Ca^{2+}-signaling complex in polarized epithelial cells. Role in apical Ca^{2+} influx. J Biol Chem 280:12908–12916

Bandyopadhyay BC, Ong HL, Lockwich TP, Liu X, Paria BC, Singh BB, Ambudkar IS (2008) TRPC3 controls agonist-stimulated intracellular Ca^{2+} release by mediating the interaction between inositol 1,4,5-trisphosphate receptor and RACK1. J Biol Chem 283:32821–32830

Bandyopadhyay BC, Swaim WD, Sarkar A, Liu X, Ambudkar IS (2012) Extracellular Ca^{2+} sensing in salivary ductal cells. J Biol Chem 287:30305–30316

Beech DJ (2005) TRPC1: store-operated channel and more. Pflugers Arch 451:53–60

Bergdahl A, Gomez MF, Dreja K, Xu SZ, Adner M, Beech DJ, Broman J, Hellstrand P, Sward K (2003) Cholesterol depletion impairs vascular reactivity to endothelin-1 by reducing store-operated Ca^{2+} entry dependent on TRPC1. Circ Res 93:839–847

Bezzerides VJ, Ramsey IS, Kotecha S, Greka A, Clapham DE (2004) Rapid vesicular translocation and insertion of TRP channels. Nat Cell Biol 6:709–720

Blair NT, Kaczmarek JS, Clapham DE (2009) Intracellular calcium strongly potentiates agonist-activated TRPC5 channels. J Gen Physiol 133:525–546

Bollimuntha S, Cornatzer E, Singh BB (2005a) Plasma membrane localization and function of TRPC1 is dependent on its interaction with beta-tubulin in retinal epithelium cells. Vis Neurosci 22:163–170

Bollimuntha S, Singh BB, Shavali S, Sharma SK, Ebadi M (2005b) TRPC1-mediated inhibition of 1-methyl-4-phenylpyridinium ion neurotoxicity in human SH-SY5Y neuroblastoma cells. J Biol Chem 280:2132–2140

Bomben VC, Turner KL, Barclay TT, Sontheimer H (2011) Transient receptor potential canonical channels are essential for chemotactic migration of human malignant gliomas. J Cell Physiol 226:1879–1888

Brazer SC, Singh BB, Liu X, Swaim W, Ambudkar IS (2003) Caveolin-1 contributes to assembly of store-operated Ca influx channels by regulating plasma membrane localization of TRPC1. J Biol Chem 278:27208–27215

Bush EW, Hood DB, Papst PJ, Chapo JA, Minobe W, Bristow MR, Olson EN, McKinsey TA (2006) Canonical transient receptor potential channels promote cardiomyocyte hypertrophy through activation of calcineurin signaling. J Biol Chem 281:33487–33496

Cayouette S, Lussier MP, Mathieu EL, Bousquet SM, Boulay G (2004) Exocytotic insertion of TRPC6 channel into the plasma membrane upon Gq protein-coupled receptor activation. J Biol Chem 279:7241–7246

Cayouette S, Bousquet SM, Francoeur N, Dupre E, Monet M, Gagnon H, Guedri YB, Lavoie C, Boulay G (2010) Involvement of Rab9 and Rab11 in the intracellular trafficking of TRPC6. Biochim Biophys Acta 1803:805–812

Chen J, Crossland RF, Noorani MM, Marrelli SP (2009) Inhibition of TRPC1/TRPC3 by PKG contributes to NO-mediated vasorelaxation. Am J Physiol Heart Circ Physiol 297:H417–H424

Cheng KT, Liu X, Ong HL, Ambudkar IS (2008) Functional requirement for Orai1 in store-operated TRPC1-STIM1 channels. J Biol Chem 283:12935–12940

Cheng KT, Liu X, Ong HL, Swaim W, Ambudkar IS (2011) Local Ca^{2+} entry via Orai1 regulates plasma membrane recruitment of TRPC1 and controls cytosolic Ca^{2+} signals required for specific cell functions. PLoS Biol 9:e1001025

Cheng KT, Ong HL, Liu X, Ambudkar IS (2013) Contribution and regulation of TRPC channels in store-operated Ca^{2+} entry. In: Prakriya M (ed) Store-operated calcium channels, vol 71. Elsevier, Amterdam, pp 150–179

Cheung KK, Yeung SS, Au SW, Lam LS, Dai ZQ, Li YH, Yeung EW (2011) Expression and association of TRPC1 with TRPC3 during skeletal myogenesis. Muscle Nerve 44:358–365

Chu X, Tong Q, Cheung JY, Wozney J, Conrad K, Mazack V, Zhang W, Stahl R, Barber DL, Miller BA (2004) Interaction of TRPC2 and TRPC6 in erythropoietin modulation of calcium influx. J Biol Chem 279:10514–10522

Cioffi DL, Wu S, Alexeyev M, Goodman SR, Zhu MX, Stevens T (2005) Activation of the endothelial store-operated I_{SOC} Ca^{2+} channel requires interaction of protein 4.1 with TRPC4. Circ Res 97:1164–1172

Cuddapah VA, Turner KL, Sontheimer H (2013) Calcium entry via TRPC1 channels activates chloride currents in human glioma cells. Cell Calcium 53:187–194

Dibattista M, Amjad A, Maurya DK, Sagheddu C, Montani G, Tirindelli R, Menini A (2012) Calcium-activated chloride channels in the apical region of mouse vomeronasal sensory neurons. J Gen Physiol 140:3–15

Dietrich A, Kalwa H, Rost BR, Gudermann T (2005) The diacylgylcerol-sensitive TRPC3/6/7 subfamily of cation channels: functional characterization and physiological relevance. Pflugers Arch 451:72–80

Fiorio Pla A, Maric D, Brazer SC, Giacobini P, Liu X, Chang YH, Ambudkar IS, Barker JL (2005) Canonical transient receptor potential 1 plays a role in basic fibroblast growth factor (bFGF)/ FGF receptor-1-induced Ca^{2+} entry and embryonic rat neural stem cell proliferation. J Neurosci 25:2687–2701

Gervasio OL, Whitehead NP, Yeung EW, Phillips WD, Allen DG (2008) TRPC1 binds to caveolin-3 and is regulated by Src kinase – role in Duchenne muscular dystrophy. J Cell Sci 121:2246–2255

Goel M, Sinkins WG, Schilling WP (2002) Selective association of TRPC channel subunits in rat brain synaptosomes. J Biol Chem 277:48303–48310

Goel M, Sinkins W, Keightley A, Kinter M, Schilling WP (2005) Proteomic analysis of TRPC5- and TRPC6-binding partners reveals interaction with the plasmalemmal Na^+/K^+-ATPase. Pflugers Arch 451:87–98

Greka A, Navarro B, Oancea E, Duggan A, Clapham DE (2003) TRPC5 is a regulator of hippocampal neurite length and growth cone morphology. Nat Neurosci 6:837–845

Gross SA, Guzman GA, Wissenbach U, Philipp SE, Zhu MX, Bruns D, Cavalie A (2009) TRPC5 is a Ca^{2+}-activated channel functionally coupled to Ca^{2+}-selective ion channels. J Biol Chem 284:34423–34432

Harada M, Luo X, Qi XY, Tadevosyan A, Maguy A, Ordog B, Ledoux J, Kato T, Naud P, Voigt N, Shi Y, Kamiya K, Murohara T, Kodama I, Tardif JC, Schotten U, Van Wagoner DR, Dobrev D, Nattel S (2012) Transient receptor potential canonical-3 channel-dependent fibroblast regulation in a trial fibrillation. Circulation 126:2051–2064

Heo DK, Chung WY, Park HW, Yuan JP, Lee MG, Kim JY (2012) Opposite regulatory effects of TRPC1 and TRPC5 on neurite outgrowth in PC12 cells. Cell Signal 24:899–906

Hirschler-Laszkiewicz I, Tong Q, Conrad K, Zhang W, Flint WW, Barber AJ, Barber DL, Cheung JY, Miller BA (2009) TRPC3 activation by erythropoietin is modulated by TRPC6. J Biol Chem 284:4567–4581

Hisatsune C, Kuroda Y, Nakamura K, Inoue T, Nakamura T, Michikawa T, Mizutani A, Mikoshiba K (2004) Regulation of TRPC6 channel activity by tyrosine phosphorylation. J Biol Chem 279:18887–18894

Hofmann T, Schaefer M, Schultz G, Gudermann T (2002) Subunit composition of mammalian transient receptor potential channels in living cells. Proc Natl Acad Sci U S A 99:7461–7466

Hogan PG, Lewis RS, Rao A (2010) Molecular basis of calcium signaling in lymphocytes: STIM and ORAI. Annu Rev Immunol 28:491–533

Hoth M, Penner R (1992) Depletion of intracellular calcium stores activates a calcium current in mast cells. Nature 355:353–356

Hoth M, Fasolato C, Penner R (1993) Ion channels and calcium signaling in mast cells. Ann NY Acad Sci 707:198–209

Hou X, Pedi L, Diver MM, Long SB (2012) Crystal structure of the calcium release-activated calcium channel Orai. Science 338:1308–1313

Hu G, Oboukhova EA, Kumar S, Sturek M, Obukhov AG (2009) Canonical transient receptor potential channels expression is elevated in a porcine model of metabolic syndrome. Mol Endocrinol 23:689–699

Huang GN, Zeng W, Kim JY, Yuan JP, Han L, Muallem S, Worley PF (2006) STIM1 carboxyl-terminus activates native SOC, I(crac) and TRPC1 channels. Nat Cell Biol 8:1003–1010

Hui H, McHugh D, Hannan M, Zeng F, Xu SZ, Khan SU, Levenson R, Beech DJ, Weiss JL (2006) Calcium-sensing mechanism in TRPC5 channels contributing to retardation of neurite outgrowth. J Physiol 572:165–172

Ji W, Xu P, Li Z, Lu J, Liu L, Zhan Y, Chen Y, Hille B, Xu T, Chen L (2008) Functional stoichiometry of the unitary calcium-release-activated calcium channel. Proc Natl Acad Sci U S A 105:13668–13673

Ju M, Shi J, Saleh SN, Albert AP, Large WA (2010) Ins(1,4,5)P3 interacts with PIP2 to regulate activation of TRPC6/C7 channels by diacylglycerol in native vascular myocytes. J Physiol 588:1419–1433

Kim JY, Saffen D (2005) Activation of M1 muscarinic acetylcholine receptors stimulates the formation of a multiprotein complex centered on TRPC6 channels. J Biol Chem 280:32035–32047

Kim JY, Zeng W, Kiselyov K, Yuan JP, Dehoff MH, Mikoshiba K, Worley PF, Muallem S (2006a) Homer 1 mediates store- and inositol 1,4,5-trisphosphate receptor-dependent translocation and retrieval of TRPC3 to the plasma membrane. J Biol Chem 281:32540–32549

Kim MT, Kim BJ, Lee JH, Kwon SC, Yeon DS, Yang DK, So I, Kim KW (2006b) Involvement of calmodulin and myosin light chain kinase in activation of mTRPC5 expressed in HEK cells. Am J Physiol Cell Physiol 290:C1031–C1040

Kiselyov K, Mignery GA, Zhu MX, Muallem S (1999) The N-terminal domain of the IP3 receptor gates store-operated hTrp3 channels. Mol Cell 4:423–429

Kiselyov K, Shin DM, Kim JY, Yuan JP, Muallem S (2007) TRPC channels: interacting proteins. In: Flockerzi V, Nilius B (eds) Transient receptor potential (TRP) channels, vol 179. Springer, New York, NY, pp 559–574

Kitajima N, Watanabe K, Morimoto S, Sato Y, Kiyonaka S, Hoshijima M, Ikeda Y, Nakaya M, Ide T, Mori Y, Kurose H, Nishida M (2011) TRPC3-mediated Ca^{2+} influx contributes to Rac1-mediated production of reactive oxygen species in MLP-deficient mouse hearts. Biochem Biophys Res Commun 409:108–113

Kwiatek AM, Minshall RD, Cool DR, Skidgel RA, Malik AB, Tiruppathi C (2006) Caveolin-1 regulates store-operated Ca^{2+} influx by binding of its scaffolding domain to transient receptor potential channel-1 in endothelial cells. Mol Pharmacol 70:1174–1183

Lee YM, Kim BJ, Kim HJ, Yang SK, Zhu MH, Lee KP, So I, Kim KW (2003) TRPC5 as a candidate for the nonselective cation channel activated by muscarinic stimulation in murine stomach. Am J Physiol 284:G604–G616

Lee KP, Yuan JP, Hong JH, So I, Worley PF, Muallem S (2010) An endoplasmic reticulum/plasma membrane junction: STIM1/Orai1/TRPCs. FEBS Lett 584:2022–2027

Li Y, Jia YC, Cui K, Li N, Zheng ZY, Wang YZ, Yuan XB (2005) Essential role of TRPC channels in the guidance of nerve growth cones by brain-derived neurotrophic factor. Nature 434:894–898

Li M, Chen C, Zhou Z, Xu S, Yu Z (2012a) A TRPC1-mediated increase in store-operated Ca^{2+} entry is required for the proliferation of adult hippocampal neural progenitor cells. Cell Calcium 51:486–496

Li W, Calfa G, Larimore J, Pozzo-Miller L (2012b) Activity-dependent BDNF release and TRPC signaling is impaired in hippocampal neurons of Mecp2 mutant mice. Proc Natl Acad Sci U S A 109:17087–17092

Liao Y, Erxleben C, Yildirim E, Abramowitz J, Armstrong DL, Birnbaumer L (2007) Orai proteins interact with TRPC channels and confer responsiveness to store depletion. Proc Natl Acad Sci U S A 104:4682–4687

Liman ER, Corey DP, Dulac C (1999) TRP2: a candidate transduction channel for mammalian pheromone sensory signalling. Proc Natl Acad Sci U S A 96:5791–5796

Liou J, Kim ML, Heo WD, Jones JT, Myers JW, Ferrell JE Jr, Meyer T (2005) STIM is a Ca^{2+} sensor essential for Ca^{2+}-store-depletion-triggered Ca^{2+} influx. Curr Biol 15:1235–1241

Liu X, Singh BB, Ambudkar IS (2003) TRPC1 is required for functional store-operated Ca^{2+} channels. Role of acidic amino acid residues in the S5-S6 region. J Biol Chem 278:11337–11343

Liu X, Groschner K, Ambudkar IS (2004) Distinct Ca^{2+}-permeable cation currents are activated by internal Ca^{2+}-store depletion in RBL-2H3 cells and human salivary gland cells, HSG and HSY. J Membr Biol 200:93–104

Liu X, Bandyopadhyay BC, Singh BB, Groschner K, Ambudkar IS (2005) Molecular analysis of a store-operated and 2-acetyl-sn-glycerol-sensitive non-selective cation channel. Heteromeric assembly of TRPC1-TRPC3. J Biol Chem 280:21600–21606

Lockwich TP, Liu X, Singh BB, Jadlowiec J, Weiland S, Ambudkar IS (2000) Assembly of Trp1 in a signaling complex associated with caveolin-scaffolding lipid raft domains. J Biol Chem 275:11934–11942

Lockwich T, Singh BB, Liu X, Ambudkar IS (2001) Stabilization of cortical actin induces internalization of transient receptor potential 3 (Trp3)-associated caveolar Ca^{2+} signaling complex and loss of Ca^{2+} influx without disruption of Trp3-inositol trisphosphate receptor association. J Biol Chem 276:42401–42408

Lockwich T, Pant J, Makusky A, Jankowska-Stephens E, Kowalak JA, Markey SP, Ambudkar IS (2008) Analysis of TRPC3-interacting proteins by tandem mass spectrometry. J Proteome Res 7:979–989

Lu M, Branstrom R, Berglund E, Hoog A, Bjorklund P, Westin G, Larsson C, Farnebo LO, Forsberg L (2010) Expression and association of TRPC subtypes with Orai1 and STIM1 in human parathyroid. J Mol Endocrinol 44:285–294

Lussier MP, Cayouette S, Lepage PK, Bernier CL, Francoeur N, St-Hilaire M, Pinard M, Boulay G (2005) MxA, a member of the dynamin superfamily, interacts with the ankyrin-like repeat domain of TRPC. J Biol Chem 280:19393–19400

Lussier MP, Lepage PK, Bousquet SM, Boulay G (2008) RNF24, a new TRPC interacting protein, causes the intracellular retention of TRPC. Cell Calcium 43:432–443

Ma X, Cao J, Luo J, Nilius B, Huang Y, Ambudkar IS, Yao X (2010) Depletion of intracellular Ca^{2+} stores stimulates the translocation of vanilloid transient receptor potential 4-c1 heteromeric channels to the plasma membrane. Arterioscler Thromb Vasc Biol 30:2249–2255

Ma X, Cheng KT, Wong CO, O'Neil RG, Birnbaumer L, Ambudkar IS, Yao X (2011) Heteromeric TRPV4-C1 channels contribute to store-operated Ca^{2+} entry in vascular endothelial cells. Cell Calcium 50:502–509

Maruyama Y, Nakanishi Y, Walsh EJ, Wilson DP, Welsh DG, Cole WC (2006) Heteromultimeric TRPC6-TRPC7 channels contribute to arginine vasopressin-induced cation current of A7r5 vascular smooth muscle cells. Circ Res 98:1520–1527

Mast TG, Brann JH, Fadool DA (2010) The TRPC2 channel forms protein-protein interactions with Homer and RTP in the rat vomeronasal organ. BMC Neurosci 11:61

McGurk JS, Shim S, Kim JY, Wen Z, Song H, Ming GL (2011) Postsynaptic TRPC1 function contributes to BDNF-induced synaptic potentiation at the developing neuromuscular junction. J Neurosci 31:14754–14762

Mehta D, Ahmmed GU, Paria BC, Holinstat M, Voyno-Yasenetskaya T, Tiruppathi C, Minshall RD, Malik AB (2003) RhoA interaction with inositol 1,4,5-triphosphate receptor and transient receptor potential channel-1 regulates Ca^{2+} entry. J Biol Chem 278:33492–33500

Mery L, Magnino F, Schmidt K, Krause KH, Dufour J-F (2001) Alternative splice variants of hTrp4 differentially interact with the C-terminal portion of the inositol 1,4,5-trisphosphate receptors. FEBS Lett 487:377–383

Miehe S, Bieberstein A, Arnould I, Ihdene O, Rutten H, Strubing C (2010) The phospholipid-binding protein SESTD1 is a novel regulator of the transient receptor potential channels TRPC4 and TRPC5. J Biol Chem 285:12426–12434

Monet M, Francoeur N, Boulay G (2012) Involvement of phosphoinositide 3-kinase and PTEN protein in mechanism of activation of TRPC6 protein in vascular smooth muscle cells. J Biol Chem 287:17672–17681

Mukerji N, Damodaran TV, Winn MP (2007) TRPC6 and FSGS: the latest TRP channelopathy. Biochim Biophys Acta 1772:859–868

Murata T, Lin MI, Stan RV, Bauer PM, Yu J, Sessa WC (2007) Genetic evidence supporting caveolae microdomain regulation of calcium entry in endothelial cells. J Biol Chem 282:16631–16643

Nagamine K, Kudoh J, Minoshima S, Kawasaki K, Asakawa S, Ito F, Shimizu N (1998) Molecular cloning of a novel putative Ca^{2+} channel protein (TRPC7) highly expressed in brain. Genomics 54:124–131

Ng LC, McCormack MD, Airey JA, Singer CA, Keller PS, Shen XM, Hume JR (2009) TRPC1 and STIM1 mediate capacitative Ca^{2+} entry in mouse pulmonary arterial smooth muscle cells. J Physiol 587:2429–2442

Nilius B, Owsianik G (2010) Channelopathies converge on TRPV4. Nat Genet 42:98–100

Noorani MM, Noel RC, Marrelli SP (2011) Upregulated TRPC3 and downregulated TRPC1 channel expression during hypertension is associated with increased vascular contractility in rat. Front Physiol 2:42

Numaga T, Wakamori M, Mori Y (2007) Trpc7. In: Flockerzi V, Nilius B (eds) Transient receptor potential (TRP) channels. Springer, New York, NY, pp 143–151

Obukhov AG, Nowycky MC (2004) TRPC5 activation kinetics are modulated by the scaffolding protein ezrin/radixin/moesin-binding phosphoprotein-50 (EBP50). J Cell Physiol 201:227–235

Odell AF, Scott JL, Van Helden DF (2005) Epidermal growth factor induces tyrosine phosphorylation, membrane insertion, and activation of transient receptor potential channel 4. J Biol Chem 280:37974–37987

Ohta T, Morishita M, Mori Y, Ito S (2004) Ca^{2+} store-independent augmentation of $[Ca^{2+}]_i$ responses to G-protein coupled receptor activation in recombinantly TRPC5-expressed rat pheochromocytoma (PC12) cells. Neurosci Lett 358:161–164

Ong HL, Ambudkar IS (2012) Role of lipid rafts in the regulation of store-operated Ca^{2+} channels. In: Levitan I, Barrantes F (eds) Cholesterol regulation of ion channels and receptors. Wiley, Hoboken, NJ, pp 69–90

Ong HL, Cheng KT, Liu X, Bandyopadhyay BC, Paria BC, Soboloff J, Pani B, Gwack Y, Srikanth S, Singh BB, Gill DL, Ambudkar IS (2007) Dynamic assembly of TRPC1-STIM1-Orai1 ternary complex is involved in store-operated calcium influx. Evidence for similarities in store-operated and calcium release-activated calcium channel components. J Biol Chem 282:9105–9116

Ong EC, Nesin V, Long CL, Bai CX, Guz JL, Ivanov IP, Abramowitz J, Birnbaumer L, Humphrey MB, Tsiokas L (2013) A TRPC1-dependent pathway regulates osteoclast formation and function. J Biol Chem 288:22219–22232

Owsianik G, Talavera K, Voets T, Nilius B (2006) Permeation and selectivity of TRP channels. Annu Rev Physiol 68:685–717

Pani B, Ong HL, Liu X, Rauser K, Ambudkar IS, Singh BB (2008) Lipid rafts determine clustering of STIM1 in endoplasmic reticulum-plasma membrane junctions and regulation of store-operated Ca^{2+} entry (SOCE). J Biol Chem 283:17333–17340

Pani B, Ong HL, Brazer SC, Liu X, Rauser K, Singh BB, Ambudkar IS (2009) Activation of TRPC1 by STIM1 in ER-PM microdomains involves release of the channel from its scaffold caveolin-1. Proc Natl Acad Sci U S A 106:20087–20092

Pani B, Liu X, Bollimuntha S, Cheng KT, Niesman IR, Zheng C, Achen VR, Patel HH, Ambudkar IS, Singh BB (2012) Impairment of TRPC1-STIM1 channel assembly and AQP5 translocation compromise agonist-stimulated fluid secretion in mice lacking caveolin1. J Cell Sci 126:667–675

Parekh AB, Penner R (1997) Store depletion and calcium influx. Physiol Rev 77:901–930

Parekh AB, Putney JW Jr (2005) Store-operated calcium channels. Physiol Rev 85:757–810

Paria BC, Vogel SM, Ahmmed GU, Alamgir S, Shroff J, Malik AB, Tiruppathi C (2004) Tumor necrosis factor-alpha-induced TRPC1 expression amplifies store-operated Ca^{2+} influx and endothelial permeability. Am J Physiol Lung Cell Mol Physiol 287:L1303–L1313

Park JY, Hwang EM, Yarishkin O, Seo JH, Kim E, Yoo J, Yi GS, Kim DG, Park N, Ha CM, La JH, Kang D, Han J, Oh U, Hong SG (2008) TRPM4b channel suppresses store-operated Ca^{2+} entry by a novel protein-protein interaction with the TRPC3 channel. Biochem Biophys Res Commun 368:677–683

Patterson RL, van Rossum DB, Ford DL, Hurt KJ, Bae SS, Suh P-G, Kurosaki T, Snyder SH, Gill DL (2002) Phospholipase C-γ is required for agonist-induced Ca^{2+} entry. Cell 111:529–541

Penna A, Demuro A, Yeromin AV, Zhang SL, Safrina O, Parker I, Cahalan MD (2008) The CRAC channel consists of a tetramer formed by Stim-induced dimerization of Orai dimers. Nature 456:116–120

Phelan KD, Shwe UT, Abramowitz J, Wu H, Rhee SW, Howell MD, Gottschall PE, Freichel M, Flockerzi V, Birnbaumer L, Zheng F (2013) Canonical transient receptor channel 5 (TRPC5) and TRPC1/4 contribute to seizure and excitotoxicity by distinct cellular mechanisms. Mol Pharmacol 83:429–438

Poteser M, Graziani A, Rosker C, Eder P, Derler I, Kahr H, Zhu MX, Romanin C, Groschner K (2006) TRPC3 and TRPC4 associate to form a redox-sensitive cation channel. Evidence for expression of native TRPC3-TRPC4 heteromeric channels in endothelial cells. J Biol Chem 281:13588–13595

Puram SV, Riccio A, Koirala S, Ikeuchi Y, Kim AH, Corfas G, Bonni A (2011) A TRPC5-regulated calcium signaling pathway controls dendrite patterning in the mammalian brain. Genes Dev 25:2659–2673

Putney JW (2005) Physiological mechanisms of TRPC activation. Pflugers Arch 451:29–34

Rao JN, Platoshyn O, Golovina VA, Liu L, Zou T, Marasa BS, Turner DJ, Yuan JX, Wang JY (2006) TRPC1 functions as a store-operated Ca^{2+} channel in intestinal epithelial cells and regulates early mucosal restitution after wounding. Am J Physiol Gastrointest Liver Physiol 290:G782–G792

Redondo PC, Harper AG, Salido GM, Pariente JA, Sage SO, Rosado JA (2004) A role for SNAP-25 but not VAMPs in store-mediated Ca^{2+} entry in human platelets. J Physiol 558:99–109
Redondo PC, Jardin I, Lopez JJ, Salido GM, Rosado JA (2008) Intracellular Ca^{2+} store depletion induces the formation of macromolecular complexes involving hTRPC1, hTRPC6, the type II IP3 receptor and SERCA3 in human platelets. Biochim Biophys Acta 1783:1163–1176
Reiser J, Polu KR, Moller CC, Kenlan P, Altintas MM, Wei C, Faul C, Herbert S, Villegas I, Avila-Casado C, McGee M, Sugimoto H, Brown D, Kalluri R, Mundel P, Smith PL, Clapham DE, Pollak MR (2005) TRPC6 is a glomerular slit diaphragm-associated channel required for normal renal function. Nat Genet 37:739–744
Riccio A, Li Y, Moon J, Kim KS, Smith KS, Rudolph U, Gapon S, Yao GL, Tsvetkov E, Rodig SJ, Van't Veer A, Meloni EG, Carlezon WA Jr, Bolshakov VY, Clapham DE (2009) Essential role for TRPC5 in amygdala function and fear-related behavior. Cell 137:761–772
Roos J, DiGregorio PJ, Yeromin AV, Ohlsen K, Lioudyno M, Zhang S, Safrina O, Kozak JA, Wagner SL, Cahalan MD, Velicelebi G, Stauderman KA (2005) STIM1, an essential and conserved component of store-operated Ca^{2+} channel function. J Cell Biol 169:435–445
Sabourin J, Lamiche C, Vandebrouck A, Magaud C, Rivet J, Cognard C, Bourmeyster N, Constantin B (2009) Regulation of TRPC1 and TRPC4 cation channels requires an alpha1-syntrophin-dependent complex in skeletal mouse myotubes. J Biol Chem 284:36248–36261
Saleh SN, Albert AP, Peppiatt-Wildman CM, Large WA (2008) Diverse properties of store-operated TRPC channels activated by protein kinase C in vascular myocytes. J Physiol 586:2463–2476
Satoh S, Tanaka H, Ueda Y, Oyama J, Sugano M, Sumimoto H, Mori Y, Makino N (2007) Transient receptor potential (TRP) protein 7 acts as a G protein-activated Ca^{2+} channel mediating angiotensin II-induced myocardial apoptosis. Mol Cell Biochem 294:205–215
Schaefer M, Plant TD, Obukhov AG, Hofmann T, Gudermann T, Schultz G (2000) Receptor-mediated regulation of the nonselective cation channels TRPC4 and TRPC5. J Biol Chem 275:17517–17526
Schindl R, Frischauf I, Kahr H, Fritsch R, Krenn M, Derndl A, Vales E, Muik M, Derler I, Groschner K, Romanin C (2008) The first ankyrin-like repeat is the minimum indispensable key structure for functional assembly of homo- and heteromeric TRPC4/TRPC5 channels. Cell Calcium 43:260–269
Schindl R, Fritsch R, Jardin I, Frischauf I, Kahr H, Muik M, Riedl MC, Groschner K, Romanin C (2012) Canonical transient receptor potential (TRPC) 1 acts as a negative regulator for vanilloid TRPV6-mediated Ca^{2+} influx. J Biol Chem 287:35612–35620
Selvaraj S, Watt JA, Singh BB (2009) TRPC1 inhibits apoptotic cell degeneration induced by dopaminergic neurotoxin MPTP/MPP^{+}. Cell Calcium 46:209–218
Selvaraj S, Sun Y, Watt JA, Wang S, Lei S, Birnbaumer L, Singh BB (2012) Neurotoxin-induced ER stress in mouse dopaminergic neurons involves downregulation of TRPC1 and inhibition of AKT/mTOR signaling. J Clin Invest 122:1354–1367
Shi J, Ju M, Abramowitz J, Large WA, Birnbaumer L, Albert AP (2012) TRPC1 proteins confer PKC and phosphoinositol activation on native heteromeric TRPC1/C5 channels in vascular smooth muscle: comparative study of wild-type and $TRPC1^{-/-}$ mice. FASEB J 26:409–419
Shimizu S, Yoshida T, Wakamori M, Ishii M, Okada T, Takahashi M, Seto M, Sakurada K, Kiuchi Y, Mori Y (2006) Ca^{2+}-calmodulin-dependent myosin light chain kinase is essential for activation of TRPC5 channels expressed in HEK293 cells. J Physiol 570:219–235
Singh BB, Liu X, Tang J, Zhu MX, Ambudkar IS (2002) Calmodulin regulates Ca^{2+}-dependent feedback inhibition of store-operated Ca^{2+} influx by interaction with a site in the C terminus of TrpC1. Mol Cell 9:739–750
Singh BB, Lockwich TP, Bandyopadhyay BC, Liu X, Bollimuntha S, Brazer SC, Combs C, Das S, Leenders AG, Sheng ZH, Knepper MA, Ambudkar SV, Ambudkar IS (2004) VAMP2-dependent exocytosis regulates plasma membrane insertion of TRPC3 channels and contributes to agonist-stimulated Ca^{2+} influx. Mol Cell 15:635–646

Song X, Zhao Y, Narcisse L, Duffy H, Kress Y, Lee S, Brosnan CF (2005) Canonical transient receptor potential channel 4 (TRPC4) co-localizes with the scaffolding protein ZO 1 in human fetal astrocytes in culture. Glia 49:418–429

Sours-Brothers S, Ding M, Graham S, Ma R (2009) Interaction between TRPC1/TRPC4 assembly and STIM1 contributes to store-operated Ca^{2+} entry in mesangial cells. Exp Biol Med 234:673–682

Stiber JA, Zhang ZS, Burch J, Eu JP, Zhang S, Truskey GA, Seth M, Yamaguchi N, Meissner G, Shah R, Worley PF, Williams RS, Rosenberg PB (2008) Mice lacking Homer 1 exhibit a skeletal myopathy characterized by abnormal transient receptor potential channel activity. Mol Cell Biol 28:2637–2647

Strubing C, Krapivinsky G, Krapivinsky L, Clapham DE (2001) TRPC1 and TRPC5 form a novel cation channels in mammalian brain. Neuron 29:645–655

Strubing C, Krapivinsky G, Krapivinsky L, Clapham DE (2003) Formation of novel TRPC channels by complex subunit interactions in embryonic brain. J Biol Chem 278:39014–39019

Sukumaran P, Lof C, Pulli I, Kemppainen K, Viitanen T, Tornquist K (2013) Significance of the transient receptor potential canonical 2 (TRPC2) channel in the regulation of rat thyroid FRTL-5 cell proliferation, migration, adhesion and invasion. Mol Cell Endocrinol 374:10–21

Sundivakkam PC, Kwiatek AM, Sharma TT, Minshall RD, Malik AB, Tiruppathi C (2009) Caveolin-1 scaffold domain interacts with TRPC1 and IP3R3 to regulate Ca^{2+} store release-induced Ca^{2+} entry in endothelial cells. Am J Physiol Cell Physiol 296:C403–C413

Sundivakkam PC, Freichel M, Singh V, Yuan JP, Vogel SM, Flockerzi V, Malik AB, Tiruppathi C (2012) The Ca^{2+} sensor stromal interaction molecule 1 (STIM1) is necessary and sufficient for the store-operated Ca^{2+} entry function of transient receptor potential canonical (TRPC) 1 and 4 channels in endothelial cells. Mol Pharmacol 81:510–526

Sutton KA, Jungnickel MK, Wang Y, Cullen K, Lambert S, Florman HM (2004) Enkurin is a novel calmodulin and TRPC channel binding protein in sperm. Dev Biol 274:426–435

Tang Y, Tang J, Chen Z, Trost C, Flockerzi V, Li M, Ramesh V, Zhu MX (2000) Association of mammalian Trp4 and phospholipase C isozymes with a PDZ domain-containing protein, NHERF. J Biol Chem 275:27559–27564

Tang J, Lin Y, Zhang Z, Tikunova S, Birnbaumer L, Zhu MX (2001) Identification of common binding sites for calmodulin and IP_3 receptors on the carboxyl-termini of TRP channels. J Biol Chem 276:21303–21310

Thebault S, Flourakis M, Vanoverberghe K, Vandermoere F, Roudbaraki M, Lehen'kyi V, Slomianny C, Beck B, Mariot P, Bonnal JL, Mauroy B, Shuba Y, Capiod T, Skryma R, Prevarskaya N (2006) Differential role of transient receptor potential channels in Ca^{2+} entry and proliferation of prostate cancer epithelial cells. Cancer Res 66:2038–2047

Tian D, Jacobo SM, Billing D, Rozkalne A, Gage SD, Anagnostou T, Pavenstadt H, Hsu HH, Schlondorff J, Ramos A, Greka A (2010) Antagonistic regulation of actin dynamics and cell motility by TRPC5 and TRPC6 channels. Sci Signal 3:ra77

Tong Q, Chu X, Cheung JY, Conrad K, Stahl R, Barber DL, Mignery G, Miller BA (2004) Erythropoietin-modulated calcium influx through TRPC2 is mediated by phospholipase Cgamma and IP3R. Am J Physiol Cell Physiol 287:C1667–C1678

Torihashi S, Fujimoto T, Trost C, Nakayama S (2002) Calcium oscillation linked to pacemaking of interstitial cells of Cajal: requirement of calcium influx and localization of TRP4 in caveolae. J Biol Chem 277:19191–19197

Tsvilovskyy VV, Zholos AV, Aberle T, Philipp SE, Dietrich A, Zhu MX, Birnbaumer L, Freichel M, Flockerzi V (2009) Deletion of TRPC4 and TRPC6 in mice impairs smooth muscle contraction and intestinal motility in vivo. Gastroenterology 137:1415–1424

Tu CL, Chang W, Bikle DD (2005) Phospholipase cgamma1 is required for activation of store-operated channels in human keratinocytes. J Invest Dermatol 124:187–197

van Rossum DB, Patterson RL, Sharma S, Barrow RK, Kornberg M, Gill DL, Snyder SH (2005) Phospholipase Cgamma1 controls surface expression of TRPC3 through an intermolecular PH domain. Nature 434:99–104

Vannier B, Peyton M, Boulay G, Brown D, Qin N, Jiang MS, Zhu X, Birnbaumer L (1999) Mouse trp2, the homologue of the human trpc2 pseudogene, encodes mTrp2, a store depletion-activated capacitative Ca^{2+} entry channel. Proc Natl Acad Sci U S A 99:2060–2064

Venkatachalam K, Montell C (2007) TRP channels. Annu Rev Biochem 76:387–417

Viitanen TM, Sukumaran P, Lof C, Tornquist K (2013) Functional coupling of TRPC2 cation channels and the calcium-activated anion channels in rat thyroid cells: implications for iodide homeostasis. J Cell Physiol 228:814–823

Wenning AS, Neblung K, Strauss B, Wolfs MJ, Sappok A, Hoth M, Schwarz EC (2011) TRP expression pattern and the functional importance of TRPC3 in primary human T-cells. Biochim Biophys Acta 1813:412–423

Wes PD, Chevesich J, Jeromin A, Rosenberg C, Stetten S, Montell C (1995) TRPC1, a human homolog of a Drosophila store-operated channel. Proc Natl Acad Sci U S A 92:9652–9656

Williams IA, Allen DG (2007) Intracellular calcium handling in ventricular myocytes from *mdx* mice. Am J Physiol Heart Circ Physiol 292:H846–H855

Wissenbach U, Schroth G, Philipp S, Flockerzi V (1998) Structure and mRNA expression of a bovine trp homologue related to mammalian trp2 transcripts. FEBS Lett 429:61–66

Woo JS, Kim do H, Allen PD, Lee EH (2008) TRPC3-interacting triadic proteins in skeletal muscle. Biochem J 411:399–405

Woo JS, Cho CH, Kim do H, Lee EH (2010) TRPC3 cation channel plays an important role in proliferation and differentiation of skeletal muscle myoblasts. Exp Mol Med 42:614–627

Woodard GE, Lopez JJ, Jardin I, Salido GM, Rosado JA (2010) TRPC3 regulates agonist-stimulated Ca^{2+} mobilization by mediating the interaction between type I inositol 1,4,5-trisphosphate receptor, RACK1, and Orai1. J Biol Chem 285:8045–8053

Worley PF, Zeng W, Huang GN, Yuan JP, Kim JY, Lee MG, Muallem S (2007) TRPC channels as STIM1-regulated store-operated channels. Cell Calcium 42:205–211

Wu X, Zagranichnaya TK, Gurda GT, Eves EM, Villereal ML (2004) A TRPC1/TRPC3-mediated increase in store-operated calcium entry is required for differentiation of H19-7 hippocampal neuronal cells. J Biol Chem 279:43392–43402

Wu X, Eder P, Chang B, Molkentin JD (2010) TRPC channels are necessary mediators of pathologic cardiac hypertrophy. Proc Natl Acad Sci U S A 107:7000–7005

Xu XZ, Li HS, Guggino WB, Montell C (1997) Coassembly of TRP and TRPL produces a distinct store-operated conductance. Cell 89:1155–1164

Xu SZ, Muraki K, Zeng F, Li J, Sukumar P, Shah S, Dedman AM, Flemming PK, McHugh D, Naylor J, Cheong A, Bateson AN, Munsch CM, Porter KE, Beech DJ (2006) A sphingosine-1-phosphate-activated calcium channel controlling vascular smooth muscle cell motility. Circ Res 98:1381–1389

Yildirim E, Dietrich A, Birnbaumer L (2003) The mouse C-type transient receptor potential 2 (TRPC2) channel: alternative splicing and calmodulin binding to its N terminus. Proc Natl Acad Sci U S A 100:2220–2225

Yuan JP, Kiselyov K, Shin DM, Chen J, Shcheynikov N, Kang SH, Dehoff MH, Schwarz MK, Seeburg PH, Muallem S, Worley PF (2003) Homer binds TRPC family channels and is required for gating of TRPC1 by IP3 receptors. Cell 114:777–789

Yuan JP, Zeng W, Huang GN, Worley PF, Muallem S (2007) STIM1 heteromultimerizes TRPC channels to determine their function as store-operated channels. Nat Cell Biol 9:636–645

Yuan JP, Zeng W, Dorwart MR, Choi YJ, Worley PF, Muallem S (2009) SOAR and the polybasic STIM1 domains gate and regulate Orai channels. Nat Cell Biol 11:337–343

Yuasa K, Matsuda T, Tsuji A (2011) Functional regulation of transient receptor potential canonical 7 by cGMP-dependent protein kinase Ialpha. Cell Signal 23:1179–1187

Zagranichnaya TK, Wu X, Villereal ML (2005) Endogenous TRPC1, TRPC3, and TRPC7 proteins combine to form native store-operated channels in HEK-293 cells. J Biol Chem 280:29559–29569

Zeng W, Yuan JP, Kim MS, Choi YJ, Huang GN, Worley PF, Muallem S (2008) STIM1 gates TRPC channels, but not Orai1, by electrostatic interaction. Mol Cell 32:439–448

Zhang Z, Tang J, Tikunova S, Johnson JD, Chen Z, Qin N, Dietrich A, Stefani E, Birnbaumer L, Zhu MX (2001) Activation of Trp3 by inositol 1, 4, 5-trisphosphate receptors through displacement of inhibitory calmodulin from a common binding domain. Proc Natl Acad Sci U S A 98:3168–3173

Zhang ZY, Pan LJ, Zhang ZM (2010) Functional interactions among STIM1, Orai1 and TRPC1 on the activation of SOCs in HL-7702 cells. Amino Acids 39:195–204

Zhu X, Chu PB, Peyton M, Birnbaumer L (1995) Molecular cloning of a widely expressed human homologue for the Drosophila trp gene. FEBS Lett 373:193–198

Zhu X, Jiang M, Peyton M, Boulay G, Hurst R, Stefani E, Birnbaumer L (1996) trp, a novel mammalian gene family essential for agonist-activated capacitative Ca^{2+} entry. Cell 85:661–671

Zimmermann K, Lennerz JK, Hein A, Link AS, Kaczmarek JS, Delling M, Uysal S, Pfeifer JD, Riccio A, Clapham DE (2011) Transient receptor potential cation channel, subfamily C, member 5 (TRPC5) is a cold-transducer in the peripheral nervous system. Proc Natl Acad Sci U S A 108:18114–18119

The TRPCs–STIM1–Orai Interaction

Seok Choi, Jozsef Maleth, Archana Jha, Kyu Pil Lee, Min Seuk Kim, Insuk So, Malini Ahuja, and Shmuel Muallem

Contents

S. Choi • J. Maleth • A. Jha • M. Ahuja
Epithelial Signaling and Transport Section, Molecular Physiology and Therapeutics Branch, National Institute of Dental and Craniofacial Research, National Institute of Health, Bethesda, MD 20892, USA

K.P. Lee
Department of Physiology, College of Veterinary Medicine, Chungnam National University, Daejeon 305-764, Republic of Korea

M.S. Kim
Department of Oral Physiology, School of Dentistry, Wonkwang University, Iksan City, Jeonbuk, Republic of Korea

I. So
Department of Physiology and Biophysics, Seoul National University College of Medicine, Seoul 110-799, Republic of Korea

S. Muallem (✉)
Epithelial Signaling and Transport Section, Molecular Physiology and Therapeutics Branch, NIDCR, NIH, Building 10, Room 1N-112, Bethesda, MD 20892, USA
e-mail: shmuel.muallem@nih.gov

B. Nilius and V. Flockerzi (eds.), *Mammalian Transient Receptor Potential (TRP) Cation Channels*, Handbook of Experimental Pharmacology 223, DOI 10.1007/978-3-319-05161-1_13,

Abstract
Ca^{2+} signaling entails receptor-stimulated Ca^{2+} release from the ER stores that serves as a signal to activate Ca^{2+} influx channels present at the plasma membrane, the store-operated Ca^{2+} channels (SOCs). The two known SOCs are the Orai and TRPC channels. The SOC-dependent Ca^{2+} influx mediates and sustains virtually all Ca^{2+}-dependent regulatory functions. The signal that transmits the Ca^{2+} content of the ER stores to the plasma membrane is the ER resident, Ca^{2+}-binding protein STIM1. STIM1 is a multidomain protein that clusters and dimerizes in response to Ca^{2+} store depletion leading to activation of Orai and TRPC channels. Activation of the Orais by STIM1 is obligatory for their function as SOCs, while TRPC channels can function as both STIM1-dependent and STIM1-independent channels. Here we discuss the different mechanisms by which STIM1 activates the Orai and TRPC channels, the emerging specific and non-overlapping physiological functions of Ca^{2+} influx mediated by the two channel types, and argue that the TRPC channels should be the preferred therapeutic target to control the toxic effect of excess Ca^{2+} influx.

Keywords
TRPC channels • STIM1 • Gating • Physiology • Pathology

1 Introduction

The receptor-evoked Ca^{2+} signal is initiated by the hydrolysis of PI(4,5)P2 either by phospholipase Cβ (G protein-coupled receptors) or phospholipase Cγ (tyrosine kinase receptors) to generate IP_3 and diacylglycerol. IP_3 releases Ca^{2+} primarily from the endoplasmic reticulum (ER), which provides a signal to activate Ca^{2+} influx channels at the plasma membrane. The ensuing increase in free cytoplasmic Ca^{2+} ($[Ca^{2+}]_i$) activates the plasma membrane Ca^{2+} ATPase (PMCA) and the sarcoplasmic reticulum Ca^{2+} ATPase (SERCA) pumps that clear Ca^{2+} from the cytosol (Kiselyov et al. 2003). At weak receptor stimulation, this cycle is periodically repeated to generate Ca^{2+} oscillations, while at intense receptor stimulation, the IP_3 receptors (IP_3Rs) remain active to keep the ER depleted of Ca^{2+}, and $[Ca^{2+}]_i$ is determined by a "pump-leak turnover" with the Ca^{2+} influx channels providing the leak and the PMCA the pump (Kiselyov et al. 2006).

The Ca^{2+} influx channels participate in all aspects of the receptor-evoked Ca^{2+} signal. They contribute to the initial increase in $[Ca^{2+}]_i$ upon receptor stimulation, they reload the stores with Ca^{2+} between the Ca^{2+} spikes during Ca^{2+} oscillations, and they remain active for many minutes at the end of the stimulation period until the stores are fully loaded with Ca^{2+} (Muallem et al. 1988; Pandol et al. 1987; Parekh and Putney 2005). In polarized cells the Ca^{2+} signal initiates at the apical pole and propagates to the basal pole in the form of Ca^{2+} waves (Kasai and Augustine 1990; Thorn et al. 1993). Ca^{2+} influx accelerates the rate of Ca^{2+} wave propagation (Lee et al. 1997). The Ca^{2+} influx channels determine practically all physiological and pathological effects of $[Ca^{2+}]_i$ by determining the size and shape of the Ca^{2+} signal. In fact, all Ca^{2+}-dependent physiological functions occurring after the first 1 min of cell stimulation stops in the absence of Ca^{2+} influx (Parekh and Putney 2005), while excessive activation of the Ca^{2+} influx channels and Ca^{2+} influx is largely responsible for the pathological effects of Ca^{2+} (Lee et al. 2010a; Petersen et al. 2006).

Two types of channels mediate the bulk of the receptor-stimulated Ca^{2+} influx, the TRPC and Orai channels (Lee et al. 2010a). The contribution and role of the TRPC channels to the receptor-stimulated Ca^{2+} influx has been extensively studied (Pandol et al. 1987; Lee et al. 1997) before the discovery of the Orai channels (Feske et al. 2006; Zhang et al. 2006; Vig et al. 2006) and of STIM1 (Roos et al. 2005; Liou et al. 2005). In this chapter we will argue that both the TRPC and Orai channels are gated by STIM1, that the contribution of both channel types is vital for cell function, that the TRPC and Orai channels sense and respond to the activity of each other, that increasing evidence suggests that each channel type may regulate separate Ca^{2+}-dependent cell functions, and that the TRPC channels may provide an important and perhaps preferred therapeutic target to guard against Ca^{2+}-dependent toxicity and pathology.

2 The STIM1 Functional Domains

2.1 The STIM1 Domains

The relationship between the ER Ca^{2+} load and the function of the Ca^{2+} influx channels was resolved with the discovery of the ER Ca^{2+} content sensor STIM1 (Roos et al. 2005; Liou et al. 2005). STIM1 is a multidomain protein, the function of which is continually being updated as more information becomes available. The currently known STIM1 domains are depicted in Fig. 1a. These include the ER lumen resident Ca^{2+} binding EF hand and SAM domains. The EF hand is the Ca^{2+}-binding site, while the SAM domain participates in STIM1 clustering (Stathopulos et al. 2008; Zheng et al. 2008). Binding of Ca^{2+} to the EF hand keeps STIM1 in a non-clustered form and restrict its access to the Ca^{2+} influx channels. The luminal domain also has reactive cysteines that appear to participate in ER redox sensing through S-glutathionylation of cysteine 56 (Hawkins et al. 2010). The SAM domain is followed by a short transmembrane domain and a long cytoplasmic domain. The cytoplasmic domain of STIM1 mediates the opening of the TRPC and Orai channels (Huang et al. 2006). The cytoplasmic domain starts with the first coiled-coil domain (CC1), which includes a short helix at its C terminus that functions as an inhibitory helix (IH) (Yu et al. 2013; Yang et al. 2012). CC1 is followed by the SOAR domain (Yuan et al. 2009), which is also known as CAD (Park et al. 2009) or CCb9 (Feske et al. 2010). SOAR is the minimal STIM1 domain needed to fully activate Orai1 (Yuan et al. 2009; Park et al. 2009; Feske et al. 2010). SOAR may also participate in the regulation of TRPC channels by STIM1 (see below). The

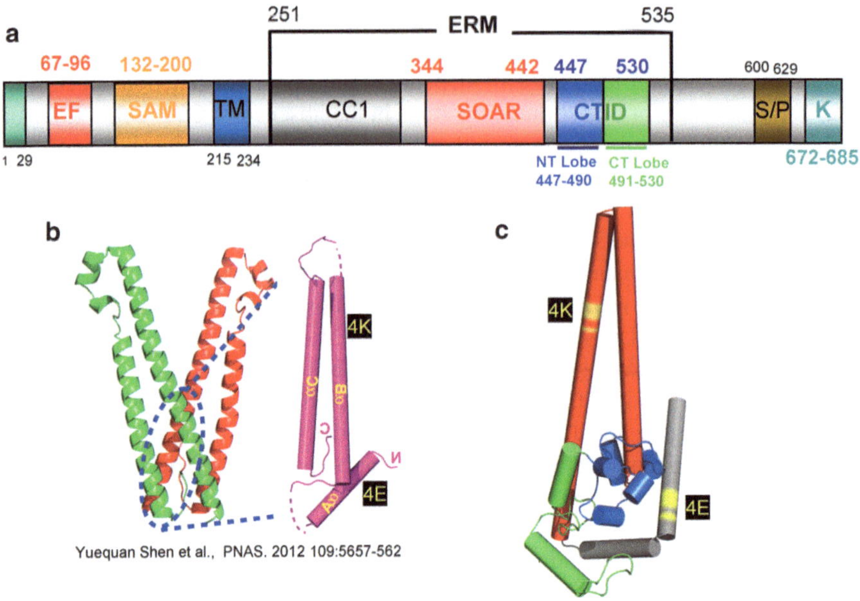

Fig. 1 The STIM1 domains. The currently known domains of STIM1 are shown in (**a**). Panel (**b**) shows the crystal structure of the SOAR dimer and monomer and the position of the four lysines (4K) and four glutamates (4E) in IH taken from Archana Jha Malini Ahuja et al. (2013). Panel (**c**) is a predicted structure of SOAR (*red*) and the N terminal (*blue*) and C terminal (*green*) lobes of CTID

crystal structure of the SOAR domain was resolved recently (Jha et al. 2013; Fig. 1b). Recent analysis identified a new highly conserved domain immediately C terminus to SOAR that we named CTID (for C terminus inhibitory domain). The predicted structure of CTID is shown in Fig. 1c (see also Jha et al. 2013). CTID is followed by a long sequence, the function of which is only beginning to emerge. Recent study reported that the function of this C terminus portion of STIM1 is to keep SOAR in an inactive state (Yu et al. 2013). The C terminal end of STIM1 is polybasic with multiple lysine residues (K-domain). The K-domain interacts with PI(4,5)P2 at the plasma membrane to stabilize the clustered STIM1 at the ER/plasma membrane microdomain (Liou et al. 2007). Most significantly, the K-domain gates open the TRPC channels (see Zeng et al. (2008) and below).

Several recent studies revealed that when the ER stores are filled with Ca^{2+}, STIM1 resumes a compact structure. Upon Ca^{2+} store depletion, STIM1 undergoes massive conformational transition to a more extended conformation (Yu et al. 2013; Korzeniowski et al. 2010; Muik et al. 2011). Sequence analysis identified four conserved lysines in SOAR (marked by 4K in Fig. 1b, c) that potentially interact either with four conserved glutamates in the IH domain in CC1 (marked by 4E in Fig. 1b, c) or with conserved glutamates in Orai1 that may fold similarly (Korzeniowski et al. 2010). Mutations of the glutamates in IH resulted in a constitutively active STIM1, while mutations of the lysines in SOAR inhibited

STIM1 function. Based on these findings, it was suggested that CC1 interacts with SOAR to keep it in inactive state (Yu et al. 2013; Korzeniowski et al. 2010; Muik et al. 2011). Direct evidence for folding and interaction of CC1 and SOAR was provided by convincing experiments using a STIM1 conformational FRET reporter (Muik et al. 2011). These studies suggested that the CC1–SOAR interaction is mediated by hydrophobic interactions. This conclusion is supported by further analysis of the role of the glutamates in the CC1 domain (Yu et al. 2013). In addition, recent crystal structure of SOAR and IH revealed that the lysines in SOAR and the glutamates in CC1 do not interact (Yang et al. 2012). However, they are consistent with the possibility that the lysines in SOAR may mediate SOAR interaction with Orai1 (Korzeniowski et al. 2010). Our studies suggest that mutation of the lysines in SOAR likely disrupts its structure to the extent that STIM1 with the lysines mutated to alanines aggregates in small intracellular inclusions (Jha et al. 2013). Another fold that occludes SOAR is present in STIM1 C terminus to SOAR (Yu et al. 2013; Jha et al. 2013). Initially, it was shown that replacing the STIM1 C terminus with GFP resulted in activation of STIM1 by making SOAR available for interaction with Orai1 (Yu et al. 2011). Subsequently, another STIM1 conformational reporter that includes the entire STIM1 cytoplasmic domain was used to show that the STIM1 C terminus folds to contact CC1, likely in the IH, and participates in the occlusion of SOAR (Yu et al. 2013). Together, the conformational reporters and SOAR structure suggest that in the basal state, STIM1 is folded in a manner that CC1 interacts with SOAR and the C terminus folds back to interact with the IH in CC1, resulting in occlusion of SOAR.

2.2 STIM1 Interactors

Additional regulation of STIM1 function is due to its interaction with other proteins. Comprehensive analysis of the STIM1 interactome is not available as yet. The two established interactors are the microtubule-plus-end-tracking protein end binding (EB1) (Grigoriev et al. 2008) and the SOCE-associated regulatory factor (SARAF) (Palty et al. 2012). Interaction of STIM1 with EB1 forms comet-like accumulations at the sites of microtubule end-ER contacts. The interaction is regulated by phosphorylation of STIM1 in serine/threonine residues in the C terminus region of STIM1 (Smyth et al. 2012). Interestingly, this interaction appears to participate in ER remodeling (Grigoriev et al. 2008; Smyth et al. 2012). In particular, STIM1 interaction with EB1 is inhibited during mitosis, resulting in exclusion of STIM1 from the mitotic spindle, and this is reversed by phosphorylation-compromised STIM1 mutant (Smyth et al. 2012). Importantly, STIM1 phosphorylation at the sites that affect EB1 interaction and spindle exclusion has no role in the regulation of Ca^{2+} influx by STIM1 (Smyth et al. 2012), suggesting that STIM1 has roles outside the regulation of Ca^{2+} influx channels. This has also been shown for the regulation of endothelial barrier function that is regulated by STIM1-mediated coupling of the thrombin receptor to activation of

RhoA, formation of actin stress fibers, and the eventual loss of cell–cell adhesion (Shinde et al. 2013).

High-throughput screen has identified SARAF as a negative regulator of STIM1 (Palty et al. 2012). SARAF is a single transmembrane span, ER resident protein with the N terminus in the ER lumen and the C terminus in the cytoplasm. SARAF interacts with STIM1 upon Ca^{2+} store depletion to facilitate inactivation of Orai1 by Ca^{2+}. It appears that the C terminus of SARAF mediates the interaction with STIM1 (Palty et al. 2012). Once activated, Orai1 undergoes two types of Ca^{2+}-mediated inactivation, fast (FCDI) and slow (SCDI) inactivation with time courses of ms and min (Parekh and Putney 2005). Prior work identified a negatively charged STIM1 sequence (^{475}DDVDDMDEE^{483}) that mediates the FCDI (Lee et al. 2009; Mullins et al. 2009; Derler et al. 2009). Before the discovery of SARAF, no information was available on the mechanism of the SCDI. In a systematic domain analysis, we identified SOAR as the STIM1 domain that interacts with SARAF (Jha et al. 2013). Moreover, we found that CTID, the highly conserved STIM1 domain C terminus to SOAR, controls the access of SARAF to SOAR to mediate the SCDI. As shown in Fig. 1c, CTID appears to have two lobes, with the N terminal lobe [STIM1(447–490)] restricting the access of SARAF to SOAR and the C terminal lobe [STIM1(491–530)] facilitating interaction of SARAF with SOAR (Jha et al. 2013). Finally, CTID appears to also have a role in the FCDI of Orai1, as modification of the interaction of SARAF with STIM1 alters the kinetics of or prevents FCDI (Jha et al. 2013).

That STIM1 has both Ca^{2+} influx-dependent and Ca^{2+} influx-independent functions strongly suggest that STIM1 interacts with many more proteins in addition to EB1 and SARAF within signaling complexes. It is likely that the STIM1 interactome will become available in the near future.

3 Gating of Ca^{2+} Influx Channels by STIM1

3.1 Gating of Orai Channels by STIM1

This topic is discussed only briefly. Further details can be found in recent comprehensive reviews (Engh et al. 2012; Shaw and Feske 2012; Muik et al. 2012). STIM1 activates the Orai channels to mediate the Ca^{2+}-release-activated Ca^{2+} (CRAC) current (Zhang et al. 2006). The Orai channels are four transmembrane span proteins with cytoplasmic N and C termini. Early biochemical and single molecule photobleaching studies suggested that the Orai channels are tetramers (Demuro et al. 2011; Penna et al. 2008; Ji et al. 2008). However, the recent crystal structure of Orai1 revealed that the channel is a hexamer with the selectivity filter composed of a ring of glutamates (Hou et al. 2012). Cytoplasmic extensions of the fourth transmembrane domain of four of the six subunits fold into coiled-coil domains. Functional (Muik et al. 2008; Frischauf et al. 2009) and the structural (Hou et al. 2012) information showed that the C terminus coiled-coil domains are essential for activation of the Orai channels by STIM1. SOAR opens all three

Orai channels by interaction with the C-terminus coiled-coil domains of the Orais (Lee et al. 2009). A second interaction of SOAR/CAD was found with the N-terminus of Orai1 downstream of residue 73 (Park et al. 2009). Recently, it was suggested that STIM1 interaction with the Orai1 C terminus mediates the binding between the proteins, while interaction of STIM1 with the N terminus of Orai1 gates the opening of Orai1 (Zheng et al. 2013). However, farther recent examination of this topic shows that interaction of STIM1 with both the C and N termini of the Orai1 participates in channel gating (McNally et al. 2013). Moreover, both interactions are required for the effect of STIM1 on Orai1 channel selectivity (McNally et al. 2012, 2013). The later findings are also compatible with the role of the Orai channels C termini in the extent and mode of channel regulation. The three Orai channels have a typical FCDI, with Orai1 showing the least prominent and Orai3 the most prominent inactivation (Lis et al. 2007). The information for the type of inactivation is coded in their C termini, as revealed by translocation of the inactivation kinetics with the C-termini when they are swapped among the Orais (Lee et al. 2009). Since removal of CTID eliminates both the FCDI and SCDI (Jha et al. 2013) and SARAF accelerates the SCDI of Orai1 (Jha et al. 2013; Palty et al. 2012), it should be of interest to determine whether and how SARAF affects the FCDI and SCDI of Orai2 and Orai3.

4 TRPC Channels and STIM1

The regulation of TRPC channels by STIM1 is poorly understood and sparsely studied and thus remains controversial to some extent. Several aspects of TRPC channels function and regulation are generally accepted. TRPC channels mediate significant portion of the receptor-stimulated Ca^{2+} influx, TRPC3, TRPC6, and TRPC7, but not TRPC1, TRPC4, and TRPC5 are activated by diacylglycerol, and Ca^{2+} influx by TRPC channels has multiple roles in virtually every cell type (Parekh and Putney 2005; Pedersen et al. 2005; Freichel et al. 2005). Biochemical interaction between TRPC channels and STIM1 is a highly consistent finding. It is also a consistent finding that knockdown or deletion of Orai1 or STIM1 eliminates the store-dependent and receptor-activated Ca^{2+} influx (Lee et al. 2010a). At the same time, numerous studies in multiple cell types consistently find that knockout or knockdown of TRPC channels significantly or markedly reduce the store-mediated and receptor-activated Ca^{2+} influx. Examples of the effect of TRPC1 and TRPC3 knockout of receptor- and store-dependent Ca^{2+} influx in pancreatic acini are illustrated in Fig. 2. Deletion of TRPC1 (green traces and columns) or TRPC3 (red traces and columns) reduces the receptor-stimulated Ca^{2+} influx (reduced plateau) and Ca^{2+} influx evoked by store depletion (Fig. 2a, c, d) and thus the frequency of Ca^{2+} oscillations (Fig. 2b). Additional examples of the effect of TRPC channels knockdown on Ca^{2+} influx are in Lee et al. (2010a), Ng et al. (2012), Antigny et al. (2013), Zhang et al. (2010), Rao et al. (2012), and Sundivakkam et al. (2012). Given the complete inhibition of Ca^{2+} influx by deletion of STIM1 and Orai1 and the substantial inhibition of Ca^{2+} influx by deletion of TRPC channels, it

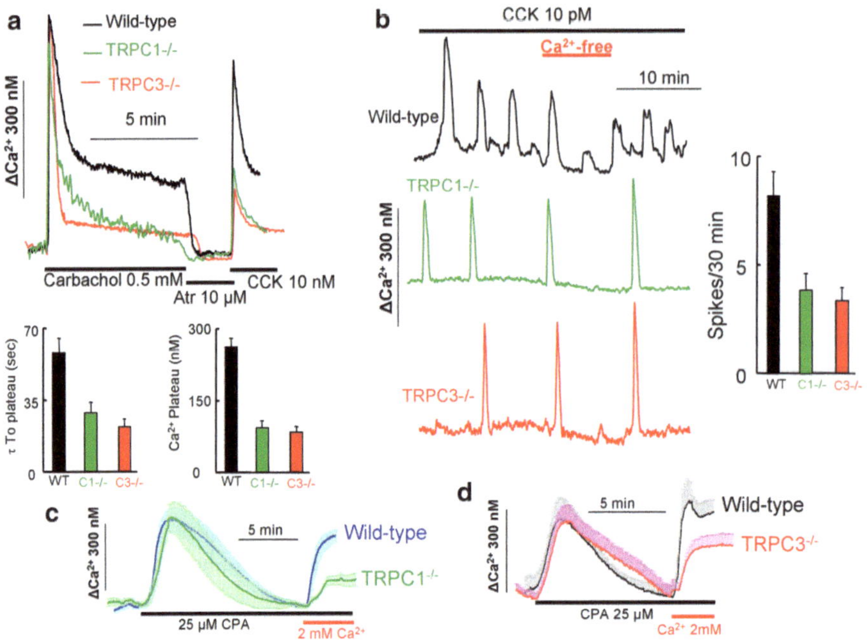

Fig. 2 TRPC1 and TRPC3 as store-operated Ca^{2+} influx channels in pancreatic acini. $[Ca^{2+}]_i$ was measured in wild-type (*black traces and columns*), $Trpc3^{-/-}$ (*red traces and columns*), and $Trpc1^{-/-}$ (*green traces and columns*) pancreatic acini stimulated with the indicated concentration of agonists to generate a sustain response (**a**) or Ca^{2+} oscillations (**b**) or treated with 25 μM CPA to deplete the stores and measure Ca^{2+} influx by Ca^{2+} re-addition (**c**, **d**). Results were reproduced from Kim et al. (2009b)

follows that TRPC channels required Orai1 function and/or Orai1 requires TRPC channels function for their activity and, moreover, that TRPC channels are regulated by STIM1.

4.1 STIM1-Dependent and STIM1-Independent Function of TRPC Channels

In spite of the findings mentioned above, regulation of TRPC channels by STIM1 remains poorly understood and is not fully accepted or appreciated. This is mostly because regulation of TRPC channels by STIM1 is not always observed when the two are co-expressed in model systems. How can this be reconciled with the knockout/knockdown results? A major reason is that TRPC channels can function as both STIM1-dependent and STIM1-independent channels while the Orai channels function is strictly dependent on STIM1. Thus, the early studies already showed that TRPC1, TRPC2, TRPC4, and TRPC5 can directly interact with STIM1, while TRPC3, TRPC6, and TRPC7 do not (Huang et al. 2006). However, it is well established that TRPC channels can heteromultimerize, including TRPC1–TRPC3 (Yuan et al. 2007; Liu et al. 2005) and TRPC4–TRPC6 (Yuan et al. 2007). It turned out that TRPC3 can function as STIM1-dependent channel only in the presence of TRPC1 and TRPC6 functions as STIM1-dependent channel only in the presence of TRPC4 (Yuan et al. 2007). Notably, in the absence of TRPC1 and TRPC4, TRPC3 and TRPC6, respectively, are fully active, but their activity is STIM1 independent. This indicates that (1) the cellular composition of TRPC channel isoforms will determine whether their function is STIM1 dependent or not and (2) a precise, physiological expression level and ratio of TRPC channels isoforms is required to observe and study their STIM1 dependence. Hence, a common problem in the field is that often marked overexpression of the channels is used to study their regulation by STIM1, conditions that ensure their STIM1-independent function.

The study of the regulation of TRPC channels by STIM1 is further complicated by the requirement of Orai1 for TRPC channels function. Undoubtedly, the Orai and TRPC channels can function independent of each other to mediate the CRAC current (Zeng et al. 2008) and nonselective, Ca^{2+}-permeable current (Pedersen et al. 2005; Freichel et al. 2005; Nilius et al. 2007) and dominate Ca^{2+} influx in particular cell types. For example, Orai1-mediated current is prominent in all blood-born cell types (Engh et al. 2012; Shaw and Feske 2012). However, in most cell types, it is difficult to record Icrac current, where TRPC channels likely mediate most of the Ca^{2+} influx. In such cells, both the Orai and TRPC channels appear to be required for SOC by affecting the activity of each other. Initial indication for this was obtained by demonstrating that deletion of Orai1 inhibits all forms of Ca^{2+} influx. Subsequently it was shown that all Orai channels interact with TRPC channels to complex with STIM1 and enhance TRPC channels store dependence (Liao et al. 2007, 2008, 2009). Additional relationship between the channels was shown by demonstrating that functional Orai1 and TRPC channels are required to restore the physiological store-mediated Ca^{2+} influx, with channel-dead Orai1 or TRPC channel mutants unable to do so (Kim et al. 2009a). The interdependence of the channels is attributed to their insertion in the plasma membrane and stabilization of the Orai1–STIM1–TRPC complexes in plasma membrane microdomain (Kim et al. 2009a; Ong et al. 2007; Cheng et al. 2011). Importantly, modulation of TRPC channels function by the Orais and reconstitution of SOCs were observed only when the Orai and TRPC channels are expressed at very low, close to physiological levels (Liao et al. 2007, 2008, 2009; Kim et al. 2009a). The latter suggest that the ratio of Orai1/TRPC channels is important for the STIM1-dependent function of the TRPC channels.

Another factor contributing to the problem in studying TRPC channels regulation by STIM1 is the activation/regulation of TRPC channels by other ligands/regulators. For example, TRPC3, TRPC6, and TRPC7 are activated by diacylglycerol (Pedersen et al. 2005; Freichel et al. 2005), TRPC4 and TRPC5 are activated by Gα subunits (Jeon et al. 2012), TRPC5 is activated by Ca^{2+} (Blair et al. 2009), and TRPC4 and TRPC5 are activated by redox compounds (Xu et al. 2008; Takahashi and Mori 2011). These forms of activation are

STIM1-independent and, when functioning, may overwhelm the STIM1-dependent activation of the channels.

The discussion above points to the caution needed when selecting the system and conditions to study regulation of TRPC channels by STIM1. However, the important point is that this topic is neglected and grossly understudied by investigators interested in TRP channels and in the role of TRPC channels in receptor-mediated Ca^{2+} influx and its functions. Biochemical interaction between STIM1 and the TRPC channels is a highly consistent observation. Therefore, understanding how STIM1 interacts with and affects the function of TRPC channels is paramount for the understanding receptor-mediated Ca^{2+} signaling.

4.2 Gating of TRPC Channels by STIM1

Probably the strongest evidence that TRPC channels are gated by STIM1 is the finding that STIM1 gates TRPC channels by electrostatic interaction (Zeng et al. 2008). Initial structure-function studies revealed that deletion of the STIM1 K-domain prevents activation of TRPC1 by STIM1, Ca^{2+} influx, and translocation of NFAT to the nucleus (Huang et al. 2006). The K-domain is predicted to fold as a helix with the positive charge of the last two lysines (K) on the surface at the end of the helix. This prompted us to search for a helix with negatively charged residues that might complement the positive charges in the STIM1 K-domain. Such two conserved residues (DD/DE/EE) were identified in all TRPC channels (Park et al. 2009). Mutation analysis revealed that the negatively charged residues (D^{639} and D^{640} in the case of TRPC1) interact with the last two lysines (K^{684} and K^{685}) in STIM1, that interaction of both lysines with the two negative charges is required, that insertion or deletion of a single glycine within 100 residues upstream of the STIM1 C terminus disrupted the gating, and, most notably, that gating of TRPC channels by STIM1 occurs whether the positive or negative charges are on STIM1 or TRPC channels, as long as the negative charges on one protein are matched with positive charges on the other protein (Zeng et al. 2008).

Subsequent studies examined gating by electrostatic interaction with other TRPC channels to show that the same mechanism operates with TRPC3, TRPC4, TRPC5, and TRPC6 (Lee et al. 2010b). These studies also resulted in an excellent tool to selectively inhibit TRPC channels independent of the inhibition of Orai1. Since the STIM1 K-domain is not required for activation of Orai1 (Yuan et al. 2009), STIM1($K^{684}E/K^{685}E$) has no effect on Orai1, but inhibits the native TRPC channels activity. This tool was used together with mutations of the conserved TRPC3($D^{697}K/D^{698}K$) to demonstrate gating of TRPC3 by electrostatic interaction and the STIM1-dependent and STIM1-independent function of TRPC3. As illustrated in Fig. 3 (taken from Lee et al. 2010b), while wild-type STIM1 increased, the mutant STIM1($K^{684}E/K^{685}E$) strongly inhibited the current of wild-type TRPC3 (Fig. 3a–c). The TRPC3($D^{697}K/D^{698}K$) mutant that is not gated by STIM1 retained full activity. Notably, the current of TRPC3($D^{697}K/D^{698}K$) was not

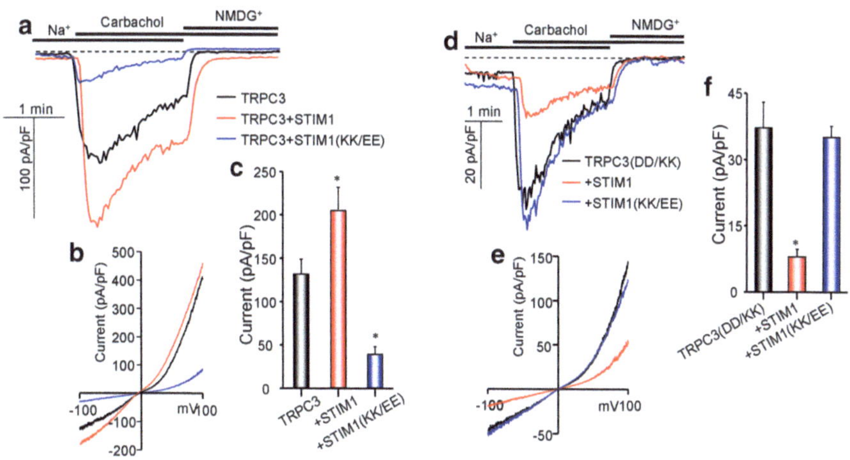

Fig. 3 Function of TRPC3 in STIM1-dependent and STIM1-independent modes. HEK cells were transfected with wild-type TRPC3 (**a–c**) or the mutant TRPC3(DD/KK) (**d–f**) and either empty vector (*black traces and columns*), STIM1 (*red traces and columns*), or STIM1(KK/EE) (*blue traces and columns*), and the carbachol-activated current was measured. Note that STIM1 activated wild-type TRPC3 but strongly inhibited the mutant TRPC3(DD/KK) while the reverse charge STIM1(KK/EE) had the opposite effects. Results were reproduced from Lee et al. (2010b) * In c,f denotes $p<0.05$

affected by the reverse charge STIM1($K^{684}E/K^{685}E$), but was fully inhibited by wild-type STIM1 (Fig. 3d–f).

An important remaining question in the gating of TRPC channels by STIM1 is their interacting modules. The STIM1 K-domain that gates the channel is not required for STIM1–TRPC channels interaction (Huang et al. 2006). The most likely STIM1 domain that interacts with the TRPC channels is SOAR. SOAR folds into a coiled-coil domain structure and crystallizes as a V-shaped dimer with lysines lining both sides of the V interior (Yang et al. 2012; Fig. 1b) that may interact with the Orai1 C terminus (Korzeniowski et al. 2010). The Orai1 C terminus folds into a coiled-coil domain structure (Hou et al. 2012), and the coiled-coil domain fold is required for activation of Orai1 by STIM1 (Muik et al. 2008; Frischauf et al. 2009). The same mechanism may hold for TRPC channels, which are predicted to have N- and C terminal coiled-coil domains. The role of the TRPC channels coiled-coil domains in interaction with STIM1 and in regulation of TRPC channels function by STIM1 was reported recently in Lee et al. 2014.

4.3 TRPC Channels in Physiology and as Therapeutic Target

The presence of two STIM1-gated Ca^{2+} influx channels in the same cells that affect the activity of each other raises the question of their expression pattern and whether Ca^{2+} influx through each of the channels mediate specific cell function. Expression

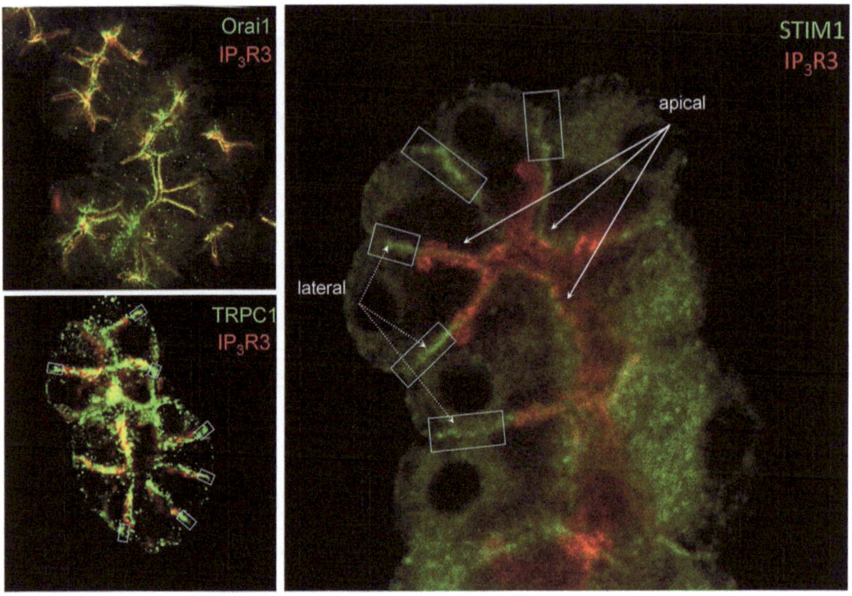

Fig. 4 Localization of the native STIM1, Orai1, and TRPC1 in polarized cells. Localization of Orai1, TRPC1, and STIM1 was determined relative to localization of IP_3Rs in the apical pole. Note that Orai1 is confined largely to the apical pole whereas TRPC1 and STIM1 are found at the apical and lateral membrane with minimal expression of the proteins in the basolateral membrane. The *turquoise squares* mark regions expressing STIM1 and TRPC1 but devoid of Orai1. Results were reproduced from Archana Jha Malini Ahuja et al. (2013)

pattern of the *native* STIM1, Orai1, and TRPC channels has been examined to a very limited extent, especially in animal tissues (rather than in cell lines). Such studies were performed in the polarized pancreatic acinar cells and an example is shown in Fig. 4 that was taken from Hong et al. (2011). Localization of the proteins was examined relative to IP_3 receptors (IP_3Rs) that are clustered at the apical pole (Lee et al. 1997; Hong et al. 2011; Yule et al. 1997). Orai1 shows perfect colocalization with IP_3Rs, while the localization of TRPC1 and STIM1 only partially overlap with the localization of IP_3Rs. This indicates that STIM1 forms several Ca^{2+} influx complexes in the same cells, Orai1–STIM1, Orai1–STIM1–TRPC1, and STIM1–TRPC1 complexes. Complexes of TRPC1–STIM1 are formed when cells are stimulated and can be readily immunoprecipitated (Hong et al. 2011).

Emerging evidence points to selective roles of the STIM1-regulated Ca^{2+} influx pathways in cellular physiology. Several of these examples are described below. Orai1-mediated Ca^{2+} influx activates the Ca^{2+}-dependent phosphatase calcineurin that dephosphorylates NFAT, resulting in translocation of NFAT to the nucleus (Shaw and Feske 2012; Darbellay et al. 2010). Selective inhibition of Ca^{2+} influx by TRPC channels was reported to have no effect on NFAT translocation to the nucleus (Cheng et al. 2011; Darbellay et al. 2010). Conversely, Ca^{2+} influx mediated by TRPC1 activates the Ca^{2+}-dependent transcription factor NFκB that is less affected by Orai1-mediated Ca^{2+} influx (Cheng et al. 2011; Darbellay et al. 2010). The TRPC1-mediated Ca^{2+} influx activates the Ca^{2+}-activated K^{+} channel in salivary glands acinar cells (Cheng et al. 2011) that are localized at the apical pole (Ong et al. 2012). This is despite the findings that Orai1 channels are present in and mediate Ca^{2+} influx across the luminal membrane (Almassy et al. 2012).

The Orai1–STIM1-mediated Ca^{2+} influx was found to be essential for postnatal human myoblast differentiation through activation of the transcription factors MEF2 and myogenin (Hong et al. 2011; Balghi et al. 2011). On the other hand, TRPC1 and TRPC4 mediate a portion of STIM1-dependent store-operated Ca^{2+} influx necessary for activation of MEF2 that mediates myoblasts fusion. Ca^{2+} influx by Orai1 did not substitute for the TRPC channel-mediated Ca^{2+} influx that facilitates the myogenic fusion (Antigny et al. 2013). Proliferation of adult neuronal progenitor cells is regulated by store-dependent Ca^{2+} influx. Knockdown of Orai1 and TRPC1 was equally effective in inhibition of the proliferation and causing cell cycle arrest (Li et al. 2012). Similarly, Ca^{2+} influx by TRPC1 is essential for the platelet-derived growth factor-BB-mediated restoration of neuronal progenitor cells proliferation that had been impaired by HIV Tat-cocaine via the cognate receptors (Yao et al. 2012). Similar effect was proposed for TRPC5 in the A2B5 neuronal progenitor cells (Shin et al. 2010).

In a Parkinson's disease model induced by neurotoxin, loss of dopaminergic neurons in the substantia nigra is associated with a decrease in TRPC1 expression and interaction of TRPC1 with STIM1. Ca^{2+} influx through the STIM1-regulated TRPC1 was essential and sufficient for activating the AKT pathway to mediate the neuroprotection (Selvaraj et al. 2012).

Vascular permeability is controlled by endothelial cells, a function regulated by TRPC channels, in particular by TRPC1 and TRPC4 when activated by STIM1 to mediate SOC-dependent Ca^{2+} influx (Di and Malik 2010). Knockdown of *Trpc4* or STIM1 inhibited vascular endothelial cells Ca^{2+} influx and caused disruption of the endothelial barrier (Sundivakkam et al. 2012). Notably, knockdown of Orai1 or Orai3 in these cells had minimal effect of Ca^{2+} influx and barrier integrity. Similarly, overexpression of Orai1 or STIM1 in *Trpc4*$^{-/-}$ cells failed to rescue Ca^{2+} influx (Sundivakkam et al. 2012), further demonstrating the essential role of TRPC channels in endothelial cell SOCs.

Another, potentially important aspect of TRPC channels relative to the Orai1 channel is their potential as therapeutic targets. Several diseases have been associated with mutations in TRPC channels. They are not discussed here, but information on this topic can be found in other chapters of this book and in Nilius and Owsianik (2011). However, the reason TRPC channels should be considered as therapeutic target is that deletion of TRPC1 (Liu et al. 2007), TRPC3 (Kim et al. 2009b, 2011; Hartmann et al. 2008), TRPC4 (Freichel et al. 2005), TRPC5 (Riccio et al. 2009), and TRPC6 (Dietrich et al. 2005) has relatively minor phenotypes, while deletion of Orai1 is embryonically lethal (Vig et al. 2006; McCarl et al. 2009). Moreover, deletion of the ubiquitous Orai1 eliminates all

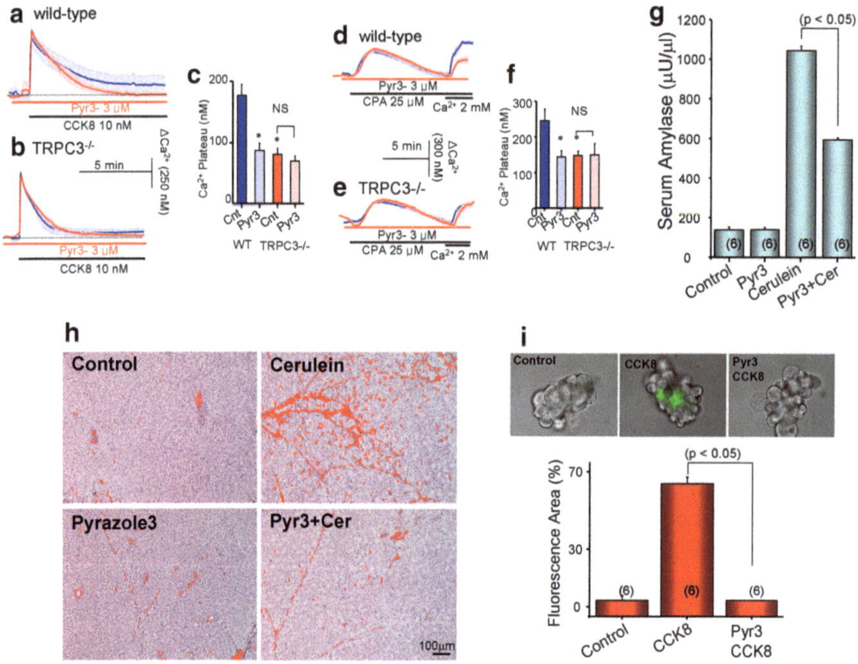

Fig. 5 Inhibition of TRPC3 protects against Ca^{2+}-dependent injury. Ca^{2+} measurements show that deletion of TRPC3 and Pyrazole 3 inhibits the same component of Ca^{2+} influx activated by receptor stimulation (**a–c**) or store depletion (**d–f**). Treating mice with Pyrazole 3 reduces the severity of cerulein-induced acute pancreatitis as assayed by the increase in serum amylase (**g**) and edema (**h**) and activation of intracellular trypsin (**i**). Results were reproduced from Smyth et al. (2012)
* In c,f denotes $p<0.05$

forms of receptor-stimulated Ca^{2+} influx. By contrast, expression of TRPC channels is more cell-specific and deletion of any of the TRPC channels results in only partial reduction in the receptor-mediated Ca^{2+} influx. Although TRPC channels also contribute to the regulation of the membrane potential by depolarizing the cells when activated, it seems that their function can be spared, at least temporarily.

Good examples for the therapeutic potential of inhibiting TRPC channels are the inhibition of TRPC3 and TRPC6 in reducing cardiac hypertrophy (Kuwahara et al. 2006; Wu et al. 2010) and of TRPC3 in ameliorating Ca^{2+}-dependent damage in the brain, pancreas, and salivary glands (Kim et al. 2009b, 2011; Hartmann et al. 2008). Analysis of patient tissues revealed increased TRPC3 and TRPC6 expression in cardiac hypertrophy that could be reproduced by overexpression of the channels in mice hearts, and hearts are protected when expression of the channels is reduced (Kuwahara et al. 2006; Wu et al. 2010; Millay et al. 2009). This suggested that pharmacological inhibition of TRPC3 should prevent cardiac hypertrophy. Pyrazole 3 is an inhibitor of TRPC3 (Kim et al. 2011; Kiyonaka et al. 2009; see Fig. 5) that is tolerated well by mice with no apparent gross side effects when infused over a 2 weeks period (Kiyonaka et al. 2009). Inhibition of TRPC3 by Pyrazole 3 prevented cardiac hypertrophy induced by aortic banding (Kiyonaka et al. 2009). Together, the genetic and pharmacological studies suggest that inhibition of TRPC3 is a promising target in treatment of cardiac hypertrophy.

It was recently shown that inhibition of TRPC3 reduces brain injury caused by hemorrhage-related brain inflammation. Intracerebral hemorrhage caused by rupture of blood vessels in the brain results in leakage of thrombin into the parenchyma and activation of neuronal and astrocytes TRPC3. Activated Ca^{2+} influx causes excessive Ca^{2+} influx that is associated with inflammation and neuronal injury (Qureshi et al. 2009). Application of Pyrazole 3 intracerebroventricularly and intraperitoneally to mice after induction of intracerebral hemorrhage to inhibit TRPC3 activation significantly reduced brain injury (Munakata et al. 2013). It should be of interest to determine whether inhibition of TRPC3 reduces brain damage due to other inflammation modalities.

Acute pancreatitis and Sjögren's syndrome are inflammatory diseases that damage the pancreas and salivary glands (Sah et al. 2012; Alevizos and Illei 2010). A common nodal point in all models of these diseases is an excess Ca^{2+} influx (Cheng et al. 2012; Petersen and Sutton 2006). Salivary glands and pancreatic acini express high level of TRPC3 (Bandyopadhyay et al. 2005; Kim et al. 2006), and TRPC3 mediates significant portion of receptor- and store-mediated Ca^{2+} influx (Kim et al. 2009b). The therapeutic potential of inhibiting TRPC3 in these diseases was tested by genetic and pharmacological inhibition of TRPC3 (Kim et al. 2011). Part of these findings is illustrated in Fig. 5 (reproduced from Kim et al. 2009b). Figure 5a–f shows similar inhibition of receptor- and store-mediated Ca^{2+} influx by deletion of TRPC3 and by Pyrazole 3 and that Pyrazole 3 has no effect on the Ca^{2+} signal in *Trpc3*$^{-/-}$ cells, indicating that in vivo Pyrazole 3 primarily inhibits TRPC3. Inhibition of TRPC3 by Pyrazole 3 reduced all indices of acute pancreatitis, including serum amylase (Fig. 5g), edema (Fig. 5h), and intracellular activation of trypsin (Fig. 5i). Pyrazole 3 similarly inhibited all Ca^{2+}-dependent acinar cells injury tested in the parotid glands (Kim et al. 2009b).

The results with Pyrazole 3 in the three model diseases illustrate the usefulness and relative safety of inhibiting TRPC channels activity as a therapeutic approach to diseases caused by excessive Ca^{2+} influx, in particular inflammatory diseases. It is hoped that as more and more potent and selective inhibitors of TRPC channels develop, the full benefit of inhibiting their function will be explored and will be used to treat human diseases.

References

Alevizos I, Illei GG (2010) MicroRNAs in Sjogren's syndrome as a prototypic autoimmune disease. Autoimmun Rev 9(9):618–621

Almassy J, Won JH, Begenisich TB, Yule DI (2012) Apical Ca2+-activated potassium channels in mouse parotid acinar cells. J Gen Physiol 139(2):121–133

Antigny F, Koenig S, Bernheim L, Frieden M (2013) During post-natal human myogenesis, normal myotube size requires TRPC1 and TRPC4 mediated Ca2+ entry. J Cell Sci 126 (Pt 11):2525–2533
Jha A, Ahuja M, Maléth J, Moreno CM, Yuan JP, Kim MS, Muallem S (2013) The STIM1 CTID domain determines access of SARAF to SOAR to regulate Orai1 channel function. J Cell Biol 202(1):71–79
Balghi H et al (2011) Enhanced Ca2+ entry due to Orai1 plasma membrane insertion increases IL-8 secretion by cystic fibrosis airways. FASEB J 25(12):4274–4291
Bandyopadhyay BC et al (2005) Apical localization of a functional TRPC3/TRPC6-Ca2+-signaling complex in polarized epithelial cells. Role in apical Ca2+ influx. J Biol Chem 280 (13):12908–12916
Blair NT, Kaczmarek JS, Clapham DE (2009) Intracellular calcium strongly potentiates agonist-activated TRPC5 channels. J Gen Physiol 133(5):525–546
Cheng KT, Liu X, Ong HL, Swaim W, Ambudkar IS (2011) Local Ca entry via Orai1 regulates plasma membrane recruitment of TRPC1 and controls cytosolic Ca signals required for specific cell functions. PLoS Biol 9(3):e1001025
Cheng KT et al (2012) STIM1 and STIM2 protein deficiency in T lymphocytes underlies development of the exocrine gland autoimmune disease, Sjogren's syndrome. Proc Natl Acad Sci U S A 109(36):14544–14549
Darbellay B et al (2010) Human muscle economy myoblast differentiation and excitation-contraction coupling use the same molecular partners, STIM1 and STIM2. J Biol Chem 285 (29):22437–22447
Demuro A et al (2011) Subunit stoichiometry of human Orai1 and Orai3 channels in closed and open states. Proc Natl Acad Sci U S A 108(43):17832–17837
Derler I et al (2009) A Ca2(+)release-activated Ca2(+) (CRAC) modulatory domain (CMD) within STIM1 mediates fast Ca2(+)-dependent inactivation of ORAI1 channels. J Biol Chem 284 (37):24933–24938
Di A, Malik AB (2010) TRP channels and the control of vascular function. Curr Opin Pharmacol 10(2):127–132
Dietrich A et al (2005) Increased vascular smooth muscle contractility in TRPC6-/- mice. Mol Cell Biol 25(16):6980–6989
Engh A, Somasundaram A, Prakriya M (2012) Permeation and gating mechanisms in store-operated CRAC channels. Front Biosci 17:1613–1626
Feske S et al (2006) A mutation in Orai1 causes immune deficiency by abrogating CRAC channel function. Nature 441(7090):179–185
Feske S, Picard C, Fischer A (2010) Immunodeficiency due to mutations in ORAI1 and STIM1. Clin Immunol 135(2):169–182
Freichel M et al (2005) Functional role of TRPC proteins in native systems: implications from knockout and knock-down studies. J Physiol 567(Pt 1):59–66
Frischauf I et al (2009) Molecular determinants of the coupling between STIM1 and Orai channels: differential activation of Orai1-3 channels by a STIM1 coiled-coil mutant. J Biol Chem 284(32):21696–21706
Grigoriev I et al (2008) STIM1 is a MT-plus-end-tracking protein involved in remodeling of the ER. Curr Biol 18(3):177–182
Hartmann J et al (2008) TRPC3 channels are required for synaptic transmission and motor coordination. Neuron 59(3):392–398
Hawkins BJ et al (2010) S-glutathionylation activates STIM1 and alters mitochondrial homeostasis. J Cell Biol 190(3):391–405
Hong JH et al (2011) Polarized but differential localization and recruitment of STIM1, Orai1 and TRPC channels in secretory cells. Traffic 12(2):232–245
Hou X, Pedi L, Diver MM, Long SB (2012) Crystal structure of the calcium release-activated calcium channel Orai. Science 338(6112):1308–1313

Huang GN et al (2006) STIM1 carboxyl-terminus activates native SOC, I(crac) and TRPC1 channels. Nat Cell Biol 8(9):1003–1010

Jeon JP et al (2012) Selective Galphai subunits as novel direct activators of transient receptor potential canonical (TRPC)4 and TRPC5 channels. J Biol Chem 287(21):17029–17039

Ji W et al (2008) Functional stoichiometry of the unitary calcium-release-activated calcium channel. Proc Natl Acad Sci U S A 105(36):13668–13673

Kasai H, Augustine GJ (1990) Cytosolic Ca2+ gradients triggering unidirectional fluid secretion from exocrine pancreas. Nature 348(6303):735–738

Kim JY et al (2006) Homer 1 mediates store- and inositol 1,4,5-trisphosphate receptor-dependent translocation and retrieval of TRPC3 to the plasma membrane. J Biol Chem 281(43):32540–32549

Kim MS et al (2009a) Native store-operated Ca2+ influx requires the channel function of Orai1 and TRPC1. J Biol Chem 284(15):9733–9741

Kim MS et al (2009b) Deletion of TRPC3 in mice reduces store-operated Ca2+ influx and the severity of acute pancreatitis. Gastroenterology 137(4):1509–1517

Kim MS et al (2011) Genetic and pharmacologic inhibition of the Ca2+ influx channel TRPC3 protects secretory epithelia from Ca2+-dependent toxicity. Gastroenterology 140(7):2107–2115, 2115.e2101–2104

Kiselyov K, Shin DM, Muallem S (2003) Signalling specificity in GPCR-dependent Ca2+ signalling. Cell Signal 15(3):243–253

Kiselyov K, Wang X, Shin DM, Zang W, Muallem S (2006) Calcium signaling complexes in microdomains of polarized secretory cells. Cell Calcium 40(5–6):451–459

Kiyonaka S et al (2009) Selective and direct inhibition of TRPC3 channels underlies biological activities of a pyrazole compound. Proc Natl Acad Sci U S A 106(13):5400–5405

Korzeniowski MK, Manjarres IM, Varnai P, Balla T (2010) Activation of STIM1-Orai1 involves an intramolecular switching mechanism. Sci Signal 3(148):ra82

Kuwahara K et al (2006) TRPC6 fulfills a calcineurin signaling circuit during pathologic cardiac remodeling. J Clin Invest 116(12):3114–3126

Lee MG et al (1997) Polarized expression of Ca2+ channels in pancreatic and salivary gland cells. Correlation with initiation and propagation of [Ca2+]i waves. J Biol Chem 272(25):15765–15770

Lee KP et al (2009) Molecular determinants of fast Ca2+-dependent inactivation and gating of the Orai channels. Proc Natl Acad Sci U S A 106(34):14687–14692

Lee KP et al (2010a) An endoplasmic reticulum/plasma membrane junction: STIM1/Orai1/TRPCs. FEBS Lett 584(10):2022–2027

Lee KP, Yuan JP, So I, Worley PF, Muallem S (2010b) STIM1-dependent and STIM1-independent function of transient receptor potential canonical (TRPC) channels tunes their store-operated mode. J Biol Chem 285(49):38666–38673

Lee KP, Choi S, Hong JH, Ahuja M, Graham S, Ma R, So I, Shin DM, Muallem S, Yuan J (2014) Molecular determinants mediating gating of transient receptor potential canonical (TRPC) channels by stromal interaction molecule 1 (STIM1). J Biol Chem [Epub ahead of print]. PMID: 24464579

Li M, Chen C, Zhou Z, Xu S, Yu Z (2012) A TRPC1-mediated increase in store-operated Ca2+ entry is required for the proliferation of adult hippocampal neural progenitor cells. Cell Calcium 51(6):486–496

Liao Y et al (2007) Orai proteins interact with TRPC channels and confer responsiveness to store depletion. Proc Natl Acad Sci U S A 104(11):4682–4687

Liao Y et al (2008) Functional interactions among Orai1, TRPCs, and STIM1 suggest a STIM-regulated heteromeric Orai/TRPC model for SOCE/Icrac channels. Proc Natl Acad Sci U S A 105(8):2895–2900

Liao Y et al (2009) A role for Orai in TRPC-mediated Ca2+ entry suggests that a TRPC:Orai complex may mediate store and receptor operated Ca2+ entry. Proc Natl Acad Sci U S A 106 (9):3202–3206

Liou J et al (2005) STIM is a Ca2+ sensor essential for Ca2+-store-depletion-triggered Ca2+ influx. Curr Biol 15(13):1235–1241
Liou J, Fivaz M, Inoue T, Meyer T (2007) Live-cell imaging reveals sequential oligomerization and local plasma membrane targeting of stromal interaction molecule 1 after Ca2+ store depletion. Proc Natl Acad Sci U S A 104(22):9301–9306
Lis A et al (2007) CRACM1, CRACM2, and CRACM3 are store-operated Ca(2+) channels with distinct functional properties. Curr Biol 17(9):794–800
Liu X, Bandyopadhyay BC, Singh BB, Groschner K, Ambudkar IS (2005) Molecular analysis of a store-operated and 2-acetyl-sn-glycerol-sensitive non-selective cation channel. Heteromeric assembly of TRPC1-TRPC3. J Biol Chem 280(22):21600–21606
Liu X et al (2007) Attenuation of store-operated Ca2+ current impairs salivary gland fluid secretion in TRPC1(-/-) mice. Proc Natl Acad Sci U S A 104(44):17542–17547
McCarl CA et al (2009) ORAI1 deficiency and lack of store-operated Ca2+ entry cause immunodeficiency, myopathy, and ectodermal dysplasia. J Allergy Clin Immunol 124(6):1311–1318, e1317
McNally BA, Somasundaram A, Yamashita M, Prakriya M (2012) Gated regulation of CRAC channel ion selectivity by STIM1. Nature 482(7384):241–245
McNally BA, Somasundaram A, Jairaman A, Yamashita M, Prakriya M (2013) The C- and N-terminal STIM1 binding sites on Orai1 are required for both trapping and gating CRAC channels. J Physiol 591(Pt 11):2833–2850
Millay DP et al (2009) Calcium influx is sufficient to induce muscular dystrophy through a TRPC-dependent mechanism. Proc Natl Acad Sci U S A 106(45):19023–19028
Muallem S, Schoeffield MS, Fimmel CJ, Pandol SJ (1988) Agonist-sensitive calcium pool in the pancreatic acinar cell. II. Characterization of reloading. Am J Physiol 255(2 Pt 1):G229–G235
Muik M et al (2008) Dynamic coupling of the putative coiled-coil domain of ORAI1 with STIM1 mediates ORAI1 channel activation. J Biol Chem 283(12):8014–8022
Muik M et al (2011) STIM1 couples to ORAI1 via an intramolecular transition into an extended conformation. EMBO J 30(9):1678–1689
Muik M, Schindl R, Fahrner M, Romanin C (2012) Ca(2+) release-activated Ca(2+) (CRAC) current, structure, and function. Cell Mol Life Sci 69(24):4163–4176
Mullins FM, Park CY, Dolmetsch RE, Lewis RS (2009) STIM1 and calmodulin interact with Orai1 to induce Ca2+-dependent inactivation of CRAC channels. Proc Natl Acad Sci U S A 106(36):15495–15500
Munakata M et al (2013) Transient receptor potential canonical 3 inhibitor Pyr3 improves outcomes and attenuates astrogliosis after intracerebral hemorrhage in mice. Stroke 44 (7):1981–1987
Ng LC et al (2012) TRPC1 and Orai1 interact with STIM1 and mediate capacitative Ca(2+) entry caused by acute hypoxia in mouse pulmonary arterial smooth muscle cells. Am J Physiol Cell Physiol 303(11):C1156–C1172
Nilius B, Owsianik G (2011) The transient receptor potential family of ion channels. Genome Biol 12(3):218
Nilius B, Owsianik G, Voets T, Peters JA (2007) Transient receptor potential cation channels in disease. Physiol Rev 87(1):165–217
Ong HL et al (2007) Dynamic assembly of TRPC1/STIM1/Orai1 ternary complex is involved in store operated calcium influx: evidence for similarities in SOC and CRAC channel components. J Biol Chem 282(12):9105–9116
Ong HL, Jang SI, Ambudkar IS (2012) Distinct contributions of Orai1 and TRPC1 to agonist-induced [Ca(2+)](i) signals determine specificity of Ca(2+)-dependent gene expression. PLoS One 7(10):e47146
Palty R, Raveh A, Kaminsky I, Meller R, Reuveny E (2012) SARAF inactivates the store operated calcium entry machinery to prevent excess calcium refilling. Cell 149(2):425–438

Pandol SJ, Schoeffield MS, Fimmel CJ, Muallem S (1987) The agonist-sensitive calcium pool in the pancreatic acinar cell. Activation of plasma membrane Ca2+ influx mechanism. J Biol Chem 262(35):16963–16968

Parekh AB, Putney JW Jr (2005) Store-operated calcium channels. Physiol Rev 85(2):757–810

Park CY et al (2009) STIM1 clusters and activates CRAC channels via direct binding of a cytosolic domain to Orai1. Cell 136(5):876–890

Pedersen SF, Owsianik G, Nilius B (2005) TRP channels: an overview. Cell Calcium 38(3–4):233–252

Penna A et al (2008) The CRAC channel consists of a tetramer formed by Stim-induced dimerization of Orai dimers. Nature 456(7218):116–120

Petersen OH, Sutton R (2006) Ca2+ signalling and pancreatitis: effects of alcohol, bile and coffee. Trends Pharmacol Sci 27(2):113–120

Petersen OH, Sutton R, Criddle DN (2006) Failure of calcium microdomain generation and pathological consequences. Cell Calcium 40(5–6):593–600

Qureshi AI, Mendelow AD, Hanley DF (2009) Intracerebral haemorrhage. Lancet 373 (9675):1632–1644

Rao JN et al (2012) Polyamines regulate intestinal epithelial restitution through TRPC1-mediated Ca(2)+ signaling by differentially modulating STIM1 and STIM2. Am J Physiol Cell Physiol 303(3):C308–C317

Riccio A et al (2009) Essential role for TRPC5 in amygdala function and fear-related behavior. Cell 137(4):761–772

Roos J et al (2005) STIM1, an essential and conserved component of store-operated Ca2+ channel function. J Cell Biol 169(3):435–445

Sah RP, Garg P, Saluja AK (2012) Pathogenic mechanisms of acute pancreatitis. Curr Opin Gastroenterol 28(5):507–515

Selvaraj S et al (2012) Neurotoxin-induced ER stress in mouse dopaminergic neurons involves downregulation of TRPC1 and inhibition of AKT/mTOR signaling. J Clin Invest 122(4):1354–1367

Shaw PJ, Feske S (2012) Physiological and pathophysiological functions of SOCE in the immune system. Front Biosci (Elite Ed) 4:2253–2268

Shin HY et al (2010) A role of canonical transient receptor potential 5 channel in neuronal differentiation from A2B5 neural progenitor cells. PLoS One 5(5):e10359

Shinde AV et al (2013) STIM1 controls endothelial barrier function independently of Orai1 and Ca2+ entry. Sci Signal 6(267):ra18

Smyth JT, Beg AM, Wu S, Putney JW Jr, Rusan NM (2012) Phosphoregulation of STIM1 leads to exclusion of the endoplasmic reticulum from the mitotic spindle. Curr Biol 22(16):1487–1493

Stathopulos PB, Zheng L, Li GY, Plevin MJ, Ikura M (2008) Structural and mechanistic insights into STIM1-mediated initiation of store-operated calcium entry. Cell 135(1):110–122

Sundivakkam PC et al (2012) The Ca(2+) sensor stromal interaction molecule 1 (STIM1) is necessary and sufficient for the store-operated Ca(2+) entry function of transient receptor potential canonical (TRPC) 1 and 4 channels in endothelial cells. Mol Pharmacol 81(4):510–526

Takahashi N, Mori Y (2011) TRP channels as sensors and signal integrators of redox status changes. Front Pharmacol 2:58

Thorn P, Lawrie AM, Smith PM, Gallacher DV, Petersen OH (1993) Local and global cytosolic Ca2+ oscillations in exocrine cells evoked by agonists and inositol trisphosphate. Cell 74 (4):661–668

Vig M et al (2006) CRACM1 is a plasma membrane protein essential for store-operated Ca2+ entry. Science 312(5777):1220–1223

Wu X, Eder P, Chang B, Molkentin JD (2010) TRPC channels are necessary mediators of pathologic cardiac hypertrophy. Proc Natl Acad Sci U S A 107(15):7000–7005

Xu SZ et al (2008) TRPC channel activation by extracellular thioredoxin. Nature 451(7174):69–72

Yang X, Jin H, Cai X, Li S, Shen Y (2012) Structural and mechanistic insights into the activation of Stromal interaction molecule 1 (STIM1). Proc Natl Acad Sci U S A 109(15):5657–5662

Yao H, Duan M, Yang L, Buch S (2012) Platelet-derived growth factor-BB restores human immunodeficiency virus Tat-cocaine-mediated impairment of neurogenesis: role of TRPC1 channels. J Neurosci 32(29):9835–9847

Yu F, Sun L, Courjaret R, Machaca K (2011) Role of the STIM1 C-terminal domain in STIM1 clustering. J Biol Chem 286(10):8375–8384

Yu F, Sun L, Hubrack S, Selvaraj S, Machaca K (2013) Intramolecular shielding maintains STIM1 in an inactive conformation. J Cell Sci 126(Pt 11):2401–2410

Yuan JP, Zeng W, Huang GN, Worley PF, Muallem S (2007) STIM1 heteromultimerizes TRPC channels to determine their function as store-operated channels. Nat Cell Biol 9(6):636–645

Yuan JP et al (2009) SOAR and the polybasic STIM1 domains gate and regulate Orai channels. Nat Cell Biol 11(3):337–343

Yule DI, Ernst SA, Ohnishi H, Wojcikiewicz RJ (1997) Evidence that zymogen granules are not a physiologically relevant calcium pool. Defining the distribution of inositol 1,4,5-trisphosphate receptors in pancreatic acinar cells. J Biol Chem 272(14):9093–9098

Zeng W et al (2008) STIM1 gates TRPC channels, but not Orai1, by electrostatic interaction. Mol Cell 32(3):439–448

Zhang SL et al (2006) Genome-wide RNAi screen of Ca(2+) influx identifies genes that regulate Ca(2+) release-activated Ca(2+) channel activity. Proc Natl Acad Sci U S A 103(24):9357–9362

Zhang ZY, Pan LJ, Zhang ZM (2010) Functional interactions among STIM1, Orai1 and TRPC1 on the activation of SOCs in HL-7702 cells. Amino Acids 39(1):195–204

Zheng L, Stathopulos PB, Li GY, Ikura M (2008) Biophysical characterization of the EF-hand and SAM domain containing Ca2+ sensory region of STIM1 and STIM2. Biochem Biophys Res Commun 369(1):240–246

Zheng H et al (2013) Differential roles of the C and N termini of Orai1 protein in interacting with stromal interaction molecule 1 (STIM1) for Ca2+ release-activated Ca2+ (CRAC) channel activation. J Biol Chem 288(16):11263–11272

The TRPC Family of TRP Channels: Roles Inferred (Mostly) from Knockout Mice and Relationship to ORAI Proteins

Yanhong Liao, Joel Abramowitz, and Lutz Birnbaumer

Contents

Abstract

Aside from entering into cells through voltage gated Ca channels and Na/Ca exchangers in those cells that express these proteins, for all cells be they excitable or non-excitable, Ca^{2+} enters through channels that are activated downstream of phosphoinositide mobilization (activation of phospholipase C, PLC) and through channels that are activated secondary to depletion of internal stores. Depletion of internal stores activates plasma membrane channels known as ORAIs. Activation of PLCs activates the canonical class of transient receptor potential channels (TRPCs), and, because this activation also causes depletion of Ca^{2+} stores, also ORAI based channels. Whereas the activation of ORAI is a

Y. Liao (✉)
Department of Anatomy, College of Medicine, Huazhong University of Science and Technology, Wuhan 430074, China
e-mail: yhliao1@gmail.com

J. Abramowitz • L. Birnbaumer
Laboratory of Neurobiology, National Institute of Environmental Health Sciences, National Institutes of Health, 111 TW Alexander Dr., Research Triangle Park, NC 27709, USA
e-mail: abramow1@niehs.nih.gov; birnbau1@niehs.nih.gov

B. Nilius and V. Flockerzi (eds.), *Mammalian Transient Receptor Potential (TRP) Cation Channels*, Handbook of Experimental Pharmacology 223, DOI 10.1007/978-3-319-05161-1_14,

well-accepted phenomenon, it appears that TRPC channels also participate in Ca^{2+} entry triggered by store depletion with or without participation of ORAI molecules. Regardless of molecular makeup of TRPC containing channels, a plethora of studies have shown TRPCs to be important both in physiologic systems as well as in pathophysiologic phenomena. Particularly important in defining roles of TRPCs, have been studies with mice with targeted disruption of their genes, i.e., with TRPC KO mice. In this chapter we first focus on TRPCs as regulators of body functions in health and disease, and then focus on the possible make-up of the channels of which they participate. A hypothesis is set forth, whereby ORAI dimers are proposed to be regulatory subunits of tetrameric TRPC channels and serve as structural units that form ORAI channels either as dimers of dimers or trimers of dimers.

Keywords
Ca^{2+} Influx • TRPC • ORAI • Icrac • SOCE • ROCE • TRPC physiology • TRPC pathophysiology • Knockout mice • Phenotype

Abbreviations

CaCaM	Ca^{2+}-calmodulin complex
DAG	Diacylglycerol
ER	Endoplasmic reticulum
GPCR	G protein coupled receptor
IP3	Inositol 1,4,5-trisphosphate
OAG	Oleyl-acetyl-glycerol
PLC	Phospholipase C
PM	Plasma membrane
PMCA	Plasma membrane Ca^{2+}-activated ATPase
ROC	Receptor operated channel
ROCE	Receptor operated Ca^{2+} entry
SERCA	Sarcoplasmic endoplasmic reticulum Ca^{2+}-activated ATPase
SOC	Store operated channel
SOCE	Store operated Ca^{2+} entry
TRPC	Transient receptor potential canonical
VSMC	Vascular smooth muscle cells

1 Introduction

Changes in intracellular Ca^{2+} concentrations constitute a fundamental mechanism by which a host of extracellular stimuli are transduced into cellular responses that can either be helpful in adapting to the extracellular milieu or be detrimental to the cell's survival—Ca^{2+} toxicity. One large class of stimuli acts by activating phospholipase C enzymes which, owing to the formation of inositol trisphosphate (IP3), promote release of Ca^{2+} from internal stores. The rises in cytosolic Ca^{2+} ($[Ca^{2+}]i$) are transient, owing to the activity of Ca^{2+} pumps that rapidly restore "resting" levels by returning Ca^{2+} to the stores (SERCA pumps) and by extrusion from the cell (PMCA pumps). Persistent stimuli elicit cyclical elevations—oscillations. But, because of each oscillation causing a net loss of cellular Ca^{2+}, signaling—good or bad—would stop if it were not that secondary to PLC activation, a Ca^{2+} influx pathway is activated. This Ca^{2+} entry pathway is commonly referred to as receptor-operated Ca^{2+} entry or ROCE. An entry pathway similar to ROCE can be activated also by mere inhibition of SERCA pumps, which causes gradual emptying of the stores and is referred to as store-operated Ca^{2+} entry or SOCE. This emptying is associated with the appearance of a highly Ca^{+2} selective, Ca^{2+} release activated current, ICRAC, the electrophysiological correlate to SOCE. This chapter deals with TRPC channels—transient receptor potential canonical channels—as structural components of ROCE channels, their functional relationship and interaction with SOCE channels formed by ORAI proteins, and the multiple roles in physiology and pathophysiology (health and disease) that TRPC channels have been found to have.

2 SOCE and ICRAC Are Mediated by Both ORAI-Based Channels and TRPC-Based Channels

With the discovery of ORAI (Vig et al. 2006; Feske et al. 2006; Zhang et al. 2006) and the finding that when expressed together with the endoplasmic reticulum (ER) Ca^{2+} sensor STIM, this gives rise to very large store depletion-activated currents with the bonafide characteristics of CRAC currents (Peinelt et al. 2006), Soboloff et al. 2006; Mercer et al. 2006), the previous interest in the roles and functioning of TRPC channels as molecular participants of SOCE and CRAC channels waned in the minds of many. The reasons for this were twofold: (1) when expressed in model cells, TRPCs had given rise to mostly nonselective cation channels with variable Ca^{2+} selectivity over Na^{+}, none more than 10:1, most less 3:1, whereas CRAC channels are highly Ca^{2+} selective, and (2) store depletion was not a good trigger, if at all, for TRPC activation.

However, one TRPC stood out, TRPC1, which, even if nonselective, is activated by store depletion. This was shown in one series of reports, in which thapsigargin was found to activate a store-operated current or channel (SOC) recorded in the whole-cell configuration of the patch clamp technique from cells transfected with the TRPC1 cDNA (Ong et al. 2007). Interestingly, TRPC1 SOC activity was augmented by expression of STIM1, and this required ORAI1 (Cheng et al. 2008), indicating that the TRPC1 machinery interacted with the ORAI machinery. In support of store depletion being a regulator of TRPC1, single-channel SOC currents with slope conductances of approximately 2pS were recorded from cell-attached and inside-out membrane patches of cells exposed to BAPTA-AM to deplete their Ca^{2+} stores, and these SOCs were absent in cells from *Trpc*$^{-/-}$ mice (Shi et al. 2012). At the biochemical level, STIM1 was discovered to gate not only ORAI channels but also TRPC1 channels, providing a link between

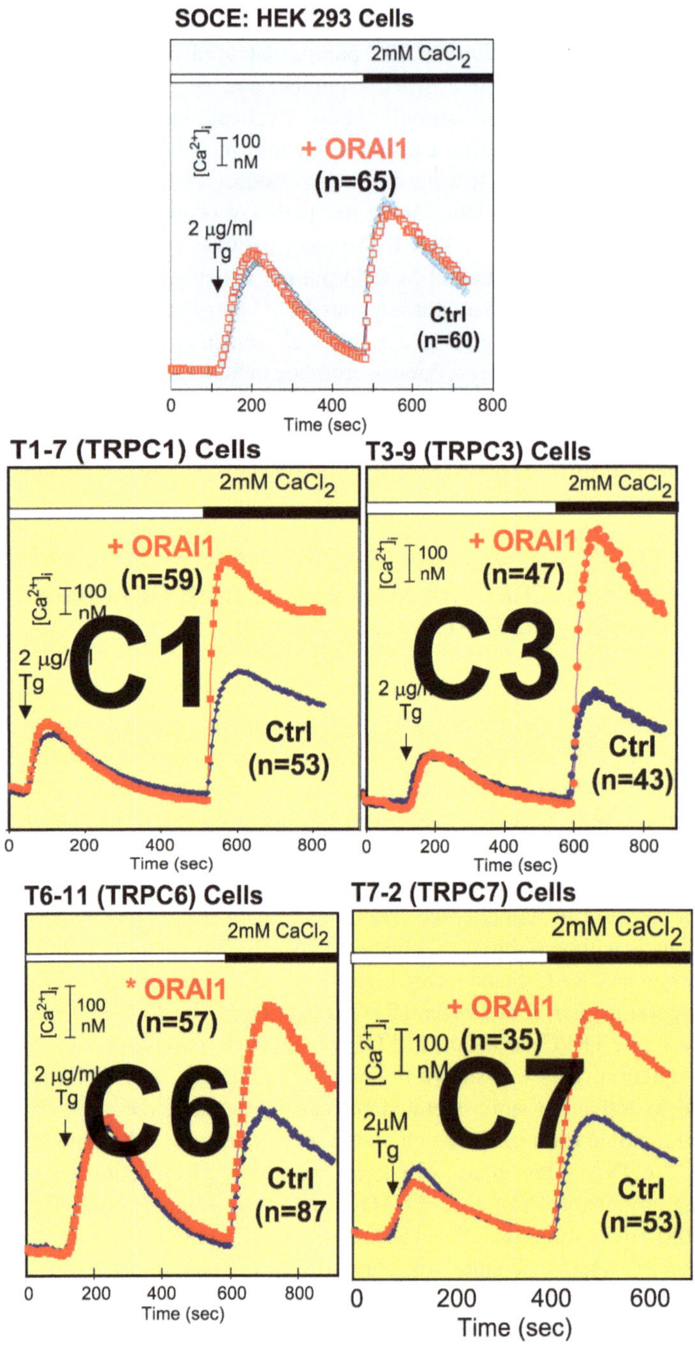

Fig. 1 Functional interaction of ORAI with TRPCs as seen in tests for SOCE. Store-operated Ca^{2+} entry is defined as the Ca^{2+} entry seen upon addition of $CaCl_2$ to the extracellular milieu of cells in which internal stores have been depleted in the absence of external Ca^{2+} by inhibition of SERCA pumps without activation of a PLC enzyme. SOCE was assessed in control HEK-293 cells and in

store depletion and TRPC1 (Zeng et al. 2008). Gating of TRPC1 and ORAI1 by STIM1 differed in that that of TRPC1 occurs by an electrostatic mechanism involving the polybasic C-terminus of STIM and the AspAsp, AspGlu or Glu/Glu (DD, DE or EE) sites next to the TRPC's EWFKAR motif (TRP box) (Zeng et al. 2008), whereas gating of ORAI channels by STIM1 occurs at ORAI sites interacting with STIM's SOAR (Stim-Orai-Activation Region, Yuan et al. 2009) or CAD [for CRAC Activation Domain, Park et al. (2009)].

Independent studies confirmed the existence of functional interactions between TRPCs and ORAI. Thus, when ORAI was expressed in cells with unpaired TRPC channels, but not in control cells not expressing an excess of TRPCs, it promoted a robust increase in store-operated Ca^{2+} entry (SOCE) (Liao et al. 2007; Fig. 1); ORAI1 expression also facilitated recording of CRAC currents (Liao et al. 2008), and under resting conditions, it silenced spontaneous activity of the unpaired TRPCs (Liao et al. 2007). These findings suggested that TRPCs and ORAIs interact at a functional level and possibly also at the physical level. This concept was then supported further by the fact that expression of Orai1[R91W] behaved as a dominant negative element (dnORAI) inhibiting not only SOCE elicited by inhibition of SERCA pumps but also Gd^{3+}-resistant ROCE and OAG-stimulated Ca^{2+} entry in cells overexpressing TRPC3, one of the OAG-activated TRPCs. SOCE channels and ICRAC, activated without participation of PLC-activating stimuli, are inhibited by 0.5–1.0 μM Gd^{3+}. As in the above-mentioned ROCE experiment, the ORAI channels were inhibited by Gd^{3+}; the ROCE channel(s) inhibited by the dnORAI1 can safely be assumed to be TRPC channels (Fig. 2). In the OAG experiment, the Ca^{2+} entry inhibited by the dnORAI1 occurred without store depletion, depended on overexpression of one of the OAG-responsive TRPCs, and was therefore also mediated by a TRPC channel (Liao et al. 2009) and not by an ORAI channel which requires, for its assembly, the molecular orchestration of STIM clusters engendered by store depletion (Luik et al. 2006; Wu et al. 2006).

3 Structural Considerations

ORAIs, of which there are three, are membrane proteins with four transmembrane (TM) domains. Biochemical studies on the structure of ORAI1 in resting cells and after activation by STIMs found ORAI1 to exist as dimers and the assembled CRAC channel to be dimers of dimers (Penna et al. 2008). Scoring of the inhibitory effect, another dnORAI1, the pore-dead mutant of ORAI1 (ORAI1[E106Q]), on the CRAC channel activity of wild-type ORAI1 transfected as concatenated dimers, trimmers, and tetramers also led to the conclusion that the ORAI1-based CRAC

Fig. 1 (continued) HEK cells that stably expressed TRPC1 (T1-7 cells), TRPC3 (T3-9 cells), TRPC6 (T6-11 cells), and TRPC7 (T7-2 cells), into which non-inhibitory amounts of myc-tagged ORAI1 expression plasmids had been transfected 48 h before imaging (adapted from Liao et al. 2007)

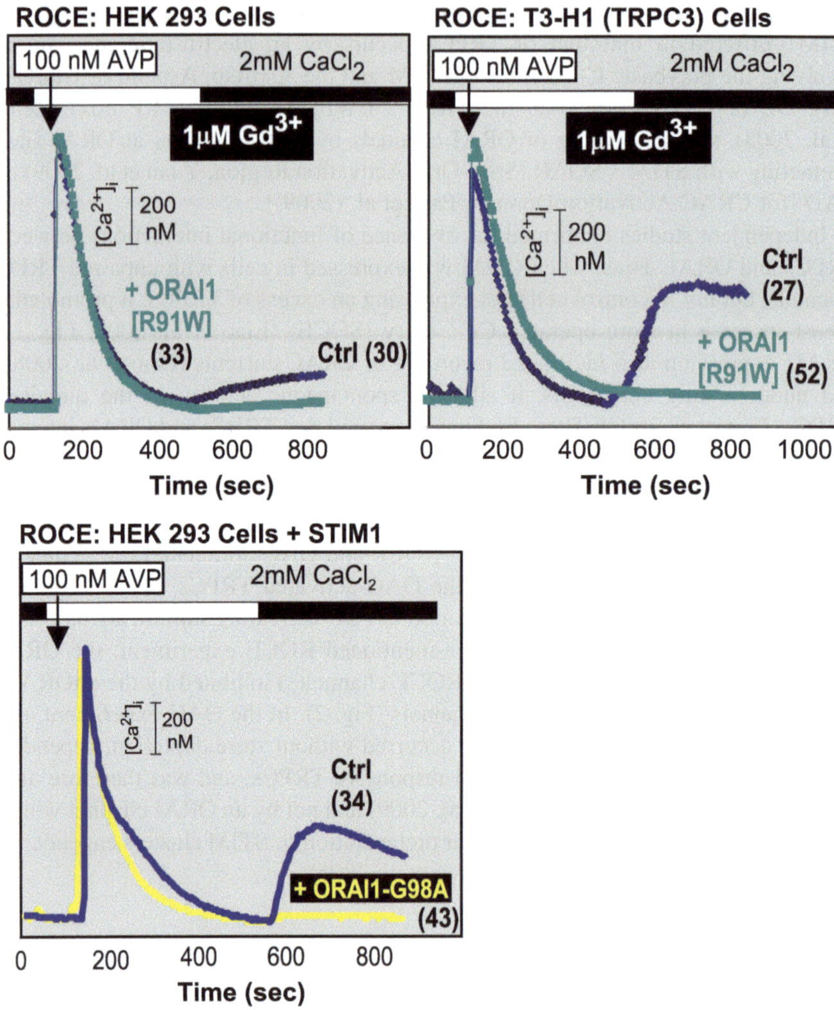

Fig. 2 Functional interaction of ORAI with TRPCs as seen in tests for ROCE. Receptor-operated Ca^{2+} entry is defined as the Ca^{2+} entry seen upon addition of Ca^{2+} to the extracellular milieu of cells in which a Gq-coupled G protein receptor agonist had been used to stimulate PLCβ activity in the absence of extracellular Ca^{2+}. The intracellular Ca^{2+} transient had been allowed to run its course—release of Ca^{2+} from the ER and reduction of cytosolic Ca by the action of PMCA and SERCA pumps. *Top two panels*: ROCE elicited by V1a receptor stimulation is inhibited by Gd^{3+} in HEK 293 cells but not in T3-H1 cells which stably overexpress TRPC3. *Bottom panel*: ROCE was assessed in HEK-293 cells transfected with the V1a receptor expression plasmid, a eYFP-STIM1 expression plasmid, with or without a plasmid driving the expression of ORAI1[G98A], a dominant negative mutant of ORAI1

channel is an ORAI1 tetramer (Mignen et al. 2008). To the surprise of many, if not all, the crystal structure of ORAI—Drosophila ORAI—showed it to be assembled as an hexamer formed of three dimers (Hou et al. 2012; Fig. 3a). ORAI1 can

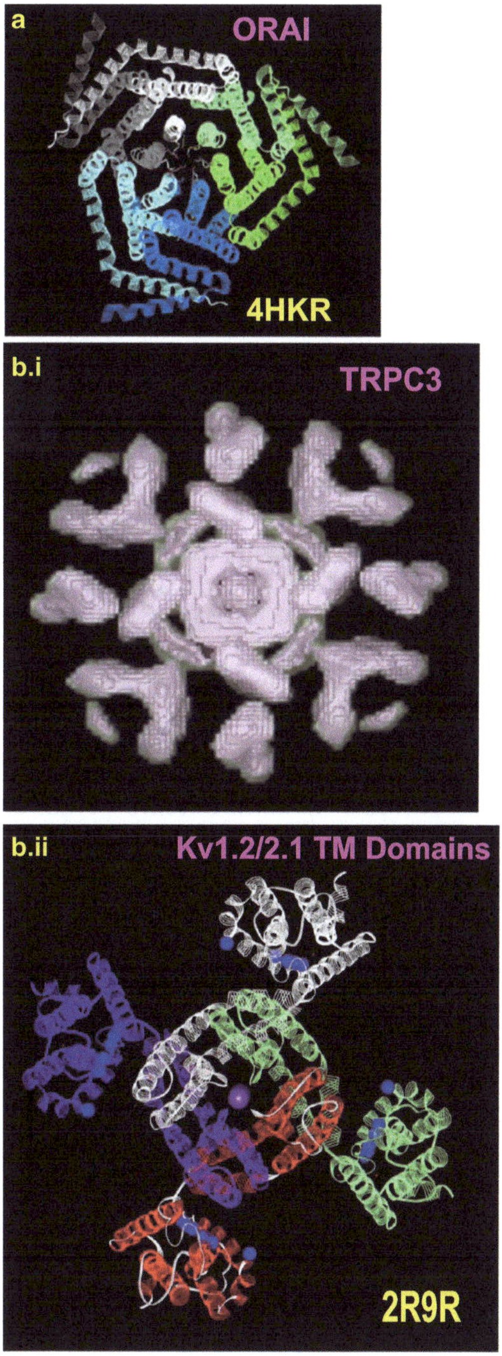

Fig. 3 Models of ORAI (**a**) and TRPCs (**b**). (**a**) Hexameric ORAI (PDB: 4HKR, Hou et al. 2012). (**b**-i), Tetrameric TRPC3 reconstructed from cryo-electron micrographs (adapted from Mio et al. 2007). (**b**-ii), Tetrameric voltage-gated K+ channel (PDB: 2R9R, Long et al. 2007) as surrogate for that of a TRPC channel

therefore be expected to coexist in three formats—dimer, tetramer, and hexamer—of which the relative abundance depends on the interaction with activated STIMs and the level of expression, with hexamers being favored in cells overexpressing ORA1, a situation that may mimic the artificially high concentrations at which proteins are set up to drive their crystallization.

Glycosylation scanning of the transmembrane topology of TRPC3 showed it to pass the membrane six times with cytosolic N- and C-termini akin to voltage-gated K^+ channels (Vannier et al. 1998) but lacking the voltage sensing Arg- and Lys-rich TM4 of voltage-gated K^+ channels. Voltage-gated K^+ channels are tetramers (e.g., Long et al. 2007; PDB: 2R9R; Fig. 3b-ii). Analysis of cryo-electron micrographs of TRPC3 expressed in HEK293 cells by Mio et al. (2007) (Fig. 3b-i) confirmed the generally held assumption that assembled TRPCs channels are tetramers. The most plausible model of a tetrameric TRPC channel is thus one resembling that of voltage-gated K^+ channel, shown in Fig. 3b-ii.

4 Complexity in the Subunit Composition of TRPCs

There are seven *TRPC* genes in the nonhuman mammalian genomes and six in Old World monkeys and human genomes in which *TRPC2* is a pseudogene. Of the seven (six) TRPCs, most cells and tissues express between three and four forming heteromeric complexes that vary with cell type and approach used to determine the subunit composition. The degree of TRPC subunit complexity was well documented by the subunit analysis of SOC channels in vascular smooth muscle cells (VSMCs) in which selective peptide-directed antibodies were added to medium bathing inside-out membrane patches in which SOCs had been activated by store depletion. As illustrated in Fig. 4, single-channel SOCs in mesenteric artery SMVCs are immunoneutralized by either anti-TRPC1 or anti-TRPC5 when derived from wild-type mice but only by anti-TRPC5 in $Trpc1^{-/-}$ mice, in which the unpaired TRPC5 channel appeared with properties that differed from those seen in the WT SOC and were those expected from TRPC5 expression in model cells (Shi et al. 2012). The likely TRPC subunit makeup of SOCs in WT mesenteric artery VSMCs is that of a heteromeric tetramer formed to TRPC1 and TRPC5. When the same analysis was applied to VSMCs from coronary artery and portal vein using four antibodies, anti-C1, anti-C5, abti-C6, and anti-C7, all SOCs were immunoneutralized by ant-C1 and anti-C5 (Fig. 4b–d), but anti-C6 immunoneutralized also coronary artery VSMC SOCs (Fig. 4b), whereas portal vein SOCs were immunoneutralized by anti-C7, but not anti-C6 (Fig. 4d). Thus, what appears to be similar cell types, vascular smooth muscle cells, show differences at the level of TRPC channel makeup. Although anti-C6 did not affect mesenteric artery VSMCs, this does not mean that the difference is due to expression or not of the *Trpc6* gene, because similar studies on single-channel currents activated by angiotensin II, instead of store depletion, showed presence of TRPC6-based cation currents in these cells (Shi et al. 2010).

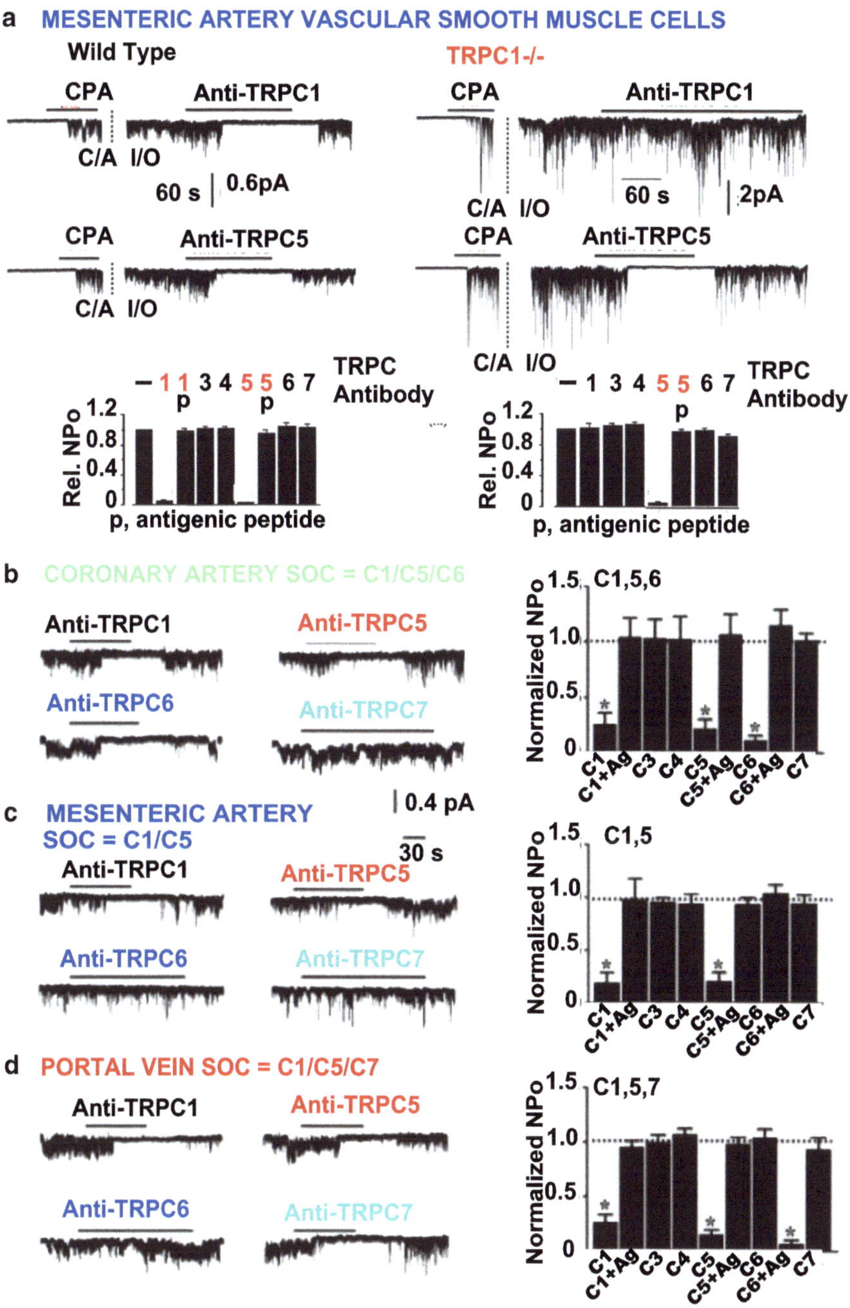

Fig. 4 Heterogeneous TRPC subunit makeup of vascular smooth muscle store-operated channels (SOCs). Vascular smooth muscle cells were patched in the cell-attached configuration of the patch clamp technique (C/A patches), their SOCs activated by treatment with the SERCA pump inhibitor cyclopiazonic acid (CPA) in the cell, and patches excised to expose the cytosolic aspect of the membrane patches [inside-out patches (I/O patches)]. The indicated antibodies were then added to test for their inhibitory activity. Washout restored activity and immunogenic peptides neutralized

Given that similar cell types assemble TRPC channels with a different subunit makeup, it stands to reason that the complexity in any given cell type cannot be predicted and has to be assumed to be unknown unless explicitly tested for. It also stands to reason that phenotypes that may develop upon single- or double-gene disruptions may differ with the cell type under study.

5 Dependence of TRPC's Channel Function on ORAI

The fact that ORAI1[R91W] inhibits OAG-activated TRPC (TRPC3 or TRPC6) raised the interesting possibility that whereas "pure" CRAC channels—activated by store depletion in the absence of PLC activation—are ORAI tetramers or hexamers, the ORAI dimers may play dual roles: structural CRAC channel subunits assembled into an active ion channel and regulatory subunits of TRPC-based channels. In this view a TRPC-based channel is formed of four TRPC molecules and four ORAI dimers. One corollary of this view is that the functioning of a TRPC channel depends on ORAI: "No ORAI, no channel-competent TRPC".

ORAI1 has been knocked out in two laboratories and it is possible to prepare *Orai1*$^{-/-}$ embryonic fibroblasts (ORAI1 KO MEFs). We used 35-cycle RT-PCR analysis of ORAI1 KO MEF RNA and found the MEFs to express both *STIM* genes (*Stim1* and *Stim2)* and the remaining *Oraii2* and *Orai3* genes. Whereas wt MEFs express *Trpc1*, *Trpc2*, and *Trpc6*, the ORAI1 KO MEFs express *Trpc1*, *Trpc2*, and *Trpc4* and therefore have an altered gene expression profile. In tests for ROCE and SOCE, we found that ORAI1 KO MEFs have essentially no SOCE or ROCE responses when challenged with thapsigargin to activate their SOCE or with a Gq-coupled GPCR agonist to activate their ROCE, respectively (Fig. 5). These experiments are in agreement with the view that TRPCs depend on an ORAI, because ROCE should not have been interfered with by the loss of ORAI1, as ORAI and TRPC channels should have been able to operate independently of each other. Consistent with this view, TRPC1 and ORAI1 co-immunoprecipitate (Ong et al. 2007).

The results with ORAI1 KO MEFs are interesting in that they indicate that endogenous levels of ORAI2 and ORAI3 cannot compensate for loss of ORAI1. Yet overexpression of ORAI2 or ORAI3 in ORAI1 KO MEFs reconstitutes SOCE, albeit to a lower extent (Fig. 5, top panel). ORAI1 is thus the dominant protein in setting SOCE and ROCE activities of a cell, and ORAI2 and ORAI3 can replace ORAI1 to varying degrees that may differ from cell to cell. In vivo studies with *Orai1* KO mice reached similar conclusions (Vig et al. 2008; Gwack et al. 2008; Kim et al. 2011).

Fig. 4 (continued) the inhibitory effects of the corresponding antibodies. The heteromeric composition of the TRPCs activated by store depletion was concluded to be C1/C5 for cells from mesenteric artery, C1/C5/C6 for cells from the coronary artery, and C1/C5/C7 for cells from the portal vein. Panel (**a**) was adapted from Shi et al. (2012), and panels (**b**), (**c**), and (**d**) were adapted from Saleh et al. (2008)

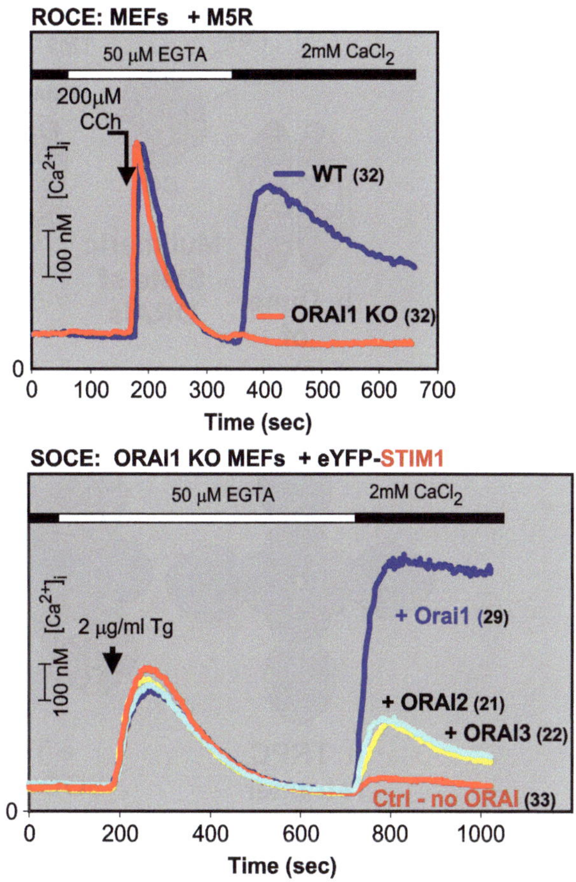

Fig. 5 ROCE and SOCE are critically dependent on ORAI. *Left Panel*: ROCE in WT and ORAI1 KO mouse embryonic fibroblasts (MEFs) activated by a transfected M5 mACh receptor. *Right panel*: SOCE in ORAI1 KO MEFs transfected with an empty pCMV expression vector (no ORAI) and with pCMV directing the expression of myc-tagged human ORAI1, ORAI2, or ORAI3

6 To What Extent Is ORAI Function Dependent on a TRPC?

Whereas TRPCs appear to depend on ORAI, the answer to the converse question is unknown. Another unknown, related to the previous, is in how many formats ORAIs may adopt. Taken together the evidence discussed so far may suggest that ORAI molecules exist in one of four configurations (Fig. 6): (1) as a non-active dimeric precursor of (2) tetrameric and (3) hexameric ORAI-only channels and (4) in association with TRPCs, where they act as regulatory subunits of TRPC channels. In turn, TRPCs may exist in one of two configurations: (a) inactive tetramers, often located within the cells and trafficking to the cell surface in response to an external stimulus (e.g., Chaudhuri and Colles 2008), and (b) active TRPC-ORAI hetero-multimers in the plasma membrane. ORAI dimers may or may not have changed conformation upon binding to TRPCs.

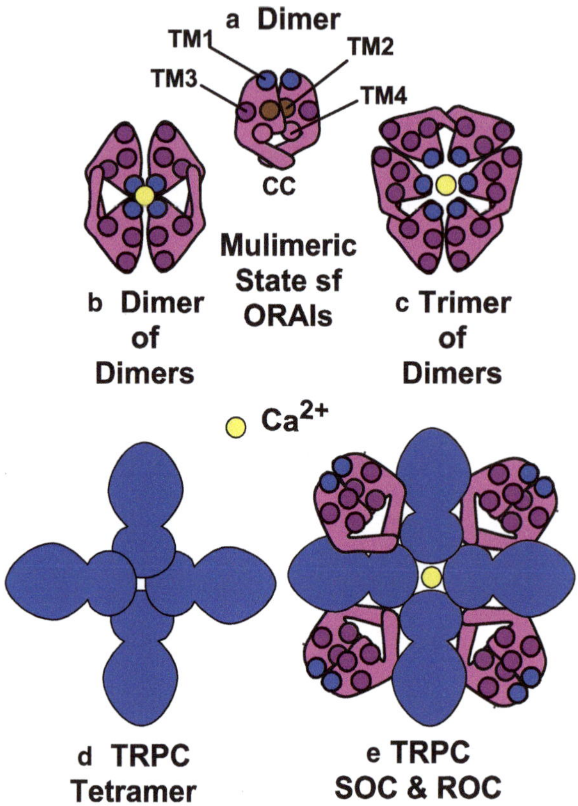

Fig. 6 Models of molecular makeup of CRAC and TRPC channels. (**a**) Cartoon of an ORAI dimer based on the crystal structure of the hexameric trimer of dimers ORAI (PDB 5HKR, Fig. 3). *Blue circle*, pore lining TM1, CC, coiled coil formed of the C-termini of the two monomers. The SOAR/CAD interacting domains located on the ORAI CC region. Mutations that disrupt the CC structure impede activation by STIM. (**b**) Hypothetical channel-competent ORAI tetramer with Ca^{2+} atom in its pore. (**c**) Cartoon of hexameric ORAI channel with Ca^{2+} in its pore. (**d**) Inactive tetrameric TRPC complex. (**e**) Channel-competent TRPC-tetramer with four ORAI dimers bound to it. It is hypothesized that ORAI tetramers and hexamers are CRAC channels and that the [TRPC]4-[ORAI2]4 complex is a CRAC-like SOC, whereas nonselective ROCs are TRPC tetramers with less than four ORAI dimers

The central roles of ORAIs in Ca^{2+} signaling downstream of PLC activation became evident early on, as a missense mutation of *Orai1*, *ORAI1[R91W]*, was found to be the molecular cause of a familial severe combined immune deficiency (Feske et al. 2006). Thus, loss of ORAI leads to strong phenotypes, which was how ORAI was discovered: RNAi-mediated suppression of *Orai1* caused loss of SOCE, and homozygous *Orai1* KO mice tend to die perinatally; do not procreate and have impaired mast cell, T cell, and B-cell function, suffer from hair loss; and have stunted growth (Vig et al. 2008; Gwack et al. 2008; Kim et al. 2011). Since ORAIs have not been expressed and/or studied in cells proven null for all seven TRPCs, the question whether perhaps the channel-competent ORAI1 tetramer may coexist and depend with a TRPC-tetramer cannot be answered at this time. In contrast, loss of any one or two of the TRPCs does not compromise normal growth nor does it interfere with procreation. The difference presumably lies in that there are only three ORAIs with one, ORAI1, being dominant, whereas there are seven TRPCs and loss of one or two can be compensated by the remaining TRPCs. This is not to say that loss of TRPCs do not bring about distinct phenotypes.

7 Phenotypic Changes in TRPC KO Mice Reveal a Broad Spectrum of Physiological and Pathophysiological Roles for TRPCs

All mouse *Trpc* genes have been disrupted as reported in Dietrich et al. (2007) for *Trpc1*, Stowers et al. (2002) for *Trpc2*, Hartmann et al. (2008) for *Trpc3*, Freichel et al. (2001) for *Trpc4*, Phelan et al. (2013) for *Trpc5*, Dietrich et al. (2005) for *Trpc6*, and Perez-Leighton et al. (2011) for *Trpc7*, and the literature is replete with functional roles for TRPCs in health and disease stemming from studies on the TRPC KO mice. As detailed in Table 1, these roles span from neuronal to smooth and skeletal muscle, the immune system, inflammation, cardiac hypertrophy, pathologic remodeling of endothelial cells in response to static stretch, wound healing, and myogenic tone and blood pressure regulation. They also include specialized systems such as vomeronasal pheromone sensing (TRPC2), light responses initiated by melanopsin in intrinsically photosensitive retinal ganglion cells (ipRGCs) that project to the suprachiasmatic nucleus to entrain circadian rhythms (TRPC6 plus TRPC7), normal touch (TRPC3 plus TRPC6), and auditory neurotransmission (TRPC3). Independent studies of human mutations and syndromes have pointed to a role of TRPC6 in normal glomerular filtration in renal kidney (Reiser et al. 2005; Winn et al. 2005) and raised the involvement of a TRPC in a host of other syndromes that are outside of the scope of this review. The phenotypic changes ascribed to loss of TRPC function listed in Table 1 are primarily those seen in knockout mice, though there are also some phenotypic changes listed as seen in cultured cardiac myocytes [TRPC3 and TRPC6; Onohara et al. (2006)] and organotypic cultures of cerebral resistance artery segments denuded of their endothelial cell layer (Welsh et al. 2003). In skeletal muscle, TRPC1 has been shown to counteract muscle fatigue (Zanou et al. 2010), to aid in muscle regeneration (Zanou et al. 2012). In a model of dopaminergic neurons, TRPC1 was shown to aid in maintenance of ER Ca^{2+} homeostasis and in so doing to counteract the effects of neurotoxin-induced ER stress and development of the unfolded protein response (Selvaraj et al. 2012), relevant in the context of Parkinson's disease development.

Gain-of-function mutations of *TRPC6* are responsible for familial forms of focal segmental glomerulosclerosis in man (Winn et al. 2005; Reiser et al. 2005), and a gain-of-function mutation of *Trpc3* in mice is responsible for the "Moonwalker" ataxia caused by loss of Purkinje cells (Becker et al. 2009). The "Moonwalker" phenotype develops in mk/+ mice; the mwk/mwk genotype is embryonic lethal.

Table 1 Examples of roles of TRPC channels (nonselective calcium-permeable cation channels) as seen from different points of view

Biochemical roles	
NFAT activation—Calcineurin (CaN) activation by Ca–calmodulin (CaCaM) (Seth et al. 2009; Poteser et al. 2011)	TRPC1; TRPC3
TLR4-CD14 signaling—MLCK (CaCaM) (Tauseef et al. 2012)	TRPC6
NFkB activation in endothelial cells (ECs)	
by GPCR—CaMKKβ (CaCaM) in ECs (Bair et al. 2009)	TRPC4
by LPS—MLCK (CaCaM) MyD88-IRAK4-MLCK complex in ECs (Tauseef et al. 2012)	TRPC6
TNFα signaling—CaMKII (CaCaM) in monocytes (Smedlund et al. 2010)	TRPC3
CaN activation by CaCaM—NFAT (Poteser et al. 2011)	TRPC3
CaMKKIIβ activation by CaCaM in activation of NFkB (Bair et al. 2009)	TRPC4
CaMKII activation by CaCaM—TNFα signaling—CaCaM-CaMKII (Tano and Vazquez 2011)	TRPC3
MLCK activation by CaCaM—activation of NFkB by LPS (Tauseef et al. 2012)	TRPC6
cGMP-independent signaling of the ANP receptor GC-A (membrane guanylyl cyclase A) (Klaiber et al. 2011)	TRPC3–C6
Physiological roles	
Agonist-induced Ca-mediated neurotransmitter release from dendrites (Munsch et al. 2003)	TRPC4
Synaptic transmission and motor control; slow EPSCs (Hartmann et al. 2008)	TRPC3
Neuronal after depolarization (Stroh et al. 2012)	TRPC1–C4[a]
Plateau potentials in hippocampal CA1 pyramidal neurons (Tai et al. 2011)	TRPC5
Control of vascular tone (Welsh et al. 2003; Dietrich et al. 2005)	TRPC6
Endothelial cell NO-EDRF (endothelium-derived relaxing factor) generation—vascular smooth muscle relaxation (Freichel et al. 2001)	TRPC4
Endothelial cell NO-independent EDH (endothelium-dependent hyperpolarization—vascular smooth muscle relaxation (Senadheera et al. 2012)	TRPC3
Endothelial cell migration—wound healing, fibroblast transdifferentiation (Davis et al. 2012)	TRPC6
Static stretch response of endothelial cells—stretch-ATR1-Gq-TRPC-Ca-ET1-ANP-GC-A-cGMP-PKG-zyxin-gene transcription (Suresh Babu et al. 2012)	TRPC3
Intestinal motility regulation by vagus (Tsvilovskyy et al. 2009)	TRPC4 + TRPC6
Cold transduction in the peripheral nervous system (Zimmermann et al. 2011)	TRPC5
Exocrine secretion (saliva) (Liu et al. 2007)	TRPC1
Efferocytosis and survival signaling in macrophages (Tano et al. 2011)	TRPC3
Normal touch (Quick et al. 2012)	TRPC3-C6[a]
Light entrainment by ipRGCs (melanopsin signaling) (Xue et al. 2011)	TRPC6-C7[a]
Innate immunity (LPS-induced NFkB activation) (Tauseef et al. 2012)	TRPC6
Platelet activation by thrombin plus collagen-related peptide—PS exposure (Harper et al. 2013)	TRPC3–C6[a]
Short-term postsynaptic memory—burst firing-induced after depolarization (Phelan et al. 2012)	TRPC1–C4[a]

(continued)

Table 1 (continued)

Pheromone signal transduction in vomeronasal sensory neurons (Stowers et al. 2002); lost in evolution between New and Old World monkeys and man (Liman and Innan 2003)	TRPC2
Sound transduction and auditory neurotransmission (Wong et al. 2013)	TRPC3
In disease (pathophysiological roles)	
Cardiac hypertrophy induced by Ang II (Onohara et al. 2006)	TRPC3 +TRPC6
Cardiac hypertrophy induced by transverse aorta constriction (TAC) (Seth et al. 2009)	TRPC1
Albuminuria associated with Ang II-induced cardiac hypertrophy (Eckel et al. 2011)	TRPC6
Epileptogenic postsynaptic regenerative plateau potentials (Phelan et al. 2012)	TRPC1–C4[a]
Calcium toxicity in secretory epithelia (Kim et al. 2011)	TRPC3
Neuronal excitotoxicity (Phelan et al. 2012)	TRPC1–C4[a]
Neurotoxin-induced ER stress response and ER calcium homeostasis (Selvaraj et al. 2012)	TRPC1 loss
Proinflammatory in murine allergic asthma (Yildirim et al. 2012)	TRPC1

Most of the roles listed here were deduced from studying phenotypes developed in the corresponding knockout mice or cells

[a]Likely operating as obligatory heteromeric channels

Gain-of-function phenotypes are rare however. An explanation of this may come from studies with gain-of-function mutations in TRPC4 and TRPC5, which when expressed in HEK cells caused their death due to Ca^{2+} toxicity. Cell death was reduced by buffering extracellular Ca^{2+} to below 1 μM (Beck et al. 2013).

An analysis of the molecular basis for the phenotypes described to loss of TRPCs points to two mechanisms (Fig. 7): (1) TRPCs cause the collapse of the membrane potential owing to their nonselective cation channel property, and (2) TRPCs directly mediate Ca^{2+} entry. In many cell types the membrane depolarization triggers additional Ca^{2+} entry either by activation of voltage-gated Ca^{2+} channels or as a consequence of Na^+ influx driving reverse mode Na^+/Ca^{2+} exchange, by the Na^+/Ca^{2+} exchangers (NCX1–3). In neurons, TRPC activation may of course trigger action potentials as is the case in ipRGCs. The opposite may also occur. In cells expressing Ca^{2+}-activated K^+ channels (SKca, IKca, BKca), the Ca^{2+} entering through the TRPC causes hyperpolarization of the cell, as happens in endothelial cells (endothelium-derived hyperpolarization or EDH), which may thus contribute to vascular smooth muscle relaxation in those vessels in which the endothelium is connected to the surrounding smooth muscle layer by myoendothelial gap junctions (Sandow et al. 2002). The EDH response to stimuli that activate the Gq-PLC pathway in endothelia sums to the production of EDRF (NO) by CaCaM-activated soluble nitric oxide synthase, a response shown to be due to Ca^{+2} entering through TRPC4, as it is impaired in TRPC4 KO mice (Table 1; Freichel et al. 2001).

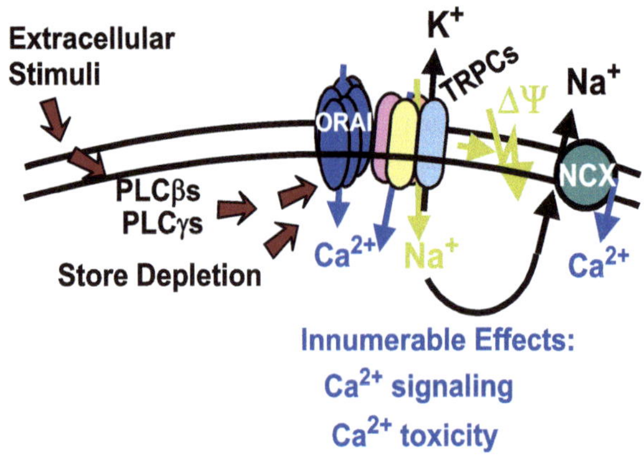

Fig. 7 Signaling initiated by TRPC and ORAI channels in response to stimuli that activate PLCs and deplete intracellular Ca^{+2} stores

Concluding Remarks

The fact that TRPC activation has a component driven by store depletion (via STIM)—an argument that was limited to TRPC1 in this article, but may also apply to other TRPCs whose presence allows for increased SOCE by fortification with ORAI (Fig. 1)—would predict that at some point, consequences of pharmacological manipulation of ORAI would overlap with consequences arising from pharmacological manipulation of TRPCs. In this context, it is interesting that loss of TRPC3 or inhibition of TRPC3 by the TRPC3-selective antagonist Pyr3 ameliorate cerulein-induced acute pancreatitis (Kim et al. 2009) as does inhibition of ORAI1 in a model of store depletion (thapsigargin)-induced pancreatitis (Gerasimenko et al. 2013). TRPC3, as TRPC1, is gated by STIM1 (Zeng et al. 2008), and both models of pancreatitis are based on prolonged increases in intracellular Ca^{2+}.

Signaling in response to ORAI and TRPC activation downstream of PLC activation and store depletion is summarized in Fig. 7. ORAI is placed next to TRPCs and does not distinguish between the regulatory roles of ORAI dimer independent of the CRAC activity of the assembled ORAI channel. The resolution of the relative roles of the different configurations of ORAI and how they regulate TRPCs awaits the results from further investigations. The dependence of a functional TRPC channel on an ORAI is supported by experimental data, several of which were presented in this review. The inverse, to what extent ORAI channels may or may not operate in the absence of a TRPC will require the creation of TRPC-null cell line. We (our laboratory) are well on the way in creating such a cell line. So far we have found that MEFs lacking TRPC1/2/3/5/6 (TRPC4 and TRPC7 remaining) have unaltered SOCE.

Acknowledgements The author thanks Yingpei Zhang for technical assistance. This research was supported by the Intramural Research Program of the NIH (project Z01-ES-101684).

References

Bair AM, Thippegowda PB, Freichel M, Cheng N, Ye RD, Vogel SM, Yu Y, Flockerzi V, Malik AB, Tiruppathi C (2009) Ca^{2+} entry via TRPC channels is necessary for thrombin-induced NF-kappaB activation in endothelial cells through AMP-activated protein kinase and protein kinase Cdelta. J Biol Chem 284:563–574. doi: 10.1074/jbc.M803984200. Epub 2008 Nov 6

Beck A, Speicher T, Stoerger C, Sell T, Dettmer V, Jusoh SA, Abdulmughni A, Cavalié A, Philipp SE, Zhu MX, Helms V, Wissenbach U, Flockerzi V (2013) Conserved gating elements in TRPC4 and TRPC5 channels. J Biol Chem 288:19471–19483

Becker EB, Oliver PL, Glitsch MD, Banks GT, Achilli F, Hardy A, Nolan PM, Fisher EM, Davies KE (2009) A point mutation in TRPC3 causes abnormal Purkinje cell development and cerebellar ataxia in moonwalker mice. Proc Natl Acad Sci U S A 106:6706–6711

Chaudhuri P, Colles SM, Bhat M, Van Wagoner DR, Birnbaumer L, Graham LM (2008) Elucidation of a TRPC6-TRPC5 channel cascade that restricts endothelial cell movement. Mol Biol Cell 19:3203–3211

Cheng KT, Liu X, Ong HL, Ambudkar IS (2008) Functional requirement for Orai1 in store-operated TRPC1-STIM1 channels. J Biol Chem 283:12935–12940

Davis J, Burr AR, Davis G, Birnbaumer L, Molkentin JD (2012) A novel TRPC6-dependent pathway for myofibroblast transdifferentiation and wound healing in vivo. Dev Cell 23:705–715

Dietrich A, Mederos Y, Schnitzler M, Gollasch M, Gross V, Storch U, Dubrovska G, Obst M, Yildirim E, Salanova B, Kalwa H, Essin K, Pinkenburg O, Luft FC, Gudermann T, Birnbaumer L (2005) Increased vascular smooth muscle contractility in $TRPC6^{-/-}$ mice. Mol Cell Biol 25:6980–6989

Dietrich A, Kalwa H, Storch U, Mederos y Schnitzler M, Salanova B, Pinkenburg O, Dubrovska G, Essin K, Gollasch M, Birnbaumer L, Gudermann T (2007) Pressure-induced and store-operated cation influx in vascular smooth muscle cells is independent of TRPC1. Pflugers Arch 455:465–477

Eckel J, Lavin PJ, Finch EA, Mukerji N, Burch J, Gbadegesin R, Wu G, Bowling B, Byrd A, Hall G, Sparks M, Zhang ZS, Homstad A, Barisoni L, Birnbaumer L, Rosenberg P, Winn MP (2011) TRPC6 enhances angiotensin II-induced albuminuria. J Am Soc Nephrol 22:526–535

Feske S, Gwack Y, Prakriya M, Srikanth S, Puppel SH, Tanasa B, Hogan PG, Lewis RS, Daly M, Rao A (2006) A mutation in Orai1 causes immune deficiency by abrogating CRAC channel function. Nature 441:179–185

Freichel M, Suh SH, Pfeifer A, Schweig U, Trost C, Weissgerber P, Biel M, Philipp S, Freise D, Droogmans G, Hofmann F, Flockerzi V, Nilius B (2001) Lack of an endothelial store-operated Ca^{2+} current impairs agonist-dependent vasorelaxation in $TRP4^{-/-}$ mice. Nat Cell Biol 3:121–127

Gerasimenko JV, Gryshchenko O, Ferdek PE, Stapleton E, Hébert TO, Bychkova S, Peng S, Begg M, Gerasimenko OV, Petersen OH (2013) Ca^{2+} release-activated Ca^{2+} channel blockade as a potential tool in antipancreatitis therapy. Proc Natl Acad Sci U S A 110:13186–13191

Gwack Y, Srikanth S, Oh-Hora M, Hogan PG, Lamperti ED, Yamashita M, Gelinas C, Neems DS, Sasaki Y, Feske S, Prakriya M, Rajewsky K, Rao A (2008) Hair loss and defective T- and B-cell function in mice lacking ORAI1. Mol Cell Biol 28:5209–5222

Harper MT, Camacho Londoño JE, Quick K, Camacho Londoño J, Flockerz V, Philipp S, Birnbaumer L, Freichel M, Poole AW (2013) Transient receptor potential channels function as a coincidence signal detector mediating phosphatidylserine exposure. Sci Signal 6:ra80

Hartmann J, Dragicevic E, Adelsberger H, Henning HA, Sumser M, Abramowitz J, Blum R, Dietrich A, Freichel M, Flockerzi V, Birnbaumer L, Konnerth A (2008) TRPC3 channels are required for synaptic transmission and motor coordination. Neuron 59:392–398

Hou X, Pedi L, Diver MM, Long SB (2012) Crystal structure of the calcium release-activated calcium channel Orai. Science 338:1308–1313

Kim MS, Zeng W, Yuan JP, Shin DM, Worley PF, Muallem S (2009) Native store-operated Ca^{2+} influx requires the channel function of Orai1 and TRPC1. J Biol Chem 284:9733–9741

Kim KD, Srikanth S, Yee MK, Mock DC, Lawson GW, Gwack Y (2011) ORAI1 deficiency impairs activated T cell death and enhances T cell survival. J Immunol 187:3620–3630

Klaiber M, Dankworth B, Kruse M, Hartmann M, Nikolaev VO, Yang RB, Völker K, Gassner B, Oberwinkler H, Feil R, Freichel M, Groschner K, Skryabin BV, Frantz S, Birnbaumer L, Pongs O, Kuhn M (2011) A cardiac pathway of cyclic GMP-independent signaling of guanylyl cyclase A, the receptor for atrial natriuretic peptide. Proc Natl Acad Sci U S A 108:18500–18505

Liao Y, Erxleben C, Yildirim E, Abramowitz J, Armstrong DL, Birnbaumer L (2007) Orai proteins interact with TRPC channels and confer responsiveness to store depletion. Proc Natl Acad Sci U S A 104:4682–4687

Liao Y, Erxleben C, Abramowitz J, Flockerzi V, Zhu MX, Armstrong DL, Birnbaumer L (2008) Functional interactions among Orai1, TRPCs, and STIM1 suggest a STIM-regulated heteromeric Orai/TRPC model for SOCE/Icrac channels. Proc Natl Acad Sci U S A 105:2895–2900

Liao Y, Plummer NW, George MD, Abramowitz J, Zhu MX, Birnbaumer L (2009) A role for Orai in TRPC-mediated Ca^{2+} entry suggests that a TRPC:Orai complex may mediate store and receptor operated Ca^{2+} entry. Proc Natl Acad Sci U S A 106:3202–3206

Liman ER, Innan H (2003) Relaxed selective pressure on an essential component of pheromone transduction in primate evolution. Proc Natl Acad Sci U S A 100:3328–3332

Liu X, Cheng KT, Bandyopadhyay BC, Pani B, Dietrich A, Paria BC, Swaim WD, Beech D, Yildrim E, Singh BB, Birnbaumer L, Ambudkar IS (2007) Attenuation of store-operated Ca^{2+} current impairs salivary gland fluid secretion in TRPC1($^{-/-}$) mice. Proc Natl Acad Sci U S A 104:17542–17547

Long SB, Tao X, Campbell EB, MacKinnon R (2007) Atomic structure of a voltage-dependent K+ channel in a lipid membrane-like environment. Nature 450:376–382

Luik RM, Wu MM, Buchanan J, Lewis RS (2006) The elementary unit of store-operated Ca^{2+} entry: local activation of CRAC channels by STIM1 at ER-plasma membrane junctions. J Cell Biol 174:815–825

Mercer JC, Dehaven WI, Smyth JT, Wedel B, Boyles RR, Bird GS, Putney JW Jr (2006) Large store-operated calcium selective currents due to co-expression of Orai1 or Orai2 with the intracellular calcium sensor, Stim1. J Biol Chem 281:24979–24990

Mignen O, Thompson JL, Shuttleworth TJ (2008) Orai1 subunit stoichiometry of the mammalian CRAC channel pore. J Physiol 586:419–425

Mio K, Ogura T, Kiyonaka S, Hiroaki Y, Tanimura Y, Fujiyoshi Y, Mori Y, Sato C (2007) The TRPC3 channel has a large internal chamber surrounded by signal sensing antennas. J Mol Biol 367(2):373–383, Epub 2006

Munsch T, Freichel M, Flockerzi V, Pape HC (2003) Contribution of transient receptor potential channels to the control of GABA release from dendrites. Proc Natl Acad Sci U S A 100:16065–16070

Ong HL, Cheng KT, Liu X, Bandyopadhyay BC, Paria BC, Soboloff J, Pani B, Gwack Y, Srikanth S, Singh BB, Gill DL, Ambudkar IS (2007) Dynamic assembly of TRPC1-STIM1-Orai1 ternary complex is involved in store-operated calcium influx. Evidence for similarities in store-operated and calcium release-activated calcium channel components. J Biol Chem 282:9105–9116

Onohara N, Nishida M, Inoue R, Kobayashi H, Sumimoto H, Sato Y, Mori Y, Nagao T, Kurose H (2006) TRPC3 and TRPC6 are essential for angiotensin II-induced cardiac hypertrophy. EMBO J 25:5305–5316

Park CY, Hoover PJ, Mullins FM, Bachhawat P, Covington ED, Raunser S, Walz T, Garcia KC, Dolmetsch RE, Lewis RS (2009) STIM1 clusters and activates CRAC channels via direct binding of a cytosolic domain to Orai1. Cell 136:876–890

Peinelt C, Vig M, Koomoa DL, Beck A, Nadler MJ, Koblan-Huberson M, Lis A, Fleig A, Penner R, Kinet JP (2006) Amplification of CRAC current by STIM1 and CRACM1 (Orai1). Nat Cell Biol 8:771–773

Penna A, Demuro A, Yeromin AV, Zhang SL, Safrina O, Parker I, Cahalan MD (2008) The CRAC channel consists of a tetramer formed by Stim-induced dimerization of Orai dimers. Nature 456:116–120

Perez-Leighton CE, Schmidt TM, Abramowitz J, Birnbaumer L, Kofuji P (2011) Intrinsic phototransduction persists in melanopsin-expressing ganglion cells lacking diacylglycerol-sensitive TRPC subunits. Eur J Neurosci 33:856–867

Phelan KD, Mock MM, Kretz O, Shwe UT, Kozhemyakin M, Greenfield LJ, Dietrich A, Birnbaumer L, Freichel M, Flockerzi V, Zheng F (2012) Heteromeric canonical transient receptor potential 1 and 4 channels play a critical role in epileptiform burst firing and seizure-induced neurodegeneration. Mol Pharmacol 81:384–392

Phelan KD, Shwe UT, Abramowitz J, Wu H, Rhee SW, Howell MD, Gottschall PE, Freichel M, Flockerzi V, Birnbaumer L, Zheng F (2013) Canonical transient receptor channel 5 (TRPC5) and TRPC1/4 contribute to seizure and excitotoxicity by distinct cellular mechanisms. Mol Pharmacol 83:429–438

Poteser M, Schleifer H, Lichtenegger M, Schernthaner M, Stockner T, Kappe CO, Glasnov TN, Romanin C, Groschner K (2011) PKC-dependent coupling of calcium permeation through transient receptor potential canonical 3 (TRPC3) to calcineurin signaling in HL-1 myocytes. Proc Natl Acad Sci U S A 108:10556–10561

Quick K, Zhao J, Eijkelkamp N, Linley JE, Rugiero F, Cox JJ, Raouf R, Gringhuis M, Sexton JE, Abramowitz J, Taylor R, Forge A, Ashmore J, Kirkwood N, Kros CJ, Richardson GP, Freichel M, Flockerzi V, Birnbaumer L, Wood JN (2012) TRPC3 and TRPC6 are essential for normal mechanotransduction in subsets of sensory neurons and cochlear hair cells. Open Biol 2:120068

Reiser J, Polu KR, Möller CC, Kenlan P, Altintas MM, Wei C, Faul C, Herbert S, Villegas I, Avila-Casado C, McGee M, Sugimoto H, Brown D, Kalluri R, Mundel P, Smith PL, Clapham DE, Pollak MR (2005) TRPC6 is a glomerular slit diaphragm-associated channel required for normal renal function. Nat Genet 37:739–744

Saleh SN, Albert AP, Peppiatt-Wildman CM, Large WA (2008) Diverse properties of store-operated TRPC channels activated by protein kinase C in vascular myocytes. J Physiol 586:2463–2476

Sandow SL, Tare M, Coleman HA, Hill CE, Parkington HC (2002) Involvement of myoendothelial gap junctions in the actions of endothelium-derived hyperpolarizing factor. Circ Res 90:1108–1113

Selvaraj S, Sun Y, Watt JA, Wang S, Lei S, Birnbaumer L, Singh BB (2012) Neurotoxin-induced ER stress in mouse dopaminergic neurons involves downregulation of TRPC1 and inhibition of AKT/mTOR signaling. J Clin Invest 122:1354–1367

Senadheera S, Kim Y, Grayson TH, Toemoe S, Kochukov MY, Abramowitz J, Housley GD, Bertrand RL, Chadha PS, Bertrand PP, Murphy TV, Tare M, Birnbaumer L, Marrelli SP, Sandow SL (2012) Transient receptor potential canonical type 3 channels facilitate endothelium-derived hyperpolarization-mediated resistance artery vasodilator activity. Cardiovasc Res 95:439–447

Seth M, Zhang ZS, Mao L, Graham V, Burch J, Stiber J, Tsiokas L, Winn M, Abramowitz J, Rockman HA, Birnbaumer L, Rosenberg P (2009) TRPC1 channels are critical for hypertrophic signaling in the heart. Circ Res 105:1023–1030

Shi J, Ju M, Saleh SN, Albert AP, Large WA (2010) TRPC6 channels stimulated by angiotensin II are inhibited by TRPC1/C5 channel activity through a Ca^{2+}– and PKC-dependent mechanism in native vascular myocytes. J Physiol 588:3671–3682

Shi J, Ju M, Abramowitz J, Large WA, Birnbaumer L, Albert AP (2012) TRPC1 proteins confer PKC and phosphoinositol activation on native heteromeric TRPC1/C5 channels in vascular smooth muscle: comparative study of wild-type and TRPC1$^{-/-}$ mice. FASEB J 26:409–419

Smedlund K, Tano JY, Vazquez G (2010) The constitutive function of native TRPC3 channels modulates vascular cell adhesion molecule-1 expression in coronary endothelial cells through nuclear factor kappaB signaling. Circ Res 106:1479–1488

Soboloff J, Spassova MA, Tang XD, Hewavitharana T, Xu W, Gill DL (2006) Orai1 and STIM reconstitute store-operated calcium channel function. J Biol Chem 281:20661–20665

Stowers L, Holy TE, Meister M, Dulac C, Koentges G (2002) Loss of sex discrimination and male-male aggression in mice deficient for TRP2. Science 295:1493–1500

Stroh O, Freichel M, Kretz O, Birnbaumer L, Hartmann J, Egger V (2012) NMDA receptor-dependent synaptic activation of TRPC channels in olfactory bulb granule cells. J Neurosci 32:5737–5746

Suresh Babu S, Wojtowicz A, Freichel M, Birnbaumer L, Hecker M, Cattaruzza M (2012) Mechanism of stretch-induced activation of the mechanotransducer zyxin in vascular endothelial cells. Sci Signal 5:ra91

Tai C, Hines DJ, Choi HB, MacVicar BA (2011) Plasma membrane insertion of TRPC5 channels contributes to the cholinergic plateau potential in hippocampal CA1 pyramidal neurons. Hippocampus 21:958–967

Tano JY, Vazquez G (2011) Requirement for non-regulated, constitutive calcium influx in macrophage survival signaling. Biochem Biophys Res Commun 407:432–437

Tano JY, Smedlund K, Lee R, Abramowitz J, Birnbaumer L, Vazquez G (2011) Impairment of survival signaling and efferocytosis in TRPC3-deficient macrophages. Biochem Biophys Res Commun 410(3):643–647. doi:10.1016/j.bbrc.2011.06.045

Tauseef M, Knezevic N, Chava KR, Smith M, Sukriti S, Gianaris S, Obukhov AG, Vogel SM, Schraufnagel SE, Dietrich A, Birnbaumer L, Malik AB, Mehta D (2012) Cation channel TRPC6 activation of TLR4 in endothelial cells mediates sepsis-induced acute lung injury. J Exp Med 209:1953–1968

Tsvilovskyy VV, Zholos AV, Aberle T, Philipp SE, Dietrich A, Zhu MX, Birnbaumer L, Freichel M, Flockerzi V (2009) Deletion of TRPC4 and TRPC6 in mice impairs smooth muscle contraction and intestinal motility in vivo. Gastroenterology 137:1415–1424

Vannier B, Zhu X, Brown D, Birnbaumer L (1998) The membrane topology of human transient receptor potential 3 as inferred from glycosylation-scanning mutagenesis and epitope immunocytochemistry. J Biol Chem 273:8675–8679

Vig M, Peinelt C, Beck A, Koomoa DL, Rabah D, Koblan-Huberson M, Kraft S, Turner H, Fleig A, Penner R, Kinet JP (2006) CRACM1 is a plasma membrane protein essential for store-operated Ca^{2+} entry. Science 312:1220–1223

Vig M, DeHaven WI, Bird GS, Billingsley JM, Wang H, Rao PE, Hutchings AB, Jouvin MH, Putney JW, Kinet JP (2008) Defective mast cell effector functions in mice lacking the CRACM1 pore subunit of store-operated calcium release-activated calcium channels. Nat Immunol 9:89–96

Welsh DG, Morielli AD, Nelson MT, Brayden JE (2003) Transient receptor potential channels regulate myogenic tone of resistance arteries. Circ Res 90:248–250

Winn MP, Conlon PJ, Lynn KL, Farrington MK, Creazzo T, Hawkins AF, Daskalakis N, Kwan SY, Ebersviller S, Burchette JL, Pericak-Vance MA, Howell DN, Vance JM, Rosenberg PB (2005) A mutation in the TRPC6 cation channel causes familial focal segmental glomerulosclerosis. Science 308:1801–1804

Wong AC, Birnbaumer L, Housley GD (2013) Canonical transient receptor potential channel subtype 3-mediated hair cell Ca^{2+} entry regulates sound transduction and auditory neurotransmission. Eur J Neurosci 37:1478–1486

Wu MM, Buchanan J, Luik RM, Lewis RS (2006) Ca^{2+} store depletion causes STIM1 to accumulate in ER regions closely associated with the plasma membrane. J Cell Biol 174:803–813

Xue T, Do MT, Riccio A, Jiang Z, Hsieh J, Wang HC, Merbs SL, Welsbie DS, Yoshioka T, Weissgerber P, Stolz S, Flockerzi V, Freichel M, Simon MI, Clapham DE, Yau KW (2011) Melanopsin signalling in mammalian iris and retina. Nature 479:67–73

Yildirim E, Carey MA, Card JW, Dietrich A, Flake GP, Zhang Y, Bradbury JA, Rebolloso Y, Germolec DR, Morgan DL, Zeldin DC, Birnbaumer L (2012) Severely blunted allergen-induced pulmonary Th2-cell response and lung hyperresponsiveness in type 1 transient receptor potential channel (TRPC1)-deficient mice. Am J Physiol Lung Cell Mol Physiol 303:L539–L549

Yuan JP, Zeng W, Dorwart MR, Choi YJ, Worley PF, Muallem S (2009) SOAR and the polybasic STIM1 domains gate and regulate Orai channels. Nat Cell Biol 11:337–343

Zanou N, Shapovalov G, Louis M, Tajeddine N, Gallo C, Van Schoor M, Anguish I, Cao ML, Schakman O, Dietrich A, Lebacq J, Ruegg U, Roulet E, Birnbaumer L, Gailly P (2010) Role of TRPC1 channel in skeletal muscle function. Am J Physiol Cell Physiol 298:C149–C162

Zanou N, Schakman O, Louis P, Ruegg UT, Dietrich A, Birnbaumer L, Gailly P (2012) Trpc1 ion channel modulates phosphatidylinositol 3-kinase/Akt pathway during myoblast differentiation and muscle regeneration. J Biol Chem 287:14524–14534

Zeng W, Yuan JP, Kim MS, Choi YJ, Huang GN, Worley PF, Muallem S (2008) STIM1 gates TRPC channels, but not Orai1, by electrostatic interaction. Mol Cell 32:439–448

Zhang SL, Yeromin AV, Zhang XH, Yu Y, Safrina O, Penna A, Roos J, Stauderman KA, Cahalan MD (2006) Genome-wide RNAi screen of Ca^{2+} influx identifies genes that regulate Ca^{2+} release-activated Ca^{2+} channel activity. Proc Natl Acad Sci U S A 103:9357–9362

Zimmermann K, Lennerz JK, Hein A, Link AS, Kaczmarek JS, Delling M, Uysal S, Pfeifer JD, Riccio A, Clapham DE (2011) Transient receptor potential cation channel, subfamily C, member 5 (TRPC5) is a cold-transducer in the peripheral nervous system. Proc Natl Acad Sci U S A 108:18114–18119

TRPs: Modulation by Drug-Like Compounds

Michael Schaefer

Contents

Abstract

Drug-like compounds that exert biological activity towards TRP channels are either being used as cell biological tools or further developed into pharmacological lead structures aiming at therapeutic use in diseased states. Although drug-likeliness is not easy to predict, common rules include a relatively low molecular weight, physicochemical constraints, and the absence of known reactive or otherwise toxic groups. Small molecules that exert a biological activity to block, activate, or modulate TRP channels are intensely sought. Such tool compounds may be useful to assign native currents to a certain TRP channel and to validate the channel as a candidate target for future pharmacological intervention. Depending on the TRP channel isotype, these activities have reached different levels, with only few TRP channels modulators already being clinically tested in humans, whereas other compounds only underwent a preliminary validation. For some TRP channels, reliable low molecular weight inhibitors are not yet available. Hence, further efforts need to be undertaken in

M. Schaefer (✉)
Rudolf-Boehm-Institut für Pharmakologie und Toxikologie, Härtelstraße 16-18, 04107 Leipzig, Germany
e-mail: michael.schaefer@medizin.uni-leipzig.de

B. Nilius and V. Flockerzi (eds.), *Mammalian Transient Receptor Potential (TRP) Cation Channels*, Handbook of Experimental Pharmacology 223, DOI 10.1007/978-3-319-05161-1_15,

order to explore the physiological impact and possible therapeutic potential of TRP channel targeting with drug-like compounds.

Keywords
Cation channel blockers • Compound screening • Hit optimization • Medicinal chemistry • Lead compound development • Chemical biology • Proof-of-principle preclinical tests • Clinical trials

1 Introduction

Owing to the plethora of cellular functions that have been demonstrated to rely on the functional activity of members of the TRP channel superfamily in health and disease, their pharmacological modulation appears highly desirable. Apart from possible future therapeutic use, isotype-specific modulators are being developed as chemical tools to identify as yet unrecognized physiological or pathophysiological functions of TRP channels at the tissue, organ, or organismic level. Therefore, both academia and the private sector have stepped up their efforts to develop potent and selective TRP channel modulators. These modulators may either activate or inhibit the channel activity in an agonist-/antagonist-like fashion by occupying the binding site of a physiological activator, or they may modulate the channel gating via allosteric mechanisms.

Like many cell surface receptors, the ligand- or second-messenger-gated TRP channels typically feature binding sites for low molecular weight compounds that bear the potential to identify competitive ligands that exert lower, no, or inverse intrinsic activity. In addition, the ion-permeable pore itself, including vestibules or fenestrations of the channel protein, offers cavities that may accommodate specifically tailored small molecules, plugging into the permeation pathway. Finally, allosteric binding sites may be located on the channel protein and prevent or enhance the channel gating by its physiological activator(s). Since many TRP channels are regulated or influenced by cellular signaling cascades, modulators of upstream receptor classes add yet another level of complexity to the possibilities of TRP channel targeting. Such indirect regulatory concepts will not be reviewed herein.

Before capitalizing on drug-like low molecular weight compounds, it should be mentioned that macromolecular modulators provide valuable tools to study TRP channel functions at least at the cellular and tissue/organ level. Antibodies that have been raised against epitopes within the third extracellular loop (E3) of TRP channels have been demonstrated to exert biological activity by blocking the respective TRP channel's ion-conducting activity (Naylor and Beech 2013; Xu et al. 2005b). Likewise, RNA interference-based knockdown of TRP channel expression, e.g., by specific siRNA, has successfully been applied in numerous studies. Since antibodies, siRNA, aptamers, antisense oligonucleotides, or heterologously expressed dominant negative TRP channel constructs do not represent drug-like compounds, these approaches are not covered herein. Also, the widespread modulation of many cation channels, including TRP channels, by lanthanides or divalent heavy metal cations should be mentioned, but is not within the focus of this contribution.

2 Historically Known TRP Channel-Modulating Compounds

Several activators of nonselective cation entry have been known long before TRP channels have been identified as the responsible molecular target. Chemically defined activators of TRPV1 as well as of some other sensory TRP channels exert or mimic specific sensory qualities and are traditionally being used as spices or are included as additives in cosmetics and ointments to treat a variety of ailments. Capsaicin or related vanilloids (e.g., N-vanillylnonanamide) are ingredients in topically applied creams or patches to relieve pain. Likewise, capsaicin is intravesically instilled to treat neurogenic bladder disorders. Last but not least, the adverse pain-inducing property of capsaicin-containing pepper spray is being utilized as a biological warfare agent in the context of self-defense or as tear gas in riot control.

Cooling compounds such as menthol, icilin, or others have been used long before it became clear that TRPM8 mediates the responses to these chemical surrogates that mimic cold temperatures. Numerous activators of the TRPA1 channel, a multivalent irritant sensor, have been known and used prior to the identification and recombinant expression of the TRPA1 gene product. These include ingredients of mustard oil (allyl isothiocyanate) or cinnamon (cinnamaldehyde) but also acrolein, a cystitis-inducing toxic metabolite of cyclophosphamide. Capsaicin and menthol were instrumental in performing expression cloning strategies and to validate the functional activity of identified candidate gene products, revealing TRPV1 and TRPM8 as the first hot- (Caterina et al. 1997) and cold-sensing (McKemy et al. 2002; Peier et al. 2002) "thermo-TRPs." Of note, several optimized compounds that have been generated later on are based on the original pharmacophore of such long known modulators.

In contrast to chemical activators of TRP channels that mimic sensory cues, channel blockers, antagonists, and allosteric inhibitors per se typically do not elicit perceivable sensations. Therefore, with only few exceptions, small molecules with drug-like properties that inhibit or block a given TRP channel with reasonable potency and specificity have not been available until recently. Only in the case of the capsaicin-sensitive cation conductance in nociceptive neurons, an open pore blocker (ruthenium red) and a competitive inhibitor (capsazepine) were recognized to counteract capsaicin-induced responses almost a decade before the molecular cloning of TRPV1 (Urban and Dray 1991; Wood et al. 1988). Since ruthenium red blocks numerous channel entities, however, it cannot be considered a selective tool or even a drug.

Table 1 Poorly selective drug-like TRP channel blockers

2-APB	10–50 μM	C1/C2/C3/C5/C6/C7 SOCE/InsP$_3$R	Bootman et al. (2002), Maruyama et al. (1997), Strübing et al. (2001), Xu et al. (2005a)
2-APB analogs	>5 μM	V6 SOCE	Hofer et al. (2013)
SKF96365	1–100 μM 0.6 μM	C1–7/V2/M8 Ca$_V$1–3	Halaszovich et al. (2000), Inoue et al. (2001), Juvin et al. (2007), Okada et al. (1999), Singh et al. (2010)
Flufenamic acid	1–100 μM	C3/C4/C5/C6(a)/C7/V1/V3/V4/M2/M3/M4/M5/A1(a)/P2	Guinamard et al. (2013), Hill et al. (2004a), Inoue et al. (2001), Lee et al. (2003)
N-(p-amylcinnamoyl)-anthranilic acid	1.7–20 μM	C3/C6/V1/M2/M8	Bari et al. (2009), Harteneck et al. (2007), Kraft et al. (2006)
KB-R7943	0.5–1.4 μM 0.09–0.12 μM	C3/C5/C6 hERG/I$_{Kr}$/NCX1	Cheng et al. (2012), Iwamoto et al. (1996), Kraft (2007)
BTP2/Pyr2	0.01–10 μM	M4 ≫ C3/C4/C5/C6/SOCE	Ishikawa et al. (2003), Kiyonaka et al. (2009), Takezawa et al. (2006)
Norgestimate	3–14 μM	C3/C5/C6	Miehe et al. (2012)
Progesterone	6–18 μM	C3/C4/C5/C6/M3	Majeed et al. (2012), Miehe et al. (2012)
W-7	5–80 μM	TRPγ > calmodulin/C6/M2/M3 > V4	Harteneck and Gollasch (2011)
Sigma-1 rec. ligands	6–100 μM	C5/M2/M3	Amer et al. (2013)

(a) augmentation/activation of channel activity

Several compounds were long known to inhibit nonselective cation currents or store-operated Ca^{2+} entry (SOCE). Most of them have neither been identified applying a molecularly defined target molecule nor have been secondarily validated as selectively acting drugs by applying an array of TRP channels. They include the poorly selective modulators SKF96365, LOE-908, 2-APB, and several fenamates (Table 1). Consequently, results that have been achieved applying such poorly validated compounds require a cautious interpretation. The pyrazole compound YM-58483 (now referred to as BTP2 or Pyr2) has initially been demonstrated to inhibit I_{CRAC} (Ishikawa et al. 2003), but later on, inhibition of TRPC3 and TRPC5 (He et al. 2005) and potentiation of TRPM4 activity (Takezawa et al. 2006) have been demonstrated to take effect at similar concentrations. Only after optimization of this compound, the successors Pyr3 (Kiyonaka et al. 2009) and Pyr10 (Schleifer et al. 2012) act as TRPC3-prevalent inhibitors. Although not identified in a hypothesis-free, unbiased manner, these optimized compounds now fulfill the requirement to be validated on an array of TRP channels and may be grouped within the first-generation TRP channel inhibitors.

3 Identification of TRP Channel Modulators by Screening Technologies

Since no crystal structures of TRP channels are available so far, rational drug design and structure-guided docking are not readily applicable. Instead, medium- or high-throughput screening technologies are indispensable for identifying novel TRP-modulating compounds in an unbiased manner. In most cases, stably transfected cell lines that either constitutively or inducibly overexpress a certain TRP channel ensure a favorable signal-to-noise ratio, which is a prerequisite for assay miniaturization and automation. Screening technologies that are applicable to TRP channels include fluorescent or luminometric indicator dyes, autofluorescent indicators, or automated patch clamp recording of ionic currents (Miller et al. 2011b). Due to the favorable assay reliability and scalability, fluorescent depolarization or Ca^{2+} indicator dyes are most frequently used in the screening of TRP channel modulators. Current technological progress with plate reader or imaging devices as well as the increasing availability of well-established cell lines and cost-saving assay miniaturization assisted by liquid handling robotics now allows scientists to embark on screening approaches not only in the commercial sector but also in academia. Equally important are new developments in the composition and availability of compound libraries in the public sector. An increasing number of chemically diverse, focused, or target-specified libraries are established either at academic screening centers or associated institutions. In addition, some companies give access to part of their libraries within the context of consortia, following private–public-partnership models to speed up the development process by obtaining initial target hit combinations that can either serve as a starting point for hit optimization and lead definition or may be instrumental for target validation.

The robustness of screening assays is critically determined by the availability of strong and reliable activators. Although indirect activation mechanisms (e.g., via receptor stimulation) may be employed, hits that are initially identified may not act at the level of the ion channel itself. For example, when stimulating TRPC channels via a muscarinic receptor, numerous drug-like molecules that bear anticholinergic activities will give rise to false-positive results and thereby increase the experimental burden during subsequent hit validation and characterization. Therefore, if available, direct channel activators are preferably being used. As an example, the direct activation of recombinantly expressed TRPC3, TRPC6, and TRPC7 by diacylglycerols provided a less error prone screening method compared to indirect stimulation via endogenous or recombinantly expressed phospholipase C-coupling receptors (Urban et al. 2012).

Despite hints that release from PIP_2-mediated block can activate TRPC5 and the TRPC4α splice variant (Otsuguro et al. 2008; Trebak et al. 2009) or that PKC- and PI3-kinase-dependent signaling pathways may contribute to the activation of TRPC1-bearing heteromeric channel complexes (Saleh et al. 2009), the direct physiological activators of TRPC1, TRPC4, and TRPC5 have not yet been unambiguously identified. Alternatively, heavy metal cations, such as La^{3+}, Gd^{3+}, Pb^{2+}, or Hg^{2+}, efficiently potentiate or stimulate TRPC5 activity (Jung et al. 2003; Sukumar and Beech 2010; Xu et al. 2012). Since the mode of channel activation may be critical for the identification of inhibitors, artificial activation modes bear the risk to overlook molecules that effectively inhibit the respective channel in more physiological settings or may yield modulators whose activity is restricted to the artificial mode of channel regulation. Accordingly, inhibition of TRPC5 by propofol has been shown to be effective upon TRPC5 stimulation with Gd^{3+}, but not when TRPC5 is activated by lysophosphatidylcholine (Bahnasi et al. 2008).

In the case of TRPM2, stimulation by hydrogen peroxide or by N-methyl-N′-nitrosoguanidine induces a PARP-1-dependent signaling cascade that results in the formation of ADP-ribose that in turn binds to and activates TRPM2 (Song et al. 2008). Consequently, screening hits will include compounds with antioxidant/antinitrosative activities as well as inhibitors of the PARP-1/PARG-pathway. Using the screening assay proposed by these authors, we have identified a large number of molecules that attenuated or blocked the hydrogen peroxide-induced and TRPM2-dependent Ca^{2+} signal, but none of them acted at the level of TRPM2 itself (Hill & Schaefer, unpublished results). In some cases, artificial channel activators have been identified that may be used in screening assays. Paradoxically, TRPV1, TRPV2, or TRPV3 can be activated by the otherwise poorly specific TRP channel inhibitor 2-APB (Chung et al. 2004; Hu et al. 2004). For TRPV4, GSK1016790A has been developed as a selective and highly potent activator (Thorneloe et al. 2008; Willette et al. 2008) that was later instrumental in screening, identifying, and optimizing TRPV4 inhibitors, e.g., in WO 2012/174340 A1 (Thorneloe et al. 2012). TRPM3 activation by pregnenolone sulfate or by nifedipine appears to be direct (Wagner et al. 2008). In TRPV5 and TRPV6, a high level of basal activity (Hoenderop et al. 1999; Peng et al. 1999) may suffice to set up a screening for inverse agonists, blockers, or allosteric inhibitors. In addition, activating channel mutations that cause calcium overload-induced cell death (Beck et al. 2013) might be rescued by pharmacological inhibitors, providing an appealing strategy for screening assays.

Taken together, TRP channel modulator screening is feasible both in the public and commercial sector, but specific attention should be paid to the design of the assay and the applied TRP channel activator. In some cases, novel direct TRP channel activators, constitutive activity, or activating channel mutations may be helpful to set up or optimize the screening methods.

4 Drug-Like TRP Channel-Modulating Compounds

4.1 Canonical TRP Channels

Compounds that are developed and validated applying molecularly defined test systems represent the first and following generations of TRP channel modulators. Despite the fact that TRPC1 was the first molecularly cloned mammalian TRP channel, no reliable blockers are available so far. In fact, it is intensely discussed whether TRPC1 alone can or cannot form an ion channel. Since TRPC1 regulation has been linked with store-operated Ca^{2+} entry (SOCE), inhibitors of SOCE such as a series of quinazoline compounds were assumed to act via TRPC1 (Wu et al. 2011), but direct evidence for inhibition of TRPC1 currents is lacking. Inhibition of ionic currents through heteromeric TRPC1:TRPC5 complexes by 2-APB has been demonstrated (Strübing et al. 2001), but this compound is neither specific nor potent. TRPC2, a cation channel that is highly expressed in the pheromone-sensing neuroepithelium of the vomeronasal organ, and functional in rodents, is also susceptible to inhibition by 2-APB (Lucas et al. 2003), but specific inhibitors of this channel are not known.

As a result of first screening projects, targeting TRPC channel isoforms, partially selective inhibitors have been identified (Table 2). In an academic high-throughput approach, ML204 (4-methyl-2-piperidinoquinoline) has been identified as a potent TRPC4-prevalent inhibitor (Miller et al. 2011a) that displays an IC_{50} of about 1 μM to block TRPC4β and a 19-fold selectivity with respect to TRPC6. TRPC5, the closest relative of TRPC4, is also inhibited by ML204, but the potency has not been exactly determined. Within a screening for bioactive molecules that act on TRPC3/TRPC6 channels, norgestimate and progesterone have been shown to inhibit TRPC3, TRPC5, and TRPC6 (Miehe et al. 2012). Progesterone also inhibits TRPC4 and TRPM3 channels (Majeed et al. 2011a, 2012; Miehe et al. 2012). In addition, the concentrations that are required to block TRPC6 are relatively high (3–100 μM), and progesterone is known to block many other channel entities as well (Kelley and Mermelstein 2011). Thus, other biological activities of progesterone-like compounds will presumably prevail when applied for longer time periods in vivo or in more complex test systems. Other TRPC channel inhibitors that exert a low potency and limited selectivity include niflumic acid; DIDS and ML204, acting on TRPC4 (Miller et al. 2011a; Walker et al. 2002); KB-R7943 (Kraft 2007); bromoenol lactone (Chakraborty et al. 2011); and amiloride (Inoue et al. 2001). Direct inhibition of TRPC5 by hydroxylated stilbenes has been demonstrated (Naylor et al. 2011), but the selectivity of this effect has not been worked out.

First TRPC6-prevalent inhibitors were identified in an academic screening, applying a library of chemically diverse drug-like compounds (Urban et al. 2012). Although the selectivity of a β-carboline compound 8009-5364 to block TRPC6 compared to its closest relative TRPC3 did not exceed a factor of 3, the finding that TRPC7, TRPV1, TRPA1, and TRPM2 were only weakly suppressed or entirely unaffected by this compound was encouraging. A 5-substituted indolinone (5408-0428) displayed an even better discrimination within the TRPC3/6/7 subfamily but also inhibited members of other TRP channel subfamilies.

First TRPC blockers that have been developed by pharmaceutical industry have recently been disclosed in patents. They include 2-(amino)-thiazole-4-carboxamide compounds as TRPC3/TRPC6 blockers developed by GlaxoSmithKline (WO 2012/037349), reaching a potency in the nanomolar range and an up to 10-fold selectivity for TRPC6 compared to TRPC3 (Washburn et al. 2013). Unfortunately, further lead

Table 2 Drug-like blockers acting on TRPC channels

Compound	EC_{50}	Selectivity	Application	References
TRPC1				
–				
TRPC2				
–				
TRPC3				
Pyr3	0.7–0.8 μM	C3/C3:C6 > C4/C5/C6/C7/M4	In vivo (i.p.)	Kiyonaka et al. (2009)
Pyr10	0.7 μM	C3 > C4/C5/C6/SOCE	In vitro	Schleifer et al. (2012)
2-(Amino)-thiazole-4-carboxamides	0.01–1 μM	C3 = C6	In vitro	WO 2012/037349 A2
TRPC4				
ML204	1 μM	C4 > C5 ≫ C6/V1/V3/M8/A1	In vitro and in vivo (oral)	Miller et al. (2011a) WO 2013/090722 A1
TRPC5				
–				
TRPC6				
8009-5364	3–5 μM	C6 > C3 > A1 ≫ C7/M2/V1	Ex vivo	Urban et al. (2012)
5408-0428	3–11 μM	C6 = M2 > A1 > V1> > C3/C7	Ex vivo	Urban et al. (2012)
Aniline-thiazoles GSK2332255B and GSK2833503A	0.01–1 μM	C3 = C6	In vitro and in vivo (s.c.)	WO 2012/037349 A2 Seo et al. (2014); Washburn et al. (2013)
Aminoindanes	12 nM	n.d.	In vitro	WO 2011/107474
TRPC7				
–				

n.d. not determined

development was abandoned due to pharmacokinetic constraints consisting in rapid metabolism, high plasma protein binding (Seo et al. 2014), and either low oral of dual TRPC3/TRPC6 blockers that have been shown to exert antifibrotic activity in a cadiac pressure overload model in vivo bioavailability or a loss of potency. A series of aminoindanes has been recently filed in a patent by Sanofi (WO 2011/107474). With a stated IC_{50} of 12 nM to block TRPC6-mediated Ca^{2+} entry, the compound 4-({(1R,2R)-2-[(3R)-3-aminopiperidin-1-yl]-2,3-dihydro-1H-inden-1-yl}oxy)-3-chlorobenzonitrile is the most potent TRPC6 blocker so far. The selectivity profile and pharmacokinetics of the Sanofi compound, however, have not been disclosed so far.

With the exception of orally applied ML204 that has been tested for a modulation of pain-related behavior after intracolonic mustard oil infusion (WO 2013/

090722 A1), biological activities of selective first-generation TRPC channel blockers have not yet been demonstrated in animal models in vivo. To further validate TRPC channels as therapeutic targets and to gain insight into possible experimental therapies, improved compounds with tolerable toxicity and reasonably well-characterized pharmacokinetic properties are urgently needed.

4.2 Vanilloid Receptor-Related TRP Channels

Phylogenetically, the TRPV family consists of the TRPV1–TRPV4 subgroup of heat-sensing, nonselective cation channels and the constitutively active channels TRPV5 and TRPV6, which display a Ca^{2+}-selective conductance. A block of inward currents by extracellularly applied ruthenium red is a common feature of all TRPV channels, although TRPV6 block requires substantially higher concentrations than that of other TRPV channels (Hoenderop et al. 2001). A highly potent but only partial (25–34 %) inhibition of TRPV6 by 27- and 13-meric peptides that have been derived from the C-terminal sequence of soricidin has been demonstrated (Bowen et al. 2013). The selectivity of the TRPV6-inhibitory macromolecules has not been probed. Thus, there is still no specific drug-like activator or inhibitor known for TRPV2, TRPV5, and TRPV6. By contrast, owing to their relevance for pain sensation, the preclinical and clinical development of TRPV1, TRPV3, and TRPV4 modulators has rapidly advanced (Table 3). Focused pharmaceutical development has provided lead structures for clinical use as TRPV1 agonists and antagonists (Table 4). In addition, potent and selective TRPV3 and TRPV4 modulators have been disclosed, providing valuable tools for probing the physiological and potential clinical relevance of these channels in vitro and in vivo.

4.2.1 TRPV1-Targeting Drug-Like Compounds

Obviously, the heat-, acid-, and capsaicin-activated TRPV1 channel has immediately attracted the interest of pharmaceutical industries. The wealth of known direct TRPV1 activators includes the pungent spices capsaicin, piperine, and gingerol, or anandamide and other lipid mediators, but also spider and jellyfish toxins (Vriens et al. 2009). The highly potent TRPV1 channel ligand resiniferatoxin was also instrumental in setting up binding assays to prove direct interaction of TRPV1 modulators with the channel protein. Interestingly, a highly potent TRPV1 antagonism by the lipid mediators resolvin D2 and resolvin E1 may provide an endogenous mechanism for tuning or terminating inflammatory hyperalgesia (Park et al. 2011). In addition to the competitive capsaicin antagonist capsazepine, which acts in the low micromolar range, tool compounds that selectively inhibit TRPV1 at submicromolar concentrations are commercially available (Kaneko and Szallasi 2013; Vriens et al. 2009). They include A 784168, ABT-102, AMG517, BCTC, JNJ 17203212, 5-iodo-resiniferatoxin, AMG 9810, SB 366791, SB 452533, and others.

TRPV1 inhibitors rapidly made their way into early clinical trials (see Table 4). Systemically applied inhibitory compounds such as AMG517 from Amgen or AZD-1386 from AstraZeneca revealed that global TRPV1 inhibition leads to a

Table 3 Drug-like TRPV channel modulators (excluding clinically tested TRPV1 inhibitors)

Compound	EC50	Selectivity	Application	References
TRPV1 activators (selection)				
Capsaicin	10–700 nM	V1 ≫ V2–V6	In vitro	Caterina et al. (1997)
Resiniferatoxin	0.16–30 nM	Mouse V1 > human V1 > rat V1	In vitro	Caterina et al. (1997)
Anandamide	1.1 μM	n.d.	In vitro	Smart et al. (2000)
5S/12S/15S-HPETE	9–10 μM	n.d.	In vitro	Hwang et al. (2000)
Piperine	38 μM	n.d.	In vitro	McNamara et al. (2005)
Olvanil	9–19 nM	n.d.	In vitro	Smart et al. (2000)
TRPV1 inhibitors (selection)				
5-Iodo-resiniferatoxin	4 nM		In vitro	Wahl et al. (2001)
Capsazepine	50–280 nM		In vitro In vivo	Caterina et al. (1997), Surowy et al. (2008)
Resolvin D2	0.1 nM	V1 = A1	In vivo (intrathecal)	Park et al. (2011)
BCTC	6–35 nM	V1 > M8/MOR (Novascreen)	In vitro In vivo	Behrendt et al. (2004), Valenzano et al. (2003)
Agatoxin AG489	0.3 μM	n.d.	In vitro	Kitaguchi and Swartz (2005)
ABT-102	5–7 nM	V1 ≫ V3/V4/M8/A1	In vitro In vivo	Surowy et al. (2008)
TRPV2 activators				
2-APB	60–190 μM	Mouse V2 > rat V2 ≫ human V2	In vitro	Hu et al. (2004), Juvin et al. (2007)
Cannabidiol	3.7 μM	n.d.	In vitro	Qin et al. (2008)
Probenecid	32 μM	V2 ≫ V1/V3/V4/M8/A1	In vitro	Bang et al. (2007)
TRPV3 activators				
2-APB	28 μM	V3 > V1/V2	In vitro	Chung et al. (2004), Hu et al. (2004)
Camphor	~2 mM	V3/M8	In vitro	Moqrich et al. (2005)
Eugenol	~1 mM		In vitro	Xu et al. (2006)
Thymol	~1 mM		In vitro	Xu et al. (2006)
Carvacrol	>500 μM	M7(i)/TRPL(i) > V3	In vitro	Xu et al. (2006)
Incensole acetate	16 μM	V3 > V1/V2/V4	In vitro	Moussaieff et al. (2008)
Farnesyl pyrophosphate	131 nM	V3 ≫ V1/V2/V4/M8/A1	In vivo (intradermal)	Bang et al. (2010)
TRPV3 inhibitors				
Hydra comp. #64	0.2-0.3 μM	V3 ≫ V1/V4/M8/A1	In vivo (i.p.)	WO 2006/122156 A2

(continued)

Table 3 (continued)

Compound	EC50	Selectivity	Application	References
Hydra comp. #82	0.2–1 μM	V3 ≫ V1/V4/M8/A1	In vivo (i.p.)	WO 2007/056124 A2
GRC 15133	0.25 μM	V3 ≫ V1/V4/M8/A1	In vivo (i.p.)	
GRC 17173	0.18 μM	V3 ≫ V1/V4/M8/A1	In vivo (oral)	
GRC 15300	?	V3 ≫ V1/V4/M8/A1	In vivo (oral)	
Cyclobutyl and cyclopentyl comp.	0.1–1 μM	?	In vivo (oral)	WO 2013/062964 A2
TRPV4 activators				
4α phorbol esters	0.2–0.4 μM	V4 ≫ V1/V2/V3	In vitro	Klausen et al. (2009), Watanabe et al. (2002)
Epoxyeicosatrienoic acids	0.1–1 μM	n.d.	In vitro	Watanabe et al. (2003)
Bisandrographolide A	1 μM	V4 ≫ V1/V2/V3	In vitro	Smith et al. (2006)
GSK1016790A	5 nM	V4 > V1	In vivo	Thorneloe et al. (2008), Willette et al. (2008)
RN-1747	0.8 μM	V4 > M8(i) ≫ V1/V3	In vitro	Vincent et al. (2009)
TRPV4 inhibitors				
GSK2193874	2–40 nM	n.d.	In vivo	Huh et al. (2012), Thorneloe et al. (2012)
HC-067047	17–133 nM	V4 > M8/hERG > C5/V1/V2/V3/V5/V6/M7	In vivo (i.v.; i.p.)	Everaerts et al. (2010)
RN-1734	2.3 μM	V4 ≫ V1/V3/V8	In vitro	Vincent et al. (2009)
GSK205	0.6 μM	n.d.	In vitro	Phan et al. (2009)
RN-9893	50–120 nM	V4 > V1/V3/M8	In vivo (oral)	Vincent and Duncton (2011)
TRPV5 inhibitors				
Econazole	1.3–300 μM	n.d.	In vitro	Landowski et al. (2011), Nilius et al. (2001)
TRPV6 inhibitors				
Econazole	190 μM	V6/V5	In vitro	Landowski et al. (2011)
Miconazole	200 μM	V4 ≫ V6 > V5	In vitro	Landowski et al. (2011), Watanabe et al. (2003)
TH-1177 and derivatives	90–900 μM	SOCE ≫ V6 = V5	In vitro	Haverstick et al. (2000), Landowski et al. (2011)
Soricidin-derived peptides SOR-C3; SOR-C27	14–65 nM	Partial 25–34 % TRPV6 inhibition; selectivity n.d.	In vivo (i.v.; i.p.)	Bowen et al. (2013)

(i) inhibitory action, *n.d.* not determined

Table 4 Selection of interventional clinical trials involving TRPV1-targeting drugs

Compound	Sponsor	Indication	Application	Study code/ reference
TRPV1 activators				
Capsaicin patch, Qutenza®	Acorda, Astellas	Postherpetic neuralgia	Dermal	Marketed
Capsaicin spray, Sinol-M®	Sinol	Allergic rhinitis	Intranasal	NCT00825656
Zucapsaicin civamide	Winston Laboratories	Osteoarthritis	Cream	NCT00077935 NCT00995306
		Postherpetic neuralgia	Patch	NCT00845923
		Episodic cluster headache	Intranasal	NCT00033839 NCT00069082 NCT01341548
Resiniferatoxin	NIDCR	Pain in advanced cancer	Intrathecal	NCT00804154
TRPV1 inhibitors				
AMG517	Amgen	Phase I/tooth extraction	Oral	Gavva et al. (2008)
AZD-1386	AstraZeneca	Osteoarthritis	Oral multiple	NCT00878501
		Tooth extraction	Oral single	NCT00672646
		Heat-/capsaicin-induced pain	Oral single	NCT00692146
		Posttraumatic and postherpetic neuralgia	Multiple oral doses	NCT00976534
SB-705498	GSK	Migraine (acute)		NCT00269022
		Tooth extraction	Oral single	NCT00281684
		Chronic cough	Oral	NCT01476098
		Rectal pain	Single dose	NCT00461682
		Allergic rhinitis	Intranasal	NCT01424397 NCT01424514
		Nonallergic rhinitis	Intranasal	NCT01439308
JNJ-39439335 (mavatrep)	Johnson & Johnson	Osteoarthritis	Oral single, oral multiple	NCT00933582 NCT01343303
		Heat-induced pain	Single oral dose	NCT01006304
DWP-05195	Daewoong	Postherpetic neuralgia	Oral	NCT01557010
XEN-D0501	Provesica	Overactive bladder		
SYL-1001	Sylentis	Ocular pain	Topical	NCT01776658
PAC-14028	Amorepacific Corporation	Heat tolerance	Single and multiple oral dosing	NCT01264224

NIDCR National Institute of Dental and Craniofacial Research, *GSK* GlaxoSmithKline

mild but undesirable increase in the body temperature (Gavva et al. 2008; Krarup et al. 2011). Since TRPV1 inhibition lowers the sensitivity to noxious heat, this state can be associated with an increasing risk of scalding injury (Brederson et al. 2013). Due to these problems, the therapeutic value of systemic TRPV1 inhibition is still under debate, and further development of AMG 517, but also of GRC 6211 (Glenmark/Eli Lilly), and of NGD-8243 (Neurogen/Merck) has been stopped or suspended. More recent data indicate that hyperthermia in humans receiving TRPV1-inhibiting drugs may be transient (Othman et al. 2013). Several clinical trials currently investigate the topical or systemic use of TRPV1 inhibitors in neuropathic pain, ocular pain, pruritus, chronic cough, and atopic dermatitis (Kaneko and Szallasi 2013; Premkumar and Abooj 2013). In addition, TRPV1-inhibitory (e.g., XEN-D0501 from Provesica) and TRPV1-activating compounds (e.g., resiniferatoxin) are clinically tested for the treatment of overactive bladder. The mechanistic concept behind the use of TRPV1 activators for relieving pain states is mostly based on the assumption that long-lasting TRPV1 activation may lead to a depolarization block of nerve terminals or even causes the ablation of peripheral nerve terminals (Bishnoi et al. 2011; Jeffry et al. 2009). The topically applied rapid delivery capsaicin patch Qutenza® (formerly NGX-4010) is the only approved TRP channel-targeting drug today. It may be used to treat neuropathic pain, including postherpetic pain and HIV-associated pain (Mou et al. 2013), presumably by inducing a long-lasting desensitization of TRPV1-bearing nerve endings (Derry et al. 2009). A TRPV1-activating cis-capsaicin (zucapsaicin or civamide) is still awaiting approval for the treatment of neuropathic pain and osteoarthritis. Additional envisaged indications include cluster headache and migraine.

4.2.2 TRPV2-Targeting Drug-Like Compounds

Chemical tools to activate TRPV2 include the promiscuous modulator 2-APB, probenecid, and cannabinoid derivatives (Bang et al. 2007; Hu et al. 2004; Qin et al. 2008). Tranilast has been described to inhibit TRPV2 (Hisanaga et al. 2009), but a detailed characterization of this effect is not available.

4.2.3 TRPV3-Targeting Drug-Like Compounds

Despite its predominant expression in keratinocytes, TRPV3 may also contribute to pain sensation or inflammation-induced hypersensitivity via the release of paracrine mediators, such as ATP, prostaglandins, or arachidonic acid (Mandadi et al. 2009). Its expression in DRG neurons at lower levels has been shown (Frederick et al. 2007). Its contribution to heat perception in mice has been proposed (Moqrich et al. 2005), but this finding was not confirmed in another study (Huang et al. 2011). Unfortunately, the currently available chemical activators of TRPV3 are either poorly selective (e.g., 2-APB) or exert a low potency of about 20 μM to 100 mM, precluding their reliable systemic use in vivo. Nonetheless, TRPV3 is increasingly recognized as a possible target for pain therapy (Grubisha et al. 2013; Premkumar and Abooj 2013). Specific blockers of TRPV3 have been identified and optimized by Hydra Biosciences, Glenmark, and Abbvie (formerly Abbott). As a result, lead compounds with nanomolar potency and good selectivity with respect to other sensory TRP channels have been achieved and tested in preclinical models of neuropathic pain. While some compounds did not seem to be orally bioavailable, lead candidates such as GRC 15300 (Glenmark, now in licensed by Sanofi Aventis) are. As to today, the important in vitro and preclinical data are only partially disclosed in the patents, and the inhibitors are not made available for the public. Therefore, important tools are missing, and independent evidence for possible beneficial and adverse effects cannot be achieved. A major concern comes from the fact that TRPV3 is also important for the regulation of epidermal keratinocyte proliferation and differentiation regulation (Cheng et al. 2010).

4.2.4 TRPV4-Targeting Drug-Like Compounds

TRPV4 is a remarkable example of how drug-like small molecules can contribute to clarifying physiological and pathophysiological roles of their target protein. The first-known modes of TRPV4 activation included not only hypoosmolaric stress and heat but also phorbol esters and epoxyeicosatrienoic acids (EET) as naturally occurring and endogenous mediators (Guler et al. 2002; Strotmann et al. 2000; Watanabe et al. 2002, 2003). A direct and selective TRPV4 activation could be achieved by several compound classes, including bisandrographolide A, GSK1016790A, and RN-1747. Originating from a screening hit that was a known cathepsin inhibitor (Vincent and Duncton 2011), the development of a highly efficient, superagonist-like TRPV4 activator GSK1016790A yielded a compound that, upon i.v. administration in mice, rats, and dogs, caused acute and lethal drop of blood pressure (Willette et al. 2008). This response was not observed in TRPV4$^{-/-}$ mice, indicating an on-target effect of the superagonist, and confirming the important role of TRPV4 in the regulation of endothelium-dependent vasorelaxation (Nilius et al. 2003). Similarly, intravesical instillation of GSK1016790A caused bladder overactivity only in wild-type, but not in TRPV4-deficient, mice (Thorneloe et al. 2008). Interestingly, further chemical modification of GSK1016790A by introducing a steric hindrance into the otherwise maintained chemotype resulted in compounds with inhibitory properties (WO 2009/111680). Although this scaffold was intensely explored by more than 300 chemical variations (Vincent and Duncton 2011), another scaffold was developed by GlaxoSmithKline (WO 2011/119701 A1) and further modified. The highly potent TRPV4 inhibitor GSK2193874 has then been demonstrated to prevent pulmonary edema and vascular permeability in acute and chronic heart failure models ex vivo and in vivo (Thorneloe et al. 2012). Considering the widespread expression of TRPV4 and its involvement in multiple channelopathies (Nilius and Voets 2013), future work will have to demonstrate whether pharmacological targeting of TRPV4 indeed opens new therapeutic avenues.

4.2.5 TRPV5- and TRPV6-Targeting Drug-Like Compounds

Modulators of the Ca^{2+}-selective channels TRPV5 and TRPV6 include the poorly selective antifungal drug econazole (Nilius et al. 2001). Attempts to design 2-APB-related compounds with TRPV6-selective activity were undertaken, but only minor improvements could be achieved (Hofer et al. 2013). Chemical variation of another SOCE-inhibitor, TH-1177 (Haverstick et al. 2000), yielded a compound that displayed a moderate potency to inhibit TRPV6 ($IC_{50} = 90$ µM) but a TRPV6:TRPV5 selectivity of about fivefold (Landowski et al. 2011).

4.3 Melastatin-Related TRP Channels

The founding member of the TRPM channel family, TRPM1 (or melastatin), is highly expressed and physiologically activated via a mGluR6-mediated cascade in on bipolar cells of the retina (Morgans et al. 2009; Shen et al. 2009). Like its closest relative TRPM3, TRPM1 can be activated by micromolar concentrations of the steroidal compound pregnenolone sulfate (Lambert et al. 2011; Wagner et al. 2008) (Table 5). At present more selective drug-like activators or inhibitors are not available for TRPM1. Like TRPM1, TRPM3 is responsive to pregnenolone sulfate but can also be activated by temperatures above 30 °C or by 10–50 µM concentrations of nifedipine (Vriens et al. 2011; Wagner et al. 2008). Activation by D-erythro-sphingosine has been described (Grimm et al. 2005) but may not reach the effectivity of the other activators (Wagner et al. 2008). Poorly selective TRPM3 inhibitors include progesterone and mefenamic acid (Klose et al. 2011; Majeed et al. 2012). Interestingly, a PPARγ receptor-independent inhibition of TRPM3 by the antidiabetic drug rosiglitazone has been shown (Majeed et al. 2011b), although it should be mentioned that this effect requires about 300-fold higher concentrations than its agonism on PPARγ. By screening a library of natural compounds, a set of citrus fruit or fabaceae-derived flavonoids has been shown to inhibit TRPM3. Their IC_{50} to inhibit TRPM3-mediated ionic currents and Ca^{2+} entry is in the range of 0.3–2 µM (Straub et al. 2013b). Compared to other sensory TRP channels, the selectivity of the flavonoid-type TRPM3 inhibitors was promising. The flavanone chemotype was further explored by identifying liquiritigenin and isosakuranetin as highly potent and selective TRPM3 blockers. With an IC_{50} of 50–120 nM (depending on the assay), isosakurantin is the most potent drug-like TRPM3 inhibitor so far, allowing the in vivo application to confirm the role of TRPM3 as a heat-sensing ion channel (Straub et al. 2013a; Vriens et al. 2011).

As mentioned above, drug-like molecules that interfere with TRPM2 are difficult to establish. The physiological activators ADP-ribose and cyclic ADP-ribose are membrane impermeable, and other direct TRPM2 activators are not available so far. When activated by hydrogen peroxide-triggered pathways, inhibition of TRPM2 can be achieved by a wealth of compounds with antioxidant properties or by drugs that impede on the PARG/PARP pathway, whose activity is critically required for the formation of the ultimately activating molecule ARD-ribose. Nonetheless, a small set of drug-like compounds is also inhibitory when TRPM2 is directly activated by intracellularly perfused ADP-ribose. They include the antimycotic drugs econazole and clotrimazole, as well as the poorly specific blockers N-(p-amylcinnamoyl)-anthranilic acid, flufenamic acid, and 2-APB. Despite the demonstration that the membrane-permeable compound 8Br-ADP-

Table 5 Drug-like modulators of TRPM channels

Compound	EC_{50}	Selectivity	Application	References
TRPM1 activator				
Pregnenolone sulfate	n.d.	M3 = M1	In vitro	Lambert et al. (2011)
TRPM2 activators				
ADP-ribose	90 μM	n.d.	In vitro	Perraud et al. (2001)
Cyclic ADP-ribose	10–100 μM	n.d.	In vitro	Kolisek et al. (2005)
TRPM2 inhibitors				
8Br-ADP-ribose	~100 μM	M4 ≫ SOCE	In vitro	Partida-Sanchez et al. (2007)
Econazole, clotrimazole	<3 μM	n.d.	In vitro	Hill et al. (2004b)
TRPM3 activators				
Pregnenolone sulfate	12–23 μM	M3 ≫ V4/V6/M2/M7/M8	In vitro	Wagner et al. (2008)
d-Erythro-sphingosine	12 μM	M3 ≫ C3/C4/C5/V4/V5/V6/M2	In vitro	Grimm et al. (2005)
Nifedipine	~30 μM	$Ca_V1.2$(i) > $Ca_V1.4$(i) ≫ M3/$K_V1.5$(i)	In vitro	Wagner et al. (2008)
TRPM3 inhibitors				
Rosiglitazone	12 μM	M3 > C5(a)/$K_{IR}6$	In vitro	Majeed et al. (2011b), Yu et al. (2012)
Mefenamic acid	6.6 μM	M3 ≫ C6/V4/M2	In vitro	Klose et al. (2011)
Naringenin, hesperetin, ononetin, eriodictyol	0.3–2 μM	M3 ≫ V1/M8/A1	In vitro	Straub et al. (2013b)
Liquiritigenin, isosakuranetin	50–500 nM	M3 ≫ V1/M8/A1	In vitro In vivo	Straub et al. (2013a)
TRPM4 inhibitors				
Glibenclamide	1 μM	M4/CFTR/SUR	In vitro In vivo	Demion et al. (2007), Schattling et al. (2012), Wei et al. (2001)
9-Phenanthrol	16–23 μM	M4 > M5	In vitro	Grand et al. (2008)
TRPM5 inhibitors				
Triphenylphosphine oxide	12–30 μM	M5 > V1/M4b/A1	In vitro	Palmer et al. (2010)
(3,4-Dimethoxy-benzylidene)hydrazides	0.4–0.6 μM	n.d.	In vitro	WO 2008/097504 A2
TRPM6 inhibitors				
Sphingosine	0.6 μM	M6/M7	In vitro	Qin et al. (2013)
TG-100-115	7.9 nM	Inhibits M6 kinase activity = PIK3C2G/PIK3CG/PIK3C2B	In vitro	Davis et al. (2011)
TRPM7 inhibitors				
Waixenicin A	0.02–7 μM	M7 ≫ M2/M4/M6/I_{CRAC}	In vitro	Zierler et al. (2011)

(continued)

Table 5 (continued)

Compound	EC_{50}	Selectivity	Application	References
NS8593	1.6–5.9 μM	$K_{Ca}2.1$–2.3 > M7 > M3	In vitro	Chubanov et al. (2012)
Sphingosine, N, N-dimethyl-D-erythro-sphingosine	0.3–0.6 μM	M6/M7	In vitro	Qin et al. (2013)
Fingolimod (FTY720)	0.7 μM	n.d.	In vitro	Qin et al. (2013)
AA861, MK886, NDGA	~10 μM	Lipoxygenase inhibitors	In vitro	Chen et al. (2010a)
Nafamostat mesilate	14–200 μM	n.d.	In vitro	Chen et al. (2010b)
Carvacrol	300 μM	M7/TRPL/V3(a)	In vitro	Parnas et al. (2009)
TRPM8 activators (selection)				
Menthol	0.02–7 μM	M8/A1	In vivo	Peier et al. (2002)
Icilin	0.2 μM	M8/A1	In vitro In vivo	Chuang et al. (2004), Ramachandran et al. (2013)
Camphor	4.5 mM	M8/V3	In vitro	Selescu et al. (2013)
Geraniol, eucalyptol, hydroxycitronellal	6–20 mM	n.d.	In vitro	Behrendt et al. (2004)
Frescolat ML	3.3 μM	n.d.	In vitro	Behrendt et al. (2004)
TRPM8 inhibitors				
Capsazepine	18 μM	V1 > M8	In vitro	Behrendt et al. (2004)
AMTB	59 nM	n.d.	In vivo (i.v.)	Lashinger et al. (2008)
Tetrahydroisoquinolines	56 nM	M8 ≫ V1/V3/V4/A1	In vivo (oral)	Tamayo et al. (2012)
Diaryl-phosphonates	64 nM	M8 ≫ V1	In vivo (i.p.)	Matthews et al. (2012)
2,5-Disubstituted benzimidazoles	3–20 nM	n.d.	In vivo (oral)	Calvo et al. (2012), Parks et al. (2011)
Arylglycines	5–20 nM	n.d.	In vivo	Zhu et al. (2013)
Carboxamide compounds	9 nM	M8 ≫ V1/V3/V4/A1	In vivo (oral)	Chaudhari et al. (2013)
2-(Benzyloxy)benz-amides	19–60 nM	n.d.	In vitro	Brown et al. (2013)
BCTC	0.8 μM	V1 > M8	In vitro	Behrendt et al. (2004)
PBMC	0.4–0.6 nM	M8 ≫ V1/A1	In vivo (s.c., i.p.)	Knowlton et al. (2011)

(a) augmentation/activation, *(i)* inhibitory activity, *n.d.* not determined

ribose can inhibit TRPM2 activation when applied at high micromolar concentrations (Partida-Sanchez et al. 2007), specifically acting and highly potent TRPM2 inhibitors are lacking.

The closely related TRPM4 and TRPM5 channels show a similar mode of Ca^{2+}-dependent activation and monovalent-selective permeability. For TRPM4, an activation by nanomolar concentrations of the pyrazole compound BTP2 (Takezawa et al. 2006) and inhibition by 9-phenanthrol (Grand et al. 2008) have been observed. In addition, TRPM4 has been reported to physically interact and form a complex with Sur1, the sulfonylurea receptor 1 (Woo et al. 2013). Accordingly, TRPM4 currents can be inhibited by the sulfonylurea receptor ligand glibenclamide when applied at 50–100 μM concentration (Demion et al. 2007), an effect which may explain biological activities of the antidiabetic drug that are unrelated to regulation of K_{IR} channels, the canonical interaction partner of Sur1. Highly potent and selective inhibitors are not available to date. Drug-like molecules that can activate TRPM5 are not known. Likewise, inhibitory compounds are poorly established. Triphenylphosphine oxide has been shown to inhibit heterologously expressed and native TRPM5 in vitro (Palmer et al. 2010) with an apparent affinity in the mid micromolar range. Highly potent TRPM5 blocker is not publicly available, although a series of (3,4-dimethoxy-benzylidene)hydrazides, displaying a claimed submicromolar potency to inhibit TRPM5 currents, have been recently published in a patent (WO 2008/097504 A2).

TRPM6 and TRPM7 are large channel proteins that additionally feature a kinase domain at their large cytosolic C-terminus. They can form heterooligomeric complexes and are activated by low intracellular Mg^{2+} or Mg-ATP concentrations. Owing to their dual domain organization, they are also referred to as chanzymes. No specific drug-like inhibitors of TRPM6 channel activity are known. However, a potent inhibition of TRPM7 and TRPM6 (or heteromeric TRPM6:TRPM7 complexes) by sphingosine has been observed (Qin et al. 2013) and may represent an endogenous regulatory mechanism. Of note, a recent large-scale kinase profiling study revealed that the phosphoinositide-3-kinase inhibitor TG-100-115 also inhibits the kinase activity of TRPM6 with an IC_{50} of 7.9 nM (Davis et al. 2011). Interestingly, the TRPM6 kinase was among the least frequently hit kinases within the screen.

Like for TRPM6, no drug-like direct activator of TRPM7 is known. Due to its widespread expression and possible importance in regulating tumor progression and cancer cell motility (Chen et al. 2013b), considerable efforts have been undertaken to identify drug-like inhibitors of TRPM7. Like TRPM6, TRPM7 can be inhibited by submicromolar concentrations of sphingosine but also by fingolimod (FTY720; $IC_{50} = 0.7$ μM) (Qin et al. 2013). Waixenicin A, a natural compound isolated from a soft coral Sarcothelia edmondsoni, has been isolated and identified as a potent inhibitor that cooperates with intracellular Mg^{2+} to block the TRPM7 pore (Zierler et al. 2011). Less potent TRPM7 inhibitors include the lipoxygenase inhibitors NDGA, AA861, and MK886, which require about 10 μM concentrations, carvacrol (300 μM), which also inhibits TRPL and activates TRPV3 (Parnas et al. 2009), and nafamostat mesilate, which exerts only a poor potency in the presence of physiological extracellular concentrations of divalent cations (Chen et al. 2010b). From a current view, waixenicin A, NS8593, and sphingosine derivatives may represent the most promising candidates for further optimization and TRPM7 inhibition in vivo.

The cold receptor TRPM8 is activated by temperatures lower than 18 °C or by compounds that confer a sensation of cold on the skin or in the mouth, such as menthol (Peier et al. 2002). TRPM8 activators include naturally occurring compounds, such as geraniol/lemonol, eucalyptol/1,8-cineole, hydroxycitronellal, and many semisynthetic or synthetic "cooling compounds" (Behrendt et al. 2004; Sherkheli et al. 2010). Notably, an oral formulation of 1,8-cineol (Soledum®) is marketed as cough reliever and considered in COPD and allergic airway disease (Worth et al. 2009; Worth and Dethlefsen 2012). The TRPM8-activating "supercooling" drug icilin, which can also activate TRPA1 (Story et al. 2003), has been shown to exert anti-inflammatory effects in mouse models of colitis (Ramachandran et al. 2013). A possible antinociceptive effect of TRPM8 activators and inhibitors is intensely studied. Like for TRPV1, topical application of TRPM8 activators may relieve pain by indirect mechanisms, and several clinical trials are currently being initiated. They include the use of menthol and L-camphor to treat osteoarthritic knee pain or shoulder pain (NCT01565070, NCT01827306) or of menthol in chemotherapy-induced peripheral neuropathy (NCT01855607). By contrast, TRPM8-inhibiting compounds have not proceeded to clinical trials so far. TRPM8 inhibitors with or without proven oral bioavailability have been developed by several companies (see Table 5). Some of them display nanomolar potency, oral bioavailability, and effects in preclinical proof-of-concept tests, e.g., in the "wet dog shake" assay, monitoring shivering responses upon systemic application of the TRPM8 activator icilin. Only few of these novel compounds are commercially available.

4.4 Drug-Like Compounds Targeting TRPA1

Being activated via covalently binding cysteine-modifying agents (Hinman et al. 2006), by lipid peroxidation products (Trevisani et al. 2007) and a plethora of irritants (Strassmaier and Bakthavatchalam 2011), TRPA1 is presumably the most promiscuous receptor within the TRP channel family. Activation of TRPA1 by cold temperatures is discussed and may be species specific (Chen et al. 2013a). In fluorometric assays applying Ca^{2+} or voltage indicator dyes, additional interference may originate from the exposure to ultraviolet or visible light, which directly or via photosensitizer-mediated generation of reactive oxygen species leads to TRPA1 activation (Bellono et al. 2013; Hill and Schaefer 2009). Tear gas components such as acrolein also strongly activate TRPA1 (Bautista et al. 2006). Thereby a wealth of electrophilic or colored compounds can directly or indirectly activate TRPA1. Known therapeutic drugs that activate TRPA1 include cannabinoids (Jordt et al. 2004), chlorpromazine (Hill and Schaefer 2007), apomorphine (Schulze et al. 2013), oxaliplatin (Zhao et al. 2012), chloroquine (Than et al. 2013), etodolac (Inoue et al. 2012), desflurane (Mutoh et al. 2013), and auranofin (Hatano et al. 2013), only to mention a few. The reference activator allyl isothiocyanate (Jordt et al. 2004) may bear some problems when used in high-throughput assays. Therefore, PF-484018, a potent ($EC_{50} = 23$ nM on human and 97 nM on rat TRPA1), non-electrophilic TRPA1 activator has been developed (Ryckmans et al. 2011).

A more challenging task is to develop pharmaceutical TRPA1 inhibitors. With an IC_{50} of 5–6 μM, the screening hit HC-030031 identified by Hydra Biosciences was the first robustly acting TRPA1 inhibitor (McNamara et al. 2007). The closely related Chembridge-5861528 compound was applied in vivo (i.p.) to demonstrate its impact on pain-related behavior in diabetic rats (Wei et al. 2009). A cinnamaldehyde-related inhibitory AP18 compound reached an IC_{50} of 3–4.5 μM and displayed selectivity for TRPA1 compared to TRPV1–TRPV4 and TRPM8 (Petrus et al. 2007). Further development of this chemotype led to A-967079, the first drug-like TRPA1 inhibitors reaching a submicromolar potency of 67 nM and 289 nM on human and rat TRPA1, respectively (McGaraughty et al. 2010). AZ465 is a species-selective inhibitor of human, but not mouse TRPA1, showing ex vivo biological activity on the stimulated CGRP release from human dental pulp (Nyman et al. 2013). First clinical trials are now being initiated with TRPA1-inhibiting compounds. GRC17536 (Glenmark) is currently being tested in a phase II study for beneficial effects in diabetic peripheral neuropathy (NCT01726413). Considering the proposed roles of the irritant-sensing TRPA1 in pain, airway diseases, cough, itch, and inflammation, more drug development activities and clinical trials will surely follow soon.

References

Amer MS, McKeown L, Tumova S, Liu R, Seymour VA, Wilson LA, Naylor J, Greenhalgh K, Hou B, Majeed Y, Turner P, Sedo A, O'Regan DJ, Li J, Bon RS, Porter KE, Beech DJ (2013) Inhibition of endothelial cell Ca^{2+} entry and transient receptor potential channels by Sigma-1 receptor ligands. Br J Pharmacol 168:1445–1455

Bahnasi YM, Wright HM, Milligan CJ, Dedman AM, Zeng F, Hopkins PM, Bateson AN, Beech DJ (2008) Modulation of TRPC5 cation channels by halothane, chloroform and propofol. Br J Pharmacol 153:1505–1512

Bang S, Kim KY, Yoo S, Lee SH, Hwang SW (2007) Transient receptor potential V2 expressed in sensory neurons is activated by probenecid. Neurosci Lett 425:120–125

Bang S, Yoo S, Yang TJ, Cho H, Hwang SW (2010) Farnesyl pyrophosphate is a novel pain-producing molecule via specific activation of TRPV3. J Biol Chem 285:19362–19371

Bari MR, Akbar S, Eweida M, Kuhn FJ, Gustafsson AJ, Luckhoff A, Islam MS (2009) H_2O_2-induced Ca^{2+} influx and its inhibition by N-(p-amylcinnamoyl) anthranilic acid in the beta-cells: involvement of TRPM2 channels. J Cell Mol Med 13:3260–3267

Bautista DM, Jordt SE, Nikai T, Tsuruda PR, Read AJ, Poblete J, Yamoah EN, Basbaum AI, Julius D (2006) TRPA1 mediates the inflammatory actions of environmental irritants and proalgesic agents. Cell 124:1269–1282

Beck A, Speicher T, Stoerger C, Sell T, Dettmer V, Jusoh SA, Abdulmughni A, Cavalie A, Philipp SE, Zhu MX, Helms V, Wissenbach U, Flockerzi V (2013) Conserved gating elements in TRPC4 and TRPC5 channels. J Biol Chem 288:19471–19483

Behrendt HJ, Germann T, Gillen C, Hatt H, Jostock R (2004) Characterization of the mouse cold-menthol receptor TRPM8 and vanilloid receptor type-1 VR1 using a fluorometric imaging plate reader (FLIPR) assay. Br J Pharmacol 141:737–745

Bellono NW, Kammel LG, Zimmerman AL, Oancea E (2013) UV light phototransduction activates transient receptor potential A1 ion channels in human melanocytes. Proc Natl Acad Sci U S A 110:2383–2388

Bishnoi M, Bosgraaf CA, Premkumar LS (2011) Preservation of acute pain and efferent functions following intrathecal resiniferatoxin-induced analgesia in rats. J Pain 12:991–1003

Bootman MD, Collins TJ, Mackenzie L, Roderick HL, Berridge MJ, Peppiatt CM (2002) 2-Aminoethoxydiphenyl borate (2-APB) is a reliable blocker of store-operated Ca^{2+} entry but an inconsistent inhibitor of $InsP_3$-induced Ca^{2+} release. FASEB J 16:1145–1150

Bowen CV, DeBay D, Ewart HS, Gallant P, Gormley S, Ilenchuk TT, Iqbal U, Lutes T, Martina M, Mealing G, Merkley N, Sperker S, Moreno MJ, Rice C, Syvitski RT, Stewart JM (2013) *In vivo* detection of human TRPV6-rich tumors with anti-cancer peptides derived from soricidin. PLoS One 8:e58866

Brederson JD, Kym PR, Szallasi A (2013) Targeting TRP channels for pain relief. Eur J Pharmacol 716:61–76

Brown A, Ellis D, Favor DA, Kirkup T, Klute W, Mackenny M, McMurray G, Stennett A (2013) Serendipity in drug-discovery: a new series of 2-(benzyloxy)benzamides as TRPM8 antagonists. Bioorg Med Chem Lett 23:6118–6122

Calvo RR, Meegalla SK, Parks DJ, Parsons WH, Ballentine SK, Lubin ML, Schneider C, Colburn RW, Flores CM, Player MR (2012) Discovery of vinylcycloalkyl-substituted benzimidazole TRPM8 antagonists effective in the treatment of cold allodynia. Bioorg Med Chem Lett 22:1903–1907

Caterina MJ, Schumacher MA, Tominaga M, Rosen TA, Levine JD, Julius D (1997) The capsaicin receptor: a heat-activated ion channel in the pain pathway. Nature 389:816–824

Chakraborty S, Berwick ZC, Bartlett PJ, Kumar S, Thomas AP, Sturek M, Tune JD, Obukhov AG (2011) Bromoenol lactone inhibits voltage-gated Ca^{2+} and transient receptor potential canonical channels. J Pharmacol Exp Ther 339:329–340

Chaudhari SS, Kadam AB, Khairatkar-Joshi N, Mukhopadhyay I, Karnik PV, Raghuram A, Rao SS, Vaiyapuri TS, Wale DP, Bhosale VM, Gudi GS, Sangana RR, Thomas A (2013) Synthesis and pharmacological evaluation of novel N-aryl-3,4-dihydro-1′H-spiro[chromene-2,4′-piperidine]-1′-carboxamides as TRPM8 antagonists. Bioorg Med Chem 21:6542–6553

Chen HC, Xie J, Zhang Z, Su LT, Yue L, Runnels LW (2010a) Blockade of TRPM7 channel activity and cell death by inhibitors of 5-lipoxygenase. PLoS One 5:e11161

Chen X, Numata T, Li M, Mori Y, Orser BA, Jackson MF, Xiong ZG, MacDonald JF (2010b) The modulation of TRPM7 currents by nafamostat mesilate depends directly upon extracellular concentrations of divalent cations. Mol Brain 3:38

Chen J, Kang D, Xu J, Lake M, Hogan JO, Sun C, Walter K, Yao B, Kim D (2013a) Species differences and molecular determinant of TRPA1 cold sensitivity. Nat Commun 4:2501

Chen YF, Chen YT, Chiu WT, Shen MR (2013b) Remodeling of calcium signaling in tumor progression. J Biomed Sci 20:23

Cheng X, Jin J, Hu L, Shen D, Dong XP, Samie MA, Knoff J, Eisinger B, Liu ML, Huang SM, Caterina MJ, Dempsey P, Michael LE, Dlugosz AA, Andrews NC, Clapham DE, Xu H (2010) TRP channel regulates EGFR signaling in hair morphogenesis and skin barrier formation. Cell 141:331–343

Cheng H, Zhang Y, Du C, Dempsey CE, Hancox JC (2012) High potency inhibition of hERG potassium channels by the sodium-calcium exchange inhibitor KB-R7943. Br J Pharmacol 165:2260–2273

Chuang HH, Neuhausser WM, Julius D (2004) The super-cooling agent icilin reveals a mechanism of coincidence detection by a temperature-sensitive TRP channel. Neuron 43:859–869

Chubanov V, Schnitzler M, Meissner M, Schäfer S, Abstiens K, Hofmann T, Gudermann T (2012) Natural and synthetic modulators of SK ($K_{Ca}2$) potassium channels inhibit magnesium-

dependent activity of the kinase-coupled cation channel TRPM7. Br J Pharmacol 166:1357–1376
Chung MK, Lee H, Mizuno A, Suzuki M, Caterina MJ (2004) 2-Aminoethoxydiphenyl borate activates and sensitizes the heat-gated ion channel TRPV3. J Neurosci 24:5177–5182
Davis MI, Hunt JP, Herrgard S, Ciceri P, Wodicka LM, Pallares G, Hocker M, Treiber DK, Zarrinkar PP (2011) Comprehensive analysis of kinase inhibitor selectivity. Nat Biotechnol 29:1046–1051
Demion M, Bois P, Launay P, Guinamard R (2007) TRPM4, a Ca^{2+}-activated nonselective cation channel in mouse sino-atrial node cells. Cardiovasc Res 73:531–538
Derry S, Lloyd R, Moore RA, McQuay HJ (2009) Topical capsaicin for chronic neuropathic pain in adults. Cochrane Database Syst Rev CD007393
Everaerts W, Zhen X, Ghosh D, Vriens J, Gevaert T, Gilbert JP, Hayward NJ, McNamara CR, Xue F, Moran MM, Strassmaier T, Uykal E, Owsianik G, Vennekens R, De RD, Nilius B, Fanger CM, Voets T (2010) Inhibition of the cation channel TRPV4 improves bladder function in mice and rats with cyclophosphamide-induced cystitis. Proc Natl Acad Sci U S A 107:19084–19089
Frederick J, Buck ME, Matson DJ, Cortright DN (2007) Increased TRPA1, TRPM8, and TRPV2 expression in dorsal root ganglia by nerve injury. Biochem Biophys Res Commun 358:1058–1064
Gavva NR, Treanor JJ, Garami A, Fang L, Surapaneni S, Akrami A, Alvarez F, Bak A, Darling M, Gore A, Jang GR, Kesslak JP, Ni L, Norman MH, Palluconi G, Rose MJ, Salfi M, Tan E, Romanovsky AA, Banfield C, Davar G (2008) Pharmacological blockade of the vanilloid receptor TRPV1 elicits marked hyperthermia in humans. Pain 136:202–210
Grand T, Demion M, Norez C, Mettey Y, Launay P, Becq F, Bois P, Guinamard R (2008) 9-Phenanthrol inhibits human TRPM4 but not TRPM5 cationic channels. Br J Pharmacol 153:1697–1705
Grimm C, Kraft R, Schultz G, Harteneck C (2005) Activation of the melastatin-related cation channel $TRPM_3$ by D-*erythro*-sphingosine. Mol Pharmacol 67:798–805
Grubisha O, Mogg AJ, Sorge JL, Ball LJ, Sanger H, Ruble CL, Folly EA, Ursu D, Broad LM (2013) Pharmacological profiling of the TRPV3 channel in recombinant and native assays. Br J Pharmacol, doi:10.1111/bph.12303
Guinamard R, Simard C, Del NC (2013) Flufenamic acid as an ion channel modulator. Pharmacol Ther 138:272–284
Guler AD, Lee H, Iida T, Shimizu I, Tominaga M, Caterina M (2002) Heat-evoked activation of the ion channel, TRPV4. J Neurosci 22:6408–6414
Halaszovich CR, Zitt C, Jüngling E, Lückhoff A (2000) Inhibition of TRP3 channels by lanthanides. Block from the cytosolic side of the plasma membrane. J Biol Chem 275:37423–37428
Harteneck C, Gollasch M (2011) Pharmacological modulation of diacylglycerol-sensitive TRPC3/6/7 channels. Curr Pharm Biotechnol 12:35–41
Harteneck C, Frenzel H, Kraft R (2007) *N*-(p-amylcinnamoyl)anthranilic acid (ACA): a phospholipase A_2 inhibitor and TRP channel blocker. Cardiovasc Drug Rev 25:61–75
Hatano N, Suzuki H, Muraki Y, Muraki K (2013) Stimulation of human TRPA1 channels by clinical concentrations of the antirheumatic drug auranofin. Am J Physiol Cell Physiol 304: C354–C361
Haverstick DM, Heady TN, Macdonald TL, Gray LS (2000) Inhibition of human prostate cancer proliferation in vitro and in a mouse model by a compound synthesized to block Ca^{2+} entry. Cancer Res 60:1002–1008
He LP, Hewavitharana T, Soboloff J, Spassova MA, Gill DL (2005) A functional link between store-operated and TRPC channels revealed by the 3,5-bis(trifluoromethyl)pyrazole derivative, BTP2. J Biol Chem 280:10997–11006
Hill K, Schaefer M (2007) TRPA1 is differentially modulated by the amphipathic molecules trinitrophenol and chlorpromazine. J Biol Chem 282:7145–7153

Hill K, Schaefer M (2009) Ultraviolet light and photosensitising agents activate TRPA1 via generation of oxidative stress. Cell Calcium 45:155–164
Hill K, Benham CD, McNulty S, Randall AD (2004a) Flufenamic acid is a pH-dependent antagonist of TRPM2 channels. Neuropharmacology 47:450–460
Hill K, McNulty S, Randall AD (2004b) Inhibition of TRPM2 channels by the antifungal agents clotrimazole and econazole. Naunyn Schmiedebergs Arch Pharmacol 370:227–237
Hinman A, Chuang HH, Bautista DM, Julius D (2006) TRP channel activation by reversible covalent modification. Proc Natl Acad Sci U S A 103:19564–19568
Hisanaga E, Nagasawa M, Ueki K, Kulkarni RN, Mori M, Kojima I (2009) Regulation of calcium-permeable TRPV2 channel by insulin in pancreatic beta-cells. Diabetes 58:174–184
Hoenderop JG, van der Kemp AW, Hartog A, van de Graaf SF, van Os CH, Willems PH, Bindels RJ (1999) Molecular identification of the apical Ca^{2+} channel in 1, 25-dihydroxyvitamin D_3-responsive epithelia. J Biol Chem 274:8375–8378
Hoenderop JG, Vennekens R, Muller D, Prenen J, Droogmans G, Bindels RJ, Nilius B (2001) Function and expression of the epithelial Ca^{2+} channel family: comparison of mammalian ECaC1 and 2. J Physiol 537:747–761
Hofer A, Kovacs G, Zappatini A, Leuenberger M, Hediger MA, Lochner M (2013) Design, synthesis and pharmacological characterization of analogs of 2-aminoethyl diphenylborinate (2-APB), a known store-operated calcium channel blocker, for inhibition of TRPV6-mediated calcium transport. Bioorg Med Chem 21:3202–3213
Hu HZ, Gu Q, Wang C, Colton CK, Tang J, Kinoshita-Kawada M, Lee LY, Wood JD, Zhu MX (2004) 2-Aminoethoxydiphenyl borate is a common activator of TRPV1, TRPV2, and TRPV3. J Biol Chem 279:35741–35748
Huang SM, Li X, Yu Y, Wang J, Caterina MJ (2011) TRPV3 and TRPV4 ion channels are not major contributors to mouse heat sensation. Mol Pain 7:37
Huh D, Leslie DC, Matthews BD, Fraser JP, Jurek S, Hamilton GA, Thorneloe KS, McAlexander MA, Ingber DE (2012) A human disease model of drug toxicity-induced pulmonary edema in a lung-on-a-chip microdevice. Sci Transl Med 4:159ra147
Hwang SW, Cho H, Kwak J, Lee SY, Kang CJ, Jung J, Cho S, Min KH, Suh YG, Kim D, Oh U (2000) Direct activation of capsaicin receptors by products of lipoxygenases: endogenous capsaicin-like substances. Proc Natl Acad Sci U S A 97:6155–6160
Inoue R, Okada T, Onoue H, Hara Y, Shimizu S, Naitoh S, Ito Y, Mori Y (2001) The transient receptor potential protein homologue TRP6 is the essential component of vascular $alpha_1$-adrenoceptor-activated Ca^{2+}-permeable cation channel. Circ Res 88:325–332
Inoue N, Ito S, Nogawa M, Tajima K, Kyoi T (2012) Etodolac blocks the allyl isothiocyanate-induced response in mouse sensory neurons by selective TRPA1 activation. Pharmacology 90:47–54
Ishikawa J, Ohga K, Yoshino T, Takezawa R, Ichikawa A, Kubota H, Yamada T (2003) A pyrazole derivative, YM-58483, potently inhibits store-operated sustained Ca^{2+} influx and IL-2 production in T lymphocytes. J Immunol 170:4441–4449
Iwamoto T, Watano T, Shigekawa M (1996) A novel isothiourea derivative selectively inhibits the reverse mode of Na^+/Ca^{2+} exchange in cells expressing NCX1. J Biol Chem 271:22391–22397
Jeffry JA, Yu SQ, Sikand P, Parihar A, Evans MS, Premkumar LS (2009) Selective targeting of TRPV1 expressing sensory nerve terminals in the spinal cord for long lasting analgesia. PLoS One 4:e7021
Jordt SE, Bautista DM, Chuang HH, McKemy DD, Zygmunt PM, Hogestatt ED, Meng ID, Julius D (2004) Mustard oils and cannabinoids excite sensory nerve fibres through the TRP channel ANKTM1. Nature 427:260–265
Jung S, Mühle A, Schaefer M, Strotmann R, Schultz G, Plant TD (2003) Lanthanides potentiate TRPC5 currents by an action at extracellular sites close to the pore mouth. J Biol Chem 278:3562–3571

Juvin V, Penna A, Chemin J, Lin YL, Rassendren FA (2007) Pharmacological characterization and molecular determinants of the activation of transient receptor potential V2 channel orthologs by 2-aminoethoxydiphenyl borate. Mol Pharmacol 72:1258–1268

Kaneko Y, Szallasi A (2013) Transient Receptor Potential (TRP) channels: a clinical perspective. Br J Pharmacol, doi:10.1111/bph.12414

Kelley BG, Mermelstein PG (2011) Progesterone blocks multiple routes of ion flux. Mol Cell Neurosci 48:137–141

Kitaguchi T, Swartz KJ (2005) An inhibitor of TRPV1 channels isolated from funnel Web spider venom. Biochemistry 44:15544–15549

Kiyonaka S, Kato K, Nishida M, Mio K, Numaga T, Sawaguchi Y, Yoshida T, Wakamori M, Mori E, Numata T, Ishii M, Takemoto H, Ojida A, Watanabe K, Uemura A, Kurose H, Morii T, Kobayashi T, Sato Y, Sato C, Hamachi I, Mori Y (2009) Selective and direct inhibition of TRPC3 channels underlies biological activities of a pyrazole compound. Proc Natl Acad Sci U S A 106:5400–5405

Klausen TK, Pagani A, Minassi A, Ech-Chahad A, Prenen J, Owsianik G, Hoffmann EK, Pedersen SF, Appendino G, Nilius B (2009) Modulation of the transient receptor potential vanilloid channel TRPV4 by 4α-phorbol esters: a structure-activity study. J Med Chem 52:2933–2939

Klose C, Straub I, Riehle M, Ranta F, Krautwurst D, Ullrich S, Meyerhof W, Harteneck C (2011) Fenamates as TRP channel blockers: mefenamic acid selectively blocks TRPM3. Br J Pharmacol 162:1757–1769

Knowlton WM, Daniels RL, Palkar R, McCoy DD, McKemy DD (2011) Pharmacological blockade of TRPM8 ion channels alters cold and cold pain responses in mice. PLoS One 6: e25894

Kolisek M, Beck A, Fleig A, Penner R (2005) Cyclic ADP-ribose and hydrogen peroxide synergize with ADP-ribose in the activation of TRPM2 channels. Mol Cell 18:61–69

Kraft R (2007) The Na^+/Ca^{2+} exchange inhibitor KB-R7943 potently blocks TRPC channels. Biochem Biophys Res Commun 361:230–236

Kraft R, Grimm C, Frenzel H, Harteneck C (2006) Inhibition of TRPM2 cation channels by *N*-(p-amylcinnamoyl)anthranilic acid. Br J Pharmacol 148:264–273

Krarup AL, Ny L, Astrand M, Bajor A, Hvid-Jensen F, Hansen MB, Simren M, Funch-Jensen P, Drewes AM (2011) Randomised clinical trial: the efficacy of a transient receptor potential vanilloid 1 antagonist AZD1386 in human oesophageal pain. Aliment Pharmacol Ther 33:1113–1122

Lambert S, Drews A, Rizun O, Wagner TF, Lis A, Mannebach S, Plant S, Portz M, Meissner M, Philipp SE, Oberwinkler J (2011) Transient receptor potential melastatin 1 (TRPM1) is an ion-conducting plasma membrane channel inhibited by zinc ions. J Biol Chem 286:12221–12233

Landowski CP, Bolanz KA, Suzuki Y, Hediger MA (2011) Chemical inhibitors of the calcium entry channel TRPV6. Pharm Res 28:322–330

Lashinger ES, Steiginga MS, Hieble JP, Leon LA, Gardner SD, Nagilla R, Davenport EA, Hoffman BE, Laping NJ, Su X (2008) AMTB, a TRPM8 channel blocker: evidence in rats for activity in overactive bladder and painful bladder syndrome. Am J Physiol Renal Physiol 295:F803–F810

Lee YM, Kim BJ, Kim HJ, Yang DK, Zhu MH, Lee KP, So I, Kim KW (2003) TRPC5 as a candidate for the nonselective cation channel activated by muscarinic stimulation in murine stomach. Am J Physiol Gastrointest Liver Physiol 284:G604–G616

Lucas P, Ukhanov K, Leinders-Zufall T, Zufall F (2003) A diacylglycerol-gated cation channel in vomeronasal neuron dendrites is impaired in TRPC2 mutant mice: mechanism of pheromone transduction. Neuron 40:551–561

Majeed Y, Amer MS, Agarwal AK, McKeown L, Porter KE, O'Regan DJ, Naylor J, Fishwick CW, Muraki K, Beech DJ (2011a) Stereo-selective inhibition of transient receptor potential TRPC5 cation channels by neuroactive steroids. Br J Pharmacol 162:1509–1520

Majeed Y, Bahnasi Y, Seymour VA, Wilson LA, Milligan CJ, Agarwal AK, Sukumar P, Naylor J, Beech DJ (2011b) Rapid and contrasting effects of rosiglitazone on transient receptor potential TRPM3 and TRPC5 channels. Mol Pharmacol 79:1023–1030

Majeed Y, Tumova S, Green BL, Seymour VA, Woods DM, Agarwal AK, Naylor J, Jiang S, Picton HM, Porter KE, O'Regan DJ, Muraki K, Fishwick CW, Beech DJ (2012) Pregnenolone sulphate-independent inhibition of TRPM3 channels by progesterone. Cell Calcium 51:1–11

Mandadi S, Sokabe T, Shibasaki K, Katanosaka K, Mizuno A, Moqrich A, Patapoutian A, Fukumi-Tominaga T, Mizumura K, Tominaga M (2009) TRPV3 in keratinocytes transmits temperature information to sensory neurons via ATP. Pflugers Arch 458:1093–1102

Maruyama T, Kanaji T, Nakade S, Kanno T, Mikoshiba K (1997) 2APB, 2-aminoethoxydiphenyl borate, a membrane-penetrable modulator of Ins(1,4,5)P_3-induced Ca^{2+} release. J Biochem 122:498–505

Matthews JM, Qin N, Colburn RW, Dax SL, Hawkins M, McNally JJ, Reany L, Youngman MA, Baker J, Hutchinson T, Liu Y, Lubin ML, Neeper M, Brandt MR, Stone DJ, Flores CM (2012) The design and synthesis of novel, phosphonate-containing transient receptor potential melastatin 8 (TRPM8) antagonists. Bioorg Med Chem Lett 22:2922–2926

McGaraughty S, Chu KL, Perner RJ, Didomenico S, Kort ME, Kym PR (2010) TRPA1 modulation of spontaneous and mechanically evoked firing of spinal neurons in uninjured, osteoarthritic, and inflamed rats. Mol Pain 6:14

McKemy DD, Neuhausser WM, Julius D (2002) Identification of a cold receptor reveals a general role for TRP channels in thermosensation. Nature 416:52–58

McNamara FN, Randall A, Gunthorpe MJ (2005) Effects of piperine, the pungent component of black pepper, at the human vanilloid receptor (TRPV1). Br J Pharmacol 144:781–790

McNamara CR, Mandel-Brehm J, Bautista DM, Siemens J, Deranian KL, Zhao M, Hayward NJ, Chong JA, Julius D, Moran MM, Fanger CM (2007) TRPA1 mediates formalin-induced pain. Proc Natl Acad Sci U S A 104:13525–13530

Miehe S, Crause P, Schmidt T, Lohn M, Kleemann HW, Licher T, Dittrich W, Rutten H, Strübing C (2012) Inhibition of diacylglycerol-sensitive TRPC channels by synthetic and natural steroids. PLoS One 7:e35393

Miller M, Shi J, Zhu Y, Kustov M, Tian JB, Stevens A, Wu M, Xu J, Long S, Yang P, Zholos AV, Salovich JM, Weaver CD, Hopkins CR, Lindsley CW, McManus O, Li M, Zhu MX (2011a) Identification of ML204, a novel potent antagonist that selectively modulates native TRPC4/C5 ion channels. J Biol Chem 286:33436–33446

Miller M, Wu M, Xu J, Weaver D, Li M, Zhu MX (2011b) High-throughput screening of TRPC channel ligands using cell-based assays. CRC, TRP Channels. Boca Raton, FL

Moqrich A, Hwang SW, Earley TJ, Petrus MJ, Murray AN, Spencer KS, Andahazy M, Story GM, Patapoutian A (2005) Impaired thermosensation in mice lacking TRPV3, a heat and camphor sensor in the skin. Science 307:1468–1472

Morgans CW, Zhang J, Jeffrey BG, Nelson SM, Burke NS, Duvoisin RM, Brown RL (2009) TRPM1 is required for the depolarizing light response in retinal ON-bipolar cells. Proc Natl Acad Sci U S A 106:19174–19178

Mou J, Paillard F, Turnbull B, Trudeau J, Stoker M, Katz NP (2013) Efficacy of Qutenza (capsaicin) 8 % patch for neuropathic pain: a meta-analysis of the Qutenza Clinical Trials Database. Pain 154:1632–1639

Moussaieff A, Rimmerman N, Bregman T, Straiker A, Felder CC, Shoham S, Kashman Y, Huang SM, Lee H, Shohami E, Mackie K, Caterina MJ, Walker JM, Fride E, Mechoulam R (2008) Incensole acetate, an incense component, elicits psychoactivity by activating TRPV3 channels in the brain. FASEB J 22:3024–3034

Mutoh T, Taki Y, Tsubone H (2013) Desflurane but not sevoflurane augments laryngeal C-fiber inputs to nucleus tractus solitarii neurons by activating transient receptor potential-A1. Life Sci 92:821–828

Naylor J, Beech DJ (2013) Generation of antibodies that are externally acting isoform-specific inhibitors of ion channels. Methods Mol Biol 998:245–256

Naylor J, Al-Shawaf E, McKeown L, Manna PT, Porter KE, O'Regan D, Muraki K, Beech DJ (2011) TRPC5 channel sensitivities to antioxidants and hydroxylated stilbenes. J Biol Chem 286:5078–5086

Nilius B, Voets T (2013) The puzzle of TRPV4 channelopathies. EMBO Rep 14:152–163

Nilius B, Prenen J, Vennekens R, Hoenderop JG, Bindels RJ, Droogmans G (2001) Pharmacological modulation of monovalent cation currents through the epithelial Ca^{2+} channel ECaC1. Br J Pharmacol 134:453–462

Nilius B, Droogmans G, Wondergem R (2003) Transient receptor potential channels in endothelium: solving the calcium entry puzzle? Endothelium 10:5–15

Nyman E, Franzen B, Nolting A, Klement G, Liu G, Nilsson M, Rosen A, Bjork C, Weigelt D, Wollberg P, Karila P, Raboisson P (2013) In vitro pharmacological characterization of a novel TRPA1 antagonist and proof of mechanism in a human dental pulp model. J Pain Res 6:59–70

Okada T, Inoue R, Yamazaki K, Maeda A, Kurosaki T, Yamakuni T, Tanaka I, Shimizu S, Ikenaka K, Imoto K, Mori Y (1999) Molecular and functional characterization of a novel mouse transient receptor potential protein homologue TRP7. J Biol Chem 274:27359–27370

Othman AA, Nothaft W, Awni WM, Dutta S (2013) Effects of the TRPV1 antagonist ABT-102 on body temperature in healthy volunteers: pharmacokinetic/pharmacodynamic analysis of three phase 1 trials. Br J Clin Pharmacol 75:1029–1040

Otsuguro K, Tang J, Tang Y, Xiao R, Freichel M, Tsvilovskyy V, Ito S, Flockerzi V, Zhu MX, Zholos AV (2008) Isoform-specific inhibition of TRPC4 channel by phosphatidylinositol 4,5-bisphosphate. J Biol Chem 283:10026–10036

Palmer RK, Atwal K, Bakaj I, Carlucci-Derbyshire S, Buber MT, Cerne R, Cortes RY, Devantier HR, Jorgensen V, Pawlyk A, Lee SP, Sprous DG, Zhang Z, Bryant R (2010) Triphenylphosphine oxide is a potent and selective inhibitor of the transient receptor potential melastatin-5 ion channel. Assay Drug Dev Technol 8:703–713

Park CK, Xu ZZ, Liu T, Lu N, Serhan CN, Ji RR (2011) Resolvin D2 is a potent endogenous inhibitor for transient receptor potential subtype V1/A1, inflammatory pain, and spinal cord synaptic plasticity in mice: distinct roles of resolvin D1, D2, and E1. J Neurosci 31:18433–18438

Parks DJ, Parsons WH, Colburn RW, Meegalla SK, Ballentine SK, Illig CR, Qin N, Liu Y, Hutchinson TL, Lubin ML, Stone DJ Jr, Baker JF, Schneider CR, Ma J, Damiano BP, Flores CM, Player MR (2011) Design and optimization of benzimidazole-containing transient receptor potential melastatin 8 (TRPM8) antagonists. J Med Chem 54:233–247

Parnas M, Peters M, Dadon D, Lev S, Vertkin I, Slutsky I, Minke B (2009) Carvacrol is a novel inhibitor of Drosophila TRPL and mammalian TRPM7 channels. Cell Calcium 45:300–309

Partida-Sanchez S, Gasser A, Fliegert R, Siebrands CC, Dammermann W, Shi G, Mousseau BJ, Sumoza-Toledo A, Bhagat H, Walseth TF, Guse AH, Lund FE (2007) Chemotaxis of mouse bone marrow neutrophils and dendritic cells is controlled by adp-ribose, the major product generated by the CD38 enzyme reaction. J Immunol 179:7827–7839

Peier AM, Moqrich A, Hergarden AC, Reeve AJ, Andersson DA, Story GM, Earley TJ, Dragoni I, McIntyre P, Bevan S, Patapoutian A (2002) A TRP channel that senses cold stimuli and menthol. Cell 108:705–715

Peng JB, Chen XZ, Berger UV, Vassilev PM, Tsukaguchi H, Brown EM, Hediger MA (1999) Molecular cloning and characterization of a channel-like transporter mediating intestinal calcium absorption. J Biol Chem 274:22739–22746

Perraud AL, Fleig A, Dunn CA, Bagley LA, Launay P, Schmitz C, Stokes AJ, Zhu Q, Bessman MJ, Penner R, Kinet JP, Scharenberg AM (2001) ADP-ribose gating of the calcium-permeable LTRPC2 channel revealed by Nudix motif homology. Nature 411:595–599

Petrus M, Peier AM, Bandell M, Hwang SW, Huynh T, Olney N, Jegla T, Patapoutian A (2007) A role of TRPA1 in mechanical hyperalgesia is revealed by pharmacological inhibition. Mol Pain 3:40

Phan MN, Leddy HA, Votta BJ, Kumar S, Levy DS, Lipshutz DB, Lee SH, Liedtke W, Guilak F (2009) Functional characterization of TRPV4 as an osmotically sensitive ion channel in porcine articular chondrocytes. Arthritis Rheum 60:3028–3037

Premkumar LS, Abooj M (2013) TRP channels and analgesia. Life Sci 92:415–424

Qin N, Neeper MP, Liu Y, Hutchinson TL, Lubin ML, Flores CM (2008) TRPV2 is activated by cannabidiol and mediates CGRP release in cultured rat dorsal root ganglion neurons. J Neurosci 28:6231–6238

Qin X, Yue Z, Sun B, Yang W, Xie J, Ni E, Feng Y, Mahmood R, Zhang Y, Yue L (2013) Sphingosine and FTY720 are potent inhibitors of the transient receptor potential melastatin 7 (TRPM7) channels. Br J Pharmacol 168:1294–1312

Ramachandran R, Hyun E, Zhao L, Lapointe TK, Chapman K, Hirota CL, Ghosh S, McKemy DD, Vergnolle N, Beck PL, Altier C, Hollenberg MD (2013) TRPM8 activation attenuates inflammatory responses in mouse models of colitis. Proc Natl Acad Sci U S A 110:7476–7481

Ryckmans T, Aubdool AA, Bodkin JV, Cox P, Brain SD, Dupont T, Fairman E, Hashizume Y, Ishii N, Kato T, Kitching L, Newman J, Omoto K, Rawson D, Strover J (2011) Design and pharmacological evaluation of PF-4840154, a non-electrophilic reference agonist of the TrpA1 channel. Bioorg Med Chem Lett 21:4857–4859

Saleh SN, Albert AP, Large WA (2009) Activation of native TRPC1/C5/C6 channels by endothelin-1 is mediated by both PIP_3 and PIP_2 in rabbit coronary artery myocytes. J Physiol 587:5361–5375

Schattling B, Steinbach K, Thies E, Kruse M, Menigoz A, Ufer F, Flockerzi V, Bruck W, Pongs O, Vennekens R, Kneussel M, Freichel M, Merkler D, Friese MA (2012) TRPM4 cation channel mediates axonal and neuronal degeneration in experimental autoimmune encephalomyelitis and multiple sclerosis. Nat Med 18:1805–1811

Schleifer H, Doleschal B, Lichtenegger M, Oppenrieder R, Derler I, Frischauf I, Glasnov TN, Kappe CO, Romanin C, Groschner K (2012) Novel pyrazole compounds for pharmacological discrimination between receptor-operated and store-operated Ca^{2+} entry pathways. Br J Pharmacol 167:1712–1722

Schulze A, Oehler B, Urban N, Schaefer M, Hill K (2013) Apomorphine is a bimodal modulator of TRPA1 channels. Mol Pharmacol 83:542–551

Selescu T, Ciobanu AC, Dobre C, Reid G, Babes A (2013) Camphor activates and sensitizes transient receptor potential melastatin 8 (TRPM8) to cooling and icilin. Chem Senses 38:563–575

Seo K, Rainer PP, Shalkey Hahn V, Lee DI, Jo SH, Andersen A, Liu T, Xu X, Willette RN, Lepore JJ, Marino JP Jr, Birnbaumer L, Schnackenberg CG, Kass DA (2014) Combined TRPC3 and TRPC6 blockade by selective small-molecule or genetic deletion inhibits pathological cardiac hypertrophy. Proc Natl Acad Sci USA 111:1551–1556, http://www.ncbi.nlm.nih.gov/pubmed/24453217

Shen Y, Heimel JA, Kamermans M, Peachey NS, Gregg RG, Nawy S (2009) A transient receptor potential-like channel mediates synaptic transmission in rod bipolar cells. J Neurosci 29:6088–6093

Sherkheli MA, Vogt-Eisele AK, Bura D, Beltran Marques LR, Gisselmann G, Hatt H (2010) Characterization of selective TRPM8 ligands and their structure activity response (S.A.R) relationship. J Pharm Pharm Sci 13:242–253

Singh A, Hildebrand ME, Garcia E, Snutch TP (2010) The transient receptor potential channel antagonist SKF96365 is a potent blocker of low-voltage-activated T-type calcium channels. Br J Pharmacol 160:1464–1475

Smart D, Gunthorpe MJ, Jerman JC, Nasir S, Gray J, Muir AI, Chambers JK, Randall AD, Davis JB (2000) The endogenous lipid anandamide is a full agonist at the human vanilloid receptor (hVR1). Br J Pharmacol 129:227–230

Smith PL, Maloney KN, Pothen RG, Clardy J, Clapham DE (2006) Bisandrographolide from *Andrographis paniculata* activates TRPV4 channels. J Biol Chem 281:29897–29904

Song Y, Buelow B, Perraud AL, Scharenberg AM (2008) Development and validation of a cell-based high-throughput screening assay for TRPM2 channel modulators. J Biomol Screen 13:54–61

Story GM, Peier AM, Reeve AJ, Eid SR, Mosbacher J, Hricik TR, Earley TJ, Hergarden AC, Andersson DA, Hwang SW, McIntyre P, Jegla T, Bevan S, Patapoutian A (2003) ANKTM1, a TRP-like channel expressed in nociceptive neurons, is activated by cold temperatures. Cell 112:819–829

Strassmaier T, Bakthavatchalam R (2011) Transient receptor potential A1 modulators. Curr Top Med Chem 11:2227–2236

Straub I, Krügel U, Mohr F, Teichert J, Rizun O, Konrad M, Oberwinkler J, Schaefer M (2013a) Flavanones that selectively inhibit TRPM3 attenuate thermal nociception *in vivo*. Mol Pharmacol 84:736–750

Straub I, Mohr F, Stab J, Konrad M, Philipp SE, Oberwinkler J, Schaefer M (2013b) Citrus fruit and fabaceae secondary metabolites potently and selectively block TRPM3. Br J Pharmacol 168:1835–1850

Strotmann R, Harteneck C, Nunnenmacher K, Schultz G, Plant TD (2000) OTRPC4, a nonselective cation channel that confers sensitivity to extracellular osmolarity. Nat Cell Biol 2:695–702

Strübing C, Krapivinsky G, Krapivinsky L, Clapham DE (2001) TRPC1 and TRPC5 form a novel cation channel in mammalian brain. Neuron 29:645–655

Sukumar P, Beech DJ (2010) Stimulation of TRPC5 cationic channels by low micromolar concentrations of lead ions (Pb^{2+}). Biochem Biophys Res Commun 393:50–54

Surowy CS, Neelands TR, Bianchi BR, McGaraughty S, El KR, Han P, Chu KL, McDonald HA, Vos M, Niforatos W, Bayburt EK, Gomtsyan A, Lee CH, Honore P, Sullivan JP, Jarvis MF, Faltynek CR (2008) (R)-(5-tert-butyl-2,3-dihydro-1H-inden-1-yl)-3-(1H-indazol-4-yl)-urea (ABT-102) blocks polymodal activation of transient receptor potential vanilloid 1 receptors in vitro and heat-evoked firing of spinal dorsal horn neurons in vivo. J Pharmacol Exp Ther 326:879–888

Takezawa R, Cheng H, Beck A, Ishikawa J, Launay P, Kubota H, Kinet JP, Fleig A, Yamada T, Penner R (2006) A pyrazole derivative potently inhibits lymphocyte Ca^{2+} influx and cytokine production by facilitating transient receptor potential melastatin 4 channel activity. Mol Pharmacol 69:1413–1420

Tamayo NA, Bo Y, Gore V, Ma V, Nishimura N, Tang P, Deng H, Klionsky L, Lehto SG, Wang W, Youngblood B, Chen J, Correll TL, Bartberger MD, Gavva NR, Norman MH (2012) Fused piperidines as a novel class of potent and orally available transient receptor potential melastatin type 8 (TRPM8) antagonists. J Med Chem 55:1593–1611

Than JY, Li L, Hasan R, Zhang X (2013) Excitation and modulation of TRPA1, TRPV1, and TRPM8 channel-expressing sensory neurons by the pruritogen chloroquine. J Biol Chem 288:12818–12827

Thorneloe KS, Sulpizio AC, Lin Z, Figueroa DJ, Clouse AK, McCafferty GP, Chendrimada TP, Lashinger ES, Gordon E, Evans L, Misajet BA, Demarini DJ, Nation JH, Casillas LN, Marquis RW, Votta BJ, Sheardown SA, Xu X, Brooks DP, Laping NJ, Westfall TD (2008) N-((1S)-1-{[4-((2S)-2-{[(2,4-dichlorophenyl)sulfonyl]amino}-3-hydroxypropanoyl)-1-piperazinyl]carbonyl}-3-methylbutyl)-1-benzothiophene-2-carboxamide (GSK1016790A), a novel and potent transient receptor potential vanilloid 4 channel agonist induces urinary bladder contraction and hyperactivity: Part I. J Pharmacol Exp Ther 326:432–442

Thorneloe KS, Cheung M, Bao W, Alsaid H, Lenhard S, Jian MY, Costell M, Maniscalco-Hauk K, Krawiec JA, Olzinski A, Gordon E, Lozinskaya I, Elefante L, Qin P, Matasic DS, James C, Tunstead J, Donovan B, Kallal L, Waszkiewicz A, Vaidya K, Davenport EA, Larkin J, Burgert M, Casillas LN, Marquis RW, Ye G, Eidam HS, Goodman KB, Toomey JR, Roethke TJ, Jucker BM, Schnackenberg CG, Townsley MI, Lepore JJ, Willette RN (2012) An orally active TRPV4 channel blocker prevents and resolves pulmonary edema induced by heart failure. Sci Transl Med 4:159ra148

Trebak M, Lemonnier L, DeHaven WI, Wedel BJ, Bird GS, Putney JW Jr (2009) Complex functions of phosphatidylinositol 4,5-bisphosphate in regulation of TRPC5 cation channels. Pflugers Arch 457:757–769

Trevisani M, Siemens J, Materazzi S, Bautista DM, Nassini R, Campi B, Imamachi N, Andre E, Patacchini R, Cottrell GS, Gatti R, Basbaum AI, Bunnett NW, Julius D, Geppetti P (2007) 4-Hydroxynonenal, an endogenous aldehyde, causes pain and neurogenic inflammation through activation of the irritant receptor TRPA1. Proc Natl Acad Sci U S A 104:13519–13524

Urban L, Dray A (1991) Capsazepine, a novel capsaicin antagonist, selectively antagonises the effects of capsaicin in the mouse spinal cord *in vitro*. Neurosci Lett 134:9–11

Urban N, Hill K, Wang L, Kuebler WM, Schaefer M (2012) Novel pharmacological TRPC inhibitors block hypoxia-induced vasoconstriction. Cell Calcium 51:194–206

Valenzano KJ, Grant ER, Wu G, Hachicha M, Schmid L, Tafesse L, Sun Q, Rotshteyn Y, Francis J, Limberis J, Malik S, Whittemore ER, Hodges D (2003) N-(4-tertiarybutylphenyl)-4-(3-chloropyridin-2-yl)tetrahydropyrazine-1(2H)-carboxamide (BCTC), a novel, orally effective vanilloid receptor 1 antagonist with analgesic properties: I. *In vitro* characterization and pharmacokinetic properties. J Pharmacol Exp Ther 306:377–386

Vincent F, Duncton MA (2011) TRPV4 agonists and antagonists. Curr Top Med Chem 11:2216–2226

Vincent F, Acevedo A, Nguyen MT, Dourado M, DeFalco J, Gustafson A, Spiro P, Emerling DE, Kelly MG, Duncton MA (2009) Identification and characterization of novel TRPV4 modulators. Biochem Biophys Res Commun 389:490–494

Vriens J, Appendino G, Nilius B (2009) Pharmacology of vanilloid transient receptor potential cation channels. Mol Pharmacol 75:1262–1279

Vriens J, Owsianik G, Hofmann T, Philipp SE, Stab J, Chen X, Benoit M, Xue F, Janssens A, Kerselaers S, Oberwinkler J, Vennekens R, Gudermann T, Nilius B, Voets T (2011) TRPM3 is a nociceptor channel involved in the detection of noxious heat. Neuron 70:482–494

Wagner TF, Loch S, Lambert S, Straub I, Mannebach S, Mathar I, Dufer M, Lis A, Flockerzi V, Philipp SE, Oberwinkler J (2008) Transient receptor potential M3 channels are ionotropic steroid receptors in pancreatic beta cells. Nat Cell Biol 10:1421–1430

Wahl P, Foged C, Tullin S, Thomsen C (2001) Iodo-resiniferatoxin, a new potent vanilloid receptor antagonist. Mol Pharmacol 59:9–15

Walker RL, Koh SD, Sergeant GP, Sanders KM, Horowitz B (2002) TRPC4 currents have properties similar to the pacemaker current in interstitial cells of Cajal. Am J Physiol Cell Physiol 283:C1637–C1645

Washburn DG, Holt DA, Dodson J, McAtee JJ, Terrell LR, Barton L, Manns S, Waszkiewicz A, Pritchard C, Gillie DJ, Morrow DM, Davenport EA, Lozinskaya IM, Guss J, Basilla JB, Negron LK, Klein M, Willette RN, Fries RE, Jensen TC, Xu X, Schnackenberg CG, Marino JP Jr (2013) The discovery of potent blockers of the canonical transient receptor channels, TRPC3 and TRPC6, based on an anilino-thiazole pharmacophore. Bioorg Med Chem Lett 23:4979–4984

Watanabe H, Davis JB, Smart D, Jerman JC, Smith GD, Hayes P, Vriens J, Cairns W, Wissenbach U, Prenen J, Flockerzi V, Droogmans G, Benham CD, Nilius B (2002) Activation of TRPV4 channels (hVRL-2/mTRP12) by phorbol derivatives. J Biol Chem 277:13569–13577

Watanabe H, Vriens J, Prenen J, Droogmans G, Voets T, Nilius B (2003) Anandamide and arachidonic acid use epoxyeicosatrienoic acids to activate TRPV4 channels. Nature 424:434–438

Wei L, Freichel M, Jaspers M, Cuppens H, Cassiman JJ, Droogmans G, Flockerzi V, Nilius B (2001) Functional interaction between TRP4 and CFTR in mouse aorta endothelial cells. BMC Physiol 1:3

Wei H, Hamalainen MM, Saarnilehto M, Koivisto A, Pertovaara A (2009) Attenuation of mechanical hypersensitivity by an antagonist of the TRPA1 ion channel in diabetic animals. Anesthesiology 111:147–154

Willette RN, Bao W, Nerurkar S, Yue TL, Doe CP, Stankus G, Turner GH, Ju H, Thomas H, Fishman CE, Sulpizio A, Behm DJ, Hoffman S, Lin Z, Lozinskaya I, Casillas LN, Lin M, Trout RE, Votta BJ, Thorneloe K, Lashinger ES, Figueroa DJ, Marquis R, Xu X (2008) Systemic activation of the transient receptor potential vanilloid subtype 4 channel causes endothelial failure and circulatory collapse: Part 2. J Pharmacol Exp Ther 326:443–452

Woo SK, Kwon MS, Ivanov A, Gerzanich V, Simard JM (2013) The sulfonylurea receptor 1 (Sur1)-transient receptor potential melastatin 4 (Trpm4) channel. J Biol Chem 288:3655–3667

Wood JN, Winter J, James IF, Rang HP, Yeats J, Bevan S (1988) Capsaicin-induced ion fluxes in dorsal root ganglion cells in culture. J Neurosci 8:3208–3220

Worth H, Dethlefsen U (2012) Patients with asthma benefit from concomitant therapy with cineole: a placebo-controlled, double-blind trial. J Asthma 49:849–853

Worth H, Schacher C, Dethlefsen U (2009) Concomitant therapy with Cineole (Eucalyptole) reduces exacerbations in COPD: a placebo-controlled double-blind trial. Respir Res 10:69

Wu J, Shih HP, Vigont V, Hrdlicka L, Diggins L, Singh C, Mahoney M, Chesworth R, Shapiro G, Zimina O, Chen X, Wu Q, Glushankova L, Ahlijanian M, Koenig G, Mozhayeva GN, Kaznacheyeva E, Bezprozvanny I (2011) Neuronal store-operated calcium entry pathway as a novel therapeutic target for Huntington's disease treatment. Chem Biol 18:777–793

Xu SZ, Zeng F, Boulay G, Grimm C, Harteneck C, Beech DJ (2005a) Block of TRPC5 channels by 2-aminoethoxydiphenyl borate: a differential, extracellular and voltage-dependent effect. Br J Pharmacol 145:405–414

Xu SZ, Zeng F, Lei M, Li J, Gao B, Xiong C, Sivaprasadarao A, Beech DJ (2005b) Generation of functional ion-channel tools by E3 targeting. Nat Biotechnol 23:1289–1293

Xu H, Delling M, Jun JC, Clapham DE (2006) Oregano, thyme and clove-derived flavors and skin sensitizers activate specific TRP channels. Nat Neurosci 9:628–635

Xu SZ, Zeng B, Daskoulidou N, Chen GL, Atkin SL, Lukhele B (2012) Activation of TRPC cationic channels by mercurial compounds confers the cytotoxicity of mercury exposure. Toxicol Sci 125:56–68

Yu L, Jin X, Cui N, Wu Y, Shi Z, Zhu D, Jiang C (2012) Rosiglitazone selectively inhibits K_{ATP} channels by acting on the K_{IR} 6 subunit. Br J Pharmacol 167:26–36

Zhao M, Isami K, Nakamura S, Shirakawa H, Nakagawa T, Kaneko S (2012) Acute cold hypersensitivity characteristically induced by oxaliplatin is caused by the enhanced responsiveness of TRPA1 in mice. Mol Pain 8:55

Zhu B, Xia M, Xu X, Ludovici DW, Tennakoon M, Youngman MA, Matthews JM, Dax SL, Colburn RW, Qin N, Hutchinson TL, Lubin ML, Brandt MR, Stone DJ, Flores CM, Macielag MJ (2013) Arylglycine derivatives as potent transient receptor potential melastatin 8 (TRPM8) antagonists. Bioorg Med Chem Lett 23:2234–2237

Zierler S, Yao G, Zhang Z, Kuo WC, Porzgen P, Penner R, Horgen FD, Fleig A (2011) Waixenicin A inhibits cell proliferation through magnesium-dependent block of transient receptor potential melastatin 7 (TRPM7) channels. J Biol Chem 286:39328–39335

TRP Channels in Reproductive (Neuro)Endocrinology

Trese Leinders-Zufall and Ulrich Boehm

Contents

Abstract

Transient receptor potential (TRP) ion channels have been detected in neurons that are part of the neural network controlling reproductive physiology and behavior. In this chapter we will primarily take a look at the classical/canonical TRP (TRPC) channels but will also examine some other members of the TRP channel superfamily in reproductive (neuro)endocrinology. The referenced data suggest that different TRP proteins could play functional roles at different levels of the reproductive pathway. Still, our understanding of TRP channel involvement in (neuro)endocrinology is quite limited. Due to their mechanism of activation and complex regulation, these channels are however ideally suited to be part of the transduction machinery of hormone-secreting cells.

T. Leinders-Zufall (✉)
Department of Physiology, University of Saarland School of Medicine, 66421 Homburg, Germany
e-mail: trese.leinders@uks.eu

U. Boehm
Department of Pharmacology and Toxicology, University of Saarland School of Medicine, 66421 Homburg, Germany
e-mail: ulrich.boehm@uks.eu

B. Nilius and V. Flockerzi (eds.), *Mammalian Transient Receptor Potential (TRP) Cation Channels*, Handbook of Experimental Pharmacology 223, DOI 10.1007/978-3-319-05161-1_16,

Keywords
GnRH • Kisspeptin • Hormone • Reproduction • Hypothalamus • Pituitary

1 Introduction

The hypothalamus–pituitary–gonadal (hpg) axis ensures central control of reproduction in vertebrates (Gore 2002). Starting with the onset of puberty, gonadotropin-releasing hormone (GnRH) neurons in the anterior hypothalamus release GnRH in short pulses into the hypophyseal portal vasculature from axon terminals at the median eminence. Pulsatile GnRH release increases in frequency and amplitude during the female reproductive cycle (Sisk et al. 2001). GnRH binds to the GnRH receptor (GnRHR) specifically expressed on gonadotrope cells in the anterior pituitary gland to regulate production and release of the two gonadotropins, luteinizing hormone (LH) and follicle-stimulating hormone (FSH) (Jeong and Kaiser 2006). LH and FSH in turn act on the gonads to regulate steroidogenesis and reproductive function. Steroid hormones, e.g., estradiol, progesterone, and testosterone, produced by the gonads feed back onto both the pituitary and the hypothalamus.

To ensure reproductive success, GnRH neurons need to integrate different inputs indicating optimal conditions for reproduction (Fig. 1). GnRH neurons are part of a complex neural network and are connected to many different neurons in a variety of brain areas (Boehm et al. 2005; Yoon et al. 2005; Herbison 2006). Signals known to modulate the reproductive axis range from external chemosensory cues such as pheromones (Chamero et al. 2012) to metabolic hormones such as leptin (Chehab 1996) and insulin (Bruning et al. 2000). How GnRH neurons integrate these different inputs is not well understood. Specifically the molecular and ionic mechanisms underlying the physiological activation of GnRH neurons and their adjustment to different hormonal stages are just beginning to emerge. Recently, transient receptor potential (TRP) ion channels have been identified in GnRH neurons and in other cells of the neural network controlling reproductive physiology and behavior. Canonical or classic TRP channels (TRPC channels) are one of the three TRP subfamilies known to mediate receptor-operated and store-operated Ca^{2+} entry and are divided into four subfamilies (TRPC1, TRPC2, TRPC4/5, and TRPC3/6/7, respectively) on the basis of sequence homology and functional similarities (Venkatachalam and Montell 2007; Clapham et al. 2001). These channels are ideally suited to be part of the transduction machinery of hormone-secreting cells due to their mechanism of activation and complex regulation.

2 TRPC Channel Distribution in the Brain

TRPC channel expression in the central nervous system (CNS) is generally not well documented and even less so for specific nuclei, let alone individual neuronal populations in heterogeneous brain regions like the hypothalamus. This lack of

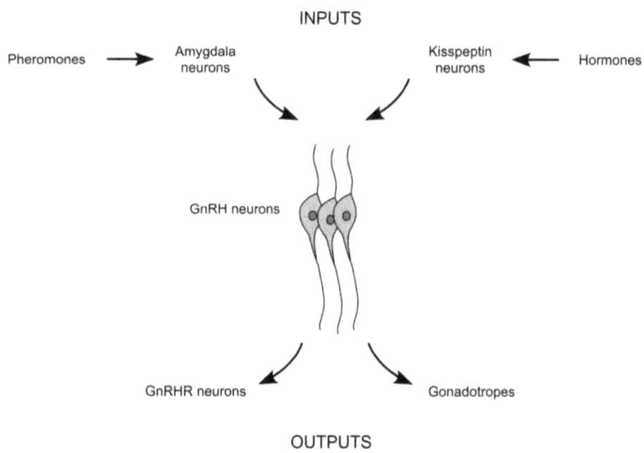

Fig. 1 GnRH neural circuitry. GnRH neurons integrate different inputs to optimize reproductive success. Input signals range from external chemosensory cues such as pheromones to internal hormonal cues. Secreted GnRH binds to the GnRH receptor (GnRHR) expressed on gonadotropes in the anterior pituitary gland to regulate production and release of gonadotropins. In addition to impinging onto the hypothalamus-pituitary-gonadal axis, GnRH-secreting fibers contact GnRH receptor neurons (GnRHR neurons) in various different brain areas (Wen et al. 2011)

information is partly due to low expression levels and to the scarcity of available antisera specific for these channels. Furthermore, most of the commercial antisera against TRPC channels have not yet been validated in tissues prepared from the corresponding TRPC knockout mice. Using quantitative RT-PCR, human TRPC (hTRPC) mRNA expression has been observed in discrete brain areas (Riccio et al. 2002). *hTRPC1* mRNA was found to be ubiquitously expressed with the highest expression levels reported in the cerebellum and also in the pituitary gland. *hTRPC3* mRNA was mainly localized to the basal ganglia as well as to the pituitary. *hTRPC4* and *hTRPC6* mRNA was detected at low levels all over the brain, whereas *hTRPC5* mRNA was found to be highly expressed in all brain areas except for the basal ganglia. *hTRPC7* mRNA expression seemed to occur mainly in the pituitary. In rat, TRPC4 and TRPC5 channels seem to be the most prominently expressed TRP channels in the brain (Fowler et al. 2007). TRPC5 expression in the hypothalamus was also confirmed in mice (Riccio et al. 2009) and is independently documented in the Allen Brain Atlas (http://mouse.brain-map.org). Thus far, only one study demonstrating TRPC5 immunostaining in the lateral amygdala used corresponding tissue from a TRPC5-deficient mouse strain as a control for the specificity of the immunocytochemistry (Riccio et al. 2009). Information about the cellular resolution of TRPC mRNA and protein expression is essentially missing in non-layered multifunctional brain areas such as the hypothalamus, where different nuclei contain multiple distinct hormone-secreting neuronal populations. In addition, it is often unclear which cell type expresses the TRP channels: glia, neurons, tanycytes, or even elements of the CNS vascular system. Therefore, determining the distribution of TRP channels with single cell resolution, in particular in brain areas, which are of pivotal importance to the homeostatic regulation of different body functions via the neuroendocrine system, could be a first but fundamental step in understanding their physiological function.

3 TRP Channel Expression in (Neuro)Endocrine Cancer Cell Lines

The presence of TRP channels in several immortalized cell lines commonly used in reproductive physiology research has been described. A major limitation of these cell lines is that the available models are almost invariably poor representations of cells found in intact tissues. Therefore, it is not always clear whether TRP channel expression in these cells is indeed real and reflects a physiological role in vivo or whether they express TRP channels because of their genetic reprogramming causing an infinite number of cell divisions. In addition, cancer-derived cells contain various genomic rearrangements possibly leading to chromosome instability (Landry et al. 2013). The characteristics of immortalized and cancer-derived cells can thus change with every passage, but most authors do not indicate this information in their publication, which may cause problems in comparing or verifying the obtained results. In addition to being quite valuable in cancer research, the major benefit of using cell lines is that they offer an infinite supply of cells that are capable of self-replication in standard cell culture medium compared to finding a limited number of neurons which are often hard to identify in multifunctional organs.

All investigated TRPC channels (i.e., TRPC1, C3, C4, C5, and C6) have been detected in both the GN11 and the GT1-7 cell lines (Dalmazzo et al. 2008; Krsmanovic et al. 1999; Ariano et al. 2011). Both cell lines have been used as models to investigate GnRH neuron cell biology. While GT1-7 cells are thought to represent mature hypothalamic GnRH-secreting neurons, GN11 cells are more immature and display some migratory activity. Perhaps not surprisingly, these cell line data are in stark contrast to a study analyzing TRP channel expression in genetically tagged primary GnRH neurons prepared from a transgenic mouse strain, in which these neurons can be identified due to their simultaneous expression of GFP. Primary GnRH neurons were found to contain different combinations of TRPC channel transcripts. These data suggest that different subpopulations of GnRH neurons express individual combinations of TRPC channels (e.g., C1/C4/C6 or C3/C4/C7) (Zhang et al. 2008). Similar combinatorial TRP channel expression patterns were also found in other native neurons (Sergeeva et al. 2003; Qiu et al. 2010, 2011). Unfortunately, TRPC2 expression in GnRH neurons was not analyzed in this study (see below).

Further down in the reproductive pathway, Ca^{2+}-activated nonselective cationic currents that could be mediated by TRP channels have also been described in gonadotrope cells of the anterior pituitary gland (Vergara et al. 1997). Information about signaling in primary gonadotropes is still somewhat limited. Previous experiments to characterize GnRH-/GnRHR-induced signal transduction mechanisms were largely carried out in gonadotrope-like carcinoma cell lines, since it was not possible to study living primary gonadotrope cells in situ until recently (Wen et al. 2008). These account for only <15 % of the hormone-secreting cells of the anterior pituitary and are morphologically not easy to distinguish from other cell types. The mRNA for TRPC1, TRPC3, TRPC5, and TRPC7 has been identified in human pituitary cells (Riccio et al. 2002). TRP channel expression in primary gonadotropes, which have become amenable for isolation via FACs (Hoivik et al. 2011), awaits experimental analysis at different hormonal stages.

4 TRPC Function in the Central Control of Mammalian Reproduction

GnRH neurons receive information about olfactory signals both generated in the vomeronasal organ (VNO) and in main olfactory epithelium (Boehm et al. 2005; Yoon et al. 2005). Processing of chemosensory cues in the nose relies on TRP channel function at various levels (Munger et al. 2009, Zufall 2014). Of particular interest is a mouse strain deficient for TRPC2, which is predominately expressed in sensory neurons of the VNO (Leypold et al. 2002; Stowers et al. 2002; Liman et al. 1999). These cells are thought to be specialized in the detection of pheromones. *Trpc2−/−* mice show some striking alterations in a range of social behaviors including loss of male-male and maternal aggression, defective territorial marking, and enhanced courtship behaviors of mutant males toward other males (Stowers et al. 2002; Leypold et al. 2002). Furthermore, TRPC2 deletion in female mice caused male-like sexual and courtship behaviors (Kimchi et al. 2007). These data have been interpreted to indicate that chemosensory inputs act in wild-type females to repress male behavior and to suggest the existence of specialized neural circuits underlying male-specific behaviors in the normal female mouse brain (Kimchi et al. 2007; Spors and Sobel 2007; Shah and Breedlove 2007). However, the various behavioral abnormalities described in these genetically modified animals differ strikingly from those described after surgical disruption of vomeronasal function (VNX) (Keverne 2002). What could account for these behavioral differences? First, the two experimental approaches differ in the time point of the functional VNO removal. An alternative and perhaps more simple explanation is that TRPC2 may be expressed in a subset of neurons controlling reproductive physiology and behavior. While male-typical sexual behavior of female *Trpc2−/−* mice was reported to be independent of the adult female endocrine status, testosterone levels in ovary-intact *Trpc2−/−* females were significantly elevated (Kimchi et al. 2007). In another study adult testosterone treatment rather than VNO signaling was observed to enhance male-typical sexual behavior in female mice (Martel and Baum 2009), raising the possibility that TRPC2 deficiency may have influenced the reproductive axis at another level of the neural circuitry.

In addition to external input onto the reproductive axis, reproductive function is modulated by an array of neuronal inputs, which ultimately adjust the neuronal activity of the GnRH secreting terminals in the brain. For example, a variable set of hormones is involved in the control of reproduction that are part of either the reproductive axis itself, like the gonadal hormones (estradiol), or signaling pathways that determine energy balance (leptin, insulin, nutrition) and stress (corticotropins). However, GnRH neurons themselves do express receptors neither for estrogen (Radovick et al. 2012) nor for leptin or insulin (Divall et al. 2010; Elias 2012), highlighting the importance of other neurons in the neural circuits controlling reproduction in integrating these signals. One pivotal player in this is a population of hypothalamic neurons that produce the neuropeptide kisspeptin, a potent activator of GnRH neurons (Han et al. 2005). These cells have recently been shown to relay information about estradiol to GnRH neurons during maturation of the female reproductive axis (Mayer et al. 2010). An exciting alternative to reproductive axis modulation via hormone receptor activation has been brought up in a recent study reporting that TRPC channels may be directly regulated by steroid hormones. TRPC3, C5, and C6 activity can be inhibited by synthetic and natural steroids (Miehe et al. 2012), suggesting common features between TRP channels that might be susceptible for direct steroid action and therefore interfere with or adjust reproductive physiology depending on the hormonal status of the animal.

5 Mechanisms of Activation and Complexity in Regulation

TRPC channels can be activated by G-protein-coupled receptors (GPCRs) via phospholipase C (PLC). However, there is variability in the mechanisms coupling PLC activity to channel stimulation. TRPC channels may be activated by either (1) Ca^{2+} from internal stores via IP_3 activity, (2) diacylglycerol (DAG), or (3) its metabolites, polyunsaturated fatty acids (PUFAs). The dependence of TRPC activity on store depletion is still controversial in some cases. Of the various TRPC channels, TRPC1 is the best characterized channel to contribute to store-operated Ca^{2+} entry (Cheng et al. 2011). TRPC5 has been reported to bind to STIM1 and suggested to function as a store-operated channel (Worley et al. 2007), but has also been shown to be activated independent of store depletion (Schaefer et al. 2000; Okada et al. 1998). The majority of the evidence regarding activation and regulation relies unfortunately on deletion of TRPC isoforms or their silencing by antisense or siRNA in heterologous expression systems. Conditional gene targeting experiments will be needed to confirm these results in primary cells in vivo. Native TRPC2 channels present in the mouse olfactory system are activated exclusively by DAG and not by either IP_3, or Ca^{2+} stores or the PUFA arachidonic acid, supporting a role for PLC activity and DAG in vivo (Lucas et al. 2003).

On the other hand, the signaling pathway suggested in kisspeptin receptor (GPR54)-expressing GnRH neurons seems to involve the activation of a cSrc tyrosine kinase, which is proposed to directly modulate TRPC4 channel activity (Zhang et al. 2013). Binding of kisspeptin to the kisspeptin receptor GPR54 leads to activation of Gq in these cells (Zhang et al. 2013). The phytoestrogen genistein, a general tyrosine kinase inhibitor (Akiyama et al. 1987), as well as the more specific kinase inhibitor PP2 suppressed the kisspeptin-induced TRPC current in GnRH neurons. Interestingly, genistein on itself can directly interact with estrogen receptors and causes effects similar to those of the hormone estrogen (Kuiper et al. 1998). The reported genistein effects could therefore depend on the presence of estrogen receptors in these neuroendocrine cells. Similar effects of PP2 on estrogen receptors have not been reported. In contrast, genistein enhances TRPC5-mediated Ca^{2+} influx in TRPC5-overexpressing HEK cells (Wong et al. 2010). The potentiation of this Ca^{2+} influx was shown to be independent of estrogen receptors or PLC. It thus seems that tyrosine phosphorylation of TRP family members can result in altered channel activity; however, the link between these two events is still unclear. In favor of the idea is a study showing that two tyrosine residues in the C-terminus of hTRPC4 can be phosphorylated following epidermal growth factor receptor stimulation in cell lines (Odell et al. 2005). Another alternative pathway may involve Src-family tyrosine kinase phosphorylation of PLCγ (Liao et al. 1993; Sato et al. 2003), which may then again hydrolyse PIP2 into DAG and IP_3. Moreover, TRPC5 has been labeled a lipid ionotropic receptor as well as a non-genomic steroid sensor due to the regulation of TRPC5 activity by lysophosphatidylcholine and its modulation of channel activity by steroids, respectively (Majeed et al. 2011; Flemming et al. 2006). TRPC1 and TRPC3 have also both been indicated to be modulated by Src-family tyrosine kinase (Vazquez et al. 2004; Gervasio et al. 2008).

Understanding of the endogenous activators and function of many TRPC channels is still limited and may depend on the subunit composition or complex signaling environment (Strubing et al. 2001; Beech 2007). TRPC5 can form heterotetrameric complexes with TRPC1 or TRPC4 (Strubing et al. 2001; Hofmann et al. 2002; Meis et al. 2007) and the presence of these subunits has been described in the amygdala and hypothalamic areas (Fowler et al. 2007; Qiu et al. 2011; Meis et al. 2007). It is however not yet clear if these TRPC channels are found within the same neurons. Single cell RT-PCR analyses of genetically labeled GnRH neurons suggest combinatorial TRPC channel expression (Zhang et al. 2008). Therefore, the data imply that subunit co-expression patterns might be regional and cell-type specific. Furthermore, the dependence of neuroendocrine cells on the hormonal state of an individual may even alter the presence or abundance of TRP channel expression.

6 What Might Be the Function of TRPC Channels in Neuroendocrine Cells?

Neurons typically fire action potentials having a variable range of either single spikes (tonic firing) or high-frequency bursting. Dependent on the resting membrane potential, a persistent inward current induces a more depolarized membrane potential and can cause either an increase in the occurrence or prevent the generation of action potentials. TRPC activity- stimulating hormones such as leptin and kisspeptin have been proposed to depolarize their target cells in the CNS ensuring a sustained activation of the neurons (Zhang et al. 2008, 2013). These induced plateau potentials can drive burst-firing behavior, which could depend on TRPC activity (Phelan et al. 2013). Neurons firing action potentials in bursts are proposed to increase the reliability of communication between neurons. Decreasing the interval between the action potentials within a burst will enlarge the combined postsynaptic potential (Izhikevich et al. 2003; Lisman 1997) and therefore facilitate neurotransmitter release. Modulating the firing behavior via changed TRPC activity could thus enhance or decrease communication between neurons, possibly interfering with the physiological activity of the neuroendocrine network. Unfortunately, information on potential perturbations of the reproductive axis in the different TRPC-deficient mouse strains generated this far (see Birnbaumer 2014) is quite limited. Ultimately, TRPC channel function in individual neuroendocrine cells will need to be addressed via conditional gene targeting.

7 What About Other TRP Channels?

This question is still hard to answer since not many have investigated the various TRP channels in relation to reproductive physiology. Interesting other candidates next to the TRPC channels are TRPM3, TRPM8, TRPV4, and TRPV6. These channels have been either proposed or described to be regulated by steroids or steroid receptor activity and could be up- or downregulated to influence reproductive physiology. Steroid hormones affect TRPM8, TRPV4, and TRPV6 expression in prostate epithelial cells (Bidaux et al. 2005) and mammary gland epithelial cells (Jung et al. 2009; Bolanz et al. 2008), respectively. Changes in neuronal activity due to changes in the expression of ion channels are known to occur during the female reproductive cycle and might take place during stress-induced aggressive behaviors affecting the male (Arbogast 2008; Roepke et al. 2007).

TRPM3 is expressed in several brain areas, mainly in oligodendrocytes and during development in some neurons (Hoffmann et al. 2010). Using calcium imaging techniques and electrophysiological whole-cell recordings, TRPM3 is described to be directly activated by pregnenolone sulfate (Wagner et al. 2008). Moreover, this channel can be modulated by various other steroid hormones, including progesterone, 17ß-estradiol, and dihydrotestosterone (Majeed et al. 2011). Other TRP channels, like TRPM4/M5 and TRPV1 immunoreactivity, have been documented in the supraoptic and paraventricular nucleus (Teruyama et al. 2011; Sharif Naeini et al. 2006). In this latter region, GnRH target neurons, which express the GnRH receptor (GnRHR), are present (Wen et al. 2011), suggesting additional promising TRP channel candidates for manipulating reproductive physiology and behavior in vertebrates.

References

Akiyama T, Ishida J, Nakagawa S, Ogawara H, Watanabe S, Itoh N, Shibuya M, Fukami Y (1987) Genistein, a specific inhibitor of tyrosine-specific protein kinases. J Biol Chem 262 (12):5592–5595

Arbogast LA (2008) Estrogen genomic and membrane actions at an intersection. Trends Endocrinol Metab 19(1):1–2

Ariano P, Dalmazzo S, Owsianik G, Nilius B, Lovisolo D (2011) TRPC channels are involved in calcium-dependent migration and proliferation in immortalized GnRH neurons. Cell Calcium 49(6):387–394

Beech DJ (2007) Canonical transient receptor potential 5. Handb Exp Pharmacol 179:109–123

Bidaux G, Roudbaraki M, Merle C, Crepin A, Delcourt P, Slomianny C, Thebault S, Bonnal JL, Benahmed M, Cabon F, Mauroy B, Prevarskaya N (2005) Evidence for specific TRPM8 expression in human prostate secretory epithelial cells: functional androgen receptor requirement. Endocr Relat Cancer 12(2):367–382

Birnbaumer L (2014) The TRPC family of TRP channels: Regulators of Ca^{2+} influx, relationship to ORAI proteins, and their multiple roles in physiology and pathophysiology as deduced primarily from phenotypes that developed in knockout mice. In: Nilius B, Flockerzi V (eds) Mammalian transient receptor potential (TRP) cation channels. Handbook of experimental pharmacology. Springer, Heidelberg

Boehm U, Zou Z, Buck LB (2005) Feedback loops link odor and pheromone signaling with reproduction. Cell 123(4):683–695

Bolanz KA, Hediger MA, Landowski CP (2008) The role of TRPV6 in breast carcinogenesis. Mol Cancer Ther 7(2):271–279

Bruning JC, Gautam D, Burks DJ, Gillette J, Schubert M, Orban PC, Klein R, Krone W, Muller-Wieland D, Kahn CR (2000) Role of brain insulin receptor in control of body weight and reproduction. Science 289(5487):2122–2125

Chamero P, Leinders-Zufall T, Zufall F (2012) From genes to social communication: molecular sensing by the vomeronasal organ. Trends Neurosci 35:597–606

Chehab FF (1996) A broader role for leptin. Nat Med 2(7):723–724

Cheng KT, Ong HL, Liu X, Ambudkar IS (2011) Contribution of TRPC1 and Orai1 to Ca(2+) entry activated by store depletion. Adv Exp Med Biol 704:435–449

Clapham DE, Runnels LW, Strubing C (2001) The TRP ion channel family. Nat Rev Neurosci 2(6):387–396

Dalmazzo S, Antoniotti S, Ariano P, Gilardino A, Lovisolo D (2008) Expression and localisation of TRPC channels in immortalised GnRH neurons. Brain Res 1230:27–36

Divall SA, Williams TR, Carver SE, Koch L, Bruning JC, Kahn CR, Wondisford F, Radovick S, Wolfe A (2010) Divergent roles of growth factors in the GnRH regulation of puberty in mice. J Clin Invest 120(8):2900–2909

Elias CF (2012) Leptin action in pubertal development: recent advances and unanswered questions. Trends Endocrinol Metab 23(1):9–15

Flemming PK, Dedman AM, Xu SZ, Li J, Zeng F, Naylor J, Benham CD, Bateson AN, Muraki K, Beech DJ (2006) Sensing of lysophospholipids by TRPC5 calcium channel. J Biol Chem 281 (8):4977–4982

Fowler MA, Sidiropoulou K, Ozkan ED, Phillips CW, Cooper DC (2007) Corticolimbic expression of TRPC4 and TRPC5 channels in the rodent brain. PLos One 2(6):e573

Gervasio OL, Whitehead NP, Yeung EW, Phillips WD, Allen DG (2008) TRPC1 binds to caveolin-3 and is regulated by Src kinase - role in Duchenne muscular dystrophy. J Cell Sci 121(Pt 13):2246–2255

Gore AC (2002) GnRH: The master molecule of reproduction. Kluwer, Norwell, MA

Han SK, Gottsch ML, Lee KJ, Popa SM, Smith JT, Jakawich SK, Clifton DK, Steiner RA, Herbison AE (2005) Activation of gonadotropin-releasing hormone neurons by kisspeptin as a neuroendocrine switch for the onset of puberty. J Neurosci 25(49):11349–11356

Herbison AE (2006) Physiology of the gonadotropin-releasing hormone neuronal network. In: Neill JD (ed) Knobil and Neill's physiology of reproduction, 3rd edn. Elsevier Academic, St. Louis, pp 1415–1482

Hoffmann A, Grimm C, Kraft R, Goldbaum O, Wrede A, Nolte C, Hanisch UK, Richter-Landsberg C, Bruck W, Kettenmann H, Harteneck C (2010) TRPM3 is expressed in sphingosine-responsive myelinating oligodendrocytes. J Neurochem 114(3):654–665
Hofmann T, Schaefer M, Schultz G, Gudermann T (2002) Subunit composition of mammalian transient receptor potential channels in living cells. Proc Natl Acad Sci U S A 99(11):7461–7466
Hoivik EA, Bjanesoy TE, Mai O, Okamoto S, Minokoshi Y, Shima Y, Morohashi K, Boehm U, Bakke M (2011) DNA methylation of intronic enhancers directs tissue-specific expression of steroidogenic factor 1/adrenal 4 binding protein (SF-1/Ad4BP). Endocrinology 152(5):2100–2112
Izhikevich EM, Desai NS, Walcott EC, Hoppensteadt FC (2003) Bursts as a unit of neural information: selective communication via resonance. Trends Neurosci 26(3):161–167
Jeong K-H, Kaiser UB (2006) Gonadotropin-releasing hormone regulation of gonadotropin biosynthesis and secretion. In: Knobil E, Neill JD (eds) Physiology of reproduction. Elsevier Academic, St. Louis, pp 1635–1726
Jung C, Fandos C, Lorenzo IM, Plata C, Fernandes J, Gene GG, Vazquez E, Valverde MA (2009) The progesterone receptor regulates the expression of TRPV4 channel. Pflugers Arch 459 (1):105–113
Keverne EB (2002) Mammalian pheromones: from genes to behaviour. Curr Biol 12(23):R807–809
Kimchi T, Xu J, Dulac C (2007) A functional circuit underlying male sexual behaviour in the female mouse brain. Nature 448(7157):1009–1014
Krsmanovic LZ, Martinez-Fuentes AJ, Arora KK, Mores N, Navarro CE, Chen HC, Stojilkovic SS, Catt KJ (1999) Autocrine regulation of gonadotropin-releasing hormone secretion in cultured hypothalamic neurons. Endocrinology 140(3):1423–1431
Kuiper GG, Lemmen JG, Carlsson B, Corton JC, Safe SH, van der Saag PT, van der Burg B, Gustafsson JA (1998) Interaction of estrogenic chemicals and phytoestrogens with estrogen receptor beta. Endocrinology 139(10):4252–4263
Landry JJ, Pyl PT, Rausch T, Zichner T, Tekkedil MM, Stutz AM, Jauch A, Aiyar RS, Pau G, Delhomme N, Gagneur J, Korbel JO, Huber W, Steinmetz LM (2013) The genomic and transcriptomic landscape of a HeLa cell line. G3 (Bethesda) 3(8):1213–1224
Leypold BG, Yu CR, Leinders-Zufall T, Kim MM, Zufall F, Axel R (2002) Altered sexual and social behaviors in trp2 mutant mice. Proc Natl Acad Sci U S A 99(9):6376–6381
Liao F, Shin HS, Rhee SG (1993) In vitro tyrosine phosphorylation of PLC-gamma 1 and PLC-gamma 2 by src-family protein tyrosine kinases. Biochem Biophys Res Commun 191 (3):1028–1033
Liman ER, Corey DP, Dulac C (1999) TRP2: a candidate transduction channel for mammalian pheromone sensory signaling. Proc Natl Acad Sci U S A 96(10):5791–5796
Lisman JE (1997) Bursts as a unit of neural information: making unreliable synapses reliable. Trends Neurosci 20(1):38–43
Lucas P, Ukhanov K, Leinders-Zufall T, Zufall F (2003) A diacylglycerol-gated cation channel in vomeronasal neuron dendrites is impaired in TRPC2 mutant mice: mechanism of pheromone transduction. Neuron 40(3):551–561
Majeed Y, Amer MS, Agarwal AK, McKeown L, Porter KE, O'Regan DJ, Naylor J, Fishwick CW, Muraki K, Beech DJ (2011) Stereo-selective inhibition of transient receptor potential TRPC5 cation channels by neuroactive steroids. Br J Pharmacol 162(7):1509–1520
Martel KL, Baum MJ (2009) Adult testosterone treatment but not surgical disruption of vomeronasal function augments male-typical sexual behavior in female mice. J Neurosci 29 (24):7658–7666
Mayer C, Acosta-Martinez M, Dubois SL, Wolfe A, Radovick S, Boehm U, Levine JE (2010) Timing and completion of puberty in female mice depend on estrogen receptor alpha-signaling in kisspeptin neurons. Proc Natl Acad Sci U S A 107(52):22693–22698

Meis S, Munsch T, Sosulina L, Pape HC (2007) Postsynaptic mechanisms underlying responsiveness of amygdaloid neurons to cholecystokinin are mediated by a transient receptor potential-like current. Mol Cell Neurosci 35(2):356–367
Miehe S, Crause P, Schmidt T, Lohn M, Kleemann HW, Licher T, Dittrich W, Rutten H, Strubing C (2012) Inhibition of diacylglycerol-sensitive TRPC channels by synthetic and natural steroids. PLoS One 7(4):e35393
Munger SD, Leinders-Zufall T, Zufall F (2009) Subsystem organization of the mammalian sense of smell. Annu Rev Physiol 71:115–140
Odell AF, Scott JL, Van Helden DF (2005) Epidermal growth factor induces tyrosine phosphorylation, membrane insertion, and activation of transient receptor potential channel 4. J Biol Chem 280(45):37974–37987
Okada T, Shimizu S, Wakamori M, Maeda A, Kurosaki T, Takada N, Imoto K, Mori Y (1998) Molecular cloning and functional characterization of a novel receptor-activated TRP Ca^{2+} channel from mouse brain. J Biol Chem 273(17):10279–10287
Phelan KD, Shwe UT, Abramowitz J, Wu H, Rhee SW, Howell MD, Gottschall PE, Freichel M, Flockerzi V, Birnbaumer L, Zheng F (2013) Canonical transient receptor channel 5 (TRPC5) and TRPC1/4 contribute to seizure and excitotoxicity by distinct cellular mechanisms. Mol Pharmacol 83(2):429–438
Qiu J, Fang Y, Ronnekleiv OK, Kelly MJ (2010) Leptin excites proopiomelanocortin neurons via activation of TRPC channels. J Neurosci 30(4):1560–1565
Qiu J, Fang Y, Bosch MA, Ronnekleiv OK, Kelly MJ (2011) Guinea pig kisspeptin neurons are depolarized by leptin via activation of TRPC channels. Endocrinology 152(4):1503–1514
Radovick S, Levine JE, Wolfe A (2012) Estrogenic regulation of the GnRH neuron. Front Endocrinol (Lausanne) 3:52
Riccio A, Medhurst AD, Mattei C, Kelsell RE, Calver AR, Randall AD, Benham CD, Pangalos MN (2002) mRNA distribution analysis of human TRPC family in CNS and peripheral tissues. Brain Res Mol Brain Res 109(1–2):95–104
Riccio A, Li Y, Moon J, Kim KS, Smith KS, Rudolph U, Gapon S, Yao GL, Tsvetkov E, Rodig SJ, Van't Veer A, Meloni EG, Carlezon WA Jr, Bolshakov VY, Clapham DE (2009) Essential role for TRPC5 in amygdala function and fear-related behavior. Cell 137(4):761–772
Roepke TA, Malyala A, Bosch MA, Kelly MJ, Ronnekleiv OK (2007) Estrogen regulation of genes important for K^+ channel signaling in the arcuate nucleus. Endocrinology 148(10):4937–4951
Sato K, Tokmakov AA, He CL, Kurokawa M, Iwasaki T, Shirouzu M, Fissore RA, Yokoyama S, Fukami Y (2003) Reconstitution of Src-dependent phospholipase Cgamma phosphorylation and transient calcium release by using membrane rafts and cell-free extracts from Xenopus eggs. J Biol Chem 278(40):38413–38420
Schaefer M, Plant TD, Obukhov AG, Hofmann T, Gudermann T, Schultz G (2000) Receptor-mediated regulation of the nonselective cation channels TRPC4 and TRPC5. J Biol Chem 275 (23):17517–17526
Sergeeva OA, Korotkova TM, Scherer A, Brown RE, Haas HL (2003) Co-expression of non-selective cation channels of the transient receptor potential canonical family in central aminergic neurones. J Neurochem 85(6):1547–1552
Shah NM, Breedlove SM (2007) Behavioural neurobiology: females can also be from Mars. Nature 448(7157):999–1000
Sharif Naeini R, Witty MF, Seguela P, Bourque CW (2006) An N-terminal variant of Trpv1 channel is required for osmosensory transduction. Nat Neurosci 9(1):93–98
Sisk CL, Richardson HN, Chappell PE, Levine JE (2001) In vivo gonadotropin-releasing hormone secretion in female rats during peripubertal development and on proestrus. Endocrinology 142 (7):2929–2936
Spors H, Sobel N (2007) Male behavior by knockout. Neuron 55(5):689–693
Stowers L, Holy TE, Meister M, Dulac C, Koentges G (2002) Loss of sex discrimination and male-male aggression in mice deficient for TRP2. Science 295(5559):1493–1500

Strubing C, Krapivinsky G, Krapivinsky L, Clapham DE (2001) TRPC1 and TRPC5 form a novel cation channel in mammalian brain. Neuron 29(3):645–655

Teruyama R, Sakuraba M, Kurotaki H, Armstrong WE (2011) Transient receptor potential channel m4 and m5 in magnocellular cells in rat supraoptic and paraventricular nuclei. J Neuroendocrinol 23(12):1204–1213

Vazquez G, Wedel BJ, Kawasaki BT, Bird GS, Putney JW Jr (2004) Obligatory role of Src kinase in the signaling mechanism for TRPC3 cation channels. J Biol Chem 279(39):40521–40528

Venkatachalam K, Montell C (2007) TRP channels. Annu Rev Biochem 76:387–417

Vergara L, Rojas E, Stojilkovic SS (1997) A novel calcium-activated apamin-insensitive potassium current in pituitary gonadotrophs. Endocrinology 138(7):2658–2664

Wagner TF, Loch S, Lambert S, Straub I, Mannebach S, Mathar I, Dufer M, Lis A, Flockerzi V, Philipp SE, Oberwinkler J (2008) Transient receptor potential M3 channels are ionotropic steroid receptors in pancreatic beta cells. Nat Cell Biol 10(12):1421–1430

Wen S, Schwarz JR, Niculescu D, Dinu C, Bauer CK, Hirdes W, Boehm U (2008) Functional characterization of genetically labeled gonadotropes. Endocrinology 149(6):2701–2711

Wen S, Gotze IN, Mai O, Schauer C, Leinders-Zufall T, Boehm U (2011) Genetic identification of GnRH receptor neurons: a new model for studying neural circuits underlying reproductive physiology in the mouse brain. Endocrinology 152(4):1515–1526

Wong CO, Huang Y, Yao X (2010) Genistein potentiates activity of the cation channel TRPC5 independently of tyrosine kinases. Br J Pharmacol 159(7):1486–1496

Worley PF, Zeng W, Huang GN, Yuan JP, Kim JY, Lee MG, Muallem S (2007) TRPC channels as STIM1-regulated store-operated channels. Cell Calcium 42(2):205–211

Yoon H, Enquist LW, Dulac C (2005) Olfactory inputs to hypothalamic neurons controlling reproduction and fertility. Cell 123(4):669–682

Zhang C, Roepke TA, Kelly MJ, Ronnekleiv OK (2008) Kisspeptin depolarizes gonadotropin-releasing hormone neurons through activation of TRPC-like cationic channels. J Neurosci 28 (17):4423–4434

Zhang C, Bosch MA, Ronnekleiv OK, Kelly MJ (2013) Kisspeptin Activation of TRPC4 Channels in Female GnRH Neurons Requires PIP2 Depletion and cSrc Kinase Activation. Endocrinology 154(8):2772–2783

Zufall F (2014) TRPs in olfaction. In: Nilius B, Flockerzi V (eds) Mammalian transient receptor potential (TRP) cation channels, Handbook of experimental pharmacology. Springer, Heidelberg

Modulation of TRP Ion Channels by Venomous Toxins

Jan Siemens and Christina Hanack

Contents

Abstract

Venoms are evolutionarily fine-tuned mixtures of small molecules, peptides, and proteins—referred to as toxins—that have evolved to specifically modulate and interfere with the function of diverse molecular targets within the envenomated animal. Many of the identified toxin targets are membrane receptors and ion channels. Due to their high specificity, toxins have emerged as an invaluable tool set for the molecular characterization of ion channels, and a selected group of toxins even have been developed into therapeutics. More recently, TRP ion channels have been included as targets for venomous toxins. In particular, a number of apparently unrelated peptide toxins target the capsaicin receptor TRPV1 to produce inflammatory pain. These toxins have turned out to be invaluable for structural and functional characterizations of the capsaicin

J. Siemens (✉) • C. Hanack
Department of Pharmacology, University Clinic Heidelberg, Im Neuenheimer Feld 366, 69120 Heidelberg, Germany
e-mail: jan.siemens@pharma.uni-heidelberg.de

B. Nilius and V. Flockerzi (eds.), *Mammalian Transient Receptor Potential (TRP) Cation Channels*, Handbook of Experimental Pharmacology 223, DOI 10.1007/978-3-319-05161-1_17,

receptor. If toxins will serve similar roles for other TRP ion channels, only future will tell.

Keywords
Toxin • Vanillotoxin • DkTx • Capsaicin receptor • TRPV1 • TRP ion channel

1 Introduction: Venom Biology

In the animal kingdom, venoms are chiefly used either as part of a predatory strategy to paralyze, capture, and kill prey or as a defense mechanism to ward of predators and/or competitors (Evans and Schmidt 1990).

Additional functions have been proposed, and some insect venom ingredients may serve chemical communication (Speed et al. 2012).

Venoms from spiders, snakes, cone snails, fish, and scorpions are complex cocktails of salts, nucleotides (e.g., ATP), free amino acids, neurotransmitters, polyamines, peptides and proteins. From a pharmacological viewpoint, the latter two toxin groups, peptides, and proteins have attracted the most interest and have been invaluable to dissect and probe the biology of their targeted receptors. A few selected toxins have also inspired drug development (described in the following sections), and there is likely more to be expected in the future.

The astounding specificity, selectivity, and potency of many of the identified toxins for their molecular targets are owed to sophisticated adaptive mechanisms of diversification and expansion of selected toxins and toxin families (Blumenthal and Seibert 2003; Han et al. 2008; Lynch et al. 2006; Terlau and Olivera 2004; Lewis and Garcia 2003; Phui Yee et al. 2004; Twede et al. 2009).

The toxin profiles of different venoms are highly diverse. Diversity is not only found when comparing venoms of different species but a great deal of variability can also be observed when comparing individuals of a given species (Speed et al. 2012). The utility of such variability within a given species is not entirely clear and is not understood if this form of variability represents "ecological noise" or has adaptive evolutionary significance, potentially reflecting adaptation to different habitats.

Despite peptide toxin variability and complexity, there are common principles dictating the nature of structural building blocks and scaffolds. One structural motif, which is shared by many spider, scorpion and cone snail toxins, is the so-called inhibitor cysteine knot (ICK) fold of peptide toxins that are typically comprised of a total of 25–50 residues (Daly and Craik 2011). A hallmark of ICK toxins are the presence of 6 cysteine residues that form intramolecular disulfide bridges, constraining the toxin into a knot-like structure. The ICK fold is among the most abundant toxin motifs, and ICK toxins are estimated to account for 10^5–10^6 unique toxin sequences (Craik et al. 2001; Zhu et al. 2003). Thus, the ICK motif can be considered an "evolutionary vetted" scaffold that provides stability and protection against proteolytic cleavage and reducing environments when injected into another animal. Additionally, the ICK fold allows for high variability in the inter-cysteine loops that govern target receptor-specific interaction sites (Craik et al. 2001).

Spacing of cysteine residues appears to be important for productive folding, and therefore the number of naturally occurring cysteine patterns found in toxins is limited. The loops in between cysteine residues comprise regions of highly variable amino acid composition, which confer specificity to discrete target ion channel and receptor types. Given the relatively conserved nature of cysteine-residue position, it is suggested that they share a common evolutionary origin (Conticello et al. 2001). This raises the question of how the hypervariable regions in between cysteine residues are generated. A few studies, mainly carried out on fish-hunting cone snails, the so-called conotoxins, have been conducted. These studies suggest that a highly mutagenic molecular replication mechanism is at play, similar to the DNA-Polymerase V in bacterial stress-response mutagenesis (Conticello et al. 2000). At the same time macromolecules binding to cysteine-encoding codon triplets have been postulated to protect and conserve cysteine patterns of the toxin protein product (Conticello et al. 2000; Duda and Palumbi 1999; Olivera et al. 1999). While such a scenario is attractive and could explain the high diversity of cysteine-knot toxins, components of such a mutagenic replication machinery or molecules binding to specific DNA or RNA triplets have not been identified from venom ducts. Future studies are required to unveil and characterize this putative mechanism and reveal whether it can be exploited to generate pharmacologically relevant toxin libraries in vitro.

It has been proposed that the ICK motif has emerged from a simpler, ancestral toxin fold, encompassing only two disulfide bonds, the so-called disulfide-directed β-hairpin (DDH) (Escoubas and Rash 2004; Smith et al. 2011; Wang et al. 2000).

While the DDH motif is found in many proteins and in most phyla, direct evidence for molecular evolution from DDH to ICK motif was lacking until recently, when a scorpion toxin was identified and shown to adopt to this previously hypothetical DDH fold (Smith et al. 2011). By comparisons of ICK structures with that of the newly identified DDH toxin, it has been found that the two central disulfide bridges of ICK toxins nicely align with the DDH disulfides, suggesting that the "outer" solvent exposed additional disulfide, that closes the ring of the ICK motif, has emerged later during evolution (Fig. 1).

While many venoms can cause inflammatory pain as a consequence of severe tissue damage, some venoms can produce a robust perception of pain without eliciting appreciable trauma or damage. As illustrated below, venoms of the latter category have been found to directly engage and hijack ion channels and receptors of nociceptive sensory neurons, among them TRP ion channels. Likely, this serves to deter and avert predators and competitors by inflicting a painful and memorable experience. This parallels the mechanisms adopted by numerous plant species to deter predatory mammals through the production of chemical irritants (such as capsaicin or isothiocyanates) that also target TRP channels on sensory neurons of the pain pathway (Basbaum et al. 2009).

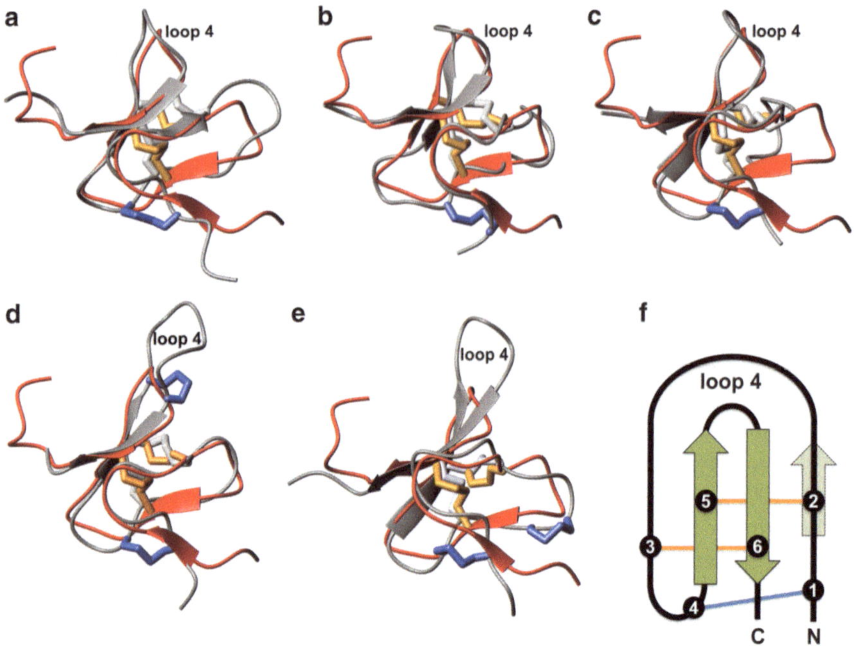

Fig. 1 Overlay of the newly identified ancestral DDH toxin (U1-LITX-Lw1a, *red*) with homologous ICK toxins (*gray*). Homologous spider ICK toxins include (**a**) guangxitoxin, (**b**) GsMTX-4, (**c**) Hainantoxin-I, (**d**) Purotoxin, and (**e**) κ-hexatoxin-Hv1c. The two central disulfides of the ICK toxins shown in *gray* tubes overlap with the two disulfides of DDH (*gold tubes*). Additional ICK disulfides are shown in *blue*. (**f**) Graphical representation of the DDH and ICK motifs, with the disulfides and β-sheet of the DDH motif shown in *orange* and *green*, respectively. The third disulfide necessary for the formation of the ICK motif is shown in *blue*. (Adopted with permission from Smith JJ. et al. 2011, *PNAS*, Vol. 108, 26, 10478–10483.)

2 Toxins of Venomous Organisms in Ion Channel Research and Medical Therapy

A lot of the knowledge on ion channel structure and function (ligand or voltage-gated) was elucidated using specific purified or synthesized peptide toxins (Catterall 1986; Norton and Olivera 2006; Terlau and Olivera 2004; Tsetlin 1999).

The Tarantula toxin, Hanatoxin, for instance, helped to pave the way for structure-function analysis of potassium and sodium ion channels. Other known examples include tetrodotoxin (TTX) from tetraodon pufferfish and saxitoxin (STX) from shellfish (Llewellyn 2006) for blocking sodium channels. Both toxins have revealed valuable information about the way neurons communicate with each other. α-bungarotoxin and other α-neurotoxins from *Elapidae* and *Hydrophidae* snakes were used to provide the first identification of nAChRs (Hucho et al. 1996; Lena and Changeux 1998), and α-bungarotoxin is still used as the most reliable pharmacological tool to study neuromuscular blockage. Furthermore, α- and β-neurotoxins from scorpions have been used to modulate sodium channels by delaying inactivation (α), or shifting the membrane potential dependence (β) (Bosmans and Tytgat 2007; Zuo and Ji 2004). The high degree of specificity with which venom peptides bind to different ion channel families as well as to other membrane receptors (including voltage-gated sodium (Nav), calcium (Cav), and potassium (Kv) ion channels and ligand-gated receptors such as nicotinic acetylcholine receptors (nAChRs), *N*-methyl-D-aspartate (NMDA), and G-protein coupled receptors (GPCRs)) make them ideal tools to study biophysical properties of target receptors as well as their physiological functions in vivo. Additionally, their high degree of specificity makes them suitable for manipulating the activity of selective cell subtypes. These properties (specificity/selectivity and high biological activity) that make toxins effective venoms are also what make them so valuable for medical applications. Many venom toxins target the same molecules that need to be controlled during diseases. A specific therapeutic area in which peptide toxins have already proven their potential includes, for instance, the treatment of chronic pain in humans by ω-conotoxin MVIIA (commercialized as Prialt) (Miljanich 2004; Staats et al. 2004; Zamponi et al. 2009). Snail venom peptides called conantokins are being tested with some success against epileptic seizures (Jimenez et al. 2002). Both, conotoxins and conantokins, are evaluated in studies of Alzheimer's and Parkinson's diseases, depression, and nicotine addiction (Moriguchi et al. 2012; Ragnarsson et al. 2002). In 2005, a venom component of Gila monster, called Exenatide (Byetta), was approved by the US Food and Drug Administration for the management of type 2 diabetes mellitus (Bond 2006). This peptide stimulates the secretion of insulin in the presence of elevated blood glucose levels. The sea anemone holds great potential for treating autoimmune diseases such as multiple sclerosis, rheumatoid arthritis, psoriasis, and lupus (Chi et al. 2012). For instance, Stichodactyla toxin (ShK) blocks Kv1.1 and Kv1.3 channels and is currently in phase 1 human trials for the treatment of autoimmune diseases. Eastern green mambas possess toxins that impair blood circulation and could be life saving to heart patients; Cenderitide, a fusion of a key peptide of the venom with a peptide from cells of human blood vessels, is intended to lower blood pressure and reduce fibrosis in a failing heart (Martin et al. 2012; von Lueder et al. 2013). Interestingly, ASIC channels are important targets by various venomous toxins. These include the sea anemone toxin, APETx2, that diminishes postoperative pain (Diochot et al. 2004), as well as psalmotoxin 1 (PcTx1) of the Trinidad chevron tarantula and Black mamba toxin peptides (Mambalgins) targeting ASIC channels with great potential to be effective new analgesics in the treatment of chronic pain (Diochot et al. 2012; Mazzuca et al. 2007). Moreover, a drug based on an anticoagulant toxin in the saliva of the vampire bat is now in clinical trials to test its potential to dissolve blood clots of stroke patients. It was found that the thrombolytic activity of the toxin exceeds that of conventional treatments, extending the time window for treatment from currently 4 to potentially 9 h (Liberatore et al. 2003; Medcalf 2012; Reddrop et al. 2005).

3 Toxins: Mode of Action

The mode of action of toxins will be exemplified on Kv channels, which have been most extensively studied. A number of different mechanisms have evolved for toxin-mediated modulation of ion channel activity. Among the best studied examples are two major groups: The "pore-pluggers" and the "voltage-sensor modulators."

"Pore-pluggers" bind to the outer vestibule of the ion conduction pore and physically block the flow of ions. Well-characterized examples are the scorpion toxins, charybodotoxin and agitoxin (Miller 1995), which block potassium channels and were used to identify and probe their pore region (MacKinnon and Miller 1989). The pore domain determines K^+ ion selectivity and contains the region that pore blocking toxins target. In the case of charybdotoxin, a Lys residue on the active surface of the toxin is "plugged" into the pore of the channel and interacts with potassium ions bound within the selectivity filter (Anderson et al. 1988; Banerjee et al. 2013; MacKinnon and Miller 1988; Park and Miller 1992; Rodriguez de la Vega et al. 2003).

The voltage-sensing domain is a second region of ion channels that is widely targeted by toxins. The four external arginine (or lysine) residues in S4 are positively charged and carry most of the gating charge (Aggarwal and MacKinnon 1996; Seoh et al. 1996). Binding to this region induces conformational changes during gating and thereby influences the gating mechanism by altering the stability of the closed, open, or inactivated states. Toxins targeting the voltage sensing domains of Nav channels can block or facilitate channel opening (Bosmans and Swartz 2010). Extensive studies on the voltage sensor were carried out in Kv channels. Hanatoxin, a 35-aa three-disulfide bond peptide, is the best-studied tarantula toxin that targets the voltage sensor and inhibits opening of the Kv channel by stabilizing the resting conformation of the voltage sensor (Phillips et al. 2005; Swartz and MacKinnon 1997a, b). As illustrated in Fig. 2 the region where hanatoxin binds is called the voltage sensor paddle motif (Jiang et al. 2003b), a helix-turn-helix motif composed of the C-terminal portion of S3, (S3b), and the S4 helix (Alabi et al. 2007; Jiang et al. 2003a; Ruta et al. 2005; Swartz 2008). This is a mobile region in the voltage-sensing domain that moves in response to changes in membrane voltage (Alabi et al. 2007; Jiang et al. 2003a, b). It was shown that paddle motifs are modular components that can be swapped between ion channels with a voltage-sensing domain without losing their functional properties (Alabi et al. 2007; Bosmans et al. 2008). The paddle motif is an important pharmacological target in ion channels because various tarantula toxins were shown to interact with this region (Bosmans et al. 2008; Milescu et al. 2009; Phillips et al. 2005; Lee et al. 2004). In addition, it was postulated that toxins like hanatoxin partition into the lipid membrane and bind to the paddle motif from within the bilayer (Lee and MacKinnon 2004; Milescu et al. 2007, 2009) (See also Fig. 2). This suggests that the effect of toxins and their interaction with the ion channel is not determined by the protein alone but by the lipids in the surrounding membrane.

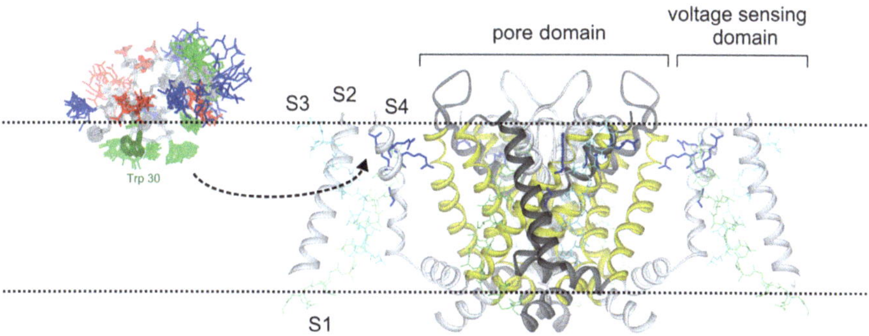

Fig. 2 Illustration of hanatoxin binding to the voltage sensor of a Kv channel. Hanatoxin, a 35-aa three-disulfide bond peptide, is among the best-studied tarantula toxin that targets the voltage sensor domain and inhibits opening of the Kv channel by stabilizing the resting conformation of the ion channel. The region where hanatoxin binds is the so-called voltage sensor paddle motif, a helix-turn-helix motif composed of the C-terminal portion of S3, (S3b) and the S4 helix. This mobile region in the voltage-sensing domain moves in response to changes in membrane voltage. It has been suggested that Hanatoxin partitions into the lipid membrane and binds to the paddle motif from within the bilayer. (Adopted with permission from Swartz 2007, Toxicon, Vol. 49, Issue 2, 213-230)

4 Toxins Modulating TRPV1 Activity

4.1 Vanillotoxins

TRPV1 is predominantly expressed in primary afferent sensory fibers that innervate the skin and viscera. Originally cloned as the receptor for capsaicin, the pungent ingredient in red chili peppers (Caterina et al. 1997), TRPV1 has subsequently emerged as a prototypical molecular nociceptor, participating in the detection of a wide range of noxious stimuli such as low pH, high temperature, and inflammatory signaling molecules (Basbaum et al. 2009). Biophysical properties as well as physiological functions of TRPV1 are described in detail in a previous chapter of this book.

Pain serves as an important warning system signaling tissue damage and distress triggering protective reflexes that allow the organism to respond to and escape from potentially dangerous environments.

Similar to selected plant species, venoms of some animals inflict pain by directly activating primary afferent sensory fibers, presumably hijacking the pain system, to deter predators and competitors by causing a memorable discomfort and distress.

Since TRPV1 and its related ion channels TRPA1 and TRPM8 are central to the pain pathway, a screen was conducted to test whether venoms contain toxins specifically activating these receptors thus explaining a pain-producing effect of bites and stings by some venomous species. Out of 22 spider and scorpion venoms, two distantly related tarantula species, *Psalmopoeus cambdridgei* and

Ornithoctonus huwena (the latter also known as "Earth Tiger"), robustly and specifically activated TRPV1 (Siemens et al. 2006). Reiterative reversed-phase chromatography of *Psalmopoeus cambdridgei* venom yielded three related peptide toxins each of which individually promoted TRPV1 activation with different efficacies in calcium imaging experiments and electrophysiological recordings of overexpressed or native vanilloid receptor TRPV1. Accordingly, the peptide toxins were named vanillotoxins (VaTx 1–3). Edman peptide sequencing combined with mass spectrometry revealed that vanillotoxins consist of 34–35 amino acids that belong to the extended family of inhibitory cysteine knot (ICK) peptides. Vatxs were also confirmed to elicit nocifensive behavior in a TRPV1-dependent fashion and to be present in sufficiently high enough concentrations in the crude venom (100–1,000 Fold their EC50 values), suggesting that a TRPV1-mediated pain response can be triggered in envenomated animals in the wild.

ICK toxins are widely recognized as inhibitors of cationic channels and transmembrane receptors, exemplified by the detailed analysis of Hanatoxin's (Hatx1 and 2) interaction with voltage-gated potassium (Kv-) channels (Swartz 2007). Vanillotoxins were among the first ICKs identified that act as positive modulators of a cationic channel rather than serving as inhibitors.

To test whether any of the VaTxs additionally exhibit a "classic" inhibitory function, they were tested on a set of Kv channels. Indeed, it was found that Vatx1 is equally potent as a TRPV1 activator and as a Kv2.1 inhibitor (Fig. 3). Vatx 2 and 3, that are more potent TRPV1 activators than VaTx1, had only minor or no appreciable activity on any of the tested Kv channels (Siemens et al. 2006). The diversification of vanillotoxins serves as a rare illustration how one of the most commonly found scaffolds in venom toxins, the ICK motif, can be adapted to novel functions. As described above, the Cysteine knot is a very robust, yet at the same time highly versatile structural scaffold, serving many functions that are acquired by gene duplication and rapid diversification (Conticello et al. 2001; Sollod et al. 2005; Kordis and Gubensek 2000). Here, Vatx 1-3 exemplify apparent evolutionary transition states with graded selectivity for Kv2.1 and TRPV1 ion channels (Bohlen and Julius 2012). This functional diversity is also reflected in sequence differences: Among the 3 vanillotoxins, promiscuous Vatx1 is most similar to Kv inhibitor toxins from related tarantulas (such as the Heteroscodra maculate toxin HmTx1) while Vatx2 and Vatx3 are progressively dissimilar to Vatx1 (and for that matter to HmTx1) and at the same time more specific and potent activators for TRPV1.

Both channel types, Kv and TRP ion channels, are believed to be structurally related and share similar membrane topology, tetrameric organization, and to a limited degree also voltage sensitivity. Given this similarity, one would intuitively speculate that VaTx1 is likely to bind to a similar region within Kv2.1 and TRPV1 ion channel polypetides. Surprisingly, directed mutagenesis as well as use of chimeric TRPV1 channels encompassing domains of toxin-insensitive frog TRPV1 and toxin-sensitive rat TRPV1 showed that this is not the case (Bohlen et al. 2010). While VaTx1 inhibits Kv channels by binding to the voltage sensor domain (the third and fourth transmembrane helices) (Fig. 2) much like other

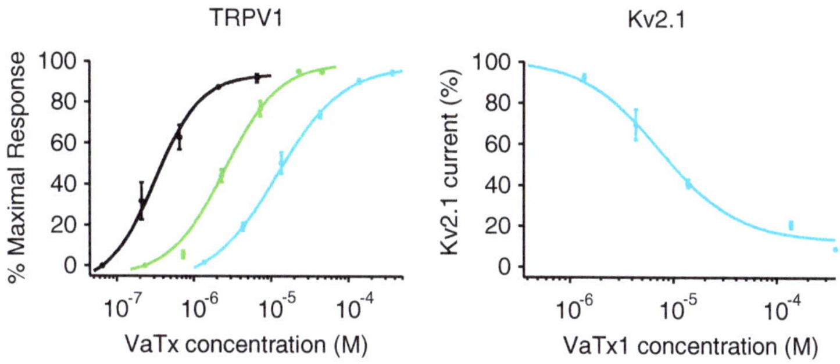

Fig. 3 Vatx1 can act both, as an activator of TRPV1 and an inhibitor of Kv2.1. *Left*: dose–response analysis by whole-cell voltage-clamp recording of TRPV1-expressing oocytes (180 mV) revealed EC50 values of 11.9 ± 1.4, 2.53 ± 0.02 and 0.32 ± 0.09 μM for VaTx1 (*blue*), VaTx2 (*green*), and VaTx3 (*black*), respectively. *Right*: toxin-mediated inhibition of Kv2.1 (at 0 mV) was assessed in oocytes. The full dose–response is shown for VaTx1 ($IC_{50} = 7.4 \pm 1.9$ μM). Error bars represent s.e.m.; $n > 3$ trials for each toxin concentration. (Adopted with permission from Siemens, J. et al. 2006, *Nature*, Vol 444)

extensively studied ICK toxins (Catterall et al. 2007; Swartz 2007), the toxin instead recognizes the pore domain (the fifth and sixth transmembrane helices) within TRPV1 protein (Bohlen et al. 2010) (Fig. 4). Opposite to Vatx1, it appears that VaTx2 and 3 have evolutionary refined their binding specificity in favor of a TRPV1 interaction.

There are several examples of toxins that have acquired new functionalities over time and others appear to be in the middle of such a transitioning process having adopted bifunctionality, such as huwentoxin-XI, which has acquired both trypsin and Kv channels' inhibitory activity (Fry et al. 2009; Yuan et al. 2008).

As such, the different vanillotoxins could be considered evolutionary intermediates transitioning between paralyzing (Kv channel inhibiting) and pain-inducing (TRPV1 activating) toxin functionalities.

4.2 Double-Knot Toxin

Although the chromatographic profile of the active ingredient in *Ornithoctonus huwena* venom suggested that it was also proteinacious in nature, with a mass of 8,522 Da it appeared unusually large for an ICK peptide.

Due to paucity of material and its rather large size, a different strategy was employed to disclose its nature. A partial sequence was obtained by de novo peptide sequencing. The peptide sequence information was used to design degenerate primers for subsequent PCR cloning of the full-length cDNA encoding the mature toxin from venom gland tissue (Bohlen et al. 2010). The *O. huwena* toxin had no apparent homology to any of the three vanillotoxins, suggesting that the TRPV1-activating toxins of the two tarantula species had independently evolved through convergent evolution. Closer analysis revealed that, most strikingly, the toxin

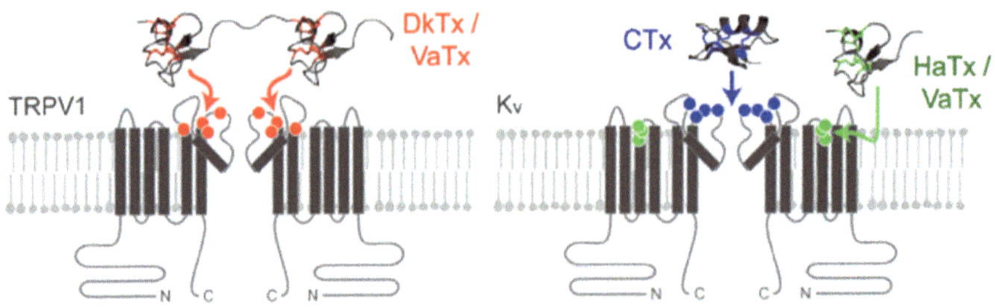

Fig. 4 Model of Toxin binding to TRPV1 and Kv channels. *Gray* bars represent transmembrane helices, and *red* dots highlights residues that are crucial for double-knot toxin (DkTx) activation. In the simplest scenario, the two knots of DkTx bind to two equivalent sites on multiple subunits of the same channel. Kv channels likely possess the same overall transmembrane topology as TRPV1 but interact with ICK toxins in different ways. For example, charybdotoxin (CTx) binds within the ion permeation path to block ion flux, and voltage-modulator toxins, such as hanatoxin (HaTx), target the voltage sensor to modify gating properties (*blue* and *green* dots represent mutations that attenuate CTx and HaTx inhibition, respectively). The single-knot vanillotoxins (VaTx) also appear to target the S3–S4 helices of Kv channels, but they activate TRPV1 through the pore region. (Adopted with permission from Bohlen CJ. et al. 2010, *Cell*, Vol. 141, p. 834–845)

consisted of tandem repeats of two individual ICK motifs and was therefore named "double knot toxin" (DkTx, Fig. 4). Presumably, DkTx has evolved from a gene duplication event that conjoined the two individual ICK motifs into a single open reading frame.

This unusual structural arrangement is paralleled by nearly irreversible activation of its target receptor TRPV1. A series of experiments led to the conclusion that the two covalently connected ICK lobes have bivalent properties, resulting in exceptionally high avidity for its multimeric target TRPV1, similar to antibody-like binding properties. Indeed, robust and long-lasting activation of TRPV1 correlated with specific, nearly irreversible direct physical interaction of DkTx with the ion channel, as deduced from binding studies (Bohlen et al. 2010).

Multimerizing ligands as a means to increase their effectiveness is a well studied and exploited phenomenon in pharmaceutical chemistry. There are numerous examples were multivalent strategies have proven extremely useful and result in enhanced kinetic parameters and drastically increased affinities of pharmacophores (Bohlen and Julius 2012). The mode of action of such multivalent ligands can be categorized into three different modes: polyvalency (targeting multiple receptors distributed over a larger surface), hetero-multivalency (targeting discrete epitopes of a given receptor), and homo-multivalency (targeting identical binding sites on different subunits of a multimeric complex). Initial estimations accounting for DkTx linker length estimations accounting for DkTx linker length as well as the presumed dimensions of homo tetrameric TRPV1 channels suggest that homo-multivalency is the likeliest scenario for a TRPV1–DkTx mode of interaction. Theses findings are corroborated by subsequent structural analysis of a DkTx-TRPV1 protein complex (Cao et al. 2013). Single channel recordings of TRPV1 in the presence of DkTx are in agreement with this interpretation: irreversible activity of DkTx is not only observed in macroscopic currents deduced from whole-cell patch-clamp recordings but have been also recapitulated in single channel recordings of TRPV1 (Bohlen et al. 2010).

While multivalency is a widespread phenomenon in nature as well as pharmaceutical chemistry, it has thus far been rarely observed for toxins. Few reported examples include sarafotoxin as well as a newly identified tandem ICK toxin from the yellow sac spider as well as a few selected other examples (Ducancel et al. 1993; Vassilevski et al. 2010). Given the widespread presence of longer peptide toxins (>6,000 Da) in scorpion, cone snail, and spider venom (Escoubas and Rash 2004), it is likely that more tandem repeat ICK toxins will be discovered in the future.

Other non-covalently linked multimeric toxins have been identified that also have multivalent properties, such as the highly potent two-component toxin from a coral snake that activates acid-sensing ion channels (ASICs) (Bohlen et al. 2011). Interestingly, ASIC activation by MitTx toxin complex also induces pain and inflammation, again pointing to an adaptation selecting for exploitation of the pain pathway of a variety of venomous animals inhabiting different environmental niches.

While the TRPV1 activator DkTx is not the only multivalent toxin, it represents thus far the clearest case that bivalency can tremendously increase potency as well as duration of activity compared to individual, monovalent vanillotoxins. DkTx can be produced in good yields, and researchers are not required to rely on miniature amounts that can be harvested from spiders (Bae et al. 2012). These properties of DkTx not only aided in probing biophysical properties and physiological functions of TRPV1, it also helped to obtain the first high resolution structures in the open conformation of the capsaicin receptor, or for that matter, of any TRP ion channel (Cao et al. 2013; see also a previous chapter in this book, highlighting TRPV1 protein structures).

Since natural toxins target a number of medically relevant receptors with high specificity and have been invaluable tools in basic research that have additionally also entered the therapeutic stage, a combinatorial approach to generate multivalent toxins may greatly improve toxin effectiveness and should be explored in the future.

5 Additional TRPV1-Mediated Toxin Effects

A number of venoms have been found to cause inflammation and pain, either locally but sometimes also more widespread and distal to the site of the bite or sting. Certainly, not all of these effects can be attributed to TRPV1 activation, and other molecular toxin targets of the somatosensory system have been identified that can mediate painful responses (Bohlen and Julius 2012). Nevertheless, some additional toxins have been found to target TRPV1, either directly or indirectly. Among them are a group of polycyclic ether toxins that are produced by marine dinoflagellates and that are carried by certain reef- and shellfish. Consumption of such infested tropical fish can lead to ciguatera fish poisoning (CFP) and neurotoxic shellfish poisoning (NSP). Common symptoms of acute intoxication include tingling of lips, hands and feet, unusual temperature perception, and burning mouth syndrome (Cameron and Capra 1993; Heir 2005). Polycyclic ether toxins gambierol and brevetoxin are derived from CFP and NSP, respectively. Cuypers et al. found that both toxins can activate TRPV1 (Cuypers et al. 2007). Both toxins appear to be allosteric modulators of TRPV1 that cannot activate the ion channel on their own (at least not at the concentrations tested) but instead enhance the effect of capsaicin or other TRPV1 agonists. Similar results were found with extracts from jellyfish tentacles that are known to cause a burning pain sensation (Cuypers et al. 2006). While it cannot be ruled out that these toxins have other molecular targets, it is likely that they act directly on TRPV1. Interestingly, similar to Vatx 1, Gambierol not only promoted activation of TRPV1 but also has inhibitory activity on Kv channels (Cuypers et al. 2008). However, the inhibitory mechanism of Gambierol-mediated Kv-channel inhibition appears to be different from ICK toxins such as HaTx or VaTx, and the site of action resides close to the Kv-ion channel pore and not within the voltage-sensor domain (Kopljar et al. 2009). The site of Gambierol's action on TRPV1 has not been determined yet. If Gambierol also constitutes an adaptive transition state of a toxin that acquired a new function, we can only speculate. It may turn out informative to test-related polycyclic ether toxins for their ion channel specificity and to determine structural elements within the polymer that dictate functionality.

Within the pain pathway, TRPV1 cannot only be activated directly by heat or by interactions with agonists such as endogenous produced lipid intermediates and protons, but it is also target of several signaling cascades that activate or sensitize the receptor (Fig. 5). Thus, TRPV1 serves as a molecular integrator (also referred to as polymodal nociceptor) of inflammatory and noxious signals (Julius and Basbaum 2001; Nieto-Posadas et al. 2011; Tominaga and Tominaga 2005). Intriguingly, venomous creatures appear to likewise exploit indirect activation and sensitization mechanisms of TRPV1 to produce a noxious sensation. Bv8, a protein toxin found in skin secretions of the yellow-bellied frog, *Bombina variagata*, is a potent activator of the prokineticin receptor 1 and 2 (PKR 1 and 2). Injection of Bv8 caused a PKR-mediated hyperalgesia that was largely dependent on TRPV1 (Negri et al. 2006; Vellani et al. 2006). Similar toxins have been identified from other reptiles as well as spiders, suggesting that nociceptor sensitization through the manipulation of signaling cascades is more widespread (Fry et al. 2006; Negri et al. 2007; Schweitz et al. 1990; Szeto et al. 2000; Wen et al. 2005) (Fig. 5).

TRPV1 inhibitory activities have also been found in venoms. Partial TRPV1 inhibition has been observed from tropical sea anemone *Heteractis crispa* venom. The polypeptide toxin APHC1 antagonized TRPV1 currents and capsaicin-induced nocifensive responses in animal models (Andreev et al. 2008).

Two non-peptidic toxins, AG489 and AG505, which attenuate TRPV1 activity have been isolated from the funnel web spider, *Agelenopsis aperta*. However, these toxins appear to inhibit a broad range of different receptors and can also lead to

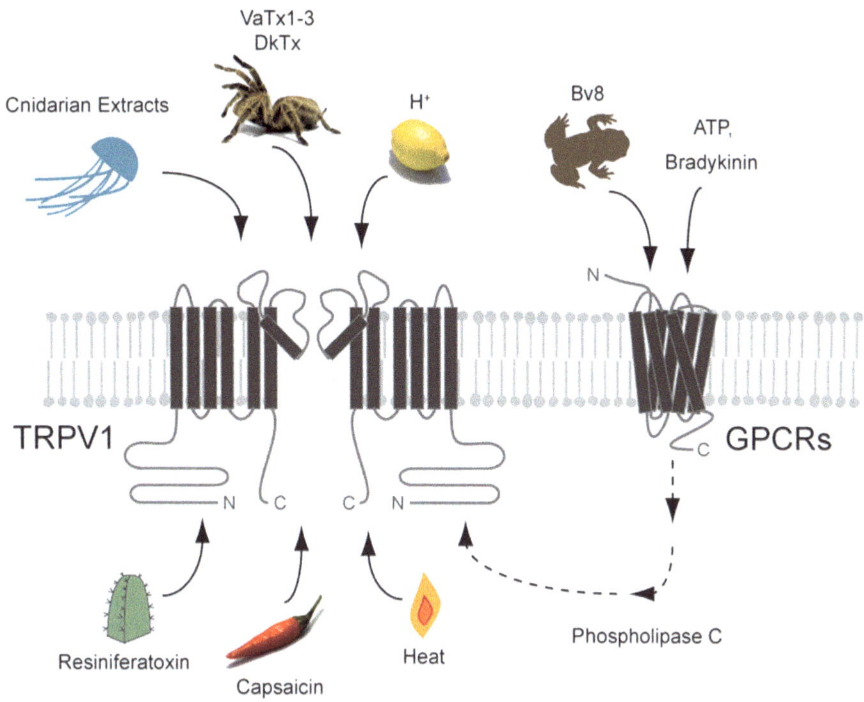

Fig. 5 Toxins targeting TRPV1. TRPV1 detects physical and chemical signals from the environment, including acidic pH and hot temperatures. Plant-derived irritants, cnidarian extracts, and toxins from tarantula venom mimic these harmful stimuli by promoting TRPV1 activation. TRPV1 is sensitized by phospoholipase C activation triggered by inflammatory signaling molecules such as bradykinin or ATP that are released downstream from venom lipases, proteases, and kallikreins. Other toxins, such as Bv8 from frog skin, produce TRPV1 sensitization through direct activation of GPCRs. (Adopted with permission from Bohlen CJ., Toxicon, 2012, Vol. 60, p. 254–264)

TRPV1-independent paralysis of envenomated species (Kitaguchi and Swartz 2005; Skinner et al. 1989).

6 Toxins Affecting Other TRPs

So far, there is only very little information about animal toxins targeting TRP channels other than TRPV1. However, a few other examples exist and the latter two illustrate that toxins indirectly affecting TRP channel function can also help to decipher their physiological roles.

6.1 TRPV6

TRPV6 is expressed in apical membranes of various tissues including kidney, intestine, pancreas, and prostate (Nijenhuis et al. 2003a, b; Zhuang et al. 2002). In the gastrointestinal tract, it is involved in apical entry of calcium ions (Hoenderop et al. 2005; Zhuang et al. 2002). TRPV6 has been implicated in tumor development and progression in a number of cancers (Bodding 2007; Nilius et al. 2007; Santoni et al. 2011). Upregulation of TRPV6 protein levels has been reported in tumors of ovary, breast, colon, and prostate cancer (Wissenbach et al. 2001; Zhuang et al. 2002). This indicates that TRPV6 could be a novel therapeutic target for the treatment of tumors over-expressing the receptor. To elucidate this further, a recent study made use of the toxin Soricidin isolated from the saliva glands of the northern short-tailed shrew (*Blarina brevicauda*) (Peters et al. 2012). Soricidin is a novel peptide that has been shown to inhibit calcium uptake via TRPV6 channels. Furthermore, it was shown that two peptide sequences from the c-terminal domain of soricidin have a high binding affinity for TRPV6 in ovarian cancer. In the study of Bowen et al. (2013), these two peptide sequences were conjugated with a fluorescent dye or an magnetic resonance imaging (MRI) contrast agent which enabled them to monitor the distribution of the peptides in vivo (Bowen et al. 2013). Imaging tumors over-expressing TRPV6 provides opportunities for early detection of ovarian cancer. This is especially important because ovarian cancers are difficult to detect in their early stages, and early detection could be lifesaving.

6.2 TRPA1

TRPA1 is another TRP channel family member that is indirectly modulated by a venomous compound. TRPA1 channels are expressed in nociceptive DRG neurons and are important transducers of pungent or irritating environmental stimuli and thus play a crucial role in nociception. Similar to TRPV1, the receptor has recently been implicated in the development of ciguatoxin-induced symptoms (Vetter et al. 2012). As stated above, Ciguatera fish poisoning is an acute intoxication resulting from the consumption of tropical reef fishes (Lewis 2006). Symptoms of this poisoning also include major peripheral sensory disturbances, including a hypersensitivity to cold (allodynia), which is characterized by intense stabbing and burning pain in response to mild cooling. Of the different ciguatoxin variants that exist, P-CTX (Pacific Ocean CTX) is the most potent and thought to be responsible for the majority of neurological symptoms (Lewis 2001). Interestingly, Vetter et al. now discovered that Ciguatoxin-induced allodynia is due to indirect activation of TRPA1 ion channels. Evidence for this is based on heterologously expressed TRPA1 which could only be activated by P-CTX in the presence of Na (v) channels, not if expressed individually. Importantly, mice lacking TRPA1 showed a reduction of the effect of CTX on C-fibers and on ciguatoxin-induced cold allodynia. Furthermore, functional MRI studies revealed that ciguatoxin-induced cold allodynia enhanced the BOLD (Blood Oxygenation Level Dependent) signal, an effect that did not occur in TRPA1-knockout mice. As a consequence of ciguatoxin-induced activation of voltage-gated sodium channels, TRPA1 channels are activated at a temperature where they usually would not be activated. Thus, when ciguatoxins are present, a temperature decrease can lead to activation of TRPA1 channels, leading to the inappropriate sensation of painful burning. This may serve as a warning device to alert the body of danger (Voets 2012).

6.3 TRPCs

Bee stings cause persistent pain and swelling around the sting area. The main peptide of bee venom responsible for pain is the toxin melittin. This toxin activates different subpopulations of primary nociceptive neurons. It was recently shown that TRPC channels may represent a possible target by melittin (Ding et al. 2012).

TRPCs are the closest homologs to drosophila TRP channels and are expressed in the central and peripheral nervous system (Chung et al. 2006; Riccio et al. 2002). In the DRG, TRPC1, TRPC3, TRPC4, and TRPC6 are most abundant (Kress et al. 2008). To date, the role of TRPC channels in the somatosensory neurons is not very well established. Ding et al. (2012) recently postulated a possible role for this subgroup of TRP channels in somatosensation. For their study they used the melittin model to induce inflammatory pain (Chen and Lariviere 2010) in rats. Subcutaneous melittin injection causes persistent pain and pronounced thermal and mechanical hyperalgesia (Chen et al. 2006; Li and Chen 2004). Melittin was already shown to indirectly activate TRPV1 (Du et al. 2011). Interestingly, a subpopulation of neurons positive for TRPC1 and TRPC3 does not express TRPV1 but is sensitive to melittin (Ding et al. 2011; Kress et al. 2008), suggesting that TRPV1 is not the only target of this toxin. For this reason, Ding et al. investigated melittin-induced pain behaviors in the presence of the nonselective TRPC antagonist SKF-96365. In fact, administration of this blocker prior or after subcutaneous melittin injection reduced the development of persistent spontaneous pain, suggesting that TRPC channels may be involved in mediating the pathophysiological processes of a bee sting. However, additional experimentation will be needed to fully elucidate the mechanism of melittin-evoked nociception. Due to nonselective properties of SKF-96365, it cannot be ruled out that channels other than TRPCs are at play here.

7 Discussion and Outlook

While it is appealing and intuitive to assume that pain-inflicting toxins provide an adaptive advantage by averting predatory threats, it is difficult to assess their physiological role directly in the wild. As a starting point for addressing this evolutionary aspect, it may be helpful to categorize venomous species into ones that utilize their venom mainly to capture and kill prey and others that employ their toxin cocktails foremost as a defense mechanism. The assumption would be that venoms used as a defense mechanism may contain pain-inflicting "nocitoxins," while predatory species have less or no nocitoxins since it would not provide an advantage to them. In fact, it may even be disadvantageous to produce a pain response when prey paralysis is the main purpose of the toxin cocktail. In agreement with this notion, toxin combinations have been found in predators that suppress the fight-or-flight response in prey (Olivera et al. 1999).

For some species such as spiders, the venom serves multiple functions and is optimized to encompass both goals of prey capture and chemical defense. In fact, some components of certain insect venoms may even serve less appreciated roles as pheromones and in chemical communication (Speed et al. 2012). But clearly, predatory cone snails chiefly (if not exclusively) use their venom to paralyze and kill prey and thus may have no adaptive need for nocitoxins. In this regard it is noteworthy that a cellular screen similar to that published in Siemens et al. (2006) has also been conducted with venoms of a dozen different cone snail species. In opposite to spider venoms, this screen did not yield any agonist activity on any of the three somatosensory TRP ion channels tested, namely TRPV1, TRPA1, and TRPM8 (David Julius and Jan Siemens—unpublished data). While more comprehensive studies are required to obtain conclusive results for the two categories (prey-capture venoms vs. defensive venoms), these preliminary result argues that in regard of inflicting a painful sensation, there may be adaptive differences.

For the elaboration of this hypothesis, it is also important to consider the perspective of the other party involved, that of the predator or competitor. One particular example illustrates how an adaptive advantage of a pain-inflicting toxin from certain scorpion species, the Bark scorpions, has been negated by a subsequent evolutionary adaption of its predator, the Grasshopper mouse. The venom of Bark scorpions inflicts pain in many mammals such as house mice, rats, and humans by directly activating voltage-gated Na^+ channels. Grasshopper mice express a Na^+ channel variant that is inhibited by the scorpion toxin, thereby blocking action potential propagation and a painful sensation (Rowe et al. 2013). This is an example of the adaptive "arms race" and illustrates the intricate predator—prey relationship, even at the level of somatosensory processing.

It may not be surprising that a number of different venom components target TRPV1, thereby causing a noxious sensation. After all, this is what venoms of many species have evolved for: to deter predators and competitors. Apparently, this can easily be achieved by injecting (through bites and stings) a TRPV1 agonist. ASIC channels are also included as venom targets to produce a noxious sensation (Bohlen et al. 2011) confirming that nocitoxins are not limited to exclusively targeting sensory TRP ion channels. On the other hand, it is somewhat surprising that not more pro-nociceptive toxins have been identified from venomous species that target other receptors relevant to the pain pathway. A number of plant-derived products have been found to activate other sensory receptors of the TRP ion channel family besides TRPV1 (Vriens et al. 2008); however, no additional toxins have been described for these thus far.

Why toxins specific for additional TRPs have not yet been identified, we can only speculate. One trivial explanation is that people have either not searched for such toxins or no robust cellular assays are available for certain TRPs to test toxin activity. Another, more provocative, possibility is that either activation or inhibition of other TRP ion channels does not provide an adaptive advantage to the venom-producing species since it would neither lead to paralysis/death nor a memorable nocifensive response. Thus, toxins affecting other TRP channels may not have been evolutionary selected for.

Yet another possibility is that the gating mechanism of other TRPs may differ from that of TRPV1 and may not be compatible with toxin bioactivity. Most TRP ion channels harbor only relatively small extracellular domains that can be accessed by peptide toxins, which are not able to freely permeate into the cytosol. Binding sites for TRP channel agonists are primarily located intracellularly, which may explain why these receptors are only weakly influenced by extracellular peptides.

In this regard the irritant receptor TRPA1 may pose a particular example. This receptor can be activated by a multitude of different endogenous and exogenous ligands, including several plant products. Yet, no toxin of a venomous animal has been described to our knowledge that is either activating or inhibiting this promiscuous receptor of the pain pathway. Different to TRPV1, TRPA1 is activated by a number of reactive substances that covalently modify the channel protein, thereby resulting in channel opening (Hinman et al. 2006; Macpherson et al. 2007). Reactive molecules, that only have a short half-life, may not be compatible with venoms that have to be produced and stored for longer periods of time before use and thus may lose activity fairly easily, particularly in complex venom mixtures. This is reflected in ICK peptide structures that are optimized for stability even under harsh conditions, while at the same time binding with high specificity to a range of different (membrane) receptors and ion channels.

Although ICK toxins contain a number of paired cysteines that potentially could lead to covalent modification of target receptors such as TRPA1, their oxidative strength is likely very limited, reflecting the overall ICK stability. More importantly, peptide toxins have no access to intracellular redox-sensing cysteines on TRPA1.

Whether other more reactive toxin species exist, future work will show. In this regard it is worth noting that crude venom of the platypus (*Ornithorhynchus anatinus*) harbors ingredients that can activate TRPA1 in cellular assays (David Julius and Jan Siemens unpublished observation). The venom, which is delivered by spurs located at the platypus hind limbs, is known to cause excruciating pain (Mebs 2002). In the same assay, the venom did not activate TRPV1 or TRPM8. Due to paucity of material and problematic behavior in chromatographic purification approaches—potentially reflecting reactivity of the putative compound—the venom ingredient causing TRPA1 activation was never isolated and further characterized. Whether this venom entity belongs to the peptidic toxin fraction or—more likely—to the fraction of small toxin molecules, it is thus not known to date and will require further exploration.

While chemical reactivity as a mechanism for toxin function is speculative, it is known that peptide toxins carry a vast array of modifications and their contribution to receptor specificity, avidity, as well as other toxin properties are largely unexplored.

Toxins, such as HaTx, have been instrumental in understanding biophysical parameters and helped paving the way to obtain high-resolution structural information of Kv ion channels.

For the first TRP ion channel, the capsaicin receptor TRPV1, detailed structural information is now available in the closed and the open state (Liao et al. 2013; Cao et al. 2013). For stabilizing the open conformation and deriving structural details of the gating mechanisms via cryo electron microscopy, DkTx has played a noteworthy role (see also a previous chapter in this book highlighting the TRPV1 structures).

Clearly, based on the success in the biophysical characterization of the capsaicin receptor and other ion channels, it would be highly beneficial for the TRP channel field to identify toxins with specificity for other TRP members.

References

Aggarwal SK, MacKinnon R (1996) Contribution of the S4 segment to gating charge in the Shaker K+ channel. Neuron 16(6):1169–1177

Alabi AA, Bahamonde MI, Jung HJ, Kim JI, Swartz KJ (2007) Portability of paddle motif function and pharmacology in voltage sensors. Nature 450(7168):370–375

Anderson CS, MacKinnon R, Smith C, Miller C (1988) Charybdotoxin block of single Ca2+-activated K+ channels. Effects of channel gating, voltage, and ionic strength. J Gen Physiol 91 (3):317–333

Andreev YA, Kozlov SA, Koshelev SG, Ivanova EA, Monastyrnaya MM, Kozlovskaya EP, Grishin EV (2008) Analgesic compound from sea anemone Heteractis crispa is the first polypeptide inhibitor of vanilloid receptor 1 (TRPV1). J Biol Chem 283(35):23914–23921

Bae C, Kalia J, Song I, Yu J, Kim HH, Swartz KJ, Kim JI (2012) High yield production and refolding of the double-knot toxin, an activator of TRPV1 channels. PloS One 7(12)

Banerjee A, Lee A, Campbell E, Mackinnon R (2013) Structure of a pore-blocking toxin in complex with a eukaryotic voltage-dependent K(+) channel. Elife 2:e00594

Basbaum AI, Bautista DM, Scherrer G, Julius D (2009) Cellular and molecular mechanisms of pain. Cell 139(2):267–284

Blumenthal KM, Seibert AL (2003) Voltage-gated sodium channel toxins: poisons, probes, and future promise. Cell Biochem Biophys 38(2):215–238

Bodding M (2007) TRP proteins and cancer. Cell Signal 19(3):617–624

Bohlen CJ, Julius D (2012) Receptor-targeting mechanisms of pain-causing toxins: how ow? Toxicon 60(3):254–264

Bohlen CJ, Priel A, Zhou S, King D, Siemens J, Julius D (2010) A bivalent tarantula toxin activates the capsaicin receptor, TRPV1, by targeting the outer pore domain. Cell 141 (5):834–845

Bohlen CJ, Chesler AT, Sharif-Naeini R, Medzihradszky KF, Zhou S, King D, Sanchez EE, Burlingame AL, Basbaum AI, Julius D (2011) A heteromeric Texas coral snake toxin targets acid-sensing ion channels to produce pain. Nature 479(7373):410–414

Bond A (2006) Exenatide (Byetta) as a novel treatment option for type 2 diabetes mellitus. Proc (Bayl Univ Med Cent) 19(3):281–284

Bosmans F, Swartz KJ (2010) Targeting voltage sensors in sodium channels with spider toxins. Trends Pharmacol Sci 31(4):175–182
Bosmans F, Tytgat J (2007) Voltage-gated sodium channel modulation by scorpion alpha-toxins. Toxicon 49(2):142–158
Bosmans F, Martin-Eauclaire MF, Swartz KJ (2008) Deconstructing voltage sensor function and pharmacology in sodium channels. Nature 456(7219):202–208
Bowen CV, DeBay D, Ewart HS, Gallant P, Gormley S, Ilenchuk TT, Iqbal U, Lutes T, Martina M, Mealing G, Merkley N, Sperker S, Moreno MJ, Rice C, Syvitski RT, Stewart JM (2013) In vivo detection of human TRPV6-rich tumors with anti-cancer peptides derived from soricidin. PLoS One 8(3):e58866
Cameron J, Capra MF (1993) The basis of the paradoxical disturbance of temperature perception in ciguatera poisoning. J Toxicol Clin Toxicol 31(4):571–579
Cao E, Liao M, Cheng Y, Julius D (2013) TRPV1 structures in distinct conformations reveal activation mechanisms. Nature 504(7478):113–118
Caterina MJ, Schumacher MA, Tominaga M, Rosen TA, Levine JD, Julius D (1997) The capsaicin receptor: a heat-activated ion channel in the pain pathway. Nature 389(6653):816–824
Catterall WA (1986) Molecular properties of voltage-sensitive sodium channels. Annu Rev Biochem 55:953–985
Catterall WA, Cestele S, Yarov-Yarovoy V, Yu FH, Konoki K, Scheuer T (2007) Voltage-gated ion channels and gating modifier toxins. Toxicon 49(2):124–141
Chen J, Lariviere WR (2010) The nociceptive and anti-nociceptive effects of bee venom injection and therapy: a double-edged sword. Prog Neurobiol 92(2):151–183
Chen YN, Li KC, Li Z, Shang GW, Liu DN, Lu ZM, Zhang JW, Ji YH, Gao GD, Chen J (2006) Effects of bee venom peptidergic components on rat pain-related behaviors and inflammation. Neuroscience 138(2):631–640
Chi V, Pennington MW, Norton RS, Tarcha EJ, Londono LM, Sims-Fahey B, Upadhyay SK, Lakey JT, Iadonato S, Wulff H, Beeton C, Chandy KG (2012) Development of a sea anemone toxin as an immunomodulator for therapy of autoimmune diseases. Toxicon 59(4):529–546
Chung YH, Sun Ahn H, Kim D, Hoon Shin D, Su Kim S, Yong Kim K, Bok Lee W, Ik Cha C (2006) Immunohistochemical study on the distribution of TRPC channels in the rat hippocampus. Brain Res 1085(1):132–137
Conticello SG, Pilpel Y, Glusman G, Fainzilber M (2000) Position-specific codon conservation in hypervariable gene families. Trends Genet 16(2):57–59
Conticello SG, Gilad Y, Avidan N, Ben-Asher E, Levy Z, Fainzilber M (2001) Mechanisms for evolving hypervariability: the case of conopeptides. Mol Biol Evol 18(2):120–131
Craik DJ, Daly NL, Waine C (2001) The cystine knot motif in toxins and implications for drug design. Toxicon 39(1):43–60
Cuypers E, Yanagihara A, Karlsson E, Tytgat J (2006) Jellyfish and other cnidarian envenomations cause pain by affecting TRPV1 channels. FEBS Lett 580(24):5728–5732
Cuypers E, Yanagihara A, Rainier JD, Tytgat J (2007) TRPV1 as a key determinant in ciguatera and neurotoxic shellfish poisoning. Biochem Biophys Res Commun 361(1):214–217
Cuypers E, Abdel-Mottaleb Y, Kopljar I, Rainier JD, Raes AL, Snyders DJ, Tytgat J (2008) Gambierol, a toxin produced by the dinoflagellate Gambierdiscus toxicus, is a potent blocker of voltage-gated potassium channels. Toxicon 51(6):974–983
Daly NL, Craik DJ (2011) Bioactive cystine knot proteins. Curr Opin Chem Biol 15(3):362–368
Ding J, Xiao Y, Lu D, Du YR, Cui XY, Chen J (2011) Effects of SKF-96365, a TRPC inhibitor, on melittin-induced inward current and intracellular Ca2+ rise in primary sensory cells. Neurosci Bull 27(3):135–142
Ding J, Zhang JR, Wang Y, Li CL, Lu D, Guan SM, Chen J (2012) Effects of a non-selective TRPC channel blocker, SKF-96365, on melittin-induced spontaneous persistent nociception and inflammatory pain hypersensitivity. Neurosci Bull 28(2):173–181

Diochot S, Baron A, Rash LD, Deval E, Escoubas P, Scarzello S, Salinas M, Lazdunski M (2004) A new sea anemone peptide, APETx2, inhibits ASIC3, a major acid-sensitive channel in sensory neurons. EMBO J 23(7):1516–1525

Diochot S, Baron A, Salinas M, Douguet D, Scarzello S, Dabert-Gay AS, Debayle D, Friend V, Alloui A, Lazdunski M, Lingueglia E (2012) Black mamba venom peptides target acid-sensing ion channels to abolish pain. Nature 490(7421):552–555

Du YR, Xiao Y, Lu ZM, Ding J, Xie F, Fu H, Wang Y, Strong JA, Zhang JM, Chen J (2011) Melittin activates TRPV1 receptors in primary nociceptive sensory neurons via the phospholipase A2 cascade pathways. Biochem Biophys Res Commun 408(1):32–37

Ducancel F, Matre V, Dupont C, Lajeunesse E, Wollberg Z, Bdolah A, Kochva E, Boulain JC, Menez A (1993) Cloning and sequence analysis of cDNAs encoding precursors of sarafotoxins. Evidence for an unusual "rosary-type" organization. J Biol Chem 268(5):3052–3055

Duda TF Jr, Palumbi SR (1999) Molecular genetics of ecological diversification: duplication and rapid evolution of toxin genes of the venomous gastropod Conus. Proc Natl Acad Sci U S A 96 (12):6820–6823

Escoubas P, Rash L (2004) Tarantulas: eight-legged pharmacists and combinatorial chemists. Toxicon 43(5):555–574

Evans DL, Schmidt JO (1990) Insect defenses: adaptive mechanisms and strategies of prey and predators. State University of New York Press, Albany, NY

Fry BG, Vidal N, Norman JA, Vonk FJ, Scheib H, Ramjan SF, Kuruppu S, Fung K, Hedges SB, Richardson MK, Hodgson WC, Ignjatovic V, Summerhayes R, Kochva E (2006) Early evolution of the venom system in lizards and snakes. Nature 439(7076):584–588

Fry BG, Roelants K, Champagne DE, Scheib H, Tyndall JD, King GF, Nevalainen TJ, Norman JA, Lewis RJ, Norton RS, Renjifo C, de la Vega RC (2009) The toxicogenomic multiverse: convergent recruitment of proteins into animal venoms. Annu Rev Genomics Hum Genet 10:483–511

Han TS, Teichert RW, Olivera BM, Bulaj G (2008) Conus venoms - a rich source of peptide-based therapeutics. Curr Pharm Des 14(24):2462–2479

Heir GM (2005) Ciguatera neurotoxin poisoning mimicking burning mouth syndrome. Quintessence Int 36(7–8):547–550

Hinman A, Chuang HH, Bautista DM, Julius D (2006) TRP channel activation by reversible covalent modification. Proc Natl Acad Sci U S A 103(51):19564–19568

Hoenderop JG, Nilius B, Bindels RJ (2005) Calcium absorption across epithelia. Physiol Rev 85 (1):373–422

Hucho F, Tsetlin VI, Machold J (1996) The emerging three-dimensional structure of a receptor. The nicotinic acetylcholine receptor. Eur J Biochem 239(3):539–557

Jiang Y, Lee A, Chen J, Ruta V, Cadene M, Chait BT, MacKinnon R (2003a) X-ray structure of a voltage-dependent K+ channel. Nature 423(6935):33–41

Jiang Y, Ruta V, Chen J, Lee A, MacKinnon R (2003b) The principle of gating charge movement in a voltage-dependent K+ channel. Nature 423(6935):42–48

Jimenez EC, Donevan S, Walker C, Zhou LM, Nielsen J, Cruz LJ, Armstrong H, White HS, Olivera BM (2002) Conantokin-L, a new NMDA receptor antagonist: determinants for anticonvulsant potency. Epilepsy Res 51(1–2):73–80

Julius D, Basbaum AI (2001) Molecular mechanisms of nociception. Nature 413(6852):203–210

Kitaguchi T, Swartz KJ (2005) An inhibitor of TRPV1 channels isolated from funnel Web spider venom. Biochemistry 44(47):15544–15549

Kopljar I, Labro AJ, Cuypers E, Johnson HW, Rainier JD, Tytgat J, Snyders DJ (2009) A polyether biotoxin binding site on the lipid-exposed face of the pore domain of Kv channels revealed by the marine toxin gambierol. Proc Natl Acad Sci U S A 106(24):9896–9901

Kordis D, Gubensek F (2000) Adaptive evolution of animal toxin multigene families. Gene 261 (1):43–52

Kress M, Karasek J, Ferrer-Montiel AV, Scherbakov N, Haberberger RV (2008) TRPC channels and diacylglycerol dependent calcium signaling in rat sensory neurons. Histochem Cell Biol 130(4):655–667

Lee SY, MacKinnon R (2004) A membrane-access mechanism of ion channel inhibition by voltage sensor toxins from spider venom. Nature 430(6996):232–235

Lee CW, Kim S, Roh SH, Endoh H, Kodera Y, Maeda T, Kohno T, Wang JM, Swartz KJ, Kim JI (2004) Solution structure and functional characterization of SGTx1, a modifier of Kv2.1 channel gating. Biochemistry 43(4):890–897

Lena C, Changeux JP (1998) Allosteric nicotinic receptors, human pathologies. J Physiol Paris 92 (2):63–74

Lewis RJ (2001) The changing face of ciguatera. Toxicon 39(1):97–106

Lewis RJ (2006) Ciguatera: Australian perspectives on a global problem. Toxicon 48(7):799–809

Lewis RJ, Garcia ML (2003) Therapeutic potential of venom peptides. Nat Rev Drug Discov 2 (10):790–802

Li KC, Chen J (2004) Altered pain-related behaviors and spinal neuronal responses produced by s.c. injection of melittin in rats. Neuroscience 126(3):753–762

Liao M, Cao E, Julius D, Cheng Y (2013) Structure of the TRPV1 ion channel determined by electron cryo-microscopy. Nature 504(7478):107–112

Liberatore GT, Samson A, Bladin C, Schleuning WD, Medcalf RL (2003) Vampire bat salivary plasminogen activator (desmoteplase): a unique fibrinolytic enzyme that does not promote neurodegeneration. Stroke 34(2):537–543

Llewellyn LE (2006) Saxitoxin, a toxic marine natural product that targets a multitude of receptors. Nat Prod Rep 23(2):200–222

Lynch SS, Cheng CM, Yee JL (2006) Intrathecal ziconotide for refractory chronic pain. Ann Pharmacother 40(7–8):1293–1300

MacKinnon R, Miller C (1988) Mechanism of charybdotoxin block of the high-conductance, Ca2+-activated K+ channel. J Gen Physiol 91(3):335–349

MacKinnon R, Miller C (1989) Mutant potassium channels with altered binding of charybdotoxin, a pore-blocking peptide inhibitor. Science 245(4924):1382–1385

Macpherson LJ, Dubin AE, Evans MJ, Marr F, Schultz PG, Cravatt BF, Patapoutian A (2007) Noxious compounds activate TRPA1 ion channels through covalent modification of cysteines. Nature 445(7127):541–545

Martin FL, Sangaralingham SJ, Huntley BK, McKie PM, Ichiki T, Chen HH, Korinek J, Harders GE, Burnett JC Jr (2012) CD-NP: a novel engineered dual guanylyl cyclase activator with anti-fibrotic actions in the heart. PLoS One 7(12):e52422

Mazzuca M, Heurteaux C, Alloui A, Diochot S, Baron A, Voilley N, Blondeau N, Escoubas P, Gelot A, Cupo A, Zimmer A, Zimmer AM, Eschalier A, Lazdunski M (2007) A tarantula peptide against pain via ASIC1a channels and opioid mechanisms. Nat Neurosci 10(8):943–945

Mebs D (2002) Venomous and poisonous animals. Medpharm, Stuttgart, Germany

Medcalf RL (2012) Desmoteplase: discovery, insights and opportunities for ischaemic stroke. Br J Pharmacol 165(1):75–89

Milescu M, Vobecky J, Roh SH, Kim SH, Jung HJ, Kim JI, Swartz KJ (2007) Tarantula toxins interact with voltage sensors within lipid membranes. J Gen Physiol 130(5):497–511

Milescu M, Bosmans F, Lee S, Alabi AA, Kim JI, Swartz KJ (2009) Interactions between lipids and voltage sensor paddles detected with tarantula toxins. Nat Struct Mol Biol 16(10):1080–1085

Miljanich GP (2004) Ziconotide: neuronal calcium channel blocker for treating severe chronic pain. Curr Med Chem 11(23):3029–3040

Miller C (1995) The charybdotoxin family of K+ channel-blocking peptides. Neuron 15(1):5–10

Moriguchi S, Shioda N, Yamamoto Y, Tagashira H, Fukunaga K (2012) The T-type voltage-gated calcium channel as a molecular target of the novel cognitive enhancer ST101: enhancement of

long-term potentiation and CaMKII autophosphorylation in rat cortical slices. J Neurochem 121(1):44–53
Negri L, Lattanzi R, Giannini E, Colucci M, Margheriti F, Melchiorri P, Vellani V, Tian H, De Felice M, Porreca F (2006) Impaired nociception and inflammatory pain sensation in mice lacking the prokineticin receptor PKR1: focus on interaction between PKR1 and the capsaicin receptor TRPV1 in pain behavior. J Neurosci 26(25):6716–6727
Negri L, Lattanzi R, Giannini E, Melchiorri P (2007) Bv8/Prokineticin proteins and their receptors. Life Sci 81(14):1103–1116
Nieto-Posadas A, Jara-Oseguera A, Rosenbaum T (2011) TRP channel gating physiology. Curr Top Med Chem 11(17):2131–2150
Nijenhuis T, Hoenderop JG, Nilius B, Bindels RJ (2003a) (Patho)physiological implications of the novel epithelial Ca2+ channels TRPV5 and TRPV6. Pflugers Arch 446(4):401–409
Nijenhuis T, Hoenderop JG, van der Kemp AW, Bindels RJ (2003b) Localization and regulation of the epithelial Ca2+ channel TRPV6 in the kidney. J Am Soc Nephrol 14(11):2731–2740
Nilius B, Owsianik G, Voets T, Peters JA (2007) Transient receptor potential cation channels in disease. Physiol Rev 87(1):165–217
Norton RS, Olivera BM (2006) Conotoxins down under. Toxicon 48(7):780–798
Olivera BM, Walker C, Cartier GE, Hooper D, Santos AD, Schoenfeld R, Shetty R, Watkins M, Bandyopadhyay P, Hillyard DR (1999) Speciation of cone snails and interspecific hyperdivergence of their venom peptides. Potential evolutionary significance of introns. Ann N Y Acad Sci 870:223–237
Park CS, Miller C (1992) Interaction of charybdotoxin with permeant ions inside the pore of a K+ channel. Neuron 9(2):307–313
Peters AA, Simpson PT, Bassett JJ, Lee JM, Da Silva L, Reid LE, Song S, Parat MO, Lakhani SR, Kenny PA, Roberts-Thomson SJ, Monteith GR (2012) Calcium channel TRPV6 as a potential therapeutic target in estrogen receptor-negative breast cancer. Mol Cancer Ther 11(10):2158–2168
Phillips LR, Milescu M, Li-Smerin Y, Mindell JA, Kim JI, Swartz KJ (2005) Voltage-sensor activation with a tarantula toxin as cargo. Nature 436(7052):857–860
Phui Yee JS, Nanling G, Afifiyan F, Donghui M, Siew Lay P, Armugam A, Jeyaseelan K (2004) Snake postsynaptic neurotoxins: gene structure, phylogeny and applications in research and therapy. Biochimie 86(2):137–149
Ragnarsson L, Mortensen M, Dodd PR, Lewis RJ (2002) Spermine modulation of the glutamate (NMDA) receptor is differentially responsive to conantokins in normal and Alzheimer's disease human cerebral cortex. J Neurochem 81(4):765–779
Reddrop C, Moldrich RX, Beart PM, Farso M, Liberatore GT, Howells DW, Petersen KU, Schleuning WD, Medcalf RL (2005) Vampire bat salivary plasminogen activator (desmoteplase) inhibits tissue-type plasminogen activator-induced potentiation of excitotoxic injury. Stroke 36(6):1241–1246
Riccio A, Medhurst AD, Mattei C, Kelsell RE, Calver AR, Randall AD, Benham CD, Pangalos MN (2002) mRNA distribution analysis of human TRPC family in CNS and peripheral tissues. Brain Res Mol Brain Res 109(1–2):95–104
Rodriguez de la Vega RC, Merino E, Becerril B, Possani LD (2003) Novel interactions between K + channels and scorpion toxins. Trends Pharmacol Sci 24(5):222–227
Rowe AH, Xiao Y, Rowe MP, Cummins TR, Zakon HH (2013) Voltage-gated sodium channel in grasshopper mice defends against bark scorpion toxin. Science 342(6157):441–446
Ruta V, Chen J, MacKinnon R (2005) Calibrated measurement of gating-charge arginine displacement in the KvAP voltage-dependent K+ channel. Cell 123(3):463–475
Santoni G, Farfariello V, Amantini C (2011) TRPV channels in tumor growth and progression. Adv Exp Med Biol 704:947–967
Schweitz H, Bidard JN, Lazdunski M (1990) Purification and pharmacological characterization of peptide toxins from the black mamba (Dendroaspis polylepis) venom. Toxicon 28(7):847–856

Seoh SA, Sigg D, Papazian DM, Bezanilla F (1996) Voltage-sensing residues in the S2 and S4 segments of the Shaker K+ channel. Neuron 16(6):1159–1167

Siemens J, Zhou S, Piskorowski R, Nikai T, Lumpkin EA, Basbaum AI, King D, Julius D (2006) Spider toxins activate the capsaicin receptor to produce inflammatory pain. Nature 444 (7116):208–212

Skinner WS, Adams ME, Quistad GB, Kataoka H, Cesarin BJ, Enderlin FE, Schooley DA (1989) Purification and characterization of two classes of neurotoxins from the funnel web spider, Agelenopsis aperta. J Biol Chem 264(4):2150–2155

Smith JJ, Hill JM, Little MJ, Nicholson GM, King GF, Alewood PF (2011) Unique scorpion toxin with a putative ancestral fold provides insight into evolution of the inhibitor cystine knot motif. Proc Natl Acad Sci U S A 108(26):10478–10483

Sollod BL, Wilson D, Zhaxybayeva O, Gogarten JP, Drinkwater R, King GF (2005) Were arachnids the first to use combinatorial peptide libraries? Peptides 26(1):131–139

Speed MP, Ruxton GD, Mappes J, Sherratt TN (2012) Why are defensive toxins so variable? An evolutionary perspective. Biol Rev Camb Philos Soc 87(4):874–884

Staats PS, Yearwood T, Charapata SG, Presley RW, Wallace MS, Byas-Smith M, Fisher R, Bryce DA, Mangieri EA, Luther RR, Mayo M, McGuire D, Ellis D (2004) Intrathecal ziconotide in the treatment of refractory pain in patients with cancer or AIDS: a randomized controlled trial. JAMA 291(1):63–70

Swartz KJ (2007) Tarantula toxins interacting with voltage sensors in potassium channels. Toxicon 49(2):213–230

Swartz KJ (2008) Sensing voltage across lipid membranes. Nature 456(7224):891–897

Swartz KJ, MacKinnon R (1997a) Hanatoxin modifies the gating of a voltage-dependent K+ channel through multiple binding sites. Neuron 18(4):665–673

Swartz KJ, MacKinnon R (1997b) Mapping the receptor site for hanatoxin, a gating modifier of voltage-dependent K+ channels. Neuron 18(4):675–682

Szeto TH, Wang XH, Smith R, Connor M, Christie MJ, Nicholson GM, King GF (2000) Isolation of a funnel-web spider polypeptide with homology to mamba intestinal toxin 1 and the embryonic head inducer Dickkopf-1. Toxicon 38(3):429–442

Terlau H, Olivera BM (2004) Conus venoms: a rich source of novel ion channel-targeted peptides. Physiol Rev 84(1):41–68

Tominaga M, Tominaga T (2005) Structure and function of TRPV1. Pflugers Arch 451(1):143–150

Tsetlin V (1999) Snake venom alpha-neurotoxins and other 'three-finger' proteins. Eur J Biochem 264(2):281–286

Twede VD, Miljanich G, Olivera BM, Bulaj G (2009) Neuroprotective and cardioprotective conopeptides: an emerging class of drug leads. Curr Opin Drug Discov Devel 12(2):231–239

Vassilevski AA, Fedorova IM, Maleeva EE, Korolkova YV, Efimova SS, Samsonova OV, Schagina LV, Feofanov AV, Magazanik LG, Grishin EV (2010) Novel class of spider toxin: active principle from the yellow sac spider Cheiracanthium punctorium venom is a unique two-domain polypeptide. J Biol Chem 285(42):32293–32302

Vellani V, Colucci M, Lattanzi R, Giannini E, Negri L, Melchiorri P, McNaughton PA (2006) Sensitization of transient receptor potential vanilloid 1 by the prokineticin receptor agonist Bv8. J Neurosci 26(19):5109–5116

Vetter I, Touska F, Hess A, Hinsbey R, Sattler S, Lampert A, Sergejeva M, Sharov A, Collins LS, Eberhardt M, Engel M, Cabot PJ, Wood JN, Vlachova V, Reeh PW, Lewis RJ, Zimmermann K (2012) Ciguatoxins activate specific cold pain pathways to elicit burning pain from cooling. EMBO J 31(19):3795–3808

Voets T (2012) TRP channel blamed for burning cold after a tropical fish meal. EMBO J 31 (19):3785–3787

von Lueder TG, Sangaralingham SJ, Wang BH, Kompa AR, Atar D, Burnett JC Jr, Krum H (2013) Renin-angiotensin blockade combined with natriuretic peptide system augmentation: novel therapeutic concepts to combat heart failure. Circ Heart Fail 6(3):594–605

Vriens J, Nilius B, Vennekens R (2008) Herbal compounds and toxins modulating TRP channels. Curr Neuropharmacol 6(1):79–96

Wang X, Connor M, Smith R, Maciejewski MW, Howden ME, Nicholson GM, Christie MJ, King GF (2000) Discovery and characterization of a family of insecticidal neurotoxins with a rare vicinal disulfide bridge. Nat Struct Biol 7(6):505–513

Wen S, Wilson DT, Kuruppu S, Korsinczky ML, Hedrick J, Pang L, Szeto T, Hodgson WC, Alewood PF, Nicholson GM (2005) Discovery of an MIT-like atracotoxin family: spider venom peptides that share sequence homology but not pharmacological properties with AVIT family proteins. Peptides 26(12):2412–2426

Wissenbach U, Niemeyer BA, Fixemer T, Schneidewind A, Trost C, Cavalie A, Reus K, Meese E, Bonkhoff H, Flockerzi V (2001) Expression of CaT-like, a novel calcium-selective channel, correlates with the malignancy of prostate cancer. J Biol Chem 276(22):19461–19468

Yuan CH, He QY, Peng K, Diao JB, Jiang LP, Tang X, Liang SP (2008) Discovery of a distinct superfamily of Kunitz-type toxin (KTT) from tarantulas. PLoS One 3(10):e3414

Zamponi GW, Lewis RJ, Todorovic SM, Arneric SP, Snutch TP (2009) Role of voltage-gated calcium channels in ascending pain pathways. Brain Res Rev 60(1):84–89

Zhu S, Darbon H, Dyason K, Verdonck F, Tytgat J (2003) Evolutionary origin of inhibitor cystine knot peptides. FASEB J 17(12):1765–1767

Zhuang L, Peng JB, Tou L, Takanaga H, Adam RM, Hediger MA, Freeman MR (2002) Calcium-selective ion channel, CaT1, is apically localized in gastrointestinal tract epithelia and is aberrantly expressed in human malignancies. Lab Invest 82(12):1755–1764

Zuo XP, Ji YH (2004) Molecular mechanism of scorpion neurotoxins acting on sodium channels: insight into their diverse selectivity. Mol Neurobiol 30(3):265–278

Phosphoinositide Regulation of TRP Channels

Tibor Rohacs

Contents

T. Rohacs (✉)
Rutgers – New Jersey Medical School, Newark, NJ, USA
e-mail: tibor.rohacs@rutgers.edu

B. Nilius and V. Flockerzi (eds.), *Mammalian Transient Receptor Potential (TRP) Cation Channels*, Handbook of Experimental Pharmacology 223, DOI 10.1007/978-3-319-05161-1_18,

Abstract

Transient Receptor Potential (TRP) channels are activated by stimuli as diverse as heat, cold, noxious chemicals, mechanical forces, hormones, neurotransmitters, spices, and voltage. Besides their presumably similar general architecture, probably the only common factor regulating them is phosphoinositides. The regulation of TRP channels by phosphoinositides is complex. There are a large number of TRP channels where phosphatidylinositol 4,5 bisphosphate [PI(4,5)P_2 or PIP_2] acts as a positive cofactor, similarly to many other ion channels. In several cases, however, PI(4,5)P_2 inhibits TRP channel activity, sometimes even concurrently with the activating effect. This chapter will provide a comprehensive overview of the literature on regulation of TRP channels by membrane phosphoinositides.

Keywords

Phosphoinositides • PIP_2 • TRP channel • Phospholipase C • TRPV1

Abbreviations

AASt	Arachydonyl-stearyl
AKAP	A-kinase anchoring protein
DAG	Diacylglycerol
DGS-NTA	1,2-dioleoyl-sn-glycero-3-[(*N*-(5-amino-1-carboxypentyl) iminodiacetic acid)succinyl]
DRG	Dorsal root ganglion
ER	Endoplasmic reticulum
GPCR	G Protein Coupled Receptor
IP_3	Inositol-1,4,5-trisphopshate
Kir	K^+ inwardly rectifying
PI	Phosphatidylinositol
PI(4)P	Phosphatidylinositol 4-phosphate
PI(4,5)P_2	Phosphatidylinositol 4,5-bisphosphate
PI4K	Phosphatidylinositol 4-kinase
PIP5K	Phosphatidylinositol 4-phosphate 5-kinase
PI3K	Phosphoinositide 3-kinase
PLC	Phospholipase C
PG	Phosphatidyl glycerol
PKC	Protein Kinase C
PH	Pleckstrin-homology
Pirt	Phosphoinositide interacting regulator of TRP

PM	Plasma membrane
TG	Trigeminal ganglion
TPC	Two-pore channel
TRP	Transient Receptor Potential
TRPA	TRP Ankyrin
TRPC	TRP Classical
TRPL	TRP-like
TRPM	TRP Melastatin
TRPML	TRP Mucolipin
TRPP	TRP Polycystin
TRPV	TRP Vanilloid
VSP	Voltage sensitive phosphatase

1 Introduction

Membrane phospholipids had generally been viewed as passive media that ion channels are embedded in. The beginning of the end of this view was probably marked by Donald Hilgemann's finding that the cardiac Na^+/Ca^{2+} exchangers and the cardiac K_{ATP} channels require the presence of $PI(4,5)P_2$ for activity (Hilgemann and Ball 1996). There were sporadic publications before this article reported effects of $PI(4,5)P_2$ on channels or transporters (see Huang 2007 for references), but it was Hilgemann's seminal publication that sparked the explosion of interest in this lipid in the context of ion channel regulation. In 1998 high profile articles from four different laboratories extended $PI(4,5)P_2$ regulation to several inwardly rectifying K^+ (Kir) channels (Baukrowitz et al. 1998; Huang et al. 1998; Shyng and Nichols 1998; Sui et al. 1998). This was followed by an ever-growing number of publications showing that a largely unexpected number and variety of ion channels require $PI(4,5)P_2$ for activity. By now it seems that phosphoinositides are general regulators of most mammalian ion channels (Hilgemann et al. 2001; Suh and Hille 2008; Gamper and Rohacs 2012).

The first two papers on the phosphoinositide regulation of TRP channels were published in 2001. Chuang et al proposed that $PI(4,5)P_2$ tonically inhibits the heat- and capsaicin-activated TRPV1 channels, and breakdown of this lipid upon PLC activation relieves this inhibition leading to potentiation of TRPV1 activity by pro-inflammatory agents such as bradykinin (Chuang et al. 2001). Another article found that $PI(4,5)P_2$ inhibits the *drosophila* TRPL channel in excised patches (Estacion et al. 2001). Since the whole subfamily of mammalian TRPCs are activated downstream of PLC, $PI(4,5)P_2$ being a general inhibitor of TRP channels was a highly attractive hypothesis. For a while, the general view was that TRP channels are inhibited by $PI(4,5)P_2$. This started to change when paper after paper showed that $PI(4,5)P_2$ activates other TRP channels, such as TRPM7 (Runnels et al. 2002), TRPM5 (Liu and Liman 2003), TRPM8 (Liu and Qin 2005; Rohacs et al. 2005), TRPV5 (Lee et al. 2005; Rohacs et al. 2005), and TRPM4 (Nilius et al. 2006) similar to most other ion channel families. The number of TRP channels that are influenced by phosphoinositides are growing ever since. It is quite likely that at least in the 3 major families, most, if not all, members are regulated by phosphoinositides, with the majority showing a dependence on some of these lipids for activity, in most cases $PI(4,5)P_2$.

There have been several reviews on the topic published in recent years (Hardie 2007; Qin 2007; Rohacs and Nilius 2007; Nilius et al. 2008; Rohacs 2009). The field progressed considerably since; here I will give a comprehensive overview of the literature, incorporating novel findings since the last reviews, but trying to systematically describe earlier results too. Due to space restrictions, in several cases I refer to earlier reviews for details where I was not aware of new publications on $PI(4,5)P_2$ regulation. I will also have no room for extensive discussions on some of the more complex or controversial regulation themes. I will provide a brief overview of the general function of the individual channels before discussing their regulation by phosphoinositides. In some cases the function of the channel is more complicated, or debated, so these short summaries are unavoidably oversimplifications. The reader is referred to other chapters of this handbook for more detailed information on individual TRP channels.

2 Phosphoinositide Signaling

$PI(4,5)P_2$ is generated by two phosphorylation steps from phosphatidylinositol (PI) (Fig. 1). Phosphoinositide 4 kinases (PI4K) catalyze the formation of phosphatidylinositol 4-phosphate [PI(4)P], which is further phosphorylated by phosphatidylinositol 4-phosphate 5 kinases (PIP5K). Phosphoinositide 3 kinases (PI3K) add a phosphate to the 3rd position in the inositol ring and form either $PI(3,4)P_2$ or $PI(3,4,5)P_3$, which are second messengers in signaling by various growth factors. Phospholipase C (PLC) enzymes catalyze the hydrolysis of $PI(4,5)P_2$ and the formation of the two classical second messengers inositol 1,4,5 trisphosphate (IP_3) and diacylglycerol (DAG). PI(4)P is also known to be the substrate for PLC enzymes, even though most PLCs are less active in hydrolyzing this lipid than $PI(4,5)P_2$. PLCβ isoforms (PLCβ1-4) are activated by G-protein coupled receptors (GPCRs) that couple to the Gq family of heterotrimeric G-proteins. PLCγ-s (PLCγ1 and 2) are activated by receptor tyrosine kinases. PLCδ-s (PLCδ1, δ3, and δ4) do not have obvious activators; these isoforms are the most sensitive to Ca^{2+} among the three classical PLC groups, and they can be activated by Ca^{2+} influx alone (Allen et al. 1997; Lukacs et al. 2013a). In addition to the classical PLC-s, newer isoforms have also been cloned more recently (PLCη1 and η2 and PLCε); their regulation is less well understood (Fukami et al. 2010).

Besides their precursor function, and regulation of ion channels, phosphoinositides are important regulators of cytoskeletal organization, membrane traffic, and general cellular architecture (Saarikangas et al. 2010; Shewan et al. 2011).

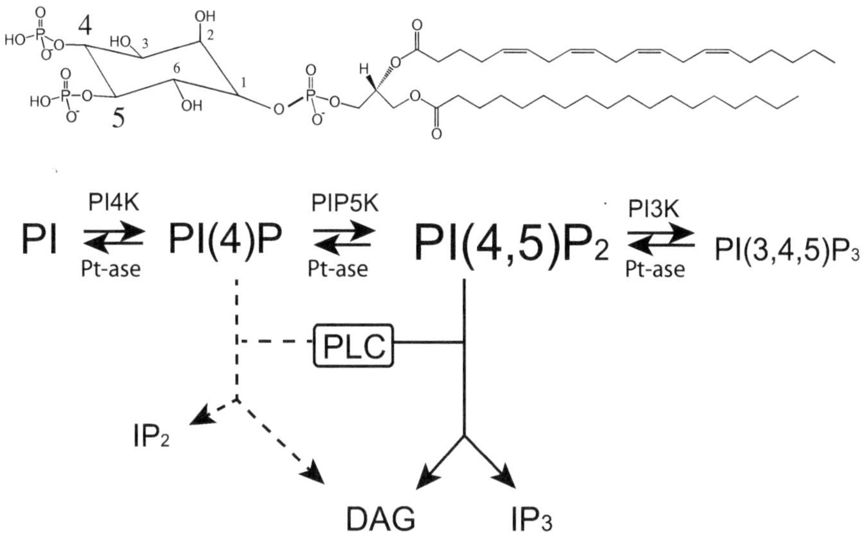

Fig. 1 Phosphoinositide metabolism

These other functions of phosphoinositides may also affect ion channel activity; see the regulation of TRPV1 and TRPC channels.

3 Methods to Study Phosphoinositide Regulation

The methods to study phosphoinositide regulation of ion channels have been reviewed extensively in several articles (Rohacs et al. 2002; Rohacs and Nilius 2007; Suh and Hille 2008; Gamper and Rohacs 2012); here I will give a brief overview of the various techniques. Figures 2 and 3 show a compilation of experimental data using a number of techniques discussed here, all demonstrating the dependence of the activity of TRPM8 on $PI(4,5)P_2$.

3.1 Excised Inside-Out Patches

Perhaps the most often used technique to demonstrate the effects of phosphoinositides is the excised inside-out patch configuration of the patch clamp technique, which allows access to the inner leaflet of the plasma membrane, where phosphoinositides are located. A common characteristic of $PI(4,5)P_2$ dependent ion channels is run-down of channel activity upon excision into an ATP free solution; see Fig. 2a for an example on TRPM8. Run-down of $PI(4,5)P_2$ dependent ion channels is caused by dephosphorylation of $PI(4,5)P_2$ and PI(4)P by lipid phosphatases associated with the patch membrane. Agents that chelate endogenous $PI(4,5)P_2$ and other phosphoinositides such as poly-Lysine (Fig. 2a) or anti-$PI(4,5)P_2$

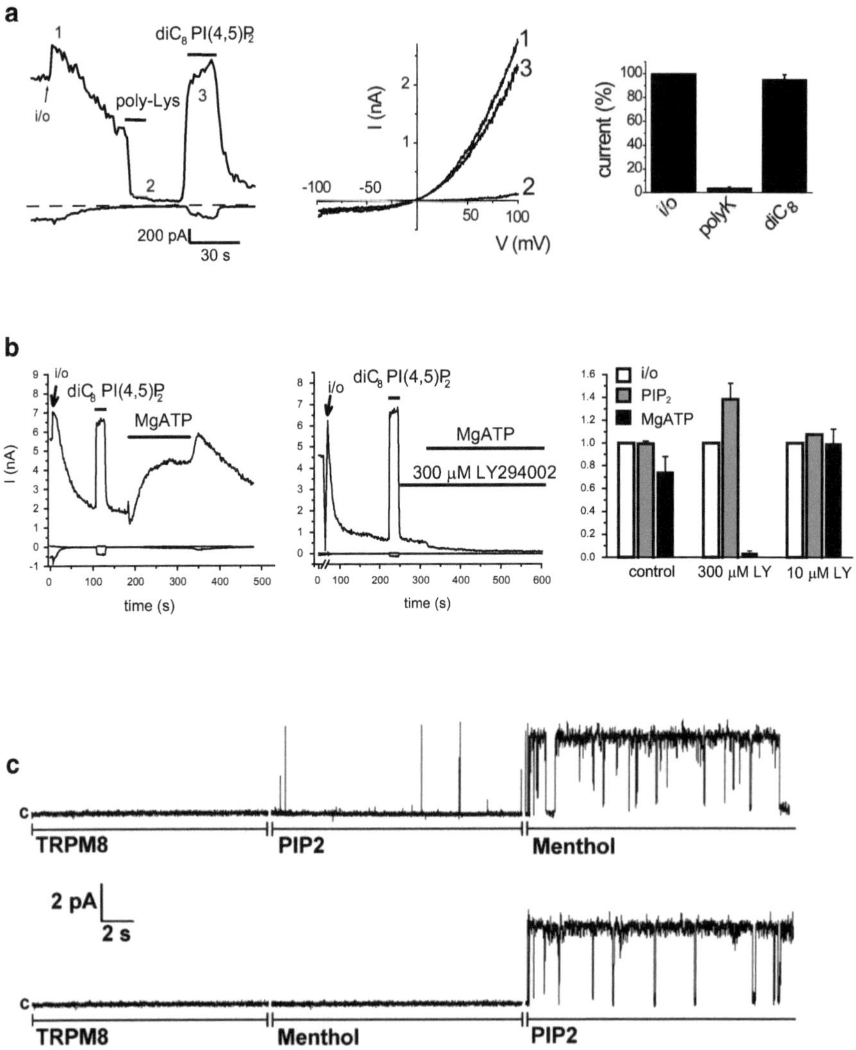

Fig. 2 Activation of TRPM8 excised inside-out patches and planar lipid bilayers. (**a**) Representative trace for recording of TRPM8 currents in large patches in *Xenopus* oocytes with 500 μM menthol in the patch pipette from (Rohacs et al. 2005). *Left panel* shows currents at −100 and +100 mV, from ramp protocol shown in the middle. Application of 30 μg/ml poly-Lysine and 50 μM diC$_8$ PI(4,5)P$_2$ are indicated with the *horizontal lines*. (**b**) Similar excised inside-out patch measurement from (Yudin et al. 2011), demonstrating the effect of 2 mM MgATP and LY294002. *Right panel* shows summary for 300 μM LY294002 that inhibits both PI4K and PI3K and for 10 μM LY294002 that selectively inhibits PI3K. (**c**) Representative measurement in planar lipid bilayers with purified TRPM8 from (Zakharian et al. 2009). In the upper trace, first, the TRPM8 protein shows no activity in the absence of menthol and PI(4,5)P$_2$, and then, diC$_8$ PI(4,5)P$_2$ is applied and then menthol in the continuous presence of PI(4,5)P$_2$. In the *bottom* trace menthol and PI(4,5)P$_2$ are applied in the reverse order

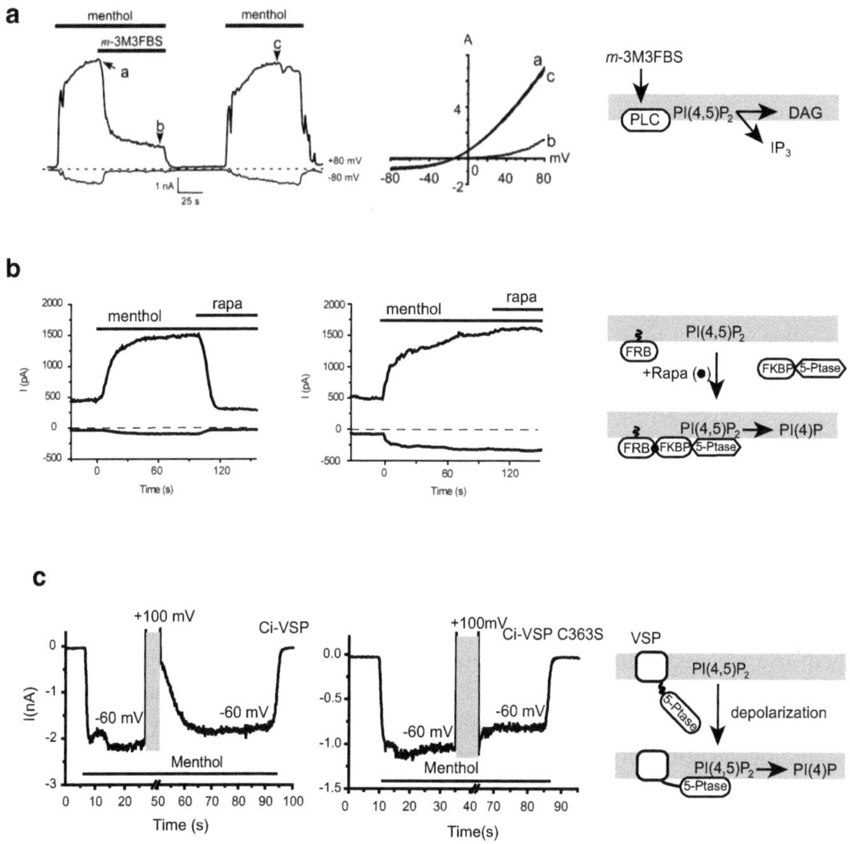

Fig. 3 Inhibition of TRPM8 by reduction of PI(4,5)P_2 in intact cells, *right panels* in **a**, **b** and **c** show cartoon of the method used. (**a**) Whole-cell patch clamp measurement from (Daniels et al. 2009) showing inhibition of menthol-induced RPM8 currents by the pharmacological PLC activator *m*-3M3FBS. (**b**) Inhibition of TRPM8 activity by the rapamycin-inducible 5-phosphatase from (Varnai et al. 2006). (**c**) Inhibition of TRPM8 activity by depolarization-induced depletion of PI(4,5)P_2 in ciVSP expressing cells from (Yudin et al. 2011). *Left panel* shows a measurement with the active ciVSP, middle panel with an inactive mutant

antibody accelerate run-down. When PI(4,5)P_2 is applied to the patch, it reactivates PI(4,5)P_2 dependent ion channels (Fig 2a, b). Natural phosphoinositides have a combination of long acyl chains, arachidonyl-stearyl (AASt) being the most common. These natural lipids or their synthetic versions such as dipalmitoyl PI(4,5)P_2 form micelles in aqueous solutions, and activate channels relatively slowly, and their effects are usually long lasting, since they accumulate in the patch membrane (Rohacs et al. 2002). Short acyl chain DiC_8 phosphoinositides are easier to handle, and their effect washes out quickly (Fig. 2a), making it possible to test multiple compounds in one patch, which can be useful for constructing dose–response curves or testing phosphoinositide specificity profiles.

Exogenously applied lipids may cause unwanted side effects (Hilgemann 2012). The effects of endogenous phosphoinositides however can also be studied in excised inside-out patches, by applying MgATP, which activates endogenous lipid kinases associated to the patch membrane (Hilgemann and Ball 1996; Sui et al. 1998). Figure 2b shows an example, where MgATP reactivated the PI(4,5)P_2 dependent TRPM8 channels, and the effect was inhibited by LY294002 at concentrations where it inhibits PI4K-kinases, but not at lower concentrations where it acts selectively on PI3-Kinases.

Applying PI(4,5)P_2 directly to the intracellular leaflet of the patch membrane is still the state of the art method to demonstrate direct activation by the lipid. Excised patches however may contain many different proteins, including the lipid phosphatases and kinases just described, and cytoskeletal elements (Sachs 2010), which may affect channel activity (Furukawa et al. 1996). Thus demonstrating a direct effect of phosphoinositides beyond doubt requires techniques based on purified proteins.

3.2 Techniques on Purified Channel Proteins

Biochemical-binding assays have also been used to demonstrate direct binding of channel proteins to PI(4,5)P_2. There is a plethora of specific techniques; most of them use isolated cytoplasmic fragments of ion channels (Rohacs 2009; Gamper and Rohacs 2012). Conceptually, if PI(4,5)P_2 acts directly on a channel, it has to bind to it. If binding of PI(4,5)P_2 is demonstrated, however, it does not prove that the biological effect of PI(4,5)P_2 happens through that binding. PI(4,5)P_2 has a very high charge density; thus, it could bind to positively charged protein surfaces, even if those do not face the plasma membrane in situ.

Perhaps the most convincing evidence for a direct effect of PI(4,5)P_2 on a channel is to study its effect on purified proteins in an artificial membrane (Zakharian et al. 2009, 2010, 2011; D'Avanzo et al. 2010; Cao et al. 2013b). Figure 2c shows an example of activation by PI(4,5)P_2 on the purified TRPM8 protein. The advantage of using reconstituted protein is that it is probably the strongest evidence for a direct functional effect of the lipid on an ion channel. The disadvantage is that the artificial membranes have different lipid compositions from that of the plasma membrane, and other lipids may modify the effects of phosphoinositides (Cheng et al. 2011).

3.3 Pharmacological Tools in Intact Cells or Whole Cell Patch Clamp

The major pathway most likely to decrease PI(4,5)P_2 levels in intact cells is activation of PLC, which often inhibits PI(4,5)P_2 sensitive channels. PLC can be activated by G-protein coupled receptors (PLCβ), Receptor tyrosine kinases (PLCγ), Ca^{2+} influx through the channel itself (PLCδ), or a chemical activator m-3M3FBS (Fig. 3a). Upon activation of PLC it is often hard to differentiate if the inhibitory effect is due to $PI(4,5)P_2$ depletion or other downstream pathways such as PKC.

PI4K can be inhibited by high concentrations of PI3K inhibitors such as LY294002 or wortmannin. These compounds slowly deplete PI(4)P and consequently $PI(4,5)P_2$ and may inhibit phosphoinositide sensitive ion channels (Rohacs et al. 2005). Since long treatments are necessary to significantly reduce $PI(4,5)P_2$ levels, many cellular processes may be affected, compounding interpretation. PI4K inhibitors have also been used to inhibit recovery from channel inhibition after PLC activation (Suh and Hille 2002; Liu et al. 2005). New more specific PI4K inhibitors are being developed, but not yet commercially available (Bojjireddy et al. 2014).

3.4 Inducible Phosphatases in Intact Cells or Whole Cell Patch Clamp

Over-expression of constitutively active lipid kinases and phosphatases can be used to modify phosphoinositide levels, but those again exert long-term effects on many cellular functions. For this reason various inducible phosphatases have been developed to rapidly and specifically dephosphorylate $PI(4,5)P_2$. One group of these tools is based on the rapamycin-induced heteromerization of FKBP12 and the mammalian target of rapamycin (mTOR). Using this system, a lipid phosphatase located in the cytoplasm can be translocated to the plasma membrane with rapamycin, where it dephosphorylates $PI(4,5)P_2$ to form PI(4)P (Suh et al. 2006; Varnai et al. 2006). If the activity of an ion channel depends on $PI(4,5)P_2$, but PI(4)P cannot activate it, translocation of the 5-phosphatase is expected to inhibit it (Fig. 3b). A novel version of this system was recently developed, where both 5 and 4 phosphatases were linked to FKBP12. This construct, pseudojanin, dephosphorylates both $PI(4,5)P_2$ and PI(4)P in response to rapamycin (Hammond et al. 2012). With the combination of these tools, the effects of $PI(4,5)P_2$ and PI(4)P can be differentiated.

Voltage sensitive phosphatases from *ciona intestinalis* (ciVSP) or *danio rerio* (drVSP) are alternative tools to dephosphorylate $PI(4,5)P_2$ (Okamura et al. 2009). These naturally occurring membrane proteins remove the 5 phosphate from $PI(4,5)P_2$ in response to depolarizing voltages. They also dephosphorylate $PI(3,4,5)P_3$, but the concentration of this lipid in non-stimulated cells is essentially zero; thus these tools can be assumed to selectively deplete $PI(4,5)P_2$ in most cases. The advantage of voltage sensitive phosphatases over the rapamycin-inducible systems is that $PI(4,5)P_2$ depletion is quickly reversible, and it is much easier to induce graded responses either with smaller or shorter depolarizing pulses. Generally, the inducible phosphatases are considered the most specific tools in intact cells to selectively decrease $PI(4,5)P_2$ levels, and they have been used to systematically clarify the role of phosphoinositides in regulation of various mammalian ion channel families (Suh et al. 2010; Kruse et al. 2012). The inducible phosphatases are not physiological however.

Overall, all of the tools have advantages and disadvantages; only the combined use of them can bring meaningful conclusion. For Kir and KCNQ channels and some TRP channels, such as TRPM8 (Figs. 2 and 3), essentially all tools support the same conclusion, channel activity depends on $PI(4,5)P_2$. There are several TRP channels however where the different tools support seemingly opposite conclusions. In some cases most data can be explained with a congruent model, in some other cases it cannot. Regulation of TRP channels by phosphoinositides is more complex than Kir and KCNQ channels, and in many cases future work is required to clarify the picture.

4 General Principles of Phosphoinositide Regulation

We have extensively discussed this topic in two previous reviews in the context of TRP channels (Rohacs 2007, 2009). The following is a brief summary of the key concepts, some of which were worked out on the best studied $PI(4,5)P_2$ sensitive ion channel families, Kir channels and voltage gated KCNQ K^+ channels (Kv7.x).

4.1 How Do Phosphoinositides Interact with Ion Channels?

Kir channels were the first ion channel family where $PI(4,5)P_2$ dependence was demonstrated. The molecular mechanism of $PI(4,5)P_2$ activation is best understood on these channels. The general view from the beginning was that the negatively charged head group of $PI(4,5)P_2$ interacts with positively charged residues in the cytoplasmic domains of the channels (Fan and Makielski 1997; Huang et al. 1998). Thorough mutagenesis studies later mapped the positively charged residues mainly to the larger C-terminal domain, but $PI(4,5)P_2$ interacting residues in the smaller N-terminus were also found (Lopes et al. 2002). Several homology models have been built based on partial crystal structures (Rosenhouse-Dantsker and Logothetis 2007), when finally in 2011 two co-crystal structures with $PI(4,5)P_2$ were published (Hansen et al. 2011; Whorton and MacKinnon 2011). Quite remarkably, the original idea that the interaction of the lipid's head group with the cytoplasmic domain(s) of the channel leads to a conformation change opening the channel (Fan and Makielski 1997) was essentially confirmed, needless to say at a much higher molecular resolution (Hansen et al. 2011; Whorton and MacKinnon 2011). Since the overall homology of TRP channels between subfamilies is negligible outside the transmembrane domains, it is unlikely that such a conserved $PI(4,5)P_2$ interaction-binding site is responsible for the effects of $PI(4,5)P_2$ on TRP channels. Accordingly, a variety of C- or N-terminal segments have been proposed as $PI(4,5)P_2$ interacting sites in different TRP channels (Rohacs et al. 2005; Nilius et al. 2006; Brauchi et al. 2007; Klein et al. 2008; Garcia-Elias et al. 2013).

4.2 Phosphoinositide Specificity

Kir channels have diverse specificities for phosphoinositides (Rohacs et al. 2003). Similarly, some TRP channels such as TRPM8 (Rohacs et al. 2005) or TRPV6 (Thyagarajan et al. 2008) are activated relatively specifically by $PI(4,5)P_2$, whereas others, like TRPV1, are activated by many different phosphoinositides (Lukacs et al. 2007).

To put the effects of various phosphoinositides in context, we need to consider their in situ concentrations in the inner leaflet of the plasma membrane. $PI(4,5)P_2$ constitutes up to 1 % of the phospholipids in the plasma membrane; PI(4)P is found in comparable quantities (Fruman et al. 1998). Their precursor PI constitutes up to 10 % of membrane lipids (Fruman et al. 1998), but it has no effect on most $PI(4,5)P_2$ sensitive channels, and its concentration is not expected to change significantly upon PLC activation. $PI(3,4,5)P_2$ and $PI(3,4)P_2$, the products of PI3K, may also activate some $PI(4,5)P_2$ sensitive ion channels, but their concentrations in the plasma membrane do not reach higher than 0.1 % even in stimulated cells (Fruman et al. 1998); thus, their effect is likely to be overridden by the much higher concentration of $PI(4,5)P_2$. Therefore, based on their concentrations the two most likely phosphoinositides regulating ion channels are $PI(4,5)P_2$ and PI(4)P; both these lipids are substrates for PLC. In most cases however, $PI(4,5)P_2$ which has a higher charge density is much more potent than PI(4)P. Some intracellular channels, such as TRPML1, may be specifically activated by $PI(3,5)P_2$, which is found mainly in intracellular membranes; see later.

4.3 Relation to Other Channel Regulators

Some TRP channels, such as TRPV5 and 6, are constitutively active, but most require another stimuli to open. In the case of $PI(4,5)P_2$ dependent TRP channels, opening happens when both $PI(4,5)P_2$ and the stimulus, such as menthol or cold for TRPM8, are present. These stimuli often modify the effect of $PI(4,5)P_2$; see below.

4.4 The Role of Apparent Affinity

The resting concentration of $PI(4,5)P_2$ in the plasma membrane is relatively high. Thus, channels with high apparent affinity for this lipid may be oversaturated, and physiological decreases in $PI(4,5)P_2$ will not affect their activity significantly. Channels with lower affinity on the other hand are more likely to be regulated by physiological changes in $PI(4,5)P_2$ levels. The apparent affinity for $PI(4,5)P_2$ is not static for several TRP channels; their chemical ligands such as capsaicin or menthol can increase $PI(4,5)P_2$ affinity (Rohacs et al. 2005; Lukacs et al. 2007).

4.5 Potential Indirect Effects of Phosphoinositides

The regulation of TRP channels by $PI(4,5)P_2$ is complex, in many cases opposing effects of the lipid are reported even on the same channel. Phosphoinositides regulate many cellular processes, including the organization of the cytoskeleton and membrane traffic (Saarikangas et al. 2010; Shewan et al. 2011). These effects are mediated by a large number and a variety of $PI(4,5)P_2$-binding proteins (DiNitto et al. 2003). It is quite likely that some or most of the complexity originates from other phosphoinositide-binding proteins influencing channel activity (Rohacs 2009). See TRPC and TRPV1 channels for some specific details.

5 TRPM Channels

TRPM channels are the most diverse group in terms of activation mechanisms and biological function. When their phosphoinositide regulation is concerned, however, we probably have the clearest picture here. In short, it is quite likely that $PI(4,5)P_2$ is a common cofactor for members of this group, necessary for activity. There are variations to this theme, but there is a clear agreement in the literature on the overall direction of the effect of phosphoinositides. Six out of the eight members have been reported to be positively regulated by $PI(4,5)P_2$ by now (Table 1).

5.1 TRPM8

The literature on phosphoinositide regulation is most extensive on the cold and menthol sensitive TRPM8. These channels are expressed in sensory neurons of the dorsal root ganglia (DRG) and trigeminal ganglia (TG). Their genetic deletion in mice significantly impaired detection of moderately cold temperatures (Bautista et al. 2007; Dhaka et al. 2007).

TRPM8 was reported by two different laboratories to be reactivated by $PI(4,5)P_2$ after run-down in excised patches and inhibited by either poly-lysine (Fig. 2a) or anti $PI(4,5)P_2$ antibody (Liu and Qin 2005; Rohacs et al. 2005). Mutations of positively charged residues in the highly conserved proximal C-terminal TRP domain shifted $PI(4,5)P_2$ dose–response to the left, and accordingly, made the channels more sensitive to inhibition by PLC, as expected from reduced apparent affinity for $PI(4,5)P_2$ (Rohacs et al. 2005). Later it was also reported that TRPM8 is reactivated by MgATP in excised patches, and this was prevented by inhibiting PI4K (Yudin et al. 2011); see Fig. 2b. TRPM8 was the first TRP channel that was reported to be inhibited by selective reduction of $PI(4,5)P_2$ by rapamycin-induced translocation of a 5-phosphatase (Varnai et al. 2006); see Fig. 3b, a finding confirmed by other laboratories (Daniels et al. 2009); see also supplemental material in (Wang et al. 2008). Later the voltage sensitive phosphatase ciVSP was also shown to inhibit the channel (Yudin et al. 2011); see also Fig. 3c. Consistent with the robust inhibition by two different 5-phosphatases, PI(4)P had very little effect

Table 1 Mammalian TRPM channels

Name	Regulation/function	Phosphoinositide effects
TRPM1	Mutation causes night blindness in humans	No phosphoinositide effect reported yet
TRPM2	Activated by ADP ribose	Poly-lysine inhibits, and $PI(4,5)P_2$ re-activates in excised patches (Toth and Csanady 2012)
TRPM3	Heat, pregnenolon sulfate	No phosphoinositide effect reported yet
TRPM4	Intracellular Ca^{2+} activates, nonselective Ca^{2+} impermeable cation channel	$PI(4,5)P_2$ activates in excised patches (Zhang et al. 2005; Nilius et al. 2006)
TRPM5	Intracellular Ca^{2+} activates nonselective Ca^{2+} impermeable cation channel	$PI(4,5)P_2$ activates in excised patches (Liu and Liman 2003)
TRPM6	Mg^{2+} transporter, mutation causes human disease	$PI(4,5)P_2$ activates in excised patches, rapamycin-inducible 5 phosphatase, and ciVSP inhibits (Xie et al. 2011)
TRPM7	cAMP, shear stress, plays important roles in development	$PI(4,5)P_2$ activates in excised patches, activation of PLC inhibits (Runnels et al. 2002). $PI(4,5)P_2$ is needed for the activity of the cardiac magnesium-inhibited TRPM7-like channels (Gwanyanya et al. 2006) Role of PLC mediated inhibition is challenged (Takezawa et al. 2004; Langeslag et al. 2007)
TRPM8	Cold, Menthol	$PI(4,5)P_2$ activates in excised patches (Liu and Qin 2005; Rohacs et al. 2005; Yudin et al. 2011) $PI(4,5)P_2$ depletion inhibits (rapamycin, ciVSP), plays a role in desensitization (Rohacs et al. 2005; Daniels et al. 2009; Yudin et al. 2011) $PI(4,5)P_2$ but not PI(4)P activates in planar lipid bilayers (Zakharian et al. 2009, 2010) $PI(4,5)P_2$ regulates temperature threshold (Fujita et al. 2013)

on TRPM8 in excised patches (Rohacs et al. 2005). The channel protein was also purified and reconstituted in planar lipid bilayers (Fig. 2c), where activation by both menthol and cold depended on the presence of $PI(4,5)P_2$, and PI(4)P could not support channel activity (Zakharian et al. 2009, 2010). $PI(3,4,5)P_3$ and $PI(3,4)P_2$ were also much less effective both in excised patches (Rohacs et al. 2005) and in lipid bilayers (Zakharian et al. 2010). These data show that $PI(4,5)P_2$ is a direct specific activator of TRPM8.

Both menthol- and cold-induced currents display a time dependent decrease, in the presence of extracellular Ca^{2+}, a phenomenon termed adaptation or desensitization. It has been shown that both menthol (Rohacs et al. 2005; Daniels et al. 2009) and cold (Yudin et al. 2011) decrease cellular $PI(4,5)P_2$ levels via Ca^{2+}-induced activation of PLCδ isoforms. Dialysis of $PI(4,5)P_2$ but not PI(4)P through the whole-cell patch pipette inhibited desensitization of menthol-induced currents

(Yudin et al. 2011). These data together with the clear dependence of channel activity on PI(4,5)P_2 support a model in which Ca^{2+} influx through TRPM8 induces depletion of PI(4,5)P_2, which limits channel activity, leading to desensitization. A more recent paper confirmed the role of PI(4,5)P_2 depletion in tachyphylaxis, the decrease in amplitude after repeated agonist applications, but proposed a role for Calmodulin (CaM) in acute desensitization (Sarria et al. 2011). A recent paper also implied that PI(4,5)P_2 is involved in setting the threshold temperature for TRPM8 (Fujita et al. 2013).

GPCRs activating PLC have also been shown to inhibit TRPM8 (Liu and Qin 2005), but the involvement of PI(4,5)P_2 depletion here is much less clear. Both PKC (Premkumar et al. 2005) and direct inhibition by Gq-alpha (Zhang et al. 2012b) have been implicated.

In conclusion, it is universally accepted now that the activity of TRPM8 depends on PI(4,5)P_2; see Figs. 2 and 3 for supporting data using a variety of techniques. It is also clear that activation of PLC either by GPCRs or by Ca^{2+} influx through the channel inhibits TRPM8. How much PI(4,5)P_2 depletion and other signaling pathways contribute to this inhibition is less clear; we refer to our recent review for further details (Yudin and Rohacs 2011).

5.2 TRPM6 and TRPM7

TRPM6 and TRPM7 are closely related to each other; both channels have an atypical kinase domain at their C-terminus. Loss of function mutations in TRPM6 cause familial hypomagnesemia with secondary hypocalcemia in humans (Walder et al. 2002). TRPM7 has been proposed to have a variety of functions, but from knockout studies it appears that this channel has important roles in development (Jin et al. 2012). Interestingly, despite the relatively restricted disease its mutation causes in humans, genetic deletion of TRPM6 in mice is also embryonic lethal (Walder et al. 2009).

TRPM7 was reported to require PI(4,5)P_2 for activity (Runnels et al. 2002; Gwanyanya et al. 2006), and PLC activation was shown to inhibit it (Runnels et al. 2002). Two articles however challenged the inhibitory effect of PLC activation (Takezawa et al. 2004; Langeslag et al. 2007). We have discussed this controversy in more detail previously (Rohacs 2009). TRPM6 was also recently reported to require PI(4,5)P_2 for activity and inhibited by PI(4,5)P_2 depletion with inducible phosphatases (Xie et al. 2011).

5.3 TRPM4 and TRPM5

Both TRPM5 and TRPM4 are Ca^{2+}-activated nonselective cation channels that are not permeable to Ca^{2+}. TRPM5 is mainly expressed in taste and related tissues, whereas TRPM4 is more widely expressed. Both channels have been shown to undergo desensitization to Ca^{2+} in excised patches, which was restored by the application of $PI(4,5)P_2$ (Liu and Liman 2003; Zhang et al. 2005; Nilius et al. 2006). On TRPM4, the $PI(4,5)P_2$ interaction site was proposed to be a distal C-terminal cluster of positively charged residues, termed PH-like domain (Nilius et al. 2006). Regions conforming to the K/R-X_{3-11}-K/R-X-K/R consensus sequence of this region can be found in most TRP channels either in the C- or the N-terminal cytoplasmic domains (Nilius et al. 2008). Interestingly, the N-terminal $PI(4,5)P_2$ interacting region recently proposed in TRPV4 also conforms to this consensus sequence (Garcia-Elias et al. 2013); see later. The role of this sequence motif in $PI(4,5)P_2$ regulation of other TRP channels remains to be determined.

5.4 TRPM2

TRPM2 is another TRP channel that has an enzyme domain on its C-terminus, an ADP ribose hydrolase; accordingly, the channel can be activated in excised patches by ADP ribose. TRPM2 has been implicated in a variety of physiological functions, such as insulin secretion and immunological responses, but the overall phenotype of the knockout mouse is quite moderate (Yamamoto et al. 2008). To our knowledge there is only one publication of examining potential effects of $PI(4,5)P_2$ on TRPM2 (Toth and Csanady 2012). In this work the authors found that run-down of TRPM2 activity in excised patches is not caused by loss of $PI(4,5)P_2$ from the patch membrane, but rather a collapse of the pore. When poly-Lysine was applied to excised inside-out patches, however, channel activity was inhibited, and this could be relieved by application of $PI(4,5)P_2$. It is quite possible that TRPM2 also requires $PI(4,5)P_2$ for activity, but its affinity to the lipid is quite high; thus loss of the lipid does not contribute to run-down at high agonist (ADP ribose) concentrations.

6 TRPV Channels

TRPV channels can be divided into two groups, TRPV1–4 are all outwardly rectifying nonselective Ca^{2+} permeable cation channels that can be activated by heat with various thresholds. TRPV5 and 6 on the other hand are Ca^{2+}-selective inwardly rectifying ion channels that play roles in epithelial Ca^{2+} transport. Phosphoinositides have been implicated in regulation of all members of this subgroup; in most cases $PI(4,5)P_2$ has a positive regulatory role, but inhibition by $PI(4,5)P_2$ has also been described with or without the presence of concurrent activating effect (Table 2).

6.1 TRPV1

TRPV1 is activated by heat, capsaicin, low extracellular pH, and a plethora of other pain-producing agents in sensory neurons. TRPV1 is perhaps the most studied member of the TRP superfamily; its phosphoinositide regulation is no exception.

Table 2 Mammalian TRPV channels

Name	Regulation/function	Phosphoinositide effects
TRPV1	Heat, capsaicin, low pH, involved in nociception	*Positive effects of phosphoinositides:* $PI(4,5)P_2$ activates in excised patches (Stein et al. 2006; Lukacs et al. 2007; Kim et al. 2008c; Klein et al. 2008; Ufret-Vincenty et al. 2011; Lukacs et al. 2013b) Channel activity runs down in excised patches, MgATP restores activity in a PI4K dependent manner (Lukacs et al. 2013b) In addition to phosphoinositides, many other negatively charged lipids including phosphatidylglycerol and oleoyl-CoA also activate TRPV1 in excised patches (Lukacs et al. 2013b) $PI(4,5)P_2$ inhibits desensitization in intact cells (Liu and Qin 2005; Lishko et al. 2007; Lukacs et al. 2007, 2013a) Channel activity is inhibited by combined depletion of $PI(4,5)P_2$ and PI(4)P using a rapamycin-inducible dual-phosphatase pseudojanin (Hammond et al. 2012; Lukacs et al. 2013a) TRPV1 is inhibited by $PI(4,5)P_2$ depletion with a rapamycin-inducible 5 phosphatase (Klein et al. 2008; Yao and Qin 2009). $PI(4,5)P_2$ enhances thermosensation and thermal hyperalgesia. Depletion of $PI(4,5)P_2$ by prostatic acid phosphatase-induced PLC activation inhibits thermal sensitivity (Sowa et al. 2010) $PI(4,5)P_2$ sensitivity was proposed to be mediated via a $PI(4,5)P_2$-binding protein Pirt (Kim et al. 2008a), was challenged (Ufret-Vincenty et al. 2011) *Negative effects of phosphoinositides:* $PI(4,5)P_2$ may partially inhibit in intact cells (Chuang et al. 2001; Prescott and Julius 2003; Lukacs et al. 2007, 2013a; Patil et al. 2011) $PI(4,5)P_2$ inhibits in lipid vesicles (Cao et al. 2013b) $PI(4,5)P_2$ inhibits via competing for AKAP (Jeske et al. 2011)
TRPV2	Growth factors activate, noxious heat activates,	$PI(4,5)P_2$ activates in excised patches (Mercado et al. 2010)
TRPV3	Heat activates in keratinocytes	$PI(4,5)P_2$ inhibits in excised patches (Doerner et al. 2011)
TRPV4	Heat, hyposmosis activates	$PI(4,5)P_2$ is required for osmotic and heat activation (Garcia-Elias et al. 2013)
TRPV5	Constitutively active epithelial Ca^{2+} channel	$PI(4,5)P_2$ activates in excised patches (Lee et al. 2005; Rohacs et al. 2005)
TRPV6	Constitutively active epithelial Ca^{2+} channel	$PI(4,5)P_2$ activates in excised patches and planar lipid bilayers (Thyagarajan et al. 2008; Zakharian et al. 2011; Cao et al. 2013a) Rapamycin-inducible 5 phosphatase inhibits, $PI(4,5)P_2$ depletion plays a role in Ca^{2+}-induced inactivation (Thyagarajan et al. 2008)

Compared to the relatively simple picture with the TRPM family, the regulation of TRPV1 by phosphoinositides is quite complex and controversial.

There is full agreement in the literature that TRPV1 is activated by phosphoinositides in excised patches. First it was shown that poly-Lysine inhibits, and diC_8 $PI(4,5)P_2$ activates both native TRPV1 currents in DRG neurons and recombinant TRPV1 expressed in F11 cells (Stein et al. 2006). A following article confirmed these results in recombinant TRPV1 expressed in *Xenopus* oocytes using both DiC_8 and AASt $PI(4,5)P_2$ (Lukacs et al. 2007). The apparent affinity for $PI(4,5)P_2$ was increased by higher capsaicin concentrations, and accordingly, the velocity of current run-down was inversely correlated with the concentration of capsaicin in the patch pipette. The same article also found that all other phosphoinositides tested, $PI(3,4,5)P_2$, $PI(3,4)P_2$, $PI(3,5)P_2$, PI(5)P, and PI(4)P activated TRPV1. PI(4)P was less potent than $PI(4,5)P_2$, i.e., its dose–response was shifted to the right, but the maximal effect was similar to that induced by $PI(4,5)P_2$ (Lukacs et al. 2007). Similar results have been obtained in excised patches with PI(4)P in two additional publications (Klein et al. 2008; Ufret-Vincenty et al. 2011). Finally, an article from a different laboratory studying mainly TRPA1 channels confirmed the activating effect of $PI(4,5)P_2$ on TRPV1 in excised patches (Kim et al. 2008c).

Results with inducible phosphatases also generally confirm the positive regulatory role of $PI(4,5)P_2$, but there is a debate on weather or not PI(4)P contributes to channel activity in intact cells. Shortly after the first publication with the rapamycin-inducible 5-phosphatase, it was shown that using this construct, depletion of $PI(4,5)P_2$ via conversion to PI(4)P did not inhibit TRPV1 activity at saturating capsaicin concentrations, whereas it inhibited TRPM8 (Lukacs et al. 2007). The lack of inhibition of TRPV1 was explained by the fact that PI(4)P also activates these channels, whereas TRPM8 is specifically activated by $PI(4,5)P_2$. Two recent publications using the rapamycin-inducible dual specificity phosphatase pseudojanin supported this conclusion. First it was shown that capsaicin-induced TRPV1 currents were inhibited by rapamycin-induced activation of pseudojanin, which dephosphorylates both $PI(4,5)P_2$ and PI(4)P, but not when either the 5 or the 4 phosphatase activity was eliminated by mutations (Hammond et al. 2012). Another article confirmed these results and showed that TRPV1 currents induced by low pH were inhibited by rapamycin in pseudojanin expressing cells, but they were not inhibited by the specific 5-phosphatase drVSP (Lukacs et al. 2013a). In contrast to these publications, two articles found that a rapamycin-inducible 5-phosphatase inhibited capsaicin-induced TRPV1 currents (Klein et al. 2008; Yao and Qin 2009). The reason for the discrepancy between these and earlier results is not clear, but the rapamycin-inducible system in these studies used a different 5-phosphatase from the one used by (Lukacs et al. 2007). Despite discrepancies on the specific roles of $PI(4,5)P_2$ and PI(4)P, data using excised patches and inducible phosphatases are consistent with the idea that phosphoinositides are positive regulators or cofactors of TRPV1.

Data in the literature support the idea that the positive regulation by phosphoinositides plays a role in desensitization of TRPV1 currents, in a similar fashion that was described earlier for TRPM8 (Rohacs et al. 2005). It was shown that desensitization of recombinant (Lishko et al. 2007; Lukacs et al. 2007) and native TRPV1 (Lukacs et al. 2013a) is inhibited by dialysis of either $PI(4,5)P_2$ or PI(4)P through the whole-cell patch pipette. It was also shown that activation of recombinant (Liu et al. 2005; Lukacs et al. 2007; Yao and Qin 2009) and native TRPV1 channels in DRG neurons (Lukacs et al. 2013a) leads to a depletion of $PI(4,5)P_2$ and PI(4)P presumably via activation of PLCδ isoforms by the massive Ca^{2+} influx through the channels. Phosphoinositide depletion is probably not the only mechanism of desensitization, since neither $PI(4,5)P_2$ nor PI(4)P could fully inhibit desensitization. Consistent with this, several other mechanisms such as CaM, calcineurin, and direct binding of ATP to the channel have been proposed to play roles in desensitization (Mohapatra and Nau 2005; Lishko et al. 2007; Rohacs 2013).

As mentioned earlier however, TRPV1 was first proposed to be inhibited by $PI(4,5)P_2$ (Chuang et al. 2001) via direct binding to a distal C-terminal lipid interacting site (Prescott and Julius 2003). It was also proposed that pro-inflammatory mediators such as bradykinin sensitize TRPV1 to moderate stimuli via removing this inhibitory effect by PLC mediated $PI(4,5)P_2$ hydrolysis. This model was mainly based on indirect evidence, and the effects of phosphoinositides were not tested in excised patches. To lend further support to the idea that $PI(4,5)P_2$ inhibits TRPV1, the same laboratory recently purified the TRPV1 protein and incorporated it into artificial liposomes. By patch clamping those liposomes, they showed that the channel was fully active in the absence of phosphoinositides and that incorporation of $PI(4,5)P_2$, PI(4)P, or even PI inhibited TRPV1 activity or, more precisely, shifted the capsaicin dose–response curve to the right. Removal of the distal C-terminal phosphoinositide-binding domain eliminated inhibition by $PI(4,5)P_2$ (Cao et al. 2013b).

How do we reconcile these two seemingly incompatible views? First of all, the fact that TRPV1 showed both heat and capsaicin activation in an artificial membrane devoid of phosphoinositides makes it very clear that these lipids are not necessary for channel activity *per se*. As mentioned earlier, activation of TRPV1 by phosphoinositides was not specific to $PI(4,5)P_2$; any other phosphoinositide could exert similar effects in excised patches (Lukacs et al. 2007). Among Kir channels, Kir6.2 (K_{ATP}) has a similar nonspecific activation profile for phosphoinositides (Rohacs et al. 2003). Those channels can be activated not only by phosphoinositides but also by many other negatively charged lipids such as oleoyl-CoA (Rohacs et al. 2003), phosphatidylserine at high concentrations (Fan and Makielski 1997), and even the artificial lipid DGS-NTA (Krauter et al. 2001). The liposomes used by (Cao et al. 2013b) contained high concentrations (~25 %) of phosphatidyl glycerol (PG), which has a negative charge. One possible explanation for the full activity of TRPV1 in the complete absence of phosphoinositides is that the high concentrations of PG satisfied the requirement for negatively charged lipids for the activity of the channel. Indeed, when PG was tested in excised inside-out patches, it reactivated TRPV1 currents after run-down when applied at very high concentrations (Lukacs et al. 2013b). Other phospholipids with single negative charges, phosphatidylserine (PS) and phosphatidylinositol (PI), could also support

TRPV1 activity at high concentrations. Similar to K_{ATP} channels, oleoyl-CoA and DGS-NTA also activated TRPV1 (Lukacs et al. 2013b). When the purified channel was incorporated in planar lipid bilayers consisting of neutral lipids, capsaicin activation depended on the presence of PI(4,5)P_2 (Lukacs et al. 2013b). TRPV1 activity showed ~90 % run-down in excised patches after 5 min in an ATP free solution, and MgATP reactivated the channel in a PI4K dependent manner. This shows that phospholipids with single negative charges (PS, PI, and PG) that do not become dephosphorylated in excised patches contribute to TRPV1 activity 10 % or less. In conclusion, despite its promiscuous lipid specificity profile, in the context of the plasma membrane, the major intracellular lipids supporting TRPV1 activity are PI(4,5)P_2 and potentially PI(4)P (Lukacs et al. 2013b).

As to the inhibitory effect of PI(4,5)P_2, there are many other data supporting this idea, most are based on measurements in intact cells or whole-cell patch clamp. We have found earlier that at low stimulation levels with capsaicin or heat, the rapamycin-inducible 5-phoshatase potentiated TRPV1 currents both in HEK cell and in Xenopus oocytes (Lukacs et al. 2007). More recently we found that depleting PI(4,5)P_2 with ciVSP potentiated TRPV1 currents induced by low capsaicin in Xenopus oocytes (unpublished observation), but neither ciVSP nor drVSP potentiated TRPV1 currents in a mammalian expression system either with low capsaicin or low pH (Lukacs et al. 2013a). On the other hand, selective depletion of PI(4,5)P_2 by drVSP potentiated the sensitizing effect of subthreshold and submaximal concentrations of the PKC agonist OAG (Lukacs et al. 2013a). Dialysis of PI(4,5)P_2, but not PI(4)P, via the whole-cell patch pipette inhibited bradykinin-induced sensitization, which again is compatible with PI(4,5)P_2 being inhibitory. Based on this and other experiments, we have proposed a model in which GPCR activation leads to a selective moderate decrease in PI(4,5)P_2 levels, which potentiates the well-known sensitizing effect of PKC (Lukacs et al. 2013a).

The fact that inhibition by PI(4,5)P_2 was not detected in excised inside-out patches by any laboratory argued that any inhibition of TRPV1 by PI(4,5)P_2 is likely to be indirect, i.e., mediated by other cellular components or PI(4,5)P_2-binding proteins. The A-Kinase anchoring protein AKAP150, for example, was proposed to mediate the inhibitory effect of PI(4,5)P_2 (Jeske et al. 2011). Another PI(4,5)P_2-binding protein that was proposed to influence TRPV1 activity is Pirt (Kim et al. 2008a). This protein, however, was proposed to mediate positive effects of PI(4,5)P_2, and its role has been recently challenged (Ufret-Vincenty et al. 2011). The finding that PI(4,5)P_2 inhibited the purified TRPV1 in artificial membranes provided strong support for a direct inhibition by phosphoinositides (Cao et al. 2013b). This, however, may be explained by the recent finding that high concentrations of PI(4,5)P_2 inhibit TRPV1 activity when applied to the extracellular leaflet in outside-out patches, which is not the physiological location of phosphoinositides (Senning et al. 2014). Clearly, more work is needed at this point to understand the mechanism of negative modulation of TRPV1 by phosphoinositides.

Overall, the author's view is that TRPV1 requires negatively charged lipids for activity, and in the cellular context these lipids are PI(4,5)P_2 and probably PI(4)P.

Removal of these phosphoinositides contribute to desensitization upon maximal pharmacological activation. A concurrent inhibitory effect is also quite likely, and smaller selective decreases in $PI(4,5)P_2$ levels contribute to sensitization by pro-inflammatory mediators such as bradykinin (Lukacs et al. 2013a, b; Rohacs 2013).

6.2 TRPV2

TRPV2 was originally proposed to be a noxious heat sensor, with activation temperatures over 50 °C, but subsequent studies showed that $TRPV2^{-/-}$ mice have no temperature sensation deficit (Park et al. 2011). At the same time, the knockout animals had impaired phagocytosis by macrophages (Link et al. 2010). As opposed to the large number of papers on TRPV1, there is only one publication on TRPV2 in connection to $PI(4,5)P_2$ (Mercado et al. 2010). This article showed that TRPV2 is inhibited by poly-Lys and reactivated by DiC_8 $PI(4,5)P_2$ in excised patches. It was also shown that activation of the channel with its chemical agonist 2-APB induced a depletion of $PI(4,5)P_2$ resulting in desensitization of the channel.

6.3 TRPV3

TRPV3 is activated by moderate heat in keratinocytes. Gain of function mutations in TRPV3 cause Olmsted syndrome, a rare congenital disorder characterized by palmoplantar and periorificial keratoderma, alopecia, and severe itching (Lin et al. 2012). This channel was reported to be inhibited by $PI(4,5)P_2$ in excised patches and its voltage and temperature dependent gating was potentiated by hydrolysis of the lipid (Doerner et al. 2011).

6.4 TRPV4

TRPV4 is another temperature sensitive TRP channel, which also acts as an osmosensor. Mutations in this channel are associated with a variety of human diseases (Nilius and Voets 2013). A recent report (Garcia-Elias et al. 2013) showed that activation of this channel in excised patches by heat required $PI(4,5)P_2$. In intact cells, both osmotic and heat-induced activation was eliminated if $PI(4,5)P_2$ was depleted using a rapamycin-inducible $PI(4,5)P_2$ phosphatase (Garcia-Elias et al. 2013). Activation by a direct chemical agonist 4α-phorbol 12,13-didecanoate was not affected by $PI(4,5)P_2$ depletion. Positively charged residues in the N-terminal cytoplasmic domain (^{121}KRWRK^{125}), located before the ankyrin repeats and the proline-rich domain, were implied in the effect of $PI(4,5)P_2$. $PI(4,5)P_2$ depletion increased the distance between the C-terminal ends of the channel as determined by fluorescence resonance energy transfer, indicating a structural rearrangement of the channel upon $PI(4,5)P_2$ binding (Garcia-Elias et al. 2013).

6.5 TRPV5 and TRPV6

These two channels are highly homologous to each other, but less to other members of the TRPV family. Unlike any other TRP channels, TRPV5 and 6 are Ca^{2+} selective, and they display inward rectification. Both are implicated in epithelial Ca^{2+} transport, TRPV5 in the kidney (Hoenderop et al. 2003) whereas TRPV6 in the duodenum (Bianco et al. 2007) and in epididymal epithelia (Weissgerber et al. 2011).

The activity of both TRPV5 (Lee et al. 2005) and TRPV6 (Thyagarajan et al. 2008) runs down in excised patches, and they are reactivated by $PI(4,5)P_2$ but not PI(4)P. TRPV6 has also been shown to be reactivated by MgATP in excised patches, and this effect was inhibited by three structurally different compounds that inhibit PI4K activity, showing that MgATP in this system acted by providing substrate for lipid kinases (Zakharian et al. 2011). TRPV6 was also shown to be activated by $PI(4,5)P_2$ but not PI(4)P in planar lipid bilayers, demonstrating direct effect on the channel (Zakharian et al. 2011). Consistent with the lack of effect of PI(4)P, the channel is inhibited by the rapamycin-inducible 5-phosphatase (Thyagarajan et al. 2008). Ca^{2+} influx through TRPV6 was also shown to activate PLC, contributing to channel inactivation (Thyagarajan et al. 2009). CaM has also been implicated in Ca^{2+}-induced inactivation of both TRPV6 (Niemeyer et al. 2001; Derler et al. 2006) and TRPV5 (de Groot et al. 2011). CaM was shown to inhibit TRPV6 in excised patches, which could be alleviated, but not prevented by excess $PI(4,5)P_2$ (Cao et al. 2013a). CaM and $PI(4,5)P_2$ thus compete with each other, even though it probably does happen through direct competition, as proposed for TRPC6 (Kwon et al. 2007) since we could not observe competition in biochemical-binding experiments. Overall, it is quite likely that $PI(4,5)P_2$ depletion and CaM cooperatively produce Ca^{2+}-induced inactivation of TRPV6 (Cao et al. 2013a).

7 TRPC Channels

Mammals have 6 or 7 TRPC channels, depending on the species; in humans TRPC2 is a pseudogene. From the functional point of view, this group is the most homogenous; all members are activated downstream of PLC. There are certainly flavors to this general theme, but at least on the textbook level we can make this generalization. TRP channels were discovered in *drosophila*, where they play a role in visual transduction. The channel complex that generates the receptor potential in the insect eye consists of the TRP and the TRPL channels; both are homologues of the mammalian TRPC channel, thus they will be discussed here.

Table 3 Mammalian TRPC channels and *drosophila* orthologues

Name	Regulation/function	Phosphoinositide effects
dTRPL	Activated downstream of PLC, *drosophila* vision	$PI(4,5)P_2$ inhibits in excised patches (Estacion et al. 2001) $PI(4,5)P_2$ activates but PI(4)P and PI inhibits in excised patches (Huang et al. 2010)
TRPC1	Activated downstream of PLC	$PI(4,5)P_2$ activates in excised patches, native cells (Saleh et al. 2009a, b; Shi et al. 2012)
TRPC3	Activated downstream of PLC, DAG activates	$PI(4,5)P_2$ activates in excised patches, expression system (Lemonnier et al. 2008) VSP inhibits (Imai et al. 2012; Itsuki et al. 2012)
TRPC4	Activated downstream of PLC	TRPC4α but not TRPC4β is inhibited by $PI(4,5)P_2$, whole-cell patch clamp (Otsuguro et al. 2008) TRPC4β is inhibited by $PI(4,5)P_2$ depletion (rapamycin-inducible phosphatase) (Kim et al. 2013)
TRPC5	Activated downstream of PLC	$PI(4,5)P_2$ activates in excised patches, but inhibits in whole cell, $PI(4,5)P_2$ depletion may inhibit or activate it (Trebak et al. 2009) $PI(4,5)P_2$ inhibits desensitization in whole-cell patch clamp (Kim et al. 2008b) $PI(4,5)P_2$ activates in excised inside-out patches in complex with TRPC1 in native cells, but inhibits in the absence of TRPC1 (Shi et al. 2012)
TRPC6	Activated downstream of PLC, DAG activates	$PI(4,5)P_2$ activates in excised patches (expression system) (Lemonnier et al. 2008) VSP inhibits (Imai et al. 2012; Itsuki et al. 2012) $PI(4,5)P_2$ inhibits in excised patches (native smooth muscle cells) (Albert et al. 2008; Ju et al. 2010) Extracellular $PI(4,5)P_2$ enhances its activity in platelets (Jardin et al. 2008) Calmodulin inhibits by displacing $PI(3,4,5)P_3$ (Kwon et al. 2007)
TRPC7	Activated downstream of PLC, DAG activates	$PI(4,5)P_2$ activates in excised patches, expression system (Lemonnier et al. 2008) VSP inhibits (Imai et al. 2012; Itsuki et al. 2012) $PI(4,5)P_2$ inhibits in excised patches, native channels (Ju et al. 2010)

When it comes to their phosphoinositide regulation, this group is probably the most complicated among the three major families (Table 3). Inhibition by $PI(4,5)P_2$ and relief from inhibition by depletion of this lipid upon PLC activation were proposed originally as a mechanism playing an important role in activation of TRPL channels (Estacion et al. 2001). This notion however is controversial; it may contribute to activation in some cases, but it is certainly not a general mechanism of opening. What activates TRPCs is not elucidated yet; activation of TRPC3, 6, and 7 channels by DAG is very well accepted, but there is less agreement on how other TRPCs are activated. It is quite likely that multiple and diverse mechanisms converge on most TRPC channels (Putney and Tomita 2011; Rohacs 2013).

7.1 Excised Patch Data

The first publication implying $PI(4,5)P_2$ in this channel family showed that the *drosophila* TRPL channel expressed in SF9 cells was inhibited by $PI(4,5)P_2$ in excised patches (Estacion et al. 2001). The authors proposed that $PI(4,5)P_2$ depletion together with DAG activates dTRPL. This finding fits very well the hypothesis that $PI(4,5)P_2$ is a general inhibitor of TRP channels and would also provide a logical activation mechanism for TRPC channels that are also activated downstream of PLC. Several years later, however, this finding was challenged, and it was shown that DiC_8 $PI(4,5)P_2$ activated TRPL, whereas PI(4)P and PI inhibited it (Huang et al. 2010).

Recombinant TRPC channels are often used as models to study channel regulation, since endogenous TRPC currents are often too small to measure reliably. Recombinant TRPC5 (Trebak et al. 2009) as well as TRPC3, 6, and 7 channels (Lemonnier et al. 2008) have been shown to be activated by $PI(4,5)P_2$ in excised patches.

Significant amount of work has been performed in native vascular smooth muscle cells on phosphoinositide regulation of TRPC channels, reviewed in (Large et al. 2009). TRPC1 channels were shown to be activated in excised patches either alone (Saleh et al. 2009b) or in complex with TRPC5 channels (Shi et al. 2012). These results are consistent with the general activation of recombinant TRPC channels. On the other hand, native TRPC5 (Shi et al. 2012), TRPC6 (Albert et al. 2008; Ju et al. 2010), and TRPC7 channels were shown to be inhibited by $PI(4,5)P_2$ in excised inside-out patches (Ju et al. 2010). These results contradict the activating effect of $PI(4,5)P_2$ on the same channels studied in expression systems.

7.2 Whole-Cell Patch Clamp Experiments

Recombinant TRPC5 is activated by PI4K inhibition and its activity is reduced by inclusion of $PI(4,5)P_2$ or PI(4)P in the patch pipette (Trebak et al. 2009). This is consistent with $PI(4,5)P_2$ depletion contributing to activation. As mentioned earlier however, the same paper showed that TRPC5 is activated in excised patches; furthermore, $PI(4,5)P_2$ depletion with a rapamycin-inducible 5-phosphatase inhibited the channel (Trebak et al. 2009). A general feature of TRPC channel activation by GPCRs is that after an initial peak, current amplitudes decline (desensitization). It was shown that inclusion of $PI(4,5)P_2$ in the patch pipette inhibited desensitization of TRPC5 (Kim et al. 2008b). This is compatible with the idea that hydrolysis of $PI(4,5)P_2$ initiates TRPC5 channel activation, but the loss of the lipid limits channel activity on a longer time course.

TRPC4α but not TRPC4β was shown to be inhibited by intracellular dialysis of $PI(4,5)P_2$. This effect depended on the actin cytoskeleton and was prevented by the deletion of the C-terminal PDZ-binding motif that links TRPC4 α to F-actin through the sodium-hydrogen exchanger regulatory factor and ezrin (Otsuguro et al. 2008). Intriguingly, TRPC4β was inhibited by a rapamycin-inducible 5 phosphatase, and desensitization was inhibited by both $PI(4,5)P_2$ and its non-hydrolyzable analogue (Kim et al. 2013).

Depletion of $PI(4,5)P_2$ using the voltage sensitive phosphatase drVSP inhibited recombinant TRPC3, 6, and 7 channels (Imai et al. 2012; Itsuki et al. 2012). The same study showed that channels desensitized less if endogenous muscarinic receptors were stimulated than if similar receptors were over-expressed. They also showed that stimulation of over-expressed receptors lead to a much stronger decrease in $PI(4,5)P_2$ levels than that induced by endogenous receptors. This is consistent with the idea that loss of $PI(4,5)P_2$ is a major factor in desensitization during PLC activation. Accordingly, intracellular dialysis of $PI(4,5)P_2$ slowed the desensitization of recombinant TRPC3, 6, and 7 channels and endogenous TRPCs in A7r5 vascular smooth muscle cells (Imai et al. 2012).

7.3 Conclusions

Many TRPC channels have been shown to be activated by $PI(4,5)P_2$ in excised patches and inhibited by specific inducible phosphatases in intact cells. Depletion of $PI(4,5)P_2$ may contribute to desensitization of several of them during PLC stimulation. In principle, loss of $PI(4,5)P_2$ may limit channel activity in two different ways. Since $PI(4,5)P_2$ activates many TRPCs in excised patches, it is possible that it functions as a cofactor needed for channel activity. It is also possible, however, that loss of $PI(4,5)P_2$ is limiting channel activity, because there is less substrate for PLC, thus less activating messenger, for example, DAG, is generated. PI(4)P however is also a substrate for PLC, even though generally most isoforms are more active in hydrolyzing $PI(4,5)P_2$ (Fukami et al. 2010). It also has to be noted that much of the data supporting this desensitization model is performed in expression systems, and some native channels may behave differently.

An opposing model, in which $PI(4,5)P_2$ is inhibitory, and its depletion contributes to activation, may be valid for some TRPCs, but it is unlikely to be a general mechanism. Overall, there are several discrepancies in the literature, especially between excised patch data on native TRPCs and recombinant ones (see Table 3). It is possible that endogenous channels assemble with accessory proteins that modify their function. Given that drosophila visual TRP-s function in a highly organized local complex (Montell 2012), it would not be surprising if endogenous mammalian TRPC functioned similarly. TRPC4 and TRPC5 channels for example are shown to be associated with the phospholipid-binding protein SESTD1 (Miehe et al. 2010). It has to be noted that when studying endogenous TRPC channels, it may be difficult to unambiguously identify individual TRPC channel subtypes in native tissues. Clearly, more work is needed to understand how phosphoinositides regulate TRPC channels.

Table 4 Other TRP channels

Name	Regulation/function	Phosphoinositide effects
TRPA1	Mustard oil and other noxious chemicals	$PI(4,5)P_2$ inhibits heterologous desensitization by capsaicin (Akopian et al. 2007) $PI(4,5)P_2$ activates in excised patches, inhibits desensitization in whole cell (Karashima et al. 2008) $PI(4,5)P_2$ inhibits sensitization by PAR in whole cell (Dai et al. 2007) $PI(4,5)P_2$ inhibits in excised patches in the presence of PPPi, no effect w/o PPPi (Kim and Cavanaugh 2007; Kim et al. 2008c) Depletion of $PI(4,5)P_2$ with rapamycin-inducible phosphatase has no effect (Wang et al. 2008)
TRPML1	Intracellular channel, mutation causes mucolipidosis	Specifically activated by $PI(3,5)P_2$ (Dong et al. 2010; Zhang et al. 2012a)
TRPP2	Mutated in polycystic kidney disease, mechanosensor?	$PI(4,5)P_2$ inhibits, depletion of $PI(4,5)P_2$ by EGF activates (Ma et al. 2005)

8 Other TRP Channels

8.1 TRPA1

TRPA1 was originally proposed to be a noxious cold sensor. Its activation by cold is controversial; there are numerous articles claiming that cold temperatures activate the channel and similar number claiming that cold has no effect. There is full agreement in the literature that noxious chemicals, such as mustard oil, formaldehyde, and compounds in tear gases, activate this channel. TRPA1 is especially important in the respiratory system for avoiding inhalation of harmful chemicals (Nilius et al. 2012). The involvement of TRPA1 in pain in humans is demonstrated by the finding that gain of function mutations of these channels cause Familial Episodic Pain Syndrome (Kremeyer et al. 2010).

The regulation of TRPA1 by phosphoinositides is complicated (Table 4). The activity of recombinant TRPA1 channels was found to run down in excised patches and could be reactivated by $PI(4,5)P_2$ and by MgATP. Run down however became irreversible after >2 min in ATP free conditions (Karashima et al. 2008). This behavior of TRPA1 in excised patches was quite similar in our hands (T.R. unpublished observation). The same study by Karashima et al reported that mustard oil-induced currents in DRG neurons were inhibited by high concentrations of wortmannin, where the drug inhibits PI4K. Rapid desensitization in response to mustard oil was inhibited by supplying $PI(4,5)P_2$ through the patch pipette, whereas neomycin, which chelates phosphoinositides, accelerated it (Karashima et al. 2008). These data support a model in which TRPA1 activity depends on $PI(4,5)P_2$, and its depletion plays a role in desensitization. Consistent with the proposed dependence of TRPA1 activity on $PI(4,5)P_2$, another study found that capsaicin-induced cross desensitization of TRPA1 was prevented by dialysis of

$PI(4,5)P_2$ through the patch pipette (Akopian et al. 2007). Contrary to these findings however, two articles reported no effect of $PI(4,5)P_2$ in excised patches (Kim and Cavanaugh 2007) or inhibition in the presence of inorganic poly-phosphate (Kim et al. 2008c), which is an intracellular activator of these channels. Another study found that depletion of $PI(4,5)P_2$ with a rapamycin-inducible 5-phosphatase did not inhibit TRPA1, while it inhibited TRPM8 (Wang et al. 2008). In conclusion there are indications that TRPA1 requires $PI(4,5)P_2$ for activity, but there are also contrary data in the literature, and the regulation of these channels by phosphoinositides may be more complex than a simple dependence on $PI(4,5)P_2$.

8.2 TRPML1

Members of the TRPML subfamily are Ca^{2+} and Fe^{2+} permeable nonselective cation channels located in intracellular membranes. Functional TRPML1 channels are located in the endolysosomal membranes. Mutations in this channel cause type IV mucolipidosis in humans. TRPML1 requires the presence of $PI(3,5)P_2$, which is also specifically located in endolysosomal membranes. Activation by $PI(3,5)P_2$ is highly specific, and the channels have a very high apparent affinity for this lipid; the EC_{50} of TRPML1 for $PI(3,5)P_2$ is 48 nM (Dong et al. 2010). For comparison, the highest affinity mammalian plasma membrane channels have an EC_{50} of 5 μM for Kir2.1 (Rohacs et al. 2002) and 1.3–4.9 μM for TRPV1 depending on the conditions (Lukacs et al. 2007; Klein et al. 2008). On the low affinity end, KCNQ2/3 channels have an EC_{50} of 40-80 μM (Zhang et al. 2003; Li et al. 2005). Intriguingly, a similar high specificity activation by $PI(3,5)P_2$ was also described for another intracellular ion channel family, the two-pore TPC1 and TPC2 channels, very distant homologues of TRP channels (Wang et al. 2012).

One very important function of phosphoinositides is to serve as markers of identity of various membrane compartments, such as the plasma membrane (PM), endoplasmic reticulum (ER), Golgi, or endosomes (Shewan et al. 2011). It was proposed by Hilgemann that a major function of the general dependence of ion channels on $PI(4,5)P_2$ is to avoid their opening during their passage through the ER and Golgi, which are devoid of $PI(4,5)P_2$, and selectively open on the arrival in the PM, which contains this lipid (Hilgemann et al. 2001). An "inverted" version of this theme was proposed for TRPML1 channels; it was shown that TRMPL1 is inhibited by $PI(4,5)P_2$, and thus the channels that "accidentally" reached the plasma membrane are inactivated by $PI(4,5)P_2$. PM channels could be activated by either depleting $PI(4,5)P_2$ or adding excess $PI(3,5)P_2$ (Zhang et al. 2012a).

8.3 TRPP1

TRPP1 (PKD2, Polycystin 2, also called TRPP2 in older nomenclature) is a member of the TRPP family (Wu et al. 2010). It co-assembles with PKD1, and loss of function mutations of either proteins leads to autosomal dominant polycystic kidney disease. Despite its well-established role in the pathophysiology of this quite common genetic disorder, its physiological roles and regulation is not very well understood. TRPP2 is activated by epidermal growth factor, and this was proposed to proceed via relief from PI(4,5)P_2 inhibition upon PLC activation (Ma et al. 2005).

Conclusions and Future Questions

PI(4,5)P_2 modulates many different ion channels. For Kir and KCNQ channels it has been demonstrated beyond doubt that PI(4,5)P_2 is an obligatory cofactor for the activity of all members of the respective ion channel families. This is based on results from many different techniques, including excised patches and specific inducible lipid phosphatases in intact cells, and the results with these channels are largely in agreement between different laboratories. For Kir channels high-resolution co-crystal structures of channels with and without PI(4,5)P_2 have recently become available.

By now, there are data in the literature for almost all members of the three major TRP channel families showing some form of modulation by PI(4,5)P_2. On these channels however, we are very far from a clear picture. In most cases TRP channel activity depends on PI(4,5)P_2, just like for Kirs and KCNQs. It is safe to say that activation by PI(4,5)P_2 via direct binding of the lipid to the channel is a conserved feature among many members of the three major subfamilies. The dependence of channel activity on PI(4,5)P_2 seems to play a role in desensitization via PLC-induced depletion of the lipid for a large number of TRP channels including some members of all three major subfamilies.

For several TRP channels, inhibitory effects of PI(4,5)P_2 were published, often in contradiction with the presence of an activating effect on the same channel. Often the different techniques we described gave seemingly noncoherent results. Since this is a recurring theme, it is hard to dismiss them as unreliable data. Apparently the phosphoinositide regulation of many TRP channels, TRPV1 and TRPCs for example, is quite complex, and it is likely that some of the effects of phosphoinositide effects are mediated by other PI(4,5)P_2-binding proteins. Clearly, more work is needed to solve the puzzle of how TRP channels are regulated by phosphoinositides.

Much is needed also on the structure–function front. Since direct activation by PI(4,5)P_2 was shown for several TRP channels, it is quite likely that the activating effect of the lipid results from direct binding. From the sporadic mutations affecting PI(4,5)P_2 regulation, no coherent picture is arising about the binding site. With the recently published side-chain resolution electron cryomicroscopy structure of TRPV1 however, we can be optimistic about future progress in understanding the molecular mechanism of phosphoinositide regulation of TRP channels (Liao et al. 2013).

References

Akopian AN, Ruparel NB, Jeske NA, Hargreaves KM (2007) Transient receptor potential TRPA1 channel desensitization in sensory neurons is agonist dependent and regulated by TRPV1-directed internalization. J Physiol 583:175–193

Albert AP, Saleh SN, Large WA (2008) Inhibition of native TRPC6 channel activity by phosphatidylinositol 4,5-bisphosphate in mesenteric artery myocytes. J Physiol 586:3087–3095

Allen V, Swigart P, Cheung R, Cockcroft S, Katan M (1997) Regulation of inositol lipid-specific phospholipase C-delta by changes in Ca^{2+} ion concentrations. Biochem J 327:545–552

Baukrowitz T, Schulte U, Oliver D, Herlitze S, Krauter T, Tucker SJ, Ruppersberg JP, Fakler B (1998) PIP_2 and PIP as determinants for ATP inhibition of K_{ATP} channels. Science 282:1141–1144

Bautista DM, Siemens J, Glazer JM, Tsuruda PR, Basbaum AI, Stucky CL, Jordt SE, Julius D (2007) The menthol receptor TRPM8 is the principal detector of environmental cold. Nature 448:204–208

Bianco SD, Peng JB, Takanaga H, Suzuki Y, Crescenzi A, Kos CH, Zhuang L, Freeman MR, Gouveia CH, Wu J, Luo H, Mauro T, Brown EM, Hediger MA (2007) Marked disturbance of calcium homeostasis in mice with targeted disruption of the Trpv6 calcium channel gene. J Bone Miner Res 22:274–285

Bojjireddy N, Botyanszki J, Hammond G, Creech D, Peterson R, Kemp DC, Snead M, Brown R, Morrison A, Wilson S, Harrison S, Moore C, Balla T (2014) Pharmacological and Genetic Targeting of the PI4KA Enzyme Reveals Its Important Role in Maintaining Plasma Membrane Phosphatidylinositol 4-Phosphate and Phosphatidylinositol 4,5-Bisphosphate Levels. J Biol Chem 289:6120–6132

Brauchi S, Orta G, Mascayano C, Salazar M, Raddatz N, Urbina H, Rosenmann E, Gonzalez-Nilo F, Latorre R (2007) Dissection of the components for PIP_2 activation and thermosensation in TRP channels. Proc Natl Acad Sci U S A 104:10246–10251

Cao C, Zakharian E, Borbiro I, Rohacs T (2013a) Interplay between calmodulin and phosphatidylinositol 4,5-bisphosphate in Ca^{2+}-induced inactivation of transient receptor potential vanilloid 6 channels. J Biol Chem 288:5278–5290

Cao E, Cordero-Morales JF, Liu B, Qin F, Julius D (2013b) TRPV1 channels are intrinsically heat sensitive and negatively regulated by phosphoinositide lipids. Neuron 77:667–679

Cheng WW, D'Avanzo N, Doyle DA, Nichols CG (2011) Dual-mode phospholipid regulation of human inward rectifying potassium channels. Biophys J 100:620–628

Chuang HH, Prescott ED, Kong H, Shields S, Jordt SE, Basbaum AI, Chao MV, Julius D (2001) Bradykinin and nerve growth factor release the capsaicin receptor from PtdIns(4,5)P_2-mediated inhibition. Nature 411:957–962

D'Avanzo N, Cheng WW, Doyle DA, Nichols CG (2010) Direct and specific activation of human inward rectifier K^+ channels by membrane phosphatidylinositol 4,5-bisphosphate. J Biol Chem 285:37129–37132

Dai Y, Wang S, Tominaga M, Yamamoto S, Fukuoka T, Higashi T, Kobayashi K, Obata K, Yamanaka H, Noguchi K (2007) Sensitization of TRPA1 by PAR2 contributes to the sensation of inflammatory pain. J Clin Invest 117:1979–1987

Daniels RL, Takashima Y, McKemy DD (2009) Activity of the neuronal cold sensor TRPM8 is regulated by phospholipase C via the phospholipid phosphoinositol 4,5-bisphosphate. J Biol Chem 284:1570–1582

de Groot T, Kovalevskaya NV, Verkaart S, Schilderink N, Felici M, van der Hagen EA, Bindels RJ, Vuister GW, Hoenderop JG (2011) Molecular mechanisms of calmodulin action on TRPV5 and modulation by parathyroid hormone. Mol Cell Biol 31:2845–2853

Derler I, Hofbauer M, Kahr H, Fritsch R, Muik M, Kepplinger K, Hack ME, Moritz S, Schindl R, Groschner K, Romanin C (2006) Dynamic but not constitutive association of calmodulin with rat TRPV6 channels enables fine tuning of Ca^{2+}-dependent inactivation. J Physiol 577:31–44

Dhaka A, Murray AN, Mathur J, Earley TJ, Petrus MJ, Patapoutian A (2007) TRPM8 is required for cold sensation in mice. Neuron 54:371–378

DiNitto JP, Cronin TC, Lambright DG (2003) Membrane recognition and targeting by lipid-binding domains. Sci STKE 2003:re16

Doerner JF, Hatt H, Ramsey IS (2011) Voltage- and temperature-dependent activation of TRPV3 channels is potentiated by receptor-mediated PI(4,5)P_2 hydrolysis. J Gen Physiol 137:271–288

Dong XP, Shen D, Wang X, Dawson T, Li X, Zhang Q, Cheng X, Zhang Y, Weisman LS, Delling M, Xu H (2010) PI(3,5)P_2 controls membrane trafficking by direct activation of mucolipin Ca^{2+} release channels in the endolysosome. Nat Commun 1:38

Estacion M, Sinkins WG, Schilling WP (2001) Regulation of Drosophila transient receptor potential-like (TrpL) channels by phospholipase C-dependent mechanisms. J Physiol 530:1–19

Fan Z, Makielski JC (1997) Anionic phospholipids activate ATP-sensitive potassium channels. J Biol Chem 272:5388–5395

Fruman DA, Meyers RE, Cantley LC (1998) Phosphoinositide kinases. Annu Rev Biochem 67:481–507

Fujita F, Uchida K, Takaishi M, Sokabe T, Tominaga M (2013) Ambient temperature affects the temperature threshold for TRPM8 activation through interaction of phosphatidylinositol 4,5-bisphosphate. J Neurosci 33:6154–6159

Fukami K, Inanobe S, Kanemaru K, Nakamura Y (2010) Phospholipase C is a key enzyme regulating intracellular calcium and modulating the phosphoinositide balance. Prog Lipid Res 49:429–437

Furukawa T, Yamane T, Terai T, Katayama Y, Hiraoka M (1996) Functional linkage of the cardiac ATP-sensitive K^+ channel to the actin cytoskeleton. Pflugers Arch 431:504–512

Gamper N, Rohacs T (2012) Phosphoinositide sensitivity of ion channels, a functional perspective. Subcell Biochem 59:289–333

Garcia-Elias A, Mrkonjic S, Pardo-Pastor C, Inada H, Hellmich UA, Rubio-Moscardo F, Plata C, Gaudet R, Vicente R, Valverde MA (2013) Phosphatidylinositol-4,5-biphosphate-dependent rearrangement of TRPV4 cytosolic tails enables channel activation by physiological stimuli. Proc Natl Acad Sci U S A 110:9553–9558

Gwanyanya A, Sipido KR, Vereecke J, Mubagwa K (2006) ATP and PIP_2 dependence of the magnesium-inhibited, TRPM7-like cation channel in cardiac myocytes. Am J Physiol Cell Physiol 291:C627–635

Hammond GR, Fischer MJ, Anderson KE, Holdich J, Koteci A, Balla T, Irvine RF (2012) PI4P and PI(4,5)P_2 are essential but independent lipid determinants of membrane identity. Science 337:727–730

Hansen SB, Tao X, MacKinnon R (2011) Structural basis of PIP_2 activation of the classical inward rectifier K^+ channel Kir2.2. Nature 477:495–498

Hardie RC (2007) TRP channels and lipids: from Drosophila to mammalian physiology. J Physiol 578:9–24

Hilgemann DW (2012) Fitting K(V) potassium channels into the PIP_2 puzzle: Hille group connects dots between illustrious HH groups. J Gen Physiol 140:245–248

Hilgemann DW, Ball R (1996) Regulation of cardiac Na^+, Ca^{2+} exchange and K_{ATP} potassium channels by PIP_2. Science 273:956–959

Hilgemann DW, Feng S, Nasuhoglu C (2001) The complex and intriguing lives of PIP_2 with ion channels and transporters. Sci STKE 2001:re19

Hoenderop JG, van Leeuwen JP, van der Eerden BC, Kersten FF, van der Kemp AW, Merillat AM, Waarsing JH, Rossier BC, Vallon V, Hummler E, Bindels RJ (2003) Renal Ca^{2+} wasting, hyperabsorption, and reduced bone thickness in mice lacking TRPV5. J Clin Invest 112:1906–1914

Huang CL (2007) Complex roles of PIP_2 in the regulation of ion channels and transporters. Am J Physiol Renal Physiol 293:F1761–1765

Huang CL, Feng S, Hilgemann DW (1998) Direct activation of inward rectifier potassium channels by PIP_2 and its stabilization by Gbetagamma. Nature 391:803–806

Huang J, Liu CH, Hughes SA, Postma M, Schwiening CJ, Hardie RC (2010) Activation of TRP channels by protons and phosphoinositide depletion in Drosophila photoreceptors. Curr Biol 20:189–197

Imai Y, Itsuki K, Okamura Y, Inoue R, Mori MX (2012) A self-limiting regulation of vasoconstrictor-activated TRPC3/C6/C7 channels coupled to PI(4,5)P_2-diacylglycerol signalling. J Physiol 590:1101–1119

Itsuki K, Imai Y, Okamura Y, Abe K, Inoue R, Mori MX (2012) Voltage-sensing phosphatase reveals temporal regulation of TRPC3/C6/C7 channels by membrane phosphoinositides. Channels (Austin) 6:206–209

Jardin I, Redondo PC, Salido GM, Rosado JA (2008) Phosphatidylinositol 4,5-bisphosphate enhances store-operated calcium entry through hTRPC6 channel in human platelets. Biochim Biophys Acta 1783:84–97

Jeske NA, Por ED, Belugin S, Chaudhury S, Berg KA, Akopian AN, Henry MA, Gomez R (2011) A-kinase anchoring protein 150 mediates transient receptor potential family V type 1 sensitivity to phosphatidylinositol-4,5-bisphosphate. J Neurosci 31:8681–8688

Jin J, Wu LJ, Jun J, Cheng X, Xu H, Andrews NC, Clapham DE (2012) The channel kinase, TRPM7, is required for early embryonic development. Proc Natl Acad Sci U S A 109:E225–233

Ju M, Shi J, Saleh SN, Albert AP, Large WA (2010) Ins(1,4,5)P_3 interacts with PIP_2 to regulate activation of TRPC6/C7 channels by diacylglycerol in native vascular myocytes. J Physiol 588:1419–1433

Karashima Y, Prenen J, Meseguer V, Owsianik G, Voets T, Nilius B (2008) Modulation of the transient receptor potential channel TRPA1 by phosphatidylinositol 4,5-biphosphate manipulators. Pflugers Arch 457:77–89

Kim D, Cavanaugh EJ (2007) Requirement of a soluble intracellular factor for activation of transient receptor potential A1 by pungent chemicals: role of inorganic polyphosphates. J Neurosci 27:6500–6509

Kim AY, Tang Z, Liu Q, Patel KN, Maag D, Geng Y, Dong X (2008a) Pirt, a phosphoinositide-binding protein, functions as a regulatory subunit of TRPV1. Cell 133:475–485

Kim BJ, Kim MT, Jeon JH, Kim SJ, So I (2008b) Involvement of phosphatidylinositol 4,5-bisphosphate in the desensitization of canonical transient receptor potential 5. Biol Pharm Bull 31:1733–1738

Kim D, Cavanaugh EJ, Simkin D (2008c) Inhibition of transient receptor potential A1 channel by phosphatidylinositol-4,5-bisphosphate. Am J Physiol Cell Physiol 295:C92–99

Kim H, Jeon JP, Hong C, Kim J, Myeong J, Jeon JH, So I (2013) An essential role of PI(4,5)P_2 for maintaining the activity of the transient receptor potential canonical (TRPC)4beta. Pflugers Arch 465:1011–1021

Klein RM, Ufret-Vincenty CA, Hua L, Gordon SE (2008) Determinants of molecular specificity in phosphoinositide regulation. Phosphatidylinositol (4,5)-bisphosphate (PI(4,5)P_2) is the endogenous lipid regulating TRPV1. J Biol Chem 283:26208–26216

Krauter T, Ruppersberg JP, Baukrowitz T (2001) Phospholipids as modulators of K_{ATP} channels: distinct mechanisms for control of sensitivity to sulphonylureas, K^+ channel openers, and ATP. Mol Pharmacol 59:1086–1093

Kremeyer B et al (2010) A gain-of-function mutation in TRPA1 causes familial episodic pain syndrome. Neuron 66:671–680

Kruse M, Hammond GR, Hille B (2012) Regulation of voltage-gated potassium channels by PI(4,5)P_2. J Gen Physiol 140:189–205

Kwon Y, Hofmann T, Montell C (2007) Integration of phosphoinositide- and calmodulin-mediated regulation of TRPC6. Mol Cell 25:491–503

Langeslag M, Clark K, Moolenaar WH, van Leeuwen FN, Jalink K (2007) Activation of TRPM7 channels by phospholipase C-coupled receptor agonists. J Biol Chem 282:232–239

Large WA, Saleh SN, Albert AP (2009) Role of phosphoinositol 4,5-bisphosphate and diacylglycerol in regulating native TRPC channel proteins in vascular smooth muscle. Cell Calcium 45:574–582

Lee J, Cha SK, Sun TJ, Huang CL (2005) PIP_2 activates TRPV5 and releases its inhibition by intracellular Mg^{2+}. J Gen Physiol 126:439–451

Lemonnier L, Trebak M, Putney JW Jr (2008) Complex regulation of the TRPC3, 6 and 7 channel subfamily by diacylglycerol and phosphatidylinositol-4,5-bisphosphate. Cell Calcium 43:506–514

Li Y, Gamper N, Hilgemann DW, Shapiro MS (2005) Regulation of Kv7 (KCNQ) K^+ channel open probability by phosphatidylinositol 4,5-bisphosphate. J Neurosci 25:9825–9835

Liao M, Cao E, Julius D, Cheng Y (2013) Structure of the TRPV1 ion channel determined by electron cryo-microscopy. Nature 504:107–112

Lin Z et al (2012) Exome sequencing reveals mutations in TRPV3 as a cause of Olmsted syndrome. Am J Hum Genet 90:558–564

Link TM, Park U, Vonakis BM, Raben DM, Soloski MJ, Caterina MJ (2010) TRPV2 has a pivotal role in macrophage particle binding and phagocytosis. Nat Immunol 11:232–239

Lishko PV, Procko E, Jin X, Phelps CB, Gaudet R (2007) The ankyrin repeats of TRPV1 bind multiple ligands and modulate channel sensitivity. Neuron 54:905–918

Liu D, Liman ER (2003) Intracellular Ca^{2+} and the phospholipid PIP_2 regulate the taste transduction ion channel TRPM5. Proc Natl Acad Sci U S A 100:15160–15165

Liu B, Qin F (2005) Functional control of cold- and menthol-sensitive TRPM8 ion channels by phosphatidylinositol 4,5-bisphosphate. J Neurosci 25:1674–1681

Liu B, Zhang C, Qin F (2005) Functional recovery from desensitization of vanilloid receptor TRPV1 requires resynthesis of phosphatidylinositol 4,5-bisphosphate. J Neurosci 25:4835–4843

Lopes CM, Zhang H, Rohacs T, Jin T, Yang J, Logothetis DE (2002) Alterations in conserved Kir channel-PIP_2 interactions underlie channelopathies. Neuron 34:933–944

Lukacs V, Thyagarajan B, Varnai P, Balla A, Balla T, Rohacs T (2007) Dual regulation of TRPV1 by phosphoinositides. J Neurosci 27:7070–7080

Lukacs V, Yudin Y, Hammond GR, Sharma E, Fukami K, Rohacs T (2013a) Distinctive changes in plasma membrane phosphoinositides underlie differential regulation of TRPV1 in nociceptive neurons. J Neurosci 33:11451–11463

Lukacs V, Rives JM, Sun X, Zakharian E, Rohacs T (2013b) Promiscuous activation of transient receptor potential vanilloid 1 channels by negatively charged intracellular lipids, the key role of endogenous phosphoinositides in maintaining channel activity. J Biol Chem 288:35003–35013

Ma R, Li WP, Rundle D, Kong J, Akbarali HI, Tsiokas L (2005) PKD2 functions as an epidermal growth factor-activated plasma membrane channel. Mol Cell Biol 25:8285–8298

Mercado J, Gordon-Shaag A, Zagotta WN, Gordon SE (2010) Ca^{2+}-dependent desensitization of TRPV2 channels is mediated by hydrolysis of phosphatidylinositol 4,5-bisphosphate. J Neurosci 30:13338–13347

Miehe S, Bieberstein A, Arnould I, Ihdene O, Rutten H, Strubing C (2010) The phospholipid-binding protein SESTD1 is a novel regulator of the transient receptor potential channels TRPC4 and TRPC5. J Biol Chem 285:12426–12434

Mohapatra DP, Nau C (2005) Regulation of Ca^{2+}-dependent desensitization in the vanilloid receptor TRPV1 by calcineurin and cAMP-dependent protein kinase. J Biol Chem 280:13424–13432

Montell C (2012) Drosophila visual transduction. Trends Neurosci 35:356–363

Niemeyer BA, Bergs C, Wissenbach U, Flockerzi V, Trost C (2001) Competitive regulation of CaT-like-mediated Ca^{2+} entry by protein kinase C and calmodulin. Proc Natl Acad Sci U S A 98:3600–3605

Nilius B, Voets T (2013) The puzzle of TRPV4 channelopathies. EMBO Rep 14:152–163

Nilius B, Mahieu F, Prenen J, Janssens A, Owsianik G, Vennekens R, Voets T (2006) The Ca^{2+}-activated cation channel TRPM4 is regulated by phosphatidylinositol 4,5-biphosphate. EMBO J 25:467–478
Nilius B, Owsianik G, Voets T (2008) Transient receptor potential channels meet phosphoinositides. EMBO J 27:2809–2816
Nilius B, Appendino G, Owsianik G (2012) The transient receptor potential channel TRPA1: from gene to pathophysiology. Pflugers Arch 464:425–458
Okamura Y, Murata Y, Iwasaki H (2009) Voltage-sensing phosphatase: actions and potentials. J Physiol 587:513–520
Otsuguro K, Tang J, Tang Y, Xiao R, Freichel M, Tsvilovskyy V, Ito S, Flockerzi V, Zhu MX, Zholos AV (2008) Isoform-specific inhibition of TRPC4 channel by phosphatidylinositol 4,5-bisphosphate. J Biol Chem 283:10026–10036
Park U, Vastani N, Guan Y, Raja SN, Koltzenburg M, Caterina MJ (2011) TRP vanilloid 2 knock-out mice are susceptible to perinatal lethality but display normal thermal and mechanical nociception. J Neurosci 31:11425–11436
Patil MJ, Belugin S, Akopian AN (2011) Chronic alteration in phosphatidylinositol 4,5-biphosphate levels regulates capsaicin and mustard oil responses. J Neurosci Res 89:945–954
Premkumar LS, Raisinghani M, Pingle SC, Long C, Pimentel F (2005) Downregulation of transient receptor potential melastatin 8 by protein kinase C-mediated dephosphorylation. J Neurosci 25:11322–11329
Prescott ED, Julius D (2003) A modular PIP_2 binding site as a determinant of capsaicin receptor sensitivity. Science 300:1284–1288
Putney JW, Tomita T (2011) Phospholipase C signaling and calcium influx. Adv Enzyme Regul 52:152–164
Qin F (2007) Regulation of TRP ion channels by phosphatidylinositol-4,5-bisphosphate. Handb Exp Pharmacol 509–525
Rohacs T (2007) Regulation of TRP channels by PIP_2. Pflugers Arch 453:753–762
Rohacs T (2009) Phosphoinositide regulation of non-canonical transient receptor potential channels. Cell Calcium 45:554–565
Rohacs T (2013) Regulation of transient receptor potential channels by the phospholipase C pathway. Adv Biol Regul 53:341–355
Rohacs T, Nilius B (2007) Regulation of transient receptor potential (TRP) channels by phosphoinositides. Pflugers Arch 455:157–168
Rohacs T, Lopes C, Mirshahi T, Jin T, Zhang H, Logothetis DE (2002) Assaying phosphatidylinositol bisphosphate regulation of potassium channels. Methods Enzymol 345:71–92
Rohacs T, Lopes CM, Jin T, Ramdya PP, Molnar Z, Logothetis DE (2003) Specificity of activation by phosphoinositides determines lipid regulation of Kir channels. Proc Natl Acad Sci U S A 100:745–750
Rohacs T, Lopes CM, Michailidis I, Logothetis DE (2005) $PI(4,5)P_2$ regulates the activation and desensitization of TRPM8 channels through the TRP domain. Nat Neurosci 8:626–634
Rosenhouse-Dantsker A, Logothetis DE (2007) Molecular characteristics of phosphoinositide binding. Pflugers Arch 455:45–53
Runnels LW, Yue L, Clapham DE (2002) The TRPM7 channel is inactivated by PIP_2 hydrolysis. Nat Cell Biol 4:329–336
Saarikangas J, Zhao H, Lappalainen P (2010) Regulation of the actin cytoskeleton-plasma membrane interplay by phosphoinositides. Physiol Rev 90:259–289
Sachs F (2010) Stretch-activated ion channels: what are they? Physiology (Bethesda) 25:50–56
Saleh SN, Albert AP, Large WA (2009a) Activation of native TRPC1/C5/C6 channels by endothelin-1 is mediated by both PIP3 and PIP_2 in rabbit coronary artery myocytes. J Physiol 587:5361–5375

Saleh SN, Albert AP, Large WA (2009b) Obligatory role for phosphatidylinositol 4,5-bisphosphate in activation of native TRPC1 store-operated channels in vascular myocytes. J Physiol 587:531–540
Sarria I, Ling J, Zhu MX, Gu JG (2011) TRPM8 acute desensitization is mediated by calmodulin and requires PIP_2: distinction from tachyphylaxis. J Neurophysiol 106:3056–3066
Senning EN, Collins MD, Stratiievska A, Ufret-Vincenty CA, Gordon SE (2014) Regulation of TRPV1 by phosphoinositide (4,5)-bisphosphate: role of membrane asymmetry. J Biol Chem (in press)
Shewan A, Eastburn DJ, Mostov K (2011) Phosphoinositides in cell architecture. Cold Spring Harb Perspect Biol 3:a004796
Shi J, Ju M, Abramowitz J, Large WA, Birnbaumer L, Albert AP (2012) TRPC1 proteins confer PKC and phosphoinositol activation on native heteromeric TRPC1/C5 channels in vascular smooth muscle: comparative study of wild-type and $TRPC1^{-/-}$ mice. FASEB J 26:409–419
Shyng SL, Nichols CG (1998) Membrane phospholipid control of nucleotide sensitivity of K_{ATP} channels. Science 282:1138–1141
Sowa NA, Street SE, Vihko P, Zylka MJ (2010) Prostatic acid phosphatase reduces thermal sensitivity and chronic pain sensitization by depleting phosphatidylinositol 4,5-bisphosphate. J Neurosci 30:10282–10293
Stein AT, Ufret-Vincenty CA, Hua L, Santana LF, Gordon SE (2006) Phosphoinositide 3-kinase binds to TRPV1 and mediates NGF-stimulated TRPV1 trafficking to the plasma membrane. J Gen Physiol 128:509–522
Suh BC, Hille B (2002) Recovery from muscarinic modulation of M current channels requires phosphatidylinositol 4,5-bisphosphate synthesis. Neuron 35:507–520
Suh BC, Hille B (2008) PIP_2 is a necessary cofactor for ion channel function: how and why? Annu Rev Biophys 37:175–195
Suh BC, Inoue T, Meyer T, Hille B (2006) Rapid chemically induced changes of PtdIns(4,5)P_2 gate KCNQ ion channels. Science 314:1454–1457
Suh BC, Leal K, Hille B (2010) Modulation of high-voltage activated Ca^{2+} channels by membrane phosphatidylinositol 4,5-bisphosphate. Neuron 67:224–238
Sui JL, Petit-Jacques J, Logothetis DE (1998) Activation of the atrial K_{ACh} channel by the betagamma subunits of G proteins or intracellular Na^+ ions depends on the presence of phosphatidylinositol phosphates. Proc Natl Acad Sci U S A 95:1307–1312
Takezawa R, Schmitz C, Demeuse P, Scharenberg AM, Penner R, Fleig A (2004) Receptor-mediated regulation of the TRPM7 channel through its endogenous protein kinase domain. Proc Natl Acad Sci U S A 101:6009–6014
Thyagarajan B, Lukacs V, Rohacs T (2008) Hydrolysis of phosphatidylinositol 4,5-bisphosphate mediates calcium-induced inactivation of TRPV6 channels. J Biol Chem 283:14980–14987
Thyagarajan B, Benn BS, Christakos S, Rohacs T (2009) Phospholipase C-mediated regulation of transient receptor potential vanilloid 6 channels: implications in active intestinal Ca^{2+} transport. Mol Pharmacol 75:608–616
Toth B, Csanady L (2012) Pore collapse underlies irreversible inactivation of TRPM2 cation channel currents. Proc Natl Acad Sci U S A 109:13440–13445
Trebak M, Lemonnier L, DeHaven WI, Wedel BJ, Bird GS, Putney JW Jr (2009) Complex functions of phosphatidylinositol 4,5-bisphosphate in regulation of TRPC5 cation channels. Pflugers Arch 457:757–769
Ufret-Vincenty CA, Klein RM, Hua L, Angueyra J, Gordon SE (2011) Localization of the PIP_2 sensor of TRPV1 ion channels. J Biol Chem 286:9688–9698
Varnai P, Thyagarajan B, Rohacs T, Balla T (2006) Rapidly inducible changes in phosphatidylinositol 4,5-bisphosphate levels influence multiple regulatory functions of the lipid in intact living cells. J Cell Biol 175:377–382
Walder RY, Landau D, Meyer P, Shalev H, Tsolia M, Borochowitz Z, Boettger MB, Beck GE, Englehardt RK, Carmi R, Sheffield VC (2002) Mutation of TRPM6 causes familial hypomagnesemia with secondary hypocalcemia. Nat Genet 31:171–174

Walder RY, Yang B, Stokes JB, Kirby PA, Cao X, Shi P, Searby CC, Husted RF, Sheffield VC (2009) Mice defective in Trpm6 show embryonic mortality and neural tube defects. Hum Mol Genet 18:4367–4375

Wang YY, Chang RB, Waters HN, McKemy DD, Liman ER (2008) The nociceptor ion channel TRPA1 is potentiated and inactivated by permeating calcium ions. J Biol Chem 283:32691–32703

Wang X, Zhang X, Dong XP, Samie M, Li X, Cheng X, Goschka A, Shen D, Zhou Y, Harlow J, Zhu MX, Clapham DE, Ren D, Xu H (2012) TPC proteins are phosphoinositide-activated sodium-selective ion channels in endosomes and lysosomes. Cell 151:372–383

Weissgerber P, Kriebs U, Tsvilovskyy V, Olausson J, Kretz O, Stoerger C, Vennekens R, Wissenbach U, Middendorff R, Flockerzi V, Freichel M (2011) Male fertility depends on Ca^{2+} absorption by TRPV6 in epididymal epithelia. Sci Signal 4:ra27

Whorton MR, MacKinnon R (2011) Crystal structure of the mammalian GIRK2 K^+ channel and gating regulation by G proteins, PIP_2, and sodium. Cell 147:199–208

Wu LJ, Sweet TB, Clapham DE (2010) International Union of Basic and Clinical Pharmacology. LXXVI. Current progress in the mammalian TRP ion channel family. Pharmacol Rev 62:381–404

Xie J, Sun B, Du J, Yang W, Chen HC, Overton JD, Runnels LW, Yue L (2011) Phosphatidylinositol 4,5-bisphosphate (PIP_2) controls magnesium gatekeeper TRPM6 activity. Sci Rep 1:146

Yamamoto S, Shimizu S, Kiyonaka S, Takahashi N, Wajima T, Hara Y, Negoro T, Hiroi T, Kiuchi Y, Okada T, Kaneko S, Lange I, Fleig A, Penner R, Nishi M, Takeshima H, Mori Y (2008) TRPM2-mediated Ca^{2+} influx induces chemokine production in monocytes that aggravates inflammatory neutrophil infiltration. Nat Med 14:738–747

Yao J, Qin F (2009) Interaction with phosphoinositides confers adaptation onto the TRPV1 pain receptor. PLoS Biol 7:e46

Yudin Y, Rohacs T (2011) Regulation of TRPM8 channel activity. Mol Cell Endocrinol 353:68–74

Yudin Y, Lukacs V, Cao C, Rohacs T (2011) Decrease in phosphatidylinositol 4,5-bisphosphate levels mediates desensitization of the cold sensor TRPM8 channels. J Physiol 589:6007–6027

Zakharian E, Thyagarajan B, French RJ, Pavlov E, Rohacs T (2009) Inorganic polyphosphate modulates TRPM8 channels. PLoS One 4:e5404

Zakharian E, Cao C, Rohacs T (2010) Gating of transient receptor potential melastatin 8 (TRPM8) channels activated by cold and chemical agonists in planar lipid bilayers. J Neurosci 30:12526–12534

Zakharian E, Cao C, Rohacs T (2011) Intracellular ATP supports TRPV6 activity via lipid kinases and the generation of $PtdIns(4,5)P_2$. FASEB J 25:3915–3928

Zhang H, Craciun LC, Mirshahi T, Rohacs T, Lopes CM, Jin T, Logothetis DE (2003) PIP_2 activates KCNQ channels, and its hydrolysis underlies receptor-mediated inhibition of M currents. Neuron 37:963–975

Zhang Z, Okawa H, Wang Y, Liman ER (2005) Phosphatidylinositol 4,5-bisphosphate rescues TRPM4 channels from desensitization. J Biol Chem 280:39185–39192

Zhang X, Li X, Xu H (2012a) Phosphoinositide isoforms determine compartment-specific ion channel activity. Proc Natl Acad Sci U S A 109:11384–11389

Zhang X, Mak S, Li L, Parra A, Denlinger B, Belmonte C, McNaughton PA (2012b) Direct inhibition of the cold-activated TRPM8 ion channel by Galpha(q). Nat Cell Biol 14:851–858

TRP Modulation by Natural Compounds

Flavia Carla Meotti, Edinéia Lemos de Andrade, and João B. Calixto

Contents

F.C. Meotti
Departamento de Bioquímica, Instituto de Química, Universidade de São Paulo, 05508-000 São Paulo, SP, Brazil

E. Lemos de Andrade • J.B. Calixto (✉)
Departamento de Farmacologia, Centro de Ciências Biológicas, Universidade Federal de Santa Catarina, 88049-900 Florianópolis, Santa Catarina, Brazil

Center of Innovation and Preclinical Studies- CIEnP, Av. Luiz Boiteux Piazza, 1302, Cachoeira do Bom Jesus, 88056-000 Florianópolis, SC, Brazil
e-mail: calixto@farmaco.ufsc.br; joao.calixto@cienp.org.br

B. Nilius and V. Flockerzi (eds.), *Mammalian Transient Receptor Potential (TRP) Cation Channels*, Handbook of Experimental Pharmacology 223, DOI 10.1007/978-3-319-05161-1_19,

Abstract

The use of medicinal plants or other naturally derived products to relieve illness can be traced back over several millennia, and these natural products are still extensively used nowadays. Studies on natural products have, over the years, enormously contributed to the development of therapeutic drugs used in modern medicine. By means of the use of these substances as selective agonists, antagonists, enzyme inhibitors or activators, it has been possible to understand the complex function of many relevant targets. For instance, in an attempt to understand how pepper species evoke hot and painful actions, the pungent and active constituent capsaicin (from *Capsicum sp.*) was isolated in 1846 and the receptor for the biological actions of capsaicin was cloned in 1997, which is now known as TRPV1 (transient receptor potential vanilloid 1). Thus, TRPV1 agonists and antagonists have currently been tested in order to find new drug classes to treat different disorders. Indeed, the transient receptor potential (TRP) proteins are targets for several natural compounds, and antagonists of TRPs have been synthesised based on the knowledge of naturally derived products. In this context, this chapter focuses on naturally derived compounds (from plants and animals) that are reported to be able to modulate TRP channels. To clarify and make the understanding of the modulatory effects of natural compounds on TRPs easier, this chapter is divided into groups according to TRP subfamilies: TRPV (TRP vanilloid), TRPA (TRP ankyrin), TRPM (TRP melastatin), TRPC (TRP canonical) and TRPP (TRP polycystin). A general overview on the naturally derived compounds that modulate TRPs is depicted in Table 1.

Keywords

Transient receptor potential agonists • Antagonists • Modulators and Natural compounds

Abbreviations

ADP	Adenosine monophosphate
ADPDK	Polycystic kidney disease
Akt	v-*Akt* murine thymoma viral oncogene
CAMKIV	Calcium/calmodulin-dependent protein kinase type IV
CGRP	Calcitonin gene-related peptide
CHO	Chinese hamster ovary
CREB	cAMP-responsive element binding protein
DRG	Dorsal root ganglion

HEK293	Human embryonic kidney 293
hTM	Human transmembrane domain
mTM	Mouse transmembrane domain
ICK	Inhibitor cysteine knot
nAChRs	Nicotinic acetylcholine receptor
PI3K	Phosphoinositide 3-Kinase
PKD	Protein kinase D
PLC	Phospholipase C
TG	Trigeminal ganglion
THC	Δ^9-tetrahydrocannabinol
TRP	Transient receptor potential
TRPA	TRP ankyrin
TRPC	TRP canonical
TRPM	TRP melastatin
TRPP	TRP polycystin
TRPV	TRP vanilloid
VaTx	Vanillotoxins

1 TRPV

The TRPV subfamily is divided into four groups based on structure and function: TRPV1/TRPV2, TRPV3, TRPV4 and TRPV5/TRPV6. TRPV-1, 2, 3 and 4 are modestly permeable for calcium (Ca^{+2}), while TRPV5 and TRPV6 are the only highly Ca^{+2}-selective channels in the TRPV family. TRPV channels are widely distributed in sensory and non-sensory cells. These channels can be activated by a diverse range of stimuli (osmotic/mechanical stress, noxious heat and chemical stimuli). The basic properties and functions of TRPVs demonstrated that some members of this subfamily are important for the perception of pain, temperature sensitivity, osmotic regulation and maintenance of Ca^{+2} homeostasis (Hardie 2007; Pedersen et al. 2005; O'Neil and Brown 2003; Nilius and Owsianik 2011).

1.1 TRPV1

TRPV1 channel can be activated by a heterogeneous array of naturally derived products that include dietary compounds and some plant and animal toxins (Table 1). Of great importance is the activation of TRPV1 by capsaicin (*trans*-8-methyl-*N*-vanillyl-6-nonenamide), the active ingredient of Capsicum species known as hot chilli pepper (Caterina et al. 1997). Capsaicin belongs to the vanilloid class with a vanillyl moiety in the chemical structure that seems essential for the TRPV1 activation (Fig. 1). Early studies showed that the application of capsaicin to

Table 1 Natural products that modulate TRPs

	Compound	Origin	Target(s)	Effects
Capsaicinoids and related compounds (Fig. 1)	Capsaicin	*Capsicum annuum*	TRPV1	Agonist
	Capsiate	*Capsicum annuum* *Capsicum frutescens*	TRPV1/ TRPA1	Nonselective agonist
	Dihydrocapsiate		TRPV1	Agonist
	Nordihydrocapsiate		TRPV1	Agonist
	Piperine	*Piper nigrum* *Piper longum*	TRPV1	Agonist
	Eugenol	*Eugenia caryophyllata* *Ocimum gratissimum* *Cinnamomum zeylanicum*	TRPV1/ TRPV3/ TRPA1/ TRPM8	Nonselective agonist
	Resiniferatoxin	*Euphorbia resinífera*	TRPV1	Agonist
Ginger-derived products (Fig. 2)	[8]-Gingerol	*Zingiber officinale*	TRPV1/ TRPA1	Nonselective agonist
	[6]-Gingerol		TRPV1/ TRPA1	
	[6]-Shogaol		TRPV1	Agonist
	[10]-Shogaol		TRPV1	Agonist
	Zingerone		TRPV1	Agonist
	Paradol		TRPV1	Agonist
Phytocannabinoids (Fig. 3)	Δ^9-tetrahydrocannabinol	*Cannabis sativa*	TRPV2/ TRPA1/ TRPM8	Nonselective agonist/ antagonist
	Δ^9-tetrahydrocannabinolic acid		TRPA1/ TRPM8	Agonist/ antagonist
	Cannabidiol		TRPV1/ TRPV2/ TRPA1/ TRPM8	Agonist/ antagonist
	Cannabidiolic acid		TRPA1/ TRPM8	Agonist/ antagonist
	Cannabinol		TRPA1	Agonist
	Cannabichromene		TRPA1	Agonist
	Cannabigerol		TRPA1/ TRPV2/ TRPM8	Nonselective agonist/ antagonist
	Tetrahydrocannabivarin		TRPA1/ TRPV2	Nonselective agonist

(continued)

Table 1 (continued)

	Compound	Origin	Target(s)	Effects
Unsaturated dialdehyde terpenes (Fig. 4)	Polygodial	*Drymis winteri*	TRPV1/ TRPA1	Nonselective agonist
	Drimanial		TRPV1	Agonist
	Isovelleral	*Lactarius vellereus*	TRPV1/ TRPA1	Mixed agonist-antagonist
	Cinnamodial	*Cinnamosma fragrans*	TRPV1	Agonist
	Cinnamosmolide		TRPV1	Agonist
	Cinnamolide		TRPV1	Agonist
	Warburganal	*Warburgia ugandensis* *Warburgia stuhlmannii*	TRPV1	Agonist
	Scalaradial	Sponges	TRPV1	Agonist
	Aframodial	*Aframomum danielli*	TRPV1	Agonist
	Ancistrodial	*Ancistrotermes cavithorax*	TRPV1	Agonist
	Merulidial	Fungi	TRPV1	Agonist
	Drimenol	*Lactarius uvidus*	TRPV1	Agonist
Monoterpenes (Fig. 5)	Carvacrol	*Origanum majorana* *Origanum vulgare*	TRPV3/ TRPA1	Agonist
	Thymol	*Origanum vulgare* *Thymus vulgaris*	TRPV3/ TRPA1/ TRPM8	Agonist
	Camphor	*Cinnamomum camphora*	TRPV1/ TRPV3/ TRPA1/ TRPM8	Mixed agonist-antagonist
	Umbellulone	*Umbellularia californica*	TRPA1/ TRPM8	Mixed agonist-antagonist
	Menthol	*Mentha piperita* *Mentha arvensis*	TRPV3/ TRPA1/ TRPM8	Mixed agonist-antagonist
	1,4-cineole	*Eucalyptus polybractea*	TRPA1	Agonist
	1,8-cineole		TRPM8/ TRPA1	Agonist/ antagonist
	Carveol	Spearmint essential oil *Mentha spicata*	TRPV3	Agonist
	6-tert-butyl-m-cresol		TRPV3	Agonist
	Dihydrocarveol		TRPV3	Agonist
	Borneol		TRPV3	Agonist
	Perillaldehyde	*Perilla frutescens*	TRPA1/ TRPM8	Agonist/ antagonist
	Perillaketone		TRPA1/ TRPM8	Agonist/ antagonist

(continued)

Table 1 (continued)

	Compound	Origin	Target(s)	Effects
Triprenyl phenols (Fig. 6)	Grifolin	*Albatrellus ovinus*	TRPV1	Antagonist
	Neogrifolin		TRPV1	Antagonist
	Albaconol		TRPV1	Antagonist
	Scutigeral		TRPV1	Agonist
Chalcones (Fig. 7)	Cochinchinenin A	*Dracaena cochinchinensis*	TRPV1	Agonist
	Cochinchinenin B		TRPV1	Agonist
	Loureirin B		TRPV1	Agonist
Electrophilic modulators (Fig. 8)	Allyl isothiocyanate	*Brassica nigra*	TRPV1/ TRPA1	Agonist
	Cinnamaldehyde	*Cinnamomum cassia* *Cinnamomum zeylanicum*	TRPA1	Agonist
	Allicin	*Allium sp*	TRPA1	Agonist
	Diallyl disulphide		TRPA1	Agonist
	Ligustilide	*Angelica sinensis*	TRPA1	Agonist
	Dehydroligustilide	*Ligusticum chuanxiong* *Ligusticum porteri*	TRPA1	Agonist/ antagonist
Other modulators (Figs. 9, 10 and 11)	Alpha-spinasterol	*Vernonia tweedieana*	TRPV1	Antagonist
	Guaiacol	*Guaiacum officinale*	TRPV1	Agonist
	Methyl salicylate	*Gaultheria procumbens*	TRPA1	Agonist
	Vanillin	*Vanilla planifolia*	TRPV3	Agonist
	Methyl syringate	*Kalopanax pictus*	TRPA1	Agonist
	Resveratrol	*Vitis vinifera*	TRPV1/ TRPA1	Antagonist
	Pinosylvin methyl ether		TRPV1	Antagonist
	Salidroside	*Rhodiola rosea*	TRPM8	Antagonist
	Bisandrographolide A	*Andrographis paniculata*	TRPV4	Agonist
	Ginsenoside Rg1	*Panax notoginseng*	TRPV1	Antagonist
	Triptolide	*Tripterygium wilfordii*	TRPP	Agonist
	Evodiamine	*Evodia rutaecarpa*	TRPV1	Agonist
	Nicotine	*Nicotiana tabacum*	TRPV1/ TRPA1	Agonist
	Yohimbine	*Pausinystalia yohimbe* *Rauwolfia serpentine*	TRPV1	Antagonist
	Hydroxy-alpha-sanshool	*Zanthoxylum piperitum*	TRPV1	Agonist
	Artepillin C	*Baccharis dracunculifolia*	TRPA1	Agonist
	Thapsigargin	*Thapsia garganica*	TRPV1	Antagonist
	Ricinoleic acid	*Ricinus communis*	TRPV1	Agonist
	Citral	*Cymbopogon citratus*	TRPV3	Agonist

(continued)

Table 1 (continued)

Compound	Origin	Target(s)	Effects
Incensole acetate	*Boswellia papyrifera*	TRPV3	Agonist
Hyperforin	*Hypericum perforatum*	TRPC6	Agonist
Cnidarian envenomations	Jellyfish Sea anemone	TRPV1	Agonist
Vanillotoxins (VaTx) 1, 2, 3	*Psalmopoeus cambridgei*	TRPV1	Agonist
Quinine	*Cinchona sp.*	TRPM7	Antagonist
Waixenicin A	*Sarcothelia edmondsoni*	TRPM7	Antagonist

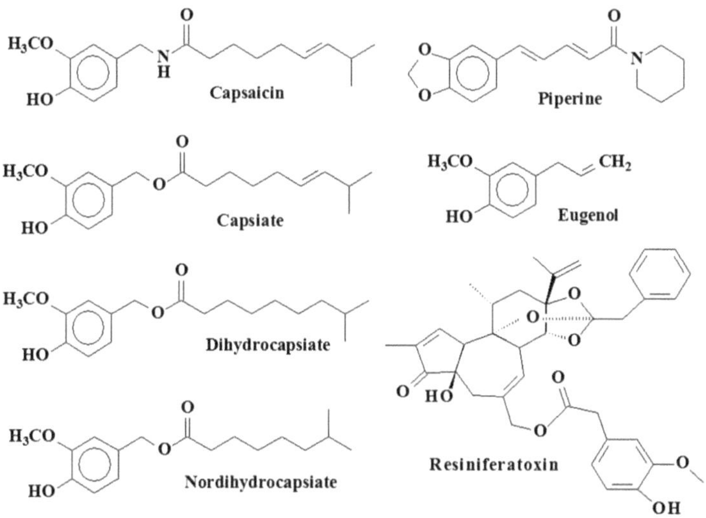

Fig. 1 Capsaicinoids and related compounds

human skin causes an early burning sensation and vasodilatation (Jancsó and Jancsó-Gabor 1949). The investigations to characterise the action of capsaicin on sensory neurons revealed that capsaicin was able to depolarise sensory fibres and cell bodies to produce an increased conductance associated with an enhancement of Na^{+} and Ca^{+2} permeability (Marsh et al. 1987). In this context, the burning pain sensation produced by capsaicin seems to be related to the influx of cations and the production of action potentials that propagate to the central nervous system (Bevan and Szolcsanyi 1990).

Notably, the initial neuronal stimulation by capsaicin of sensory neurons (perceived in humans as a burning sensation) is followed by a lasting refractory state, traditionally referred to as desensitisation or, under certain conditions, as neurotoxicity (Jancsó et al. 1977). It has been well known that following a desensitisation process by capsaicin, previously excited neurons become unresponsive not only to a repeated capsaicin challenge but also to other unrelated stimuli (Szallasi and Blumberg 1999). Capsaicin desensitisation seems to be a reversible phenomenon, since skin punch biopsies taken from volunteers exposed to high dose of capsaicin revealed a fully reversible loss of TRPV1-like immunoreactivity in the dermis (Malmberg et al. 2004; Kennedy et al. 2010). Currently, it has been reported that capsaicin desensitisation may involve additional mechanisms besides TRPV1 stimulation, because capsaicin also interferes with mitochondrial electron transport (Athanasiou et al. 2007). Thus, mitochondrial dysfunction in sensory neurons could contribute to capsaicin-induced desensitisation (Bley 2013).

1.1.1 Capsaicinoids, Capsinoids and Other Capsaicin-Related Compounds

1.1.1.1 Capsaicinoids

Capsaicin is the main active component found in chilli peppers (*Capsicum annuum*). However, this plant species contains other constituents such as dihydrocapsaicin, nordihydrocapsaicin, homodihydrocapsaicin and homocapsaicin, all of which are known as capsaicinoids (Davis et al. 2007; Mueller-Seitz et al. 2008). It is currently believed that these chemicals are synthesised by the plant as a defence mechanism against the attack of mammalian herbivores and fungi (Tewksbury et al. 2006). Capsicum species have been used in culinary and folk medicine since 7000 BC. Native Americans used *Capsicum* to cure cramps, diarrhoea and dyspepsia. The capsaicinoids were used for treating toothache in the nineteenth century. Folk medicinal uses of capsaicin include appetite stimulation, treatment of gastric ulcers and rheumatism and restoration of hair growth (Szallasi 1995). In Germany the so-called ABC patch is still used as a registered drug to apply warmth and thereby to treat lumbago, muscle tenseness and back pain. The only active ingredient nowadays is capsaicin (C), whereas originally the patches also contained arnica (A) and belladonna (B).

Currently, a great amount of evidence suggests a wide range of actions for capsaicinoids such as anticancer (Yang et al. 2010), anti-inflammatory (Choi et al. 2011) and antioxidant (Henning et al. 2011) properties. Interestingly, in vivo and in vitro antitumour activities have been attributed to the capsaicinoid compounds (Surh 2002; Oh et al. 2008; Thoennissen et al. 2010; Yang et al. 2010; Shirakawa et al. 2008; Amantini et al. 2009). In this context, capsaicin and dihydrocapsaicin have been reported to repress the growth of various immortalised or malignant cell lines via the induction of cycle arrest, apoptosis and autophagy and/or via the inhibition of cellular metabolic activation (Oh et al. 2008; Choi et al. 2010a, b; Ghosh and Basu 2010). Furthermore, capsaicin is able to inhibit both in vivo and in vitro growths of tumour cells such as adenocarcinoma, leukaemia and glioblastoma cells by inducing apoptosis (Hail and Lotan 2002; Lee et al. 2000; Kang et al. 2003; Ito et al. 2004; Qiao et al. 2005; Mori et al. 2006). In clinical studies, capsaicin inhibited the growth of leukaemic cells (Ito et al. 2004).

The apoptotic action of the capsaicin is thought to be mediated by target action, either through a direct pathway independent of TRPV1 or by an indirect pathway dependent on TRPV1 (Ziglioli et al. 2009). In the indirect pathway, capsaicin needs to interact with TRPV1, which in turn leads to an increase in intracellular Ca^{+2}, triggering early and late elements of apoptosis (Ziglioli et al. 2009). In glioma cells, capsaicin induces apoptosis, an effect that appears to be mediated by TRPV1 and involves a complex mechanism characterised by increased Ca^{+2} influx, p38 activation, phosphatidylserine exposure, mitochondrial permeability transmembrane pore opening and mitochondrial transmembrane potential dissipation, caspase 3 activation and, finally, oligonucleosomal DNA fragmentation (Amantini et al. 2007).

Several studies have investigated the effects of capsaicinoids on cardiovascular function. In this context, among the cardiovascular actions attributed to the capsaicinoids are the stimulation of the neurotransmitter release, promotion and inhibition of platelet aggregation and antioxidant properties. It is now well known that the cardiovascular system is rich in capsaicin-sensitive sensory nerves, which play an extensive role in regulating cardiovascular function through the release of multiple neurotransmitters (Peng and Li 2010; Zvara et al. 2006). Of note, capsaicin, through activation of TRPV1, was able to stimulate the release of CGRP, a neurotransmitter with a protective effect on cardiovascular function (Chai et al. 2006; Li et al. 2009; Li and Peng 2002; Zhong and Wang 2008; Zhou et al. 2010; Peng and Li 2010). However, high dose of capsaicin treatment exhausts CGRP in the nerve endings and displays detrimental effects (Peng et al. 2002).

The first evidence that capsaicin could act on platelet activation was described in 1991 (Hogaboam and Wallace 1991). In this study, capsaicin was inserted into the plasma membrane of platelets and altered membrane fluidity and/or ionic permeability. In addition, capsaicin and dihydrocapsaicin have been reported to inhibit platelet aggregation and the activity of clotting factors VIII and IX (Adams et al. 2009). These data suggest that capsaicinoids could be beneficial in the prevention or reduction of the incidence of cardiovascular diseases by reducing platelet aggregation. Otherwise, capsaicin stimulating TRPV1 localised in human platelets was able to induce the Ca^{+2} release from intracellular platelet stores and subsequently contributed to ADP and thrombin-induced platelet activation (Harper et al. 2009). Therefore, the action of capsaicinoids on platelet activation is controversial and more studies are necessary to address this issue.

There is an increased amount of evidence suggesting that capsaicinoids possess antiobesity properties (Zhang et al. 2007; Joo et al. 2010; Leung 2008; Cioffi 2007). It has been reported that both capsaicin and dihydrocapsaicin are able to inhibit obesity by enhancing energy expenditure and reducing body fat accumulation effects that were observed in experimental animal models and clinical trials, probably by altering proteins related to thermogenesis and lipid metabolism in white adipose tissue (Reinbach et al. 2009; Shin and Moritani 2007; Joo et al. 2010). Of great interest was the finding that capsaicin prevented adipogenesis and obesity by activating TRPV1. The activation of TRPV1 increased cytosolic Ca^{+2} and prevented the adipogenesis of pre-adipocytes in vitro. The effects of capsaicin on adipogenesis were attenuated by knocking down TRPV1. Furthermore, capsaicin prevented TRPV1 downregulation during adipogenesis. Finally, the administration of capsaicin prevented obesity in male wild-type mice, but not in

TRPV1 knockout mice assigned to the high-fat diet group (Zhang et al. 2007). Therefore, these data support a potential role of TRPV1 in metabolism regulation and weight reduction (Cioffi 2007; Nilius and Appendino 2011; Nilius and Owsianik 2011).

Other relevant properties of TRPV1 have been reported on the gastrointestinal system. The gastroprotective effects of capsaicinoids have been also demonstrated in different animal models of gastric mucosal injury induced by hydrochloric acid, ammonia, ethanol, aspirin or indomethacin (Mozsik et al. 2007; Szolcsanyi and Bartho 2001). These effects were attenuated by the presence of TRPV1 or CGRP-selective antagonists. In contrast, the pretreatment with capsaicin, at a single dose of 50 mg/kg, was found to attenuate the protective effects of rutaecarpine (a potential agonist of TRPV1) on acetylsalicylic acid-induced gastric mucosal injury. Otherwise, high doses of capsaicinoids may have detrimental effects on the gastrointestinal tract by promoting the exhaustion of neurotransmitters and damage to capsaicin-sensitive sensory nerves (Wang et al. 2005).

1.1.1.2 Capsinoids

Capsinoids are obtained from non-pungent red peppers (*Capsicum annuum* or *Capsicum frutescens*) named CH-19 Sweet (Kobata et al. 1999) and include capsiate, dihydrocapsiate and nordihydrocapsiate (Fig. 1) (Lang et al. 2009). These bear a similar structure to capsaicinoids, i.e. an aliphatic hydroxyl group in vanillyl alcohol with a fatty acid, but with different central linkage among the groups, yielding an amide in capsaicinoids and an ester group in capsinoids (He et al. 2009; Barbero et al. 2010; Luo et al. 2011). Although capsinoids have less pungency than capsaicin, they are also able to activate TRPV1 (Iida et al. 2003). Capsiate (4-hydroxy-3-methoxybenzyl (E)-8-methyl-6-nonenoate) was found to activate inward currents in hTRPV1-HEK293 cells with potency similar to that of capsaicin. The EC_{50} values for capsiate and capsaicin were, respectively, 290 nM (1.91, Hill coefficient) and 99.0 nM (1.46, Hill coefficient). The activation of currents by capsiate was reverted by the TRPV1 antagonist capsazepine. Moreover, capsiate administration in mice hind paw elicited pain similar to that induced by capsaicin. These data indicate that capsiate is a TRPV1 agonist and, similarly to capsaicin, is able to excite peripheral nociceptors. However, TRPV1 receptors in the skin surface, eye or oral cavity of mice are not sensitive to capsinoids. This is probably because of the high capsinoid lipophilicity and low molecule stability in aqueous solution (Iida et al. 2003). Capsiate, dihydrocapsiate and nordihydrocapsiate are also activators of TRPA1 expressed in HEK293 cells or in mouse DRG neurons, but they are more potent upon TRPV1 than upon TRPA1 (Shintaku et al. 2012). In addition, capsiate exhibited antiobesity actions, with an effect that is quite similar to that exerted by capsaicin. A direct evidence of the participation of TRPV1 in the antiobesity activity of capsiate was demonstrated by Kawabata and colleagues (Kawabata et al. 2009). In this study, the energy metabolism in wild-type mice, but not in TRPV1 knockout mice, was increased by capsiate. Currently, capsiate is a commercially available product in Japan and USA for body weight control (Luo et al. 2011).

1.1.1.3 Piperine

Piperine (piperinoyl-piperidine, Fig. 1) is an alkaloid found in plants belonging to the Piperaceae family, such as *Piper nigrum* (black pepper) and *Piper longum* (long pepper). The uses of black pepper in folk medicine include antipyretic and anti-inflammatory actions, metabolism stimulation, aiding in the absorption of nutrients and boosting the efficacy of drugs (Szallasi 2005). The initial characterisation of piperine as a TRPV1 activator was first reported by Szallasi and colleagues by demonstrating that this compound inhibited the binding of [^{3}H]-RTX in the dorsal horn of pig spinal cord (Szallasi and Blumberg 1991). Later, Liu and Simon showed that these compounds evoked similar currents in rat trigeminal ganglion neurons in vitro (Liu and Simon 1996b). In a more recent study, McNamara and co-workers concluded that piperine is not only similar to capsaicin in its effects at human TRPV1 expressed in HEK293 cells but exhibits greater efficacy than capsaicin in both events of activation and desensitisation of TRPV1 (McNamara et al. 2005). The reasons for the improved desensitisation-to-excitation ratio of piperine compared with capsaicin are unclear. An attractive hypothesis to explain the pharmacodynamic differences between capsaicin and piperine is that the latter might be a better inducer of the dephosphorylation and inactivation of TRPV1 (Vriens et al. 2009).

1.1.1.4 Eugenol

Eugenol is a phenol compound with a vanillyl moiety (Fig. 1) that can be isolated from clove or clocimum oil (*Eugenia caryophyllata* and *Ocimum gratissimum*) and cinnamon leaf (*Cinnamomum zeylanicum*). This compound shows antiseptic, astringent, hypothermic, myorelaxant, anticonvulsant and dental anaesthetic actions (Dallmeier and Carlini 1981; Evans 1996). As anaesthetic agent, eugenol, has been used to toothache, pulpitis and dentin hyperalgesia (Ohkubo and Kitamura 1997; Ohkubo and Shibata 1997). This compound shares several pharmacological actions with local anaesthetics, including the inhibition of voltage-gated sodium channels and activation of TRPV1 (Park et al. 2009). The first evidence that eugenol might be acting on ion channels emerged from electrophysiological studies showing that eugenol activated Ca^{+2}-permeable ion channels (Ohkubo and Kitamura 1997). Later, eugenol was reported to induce inward currents in HEK293 cells transfected with hTRPV1 and TG neurons as well as elevation of intracellular Ca^{+2} in the same cells (Yang et al. 2003). By means of in vivo experiments, it was demonstrated that eugenol possesses anti-nociceptive properties that are comparable to those of capsaicin (Ohkubo and Shibata 1997). Recently, it was demonstrated that the subcutaneous injection of eugenol reduced the thermal nociception and capsaicin-induced thermal hyperalgesia in a dose-dependent manner in the orofacial area (Park et al. 2009). Both the in vitro and in vivo actions of eugenol are reversed by capsazepine, demonstrating that its effects appear to be mediated by TRPV1. However, eugenol is able to active other TRP channels, such as TRPA1 and TRPM8 (Bandell et al. 2004), being classified as a nonselective TRPV1 agonist. More information about the modulatory effect of eugenol upon other TRPs will be discussed in their specific sections.

1.1.1.5 Resiniferoids

In 1975, Hecker's research group isolated the active principle of *Euphorbia resinifera*, a cactus-like plant native of Morocco (Hergenhahn et al. 1974), and named it resiniferatoxin (Fig. 1). This compound is a phorbol-related diterpene with the vanillyl moiety that shows a potent activity on TRPV1. The medicinal use of dried latex of *Euphorbia resinifera* refers to its direct application to dental cavities to mitigate toothache or to suppress chronic pain (Appendino and Szallasi 1997). Resiniferatoxin mimics most biological response characteristic of capsaicin with a hundred to several thousandfold higher potency (Szallasi and Blumberg 1990; Szallasi 1994). In this context, Maggi and co-workers found that resiniferatoxin is several thousandfold more potent than capsaicin for desensitising contractions of the isolated rat urinary bladder but similar in potency to capsaicin for contracting the isolated rat urinary bladder (Maggi et al. 1990). Another property of resiniferatoxin is the control of pulmonary chemo-reflex, which is desensitised by this compound without prior excitation but is evoked by capsaicin without subsequent desensitisation (Szolcsanyi et al. 1990). In patients with urinary bladder overactivity, intravesical resiniferatoxin improves urinary bladder function (or even restores continence) without significant irritation and/or toxicity. Therefore, resiniferatoxin is an attractive alternative to capsaicin in that it achieves lasting desensitisation without the side effects that complicate capsaicin therapy (Kissin and Szallasi 2011).

1.1.2 Ginger-Derived Products

In ginger (the rhizomes of *Zingiber officinale*), several pungent substances are found such as gingerols ([8] and [6]-gingerol) (main constituents), shogaol, zingerone and paradol (Fig. 2). Gingerols are highly similar to the shogaols and paradols with [6]-gingerol differing from 6-paradol only by a single hydroxyl group at C6 of the alkyl chain (Riera et al. 2009).

The first indication that ginger constituents might participate in sensorial processes came from the study carried out by Onogi and colleagues, which demonstrated that systemic administration of [6]-shogaol produced antinociception in rats, but in a lower intensity than that observed with capsaicin (Onogi et al. 1992). Later, electrophysiology and Ca^{+2} imaging studies showed that gingerols and shogaols increased intracellular Ca^{+2} concentrations in rats TRPV1-expressing HEK293 cells via TRPV1 and that [6]-gingerol and [8]-gingerol evoked capsaicin-like Ca^{+2} transient ion currents in DRG neurons, an effect that was blocked by the TRPV1 antagonist capsazepine (Witte et al. 2002; Iwasaki et al. 2006; Dedov et al. 2002). These data highlighted TRPV1 as a target for ginger constituents. These hypotheses were reinforced by additional findings showing that the subcutaneous administration of [10]-shogaol, [6]-shogaol or capsaicin into rats hind paw induced nociceptive responses via TRPV1. Other important findings regarding ginger constituents are that [6]- and [10]-gingerols and [6]- and [10]-shogaols promote adrenal catecholamine secretion, an event that alters energy consumption. Of interest, [10]-shogaol-induced adrenaline secretion was inhibited by the administration of capsazepine, suggesting the participation of

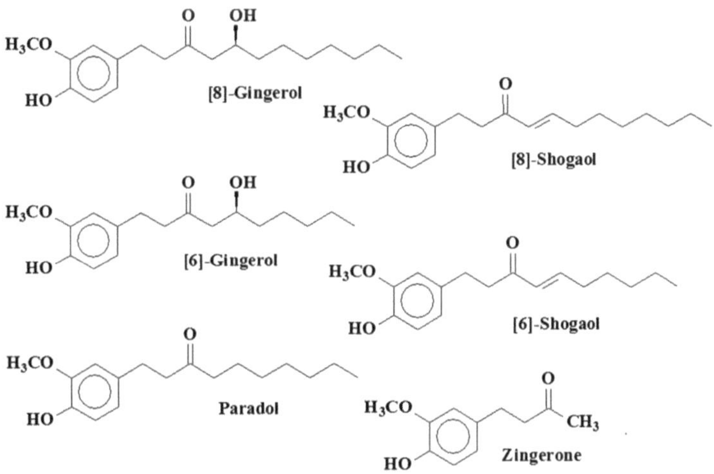

Fig. 2 Ginger-derived products

TRPV1 in this process (Iwasaki et al. 2006). However, gingerols also activate TRPA1 channel showing that these compounds are nonselective TRPV1 activators (Bandell et al. 2004).

Similarly to capsaicin, [6]-shogaol and [6]-paradol contain a vanilloid moiety and activate TRPV1 (Lee and Surh 1998; Witte et al. 2002). The activation of TRPV1 by these two compounds is thought to be the main responsible for the pungent sensation of oily extracts of Sichuan and Malagueta pepper. Zingerone [4-(4-hydroxy-3-methoxyphenyl)-2-butanone], a phenolic ketone metabolite of gingerols (Fig. 2), was suggested to activate the same receptor as capsaicin in TG neurons (Liu and Simon 1996b; Liu et al. 2000). Indeed, cultured rat TG neurons and TRPV1-Xenopus oocytes were desensitised by repeated applications of zingerone (Liu et al. 2000).

1.1.3 Phytocannabinoids

Cannabis species contain a complex mixture of substances that include 60 different cannabinoids, with the most prevalent being Δ^9-tetrahydrocannabinol (THC), cannabidiol and cannabinol (Fig. 3). Cannabidiol constitutes up to 40 % of *Cannabis sativa* extract. It is a non-psychotropic compound with possible therapeutic use as an anxiolytic, neuroprotective, anticonvulsive and anti-inflammatory agent (Mechoulam et al. 2002). Cannabidiol possesses low affinity to cannabinoid receptor, suggesting that some of the biological actions may be mediated by other receptor classes. In this context, studies have shown that cannabidiol possesses anti-inflammatory and analgesic effects similar to those of capsaicin. Reinforcing this hypothesis, cannabidiol at a micromolar concentration was able to inhibit the binding of [^{3}H]-resiniferatoxin and to increase intracellular-free Ca^{+2} in hTRPV1-HEK293 cells, behaving like a full agonist of TRPV1 (Bisogno et al. 2001).

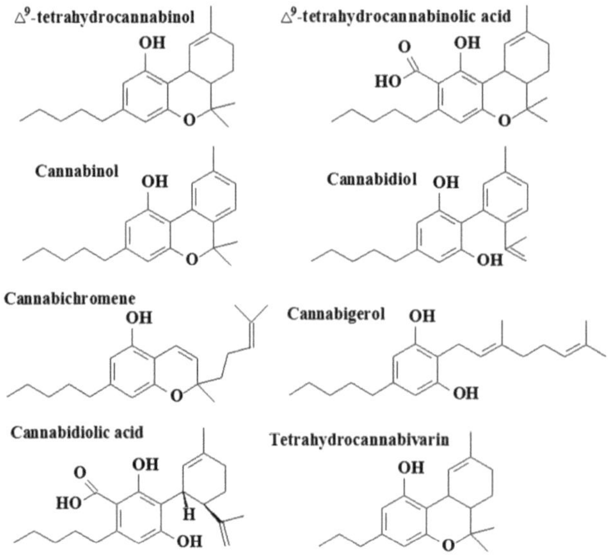

Fig. 3 Phytocannabinoids

Moreover, it was reported that the anti-nociceptive action of cannabidiol was mediated via TRPV1 (Costa et al. 2004).

1.1.4 Unsaturated Dialdehyde Terpenes

Terpenoids represent a chemical defence against environmental stress and provide a repair mechanism for wounds and injuries in plants. Interestingly, effective ingredients in several plant-derived medicinal extracts are also terpenoid compounds, such as monoterpenoid, sesquiterpenoid, diterpenoid, triterpenoid and carotenoid groups (Salminen et al. 2008). In this context, a broad spectrum of biological effects have been attributed to the terpenoids, especially to the unsaturated 1,4-dialdehydes. These effects include a decrease in serum cholesterol and inhibition of platelet aggregation (Lorenzen et al. 1994; Tanabe et al. 1993). Unsaturated 1,4-dialdeydes terpenes are structurally unrelated to capsaicin but are able to interact with TRPV1 despite of the lack vanillyl moiety in their structure (Sterner and Szallasi 1999). Importantly, not all effects attributed to unsaturated dialdehydes have a therapeutic purpose (Jonassohn and Sterner 1997).

The pungent sesquiterpenes polygodial and drimanial contain an unsaturated 1,4-dialdehyde functional group (Fig. 4). These compounds were isolated from the bark of *Drimys winteri* (Winteraceae), a native Brazilian medicinal plant used for the treatment of inflammatory diseases, in addition to being a substitute for pepper in culinary preparations (Corrêa 1978). Leaves and barks from these plants are used for analgesic and anti-inflammatory preparations to treat dental and stomach pain and other painful conditions (Andre et al. 2004). Systemic administration of polygodial or drimanial produces marked anti-nociceptive, anti-inflammatory and

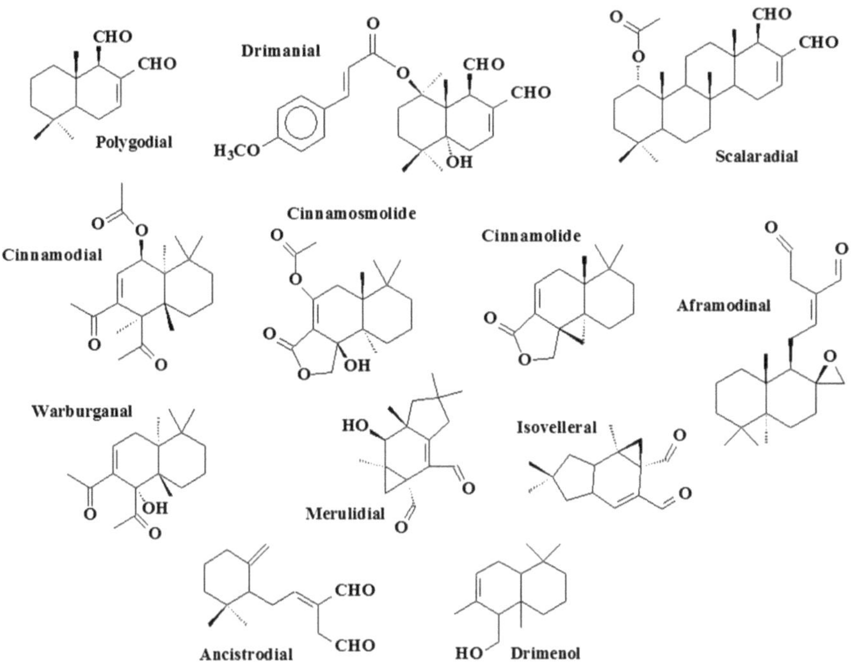

Fig. 4 Unsaturated dialdehyde terpenes

anti-allergic effects in mice (Tratsk et al. 1997; Mendes et al. 1998; da Cunha et al. 2001; Malheiros et al. 2001).

Although initially painful, polygodial acts as an analgesic through desensitisation of sensory neurons. Indeed, the neonatal treatment of rats with both polygodial and drimanial caused long-lasting and pronounced inhibition of the neurogenic nociception and thermal hyperalgesia (Andre et al. 2004), a neuronal desensitising effect similar to that described for capsaicin (Buck and Burks 1986; Holzer 1991). Moreover, sesquiterpenes produce urinary bladder contractile response through activation of a ruthenium red-sensitive channel and release of substance P or neurokinin A. However, in contrast to capsaicin, polygodial and drimanial failed to produce marked tachyphylaxis. Of note, data also show that polygodial and drimanial failed to exhibit cross-desensitisation with capsaicin. Such results suggest that although some of the actions of polygodial and drimanial in the rat urinary bladder seem to be similar to those caused by capsaicin, there are some apparent differences regarding their precise site of action (Andre et al. 2006). Corroborating with this hypothesis, ruthenium red (a nonselective blocker of TRP channels) showed an inhibitory activity that is more efficient than capsazepine (selective TRPV1 antagonist), suggesting that polygodial- or drimanial-induced responses in the urinary bladder could be mediated by other TRP channels. In this same study, André and colleagues reported that polygodial and drimanial displaced the specific binding of [^{3}H]-resiniferatoxin in rat spinal cord membranes.

In addition, polygodial and drimanial, like capsaicin, were also active in promoting [$^{45}Ca^{+2}$] uptake in the spinal cord synaptosomes and to increase the intracellular Ca^{+2} concentration in cultured rat trigeminal neurons (Andre et al. 2006). Now, it is known that polygodial activates TRPA1. Interestingly, Escalera and colleagues observed that polygodial robustly activated Ca^{+2} influx and cationic currents through TRPA1 channels expressed in primary sensory neurons and heterologous cells (Escalera et al. 2008). Therefore, the biological actions of polygodial are mediated by TRPA1, as depicted in TRPA1 section.

One of the most intensively studied defensive sesquiterpenes is isovelleral, the pungent product of the fungus *Lactarius vellereus*. Isovelleral is thought to be part of a fungal chemical defence system against predators (List and Hackenberg 1969; Magnusson et al. 1972). In humans, skin contact with the fungal milky exudate often leads to painful inflammatory responses, including eczema and blistering (Ponhold 1944; Escalera et al. 2008). Isovelleral has been proposed to act on vanilloid receptors in peripheral sensory neurons to induce pungent sensations and neurogenic inflammation. Isovelleral induced Ca^{+2} influx into cultured sensory neurons, an action which was partially abolished by the vanilloid antagonist capsazepine. In addition, isovelleral diminished binding and changed cooperativity of the high-affinity vanilloid agonist resiniferatoxin on sensory neuronal membrane preparations (Szallasi et al. 1996; Andre et al. 2006). In a similar fashion, isovelleral-activated relaxation of the mesenteric artery appeared to be partially independent of vanilloid receptors (Ralevic et al. 2003). Surprisingly, pharmacological studies failed to detect any agonist activity of isovelleral on cloned vanilloid receptors, including human, rat and porcine TRPV1 isoforms (Jerman et al. 2000; Smart et al. 2001; Ohta et al. 2005). In sharp contrast, it was found that isovelleral has TRPV1 antagonist activity, inhibiting capsaicin activation of TRPV1 at pK_b values similar to those of capsazepine and ruthenium red (7.52 ± 0.08, 6.92 ± 0.11 and 8.09 ± 0.12, respectively) (Jerman et al. 2000; Smart et al. 2001).

Other important unsaturated 1,4-dialdehydes terpenes are cinnamodial, cinnamosmolide, cinnamolide, warburganal, scalaradial, aframodial, ancistrodial, merulidial and drimenol (Fig. 4). Cinnamodial, cinnamosmolide and cinnamolide are present in the hot-tasting bark of *Cinnamosma fragrans* (Cannellaceae). These compounds induced Ca^{+2} uptake in rat DRG and inhibited specific [^{3}H]-resiniferatoxin binding in spinal cord membranes. Cinnamodial and cinnamosmolide had a bimodal effect on Ca^{+2} uptake; they increased Ca^{+2} uptake in low concentrations and inhibited it in concentrations higher than 0.3 and 3 μM, respectively (Szallasi et al. 1998). Warburganal (isolated from *Warburgia ugandensis* and *Warburgia stuhlmannii* (Cannellaceae)), scalaradial (produced by sponges), aframodial (isolated from the plant *Aframomum danielli*, Zingiberaceae), ancistrodial (isolated from the insect Ancistrotermes cavithorax), merulidial (isolated from fungi) and drimenol (isolated from *Lactarius uvidus*) were capable of inhibiting specific [^{3}H]-resiniferatoxin binding in rat spinal cord (Szallasi et al. 1996, 1998). Therefore, all these unsaturated 1,4-dialdehydes terpenes seemed to target TRPV1.

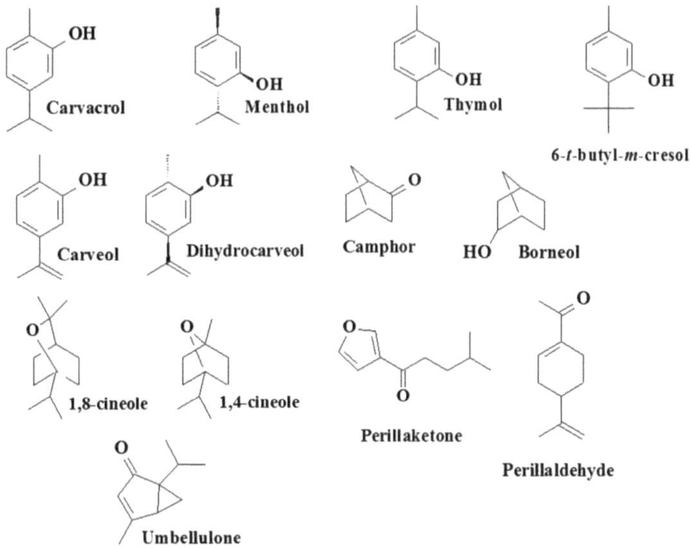

Fig. 5 Monoterpenes

1.1.5 Monoterpenes

Camphor (Fig. 5) is a naturally occurring compound that is used as an active ingredient of balms and liniments supplied as topical analgesics. Camphor activates heterologously expressed TRPV1, although this requires higher concentrations than capsaicin. Activation of TRPV1 by camphor is enhanced by phospholipase C-coupled receptor stimulation mimicking inflamed conditions. Similar camphor-activated TRPV1-like currents were observed in isolated rat DRG neurons and were strongly potentiated after the activation of protein kinase C with phorbol-12-myristate-13-acetate (PMA). The activation of rat TRPV1 by camphor was mediated by distinct channel regions from capsaicin, as indicated by experiments in the presence of the competitive TRPV1 inhibitor capsazepine and in capsaicin-insensitive point mutant. Recently, it was observed that camphor-activation pathway in TRPV1 involves the outer pore domain, particularly T633, a specific residue located in the middle of the pore helix that is also critical for direct activation of TRPV1 by protons (Marsakova et al. 2012). Although camphor activates TRPV1 less effectively, camphor application desensitised TRPV1 more rapidly and completely than capsaicin. Camphor-induced desensitisation of TRPV1 may underlie the analgesic effects of camphor (Xu et al. 2005).

1.1.6 Triprenyl Phenol

Studies have reported that the triprenyl phenols, namely, grifolin, neogrifolin, confluentin, albaconol and scutigeral (Fig. 6) obtained from *Albatrellus ovinus*, a terrestrial fungus found in Northern Europe, seem to modulate TRPV1. Scutigeral evoked $^{45}Ca^{+2}$-uptake response, which is similar in magnitude to that evoked by capsaicin and is blocked by the TRPV1 antagonist capsazepine. Under

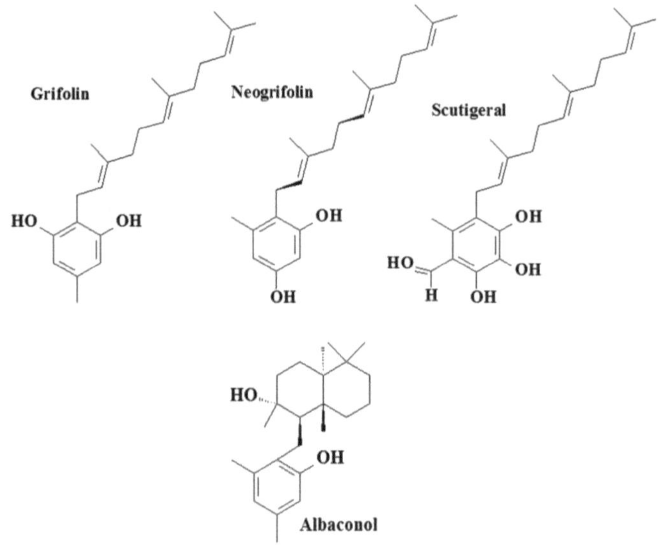

Fig. 6 Triprenyl phenols

voltage-clamp conditions, however, no scutigeral-induced inward currents could be detected. It was speculated that scutigeral gates the channel with prolonged kinetics compared to capsaicin or other pungent vanilloids. Thus, scutigeral may cause a sustained increase in intracellular calcium concentrations (Liu et al. 1997), which is not sufficient to generate action potentials but is able to block voltage-gated channels (Liu and Simon 1996a; Docherty et al. 1991). In contrast to the proposed agonist action of scutigeral, Hellwig and colleagues suggest that the prenylphenols grifolin, neogrifolin and albaconol act as weak antagonists (activity in the micromolar range) rather than exhibiting agonistic activities (Hellwig et al. 2003).

1.1.7 Chalcones

Dragon's blood from *Dracaena cochinchinensis* is one of the renowned traditional medicines. It is a multicomponent mixture with analgesic activity and has got several therapeutic uses (Gupta et al. 2008; Zhong 2010). Dragon's blood modulated tetrodotoxin-sensitive (TTX-S) and tetrodotoxin-resistant (TTX-R) sodium currents in DRG neurons. Three compounds cochinchinenin A, cochinchinenin B and loureirin B (Fig. 7) were the active components in dragon's blood interfering with pain messages (Guo et al. 2008; Liu et al. 2006). Dragon's blood solution inhibited capsaicin-induced currents in DRG with an IC_{50} value of 0.00002 ± 0.000005 % (m/v) and a Hill coefficient of 0.57 ± 0.07. The isolated compounds also inhibited capsaicin-induced currents, and the potency order was loureirin B>cochinchinenin A>cochinchinenin B. The combination of these three compounds using different proportions of each one confirmed that loureirin B was the main inhibitor of capsaicin-induced currents in DRG (Wei et al. 2013).

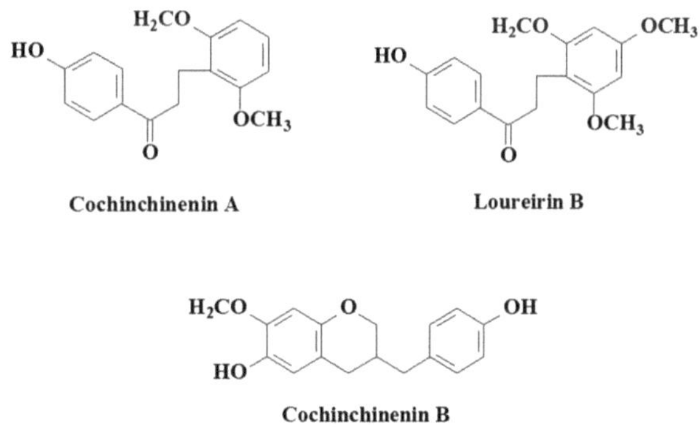

Fig. 7 Chalcones

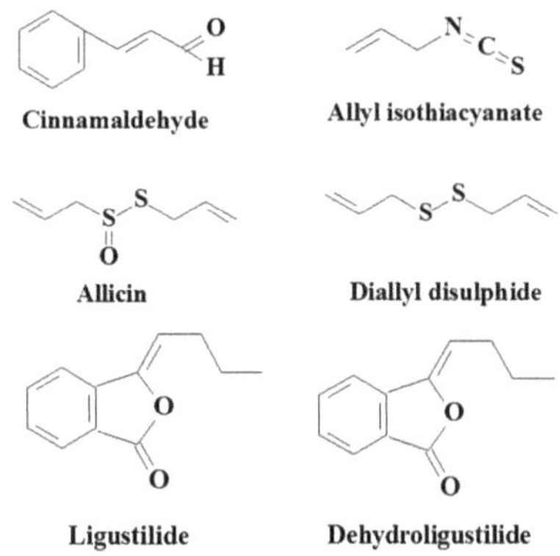

Fig. 8 Electrophilic TRP modulators

1.1.8 Other Modulators of TRPV1

1.1.8.1 Mustard Oil

Mustard oil is a pungent plant extract rich in allyl isothiocyanate (Fig. 8) obtained from mustard seed, horseradish and wasabi. TRPA1 had been purposed as the only target for mustard oil (Bandell et al. 2004; Bautista et al. 2005; Jordt et al. 2004; Macpherson et al. 2005); more details are provided in TRPA1 section. However, evidence show that mustard oil is not a selective activator of TRPA1. In this context, mustard oil induced an increase in intracellular Ca^{+2}. This event was suppressed by the TRPV1 antagonists capsazepine, ruthenium red and iodoresiniferatoxin. Mustard oil elicited outward rectified current in porcine TRPV1-expressing HEK293 cells, with a reversal potential similar to that of capsaicin. In cultured porcine DRG neurons, mustard oil elicited an intracellular calcium increase and inward current (Ohta et al. 2007). The mechanism of activation of TRPV1 by allyl isothiocyanate was largely dependent on S513, a residue that is involved in capsaicin binding. In addition, allyl isothiocyanate induced a shift of the voltage dependence of activation towards negative voltages in TRPV1, which is reminiscent of capsaicin effects, indicating that allyl isothiocyanate acts through reversible interactions with the capsaicin binding site (Gees et al. 2013). These

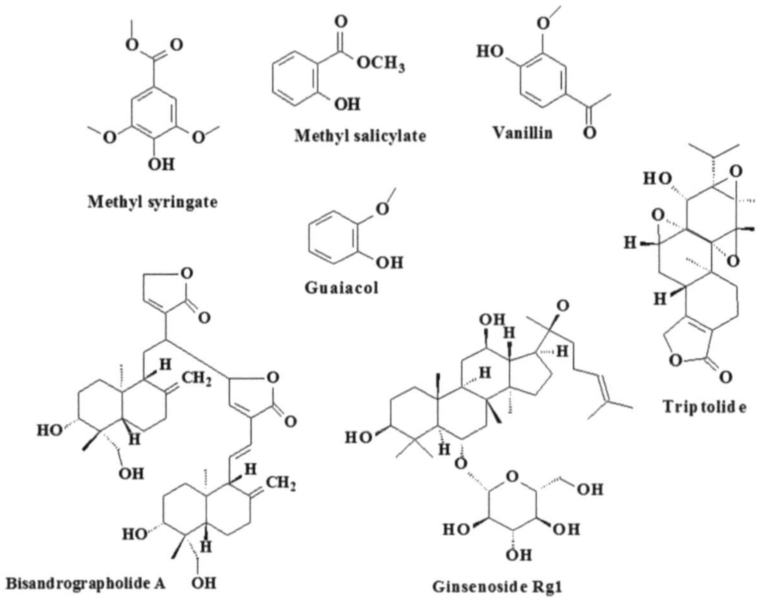

Fig. 9 TRP modulators

findings demonstrate that despite its electrophilic feature, allyl isothiocyanate activates TRPV1 in a capsaicin-like mechanism, independent of cysteine modification in TRPV1.

1.1.8.2 Ginsenosides

Ginsenosides are the main bioactive component of ginseng (*Panax ginseng*, *Panax quinquefolius*, *Panax japonicas* and *Panax notoginseng*) (Attele et al. 1999). These compounds are steroidal saponins that contain the 4-*trans*-ring rigid steroid skeleton and differ mainly by the number, type and location of their sugar moieties (Shibata et al. 1985; Sticher 1998). Ginsenosides seem also to be involved in pain modulation. Ginsenosides induced anti-nociceptive action in mice and were capable of inhibiting pain behaviour induced by capsaicin (Mogil et al. 1998; Nah et al. 2000). Corroborating these findings, ginsenosides can directly block channels activated by capsaicin, resulting in attenuation of the currents in rat sensory neurons. Ginsenoside Rg1 (Fig. 9) (bioactive component of *Panax notoginseng)* blocked Ca^{+2} influx caused by both capsaicin and proton activation in a TRPV1-dependent mechanism. Furthermore, ginsenoside Rg1 inhibited expression of the enzyme cyclooxigenase-2 (COX-2) and nuclear factor-kappa B (NF-κB) activity induced by capsaicin in keratinocytes. The inhibitory effect of ginsenoside Rg1 was quite similar to that caused by capsazepine, the antagonist of TRPV1. More importantly, ginsenoside Rg1 inhibited the secretion of prostaglandin E_2 (PGE_2) and interleukin-8 (IL-8) induced by capsaicin in human keratinocytes cells and

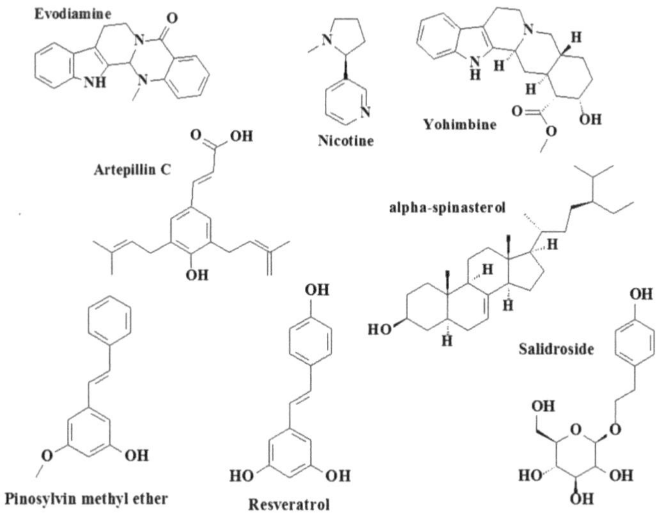

Fig. 10 TRP modulators

TRPV1-transfected HEK293 cells. On the other hand, ginsenosides also activate CHO cells expressing TRPV1, acting as agonists of the channel (Jung et al. 2001).

1.1.8.3 Guaiacol

Guaiacol (Fig. 9) is a phenolic compound isolated from the resin of the wood of *Guaiacum officinale* (Zygophyllaceae), which is similar in chemical structure to capsaicin. Traditionally, this plant has been used for the management of subacute and chronic rheumatism and rheumatoid arthritis (Newall et al. 1996). Intrathecal treatment with guaiacol or capsaicin inhibited the formalin-induced nociceptive response. Capsazepine abolished the anti-nociceptive effects of guaiacol in the acetic acid writing test. These results suggest that guaiacol may exert its anti-nociceptive effects via TRPV1 located on sensory terminals in the spinal cord (Ohkubo and Shibata 1997).

1.1.8.4 Evodia Compounds

Evodiamine (Fig. 10) is an alkaloid extracted from fruits of *Evodia rutecarpa*. The Chinese literature refers to *Evodia* fruits as a 'hot nature' herb, the same category to which chilli pepper belongs (Kobayashi et al. 2001). In vivo studies have shown that evodiamine sensitises and desensitises capsaicin-sensitive sensory afferents in mice, resulting in nociceptive and anti-nociceptive actions. The nociceptive action of intradermal evodiamine is suppressed by co-treatment with capsazepine (Kobayashi 2003). Moreover, it was reported that evodiamine inhibits [^{3}H]-resiniferatoxin binding in rTRPV1-Chinese hamster ovary (CHO) cells (Pearce et al. 2004). In spite of the similarities in the in vitro actions of evodiamine and capsaicin, evodiamine has no perceptible taste, but has a peppery hot taste that might facilitate its use in certain diseases.

1.1.8.5 Nicotine

Nicotine (Fig. 10) is an alkaloid found in tobacco leaves (*Nicotiana tabacum*, Solanaceae) and, as a result, is a component of tobacco products. When ingested by smoking, nicotine usually elicits its actions through the activation of neuronal nicotinic acetylcholine receptors (nAChRs) (Lindstrom 1997; Role and Berg 1996; Liu et al. 2004). Interestingly, some studies reported that nicotine is capable of modulating TRPV1 activity (Liu et al. 2004; Fucile et al. 2005). In this regard, nicotine induced an increase in capsaicin-activated currents in both TG neurons and in rTRPV1-CHO cells (Liu et al. 2004). This sensitising effect does not require the activation of nicotinic receptors. Additionally, nicotine contributes to the acute desensitisation, i.e. the inactivation of the current during a prolonged application of capsaicin and tachyphylaxis, i.e. the diminution of TRPV1-mediated maximal current amplitude after repetitive application of capsaicin, a phenomenon that is dependent on Ca^{+2} influx increase. Therefore, the activation of nicotinic receptors in DRG neurons by nicotine can decrease the sensitivity of TRPV1 to subsequent sensory stimuli. These data represent a possible mechanism underlying the analgesic properties of nicotinic agonists on sensory neurons (Fucile et al. 2005). Further studies are needed to understand the precise mechanisms underlying the interaction of nicotine with TRPV1.

1.1.8.6 Sanshools

Sanshools is the pungent constituent in Sichuan pepper. It is contained in the outer pod in the tiny fruit of several species of the genus Zanthoxylum (most commonly *Zanthoxylum piperitum*), which are widely grown and consumed in Asia as a spice. Despite the name, it is not related to black and chilli peppers (Sugai et al. 2005). Sanshools cause irritant, tingling and sometimes paraesthetic sensations on the tongue (Koo et al. 2007). Hydroxy-alpha-sanshool (Fig. 11) promoted Ca^{+2} influx in cells transfected with TRPV1 and evoked robust inward currents in cells transfected with TRPV1. In primary cultured sensory neurons, hydroxy-alpha-sanshool induced inward currents and Ca^{+2} influx in a capsazepine-dependent manner. Moreover, hydroxy-alpha-sanshool-induced currents and Ca^{+2} influx were greatly diminished in TRPV1 knockout mice. Hydroxy-alpha-sanshool evoked a licking behaviour when injected into a hind paw of wild-type mice, but this was reduced in TRPV1-deficient mice. These results indicate that TRPV1 is a putative molecular target of hydroxy-alpha-sanshool in sensory neurons (Koo et al. 2007).

1.1.8.7 Fatty Acids

Ricinoleic acid (Fig. 11) is a main constituent isolated from the castor oil from seeds of *Ricinus communis* (Euphorbiaceae) (Evans 1996). This fatty acid exhibits a chemical structure that is similar to some endogenous fatty acid activators of TRPV1 such as anandamide and oleoylethanolamide but different to that of capsaicin. Repeated topical and intradermal treatments with ricinoleic acid or capsaicin produced an analgesic effect when assessed in thermal and inflammatory models of pain. However, contrarily to capsaicin, ricinoleic acid does not induce pain by itself

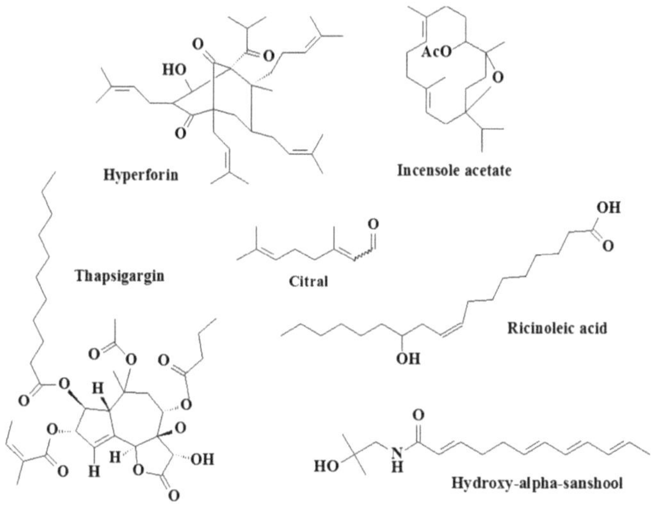

Fig. 11 TRP modulators

after acute injection in mouse paw or inward current in DRG neurons (Vieira et al. 2000, 2001). Therefore, ricinoleic acid unlikely acts as a direct agonist of TRPV1 receptor.

1.1.8.8 Cnidarian Envenomations

TRPV1 receptor was investigated as a possible target for cnidarian such as jellyfish, sea anemone and other envenomations, since that TRPV1 activation in nociceptive neurons leads to pain sensation, which is a common symptom of cnidarian stings. Interestingly, cnidarian venom induced nociception in rats, an effect that is comparable to capsaicin. Moreover, venom knocks down the desensitisation of TRPV1, evoking the typical persistent burning pain sensation (Cuypers et al. 2006).

1.1.8.9 Spider Toxins

It has been found that crude venom from *Psalmopoeus cambridgei*, a tarantula native to the Trinidad and Tobago, induces robust and reproducible signals when *was applied* to TRPV1-expressing HEK293 cells (Siemens et al. 2006). Three fractions with TRPV1 agonist activity were identified and named vanillotoxins (VaTx) 1, 2 and 3. Vanillotoxins are new members of the extended family of ICK (inhibitor cysteine knot) peptides from spiders and cone snails (Zhu et al. 2003). ICK toxins are widely recognised as blockers of cationic channels. In contrast with the predominant role of ICK toxins as channel inhibitors, the vanillotoxins function as TRPV1 agonists, providing new tools for understanding mechanisms of TRP channel gating. The in vivo studies show that the injection of VaTx3 in hind paw of mice elicits pain-related behaviours and oedema in wild-type mice, events that were attenuated in TRPV1 knockout mice (Siemens et al. 2006).

1.1.8.10 Phytosterols

Alpha-spinasterol is a phytosterol (Fig. 10) extracted from *Vernonia tweedieana* Baker, an herbaceous plant found all over the world and used as an analgesic and antitussive. Some studies have also reported the anti-nociceptive and anti-inflammatory effects of these species (Iwalewa et al. 2003; Njan et al. 2008; Risso et al. 2010; Zanon et al. 2008). Recently, it was reported that the oral administration of α-spinasterol produced anti-nociceptive effects in the noxious heat-induced nociception test and reduced the oedema and mechanical and heat hyperalgesia elicited by complete Freund's adjuvant paw injection. The evidences that α-spinasterol could be acting as a TRPV1 ligand were reported by means of in vitro studies. This compound was able to displace [^{3}H]-resiniferatoxin binding and diminish calcium influx mediated by capsaicin. In conclusion, α-spinasterol now emerges as a TRPV1 antagonist with anti-nociceptive properties (Trevisan et al. 2012).

1.1.8.11 Stilbenoids

Stilbenoids are bioactive compounds with promising beneficial effects in human health. Resveratrol (3,5,4′-trihydroxy-trans-stilbene, Fig. 10) is a widely distributed natural stilbenoid found in red wine from *Vitis vinifera* L., with reported analgesic properties against acute and chronic pain (Gentilli et al. 2001; Pham-Marcou et al. 2008; Sharma et al. 2007a, b; Torres-Lopez et al. 2002; Bertelli et al. 2008). Some of these properties of resveratrol could be attributed to the inhibition of TRPA1, a TRP family member (Yu et al. 2013). Pinosylvin methyl ether (Fig. 10), another derivative of stilbene which has a similar structure to resveratrol, blocked the capsaicin-induced currents in HEK293 cells expressing TRPV1 as well as in DRG neurons. Pretreatment with pinosylvin also decreased the nocifensive behaviour induced by capsaicin in rats. Therefore, these stilbenoids might be considered as emerging targets for anti-nociceptive and anti-inflammatory pharmacotherapy (Yu et al. 2013).

1.1.8.12 Indoles

Yohimbine (Fig. 10), an indole alkaloid obtained from the bark of the *Pausinystalia yohimbe* (Rubiaceae) tree or from the roots of *Rauwolfia serpentina* (Apocynaceae) and initially identified as an aphrodisiac, is a natural α_2-adrenoceptor antagonist and is frequently used to assess the mechanism of a drug's effect on α-adrenoceptors (MacDonald et al. 1997; Dessaint et al. 2004). This alkaloid was introduced in the treatment of erectile dysfunction more than 70 years ago (Hunner 1926) and has recently been described to inhibit the tetrodotoxin-sensitive Na^+ channels ($Na_V1.2$), the tetrodotoxin-resistant Na^+ channels $Na_V1.8$ and $Na_V1.9$ Na^+ channels and the TRPV1. The capsaicin-induced action potential firing activities in DRG neurons were eliminated by yohimbine (Dessaint et al. 2004).

1.1.8.13 Sesquiterpene Lactone

Similar to the vanilloid analogue resiniferatoxin, thapsigargin (Fig. 11) (*Thapsia garganica,* Apiaceae) is a natural product based on a tricyclic diterpene ring.

This plant is traditionally used in European and Arabian medicines for the treatment of rheumatic pain. Thapsigargin inhibited the [$^{45}Ca^{+2}$] uptake of CHO-TRVR1 cells mediated by TRPV1 and blocked [^{3}H]-resiniferatoxin binding sites in rTPV1-CHO cells. As thapsigargin inhibits Ca^{+2}-ATPases (SERCAS) and also TRPV1, further studies are necessary to interpret carefully the data of its biological effects as well as the relation with these targets (Toth et al. 2002).

1.2 TRPV2

TRPV2, another member of TRP family, was initially characterised as a promising molecule in pain processes due to its activation by noxious heat (>52 °C) and its sequence similarity to TRPV1 (Ahluwalia et al. 2002; Caterina et al. 1999; Tominaga and Caterina 2004). Nowadays, it is known that TRPV2 has a minor contribution in thermal sensing (Peralvarez-Marin et al. 2013). Cannabinoids obtained from *Cannabis sativa* have been identified as the only natural TRPV2 modulators (Table 1). In this regard, the cannabinoids Δ^9-tetrahydrocannabinol, cannabidiol and tetrahydrocannabivarin (Fig. 3) are potent TRPV2 modulators, whereas cannabinoic acids were less potent (De Petrocellis et al. 2010). However, cannabinoids are not selective for TRPV2, modulating also TRPV1, TRPV3, TRPV4, TRPA1 and TRPM8 (Akopian et al. 2008; Qin et al. 2008; De Petrocellis et al. 2008, 2010, 2012a). Both rat and human TRPV2 are activated by Δ^9-tetrahydrocannabinol and blocked by ruthenium red. Indeed, TRPV2 partially mediated the cannabidiol-evoked CGRP release in cultured rat DRG neurons (Neeper et al. 2007; Qin et al. 2008). Recently, it was demonstrated that TRPV2 activation inhibited cell proliferation of human glioblastoma multiform and reduced the resistance of the glioblastoma to the chemotherapeutic agent carmustine. The activation of TRPV2 by cannabinoids triggered Ca^{+2} influx and increased drug uptake, exerting a synergic cytotoxic effect with carmustine to induce apoptosis in glioma cells (Nabissi et al. 2013).

1.3 TRPV3

TRPV3 is a member of TRP family sensible to warm temperature (range 32–39 °C) expressed in keratinocytes and neural tissues (Facer et al. 2007). In addition to temperatures, TRPV3 is activated by monoterpenes such as carvacrol, thymol, carveol, 6-tert-butyl-m-cresol, dihydrocarveol, menthol, camphor, borneol (Table 1 and Fig. 5), eugenol (Fig. 1), cresol (a phenol from coal or wood tar), cinnamon (from the bark of *Cinnamon verum*), thujone (from the genus *Thuja*) and the acyclic monoterpenes, e.g. linalool, geraniol and propofol (Xu et al. 2006; Macpherson et al. 2006; Vogt-Eisele et al. 2007; Moqrich et al. 2005; Nilius et al. 2013; Vriens et al. 2009). Most of the monoterpenes are fragrant and form major constituents of many plant-derived essential oils (Vogt-Eisele et al. 2007). More detailed information about these compounds is provided in the TRPA1 and TRPM8 section.

Carvacrol increased intracellular Ca^{+2} in TRPV3-expressing HEK293 and epithelial cells of the tongue and skin. Additionally, heterologously expressed human TRPV3 (hTRPV3) was activated and sensitised by carvacrol and thymol in a dose-dependent manner (Xu et al. 2006).

Interestingly, cooling agents are capable of activating TRPV3. In this context, eugenol (Fig. 1) (the principal active ingredient of clove—*Caryophyllus aromaticus*) is used in dentistry as a topical analgesic. This compound activates and sensitises TRPV3 in HEK293 cells and promotes increase in intracellular Ca^{+2} levels in epithelial cells of the tongue and skin. Interestingly, the activation of TRPV3 by eugenol and the subsequent Ca^{+2} influx promoted the release of inflammatory mediator IL-1α (interleukin-1-alpha) in mouse keratinocytes (Xu et al. 2006). Camphor (*Cinnamomum camphora* tree) activated mouse cultured primary keratinocytes and this activity was abolished in TRPV3 null mice. Moreover, camphor activated and sensitised TRPV3 in CHO cells. However, camphor was found to be a weak agonist, activating TRPV3 only at higher concentrations (Moqrich et al. 2005). Of interest, TRPV3 response to camphor was completely lost in C619S mutants, which is located in the channel pore and indicates a role of the pore for gating (Sherkheli et al. 2013). Similarly, vanillin (*Vanilla planifolia*) was also characterised as a weak TRPV3 agonist (Xu et al. 2006). Menthol, a compound obtained of *Mentha piperita* activates TRPV3. However, menthol acts on other thermoTRPs: cool-activated TRMP8 and TRPA1 (Macpherson et al. 2006).

The diterpenic cembrenoid incensole acetate (Fig. 11) has also been described to modulate TRPV3 (Moussaieff et al. 2008, 2012). This compound is found in incense crude extracts from species such as *Boswellia papyrifera*. Burning of *Boswellia* resin as incense has been part of religious and cultural ceremonies for millennia (MacCulloch; Moussaieff et al. 2008; Moussaieff and Mechoulam 2009). *Boswellia* is a genus of trees known for their fragrant resin, which has many pharmacological applications particularly for inflammatory conditions. Incensole acetate has anxiolytic-like and antidepressive-like behavioural effects (Moussaieff et al. 2008; Nilius et al. 2013; Moussaieff and Mechoulam 2009; Paul and Jauch 2012). In addition, incensole acetate protects against ischaemic neuronal damage and reperfusion injury in mice, partially mediated by TRPV3 channels, as determined by studies in *Trpv3*-deficient mice and channel blockers (Moussaieff et al. 2012).

The bioactive component of lemongrass citral (*Cymbopogon citratus*) also modulates TRPV3. Citral (Fig. 11) is used as a taste enhancer, an odorant in perfumes and an insect repellent, and also for soft anxiolytic teas. In sensory neurons, citral activates TRPV3 but causes long-lasting inhibition after the initial activation (Stotz et al. 2008). Other natural compound activators of TRPV3 are the phytocannabinoids cannabidiol and tetrahydrocannabivarin (THCV). These compounds stimulate TRPV3 with high efficacy and potency, whereas the derivatives cannabigerovarin and cannabigerolic acid desensitise the channel (De Petrocellis et al. 2012b).

The investigation of TRPV3 participation in transducing signals in pain pathways suggests that TRPV3 is involved in nociceptive signalling. In addition, the identification of antagonists that are efficacious in relevant preclinical behavioural models demonstrated that TRPV3 is a putative target for pain treatment (Reilly and Kym 2011; Nilius et al. 2013).

1.4 TRPV4

TRPV4 is activated by cell swelling, moderate heat (>27 °C), endogenous epoxyeicosatrienoic acids (EETs; 5,6-EET and 8,9-EET) and synthetic phorbol ester 4α (4α-PDD) (Strotmann et al. 2000; Guler et al. 2002; Wissenbach et al. 2000; Liedtke et al. 2000; Nilius et al. 2001, 2004; Watanabe et al. 2002, 2003). The natural compound bisandrographolide A (Fig. 9), an active compound extracted from *Andrographis paniculata*, was characterised as a potent and selective agonist of TRPV4. This plant is used in traditional medicine in various parts of Asia for a wide array of ailments. Extracts are typically used as anti-inflammatory agents or immune stimulants. Indications include upper respiratory tract infections, diarrhoea, fever, tonsillitis, snakebite and many others (Smith et al. 2006). Evidences show that the transmembrane regions 3–4 of TRPV4 form a site for channel activation by bisandrographolide A, since the mutation of two hydrophobic residues in the central part of TM4 ($Leu^{584}$and Trp^{586}) caused a severe reduction of the sensitivity of the channel to 4α-PDD, bisandrographolide A and heat (Vriens et al. 2007). Functionally, bisandrographolide A activated large TRPV4-like currents in mouse keratinocytes. Bisandrographolide A also activated TRPV4 currents in HEK293 cells overexpressing TRPV4 (Smith et al. 2006).

2 TRPA1

TRPA1 is the only member of the ankyrin subfamily so far identified in mammals. Since other TRPs have only zero to eight ankyrin repeats, the fourteen ankyrin repeats in the N-terminal portion of TRPA1 make this channel an uncommon TRP member. The TRPA1 protein, originally called ANKTM1 (ankyrin-like with transmembrane domains protein 1), was first identified in human foetal lung fibroblasts as a transformation-associated gene product (Jaquemar et al. 1999). Later, studies showed that TRPA1 was expressed in DRG, nodose ganglion and TG sensory neurons (Nagata et al. 2005; Story et al. 2003), and the TRPA1 expression/activation was greatly associated to nociception (Geppetti et al. 2008; Andrade et al. 2012; Nilius et al. 2012). Nowadays, several plant-derived products are described to modulate TRPA1 channel (Table 1). A broad range of compounds can modulate TRPA1 because it contains cysteine residues in its amino-terminus that are responsible for the channel activity. Thus, carbon or sulphur-reactive electrophile compounds that form a covalent bond with thiol nucleophilic groups or oxidise it by forming intramolecular disulphide bonds can work as receptor agonists (Macpherson et al. 2007; Hinman et al. 2006; Bang and Hwang 2009). Therefore, activation of the channel depends on the reactivity of the agonist rather than the binding affinity. In parallel, TRPA1 undergoes other well-known types of activation, including a reversible interaction with the agonist within a binding pocket in the receptor (the lock and key principle of Fischer) (Xiao et al. 2008) and Ca^{+2} binding to the EF-hand domain present in the N-terminal region of the channel (Doerner et al. 2007; Zurborg et al. 2007). Taking into account all of these possibilities for activation, it is not surprising that TRPA1 would be modulated by a wide range of naturally derived compounds (Nilius and Appendino 2011, 2013; Andrade et al. 2012).

2.1 Isothiocyanate

The responsiveness of TRPA1 to natural compounds was initially demonstrated in an attempt to identify the targets responsible for the pungent effect of mustard oil (Jordt et al. 2004). Allyl isothiocyanate (Fig. 8) is the major active ingredient present in mustard oil. It can achieve more than 90 % of the volatile oil extracted from ripe seeds black mustard (*Brassica nigra*) (Evans 1996). Initially, it was only known that topical administration of mustard oil produced nociceptor excitation, burning pain, inflammation and robust hypersensitivity to both thermal and mechanical stimuli (Jancsó et al. 1968; Reeh et al. 1986; Koltzenburg et al. 1992; Jiang and Gebhart 1998), but whether this would occur through a direct action on sensory neurons or through a specific membrane receptor was still unknown at that time (Reeh et al. 1986). A breakthrough paper by Jordt and colleagues found that the pungent sensation caused by several plant-derived ingredients such as allyl, benzyl, phenylethyl, isopropyl and methyl isothiocyanates from wasabi (*Wasabia japonica*), yellow mustard (*Brassica hirta*), Brussels sprouts (*Brassica oleracea*), nasturtium seeds (*Tropaeolum majus*) and capers (*Capparis spinosa*) was due to a direct activation of TRPA1. These isolated compounds induced a direct Ca^{+2} influx in TRPA1-transfected HEK293 cells. Large membrane currents were also detected after incubating oocytes expressing TRPA1 with crude extracts from wasabi paste and brown mustard (*Brassica juncea*) (Jordt et al. 2004). Likewise, it was also demonstrated that substances present in mustard oil, wintergreen oil, clove oil, ginger and cinnamon oil, such as allyl isothiocyanate, methyl salicylate, eugenol, gingerol and cinnamaldehyde, caused a sharp increase in intracellular Ca^{+2} in mouse TRPA1-expressing CHO cells (Bandell et al. 2004). All of these compounds are known to cause a pungent burning sensation in humans, but only allyl isothiocyanate, cinnamaldehyde and methyl salicylate were selective to activate TRPA1 in this study as they did not activate TRPM8 or TRPV1 (Bandell et al. 2004). Currently, mustard oil and its main component allyl isothiocyanate have been extensively used as a tool to evaluate the role of TRPA1 in noxious somatic (Andrade et al. 2008; Fabrizi et al. 2013; Docherty et al. 2013) and visceral pain (Cattaruzza et al. 2010; O'Mullane et al. 2013), in airway mucosa irritation (Alenmyr et al. 2009), endothelial vasodilation (Earley et al. 2009; Kunkler et al. 2011), visceral smooth muscle contraction (Andrade et al. 2006), inflammation (Patil et al. 2011; Eid et al. 2008) and thermal sensation (Wang et al. 2012).

2.2 Cinnamaldehyde

Cinnamon oil is the essential oil obtained from the leaves and bark of *Cinnamomum cassia* and *Cinnamomum zeylanicum* (Lauraceae). Cinnamon oil is a yellow to brown liquid with a characteristic, aromatic odour and a sweet, pungent taste (Evans 1996). It comprises pharmacological properties including antispasmodic, carminative, orexigenic, *antidiarrhoeal* and antimicrobial. Cinnamaldehyde (Fig. 8) is the main constituent isolated from cinnamon oil (70 %) and is extensively used for flavouring purposes in foods, chewing gums and toothpastes. When orally administered (0.2 %) to human subjects, cinnamaldehyde is perceived to have a burning and tingling sensory quality (Cliff and Heymann 1992; Prescott and Swain-Campbell 2000). As TRPA1 is expressed in TG neurons that project to the tongue, TRPA1 could indeed be responsible for the burning sensory quality of cinnamaldehyde (Jordt et al. 2004). Indeed, cinnamaldehyde induced Ca^{+2} influx in TRPA1-expressing CHO cells (Bandell et al. 2004). The α,β-unsaturated bond of cinnamaldehyde is attacked by the nucleophilic mercapto group of TRPA1 cysteines in a Michael addition. Corroborating with this mechanism, a more reactive cinnamaldehyde-like Michael acceptor was a more potent activator of TRPA1 (EC_{50} 0.8 μM) than the original molecule (EC_{50} 19 μM) (Macpherson et al. 2007). Cinnamaldehyde induced spontaneous nociception, heat hyperalgesia and cold-evoked response in rodents (Tsagareli et al. 2010; Andrade et al. 2008; Fajardo et al. 2008). In humans, spontaneous pain and mechanical and heat hyperalgesia caused by cinnamaldehyde were confirmed by sensory tests and laser Doppler imaging (Namer et al. 2005; Roberts et al. 2005). Cinnamaldehyde was able to induce contractions in the rat urinary bladder (Du et al. 2007b; Andrade et al. 2006) but caused relaxation of human urethral strip preparations after phenylephrine contractions (Gratzke et al. 2009; Weinhold et al. 2010). A *protussive* effect of cinnamaldehyde through activation of TRPA1 was also reported (Andre et al. 2009). Although TRPA1 is the main target to cinnamaldehyde in pain, inflammation and smooth muscle tone control, it seems not to be responsible for the beneficial effects of cinnamaldehyde against tumour growth (Oehler et al. 2012).

2.3 Allicin

Similarly to allyl isothiocyanate, the garlic containing allicin and its diallyl sulphide by-products are reactive organosulphur substances (Fig. 8). The original substance, thiosulphinate allicin, is broken down to allicin by crushing garlic bulbs. Allicin is short lived in aqueous solution and yields diallyl sulphides, ajoene and dithiines (Block et al. 2010; Lawson and Block 1997). Allicin and diallyl sulphides account for the pungency and spicy aroma and taste of plants from the genus *Allium* (garlic, onion, leek, chives and shallots). Of interest, the application of garlic extract, purified allicin or diallyl disulphide to TG neurons caused a 30 % increase in intracellular-free Ca^{+2} (Bautista et al. 2005). Besides, garlic extracts, allicin and diallyl disulphide activated TRPA1 expressed in HEK293 cells and *Xenopus* oocytes. Voltage-clamp recordings showed that allicin produced robust membrane currents with EC_{50} (half-maximal effective concentration) 7.5 μM; the diallyl disulphide was equally efficacious but less potent (EC_{50} 192 μM) compared to allicin (Bautista et al. 2005).

2.4 Ligustilide

The electrophilic dihydrophthalide ligustilide (Fig. 8) is present in high concentrations in popular medicinal plants from the folk medicine of China [dong-quai, *Angelica sinensis* (Oliv.) Dill. (Li et al. 2006), chuanxiong, *Ligusticum chuanxiong* Hort. (Mei et al. 1991)] and North America (oshá, *Ligusticum porteri* Coult. and Rose) (Mei et al. 1991). Ligustilide is also a major constituent of the aroma of celery (*Apium graveolens* L.) (Tang et al. 1990) and lovage (*Levisticum officinale* L.) (Gijbels et al. 1982). Because of its dietary and medical relevance, ligustilide has been tested against several disorders and has been reported to induce vasodilatation (Chan et al. 2007), diminution of platelet aggregation (Zhang et al. 2009), analgesic effects (Du et al. 2007a), attenuation of lipopolysaccharide (LPS)-induced proinflammatory response, endotoxic shock (Wang et al. 2010; Shao et al. 2011), neuroprotection (Wu et al. 2011) and anti-proliferation of smooth muscle cells (Lu et al. 2006).

Ligustilide has unique electrophilicity and it is capable to trap thiol groups in a complex, multiple and reversible fashion (Beck and Stermitz 1995). This feature provided a rationale for investigating its action on TRPA1. In this context, Zhong and collaborators demonstrated that ligustilide ($EC_{50} = 44$ μM) selectively induced TRPA1 outward and inward currents in CHO cells expressing the mouse isoform of TRPA1. The oxidised product of ligustilide, dehydroligustilide (Fig. 8), also activated the channel, but with a much higher concentration, the EC_{50} value was 539 μM. Interestingly, the washout of dehydroligustilide induced a rapid increase of TRPA1 currents, and in the presence of mustard oil, dehydroligustilide induced a striking current inhibition, showing the dehydroligustilide exerts a bimodal effect upon TRPA1. Mutations of reactive cysteines in TRPA1 revealed that modulation by ligustilide was dependent on thiol groups, whereas the blocker effect of dehydroligustilide was unchanged by the substitution of cysteines by serine residues (Zhong et al. 2011b).

2.5 Unsaturated Dialdehyde Terpenes

The terpenes isovelleral and polygodial (Fig. 4) have been proposed to induce pungent sensations and neurogenic inflammation by activating TRPV1 in peripheral sensory neurons (Szallasi et al. 1996; Andre et al. 2006). A suggestion that they would activate another chemosensory was raised by using a pharmacological approach where ruthenium red, but not capsazepine, almost completely blocked polygodial-induced urinary bladder contraction (Andre et al. 2006). Likewise, isovelleral-activated relaxation of the mesenteric artery appeared to be partially independent of vanilloid receptors (Ralevic et al. 2003). Later on, it was demonstrated that both isovelleral and polygodial induced sustained currents and Ca^{+2} influx in mustard oil-sensitive neurons and in heterologous TRPA1-expressing cells (Escalera et al. 2008; Iwasaki et al. 2009). Isovelleral (EC_{50} 0.50 ± 0.13 μM) and polygodial (EC_{50} 0.4 ± 0.07 μM) also induced robust Ca^{+2} influx into HEK293 cells transiently transfected with human TRPA1 (Escalera et al. 2008). In addition, isovelleral responses were dramatically reduced in neurons from TRPA1-deficient mice. Isovelleral also induced nocifensive response (paw lifts, flicks and licks) when injected into mice hind paw. The nocifensive responses were similar among wild-type and $TRPV1^{-/-}$ mice but were dramatically reduced in $TRPA1^{-/-}$ mice (Escalera et al. 2008).

Polygodial was much more potent (0.06 μM) than allyl isothiocyanate (1.5 μM) and cinnamaldehyde (1.5 μM) to induce Ca^{+2} influx in human TRPA1-transfected cells (Iwasaki et al. 2009). The analogous terpenoids miogadial and miogatrial, found in the fresh flower buds of *Zingiber mioga* Roscoe (Iwasaki et al. 2009; Abe et al. 2002), were able to induce Ca^{+2} influx in TRPA1-expressing cells at a similar fashion to polygodial (Iwasaki et al. 2009). Even though these terpenes contain α,β-unsaturated dialdehyde moieties that are potentially capable of forming Michael adducts with thiol groups from TRPA1, the mechanism of activation was not dependent on the presence of reactive cysteines. Thus, the agonist activity was comparable to a lock key Fischer mechanism, as occurs for carvacrol (Escalera et al. 2008).

2.6 Monoterpenes

In line with the pungent effect following TRPA1 activation, carvacrol and thymol (Fig. 5), two oregano-derivative constituents, robustly activated TRPA1. Oregano and thyme are widely used as spices. The pungency of oregano might be due to the activation of TRPA1 from the free nerve endings in tongue and nose because epithelial cells from this tissue are unlikely to express TRPA1 (Xu et al. 2006). Carvacrol (3-isopropyl-6-methylphenol) represents 65 % of mass in the essential oil of oregano (Sarer et al. 1982). In HEK293 cells expressing the rat isoform of TRPA1, carvacrol (250 μM) activated TRPA1 and the current quickly desensitised in the continuous presence of agonist. Repeated stimulation resulted in much smaller responses, a phenomenon known as tachyphylaxis (Xu et al. 2006).

Thymol (2-isopropyl-5-methylphenol), a major component of oregano (*Origanum vulgare*) and thyme (*Thymus vulgaris*), potently activated membrane potential responses and intracellular Ca^{+2} increases in hTRPA1-expressing HEK293 cells (Lee et al. 2008). Thymol has medical uses in oral care products as an astringent and antibiotic (Yu et al. 2000; Burt 2004). Its distinctive sharp odour and pungent flavour are considered aversive properties. After activation by thymol, TRPA1 was desensitised to further exposure to thymol or to the agonist allyl isothiocyanate

(Lee et al. 2008). In spite of the structural similarity between thymol and carvacrol, the first presented much higher potency than the latter to induce depolarised currents in TRPA1-expressing HEK293 cells. The EC_{50} value for thymol-induced currents was 6 μM (Lee et al. 2008), whereas a concentration of 250 μM of carvacrol was necessary to induce currents (Xu et al. 2006). Of note, this comparison is limited by the species difference used in these two studies: the rat TRPA1 isoform was over-expressed for carvacrol and human TRPA1 was over-expressed for thymol studies (Xu et al. 2006; Lee et al. 2008). Confirming the low potency of carvacrol to activate TRPA1, a later report demonstrated that the vasodilation caused by carvacrol was mostly dependent on TRPV3 rather than TRPA1 (Earley et al. 2010).

The terpenoid camphor (Fig. 5) was initially believed to be an antagonist of TRPA1, since concentrations as high as 10 mM camphor completely suppressed inward TRPA1 basal currents in HEK-293 expressing rat TRPA1 and the IC_{50} occurred at a concentration of 0.66 mM (Xu et al. 2005). Therefore, the analgesic effects of camphor were associated to the inhibition of TRPA1. Camphor is isolated from the wood of the camphor laurel tree *Cinnamomum camphora* and is used as a nasal decongestant and cough suppressant (Burrow et al. 1983). Camphor is also commonly applied to the skin due to its antipruritic, analgesic and counterirritant properties (Burkhart and Burkhart 2003). The inhibitory effects of camphor upon TRPA1 were confirmed by Alpizar and co-workers who demonstrated that extracellular application of 1 mM camphor induced a decrease in basal TRPA1 currents. Interestingly, the current amplitude showed a significant overshoot upon washout, and the application of 100 μM camphor induced a threefold increase in the basal current amplitude. The subsequent application of 1 mM camphor induced current inhibition, again followed by a significant current rebound. Therefore, a bimodal effect of camphor has been clearly demonstrated, where low concentrations (300 μM or less) increase inward TRPA1 currents and high concentrations (1 mM) induce reversible inhibition (Alpizar et al. 2012). The mechanisms underlying the bimodal effect of some natural compounds on TRPA1 are not fully understood, but studies carried out with umbellulone, as we shall see, revealed some chemical features that are important for agonist/antagonist effect of these compounds upon TRPA1.

Umbellulone (Fig. 5) is the offensive principle of the so-called headache tree, California bay laurel *Umbellularia californica* (Hook. & Arn.) Nutt., which is the only species of the genus Umbellularia (Peattie 1953). Inhalation of the scent from *Umbellularia californica* leaves has been reported to cause sinus irritation, sneezing, headache and even unconsciousness (Peattie 1953; Drake and Stuhr 1935). Exposure to *Umbellularia californica* can trigger violent headache crises (Drake and Stuhr 1935; Heamy 1875; Benemei et al. 2009).

Umbellulone caused a concentration-dependent elevation of intracellular Ca^{+2} in rat and human TRPA1-transfected HEK293 cells. It was selective to TRPA1 since the Ca^{+2} intracellular elevation in trigeminal ganglia sensory neurons was inhibited by the selective TRPA1 antagonist HC-030031 but was unaffected by the TRPV1 antagonist, capsazepine. In addition, trigeminal ganglia neurons from TRPA1 null mice were insensitive to umbellulone and mustard oil (Nassini et al. 2012). Umbellulone (0.3 and 1 mM) activated whole-cell currents in transfected TRPA1-CHO cells but presented rapid rebound activation after washout. At 1 mM umbellulone reversibly inhibited the pre-activated current induced by mustard oil (Zhong et al. 2011a). An isosteric mutation of cysteine (C622S) to serine partially attenuated TRPA1 activation induced by umbellulone, indicating that activation of TRPA1 by umbellulone occurred by a dual mode, in part dependent on its electrophilic property and in part by a different unidentified mechanism. In contrast, the inhibitory effect of umbellulone upon mustard oil-induced currents was maintained in the C622S mutants, and a direct block of the channel by the compound has been suggested. Modifications at the umbellulone molecule, i.e. removing its Michael acceptor properties (nucleophilic trapping), partially decreased its ability to activate TRPA1 but greatly improved its ability to block the channel (Zhong et al. 2011a; Nassini et al. 2012). Taking these results into account, it is possible to conclude that stimulatory effect of low concentration of umbellulone on TRPA1 involves both electrophilic and non-electrophilic properties. However, the umbellulone inhibitory effect is clearly independent of reactivity with TRPA1 cysteines and is likely due to a direct blockage of the channel pore.

The cooling natural compound menthol (Fig. 5) has also been reported to have a bimodal effect upon TRPA1. At low-micromolar concentration (1 μM), menthol activated TRPA1-dependent currents, whereas higher concentrations (1 mM) led to a reversible channel block (Karashima et al. 2007; Macpherson et al. 2006). Of interest, menthol-induced TRPA1 channel opening occurred even in the absence of intracellular and extracellular Ca^{+2}, showing that the opening property of menthol does not occur by indirect Ca^{+2} binding to the channel (Karashima et al. 2007; Xiao et al. 2008). In agreement with these findings, Xiao and collaborators demonstrated a bimodal effect of menthol upon mouse TRPA1. However, this is a species-specific effect since menthol only activated or had no effect upon human and fly TRPA1, respectively (Xiao et al. 2008). To distinguish which channel domain was sensitive to menthol and was responsible for the bimodal effects of the compound, a chimera mouse-human TRPA1 was constructed. The authors found that the TM5-6 portion of the channel is responsible for the differential effect of menthol among the two species. When the TM5-6 portion of hTPRA1 was swapped with mTRPA1 TM5-6 (hTRPA1-mTM5–6), it behaved exactly like wild-type mTRPA1 in response to menthol, i.e. it was activated at low and inhibited at high concentrations of menthol. In the opposite chimera, mTRPA1-hTM5–6, menthol activated the channel with no bimodal shape. To gain further insight into its mechanism of action, the authors constructed miniature chimeras with a point mutation at specific amino acids in TM5. A total of nine amino acid substitutions (FGFATLIAM) along the TM5-pore helix TM6 of hTRPA1 greatly contributed to the phenotype switch and turned hTRPA1 into mTRPA1, with bimodal behaviour. These data demonstrate that residues at these positions govern the species-specific responses of TRPA1 and might explain the different effects of natural compounds in TRPA1 from different animal species (Klionsky et al. 2007).

Another interesting bimodal effect was described for two isomers found in eucalyptus oil, the analogous 1,4-cineole and 1,8-cineole (eucalyptol, Fig. 5); more details about these compounds are described at the TRPM8 section. These two monoterpenes have opposite effects upon TRPA1-induced inward currents. 1,4-cineole (5 mM) induced currents in TRPA1-expressing cells, whereas 1,8-cineole did not alter baseline currents but decreased allyl isothiocyanate-induced currents with an IC_{50} of 3.4 ± 0.6 mM (Takaishi et al. 2012). Further studies using these two similar compounds with opposite effects on TRPA1 might help to understand the channel dynamics.

The monoterpenes perillaldehyde and perillaketone (Fig. 5) are components of the aromatic extracts from *Perilla frutescens*, a plant widely used in Asian cuisine. Application of extracts of this plant is described in traditional Chinese medicine, for the treatment of atopic dermatitis and for other anti-inflammatory and anti-allergic properties (Yu et al. 1997). To identify the mechanisms of the somatosensory properties of this plant, Bassoli and collaborators demonstrated that perillaldehyde and perillaketone activated the cloned TRPA1 channel when expressed in heterologous cell system. In comparative study, perillaldehyde was twofold more efficacious than perillaketone to increase intracellular Ca^{+2} in HEK293 cells expressing rat TRPA1. The comparison was normalised as the percentage of the effect of allyl isothiocyanate 100 μM. By this normalisation, the concentration for half-maximal was 4.39 ± 0.09 μM for perillaldehyde and 4.71 ± 0.03 μM for perillaketone. These results confirm the role of perillaldehyde and perillaketone in the chemesthetic properties of *Perilla frutescens* (Bassoli et al. 2009).

2.7 Phytocannabinoids

A substance that targets other receptors, as the cannabinoid receptor agonist Δ^9-tetrahydrocannabinol (THC), (Fig. 3) was also described to modulate TRPA1. THC is the major cannabinoid in Cannabis species; a detailed review of this plant can be found in Hirst et al. (1998) and Calixto et al. (2000). When administered to humans, THC produces a wide range of biological effects, such as an increase in pulse rate, decreased blood pressure, muscle weakening, increased appetite and euphoria followed by drowsiness (Hirst et al. 1998). THC exerts its main effects by binding cannabinoid receptors, preferentially CB_1. However, it has been demonstrated that THC induces membrane currents in TRPA1-expressing oocytes in a similar way to that induced by mustard oil, including slight outward rectification. Noteworthy, this response was blocked by ruthenium red, a non-specific TRP antagonist (Jordt et al. 2004). Earlier studies demonstrated that THC caused hepatic and mesenteric artery relaxation in vitro via a mechanism that involved activation of capsaicin-sensitive CGRP-containing perivascular sensory nerve endings that innervate the smooth muscle. This effect was insensitive to antagonists of cannabinoid receptors but was largely blocked by ruthenium red. Since the effect was persistent in TRPV1-deficient mice, THC-induced vasorelaxation was attributed to some ruthenium red sensible target other than TRPV1 (Zygmunt et al. 2002). In agreement with these two studies, De Petrocellis and co-workers demonstrated that five phytocannabinoids structurally analogous to THC were able to increase intracellular Ca^{+2} in both TRPA1-expressing HEK293 cells and in rat DRG sensory neurons with efficacy comparable with that of mustard oil. The EC_{50} values for Ca^{+2} influx in TRPA1-expressing HEK293 cells were cannabichromene 0.06 ± 0.02 μM, cannabidiol 0.096 ± 0.012 μM, Δ^9-tetrahydrocannabinolic acid 0.24 ± 0.03 μM, cannabidiolic acid 12.0 ± 8.8 μM and cannabigerol 3.4 ± 1.0 μM. The phytocannabinoid cannabichromene also increased intracellular Ca^{+2} in a mustard oil-responsive subpopulation of DRG neurons (De Petrocellis et al. 2008). Cannabichromene, cannabidiol, cannabinol (a degradation product of THC) and tetrahydrocannabivarin (a propyl homologue of THC) are potent rat TRPA1 agonists and desensitise it (De Petrocellis et al. 2010). The activation followed by the desensitisation of TRPA1 by phytocannabinoids, endocannabinoids and synthetic cannabinoids might strongly contribute to the anti-inflammatory and analgesic effect of these compounds (De Petrocellis and Di Marzo 2010; Akopian et al. 2008).

2.8 Nicotine

TRPA1 is also a target for the alkaloid nicotine (Fig. 10). Nicotine is found in plants of the Solanaceae family. It is the agonist of the nicotinic acetylcholine receptor (nAChRs) and exerts a powerful psychoactive effect inducing addiction as consequence of tobacco consumption. The hypotheses of the nicotine targeting other receptors besides nAChRs were postulated because of the taste and smell sensations it caused, as well as strong burning, stinging and pain sensations at increasing concentrations (Thuerauf et al. 2006). Notably, all variants of smoking-cessation therapies that are based on nicotine replacement produce side effects such as local irritation (Hatsukami et al. 2008; Stead et al. 2008; Nides 2008). These effects were not consistent with the sole activation of nAChRs, because this receptor quickly desensitises under the high local concentrations of nicotine used (up to 60 mM). Taking into account these evidences, Talavera and co-workers demonstrated that application of micromolar concentrations of nicotine to mouse and human TRPA1-expressing in CHO cells induced an increase in intracellular Ca^{+2}. Cell-attached patch-clamp recordings further confirmed that nicotine activates TRPA1. Their results suggest that nicotine stabilises the open state and destabilises the closed state of the channel. In addition, nasal instillation of nicotine caused an airway constriction reflex that was dependent on TRPA1 (Talavera et al. 2009). TRPA1 is also a mediator for mucosal irritation and cough induced by smoke cigarette inhalation. However, the activation of TRPA1 in this case seems to be more associated with reactive oxygen species and reactive aldehydes that are locally formed as opposed to the component nicotine (Liu et al. 2013; Lin et al. 2010).

2.9 Artepillin C

The main pungent compound present in ethanolic extract of Brazilian propolis, artepillin C (Fig. 10), potently activated Ca^{+2} movement in hTRPA1-expressing cells (Hata et al. 2012). Propolis is a resin produced by honeybees from substances collected from tree buds and sap and use it to seal cracks in the hive and protect the bee colony against intruders, viruses, fungi and bacteria (Burdock 1998; Banskota et al. 2001). Propolis presents many pharmacological properties and is used in folk medicine. It has demonstrated antibacterial (Drago et al. 2000), antiviral (Kujumgiev et al. 1999), anti-inflammatory (Paulino et al. 2008) and anticancer effects (Aso et al. 2004; Akao et al. 2003). The chemical composition of propolis depends on the geographical origin, types of plant sources (Bankova et al. 2000) and season of the year. *Baccharis dracunculifolia* DC (Asteraceae), a plant native to Brazil, is the most important botanical source of southeastern Brazilian propolis, which is known as green propolis because of its deep green colour. Artepillin C constitutes 90 % of total solid compounds in a 50–60 % ethanol-eluted fraction, and it is believed to be the main cause of the pungent sensation tasted in this fraction (Park et al. 2002; Teixeira et al. 2005). Artepillin C potently activated Ca^{+2} influx in hTRPA1-expressing cells. The EC_{50} value of artepillin C (1.8 μM) was three times lower than allyl isothiocyanate (6.2 μM), showing that this compound is a more potent agonist than the classically described TRPA1 agonists (Hata et al. 2012).

2.10 Stilbenoids

The polyphenol stilbenoid resveratrol (3,5,4′-trihydroxy-trans-stilbene, Fig. 10) has also been described to modulate TRPA1. Resveratrol is abundant in nature, and, to date, 72 different plant species, especially grapevines, pine trees, and legumes, have been described to produce resveratrol (Soleas et al. 1997). Resveratrol was first isolated from the roots of white hellebore (*Veratrum grandiflorum* O. Loes) and from the roots of *Polygonum cuspidatum*, a plant used in traditional Chinese and Japanese medicine (Nonomura et al. 1963; Baur and Sinclair 2006). The concentration of resveratrol in grape skin is about 0.1 mg/100 g of fresh weight (Pastrana-Bonilla et al. 2003); therefore, the beneficial effects of grape beverages have been associated with the presence of resveratrol. Resveratrol has been reported to have potent anti-aging, anti-inflammatory, anticancer, antioxidant, cardio-protective and analgesic properties (Park et al. 2012; Aggarwal et al. 2004; Stojanovic et al. 2001; Bradamante et al. 2004; Pham-Marcou et al. 2008; Bertelli et al. 2008). In accordance with its analgesic properties, Yu and collaborators demonstrated that resveratrol potently inhibited inward currents induced by 100 μM allyl isothiocyanate in mTRPA1-expressing HEK293 cells. The calculated IC_{50} capable of inhibiting allyl isothiocyanate-induced currents was approximately 0.75 μM. Resveratrol also inhibited allyl isothiocyanate-induced currents in DRG (Yu et al. 2013), showing that resveratrol suppressed inward TRPA1 current in both heterologous expression systems and in sensory neurons and suggesting a direct inhibitory effect of resveratrol upon TRPA1.

2.11 Phenol Derivatives

Active constituents from *Kalopanax pictus* Nakai (Araliaceae), the phenolic compounds methyl syringate (Fig. 9) and protocatechuic acid and the saponin hederacoside C were able to induce Ca^{+2} influx in Flp-In 293 cells expressing hTRPA1. *Kalopanax pictus* is commonly known as the prickly castor oil tree, and it is mainly distributed across East Asia. Its bark and stem have long been used in Asian countries to treat rheumatoid arthritis, necrotic pain and diabetes mellitus (Lee et al. 2001; Choi et al. 2002). The first leaves of *Kalopanax pictus* which has long been used in Asian countries generally impart a somatosensation, including pungency, irritation, coldness and hotness (Hyun and Kim 2009). The ethanolic extract of *Kalopanax pictus* can activate both hTRPA1 and hTRPV1 (Son et al. 2012b). However, the isolated compound methyl syringate demonstrated to be selective to TRPA1 and was the most potent compound to activate TRPA1 among all tested compounds (Son et al. 2012a).

3 TRPM

The TRPM (long or melastatin) is characterised by the lack of ankyrin domains in the amino-terminus. The family is composed of eight mammalian members: TRPM1 to TRPM8 (Clapham 2003; Clapham et al. 2003; Jordt et al. 2004).

3.1 TRPM8

TRPM8 was identified as an mRNA encoded by the trp-p8 gene, which was up-regulated in prostatic and other cancers (Tsavaler et al. 2001). Its sensory role was recognised when it was isolated by expression cloning of a menthol receptor from trigeminal neurons (McKemy et al. 2002; Peier et al. 2002). TRPM8 is expressed in trkA positive Aδ- and C-fibre primary afferent neurons from DRG and TG neurons (Kobayashi et al. 2005). TRPM8 is widely expressed, but thought to function specifically as a thermosensor in non-expressing TRPV1 small-diameter primary sensory neurons (McKemy et al. 2002; Peier et al. 2002). It is a nonselective, outwardly rectifying channel that can be activated by cold (8–28 °C) and enhanced by 'cooling' compounds such as menthol, icilin and eucalyptol (Table 1) (Peier et al. 2002; McKemy et al. 2002; Chuang et al. 2004; Voets et al. 2004).

The monoterpenoid (−)-menthol, (1R,2S,5R)-2-isopropyl-5-methylcyclohexanol (Fig. 5), is the most studied TRPM8 agonist. Menthol is a common and popular flavouring and cooling additive in several household remedies (topical pain gels, throat lozenges and toothpaste), foods (gum, candies and teas) and cigarettes (Eccles 1994; Giovino et al. 2004). It contains local anaesthetic and counterirritant properties and is widely used to relieve minor throat irritation (Eccles 2003). It is believed that menthol may facilitate nicotine addiction because it diminishes the irritant effect of cigarette smoke constituents (Journigan and Zaveri 2013; Willis et al. 2011). Menthol is obtained from cornmint, peppermint or other mint oils along with a little menthone, the ester menthyl acetate and other compounds (Eccles 1994). *Mentha arvensis* is the primary species of mint used to make natural menthol crystals and natural menthol flakes. This species is primarily grown in the Uttar Pradesh region in India. Menthol is the natural ligand for TRPM8 (EC_{50} for hTRPM8 66.7 ± 3.3 μM) (McKemy et al. 2002) and has been described to either activate or increase cold-activated ionic currents in sensory neurons (Reid and Flonta 2002; Hui et al. 2005). Low levels of menthol elevate the threshold for TRPM8 activation to warmer temperatures, and repeated menthol stimulation rapidly desensitises TRPM8 in a Ca^{+2}-dependent manner (McKemy et al. 2002; Reid et al. 2002).

The discovery of a target protein to menthol was crucial in the development of synthetic compounds to treat cancer, inflammation, pain and mucosal irritation. TRPM8 expression is up-regulated in androgen-responsive prostate tumours (Journigan and Zaveri 2013; Zhang and Barritt 2004) and plays an important role in cellular Ca^{+2} homeostasis. In this context, menthane-based TRPM8 agonists have been patented by Dendreon Corporation because of their putative therapeutic use to treat benign prostatic hyperplasia and prostate cancer. The menthane-based TRPM8 agonists are more potent and selective than menthol; they have been described to induce apoptosis in androgen-dependent prostate tumours by increasing Ca^{+2} influx (Tolcher et al. 2010; Corey et al. 2003; Journigan and Zaveri 2013). The methane-related antagonists have also been developed, and their use has been extremely important in the comprehension of TRPM8 role in disease and nicotine addiction.

In addition to menthol, structural monoterpene analogues that also elicit a cooling sensation can modulate TRPM8. Monoterpenes are nonnutritive dietary components found in the essential oils of citrus fruits, cherry, mint and herbs. They function physiologically as chemoattractants or chemorepellents, and they are largely responsible for the distinctive fragrance of many plants (Mcgarvey and Croteau 1995). Eucalyptol (1,8-cineole) (Fig. 5) is one of the main components present in essential oils from *Eucalyptus polybractea* (Myrtaceae). This substance has a characteristic of fresh and camphoraceous fragrance and pungent taste and is used in pharmaceutical preparations as an external applicant, a nasal spray and an analgesic, as well as being used as a disinfectant or food flavouring. Moreover, eucalyptol is used to treat rhinosinusitis, bronchial asthma cough and muscular pain (Juergens et al. 2003; Kehrl et al. 2004). Like menthol, eucalyptol acts as a potent counterirritant against smoke constituents, by suppressing respiratory irritation, facilitating smoke inhalation and nicotine addiction (Willis et al. 2011). Animal studies have shown that eucalyptol possesses cardiovascular, antiulcer, anti-inflammatory and analgesic properties (Juergens et al. 2003; Santos and Rao 2000; Lahlou et al. 2002; Lima et al. 2013). Eucalyptol also elicits membrane currents and Ca^{+2} influx in TRPM8-expressing cells, but with lower efficacy and potency when compared to menthol (McKemy et al. 2002; Behrendt et al. 2004; Takaishi et al. 2012). However, the involvement of TRPM8 in the pharmacological effects of eucalyptol in vivo needs to be confirmed.

The isolated monoterpenes eugenol and menthone, the latter of which is the precursor of menthol in monoterpene biosynthesis, exhibited similar abilities to induce Ca^{+2} influx in HEK293T cells expressing hTRPM8. Interestingly, when their respective essential oils, clove and peppermint oils, were tested, peppermint induced a more pronounced Ca^{+2} influx in these cells (Takaishi et al. 2012). Noteworthy, eugenol does not specifically activate TRPM8, it also modulates other members of the TRP family channels, including TRPV1 and TRPA1 (Yang et al. 2003; Takaishi et al. 2012; Bandell et al. 2004). Other monoterpenes like umbellulone and thymol also activate TRPM8, but their effects are much more modest than menthol, representing only 30 % of the menthol capacity to induce channel activation (Zhong et al. 2011a; Lee et al. 2008). Structural modifications that removed the electrophilic property of umbellulone increased the ability to modulate TRPM8 in a bimodal manner, activating the basal state of the channel and inhibiting its activation by menthol, whereas the monoterpene camphor, at concentration as high as 10 mM, inhibited TRPM8 constitutive currents (Xu et al. 2005). The precise mechanisms through which these compounds exert their actions are currently unclear.

The monoterpenes perillaldehyde and perillaketone (Fig. 5) negatively modulated TRPM8. At a concentration of 200 μM, perillaldehyde and perillaketone inhibited by 64.3 and 39.3 %, respectively, the currents induced by icilin in TRPM8-HEK-293 cells. Of note, these compounds had no effect on basal TRPM8 currents (Bassoli et al. 2009).

Phytocannabinoids have also been described to modulate TRPM8 (Table 1). Five naturally occurring phytocannabinoids, cannabidiol, THC, cannabidiolic acid, Δ^9-tetrahydrocannabinolic acid and cannabigerol (Fig. 3), potently antagonised TRPM8-mediated Ca^{+2} elevation induced by menthol or icillin in HEK-293 cells. The IC_{50} range was 0.07–1.6 μM among the compounds (De Petrocellis et al. 2008). Taking into account the high potency of these phytocannabinoids to inhibit TRPM8 currents, the use of these compounds in in vitro studies can be suggested, but future studies are necessary to confirm the in vivo effect of these compounds upon TRPM8.

The glucoside phenol salidroside (Fig. 10) is an effective component of the traditional Chinese herb *Rhodiola rosea*, which is known to have the ability to protect individuals from cold attacks. Salidroside inhibited TRPM8-dependent Ca^{+2} influx in human bronchial epithelial (HBE16) cells when these cells were submitted to a cold stimulus (18 °C). In addition, salidroside decreased TRPM8 expression by reducing CREB activation (Li et al. 2013).

3.2 TRPM7

TRPM7 is a divalent-specific ionic channel responsible for Mg^{+2} homeostasis in cells (Schmitz et al. 2003) and thus is important in cell viability, proliferation and growth (Montell 2005). As reported by Parnas and co-workers, TRPM7 currents from both HEK293 and hippocampal culture were significantly inhibited by carvacrol (Fig. 5). Other naturally occurring compounds including menthol, borneol and cinnamaldehyde did not present such a significant ability (Parnas et al. 2009).

The plant alkaloid quinine also inhibited TRPM7 currents in cells (Chubanov et al. 2012). Quinine is an antimalarial compound that occurs naturally in the bark of *Cinchona* genus tree. The genus includes 38 species in the family Rubiaceae and is native to the tropical Andes forests of western South America (Flückiger and Hanbury 1874). It was first used to treat malaria in Rome in 1631 but had been used before by Peruvians as a muscle relaxant. It was also long used by the Quechua, who are indigenous to Peru, to halt shivering due to low temperatures (Rocco 2003). The Peruvians would mix the ground bark of cinchona trees with sweetened water to offset the bark's bitter taste, thus producing tonic water (Flückiger and Hanbury 1874).

The diterpenoid isolated from the Hawaiian soft coral *Sarcothelia edmondsoni*, waixenicin A, is a highly potent and relatively selective inhibitor for TRPM7 that effectively suppressed cell growth and proliferation. Waixenicin A was identified in a chemical library of 1,100 marine organism-derived extracts in a high-throughput assay system that measures the fluorescence quench of intracellular fura-2 by Mn^{+2} ions in HEK293 cells overexpressing murine TRPM7. Waixenicin A inhibited TRPM7-mediated Mn^{+2} quench with an IC_{50} of 12 μM. Analysis of waixenicin A in patch-clamp experiments confirmed the inhibitory effect on TRPM7, and the dose-response analysis of waixenicin A-mediated inhibition of TRPM7 revealed an IC_{50} of 7 μM (Zierler et al. 2011). More recently, the inhibitory effect of waixenicin A upon TRPM7 was described in interstitial cells of Cajal from the gastrointestinal system. In this study, waixenicin A was used as tool to understand the role of TRPM7 in these cells (Kim et al. 2013).

3.3 TRPM3

TRPM3 is prominently expressed in the kidney, liver, ovary, brain, spinal cord, pituitary, vascular smooth muscle and testis (Grimm et al. 2003; Lee et al. 2003; Naylor et al. 2010). The channel has been also identified in insulinoma cells and β-cells. The TRPM3 gene encodes many TRPM3 channel isoforms due to alternative splicing (Grimm et al. 2003; Lee et al. 2003; Oberwinkler et al. 2005; Wagner et al. 2008), which encompasses the largest number of splice variants within the TRP family of channel proteins. Interestingly, these TRPM3 isoforms differ within the pore region of the channel between the fifth and sixth transmembrane regions, and, therefore, they have different cation permeabilities (Oberwinkler et al. 2005). TRPM3 is activated by hypotonic conditions (Grimm et al. 2003) and by a series of endogenous ligands, mainly lipid metabolites (Wagner et al. 2008). In a screen of a compound library, Straub and co-workers found that citrus fruit flavanones such as naringenin and hesperetin and fabacea secondary metabolites selectively inhibited TRPM3 channel activation (Straub et al. 2013). The compounds exhibited a marked specificity for recombinant TRPM3 and blocked intracellular Ca^{+2} signals induced by the neurosteroid pregnenolone sulphate (pregS) in freshly isolated DRG neurons. The inhibitory potency of the compounds ranged from upper nanomolar to lower micromolar concentrations (Straub et al. 2013). Since physiological functions of TRPM3 channels are still poorly defined, the development and validation of potent and selective blockers is expected to contribute to the clarification of biological actions of TRPM3.

4 TRPC

The initially identified members of the TRP superfamily are referred to as the 'classical' or 'canonical' TRPs and fall into the TRPC subfamily. A common feature of the mammalian TRPC channels is that they are activated in cultured cells through pathways that engage phospholipase C (PLC) (Zhu et al. 1996; Okada et al. 1998; Boulay et al. 1997). The TRPC subfamily is unique in that its members are not only responsible for agonist-activated nonselective cation currents but they also participate in the so-called slow sustained mode of Ca^{+2} signalling, which requires sustained elevations of intracellular Ca^{+2} (Birnbaumer 2009).

Only a few natural compounds have been described to modulate TRPC channels so far. The ginseng saponin metabolite 20-*O*-β-D-glucopyranosyl-20(S)-protopanaxadiol (20-GPPD) increased Ca^{+2} entry in CT-26 cells, and this effect was suppressed by blocking TRPC channels. The activation of TRPC by 20-GPPD was thought to be responsible for the activation of AMP-activated protein kinase (AMPK) and apoptosis in these colon cancer cells (Hwang et al. 2013).

The naturally occurring monoterpene camphor (Fig. 5) has also demonstrated a modulatory effect upon TRPC5. TRPC5 is activated downstream of PLC-couple receptor activation (Schaefer et al. 2000). In HEK293 cells that express TRPC5, there are typical constitutive basal currents at positive potentials, and these currents were significantly inhibited by the application of 10 mM camphor to the cells (Xu et al. 2005). The TRPC channels are thought to be present in heteromultimeric forms (Peters et al. 2012; Li et al. 1999; Tsiokas et al. 1999); therefore, the detection of compounds that selectively target this channel family has been a challenge so far.

The active compound of St. John's wort (*Hypericum perforatum*) named hyperforin (Fig. 11), a phloroglucinol derivative, was found selectively to activate TRPC6 (Leuner et al. 2007). St. John's wort has been shown to be effective in the treatment of mild to moderate depression (Muller 2003; Linde et al. 2005, 2008). Similar to other antidepressant agents, hyperforin inhibits monoamine neurotransmitter uptake, an effect that occurs through the increase of intracellular Na^{+} concentration resulting from the activation of TRPC6 (Singer et al. 1999; Treiber et al. 2005). Otherwise, TRPC6 channels are also Ca^{+2} permeable, resulting in intracellular Ca^{+2} elevations. Some studies show that the regulation of Ca^{+2} influx in the central nervous system through the TRPC6 improves synaptic plasticity, an important phenomenon in the pathogenesis of depression. The effects on synaptic plasticity include growth cone guidance (Li et al. 2005), spine morphology changes, dendritic outgrowth (Zhou et al. 2008; Tai et al. 2008; Leuner et al. 2013) and cell survival (Du et al. 2010; Lin et al. 2013). Interestingly, hyperforin induced neurite outgrowth in PC12 cells and hippocampal neurons as well as spine morphology changes in hippocampal neurons (Leuner et al. 2007, 2013; Heiser et al. 2013). Moreover, TRPC6-dependent Ca^{2+} elevation induced by hyperforin in PC12 cells and hippocampal neurons mimics the neurotrophic effects of the nerve growth factor (NGF) such as cell differentiation and of the brain-derived neurotrophic factor (BDNF) by modifying dendritic spine morphology (Leuner et al. 2007, 2010, 2013). Different signalling pathways seem to be involved in the TRPC6-mediated neuritis outgrowth including Ras/MEK/ERK, PI3K/Akt and CAMKIV that result in CREB phosphorylation (Heiser et al. 2013), an event which is considered to be strongly involved in antidepressant response of medicines (Pittenger and Duman 2008). Therefore, these data corroborate to the understanding of St. John's wort antidepressant activity.

5 TRPP

Autosomal dominant polycystic kidney disease (ADPKD) is a major cause of end-stage renal failure. A defect in the gene product of either PKD1 or PKD2 (85 % and 15 % of all ADPKD cases, respectively) is thought to result in a loss of calcium signalling in kidney epithelial cells and a reversion to the proliferative phenotype (Nauli et al. 2003; Yamaguchi et al. 2006). The PKD2 gene encodes TRPP2 (transient receptor potential polycystin 2), which has significant homology to voltage-activated calcium and sodium TRP channels. TRPP2 is expressed in the endoplasmic/sarcoplasmic reticulum, where it modulates many cellular processes via intracellular calcium-dependent signalling pathways (Qamar et al. 2007). Extracts of *Tripterygium wilfordii* Hook. f. (leigong teng, thundergod vine) are effective in traditional Chinese medicine for the treatment of immune inflammatory diseases, including rheumatoid arthritis, systemic lupus erythematosus, nephritis and asthma. Characterisation of the terpenoids present in extracts of *Tripterygium* identified triptolide (Fig. 9), a diterpenoid triepoxide, as responsible for most of the immunosuppressive, anti-inflammatory and anti-proliferative effects observed in vitro (Qiu and Kao 2003). Triptolide binds to TRPP2 directly and induces Ca^{+2} release by a TRPP2-dependent mechanism. In murine model of ADPKD, triptolide arrests cellular proliferation and attenuates cyst formation, an effect that is attenuated in TRPP2 null mice (Leuenroth et al. 2010).

Remarkable Conclusions

The search for cellular targets of natural-occurring substances has been crucial to determine the distribution and biological functions of the TRP channels family.

Of relevance, some results have been misinterpreted because some compounds are promiscuous and target two or more TRPs, even when the channels are different in their molecular structure. In addition, a cross-sensitisation of these receptors is a common event, and, therefore, a compound that targets one subtype of TRP may indirectly modulate a second different TRP. For a while, these issues had contributed to misinterpreted conclusions about targets for natural products. Through the advance in molecular cloning, much of the misleading conclusions could be obliterated. The heterologous expression of these receptors has been crucial to determine the specificity of some TRPs agonists/antagonists from natural origin. Besides, by cloning TRPs from different animal species (flies, mice, rats, humans), it was possible to identify that TRPs from different species may present distinct sensitivity to the same compound.

The research on distribution and function of TRPs indicates a broad range of physiological and pathological roles for these channels. Some of them greatly contribute to the development and/or progression of the symptoms of diseases like neuropathic pain, overactive urinary bladder, asthma, anxiety disorders and pruritus. Indeed, gene association studies in humans indicated that single-nucleotide polymorphisms in the coding regions and/or promoters of genes that encode TRP channels are either associated with an increased risk of multifactorial diseases or they appear to be causative factors in rare heritable. Therefore, the modulation of TRPs has been purposed in therapeutic, and the use of naturally occurring agents is a promising alternative and of potential interest for the development of new therapies.

References

Abe M, Ozawa Y, Uda Y, Yamada Y, Morimitsu Y, Nakamura Y, Osawa T (2002) Labdane-type diterpene dialdehyde, pungent principle of myoga, Zingiber mioga Roscoe. Biosci Biotechnol Biochem 66(12):2698–2700

Adams MJ, Ahuja KD, Geraghty DP (2009) Effect of capsaicin and dihydrocapsaicin on in vitro blood coagulation and platelet aggregation. Thromb Res 124(6):721–723

Aggarwal BB, Bhardwaj A, Aggarwal RS, Seeram NP, Shishodia S, Takada Y (2004) Role of resveratrol in prevention and therapy of cancer: preclinical and clinical studies. Anticancer Res 24(5A):2783–2840

Ahluwalia J, Rang H, Nagy I (2002) The putative role of vanilloid receptor-like protein-1 in mediating high threshold noxious heat-sensitivity in rat cultured primary sensory neurons. Eur J Neurosci 16(8):1483–1489

Akao Y, Maruyama H, Matsumoto K, Ohguchi K, Nishizawa K, Sakamoto T, Araki Y, Mishima S, Nozawa Y (2003) Cell growth inhibitory effect of cinnamic acid derivatives from propolis on human tumor cell lines. Biol Pharm Bull 26(7):1057–1059

Akopian AN, Ruparel NB, Patwardhan A, Hargreaves KM (2008) Cannabinoids desensitize capsaicin and mustard oil responses in sensory neurons via TRPA1 activation. J Neurosci 28 (5):1064–1075

Alenmyr L, Herrmann A, Hogestatt ED, Greiff L, Zygmunt PM (2009) TRPV1 and TRPA1 stimulation induces MUC5B secretion in the human nasal airway in vivo. Clin Physiol Funct Imaging 31(6):435–444

Alpizar YA, Gees M, Sanchez A, Apetrei A, Voets T, Nilius B, Talavera K (2012) Bimodal effects of cinnamaldehyde and camphor on mouse TRPA1. Pflugers Arch 465:853–864
Amantini C, Mosca M, Nabissi M, Lucciarini R, Caprodossi S, Arcella A, Giangaspero F, Santoni G (2007) Capsaicin-induced apoptosis of glioma cells is mediated by TRPV1 vanilloid receptor and requires p38 MAPK activation. J Neurochem 102(3):977–990
Amantini C, Ballarini P, Caprodossi S, Nabissi M, Morelli MB, Lucciarini R, Cardarelli MA, Mammana G, Santoni G (2009) Triggering of transient receptor potential vanilloid type 1 (TRPV1) by capsaicin induces Fas/CD95-mediated apoptosis of urothelial cancer cells in an ATM-dependent manner. Carcinogenesis 30(8):1320–1329
Andrade EL, Ferreira J, Andre E, Calixto JB (2006) Contractile mechanisms coupled to TRPA1 receptor activation in rat urinary bladder. Biochem Pharmacol 72(1):104–114
Andrade EL, Luiz AP, Ferreira J, Calixto JB (2008) Pronociceptive response elicited by TRPA1 receptor activation in mice. Neuroscience 152(2):511–520
Andrade EL, Meotti FC, Calixto JB (2012) TRPA1 antagonists as potential analgesic drugs. Pharmacol Ther 133(2):189–204
Andre E, Ferreira J, Malheiros A, Yunes RA, Calixto JB (2004) Evidence for the involvement of vanilloid receptor in the antinociception produced by the dialdeydes unsaturated sesquiterpenes polygodial and drimanial in rats. Neuropharmacology 46(4):590–597
Andre E, Campi B, Trevisani M, Ferreira J, Malheiros A, Yunes RA, Calixto JB, Geppetti P (2006) Pharmacological characterisation of the plant sesquiterpenes polygodial and drimanial as vanilloid receptor agonists. Biochem Pharmacol 71(8):1248–1254
Andre E, Gatti R, Trevisani M, Preti D, Baraldi PG, Patacchini R, Geppetti P (2009) Transient receptor potential ankyrin receptor 1 is a novel target for pro-tussive agents. Br J Pharmacol 158(6):1621–1628
Appendino G, Szallasi A (1997) Euphorbium: modern research on its active principle, resiniferatoxin, revives an ancient medicine. Life Sci 60(10):681–696
Aso K, Kanno S, Tadano T, Satoh S, Ishikawa M (2004) Inhibitory effect of propolis on the growth of human leukemia U937. Biol Pharm Bull 27(5):727–730
Athanasiou A, Smith PA, Vakilpour S, Kumaran NM, Turner AE, Bagiokou D, Layfield R, Ray DE, Westwell AD, Alexander SP, Kendall DA, Lobo DN, Watson SA, Lophatanon A, Muir KA, Guo DA, Bates TE (2007) Vanilloid receptor agonists and antagonists are mitochondrial inhibitors: how vanilloids cause non-vanilloid receptor mediated cell death. Biochem Biophys Res Commun 354(1):50–55
Attele AS, Wu JA, Yuan CS (1999) Ginseng pharmacology: multiple constituents and multiple actions. Biochem Pharmacol 58(11):1685–1693
Bandell M, Story GM, Hwang SW, Viswanath V, Eid SR, Petrus MJ, Earley TJ, Patapoutian A (2004) Noxious cold ion channel TRPA1 is activated by pungent compounds and bradykinin. Neuron 41(6):849–857
Bang S, Hwang SW (2009) Polymodal ligand sensitivity of TRPA1 and its modes of interactions. J Gen Physiol 133(3):257–262
Bankova V, Castro S, Marcucci M (2000) Porpolis: recent advances in chemistry and plant origin. Apidologie 31:3–15
Banskota AH, Tezuka Y, Kadota S (2001) Recent progress in pharmacological research of propolis. Phytother Res 15(7):561–571
Barbero GF, Molinillo JM, Varela RM, Palma M, Macias FA, Barroso CG (2010) Application of Hansch's model to capsaicinoids and capsinoids: a study using the quantitative structure-activity relationship. A novel method for the synthesis of capsinoids. J Agric Food Chem 58 (6):3342–3349
Bassoli A, Borgonovo G, Caimi S, Scaglioni L, Morini G, Moriello AS, Di Marzo V, De Petrocellis L (2009) Taste-guided identification of high potency TRPA1 agonists from Perilla frutescens. Bioorg Med Chem 17(4):1636–1639
Baur JA, Sinclair DA (2006) Therapeutic potential of resveratrol: the in vivo evidence. Nat Rev Drug Discov 5(6):493–506

Bautista DM, Movahed P, Hinman A, Axelsson HE, Sterner O, Hogestatt ED, Julius D, Jordt SE, Zygmunt PM (2005) Pungent products from garlic activate the sensory ion channel TRPA1. Proc Natl Acad Sci U S A 102(34):12248–12252

Beck JJ, Stermitz FR (1995) Addition of methyl thioglycolate and benzylamine to (Z)-ligustilide, a bioactive unsaturated lactone constituent of several herbal medicines - an improved synthesis of (Z)-ligustilide. J Nat Prod-Lloydia 58(7):1047–1055

Behrendt HJ, Germann T, Gillen C, Hatt H, Jostock R (2004) Characterization of the mouse cold-menthol receptor TRPM8 and vanilloid receptor type-1 VR1 using a fluorometric imaging plate reader (FLIPR) assay. Br J Pharmacol 141(4):737–745

Benemei S, Appendino G, Geppetti P (2009) Pleasant natural scent with unpleasant effects: cluster headache-like attacks triggered by Umbellularia californica. Cephalalgia 30(6):744–746

Bertelli A, Falchi M, Dib B, Pini E, Mukherjee S, Das DK (2008) Analgesic resveratrol? Antioxid Redox Signal 10(3):403–404

Bevan S, Szolcsanyi J (1990) Sensory neuron-specific actions of capsaicin: mechanisms and applications. Trends Pharmacol Sci 11(8):330–333

Birnbaumer L (2009) The TRPC class of ion channels: a critical review of their roles in slow, sustained increases in intracellular $Ca^{(2+)}$ concentrations. Annu Rev Pharmacol Toxicol 49:395–426

Bisogno T, Hanus L, De Petrocellis L, Tchilibon S, Ponde DE, Brandi I, Moriello AS, Davis JB, Mechoulam R, Di Marzo V (2001) Molecular targets for cannabidiol and its synthetic analogues: effect on vanilloid VR1 receptors and on the cellular uptake and enzymatic hydrolysis of anandamide. Br J Pharmacol 134(4):845–852

Bley K (2013) Effects of topical capsaicin on cutaneous innervation: implications for pain management. Open Pain J 6:81–94

Block E, Dane AJ, Thomas S, Cody RB (2010) Applications of direct analysis in real time mass spectrometry (DART-MS) in Allium chemistry. 2-propenesulfenic and 2-propenesulfinic acids, diallyl trisulfane S-oxide, and other reactive sulfur compounds from crushed garlic and other Alliums. J Agric Food Chem 58(8):4617–4625

Boulay G, Zhu X, Peyton M, Jiang M, Hurst R, Stefani E, Birnbaumer L (1997) Cloning and expression of a novel mammalian homolog of Drosophila transient receptor potential (Trp) involved in calcium entry secondary to activation of receptors coupled by the Gq class of G protein. J Biol Chem 272(47):29672–29680

Bradamante S, Barenghi L, Villa A (2004) Cardiovascular protective effects of resveratrol. Cardiovasc Drug Rev 22(3):169–188

Buck SH, Burks TF (1986) The neuropharmacology of capsaicin: review of some recent observations. Pharmacol Rev 38(3):179–226

Burdock GA (1998) Review of the biological properties and toxicity of bee propolis (propolis). Food Chem Toxicol 36(4):347–363

Burkhart CG, Burkhart HR (2003) Contact irritant dermatitis and anti-pruritic agents: the need to address the itch. J Drugs Dermatol 2(2):143–146

Burrow A, Eccles R, Jones AS (1983) The effects of camphor, eucalyptus and menthol vapour on nasal resistance to airflow and nasal sensation. Acta Otolaryngol 96(1–2):157–161

Burt S (2004) Essential oils: their antibacterial properties and potential applications in foods–a review. Int J Food Microbiol 94(3):223–253

Calixto JB, Beirith A, Ferreira J, Santos AR, Filho VC, Yunes RA (2000) Naturally occurring antinociceptive substances from plants. Phytother Res 14(6):401–418

Caterina MJ, Schumacher MA, Tominaga M, Rosen TA, Levine JD, Julius D (1997) The capsaicin receptor: a heat-activated ion channel in the pain pathway. Nature 389(6653):816–824

Caterina MJ, Rosen TA, Tominaga M, Brake AJ, Julius D (1999) A capsaicin-receptor homologue with a high threshold for noxious heat. Nature 398(6726):436–441

Cattaruzza F, Spreadbury I, Miranda-Morales M, Grady EF, Vanner S, Bunnett NW (2010) Transient receptor potential ankyrin-1 has a major role in mediating visceral pain in mice. Am J Physiol Gastrointest Liver Physiol 298(1):G81–91

Chai W, Mehrotra S, Jan Danser AH, Schoemaker RG (2006) The role of calcitonin gene-related peptide (CGRP) in ischemic preconditioning in isolated rat hearts. Eur J Pharmacol 531(1–3):246–253

Chan SSK, Cheng TY, Lin G (2007) Relaxation effects of ligustilide and senkyunolide A, two main constituents of Ligusticum chuanxiong, in rat isolated aorta. J Ethnopharmacol 111 (3):677–680

Choi J, Huh K, Kim SH, Lee KT, Park HJ, Han YN (2002) Antinociceptive and anti-rheumatoidal effects of Kalopanax pictus extract and its saponin components in experimental animals. J Ethnopharmacol 79(2):199–204

Choi CH, Jung YK, Oh SH (2010a) Autophagy induction by capsaicin in malignant human breast cells is modulated by p38 and extracellular signal-regulated mitogen-activated protein kinases and retards cell death by suppressing endoplasmic reticulum stress-mediated apoptosis. Mol Pharmacol 78(1):114–125

Choi CH, Jung YK, Oh SH (2010b) Selective induction of catalase-mediated autophagy by dihydrocapsaicin in lung cell lines. Free Radic Biol Med 49(2):245–257

Choi SE, Kim TH, Yi SA, Hwang YC, Hwang WS, Choe SJ, Han SJ, Kim HJ, Kim DJ, Kang Y, Lee KW (2011) Capsaicin attenuates palmitate-induced expression of macrophage inflammatory protein 1 and interleukin 8 by increasing palmitate oxidation and reducing c-Jun activation in THP-1 (human acute monocytic leukemia cell) cells. Nutr Res 31(6):468–478

Chuang HH, Neuhausser WM, Julius D (2004) The super-cooling agent icilin reveals a mechanism of coincidence detection by a temperature-sensitive TRP channel. Neuron 43(6):859–869

Chubanov V, Mederos y Schnitzler M, Meissner M, Schafer S, Abstiens K, Hofmann T, Gudermann T (2012) Natural and synthetic modulators of SK (K(ca)2) potassium channels inhibit magnesium-dependent activity of the kinase-coupled cation channel TRPM7. Br J Pharmacol 166(4):1357–1376

Cioffi DL (2007) The skinny on TRPV1. Circ Res 100(7):934–936

Clapham DE (2003) TRP channels as cellular sensors. Nature 426(6966):517–524

Clapham DE, Montell C, Schultz G, Julius D (2003) International Union of Pharmacology. XLIII. Compendium of voltage-gated ion channels: transient receptor potential channels. Pharmacol Rev 55(4):591–596

Cliff M, Heymann H (1992) Descriptive analysis of oral pungency. J Sens Studies 7:279–290

Corey E, Quinn JE, Buhler KR, Nelson PS, Macoska JA, True LD, Vessella RL (2003) LuCaP 35: a new model of prostate cancer progression to androgen independence. Prostate 55(4):239–246

Corrêa MP (1978) Dicionário das plantas úteis no Brasil e das exóticas cultivadas. Imprensa Nacional, Rio de Janeiro

Costa B, Giagnoni G, Franke C, Trovato AE, Colleoni M (2004) Vanilloid TRPV1 receptor mediates the antihyperalgesic effect of the nonpsychoactive cannabinoid, cannabidiol, in a rat model of acute inflammation. Br J Pharmacol 143(2):247–250

Cuypers E, Yanagihara A, Karlsson E, Tytgat J (2006) Jellyfish and other cnidarian envenomations cause pain by affecting TRPV1 channels. FEBS Lett 580(24):5728–5732

da Cunha FM, Frode TS, Mendes GL, Malheiros A, Cechinel Filho V, Yunes RA, Calixto JB (2001) Additional evidence for the anti-inflammatory and anti-allergic properties of the sesquiterpene polygodial. Life Sci 70(2):159–169

Dallmeier K, Carlini EA (1981) Anesthetic, hypothermic, myorelaxant and anticonvulsant effects of synthetic eugenol derivatives and natural analogues. Pharmacology 22(2):113–127

Davis CB, Markey CE, Busch MA, Busch KW (2007) Determination of capsaicinoids in habanero peppers by chemometric analysis of UV spectral data. J Agric Food Chem 55(15):5925–5933

De Petrocellis L, Di Marzo V (2010) Non-CB1, non-CB2 receptors for endocannabinoids, plant cannabinoids, and synthetic cannabimimetics: focus on G-protein-coupled receptors and transient receptor potential channels. J Neuroimmune Pharmacol 5(1):103–121

De Petrocellis L, Vellani V, Schiano-Moriello A, Marini P, Magherini PC, Orlando P, Di Marzo V (2008) Plant-derived cannabinoids modulate the activity of transient receptor potential channels of ankyrin type-1 and melastatin type-8. J Pharmacol Exp Ther 325(3):1007–1015

De Petrocellis L, Ligresti A, Moriello AS, Allara M, Bisogno T, Petrosino S, Stott CG, Di Marzo V (2010) Effects of cannabinoids and cannabinoid-enriched Cannabis extracts on TRP channels and endocannabinoid metabolic enzymes. Br J Pharmacol 163(7):1479–1494

De Petrocellis L, Orlando P, Moriello AS, Aviello G, Stott C, Izzo AA, Di Marzo V (2012) Cannabinoid actions at TRPV channels: effects on TRPV3 and TRPV4 and their potential relevance to gastrointestinal inflammation. Acta Physiologica 204(2):255–266

Dedov VN, Tran VH, Duke CC, Connor M, Christie MJ, Mandadi S, Roufogalis BD (2002) Gingerols: a novel class of vanilloid receptor (VR1) agonists. Br J Pharmacol 137(6):793–798

Dessaint J, Yu W, Krause JE, Yue L (2004) Yohimbine inhibits firing activities of rat dorsal root ganglion neurons by blocking Na+ channels and vanilloid VR1 receptors. Eur J Pharmacol 485 (1–3):11–20

Docherty RJ, Robertson B, Bevan S (1991) Capsaicin causes prolonged inhibition of voltage-activated calcium currents in adult rat dorsal root ganglion neurons in culture. Neuroscience 40 (2):513–521

Docherty RJ, Ginsberg L, Jadoon S, Orrell RW, Bhattacharjee A (2013) TRPA1 insensitivity of human sural nerve axons after exposure to lidocaine. Pain 154:1569–1577

Doerner JF, Gisselmann G, Hatt H, Wetzel CH (2007) Transient receptor potential channel A1 is directly gated by calcium ions. J Biol Chem 282(18):13180–13189

Drago L, Mombelli B, De Vecchi E, Fassina MC, Tocalli L, Gismondo MR (2000) In vitro antimicrobial activity of propolis dry extract. J Chemother 12(5):390–395

Drake ME, Stuhr ET (1935) Some pharmacological and bactericidal properties of umbellulone. J Am Pharm Assoc 24:196–207

Du J, Yu Y, Ke Y, Wang CY, Zhu L, Qian ZM (2007a) Ligustilide attenuates pain behavior induced by acetic acid or formalin. J Ethnopharmacol 112(1):211–214

Du S, Araki I, Yoshiyama M, Nomura T, Takeda M (2007b) Transient receptor potential channel A1 involved in sensory transduction of rat urinary bladder through C-fiber pathway. Urology 70(4):826–831

Du W, Huang J, Yao H, Zhou K, Duan B, Wang Y (2010) Inhibition of TRPC6 degradation suppresses ischemic brain damage in rats. J Clin Invest 120(10):3480–3492

Earley S, Gonzales AL, Crnich R (2009) Endothelium-dependent cerebral artery dilation mediated by TRPA1 and Ca2+-Activated K+ channels. Circ Res 104(8):987–994

Earley S, Gonzales AL, Garcia ZI (2010) A dietary agonist of transient receptor potential cation channel V3 elicits endothelium-dependent vasodilation. Mol Pharmacol 77(4):612–620

Eccles R (1994) Menthol and related cooling compounds. J Pharm Pharmacol 46(8):618–630

Eccles R (2003) Menthol: effects on nasal sensation of airflow and the drive to breathe. Curr Allergy Asthma Rep 3(3):210–214

Eid SR, Crown ED, Moore EL, Liang HA, Choong KC, Dima S, Henze DA, Kane SA, Urban MO (2008) HC-030031, a TRPA1 selective antagonist, attenuates inflammatory- and neuropathy-induced mechanical hypersensitivity. Mol Pain 4:48–57

Escalera J, von Hehn CA, Bessac BF, Sivula M, Jordt SE (2008) TRPA1 mediates the noxious effects of natural sesquiterpene deterrents. J Biol Chem 283(35):24136–24144

Evans WC (1996) Trease and Evans' pharmacognosy. WB Saunders, London

Fabrizi L, Williams G, Lee A, Meek J, Slater R, Olhede S, Fitzgerald M (2013) Cortical activity evoked by an acute painful tissue-damaging stimulus in healthy adult volunteers. J Neurophysiol 109:2393–2403

Facer P, Casula MA, Smith GD, Benham CD, Chessell IP, Bountra C, Sinisi M, Birch R, Anand P (2007) Differential expression of the capsaicin receptor TRPV1 and related novel receptors TRPV3, TRPV4 and TRPM8 in normal human tissues and changes in traumatic and diabetic neuropathy. BMC Neurol 7:11

Fajardo O, Meseguer V, Belmonte C, Viana F (2008) TRPA1 channels mediate cold temperature sensing in mammalian vagal sensory neurons: pharmacological and genetic evidence. J Neurosci 28(31):7863–7875

Flückiger FA, Hanbury D (1874) History of quinine. In: Flückiger FA, Hanbury D (eds) Pharmacographia: a history of the principal drugs of vegetables origin, met with in Great Britain and British India. Macmillan and Co, London, pp 302–331

Fucile S, Sucapane A, Eusebi F (2005) Ca2+ permeability of nicotinic acetylcholine receptors from rat dorsal root ganglion neurones. J Physiol 565(Pt 1):219–228

Gees M, Alpizar YA, Boonen B, Sanchez A, Everaerts W, Segal A, Xue F, Janssens A, Owsianik G, Nilius B, Voets T, Talavera K (2013) Mechanisms of TRPV1 activation and sensitization by allyl isothiocyanate. Mol Pharmacol 84:325–334

Gentilli M, Mazoit JX, Bouaziz H, Fletcher D, Casper RF, Benhamou D, Savouret JF (2001) Resveratrol decreases hyperalgesia induced by carrageenan in the rat hind paw. Life Sci 68 (11):1317–1321

Geppetti P, Nassini R, Materazzi S, Benemei S (2008) The concept of neurogenic inflammation. BJU Int 101(Suppl 3):2–6

Ghosh AK, Basu S (2010) Fas-associated factor 1 is a negative regulator in capsaicin induced cancer cell apoptosis. Cancer Lett 287(2):142–149

Gijbels MJM, Scheffer JJC, Svendsen AB (1982) Phthalides in the essential oil from roots of Levisticum Officinale. Planta Med 44(4):207–211

Giovino GA, Sidney S, Gfroerer JC, O'Malley PM, Allen JA, Richter PA, Cummings KM (2004) Epidemiology of menthol cigarette use. Nicotine Tob Res 6(Suppl 1):S67–81

Gratzke C, Streng T, Waldkirch E, Sigl K, Stief C, Andersson KE, Hedlund P (2009) Transient receptor potential A1 (TRPA1) activity in the human urethra–evidence for a functional role for TRPA1 in the outflow region. Eur Urol 55(3):696–704

Grimm C, Kraft R, Sauerbruch S, Schultz G, Harteneck C (2003) Molecular and functional characterization of the melastatin-related cation channel TRPM3. J Biol Chem 278 (24):21493–21501

Guler AD, Lee HS, Iida T, Shimizu I, Tominaga M, Caterina M (2002) Heat-evoked activation of the ion channel, TRPV4. J Neurosci 22(15):6408–6414

Guo M, Chen S, Liu X (2008) Material basis for inhibition of Dragon's blood on evoked discharges of wide dynamic range neurons in spinal dorsal horn of rats. Sci China C Life Sci 51(11):1025–1038

Gupta D, Bleakley B, Gupta RK (2008) Dragon's blood: botany, chemistry and therapeutic uses. J Ethnopharmacol 115(3):361–380

Hail N Jr, Lotan R (2002) Examining the role of mitochondrial respiration in vanilloid-induced apoptosis. J Natl Cancer Inst 94(17):1281–1292

Hardie RC (2007) TRP channels and lipids: from Drosophila to mammalian physiology. J Physiol 578(Pt 1):9–24

Harper AG, Brownlow SL, Sage SO (2009) A role for TRPV1 in agonist-evoked activation of human platelets. J Thromb Haemost 7(2):330–338

Hata T, Tazawa S, Ohta S, Rhyu MR, Misaka T, Ichihara K (2012) Artepillin C, a major ingredient of Brazilian propolis, induces a pungent taste by activating TRPA1 channels. PLoS One 7(11): e48072

Hatsukami DK, Stead LF, Gupta PC (2008) Tobacco addiction. Lancet 371(9629):2027–2038

He GJ, Ye XL, Mou X, Chen Z, Li XG (2009) Synthesis and antinociceptive activity of capsinoid derivatives. Eur J Med Chem 44(8):3345–3349

Heamy JP (1875) Oreodaphne californica, Nees., Nat. Ord. Laudaceae. Am J Pharm 47:105–109

Heiser JH, Schuwald AM, Sillani G, Ye L, Muller WE, Leuner K (2013) TRPC6 channel-mediated neurite outgrowth in PC12 cells and hippocampal neurons involves activation of RAS/MEK/ ERK, PI3K and CAMKIV signaling. J Neurochem 127:303–313

Hellwig V, Nopper R, Mauler F, Freitag J, Ji-Kai L, Zhi-Hui D, Stadler M (2003) Activities of prenylphenol derivatives from fruitbodies of Albatrellus spp. on the human and rat vanilloid receptor 1 (VR1) and characterisation of the novel natural product, confluentin. Arch Pharm (Weinheim) 336(2):119–126

Henning SM, Zhang Y, Seeram NP, Lee RP, Wang P, Bowerman S, Heber D (2011) Antioxidant capacity and phytochemical content of herbs and spices in dry, fresh and blended herb paste form. Int J Food Sci Nutr 62(3):219–225
Hergenhahn M, Kusumoto S, Hecker E (1974) Diterpene esters from 'Euphorbium' and their irritant and cocarcinogenic activity. Experientia 30(12):1438–1440
Hinman A, Chuang HH, Bautista DM, Julius D (2006) TRP channel activation by reversible covalent modification. Proc Natl Acad Sci U S A 103(51):19564–19568
Hirst RA, Lambert DG, Notcutt WG (1998) Pharmacology and potential therapeutic uses of cannabis. Br J Anaesth 81(1):77–84
Hogaboam CM, Wallace JL (1991) Inhibition of platelet aggregation by capsaicin. An effect unrelated to actions on sensory afferent neurons. Eur J Pharmacol 202(1):129–131
Holzer P (1991) Capsaicin as a tool for studying sensory neuron functions. Adv Exp Med Biol 298:3–16
Hui K, Guo Y, Feng ZP (2005) Biophysical properties of menthol-activated cold receptor TRPM8 channels. Biochem Biophys Res Commun 333(2):374–382
Hunner M (1926) A practical treatise on disorders of the sexual function of the male and female. FA Davies, Philadelphia
Hwang JA, Hwang MK, Jang Y, Lee EJ, Kim JE, Oh MH, Shin DJ, Lim S, Ji G, Oh U, Bode AM, Dong Z, Lee KW, Lee HJ (2013) 20-O-beta-d-glucopyranosyl-20(S)-protopanaxadiol, a metabolite of ginseng, inhibits colon cancer growth by targeting TRPC channel-mediated calcium influx. J Nutr Biochem 24(6):1096–1104
Hyun TK, Kim JS (2009) The pharmacology and clinical properties of Kalopanax pictus. J Med Plants Res 3(9):613–620
Iida T, Moriyama T, Kobata K, Morita A, Murayama N, Hashizume S, Fushiki T, Yazawa S, Watanabe T, Tominaga M (2003) TRPV1 activation and induction of nociceptive response by a non-pungent capsaicin-like compound, capsiate. Neuropharmacology 44(7):958–967
Ito K, Nakazato T, Yamato K, Miyakawa Y, Yamada T, Hozumi N, Segawa K, Ikeda Y, Kizaki M (2004) Induction of apoptosis in leukemic cells by homovanillic acid derivative, capsaicin, through oxidative stress: implication of phosphorylation of p53 at Ser-15 residue by reactive oxygen species. Cancer Res 64(3):1071–1078
Iwalewa EO, Iwalewa OJ, Adeboye JO (2003) Analgesic, antipyretic, anti-inflammatory effects of methanol, chloroform and ether extracts of Vernonia cinerea less leaf. J Ethnopharmacol 86(2–3):229–234
Iwasaki Y, Morita A, Iwasawa T, Kobata K, Sekiwa Y, Morimitsu Y, Kubota K, Watanabe T (2006) A nonpungent component of steamed ginger–[10]-shogaol–increases adrenaline secretion via the activation of TRPV1. Nutr Neurosci 9(3–4):169–178
Iwasaki Y, Tanabe M, Kayama Y, Abe M, Kashio M, Koizumi K, Okumura Y, Morimitsu Y, Tominaga M, Ozawa Y, Watanabe T (2009) Miogadial and miogatrial with alpha, beta-unsaturated 1,4-dialdehyde moieties–novel and potent TRPA1 agonists. Life Sci 85(1–2):60–69
Jancsó N, Jancsó-Gabor A (1949) Desensitization of sensory nerve endings Kiserl Orvostud, Budapest
Jancsó N, Jancsó-Gabor A, Szolcsanyi J (1968) The role of sensory nerve endings in neurogenic inflammation induced in human skin and in the eye and paw of the rat. Br J Pharmacol Chemother 33(1):32–41
Jancsó G, Kiraly E, Jancsó-Gabor A (1977) Pharmacologically induced selective degeneration of chemosensitive primary sensory neurones. Nature 270(5639):741–743
Jaquemar D, Schenker T, Trueb B (1999) An ankyrin-like protein with transmembrane domains is specifically lost after oncogenic transformation of human fibroblasts. J Biol Chem 274 (11):7325–7333
Jerman JC, Brough SJ, Prinjha R, Harries MH, Davis JB, Smart D (2000) Characterization using FLIPR of rat vanilloid receptor (rVR1) pharmacology. Br J Pharmacol 130(4):916–922

Jiang MC, Gebhart GF (1998) Development of mustard oil-induced hyperalgesia in rats. Pain 77 (3):305–313

Jonassohn M, Sterner O (1997) Terpenoid unsaturated 1,4-di-aldehydes, occurrence and biological activities. Trends Org Chem 6:23–43

Joo JI, Kim DH, Choi JW, Yun JW (2010) Proteomic analysis for antiobesity potential of capsaicin on white adipose tissue in rats fed with a high fat diet. J Proteome Res 9(6):2977–2987

Jordt SE, Bautista DM, Chuang HH, McKemy DD, Zygmunt PM, Hogestatt ED, Meng ID, Julius D (2004) Mustard oils and cannabinoids excite sensory nerve fibres through the TRP channel ANKTM1. Nature 427(6971):260–265

Journigan VB, Zaveri NT (2013) TRPM8 ion channel ligands for new therapeutic applications and as probes to study menthol pharmacology. Life Sci 92(8–9):425–437

Juergens UR, Dethlefsen U, Steinkamp G, Gillissen A, Repges R, Vetter H (2003) Anti-inflammatory activity of 1.8-cineol (eucalyptol) in bronchial asthma: a double-blind placebo-controlled trial. Respir Med 97(3):250–256

Jung SY, Choi S, Ko YS, Park CS, Oh S, Koh SR, Oh U, Oh JW, Rhee MH, Nah SY (2001) Effects of ginsenosides on vanilloid receptor (VR1) channels expressed in Xenopus oocytes. Mol Cells 12(3):342–346

Kang HJ, Soh Y, Kim MS, Lee EJ, Surh YJ, Kim HR, Kim SH, Moon A (2003) Roles of JNK-1 and p38 in selective induction of apoptosis by capsaicin in ras-transformed human breast epithelial cells. Int J Cancer 103(4):475–482

Karashima Y, Damann N, Prenen J, Talavera K, Segal A, Voets T, Nilius B (2007) Bimodal action of menthol on the transient receptor potential channel TRPA1. J Neurosci 27(37):9874–9884

Kawabata F, Inoue N, Masamoto Y, Matsumura S, Kimura W, Kadowaki M, Higashi T, Tominaga M, Inoue K, Fushiki T (2009) Non-pungent capsaicin analogs (capsinoids) increase metabolic rate and enhance thermogenesis via gastrointestinal TRPV1 in mice. Biosci Biotechnol Biochem 73(12):2690–2697

Kehrl W, Sonnemann U, Dethlefsen U (2004) Therapy for acute nonpurulent rhinosinusitis with cineole: results of a double-blind, randomized, placebo-controlled trial. Laryngoscope 114 (4):738–742

Kennedy WR, Vanhove GF, Lu SP, Tobias J, Bley KR, Walk D, Wendelschafer-Crabb G, Simone DA, Selim MM (2010) A randomized, controlled, open-label study of the long-term effects of NGX-4010, a high-concentration capsaicin patch, on epidermal nerve fiber density and sensory function in healthy volunteers. J Pain 11(6):579–587

Kim BJ, Nam JH, Kwon YK, So I, Kim SJ (2013) The role of waixenicin A as transient receptor potential melastatin 7 blocker. Basic Clin Pharmacol Toxicol 112(2):83–89

Kissin I, Szallasi A (2011) Therapeutic targeting of TRPV1 by resiniferatoxin, from preclinical studies to clinical trials. Curr Top Med Chem 11(17):2159–2170

Klionsky L, Tamir R, Gao B, Wang W, Immke DC, Nishimura N, Gavva NR (2007) Species-specific pharmacology of Trichloro(sulfanyl)ethyl benzamides as transient receptor potential ankyrin 1 (TRPA1) antagonists. Mol Pain 3:39–48

Kobata K, Sutoh K, Todo T, Yazawa S, Iwai K, Watanabe T (1999) Nordihydrocapsiate, a new capsinoid from the fruits of a nonpungent pepper, capsicum annuum. J Nat Prod 62(2):335–336

Kobayashi Y (2003) The nociceptive and anti-nociceptive effects of evodiamine from fruits of Evodia rutaecarpa in mice. Planta Med 69(5):425–428

Kobayashi Y, Nakano Y, Kizaki M, Hoshikuma K, Yokoo Y, Kamiya T (2001) Capsaicin-like anti-obese activities of evodiamine from fruits of Evodia rutaecarpa, a vanilloid receptor agonist. Planta Med 67(7):628–633

Kobayashi K, Fukuoka T, Obata K, Yamanaka H, Dai Y, Tokunaga A, Noguchi K (2005) Distinct expression of TRPM8, TRPA1, and TRPV1 mRNAs in rat primary afferent neurons with adelta/c-fibers and colocalization with trk receptors. J Comp Neurol 493(4):596–606

Koltzenburg M, Lundberg LE, Torebjork HE (1992) Dynamic and static components of mechanical hyperalgesia in human hairy skin. Pain 51(2):207–219

Koo JY, Jang Y, Cho H, Lee CH, Jang KH, Chang YH, Shin J, Oh U (2007) Hydroxy-alpha-sanshool activates TRPV1 and TRPA1 in sensory neurons. Eur J Neurosci 26(5):1139–1147

Kujumgiev A, Tsvetkova I, Serkedjieva Y, Bankova V, Christov R, Popov S (1999) Antibacterial, antifungal and antiviral activity of propolis of different geographic origin. J Ethnopharmacol 64(3):235–240

Kunkler PE, Ballard CJ, Oxford GS, Hurley JH (2011) TRPA1 receptors mediate environmental irritant-induced meningeal vasodilatation. Pain 152(1):38–44

Lahlou S, Figueiredo AF, Magalhaes PJC, Leal-Cardoso JH (2002) Cardiovascular effects of 1,8-cineole, a terpenoid oxide present in many plant essential oils, in normotensive rats. Can J Physiol Pharmacol 80(12):1125–1131

Lang Y, Kisaka H, Sugiyama R, Nomura K, Morita A, Watanabe T, Tanaka Y, Yazawa S, Miwa T (2009) Functional loss of pAMT results in biosynthesis of capsinoids, capsaicinoid analogs, in Capsicum annuum cv. CH-19 Sweet. Plant J 59(6):953–961

Lawson LD, Block E (1997) Comments on garlic chemistry: stability of S-(2-propenyl) 2-propene-1-sulfinothioate (allicin) in blood, solvents, and simulated physiological fluids. J Agric Food Chem 45(2):542–542

Lee E, Surh YJ (1998) Induction of apoptosis in HL-60 cells by pungent vanilloids, [6]-gingerol and [6]-paradol. Cancer Lett 134(2):163–168

Lee YS, Nam DH, Kim JA (2000) Induction of apoptosis by capsaicin in A172 human glioblastoma cells. Cancer Lett 161(1):121–130

Lee EB, Li DW, Hyun JE, Kim IH, Whang WK (2001) Anti-inflammatory activity of methanol extract of Kalopanax pictus bark and its fractions. J Ethnopharmacol 77(2–3):197–201

Lee N, Chen J, Sun L, Wu S, Gray KR, Rich A, Huang M, Lin JH, Feder JN, Janovitz EB, Levesque PC, Blanar MA (2003) Expression and characterization of human transient receptor potential melastatin 3 (hTRPM3). J Biol Chem 278(23):20890–20897

Lee SP, Buber MT, Yang Q, Cerne R, Cortes RY, Sprous DG, Bryant RW (2008) Thymol and related alkyl phenols activate the hTRPA1 channel. Br J Pharmacol 153(8):1739–1749

Leuenroth SJ, Bencivenga N, Chahboune H, Hyder F, Crews CM (2010) Triptolide reduces cyst formation in a neonatal to adult transition Pkd1 model of ADPKD. Nephrol Dial Transplant 25 (7):2187–2194

Leuner K, Kazanski V, Muller M, Essin K, Henke B, Gollasch M, Harteneck C, Muller WE (2007) Hyperforin–a key constituent of St. John's wort specifically activates TRPC6 channels. FASEB J 21(14):4101–4111

Leuner K, Heiser JH, Derksen S, Mladenov MI, Fehske CJ, Schubert R, Gollasch M, Schneider G, Harteneck C, Chatterjee SS, Muller WE (2010) Simple 2,4-diacylphloroglucinols as classic transient receptor potential-6 activators-identification of a novel pharmacophore. Mol Pharmacol 77(3):368–377

Leuner K, Li W, Amaral MD, Rudolph S, Calfa G, Schuwald AM, Harteneck C, Inoue T, Pozzo-Miller L (2013) Hyperforin modulates dendritic spine morphology in hippocampal pyramidal neurons by activating Ca2+-permeable TRPC6 channels. Hippocampus 23(1):40–52

Leung FW (2008) Capsaicin-sensitive intestinal mucosal afferent mechanism and body fat distribution. Life Sci 83(1–2):1–5

Li YJ, Peng J (2002) The cardioprotection of calcitonin gene-related peptide-mediated preconditioning. Eur J Pharmacol 442(3):173–177

Li HS, Xu XZ, Montell C (1999) Activation of a TRPC3-dependent cation current through the neurotrophin BDNF. Neuron 24(1):261–273

Li Y, Jia YC, Cui K, Li N, Zheng ZY, Wang YZ, Yuan XB (2005) Essential role of TRPC channels in the guidance of nerve growth cones by brain-derived neurotrophic factor. Nature 434 (7035):894–898

Li XR, Liang YZ, Guo FQ (2006) Analysis of volatile oil in Rhizoma Ligustici Chuanxiong-Radix Paeoniae Rubra by gas chromatography-mass spectrometry and chemometric resolution. Acta Pharmacol Sin 27(4):491–498

Li D, Zhang XJ, Chen L, Yang Z, Deng HW, Peng J, Li YJ (2009) Calcitonin gene-related peptide mediates the cardioprotective effects of rutaecarpine against ischaemia-reperfusion injury in spontaneously hypertensive rats. Clin Exp Pharmacol Physiol 36(7):662–667

Li Q, Zhou XD, Kolosov VP, Perelman JM (2013) Salidroside reduces cold-induced mucin production by inhibiting TRPM8 activation. Int J Mol Med 32(3):637–646

Liedtke W, Choe Y, Marti-Renom MA, Bell AM, Denis CS, Sali A, Hudspeth AJ, Friedman JM, Heller S (2000) Vanilloid receptor-related osmotically activated channel (VR-OAC), a candidate vertebrate osmoreceptor. Cell 103(3):525–535

Lima PR, de Melo TS, Carvalho KM, de Oliveira IB, Arruda BR, de Castro Brito GA, Rao VS, Santos FA (2013) 1,8-cineole (eucalyptol) ameliorates cerulein-induced acute pancreatitis via modulation of cytokines, oxidative stress and NF-kappaB activity in mice. Life Sci 92(24–26):1195–1201

Lin YS, Hsu CC, Bien MY, Hsu HC, Weng HT, Kou YR (2010) Activations of TRPA1 and P2X receptors are important in ROS-mediated stimulation of capsaicin-sensitive lung vagal afferents by cigarette smoke in rats. J Appl Physiol 108(5):1293–1303

Lin Y, Zhang JC, Fu J, Chen F, Wang J, Wu ZL, Yuan SY (2013) Hyperforin attenuates brain damage induced by transient middle cerebral artery occlusion (MCAO) in rats via inhibition of TRPC6 channels degradation. J Cereb Blood Flow Metab 33(2):253–262

Linde K, Berner M, Egger M, Mulrow C (2005) St John's wort for depression: meta-analysis of randomised controlled trials. Br J Psychiatry 186:99–107

Linde K, Berner MM, Kriston L (2008) St John's wort for major depression. Cochrane Database Syst Rev 4, CD000448

Lindstrom J (1997) Nicotinic acetylcholine receptors in health and disease. Mol Neurobiol 15 (2):193–222

List PH, Hackenberg H (1969) Velleral and iso-velleral, acrid testing compounds from Lactarius vellereus FRIES. 17. Mushroom contents. Arch Pharm Ber Dtsch Pharm Ges 302(2):125–143

Liu L, Simon SA (1996a) Capsaicin-induced currents with distinct desensitization and Ca2+ dependence in rat trigeminal ganglion cells. J Neurophysiol 75(4):1503–1514

Liu L, Simon SA (1996b) Similarities and differences in the currents activated by capsaicin, piperine, and zingerone in rat trigeminal ganglion cells. J Neurophysiol 76(3):1858–1869

Liu L, Lo Y, Chen I, Simon SA (1997) The responses of rat trigeminal ganglion neurons to capsaicin and two nonpungent vanilloid receptor agonists, olvanil and glyceryl nonamide. J Neurosci 17(11):4101–4111

Liu L, Welch JM, Erickson RP, Reinhart PH, Simon SA (2000) Different responses to repeated applications of zingerone in behavioral studies, recordings from intact and cultured TG neurons, and from VR1 receptors. Physiol Behav 69(1–2):177–186

Liu L, Zhu W, Zhang ZS, Yang T, Grant A, Oxford G, Simon SA (2004) Nicotine inhibits voltage-dependent sodium channels and sensitizes vanilloid receptors. J Neurophysiol 91(4):1482–1491

Liu X, Chen S, Zhang Y, Zhang F (2006) Modulation of dragon's blood on tetrodotoxin-resistant sodium currents in dorsal root ganglion neurons and identification of its material basis for efficacy. Sci China C Life Sci 49(3):274–285

Liu BY, Tsai TL, Ho CY, Lu SH, Lai CJ, Kou YR (2013) Role of TRPA1 and TRPV1 in the ROS-dependent sensory irritation of superior laryngeal capsaicin-sensitive afferents by cigarette smoke in anesthetized rats. Pulm Pharmacol Ther 26(3):364–372

Lorenzen K, Anke T, Anders U, Hindermayr H, Hansske F (1994) Two inhibitors of platelet aggregation from a Panus species (Basidiomycetes). Z Naturforsch C 49(1–2):132–138

Lu Q, Qiu TQ, Yang H (2006) Ligustilide inhibits vascular smooth muscle cells proliferation. Eur J Pharmacol 542(1–3):136–140

Luo XJ, Peng J, Li YJ (2011) Recent advances in the study on capsaicinoids and capsinoids. Eur J Pharmacol 650(1):1–7

MacCulloch JA. Encyclopaedia Britannica. Encyclopaedia of religion and ethics. Charles Scribner's Sons New York, New York

MacDonald E, Kobilka BK, Scheinin M (1997) Gene targeting–homing in on alpha 2-adrenoceptor-subtype function. Trends Pharmacol Sci 18(6):211–219

Macpherson LJ, Geierstanger BH, Viswanath V, Bandell M, Eid SR, Hwang S, Patapoutian A (2005) The pungency of garlic: activation of TRPA1 and TRPV1 in response to allicin. Curr Biol 15(10):929–934

Macpherson LJ, Hwang SW, Miyamoto T, Dubin AE, Patapoutian A, Story GM (2006) More than cool: promiscuous relationships of menthol and other sensory compounds. Mol Cell Neurosci 32(4):335–343

Macpherson LJ, Dubin AE, Evans MJ, Marr F, Schultz PG, Cravatt BF, Patapoutian A (2007) Noxious compounds activate TRPA1 ion channels through covalent modification of cysteines. Nature 445(7127):541–545

Maggi CA, Patacchini R, Tramontana M, Amann R, Giuliani S, Santicioli P (1990) Similarities and differences in the action of resiniferatoxin and capsaicin on central and peripheral endings of primary sensory neurons. Neuroscience 37(2):531–539

Magnusson G, Thoren S, Wickberg G (1972) Structure of a sesquiterpene dialdehyde from Lactarius by computer stimulation of the NMR spectrum. Tetrahedron Lett 13:1105–1108

Malheiros A, Filho VC, Schmitt CB, Santos ARS, Scheidt C, Calixto JB, Monache FD, Yunes RA (2001) A sesquiterpene drimane with antinociceptive activity from Drimys winteri bark. Phytochemistry 57(1):103–107

Malmberg AB, Mizisin AP, Calcutt NA, von Stein T, Robbins WR, Bley KR (2004) Reduced heat sensitivity and epidermal nerve fiber immunostaining following single applications of a high-concentration capsaicin patch. Pain 111(3):360–367

Marsakova L, Touska F, Krusek J, Vlachova V (2012) Pore helix domain is critical to camphor sensitivity of transient receptor potential vanilloid 1 channel. Anesthesiology 116(4):903–917

Marsh SJ, Stansfeld CE, Brown DA, Davey R, McCarthy D (1987) The mechanism of action of capsaicin on sensory C-type neurons and their axons in vitro. Neuroscience 23(1):275–289

Mcgarvey DJ, Croteau R (1995) Terpenoid metabolism. Plant Cell 7(7):1015–1026

McKemy DD, Neuhausser WM, Julius D (2002) Identification of a cold receptor reveals a general role for TRP channels in thermosensation. Nature 416(6876):52–58

McNamara FN, Randall A, Gunthorpe MJ (2005) Effects of piperine, the pungent component of black pepper, at the human vanilloid receptor (TRPV1). Br J Pharmacol 144(6):781–790

Mechoulam R, Parker LA, Gallily R (2002) Cannabidiol: an overview of some pharmacological aspects. J Clin Pharmacol 42(11 Suppl):11S–19S

Mei QB, Tao JY, Cui B (1991) Advances in the pharmacological studies of Radix Angelica-Sinensis (Oliv) Diels (Chinese Danggui). Chin Med J (Engl) 104(9):776–781

Mendes GL, Santos AR, Campos MM, Tratsk KS, Yunes RA, Cechinel Filho V, Calixto JB (1998) Anti-hyperalgesic properties of the extract and of the main sesquiterpene polygodial isolated from the barks of Drymis winteri (Winteraceae). Life Sci 63(5):369–381

Mogil JS, Shin YH, McCleskey EW, Kim SC, Nah SY (1998) Ginsenoside Rf, a trace component of ginseng root, produces antinociception in mice. Brain Res 792(2):218–228

Montell C (2005) The TRP superfamily of cation channels. Sci STKE 272:3–27

Moqrich A, Hwang SW, Earley TJ, Petrus MJ, Murray AN, Spencer KS, Andahazy M, Story GM, Patapoutian A (2005) Impaired thermosensation in mice lacking TRPV3, a heat and camphor sensor in the skin. Science 307(5714):1468–1472

Mori A, Lehmann S, O'Kelly J, Kumagai T, Desmond JC, Pervan M, McBride WH, Kizaki M, Koeffler HP (2006) Capsaicin, a component of red peppers, inhibits the growth of androgen-independent, p53 mutant prostate cancer cells. Cancer Res 66(6):3222–3229

Moussaieff A, Mechoulam R (2009) Boswellia resin: from religious ceremonies to medical uses; a review of in-vitro, in-vivo and clinical trials. J Pharm Pharmacol 61(10):1281–1293

Moussaieff A, Rimmerman N, Bregman T, Straiker A, Felder CC, Shoham S, Kashman Y, Huang SM, Lee H, Shohami E, Mackie K, Caterina MJ, Walker JM, Fride E, Mechoulam R (2008) Incensole acetate, an incense component, elicits psychoactivity by activating TRPV3 channels in the brain. FASEB J 22(8):3024–3034

Moussaieff A, Yu J, Zhu H, Gattoni-Celli S, Shohami E, Kindy MS (2012) Protective effects of incensole acetate on cerebral ischemic injury. Brain Res 1443:89–97
Mozsik G, Szolcsanyi J, Domotor A (2007) Capsaicin research as a new tool to approach of the human gastrointestinal physiology, pathology and pharmacology. Inflammopharmacology 15 (6):232–245
Mueller-Seitz E, Hiepler C, Petz M (2008) Chili pepper fruits: content and pattern of capsaicinoids in single fruits of different ages. J Agric Food Chem 56(24):12114–12121
Muller WE (2003) Current St John's wort research from mode of action to clinical efficacy. Pharmacol Res 47(2):101–109
Nabissi M, Morelli MB, Santoni M, Santoni G (2013) Triggering of the TRPV2 channel by cannabidiol sensitizes glioblastoma cells to cytotoxic chemotherapeutic agents. Carcinogenesis 34(1):48–57
Nagata K, Duggan A, Kumar G, Garcia-Anoveros J (2005) Nociceptor and hair cell transducer properties of TRPA1, a channel for pain and hearing. J Neurosci 25(16):4052–4061
Nah JJ, Hahn JH, Chung S, Choi S, Kim YI, Nah SY (2000) Effect of ginsenosides, active components of ginseng, on capsaicin-induced pain-related behavior. Neuropharmacology 39 (11):2180–2184
Namer B, Seifert F, Handwerker HO, Maihofner C (2005) TRPA1 and TRPM8 activation in humans: effects of cinnamaldehyde and menthol. Neuroreport 16(9):955–959
Nassini R, Materazzi S, Vriens J, Prenen J, Benemei S, De Siena G, la Marca G, Andre E, Preti D, Avonto C, Sadofsky L, Di Marzo V, De Petrocellis L, Dussor G, Porreca F, Taglialatela-Scafati O, Appendino G, Nilius B, Geppetti P (2012) The 'headache tree' via umbellulone and TRPA1 activates the trigeminovascular system. Brain 135(Pt 2):376–390
Nauli SM, Alenghat FJ, Luo Y, Williams E, Vassilev P, Li X, Elia AE, Lu W, Brown EM, Quinn SJ, Ingber DE, Zhou J (2003) Polycystins 1 and 2 mediate mechanosensation in the primary cilium of kidney cells. Nat Genet 33(2):129–137
Naylor J, Li J, Milligan CJ, Zeng F, Sukumar P, Hou B, Sedo A, Yuldasheva N, Majeed Y, Beri D, Jiang S, Seymour VA, McKeown L, Kumar B, Harteneck C, O'Regan D, Wheatcroft SB, Kearney MT, Jones C, Porter KE, Beech DJ (2010) Pregnenolone sulphate- and cholesterol-regulated TRPM3 channels coupled to vascular smooth muscle secretion and contraction. Circ Res 106(9):1507–1515
Neeper MP, Liu Y, Hutchinson TL, Wang Y, Flores CM, Qin N (2007) Activation properties of heterologously expressed mammalian TRPV2: evidence for species dependence. J Biol Chem 282(21):15894–15902
Newall CA, Anderson LA, Phillipson JD (1996) Herbal medicines—a guide for health-care professionals. Great Britain at the University Press, Cambridge
Nides M (2008) Update on pharmacologic options for smoking cessation treatment. Am J Med 121 (4 Suppl 1):S20–31
Nilius B, Appendino G (2011) Tasty and healthy TR(i)Ps. The human quest for culinary pungency. EMBO Rep 12(11):1094–1101
Nilius B, Appendino G (2013) Spices: the savory and beneficial science of pungency. Rev Physiol Biochem Pharmacol 164:1–76
Nilius B, Owsianik G (2011) The transient receptor potential family of ion channels. Genome Biol 12(3):218
Nilius B, Prenen J, Wissenbach U, Bodding M, Droogmans G (2001) Differential activation of the volume-sensitive cation channel TRP12 (OTRPC4) and volume-regulated anion currents in HEK-293 cells. Pflugers Arch 443(2):227–233
Nilius B, Vriens J, Prenen J, Droogmans G, Voets T (2004) TRPV4 calcium entry channel: a paradigm for gating diversity. Am J Physiol Cell Physiol 286(2):C195–205
Nilius B, Appendino G, Owsianik G (2012) The transient receptor potential channel TRPA1: from gene to pathophysiology. Pflugers Arch 464(5):425–458
Nilius B, Biro T, Owsianik G (2013) TRPV3: time to decipher a poorly understood family member! J Physiol 592:295–304

Njan AA, Adzu B, Agaba AG, Byarugaba D, Diaz-Llera S, Bangsberg DR (2008) The analgesic and antiplasmodial activities and toxicology of Vernonia amygdalina. J Med Food 11(3):574–581

Nonomura S, Kanagawa H, Makimoto A (1963) Chemical constituents of polygonaceous plants. I. Studies on the components of Ko-J O-Kon. (Polygonum Cuspidatum Sieb. Et Zucc.). Yakugaku Zasshi 83:988–990

O'Mullane LM, Keast JR, Osborne PB (2013) Co-cultures provide a new tool to probe communication between adult sensory neurons and urothelium. J Urol 190:737–745

O'Neil RG, Brown RC (2003) The vanilloid receptor family of calcium-permeable channels: molecular integrators of microenvironmental stimuli. News Physiol Sci 18:226–231

Oberwinkler J, Lis A, Giehl KM, Flockerzi V, Philipp SE (2005) Alternative splicing switches the divalent cation selectivity of TRPM3 channels. J Biol Chem 280(23):22540–22548

Oehler B, Scholze A, Schaefer M, Hill K (2012) TRPA1 is functionally expressed in melanoma cells but is not critical for impaired proliferation caused by allyl isothiocyanate or cinnamaldehyde. Naunyn Schmiedebergs Arch Pharmacol 385(6):555–563

Oh SH, Kim YS, Lim SC, Hou YF, Chang IY, You HJ (2008) Dihydrocapsaicin (DHC), a saturated structural analog of capsaicin, induces autophagy in human cancer cells in a catalase-regulated manner. Autophagy 4(8):1009–1019

Ohkubo T, Kitamura K (1997) Eugenol activates Ca(2+)-permeable currents in rat dorsal root ganglion cells. J Dent Res 76(11):1737–1744

Ohkubo T, Shibata M (1997) The selective capsaicin antagonist capsazepine abolishes the antinociceptive action of eugenol and guaiacol. J Dent Res 76(4):848–851

Ohta T, Komatsu R, Imagawa T, Otsuguro K, Ito S (2005) Molecular cloning, functional characterization of the porcine transient receptor potential V1 (pTRPV1) and pharmacological comparison with endogenous pTRPV1. Biochem Pharmacol 71(1–2):173–187

Ohta T, Imagawa T, Ito S (2007) Novel agonistic action of mustard oil on recombinant and endogenous porcine transient receptor potential V1 (pTRPV1) channels. Biochem Pharmacol 73(10):1646–1656

Okada T, Shimizu S, Wakamori M, Maeda A, Kurosaki T, Takada N, Imoto K, Mori Y (1998) Molecular cloning and functional characterization of a novel receptor-activated TRP Ca2+ channel from mouse brain. J Biol Chem 273(17):10279–10287

Onogi T, Minami M, Kuraishi Y, Satoh M (1992) Capsaicin-like effect of (6)-shogaol on substance P-containing primary afferents of rats: a possible mechanism of its analgesic action. Neuropharmacology 31(11):1165–1169

Park YK, Alencar SM, Aguiar CL (2002) Botanical origin and chemical composition of Brazilian propolis. J Agric Food Chem 50(9):2502–2506

Park CK, Kim K, Jung SJ, Kim MJ, Ahn DK, Hong SD, Kim JS, Oh SB (2009) Molecular mechanism for local anesthetic action of eugenol in the rat trigeminal system. Pain 144(1–2):84–94

Park SJ, Ahmad F, Philp A, Baar K, Williams T, Luo H, Ke H, Rehmann H, Taussig R, Brown AL, Kim MK, Beaven MA, Burgin AB, Manganiello V, Chung JH (2012) Resveratrol ameliorates aging-related metabolic phenotypes by inhibiting cAMP phosphodiesterases. Cell 148(3):421–433

Parnas M, Peters M, Dadon D, Lev S, Vertkin I, Slutsky I, Minke B (2009) Carvacrol is a novel inhibitor of Drosophila TRPL and mammalian TRPM7 channels. Cell Calcium 45(3):300–309

Pastrana-Bonilla E, Akoh CC, Sellappan S, Krewer G (2003) Phenolic content and antioxidant capacity of muscadine grapes. J Agric Food Chem 51(18):5497–5503

Patil MJ, Belugin S, Akopian AN (2011) Chronic alteration in phosphatidylinositol 4,5-biphosphate levels regulates capsaicin and mustard oil responses. J Neurosci Res 89 (6):945–954

Paul M, Jauch J (2012) Efficient preparation of incensole and incensole acetate, and quantification of these bioactive diterpenes in Boswellia papyrifera by a RP-DAD-HPLC method. Nat Prod Commun 7(3):283–288

Paulino N, Abreu SR, Uto Y, Koyama D, Nagasawa H, Hori H, Dirsch VM, Vollmar AM, Scremin A, Bretz WA (2008) Anti-inflammatory effects of a bioavailable compound, Artepillin C, in Brazilian propolis. Eur J Pharmacol 587(1–3):296–301

Pearce LV, Petukhov PA, Szabo T, Kedei N, Bizik F, Kozikowski AP, Blumberg PM (2004) Evodiamine functions as an agonist for the vanilloid receptor TRPV1. Org Biomol Chem 2 (16):2281–2286

Peattie DC (1953) A natural history of western trees. Houghton Mifflin, Boston

Pedersen SF, Owsianik G, Nilius B (2005) TRP channels: an overview. Cell Calcium 38(3–4):233–252

Peier AM, Moqrich A, Hergarden AC, Reeve AJ, Andersson DA, Story GM, Earley TJ, Dragoni I, McIntyre P, Bevan S, Patapoutian A (2002) A TRP channel that senses cold stimuli and menthol. Cell 108(5):705–715

Peng J, Li YJ (2010) The vanilloid receptor TRPV1: role in cardiovascular and gastrointestinal protection. Eur J Pharmacol 627(1–3):1–7

Peng J, Lu R, Deng HW, Li YJ (2002) Involvement of alpha-calcitonin gene-related peptide in monophosphoryl lipid A-induced delayed preconditioning in rat hearts. Eur J Pharmacol 436 (1–2):89–96

Peralvarez-Marin A, Donate-Macian P, Gaudet R (2013) What do we know about the transient receptor potential vanilloid 2 (TRPV2) ion channel? FEBS J 280:5471–5487

Peters M, Trembovler V, Alexandrovich A, Parnas M, Birnbaumer L, Minke B, Shohami E (2012) Carvacrol together with TRPC1 elimination improve functional recovery after traumatic brain injury in mice. J Neurotrauma 29(18):2831–2834

Pham-Marcou TA, Beloeil H, Sun X, Gentili M, Yaici D, Benoit G, Benhamou D, Mazoit JX (2008) Antinociceptive effect of resveratrol in carrageenan-evoked hyperalgesia in rats: prolonged effect related to COX-2 expression impairment. Pain 140(2):274–283

Pittenger C, Duman RS (2008) Stress, depression, and neuroplasticity: a convergence of mechanisms. Neuropsychopharmacology 33(1):88–109

Ponhold J (1944) Schwere ekzematöse Veränderungen durch Lactarius vellereus Fries. Arch Toxicol 13:33–36

Prescott J, Swain-Campbell N (2000) Responses to repeated oral irritation by capsaicin, cinnamaldehyde and ethanol in PROP tasters and non-tasters. Chem Senses 25(3):239–246

Qamar S, Vadivelu M, Sandford R (2007) TRP channels and kidney disease: lessons from polycystic kidney disease. Biochem Soc Trans 35(Pt 1):124–128

Qiao S, Li W, Tsubouchi R, Haneda M, Murakami K, Yoshino M (2005) Involvement of peroxynitrite in capsaicin-induced apoptosis of C6 glioma cells. Neurosci Res 51(2):175–183

Qin N, Neeper MP, Liu Y, Hutchinson TL, Lubin ML, Flores CM (2008) TRPV2 is activated by cannabidiol and mediates CGRP release in cultured rat dorsal root ganglion neurons. J Neurosci 28(24):6231–6238

Qiu D, Kao PN (2003) Immunosuppressive and anti-inflammatory mechanisms of triptolide, the principal active diterpenoid from the Chinese medicinal herb Tripterygium wilfordii Hook. f. Drugs R D 4(1):1–18

Ralevic V, Jerman JC, Brough SJ, Davis JB, Egerton J, Smart D (2003) Pharmacology of vanilloids at recombinant and endogenous rat vanilloid receptors. Biochem Pharmacol 65 (1):143–151

Reeh PW, Kocher L, Jung S (1986) Does neurogenic inflammation alter the sensitivity of unmyelinated nociceptors in the rat? Brain Res 384(1):42–50

Reid G, Flonta ML (2002) Ion channels activated by cold and menthol in cultured rat dorsal root ganglion neurones. Neurosci Lett 324(2):164–168

Reid G, Babes A, Pluteanu F (2002) A cold- and menthol-activated current in rat dorsal root ganglion neurones: properties and role in cold transduction. J Physiol 545(Pt 2):595–614

Reilly RM, Kym PR (2011) Analgesic potential of TRPV3 antagonists. Curr Top Med Chem 11 (17):2210–2215

Reinbach HC, Smeets A, Martinussen T, Moller P, Westerterp-Plantenga MS (2009) Effects of capsaicin, green tea and CH-19 sweet pepper on appetite and energy intake in humans in negative and positive energy balance. Clin Nutr 28(3):260–265
Riera CE, Menozzi-Smarrito C, Affolter M, Michlig S, Munari C, Robert F, Vogel H, Simon SA, le Coutre J (2009) Compounds from Sichuan and Melegueta peppers activate, covalently and non-covalently, TRPA1 and TRPV1 channels. Br J Pharmacol 157(8):1398–1409
Risso WE, Scarminio IS, Moreira EG (2010) Antinociceptive and acute toxicity evaluation of Vernonia condensata Baker leaves extracted with different solvents and their mixtures. Indian J Exp Biol 48(8):811–816
Roberts K, Shenoy R, Anand P (2005) A novel human volunteer pain model using contact heat evoked potentials (CHEP) following topical skin application of transient receptor potential agonists capsaicin, menthol and cinnamaldehyde. J Clin Neurosci 18(7):926–932
Rocco F (2003) Quinine: malaria and the quest for a cure that changed the world. HarperCollins, New York
Role LW, Berg DK (1996) Nicotinic receptors in the development and modulation of CNS synapses. Neuron 16(6):1077–1085
Salminen A, Lehtonen M, Suuronen T, Kaarniranta K, Huuskonen J (2008) Terpenoids: natural inhibitors of NF-kappaB signaling with anti-inflammatory and anticancer potential. Cell Mol Life Sci 65(19):2979–2999
Santos FA, Rao VSN (2000) Antiinflammatory and antinociceptive effects of 1,8-cineole a terpenoid oxide present in many plant essential oils. Phytother Res 14(4):240–244
Sarer E, Scheffer JJ, Baerheim Svendsen A (1982) Monoterpenes in the essential oil of Origanum majorana. Planta Med 46(4):236–239
Schaefer M, Plant TD, Obukhov AG, Hofmann T, Gudermann T, Schultz G (2000) Receptor-mediated regulation of the nonselective cation channels TRPC4 and TRPC5. J Biol Chem 275 (23):17517–17526
Schmitz C, Perraud AL, Johnson CO, Inabe K, Smith MK, Penner R, Kurosaki T, Fleig A, Scharenberg AM (2003) Regulation of vertebrate cellular Mg2+ homeostasis by TRPM7. Cell 114(2):191–200
Shao M, Qu K, Liu K, Zhang YQ, Zhang L, Lian ZQ, Chen TT, Liu JF, Wu AL, Tang Y, Zhu HB (2011) Effects of ligustilide on lipopolysaccharide-induced endotoxic shock in rabbits. Planta Med 77(8):809–816
Sharma S, Chopra K, Kulkarni SK (2007a) Effect of insulin and its combination with resveratrol or curcumin in attenuation of diabetic neuropathic pain: participation of nitric oxide and TNF-alpha. Phytother Res 21(3):278–283
Sharma S, Kulkarni SK, Chopra K (2007b) Effect of resveratrol, a polyphenolic phytoalexin, on thermal hyperalgesia in a mouse model of diabetic neuropathic pain. Fundam Clin Pharmacol 21(1):89–94
Sherkheli MA, Vogt-Eisele AK, Weber K, Hatt H (2013) Camphor modulates TRPV3 cation channels activity by interacting with critical pore-region cysteine residues. Pak J Pharm Sci 26 (3):431–438
Shibata S, Tanaka O, Shoji J, Saito H (1985) Chemistry and pharmacology of Panax. In: Wagner HHH, Farnsworth NR (eds) Economic and medicinal plant research. Academic, London, pp 217–284
Shin KO, Moritani T (2007) Alterations of autonomic nervous activity and energy metabolism by capsaicin ingestion during aerobic exercise in healthy men. J Nutr Sci Vitaminol (Tokyo) 53 (2):124–132
Shintaku K, Uchida K, Suzuki Y, Zhou Y, Fushiki T, Watanabe T, Yazawa S, Tominaga M (2012) Activation of transient receptor potential A1 by a non-pungent capsaicin-like compound, capsiate. Br J Pharmacol 165(5):1476–1486
Shirakawa H, Yamaoka T, Sanpei K, Sasaoka H, Nakagawa T, Kaneko S (2008) TRPV1 stimulation triggers apoptotic cell death of rat cortical neurons. Biochem Biophys Res Commun 377 (4):1211–1215

Siemens J, Zhou S, Piskorowski R, Nikai T, Lumpkin EA, Basbaum AI, King D, Julius D (2006) Spider toxins activate the capsaicin receptor to produce inflammatory pain. Nature 444 (7116):208–212

Singer A, Wonnemann M, Muller WE (1999) Hyperforin, a major antidepressant constituent of St. John's Wort, inhibits serotonin uptake by elevating free intracellular Na+1. J Pharmacol Exp Ther 290(3):1363–1368

Smart D, Jerman JC, Gunthorpe MJ, Brough SJ, Ranson J, Cairns W, Hayes PD, Randall AD, Davis JB (2001) Characterisation using FLIPR of human vanilloid VR1 receptor pharmacology. Eur J Pharmacol 417(1–2):51–58

Smith PL, Maloney KN, Pothen RG, Clardy J, Clapham DE (2006) Bisandrographolide from Andrographis paniculata activates TRPV4 channels. J Biol Chem 281(40):29897–29904

Soleas GJ, Diamandis EP, Goldberg DM (1997) Resveratrol: a molecule whose time has come? And gone? Clin Biochem 30(2):91–113

Son HJ, Kim MJ, Park JH, Ishii S, Misaka T, Rhyu MR (2012a) Methyl syringate, a low-molecular-weight phenolic ester, as an activator of the chemosensory ion channel TRPA1. Arch Pharm Res 35(12):2211–2218

Son HJ, Kim Y, Misaka T, Noh BS, Rhyu MR (2012b) Activation of the chemosensory ion channels TRPA1 and TRPV1 by hydroalcohol extract of Kalopanax pictus leaves. Biomol Ther 20(6):550–555

Stead LF, Perera R, Bullen C, Mant D, Lancaster T (2008) Nicotine replacement therapy for smoking cessation. Cochrane Database Syst Rev 1, CD000146

Sterner O, Szallasi A (1999) Novel natural vanilloid receptor agonists: new therapeutic targets for drug development. Trends Pharmacol Sci 20(11):459–465

Sticher O (1998) Biochemical, pharmaceutical, and medical perspectives of ginseng. In: Lawson LDBR (ed) Phytomedicines of Europe. American Chemical Society, Washington, DC, pp 221–240

Stojanovic S, Sprinz H, Brede O (2001) Efficiency and mechanism of the antioxidant action of trans-resveratrol and its analogues in the radical liposome oxidation. Arch Biochem Biophys 391(1):79–89

Story GM, Peier AM, Reeve AJ, Eid SR, Mosbacher J, Hricik TR, Earley TJ, Hergarden AC, Andersson DA, Hwang SW, McIntyre P, Jegla T, Bevan S, Patapoutian A (2003) ANKTM1, a TRP-like channel expressed in nociceptive neurons, is activated by cold temperatures. Cell 112 (6):819–829

Stotz SC, Vriens J, Martyn D, Clardy J, Clapham DE (2008) Citral sensing by transient [corrected] receptor potential channels in dorsal root ganglion neurons. PLoS One 3(5):e2082

Straub I, Mohr F, Stab J, Konrad M, Philipp SE, Oberwinkler J, Schaefer M (2013) Citrus fruit and fabacea secondary metabolites potently and selectively block TRPM3. Br J Pharmacol 168 (8):1835–1850

Strotmann R, Harteneck C, Nunnenmacher K, Schultz G, Plant TD (2000) OTRPC4, a nonselective cation channel that confers sensitivity to extracellular osmolarity. Nat Cell Biol 2(10):695–702

Sugai E, Morimitsu Y, Kubota K (2005) Quantitative analysis of sanshool compounds in Japanese pepper (*Xanthoxylum piperitum* DC.) and their pungent characteristics. Biosci Biotechnol Biochem 69(10):1958–1962

Surh YJ (2002) More than spice: capsaicin in hot chili peppers makes tumor cells commit suicide. J Natl Cancer Inst 94(17):1263–1265

Szallasi A (1994) The vanilloid (capsaicin) receptor: receptor types and species differences. Gen Pharmacol 25(2):223–243

Szallasi A (1995) Autoradiographic visualization and pharmacological characterization of vanilloid (capsaicin) receptors in several species, including man. Acta Physiol Scand Suppl 629:1–68

Szallasi A (2005) Piperine: researchers discover new flavor in an ancient spice. Trends Pharmacol Sci 26(9):437–439

Szallasi A, Blumberg PM (1990) Resiniferatoxin and its analogs provide novel insights into the pharmacology of the vanilloid (capsaicin) receptor. Life Sci 47(16):1399–1408

Szallasi A, Blumberg PM (1991) Characterization of vanilloid receptors in the dorsal horn of pig spinal cord. Brain Res 547(2):335–338

Szallasi A, Blumberg PM (1999) Vanilloid (Capsaicin) receptors and mechanisms. Pharmacol Rev 51(2):159–212

Szallasi A, Jonassohn M, Acs G, Biro T, Acs P, Blumberg PM, Sterner O (1996) The stimulation of capsaicin-sensitive neurones in a vanilloid receptor-mediated fashion by pungent terpenoids possessing an unsaturated 1,4-dialdehyde moiety. Br J Pharmacol 119(2):283–290

Szallasi A, Biro T, Modarres S, Garlaschelli L, Petersen M, Klusch A, Vidari G, Jonassohn M, De Rosa S, Sterner O, Blumberg PM, Krause JE (1998) Dialdehyde sesquiterpenes and other terpenoids as vanilloids. Eur J Pharmacol 356(1):81–89

Szolcsanyi J, Bartho L (2001) Capsaicin-sensitive afferents and their role in gastroprotection: an update. J Physiol Paris 95(1–6):181–188

Szolcsanyi J, Szallasi A, Szallasi Z, Joo F, Blumberg PM (1990) Resiniferatoxin: an ultrapotent selective modulator of capsaicin-sensitive primary afferent neurons. J Pharmacol Exp Ther 255 (2):923–928

Tai YL, Feng SJ, Ge RL, Du WL, Zhang XX, He ZH, Wang YZ (2008) TRPC6 channels promote dendritic growth via the CaMKIV-CREB pathway. J Cell Sci 121(14):2301–2307

Takaishi M, Fujita F, Uchida K, Yamamoto S, Sawada Shimizu M, Hatai Uotsu C, Shimizu M, Tominaga M (2012) 1,8-cineole, a TRPM8 agonist, is a novel natural antagonist of human TRPA1. Mol Pain 8:86

Talavera K, Gees M, Karashima Y, Meseguer VM, Vanoirbeek JA, Damann N, Everaerts W, Benoit M, Janssens A, Vennekens R, Viana F, Nemery B, Nilius B, Voets T (2009) Nicotine activates the chemosensory cation channel TRPA1. Nat Neurosci 12(10):1293–1299

Tanabe M, Chen YD, Saito K, Kano Y (1993) Cholesterol biosynthesis inhibitory component from Zingiber officinale Roscoe. Chem Pharm Bull (Tokyo) 41(4):710–713

Tang J, Zhang YG, Hartman TG, Rosen RT, Ho CT (1990) Free and glycosidically bound volatile compounds in fresh celery (Apium-Graveolens L). J Agric Food Chem 38(10):1937–1940

Teixeira EW, Negri G, Meira RM, Message D, Salatino A (2005) Plant origin of green propolis: bee behavior, plant anatomy and chemistry. Evid Based Complement Alternat Med 2(1):85–92

Tewksbury JJ, Manchego C, Haak DC, Levey DJ (2006) Where did the chili get its spice? Biogeography of capsaicinoid production in ancestral wild chili species. J Chem Ecol 32 (3):547–564

Thoennissen NH, O'Kelly J, Lu D, Iwanski GB, La DT, Abbassi S, Leiter A, Karlan B, Mehta R, Koeffler HP (2010) Capsaicin causes cell-cycle arrest and apoptosis in ER-positive and -negative breast cancer cells by modulating the EGFR/HER-2 pathway. Oncogene 29 (2):285–296

Thuerauf N, Markovic K, Braun G, Bleich S, Reulbach U, Kornhuber J, Lunkenheimer J (2006) The influence of mecamylamine on trigeminal and olfactory chemoreception of nicotine. Neuropsychopharmacology 31(2):450–461

Tolcher A, Patnaik A, Papadopoulos K, Mays T, Stephan T, Humble DJ, Frohlich MW, Sims RB (2010) Preliminary results from a Phase 1 study of D-3263 HCl, a TRPM8 calcium channel agonist, in patients with advanced cancer. EJC Supplements 8(7):119–119

Tominaga M, Caterina MJ (2004) Thermosensation and pain. J Neurobiol 61(1):3–12

Torres-Lopez JE, Ortiz MI, Castaneda-Hernandez G, Alonso-Lopez R, Asomoza-Espinosa R, Granados-Soto V (2002) Comparison of the antinociceptive effect of celecoxib, diclofenac and resveratrol in the formalin test. Life Sci 70(14):1669–1676

Toth A, Kedei N, Szabo T, Wang Y, Blumberg PM (2002) Thapsigargin binds to and inhibits the cloned vanilloid receptor-1. Biochem Biophys Res Commun 293(2):777–782

Tratsk KS, Campos MM, Vaz ZR, Filho VC, Schlemper V, Yunes RA, Calixto JB (1997) Anti-allergic effects and oedema inhibition caused by the extract of Drymis winteri. Inflamm Res 46 (12):509–514

Treiber K, Singer A, Henke B, Muller WE (2005) Hyperforin activates nonselective cation channels (NSCCs). Br J Pharmacol 145(1):75–83

Trevisan G, Rossato MF, Walker CI, Klafke JZ, Rosa F, Oliveira SM, Tonello R, Guerra GP, Boligon AA, Zanon RB, Athayde ML, Ferreira J (2012) Identification of the plant steroid alpha-spinasterol as a novel transient receptor potential vanilloid 1 antagonist with antinociceptive properties. J Pharmacol Exp Ther 343(2):258–269

Tsagareli MG, Tsiklauri N, Zanotto KL, Carstens MI, Klein AH, Sawyer CM, Gurtskaia G, Abzianidze E, Carstens E (2010) Behavioral evidence of thermal hyperalgesia and mechanical allodynia induced by intradermal cinnamaldehyde in rats. Neurosci Lett 473(3):233–236

Tsavaler L, Shapero MH, Morkowski S, Laus R (2001) Trp-p8, a novel prostate-specific gene, is up-regulated in prostate cancer and other malignancies and shares high homology with transient receptor potential calcium channel proteins. Cancer Res 61(9):3760–3769

Tsiokas L, Arnould T, Zhu C, Kim E, Walz G, Sukhatme VP (1999) Specific association of the gene product of PKD2 with the TRPC1 channel. Proc Natl Acad Sci U S A 96(7):3934–3939

Vieira C, Evangelista S, Cirillo R, Terracciano R, Lippi A, Maggi CA, Manzini S (2000) Antinociceptive activity of ricinoleic acid, a capsaicin-like compound devoid of pungent properties. Eur J Pharmacol 407(1–2):109–116

Vieira C, Fetzer S, Sauer SK, Evangelista S, Averbeck B, Kress M, Reeh PW, Cirillo R, Lippi A, Maggi CA, Manzini S (2001) Pro- and anti-inflammatory actions of ricinoleic acid: similarities and differences with capsaicin. Naunyn Schmiedebergs Arch Pharmacol 364(2):87–95

Voets T, Droogmans G, Wissenbach U, Janssens A, Flockerzi V, Nilius B (2004) The principle of temperature-dependent gating in cold- and heat-sensitive TRP channels. Nature 430 (7001):748–754

Vogt-Eisele AK, Weber K, Sherkheli MA, Vielhaber G, Panten J, Gisselmann G, Hatt H (2007) Monoterpenoid agonists of TRPV3. Br J Pharmacol 151(4):530–540

Vriens J, Owsianik G, Janssens A, Voets T, Nilius B (2007) Determinants of 4 alpha-phorbol sensitivity in transmembrane domains 3 and 4 of the cation channel TRPV4. J Biol Chem 282 (17):12796–12803

Vriens J, Appendino G, Nilius B (2009) Pharmacology of vanilloid transient receptor potential cation channels. Mol Pharmacol 75(6):1262–1279

Wagner TF, Loch S, Lambert S, Straub I, Mannebach S, Mathar I, Dufer M, Lis A, Flockerzi V, Philipp SE, Oberwinkler J (2008) Transient receptor potential M3 channels are ionotropic steroid receptors in pancreatic beta cells. Nat Cell Biol 10(12):1421–1430

Wang L, Hu CP, Deng PY, Shen SS, Zhu HQ, Ding JS, Tan GS, Li YJ (2005) The protective effects of rutaecarpine on gastric mucosa injury in rats. Planta Med 71(5):416–419

Wang J, Du JR, Wang Y, Kuang X, Wang CY (2010) Z-ligustilide attenuates lipopolysaccharide-induced proinflammatory response via inhibiting NF-kappa B pathway in primary rat microglia. Acta Pharmacol Sin 31(7):791–797

Wang S, Lee J, Ro JY, Chung MK (2012) Warmth suppresses and desensitizes damage-sensing ion channel TRPA1. Mol Pain 8:22

Watanabe H, Davis JB, Smart D, Jerman JC, Smith GD, Hayes P, Vriens J, Cairns W, Wissenbach U, Prenen J, Flockerzi V, Droogmans G, Benham CD, Nilius B (2002) Activation of TRPV4 channels (hVRL-2/mTRP12) by phorbol derivatives. J Biol Chem 277(16):13569–13577

Watanabe H, Vriens J, Prenen J, Droogmans G, Voets T, Nilius B (2003) Anandamide and arachidonic acid use epoxyeicosatrienoic acids to activate TRPV4 channels. Nature 424 (6947):434–438

Wei LS, Chen S, Huang XJ, Yao J, Liu XM (2013) Material basis for inhibition of dragon's blood on capsaicin-induced TRPV1 receptor currents in rat dorsal root ganglion neurons. Eur J Pharmacol 702(1–3):275–284

Weinhold P, Gratzke C, Streng T, Stief C, Andersson KE, Hedlund P (2010) TRPA1 receptor induced relaxation of the human urethra involves TRPV1 and cannabinoid receptor mediated signals, and cyclooxygenase activation. J Urol 183(5):2070–2076

Willis DN, Liu B, Ha MA, Jordt SE, Morris JB (2011) Menthol attenuates respiratory irritation responses to multiple cigarette smoke irritants. FASEB J 25(12):4434–4444
Wissenbach U, Bodding M, Freichel M, Flockerzi V (2000) Trp12, a novel Trp related protein from kidney. FEBS Lett 485(2–3):127–134
Witte DG, Cassar SC, Masters JN, Esbenshade T, Hancock AA (2002) Use of a fluorescent imaging plate reader–based calcium assay to assess pharmacological differences between the human and rat vanilloid receptor. J Biomol Screen 7(5):466–475
Wu XM, Qian ZM, Zhu L, Du F, Yung WH, Gong Q, Ke Y (2011) Neuroprotective effect of ligustilide against ischaemia-reperfusion injury via up-regulation of erythropoietin and down-regulation of RTP801. Br J Pharmacol 164(2):332–343
Xiao B, Dubin AE, Bursulaya B, Viswanath V, Jegla TJ, Patapoutian A (2008) Identification of transmembrane domain 5 as a critical molecular determinant of menthol sensitivity in mammalian TRPA1 channels. J Neurosci 28(39):9640–9651
Xu H, Blair NT, Clapham DE (2005) Camphor activates and strongly desensitizes the transient receptor potential vanilloid subtype 1 channel in a vanilloid-independent mechanism. J Neurosci 25(39):8924–8937
Xu H, Delling M, Jun JC, Clapham DE (2006) Oregano, thyme and clove-derived flavors and skin sensitizers activate specific TRP channels. Nat Neurosci 9(5):628–635
Yamaguchi T, Hempson SJ, Reif GA, Hedge AM, Wallace DP (2006) Calcium restores a normal proliferation phenotype in human polycystic kidney disease epithelial cells. J Am Soc Nephrol 17(1):178–187
Yang BH, Piao ZG, Kim YB, Lee CH, Lee JK, Park K, Kim JS, Oh SB (2003) Activation of vanilloid receptor 1 (VR1) by eugenol. J Dent Res 82(10):781–785
Yang ZH, Wang XH, Wang HP, Hu LQ, Zheng XM, Li SW (2010) Capsaicin mediates cell death in bladder cancer T24 cells through reactive oxygen species production and mitochondrial depolarization. Urology 75(3):735–741
Yu H, Kosuna K, Haga M (1997) Perilla, the Genus Perilla. Taylor and Francis, London
Yu D, Pearson SK, Bowen WH, Luo D, Kohut BE, Harper DS (2000) Caries inhibition efficacy of an antiplaque/antigingivitis dentifrice. Am J Dent 13(Spec No):14C–17C
Yu L, Wang S, Kogure Y, Yamamoto S, Noguchi K, Dai Y (2013) Modulation of TRP channels by resveratrol and other stilbenoids. Mol Pain 9:3
Zanon RB, Pereira DF, Boschetti TK, Santos M, Athayde ML (2008) Phytoconstituents isolated from dichloromethane fraction of *Vernonia tweedieana* Baker leaves. Revista Brasileira de Farmacognosia 18:226–229
Zhang L, Barritt GJ (2004) Evidence that TRPM8 is an androgen-dependent Ca2+ channel required for the survival of prostate cancer cells. Cancer Res 64(22):8365–8373
Zhang LL, Yan Liu D, Ma LQ, Luo ZD, Cao TB, Zhong J, Yan ZC, Wang LJ, Zhao ZG, Zhu SJ, Schrader M, Thilo F, Zhu ZM, Tepel M (2007) Activation of transient receptor potential vanilloid type-1 channel prevents adipogenesis and obesity. Circ Res 100(7):1063–1070
Zhang L, Du JR, Wang J, Yu DK, Chen YS, He Y, Wang CY (2009) Z-ligustilide extracted from Radix Angelica Sinensis decreased platelet aggregation induced by ADP ex vivo and arterio-venous shunt thrombosis in vivo in rats. Yakugaku Zasshi-J Pharm Soc Jpn 129(7):855–859
Zhong YP (2010) Clinical application and research progress of traditional Chinese medicine dragon's blood. Mod J Integr Chin Trad Western Med 19:2469–2470
Zhong B, Wang DH (2008) N-oleoyldopamine, a novel endogenous capsaicin-like lipid, protects the heart against ischemia-reperfusion injury via activation of TRPV1. Am J Physiol Heart Circ Physiol 295(2):H728–735
Zhong J, Minassi A, Prenen J, Taglialatela-Scafati O, Appendino G, Nilius B (2011a) Umbellulone modulates TRP channels. Pflugers Arch 462(6):861–870
Zhong J, Pollastro F, Prenen J, Zhu ZM, Appendino G, Nilius B (2011b) Ligustilide: a novel TRPA1 modulator. Pflugers Archiv 462(6):841–849
Zhou J, Du W, Zhou K, Tai Y, Yao H, Jia Y, Ding Y, Wang Y (2008) Critical role of TRPC6 channels in the formation of excitatory synapses. Nat Neurosci 11(7):741–743

Zhou Z, Peng J, Wang CJ, Li D, Li TT, Hu CP, Chen XP, Li YJ (2010) Accelerated senescence of endothelial progenitor cells in hypertension is related to the reduction of calcitonin gene-related peptide. J Hypertens 28(5):931–939

Zhu X, Jiang M, Peyton M, Boulay G, Hurst R, Stefani E, Birnbaumer L (1996) trp, a novel mammalian gene family essential for agonist-activated capacitative Ca2+ entry. Cell 85 (5):661–671

Zhu S, Darbon H, Dyason K, Verdonck F, Tytgat J (2003) Evolutionary origin of inhibitor cystine knot peptides. FASEB J 17(12):1765–1767

Zierler S, Yao G, Zhang Z, Kuo WC, Porzgen P, Penner R, Horgen FD, Fleig A (2011) Waixenicin A inhibits cell proliferation through magnesium-dependent block of transient receptor potential melastatin 7 (TRPM7) channels. J Biol Chem 286(45):39328–39335

Ziglioli F, Frattini A, Maestroni U, Dinale F, Ciufifeda M, Cortellini P (2009) Vanilloid-mediated apoptosis in prostate cancer cells through a TRPV-1 dependent and a TRPV-1-independent mechanism. Acta Biomed 80(1):13–20

Zurborg S, Yurgionas B, Jira JA, Caspani O, Heppenstall PA (2007) Direct activation of the ion channel TRPA1 by Ca2+. Nat Neurosci 10(3):277–279

Zvara A, Bencsik P, Fodor G, Csont T, Hackler L Jr, Dux M, Furst S, Jancso G, Puskas LG, Ferdinandy P (2006) Capsaicin-sensitive sensory neurons regulate myocardial function and gene expression pattern of rat hearts: a DNA microarray study. FASEB J 20(1):160–162

Zygmunt PM, Andersson DA, Hogestatt ED (2002) Delta 9-tetrahydrocannabinol and cannabinol activate capsaicin-sensitive sensory nerves via a CB1 and CB2 cannabinoid receptor-independent mechanism. J Neurosci 22(11):4720–4727

What Do We Really Know and What Do We Need to Know: Some Controversies, Perspectives, and Surprises

Bernd Nilius and Veit Flockerzi

Contents

B. Nilius (✉)
Department Cell Mol Medicine, Laboratory Ion Channel Research, KU Leuven, Campus Gasthuisberg, O&N 1, Herestraat 49-Bus 802, 3000 Leuven, Belgium
e-mail: bernd.nilius@med.kuleuven.be

V. Flockerzi
Universität des Saarlandes, Institut für Experimentelle und Klinische Pharmakologie und Toxikologie, Gebäude 46, 66421 Homburg, Germany
e-mail: veit.flockerzi@uks.eu

B. Nilius and V. Flockerzi (eds.), *Mammalian Transient Receptor Potential (TRP) Cation Channels*, Handbook of Experimental Pharmacology 223, DOI 10.1007/978-3-319-05161-1_20,

Abstract

TRP channels comprise one of the most rapid growing research topics in ion channel research, in fields related to ion channels including channelopathies and translational medicine. We provide here a critical survey on our current knowledge of TRP channels and highlight some of the still open or controversial questions. This comprises questions related to evolution of TRP channels; biophysics, i.e., permeation; pore properties and gating; modulation; the still-elusive 3D structure; and channel subunits but also their role as general sensory channels and in human diseases. We will conclude that our knowledge on TRP channels is still at the very beginning of an exciting research journey.

Keywords

TRP channels • Evolution • Pores • Gating • Associated proteins • Subunits • Modulation by calcium • Sensory physiology • Phosphoinositides • Channelopathy • TRPs in networks

1 Introduction

There is no doubt: The discovery of the *trp* gene superfamily which encodes TRP (transient receptor potential) cation channels has opened a highly exciting field to understand physiological and pathophysiological basic mechanisms of cell functions which were until this time inaccessible. This book reviews many examples reaching from sensory physiology, such as vision, hearing, and touch, to homeostatic functions such as Ca^{2+} and Mg^{2+} (re)absorption and thermo-regulation and to hereditary diseases ranging from stationary night blindness to neurodegenerative disease, kidney diseases, chronic pain, and skeletal abnormalities (reviewed in Nilius and Owsianik 2010b, 2011; Nilius et al. 2007). If we date the advent of "TRP research" with the isolation of the Drosophila "trp" (transient receptor potential) mutant (Cosens and Manning 1969), approximately 20 years passed until the Drosophila *trp* gene was cloned (Montell and Rubin 1989) and a role of TRP ion channels was explicitly mentioned (Hardie and Minke 1992, 1993). Now, again 20 years later, the progress in this field is overwhelming, but there are still many open questions, controversies, and surprises. Only some of them can be addressed in such a review, and we do not claim any complete view but will provide a few snapshots.

2 The Basics: Nomenclature

As always in the beginning of a rapidly extending research field, some Babylonian language disasters happened. Some TRP clones received up to five different names, e.g., OTRPC4, VR-OAC, TRP12, VRL-2, and presently TRPV4 (Everaerts et al. 2010). Fortunately, this Babylonian disaster was at least partially solved by agreeing on a unifying nomenclature based on homology (Montell et al. 2002). All 8 subfamilies were given the name TRP (for transient receptor potential, maybe not the best decision) followed by the subfamily indicators, C, V, M, A, P, ML, and N.

However, one problem recently appeared: The polycystin family comprises 5 TRPPs (TRPP1–5). However, only three are really channel proteins. Therefore, it was suggested by a recent review from the Clapham group that this family only contains the three channel members, TRPP1 (previously known as TRPP2 or PKD2, PC2 (polycystin 2) derived from the second gene which causes autosomal dominant polycystic kidney disease (ADPKD)), TRPP2 (previously TRPP3, polycystin-like (PCL) or PKD2-like 1 (PKD2L1)), and TRPP3 (also known as TRPP5 or PKD2L2) (Wu et al. 2010). The former TRPP1 (PKD1, PC1) and TRPP4 (PKD1L1) have been eliminated because they are indeed glycoproteins which contain a large N-terminal extracellular region and multiple transmembrane domains and have no homology with ion channels and also do not function as channels. This seems to be more logical but confuses again the ADPKD community. The HUGO nomenclature still considers the old description as TRPP2, TRPP3, and TRPP5 as the TRP family polycystin members. So we stick to the HUGO nomenclature throughout this volume.

3 Some Evolutionary Puzzles

TRPs appear relatively late in the evolution. No TRPs are expressed in bacteria in contrast to most of the well-described voltage-dependent ion channels (Fig. 1). They are expressed in some algae, yeast, and other fungi. Expression occurs in the unicellular choanoflagellate *Monosiga brevicollis* and importantly in many mammalian parasites. Five metazoan TRP subfamily members (TRPA, TRPC, TRPM, TRPML, and TRPV) were identified in choanoflagellates, demonstrating that they evolved before the emergence of multicellular animals (Fig. 1). One puzzle is the complete absence in plants (see for details Cai 2008; Prole and Taylor 2011, 2012). But why make these plants drugs for TRPs, like capsaicin, tetrahydrocannabinol, nicotine, and all the other spices being effective TRP modulators (Nilius and Appendino 2011, 2013)? They make them as “secondary metabolites” which are not essential for normal growth and development of the plant but appear to function in the defense of plants against herbivores and pathogens. TRPs are widely expressed in insects which may be targeted by these plant compounds (Matsuura et al. 2009). For example, a plant can produce substances that result in adverse physiological effects in the insect, such as a bitter taste or even poisoning. Ironically, the medicinal applications and pharmacological effects of secondary metabolites in animals (insects) and humans are better understood than their functions in the plants that produce them (Liscombe and Facchini 2008).

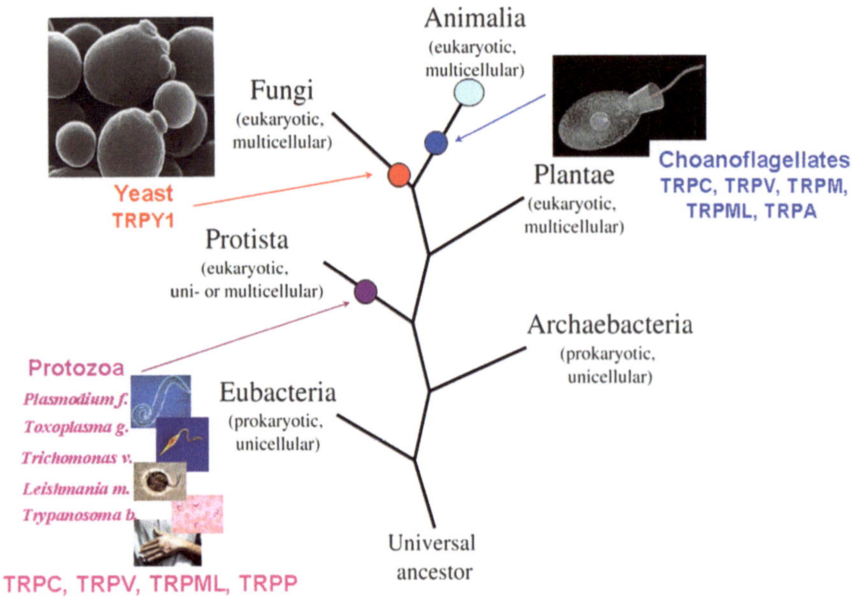

Fig. 1 TRPs in the kingdom of life and the appearance of TRPs (*circles*). See for detailed information (Cai 2008; Ihara et al. 2013; Prole and Taylor 2011, 2012; Wheeler and Brownlee 2008)

4 Do We Have the Correct Clones? Contribution of Endogenous TRPs?

There is a certain discrepancy between annotated TRP entries in protein and gene databases (e.g., UniProtKB and Ensembl Genome Browser), TRP cDNAs used for functional expression studies in vitro, and, apparently, TRP proteins expressed in vivo. For example, for human TRPC3, most in vitro studies have been performed with a cDNA encoding an 848-amino acid (aa) protein which is different to the 836-aa TRPC3 UniProtKB entry and the predicted 921-aa TRPC3 entry in the Ensembl Genome Browser. For human TRPC1, most in vitro studies have been done with the TRPC1beta variant (759 aa, Zitt et al. 1996) or with the TRPC1alfa variant (793 aa, Zhu et al. 1995), although the endogenous human TRPC1 protein may well cover an additional ~100 aa extension at its N-terminus (Ong et al. 2013). In fact, the identity of the endogenous TRP proteins is only known to a certain level of approximation: There is little direct experimental evidence to what extent the DNA sequences are translated to real proteins. This limitation refers not only to TRP channel genes but to most protein-coding genes in humans and many other organisms. In 2013, more than 95 % of protein sequences provided by the UniProt resource came from translations of coding sequences generated by gene prediction programs. Only less than 5 % of the protein entries rely on sequence data obtained by direct protein sequencing, by Edman degradation or MS/MS experiments (http://www.uniprot.org). New developments in protein analytical methods including mass spectrometry-based targeted proteomics will improve this situation in the future; but current approaches heavily rely on antibodies to pull out the protein of interest from complex biological samples. Using three independent antibodies, the endogenous protein sequence of TRPV6 could be obtained by this approach (Fecher-Trost et al. 2013). Compared to the 725-amino acid (aa) annotated sequence, the "real" TRPV6 protein contained a 40-aa extension within its N-terminus. However, adequate antibodies for most TRP proteins are not generally available (Wu et al. 2010), and therefore protein sequence information on most TRPs using mass spectrometry based has still to be shown.

Because of their convenience, many laboratories use HEK293 cells for functional expression of TRP cDNAs. *Trpc1* (Zhu et al. 1995) and the *Trpc3* cDNAs (Zhu et al. 1996) were originally cloned from HEK293 mRNA. It cannot be avoided but is hardly considered in the interpretation of data that the endogenous TRP proteins of HEK cells (TRPC1,TRPC3, and, maybe, TRPC7 and TRPM4) associate with the TRP proteins of expressed cDNAs to form functional channel complexes (Zagranichnaya et al. 2005). Considering the assumed broad expression of *Trp* genes, a similar situation will exist in most cell lines used for heterologous expression studies.

5 The Still-Elusive 3D Structure

It is agreed that all TRP consists of intracellular N- and C-termini and has six transmembrane spanning helices (TM) in which TM5 and TM6 probably form a classical channel pore. So far, only parts of TRP proteins have been crystallized, and the whole protein structure for some TRPs can be approached in electron cryomicroscopy pictures in a not very high resolution, e.g., for TRPV1 in a 19 Å resolution (Moiseenkova-Bell et al. 2008). Only 30 % of the channel protein is located in the plasma membrane (transmembrane domain TM); the other part forms a large cytosolic bulk. Both parts are connected by a fenestrated domain which might be important as an intracellular access to the TM region. Typical for the TRPV family and for TRPA1 is the N-terminal ankyrin repeat domain (ARD) (Fig. 3). TRPVs have an ARD formed by 6 ankyrin repeats (ARs) consisting of an inner and outer helix connected by unusual long "fingers" at the ARD's "concave" side (Gaudet 2008a, b) which are considered to be quite flexible. TRPA1 is proposed to have an ARD consisting of 14–19 ARs (Wang et al. 2012). It is highly speculative whether this domain is involved in gating (probably for TRPA1 via covalent cys/lys modification) or whether it is mainly required for channel trafficking and insertion in the plasma membrane (Nilius et al. 2011) (Fig. 2).

Thus, the functional role of the ARD is still unknown. It provides ATP and calmodulin binding which interferes in the TRPV family with channel activation and desensitization (inactivation) (Lishko et al. 2007; Phelps et al. 2007, 2008, 2010). Characteristically, ankyrin repeats are connected by large "fingers" which might be important for channel–protein interactions. The ARD is involved in TRPV4 dysfunction in several channelopathies (Inada et al. 2012). Interestingly,

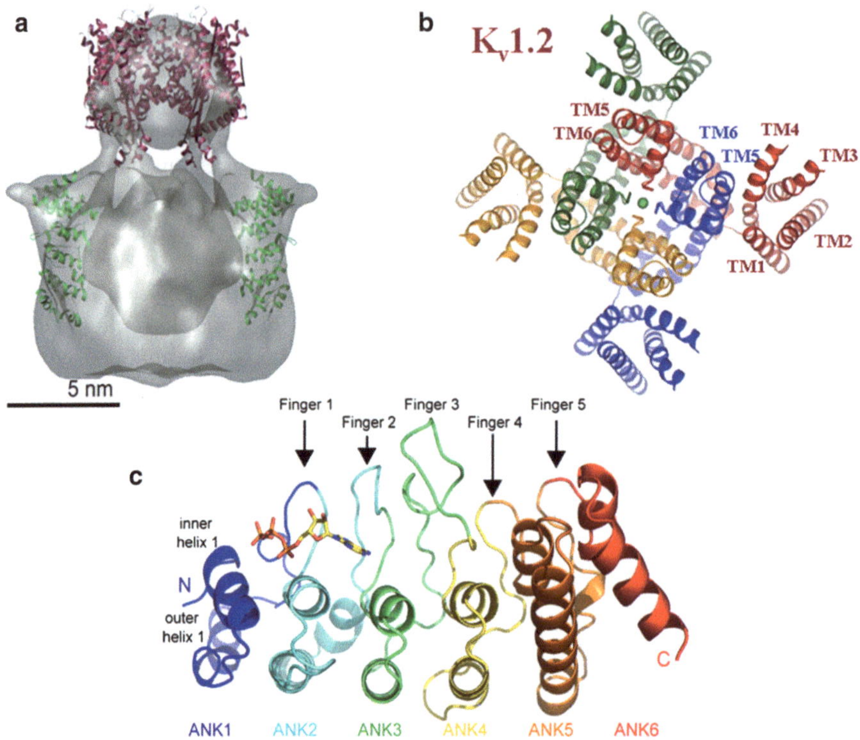

Fig. 2 Structure of TRPV1. (**a**) 3D reconstruction of TRPV1 obtained from cryo-electron microscopy. Inserted are in the plasma membrane part the atomic structures of Kv1.2 potassium channel (TM1–TM4, *pink*) and in the cytosolic part the TRPV1 ankyrin repeats (*green*). See the nice match of these structures. (**b**) High-resolution structure of Kv1.2 transmembrane domains. (**c**) TRPV1 ankyrin domains. Indicated is the ATP-binding site in TRPV1 [adapted from Lishko et al. (2007); Long et al. (2005); Moiseenkova-Bell et al. (2008) with permission]

the TM1-4 domains in TRPV1, TRPM4, and TRPM8 have some structural similarities with the TM domain of Kv1.2 including its voltage sensor (Nilius et al. 2005b). These TMs also nicely fit into the 3D structure of the transmembrane part of TRPV1 (Moiseenkova-Bell et al. 2008). It is extremely important to decipher a role of a potential voltage sensor in the TM region and the functional role of ARD in channel gating and trafficking. We refer to new and exciting data obtained from a 3D analysis of TRPV1 with a 3.4 A resolution to Chap. 11 in this book. *"High-resolution views of TRPV1 and their implications for the TRP super-family" by Ute. A. Hellmicu and Rachelle Gaudet.*

All TRPM channels have coiled-coil domains, i.e., α-helices which form helical bundles of a preferred stoichiometry. This domain was crystallized in the C-terminus of TRPM7 where it is proposed to trigger tetramerization (Fujiwara and Minor 2008; Tsuruda et al. 2006). A similar domain in TRPP2 (PKD2) has been crystallographically analyzed, and the structure provides insight into TRPP2–PKD1 heteromerization (Yu et al. 2009; Zhu et al. 2011). Also the 3D structure of the TRPM7 kinase domain was analyzed by crystallography (Yamaguchi et al. 2001). However, these are only some snapshots into the TRP structure. Thus, the high-resolution 3D structure is highly expected to solve one of the mysteries of TRP channel gating.

6 α-Subunits and Elusive β-Subunits

Most of the TRP channels are most probably tetramers as shown first for TRPV1 (Kedei et al. 2001) and later for TRPV5 and TRPV6 (Hoenderop et al. 2003). Probably, nearly all TRPs form tetrahomomers. Several heteromeric TRP channels have been described. TRPP2 forms a trimer with PKD1 (for a review see Zheng 2013). Heteromeric channels are still a matter of debate, and their functional role must be evaluated. In the classical view, these tetramers are formed from the pore-carrying α-subunit. The mechanism of co-assembly is still not solved, although several mechanisms have been proposed. N-terminal ARD domains are protein–protein interaction sites (Sedgwick and Smerdon 1999). However, its role for α-subunit assembly is very much uncertain because isolated ARDs from TRPVs do not interact (Lishko et al. 2007; Phelps et al. 2007, 2008, 2010; Zhang et al. 2011) and TRPV1 and TRPV2 crystals do not multimerize (Jin et al. 2006; Lishko et al. 2007). Also for TRPA1, evidence has been reported that the ARD probably controls membrane insertion of the channels rather than tetramerization (Nilius et al. 2011). The role of coiled-coil domains for multimerization of TRPM channels has been discussed above. It is also reported that for TRPVs a region proximal of the TRP box might be required for subunit assembly (Zhang et al. 2011; Garcia-Sanz et al. 2004). Obviously, we do not yet understand the assembly of TRP subunits to form functional channel.

Another pertinent urgent problem to solve is the existence and possible function of β-subunits, i.e., proteins which bind to the α-subunit and modulate channel function by direct interference with channel gating or regulation of membrane insertion as studied extensively for voltage-operated Ca^{2+} channels (Richards et al. 2004; Hofmann et al. 2014). Such subunits are not yet discovered for TRP channels. A putative β-subunit, although not yet confirmed in detail, might be PACSIN 3, a member of a protein family that has been implicated in synaptic vesicular membrane trafficking and regulation of dynamin-mediated endocytotic processes. PACSIN 3 binds via its SH3 domain to a site in the proline-rich domain (PRD) proximal to the ARD in TRPV4 and not only increases channel insertion in the plasma membrane but also selects the activation mode of this channel, i.e., it suppressed activation by cell swelling which is thought to be an EET-dependent process (Cuajungco et al. 2006; D'hoedt et al. 2007).

A pertinent question is not yet answered: Is TRPC1 α-subunit? Originally described as a cation channel activated by store depletion (Mori et al. 2002; Zitt et al. 1996), it was later shown that TRPC1 proteins do not form functional channels by themselves (Storch et al. 2012; Strubing et al. 2001) but contribute to heterotetrameric TRPC channels, preferably with TRPC4 and TRPC5 (Hofmann et al. 2002; Strubing et al. 2001). Apparently, TRPC4 and TRPC5 but also TRPP2 (Tsiokas et al. 1999) and TRPV4 (Ma et al. 2010, 2011) pick up TRPC1 from the ER to the plasma membrane and the primary cilium (Bai et al. 2008), whereas in their absence TRPC1 may act as a leak channel of the ER (Berbey et al. 2009). Recently, it was shown that the full-length endogenous TRPC1 amplifies Ca^{2+} release-activated Ca^{2+} currents mediated by Orai channels (Ong et al. 2013) by direct interactions with components of the store-operated Ca^{2+} entry machinery (Liao et al. 2008; Singh et al. 2002; Yuan et al. 2007). The *Trpc1* gene is supposed to be ubiquitously expressed, but how can one tell considering that there are no TRPC1 ion currents nor appropriate antibodies?

Another surprising example concerns the *Xenopus* TRPV6 homologue. xTRPC1 interacts with xTRPV6, thereby inhibiting the channel. TRPC1 is probably not part of the channel but regulates the activity of TRPV6 under physiological conditions (Schindl et al. 2012; Courjaret et al. 2013). Obviously, the search for TRP β-subunits is another high priority in TRP channel research.

Recently, it has been shown that an alternative splice variant of the mouse *Trpa1* gene (TRPA1b) can physically interact with the main variant *Trpa1* and increases the expression of TRPA1a in the plasma membrane. TRPA1a and TRPA1b co-expression significantly increases current density in response to different agonists without affecting their single-channel conductance. TRPA1 is obviously regulated through alternative splicing under physiological and also pathological conditions (Zhou et al. 2013). Thus, can such a variant in fact be considered as a β-subunit?

Another example came as a surprise. TRPM4 interacts with the sulfonylurea receptor 1 SUR1 and may confer the sensitivity of glibenclamide to this channel (Sala-Rabanal et al. 2012; Woo et al. 2013).

If a β-subunit is mainly defined as non-pore forming but with a positive effect on plasma membrane insertion of the α-subunit, then a β-subunit of voltage-dependent K^+ channels, Kvβ2, must be considered as a novel TRP β-subunit. Kvβ2 interacts with TRPV1, increases the cell surface expression levels of this channel, and results in a significant increase of TRPV1 sensitivity to capsaicin, as functionally measured with patch clamp. Thus, Kvβ2 plays a role in TRPV1 channel trafficking to the plasma membrane (Bavassano et al. 2013). It is now intriguing to ask whether β-subunits among several channel families have a certain degree of promiscuity.

7 The Correct Expression Pattern

This is one of the main but still not completely solved problems. To solve this problem, specific TRP antibodies must be available, and transgenic models and reporter strategies must be employed to identify the correct cellular localization of TRP. The poor quality of available antibodies for TRPs has caused considerable frustration and may have led to publication and perpetuation of erroneous research results. It is true that to generate an antibody for a low abundant membrane protein like a TRP is a painstaking effort (Meissner et al. 2011) requiring months of laborious benchwork and money for synthesizing, expressing, and purifying the antigen, for hosting the suitable animals to be immunized, for generating and selecting hybridoma cells, and for purifying and characterizing the final antibodies. Especially the analysis of the antibody specificity is crucial. Only in very few cases, this major ordeal is accomplished by the many commercial suppliers that have proliferated and sell antibodies directed at a wide range of proteins including TRPs (Flockerzi et al. 2005; Ong et al. 2002, 2013; Wu et al. 2010).

In addition, we need to know the subcellular, e.g., intra-organelle, localization because intracellular TRPs are important functional players in cell organelles such as the endosomes, lysosomes, SR and ER, and Golgi apparatus (for a reviews see Cheng et al. 2010; Gees et al. 2010). Probably, also mitochondria may have TRP channels. TRPC3 is probably localized in mitochondria and carries a significant fraction of mitochondrial Ca^{2+} uptake that relies on extramitochondrial Ca^{2+} concentration. Up- and downregulation of TRPC3 expression influences also the mitochondrial membrane potential (Feng et al. 2013). It has to be noted that substantial discrepancies exist in the literature concerning TRP channel distribution. The use of mRNA detection is an indicator but has several drawbacks (Anderson and Seilhamer 1997; Anderson and Anderson 1998; Gygi et al. 1999; Pradet-Balade et al. 2001; Tew et al. 1996).

Immunohistochemistry relies on selective antibodies, which might be different from the ones used for protein detection in Western blotting. Radioactive ligands might be a complementary tool for detection of TRP channels. Now, promising is the genetic modification of a TRP channel locus which allows introduction of reporter genes such as *lacZ* or *fluorescent proteins* into living animals. Expression of these genes reflects the TRP distribution. Just as an example, many functions have been attributed to TRPV1 in the central nervous system, e.g., the development of cerebellar Purkinje cells and hippocampal pyramidal neurons and its contribution to glutamatergic synapse stabilization as important player in the neonatal cerebellar cortex and in the refinement of synaptic plasticity (Vennekens et al. 2012). It came as a big surprise that TRPV1 is probably even not expressed in these brain areas (Cavanaugh et al. 2011a, b). This causes another problem. Do we really know where the TRP channels that we wish to target are really located? A crucial test is always the detection of ion current through identified TRP channels (see below).

8 News About the Pore

As required for a channel protein, we do not understand yet the regulation of permeation. Is the pore of many TRPs in fact a dynamic structure contributing to gating much more than the known α-pore of classical channels (between TM5 and TM6)?

One of the most important functional features of many TRP pores is their permeability for Mg^{2+} and other heavy metals. The Mg^{2+} permeation is still a biophysical puzzle. So far, all ion channels refuse Mg^{2+} permeation because of its very high dehydration energy. TRP channels therefore provide a unique Mg^{2+} entry pathway. However, the biophysics behind is not yet unraveled (for a review see Owsianik et al. 2006).

There is now growing evidence that upon agonist stimulation some TRP channels such as TRPV1 (Chung et al. 2008) and TRPA1 (Karashima et al. 2010) show a dynamic pore behavior, i.e., a dilatation of the pore upon agonist binding. This dilatation is in the range of several Å. Mechanistically important seems to be the fact that this dilation of the TRPV1 pore decreases after cholesterol depletion of the plasma membrane (Jansson et al. 2013). In TRPA1, it causes an increase in divalent cation selectivity and fractional Ca^{2+} current and is therefore functionally very important. How does an agonist dilate the pore? Pore dilation has been also described in the TRPC family. Activation of TRPC4β with the G protein $G\alpha_{i2}$ as compared with activation by GTPγS causes a small but significant dilation of the channel's pore and increases Ca^{2+} permeability like the dilation of the TRPA1 pore (Jeon et al. 2013). This dynamic pore behavior needs to be explained soon because it will shed some light in the gating behavior of TRP channels. Of note, a pore dilation from 3 to 12 Å occurs by binding of G proteins to open the G-loop gate in GIRK2 potassium channels, a mechanism that has been explained by changes of the pore structure (Whorton and MacKinnon 2011).

Another route to follow is even more spectacular. TRP channels might posses and exploit under certain circumstances other pathways for ion conduction as shown for voltage-activated potassium and sodium channels. This pathways comprise a permeation gate through the voltage-sensing domain (TM1–TM4) of the channel, the ω-pore (Starace and Bezanilla 2004; Tombola et al. 2007; Gamal El-Din et al. 2011; Jurkat-Rott et al. 2010; Prütting and Grissmer 2011). It might be possible that TRP channels have a similar pathway which is active during distinct modes of activation (Vriens et al. 2014).

9 Fractional Ca^{2+} Currents in TRPs

Importantly, most of the TRP channels have in fact a small Ca^{2+} selectivity, and the "fractional Ca^{2+} current" is often less than 5 %. However, for understanding the functional impact of TRP channels, we need to know which partition of the current under physiological condition is carried by Ca^{2+} compared with monovalent cation. We need to measure this "fraction" to check which amount of Ca^{2+} is really contributing to the current (Karashima et al. 2010; Samways and Egan 2011). Unfortunately, such measurements are rare (done for TRPV1, TRPA1, TRPM3, and TRPM8). TRPV5 and TRPV6 are the only "real" Ca^{2+} channels with fractional Ca^{2+} currents of ~100 %. TRPA1 and TRPM3, with a fractional Ca^{2+} current of ~20 %, can be considered as relatively efficient Ca^{2+} channels. Probably, all other TRP channels have fractional Ca^{2+} currents of less than 10 % (Gees et al. 2010). Importantly, TRPV1 is known as a TRP channel with a relatively high Ca^{2+} permeability ($P_{Ca}/P_{Na} \sim 10$) (Owsianik et al. 2006). However, its fractional Ca^{2+} current is only in the range of 5 % of the total current. High fractional Ca^{2+} current so far is only measured for TRPV5, TRPV6, TRPM3, and TRPA1 (reviewed in Gees et al. 2010). Even for the Drosophila TRP and TRPL channel, which are regarded as highly Ca^{2+}-selective channels, the fractional Ca^{2+} current is only in the range of 29 % and 17 %, respectively (Chu et al. 2013).

Knowing the fractional Ca^{2+} current will also at least partially answer the questions: which increases in $[Ca^{2+}]_i$ can be expected under physiological conditions and whether it is likely that the TRP-induced depolarization is essential rather than the Ca^{2+} influx. Obviously, TRP channels are functionally coupled with Ca^{2+}-sensing effector proteins and may therefore achieve a high signal gain. It is therefore probably an important feature of TRP channels to provide only a low fractional Ca^{2+} current or in the case of high fractional Ca^{2+} current have only a low-density expression.

Obviously, these issues have to be urgently addressed for TRPC channels.

10 Gating

TRP channels have been widely described as receptor-activated channels, ligand-activated channels, and "directly" gated channels. However, some TRP channels are constitutively open, at least in several heterologous expression systems (e.g., TRPV5, TRPV6, TRPC3, TRPA1, and others) raising the question whether a crucial gating partner is missing in the expression system. This question of constitutive open TRP channels seems to be especially intriguing in the brain: TRPV4 is active at physiological brain temperature in hippocampal neurons and thereby controls their excitability in vitro. Local changes in temperature, which occur in the brain, cause modulation of brain activity, e.g., local cooling might reduce neuronal electric excitability via a decreased TRPV4 activity and thereby contribute to brain functions (Shibasaki et al. 2007; Tominaga 2013, personal communication).

The question whether TRP channels like some K_2P channels are responsible for background currents is unanswered.

Receptor activation comprises G protein-coupled receptors (GPCRs) and receptor tyrosine kinases that activate phospholipases C (PLCs) which gate/modulate TRP channel activity by hydrolysis of phosphatidylinositol 4,5-bisphosphate (PIP2), production of diacylglycerol (DAG) or inositol 1,4,5-trisphosphate (IP3), and subsequent liberation of Ca^{2+} from intracellular stores. Ligand activation is due to exogenous small organic molecules, e.g., capsaicin, endogenous lipids, or products of lipid metabolism (diacylglycerols, phosphoinositides, eicosanoids, anandamide), and purine nucleotides and their metabolites (adenosine diphosphoribose (ADP-ribose), βNAD) but also by proteins (like G proteins for TRPCs).

Direct activation can occur via changes of temperature, changes in the membrane potential (voltage activation), and maybe also mechanical forces (Nilius and Owsianik 2011; Ramsey et al. 2006; Wu et al. 2010).

Thus, TRP channels are polymodal channels, opposite to voltage-gated channels, which only sense changes in the transmembrane electric field. TRP channel might be considered as integrators of multiple cellular signals and might be preferred coincidence detectors (Ramsey et al. 2006). However, this makes the identification of a basic gating mechanism difficult if not impossible, and probably each channel exploits multiple mechanisms to open the gate. So far, we do not have any structural insight how the gating process is initiated (Nieto-Posadas et al. 2011).

Let's first consider possibilities of direct gating. Maybe the most intriguing mechanism is TRP gating by depolarization. In this view, gating would share some similarities with the activation mechanism of classical voltage-dependent channels. However, half maximal activation occurs at very positive membrane potentials, which are physiologically irrelevant. However, several signals, like temperature changes and agonist binding, cause a shift in the activation curve toward more negative potentials, thereby increasing the fraction of open channels under steady state conditions (Nilius et al. 2005b; Voets et al. 2004). This paradigm "gating by shift" has been described for several channel types (TRPV1, TRPA1, TRPM4, TRPM8, etc.) but is mechanistically not yet understood. The shift is supported by a small apparent gating charge of TRP channels which is in the range of 0.5–0.8e (Nilius et al. 2005b; Voets et al. 2007; Zheng 2013). A proposed voltage-sensing role of basic residues in TM4 and the TM4–TM5 linker of TRPM8 might be considered as a first hint (Voets et al. 2007). Although the architectural and pharmacological similarities exist between TRP channels and Shaker K^+ channels, replacing previously identified domains and critical structural motifs of the membrane-spanning portions of Kv2.1 with corresponding regions of two TRP channels, TRPM8 and TRPV1, did not allow the formation of hybrid channels, i.e., the TM3b–TM4 paddle motif of Kv2.1 can be replaced by the analogous regions of both TRP channels without abolishing voltage activation, but swapping putative voltage-sensing TRP channel regions with the Kv2.1 voltage sensor yields non-functional channels (Kalia and Swartz 2013). Obviously, such studies should be extended. Obviously, the true voltage sensors need to be pinned down. Probably, the measurement of gating currents should be envisaged. Because of the small gating charge of TRP channels [0.4–0.9e (Nilius et al. 2005b)], better methods must be applied, e.g., the *limited slope method* (Almers 1978; Sigg and Bezanilla 1997), to measure in a low open probability range the effective gating charge or changes in gating charge caused by mutations in a putative voltage sensor (see Voets et al. 2007). From the *voltage dependence* point of view, temperature and agonists are modulators of the voltage-dependent opening of the channel gate. The true nature of this shift is still elusive. However, the existence of the open probability shift is a clear fact that needs to be explained (for a critical discussion and challenges of this simplified view see, e.g., Nieto-Posadas et al. 2011; Zheng 2013).

Direct channel gating is caused by the covalent modification of cysteine and lysine residues. Such covalent modifications of Cys and Lys residues in the N-terminus of TRPA1 by electrophiles have been extensively described (for a comprehensive review, see Nilius et al. 2012). However, there are many issues to consider. First, covalent modification by electrophilic compounds may result in the

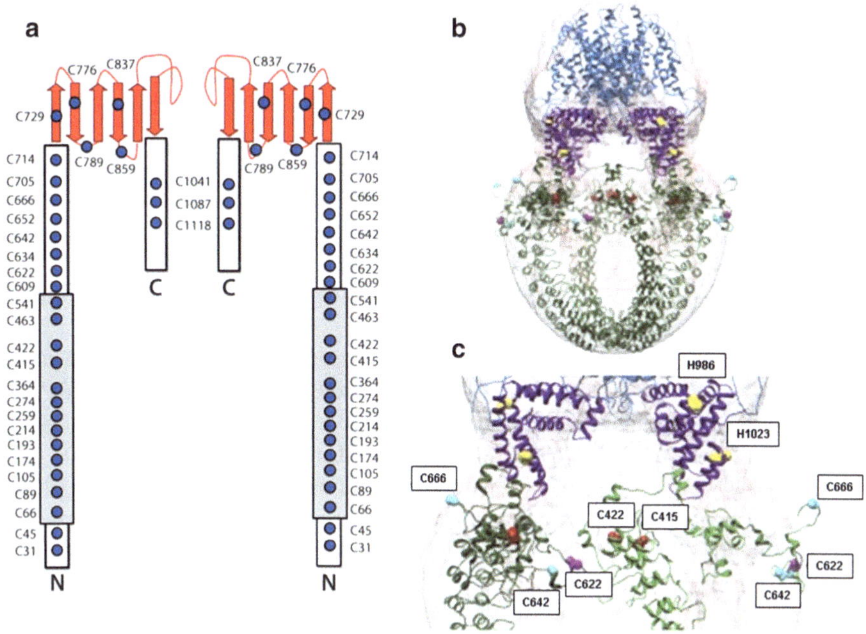

Fig. 3 (**a**) Two-dimensional representation of a TRPA1 dimer indicating the reactive cysteine residues. The cytoplasmic N- and C-termini are separated by six transmembrane helices. Each monomer contains 31 cysteine residues (*blue circles*), and the N-terminus has an extended ankyrin repeat domain (*box*). (**b**) TRPA1 is shown in the cryo-electron microscopic 3D structure. The modeled TRPV1 TM domain fits into the TRPA1 structure (*blue ribbon*). The TRPA1 N-terminus (*green*) and the C-terminus are shown. (**c**) Conserved cysteines involved in electrophilic activation in the mouse (Cys-415, Cys-422, Cys-622), human (Cys-622, Cys-642, Cys-666), and *Drosophila* (Cys-622, Cys-642) homologues are highlighted in the zoomed-in view (ball-and-stick structures and labeled, *right panel*). Histidines involved in zinc activation are also indicated in *yellow* (adapted from Cvetkov et al. (2011) with permission)

formation of Michael adducts, thiocarbamates, alkylation products of cysteine, and disulfides. The chemistry of these modified residues is not known, i.e., the subsequent intracellular modification to make the channel again available. Second, it is not yet completely clear which cysteines are really required for gating. Critical in mouse TRPA1 are C415, C422 (in the ARD), and C622 (between ARD and TM1) and in human C622, C642, and C666 (between ARD and TM1); in addition, modification of K710 is involved. However, 30 from the 31 cysteine are reactive 174, 193, 415, 422, 463, 609, 622, 634, 642, 666, and 859, and in addition four sulfide intra-chain bridges are formed between cys666-662, cys666-463, cys622-609, and cys666-193. Eleven cysteines have been identified to be possibly involved in gating (Fig. 3).

This disulfide bonding might be also involved in channel gating via an unknown mechanism. Probably, critical N-terminal cysteine residues involved in electrophilic activation are located at the interface between neighboring subunits, and conformational changes at this region might lead to unknown conformational changes resulting in channel activation (Cvetkov et al. 2011; Wang et al. 2012). How the covalent modification of the intracellular core signals to the channel gate is completely unknown. Third, surprisingly, that activation of the channel in inside-out patches is difficult and requires the presence of MgATP or polyphosphates (Kim and Cavanaugh 2007; Karashima et al. 2008). However, channel gating by non-electrophilic agonists such as Δ9-THC and menthol is still possible (Cavanaugh et al. 2008). It seems, yet not at all understood, that some intracellular factors, such as polyphosphates, may act as a scaffold to stabilize the channel in an available conformation.

Another direct form of gating remains mechanistically also enigmatic and came as a surprise. Heavy metals such as Zn^{2+}, Cd^{2+}, and Cu^{2+} can directly gate TRPA1. Activating Zn^{2+} binds to the C-terminal His983 and Cys1021 (Andersson et al. 2009; Banke and Wickenden 2009; Gu and Lin 2010; Hu et al. 2009b).

Another gating puzzle emerged with the identification of an obvious role of the pore region as part of the gating machinery. Activation of TRPV1 occurs by low pH and requires the pore residues E648 (E600 for potentiation of TRPV1 activity by low pH). Surprisingly, the non-charged T633 is also required for proton activation but not for activation by agonists (Ryu et al. 2007). In addition, TRPV1 is also activated by the double-knot toxin DkTx from the Earth tiger tarantula. This activation requires binding of the toxin to residues in the outer pore (Bohlen et al. 2010). This binding probably causes conformational changes in the pore region, thereby opening the channel gate (Bright and Sansom 2004). Also the inner pore of some TRP channels, TRPV1, TRPV3, TRPV4, and TRPA1, seems to be involved in direct channel gating as evidenced from mutagenesis experiments. TM6 in TRPA1 may possess two points of flexibility which could function as a gating hinge similar to Kv channels (Benedikt et al. 2009; Bright and Sansom 2004). Specific point mutations within the cytosolic TM4–TM5 linker of TRPA1 (N885S) (Kremeyer et al. 2010) or TRPV3 (G573C or G573S) (Asakawa et al. 2006; Xiao et al. 2008c) or the introduced G503S (TRPC4) and G504S (TRPC5) mutations in the corresponding sequences of TRPC4 and TRPC5 (Beck et al. 2013) lead to constitutively active channels indicating that the non-mutated amino acid residues are essential to keep the channels in a gateable configuration.

Obviously, many TRP channels are activated by binding of agonists. In this view, they can be considered as "ligand"-activated channels. Such ligands function endogenously or are externally applied. The most studied ligand is certainly capsaicin, a TRPV1 agonist. Capsaicin activates TRPV1 with $K_{1/2}$ values varying with temperature and voltage. Capsaicin binding occurs to the closed channel state probably at several binding sites mapped in the TM2–TM4 region (Jordt and Julius 2002), but other sites have been identified at the TRP box, the pore domain, and the N- and C-termini TRPV1. The docking of the capsaicin molecule to the interface between two channel monomers in the TM3/TM4 region, i.e., the probably voltage-sensing domain, has been analyzed in detail. The vanillyl moiety forms a $\pi-\pi$ stacking and hydrophobic interactions with Tyr511 and H-bonding with Ser512. In addition, the carbonyl group (B region) made H-bonding interactions with Tyr511 and Lys571. This binding favors an open-channel configuration (Lee et al. 2011). Although this structural modeling is intriguing, we still have no clue, even for this most advanced case, how the channel gates by ligand binding. The existence of probably more binding sites with positive cooperativity makes the underlying mechanism of channel opening still more uncertain not to speak of the still-elusive docking sites for endogenous agonists like the endocannabinoid anandamide and 2-arachidonoylglycerol (2-AG) and the lipid 12-HPETE. The situation for other TRP channels is still much less clear. Certainly, we have to fill the gap how ligand binding gates a TRP channel and binding studies have to be connected with functional data.

The situation of pinning down clear-cut gating mechanisms for TRPCs seems especially difficult. What do we really understand of TRPC gating? Obviously, Gi,o and Gq,11 are required at the same time to activate TRPC4/TRPC5. Simultaneously, an elevation of $[Ca^{2+}]_i$ is facilitating. Activation is also voltage dependent, and PI(4,5)P2 is a unique negative gating modulator (Otsuguro et al. 2008; Tsvilovskyy et al. 2009; Zholos et al. 2004). The complexity of these gating mechanisms is striking and puzzling. In addition, it has also been shown that TRPC4/TRPC5 are activated by Gαi rather than Gq/11. But what about the endogenous environment of the cell systems used? Gi,o and Gq,11 appear to be present in most cell lines used for *Trp* cDNA expression, but what about the receptors activating these G proteins (http://www.uni-leipzig.de/~strotm/molbio/endogenous_gpcr.htm)? In another approach, interaction of C-terminal SEC14-like and spectrin-type domain SESTD with Gαi was required for gating. What is the mechanism (Jeon et al. 2008, 2012, 2013)?

11 Positive and Negative Ca^{2+} Feedback

Probably all TRP channels are modulated by Ca^{2+}. Most TRPs show an activity-dependent inactivation which is mediated by Ca^{2+} and involves as possible mechanisms Ca^{2+}-dependent kinases, Ca^{2+}-dependent phosphatases, Ca^{2+}-regulated PLCs, and especially calmodulin. The mechanisms of this modulation is not clear (Gordon-Shaag et al. 2008); often Ca^{2+}-dependent PI(4,5)P_2 depletion is involved (Rohacs 2013; Rohacs and Nilius 2007). In many cases, this modulation is bimodal, e.g., comprises activation and negative feedback inhibition. As for all Ca^{2+}-permeable channels, negative feedback inhibition is extremely useful to avoid cellular Ca^{2+} overload. But do we really understand this mechanism? Some TRP channels such as TRPM4 and TRPM5 are directly activated by Ca^{2+}. Indeed, an increase in $[Ca^{2+}]_i$ is required to activate these channels in a voltage-dependent manner. Without Ca^{2+}, even very large depolarizations cannot activate the channel. Thus, this situation is different from, e.g., BK_{Ca} channels. Unfortunately, an activating binding site for Ca^{2+} has not been reliably determined. However, it has been shown that the presence of calmodulin which binds to probably several C-terminal sites shifts the voltage-dependent activation toward much more negative potentials (Nilius et al. 2005a). Again, the gating mechanism remains unclear. Both Ca^{2+}-activated channels desensitize/inactivate upon prolonged exposure to elevated $[Ca^{2+}]_i$. This effect can be nicely explained by a Ca^{2+}-dependent $PI(4,5)P_2$ depletion via activation of a Ca^{2+}-activated PLCβ4 (Nilius et al. 2006, 2008; Rohacs and Nilius 2007). A pleckstrin-like homology domain is required for $PI(4,5)P_2$ binding. Another intriguing example for the dual regulation by Ca^{2+} is TRPA1. Channel activity, if it is activated, is strongly potentiated by extracellular Ca^{2+} followed by channel desensitization. Thus, Ca^{2+} is one of the most important endogenous modulators of TRPA1. Both mechanisms are not understood. TRPA1 is affected by $[Ca^{2+}]_e$ due to entry through TRPA1 and subsequent elevation of intracellular calcium. The mutation of Asp918 in the putative TRPA1 pore greatly reduces Ca^{2+} permeability. Extracellular Ca^{2+} alone produced neither potentiation nor inactivation, e.g., channel activation is required. Application of Ca^{2+} to the cytosolic face of excised patches is sufficient to produce both potentiation and inactivation of TRPA1 channels. Moreover, in whole-cell recordings, elevation of intracellular Ca^{2+} potentiated, but did not inactivate TRPA1. Thus, potentiation and inactivation are two independent mechanistically not understood processes (Nilius et al. 2011, 2012; Wang et al. 2008). In this case, desensitization is probably much less attributed to an increased $PI(4,5)P_2$ breakdown as for other channels such as TRPM4, TRPM8, TRPV5, and TRPV6 (Karashima et al. 2008; Rohacs and Nilius 2007) or is even opposite (Kim et al. 2009). These are only examples for the probably universal modulatory activity of Ca^{2+} on TRP channels. Because of the bimodal effect of even extracellular Ca^{2+} on TRPA1, the use of a TRP-current readout in a Ca^{2+}-free solutions screening of channel modulators can easily generate false-negative (missing potentiation) or false-positive (lacking inactivation) results.

Because of the extremely important regulation of many TRP channels by Ca^{2+}, it is indicated to search downstream of Ca^{2+} for effects of Ca^{2+}-binding signaling proteins, e.g., calmodulin (CaM). CaM in most cases exerts an inhibitory effect on the channel gating. This negative feedback loop works locally and rapidly to regulate the level of channel activity suitable for cellular physiology. Activation of CaM by Ca^{2+} influx following TRP channel activation can also affect channel function indirectly, e.g., by binding of CaM to other proteins such as CaM-sensitive protein kinases. Given the diversity of TRP channels, it is perhaps not surprising that the sites of CaM binding are also quite diverse and multiple sites have been detected on many TRP channels at the N- and C-terminus (Zhu 2005). Because of this diversity, clear-cut functional data related to CaM should always substantiate such binding studies. As one striking example, the crystal form of the ARD of TRPV1 exists as a complex with CaM (Lishko et al. 2007). Thus, the ARD of TRPV1 (TRPV3, TRPV4) serves as a CaM-binding site. ATP was also found to bind to the CaM-binding site of TRPV1 and competes. CaM binding occurs in the absence of Ca^{2+}. The functional impact of the ATP–CaM competition is accelerated channel desensitization in the presence of CaM and a delay in the presence of ATP and absence of CaM. The diversity of CaM action is also shown in the surprising findings that Ca^{2+} binding to CaM may underlie the Ca^{2+}-dependent activation of

TRPM4 (Nilius et al. 2005a) and the Ca^{2+}-dependent potentiation of TRPV3 (Xiao et al. 2008b) and TRPV4 (Strotmann et al. 2003, 2010). Again, this diversity of CaM action needs to be understood and underpinned by more functional and structural analysis.

12 The PI(4,5)P_2 Puzzle

One exciting aspect of TRP channel regulation is the interaction with and modulation by plasma membrane phosphatidylinositol phosphates (PIPs) and in particular by phosphatidylinositol 4,5-bisphosphate (PI(4,5)P2). Probably all TRPs are regulated by PIPs. PI(4,5)P2-binding sites in nearly all PIP-modulated TRPs have been proposed and functionally tested via mutagenesis (Nilius et al. 2008). PI(4,5) P_2 positively modulates the activity of TRPM4, TRPM5, TRPM7, TRPM8, TRPV5, and TRPV6 but inhibits TRPC4α. Effects on TRPA1 are still somewhat controversial (reviewed in Rohacs 2013; Rohacs and Nilius 2007). Probably the most intriguing yet important problem to solve refers to a complex regulation of TRPV1 by PI(4,5)P2. It has been first shown that nerve growth factor (NGF) and bradykinin activate PLC, which hydrolyzes PI(4,5)P_2 and subsequently sensitizes the channel through interactions with a region in the C-terminus (Chuang et al. 2001; Prescott and Julius 2003). The opposite effect has been later described. Indeed, depletion of PI(4,5)P2 in inside-out patches leads to channel inhibition, and direct application of PI(4,5)P2 activated the channel (Klein et al. 2008). Again somewhat later, a new transmembrane protein has been found, Pirt (phosphoinositide interacting regulator of TRP), that interacts with both PI(4,5)P_2 and TRPV1, thereby inducing an activating effect on TRPV1. Thus, PI(4,5)P2 only activated TRPV1 in the presence of Pirt (Kim et al. 2008). However, this mechanism has been recently challenged. Pirt is probably not required for PI(4,5)P2 modulation of TRPV1. Pirt does not alter the phosphoinositide sensitivity of TRPV1 in HEK293 cells. Dorsal root ganglion neurons from Pirt knockout mice have an apparent affinity for PI(4,5)P_2 indistinguishable from that of their wild-type littermates. A proximal C-terminal region of TRPV1 is sufficient for PI(4,5)P_2 binding. This proximal C-terminal region of TRPV1 can interact directly with PI (4,5)P_2 and may play a key role in PIP2 regulation of the channel. Importantly, Pirt does probably not bind to TRPV1 (Ufret-Vincenty et al. 2011). However, Pirt might be an accessory protein for other channels, e.g., TRPM8, and may act as a positive modulator under conditions of channel activation (Tang et al. 2013). In another approach, PLC activation by bradykinin leads to a moderate decrease in PI(4,5)P2), but no sustained change in the levels of its precursor PI(4)P. Preventing this selective decrease in PI(4,5)P2 inhibited TRPV1 sensitization, while selectively decreasing PI(4,5)P2 independently of PLC potentiated the sensitizing effect of protein kinase C (PKC) on the channel, thereby inducing increased TRPV1 responsiveness. Maximal pharmacological TRPV1 stimulation leads to a large decrease of both PI(4,5)P2 and its precursor PI(4)P. Attenuating the decrease of either lipid significantly reduced desensitization, and simultaneous reduction of PI

(4,5)P2 and PI(4)P independently of PLC inhibited TRPV1. In this view, differential changes in phosphoinositide levels are mediated by probably distinct PLC isoforms which can result in opposite effects on TRPV1 (Lukacs et al. 2013). Recently, TRPV1 was inserted into artificial liposomes. It can be activated by capsaicin, protons, and heat. Under these conditions, TRPV1 is fully functional in the absence of phosphoinositides, i.e., does not require PI(4,5)P2 for activation. PI (4,5)P2, PI(4)P, and phosphatidylinositol inhibit TRPV1 (Cao et al. 2013). Obviously, there is still no unifying view how PIPs act on the probably best studied TRP channel, TRPV1. However, this modulation is critical for the understanding of TRPV1 function under conditions of inflammation and other disease-causing influences. In addition, we are still at the beginning of the beginning to understand specific actions of different PIPs on TRP channels. For the TRPML family, a completely different modulation pattern as for plasma membrane TRPs has been described. In endolysosome, PI(3)P and specifically PI(3,5)P2 are positive regulators for TRPML1–3. These lipids are also required for the correct channel trafficking. In late endosomes, PI(3,5)P2 is the positive regulator, whereas PI(4,5)P_2 is a negative regulator, and its depletion activated TRPML1 (Cheng et al. 2010; Zhang et al. 2012; Dong et al. 2010) indicating a probably highly diverse modulation pattern of TRPs by PIPs.

13 Understanding Networks

The last example on modulation of TRPV1 by PI(4,5)P_2 points to an integrated action of Ca^{2+} influx via the channel and activation of different isoforms of PLC (PLCδ, PLCβ4) and PKCε. Obviously, this is an example for a functional network which causes extremely variable channel reaction. Submaximal activation via bradykinin receptors (coupled to PLCβ4) causes a restricted PI(4,5)P_2 depletion and no change in PI(4)P; activation of TRPV1 by PKCε activation via the Ca^{2+} influx overwrites the negative effect of PI(4,5)P_2 depletion. On the other hand, maximal TRPV1 activation, e.g., by high concentrations of capsaicin, causes a large PI(4,5)P_2 and PI(4)P depletion via PLCδ activation which overwrites activating effects of PKCε (Lukacs et al. 2013).

Another striking example for a network modulating function of TRPC1 has been recently published and is discussed here as another example. Protein kinase C (PKC) and PI(4,5)P_2 are obligatory for the activation of TRPC1 in vascular smooth muscle cells (VSMCs). This activation is coordinated by myristoylated alanine-rich C kinase substrate (MARCKS) which is required for the TRPC1 channel's activation by PKC and PI(4,5)P_2. TRPC1 channels and MARCKS form signaling complexes. If PI(4,5)P_2 is bound to MARCKS, TRPC1 is closed. Activators of TRPC1 channels induce PKC phosphorylation of TRPC1 proteins, which causes dissociation of TRPC1 subunits from MARCKS and a release of the bound PI(4,5) P_2. The released PI(4,5)P_2 may now bind to TRPC1 and causes channel opening. Thus, MARCKS regulates the native TRPC1 in VSMCs by acting as a reversible PI (4,5)P_2 buffer, which in turn is regulated by PKC-mediated TRPC1

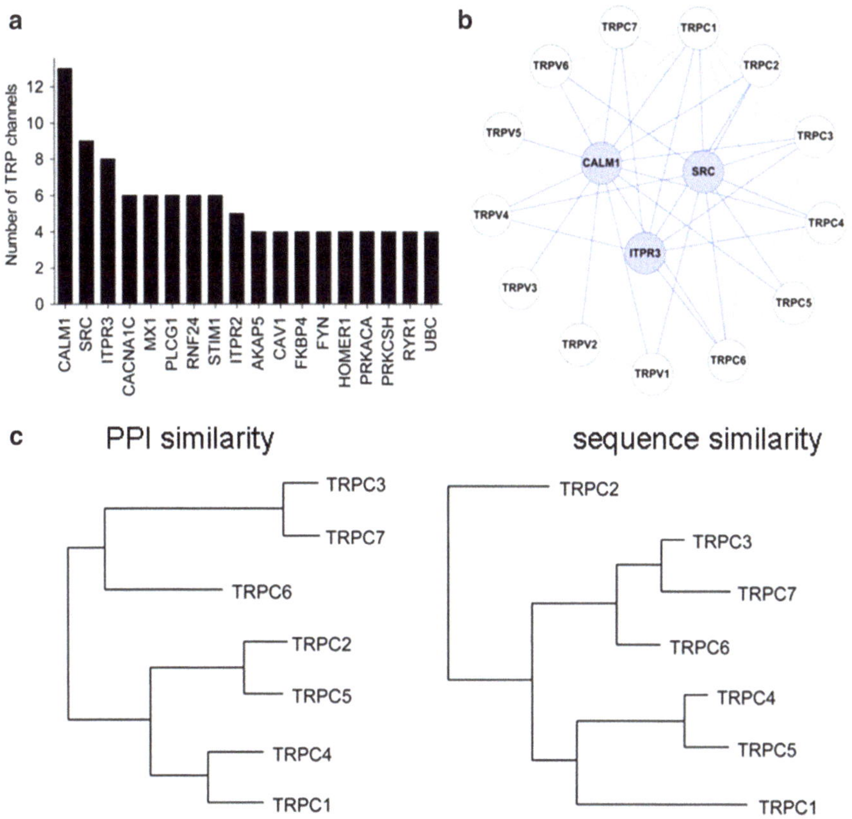

Fig. 4 (**a**) TRP channel-interacting proteins published to bind to multiple TRP channels. (**b**) CALM1, SRC, and ITPR3 interact with at least eight TRP channel isotypes within or across the subfamilies (*light blue* edges). *Gray line* marks the interactions among TRP channels. (**c**) Discordance between protein–protein interactions (PPI) and sequence similarities. PPI similarity denotes here how many interacting proteins are shared among TRPC channels. Sequence similarity is calculated using a standard protein BLAST tool (from Chun et al. (2013) with permission)

phosphorylation, and $PI(4,5)P_2$ acts finally as gating ligand of TRPC1 (Shi et al. 2013a). Because of the complexity of such a mechanism in native cells, it is important to verify similar mechanisms in other TRPC1 expressing cell types!

These are only a very simple example that TRPs function in complex networks. Probably, network-based approaches may facilitate the understanding of TRP channel biology and their function in disease. Using databases of protein–protein interaction (PPI) (Chun et al. 2013), the multifunctionality of TRP channels might be better understandable. An example is given in Fig. 4a showing the number of interacting TRP channels with several proteins and in Fig. 4b representing the network of the interfering interaction of TRP channels with 3 proteins (ITPR3, inositol 1,4,5-trisphosphate receptor, type 3; CALM1, calmodulin 1; and the SRC tyrosine-protein kinase). Interestingly, comparing the similarity of TRP subfamilies based on PPI or sequence data may give additional information (Fig. 4c). We refer to a review on network-based approaches in biomedical science, which describes the current state of TRP channel network biology and which discusses a promising future direction of TRP channel research (Chun et al. 2013). However, as always with conclusions drawn from databases, they heavily rely on the data present in the databases. So far, no rigorous and comprehensive targeted high-resolution proteomic study is available on the nano-environment of any endogenous TRP channel. Such studies are available for other ion channels such as voltage-activated Cav2 channels (Muller et al. 2010) or AMPA receptors (Schwenk et al. 2012). The very few published studies on TRP proteomes even failed to identify the "targeted" TRP. As long as we do not know details of the specific makeup of endogenous TRP channel complexes, conclusions on networks may be premature.

14 The Quest for Selective Tools

The pharmaceutical industry focuses on generation of small molecules, which are synthesized in vitro and screened with high-throughput methods. Although there is now considerable progress in the synthesis of highly selective and also reversible TRP channel modulators (like for TRPV1, TRPV4, TRPA1), this is not yet the case for all TRP channels. It is highly desirable to ask the pharmaceutical industry for distributing such selective tools especially if they are not golden bullets for the therapeutically useful TRP targets. Obviously, the progress in the research on TRPV4 was mainly triggered by the early release of the relatively selective agonist 4αPDD from Glaxo (Watanabe et al. 2002).

15 The Use of Natural Compounds

Natural compounds teach us a lot about TRP channels. The problem seems to be that they are difficult to synthesize, but is there not a huge intellectual quest for natural compounds? Since probably more than 5,000 years, mankind fights for spices added to our food. Many of them are just efficient TRP channel modulators (Nilius and Appendino 2011, 2013). Interestingly, some natural compounds have been used in traditional medicine, e.g., as antidepressants have TRP channels as targets. One of these compounds is incensole acetate which is released by the burning of resin from the *Boswellia* plant and has been used for religious and cultural ceremonies for millennia. It activates TRPV3, which is expressed in the brain and causes anxiolytic-like and antidepressant-like behavioral effects (Moussaieff and Mechoulam 2009; Moussaieff et al. 2008). St. John's wort has been used medicinally for over 5,000 years. Relatively recently, one of its phloroglucinol derivatives, hyperforin, an antidepressant compound, has been identified as effective activator of TRPC6 (Leuner et al. 2007, 2010). Hyperforin has been shown to have cognitive-enhancing, memory-facilitating properties and has probably neuroprotective effects (Griffith et al. 2010). This example is only mentioned to open the interest of TRPists for natural compounds. Estimations are done that less than 1 % of all blooming plants have never been tested for natural compounds which might be useful as medical compounds. One has to remember that these molecules were evaluated during more than 3.5 billions of years in evolution! A striking example how optimized screening natural compounds, i.e., of flavanone derivatives from citrus plants, has been successfully used to identify highly potent blockers of TRPM3, a channel with a rather elusive and intractable pharmacology (Straub et al. 2013a, b).

16 Why Is It So Difficult to Measure TRP Currents in Native Cells?

TRP proteins form channels. Obviously, the most direct evidence that such a channel generates physiological signals is the detection of a current via TRP channels in native cells. It turned out that this is a difficult task. First, for reasons explained above, TRP channel expression might be low, and especially only few channels are located in the plasma membrane where they form available ion channels. Second, although the single-channel conductance of several TRPs might be large under conditions of permeation of monovalent ions, under physiological conditions in the presence of Ca^{2+}, this conductance might be low. However, because of all difficulties with an exact localization of TRP channels in the plasma membrane, such a functional readout is unavoidable and necessary. As a striking example, when TRPV5 and TRPV6 were first described as Ca^{2+}-selective ion channels (Nilius et al. 2000), in heterologous expression systems, whole-cell currents and channel current (in the absence of Ca^{2+}) could be measured but not in the physiological most relevant native cell system, the Ca^{2+}-reabsorbing cells in the rabbit kidney connecting tubule and cortical collecting duct epithelial cells from which the *trpv5* gene was cloned (Vennekens et al. 2000); also TRPV5-dependent $^{45}Ca^{2+}$ fluxes could be measured. This is probably due to the low single-channel conductance if Ca^{2+} is the charge carrier (in contrast to a high conductance of 77pS if monovalent cations are the charge carrier). Currents are probably too small to be detected indicating the need of using more refined methods such as single-channel measurements and fluctuation analysis. On the other hand, the amount of eGFP-tagged TRPV5 (den Dekker et al. 2005) or TRPV6 protein (Fecher-Trost et al. 2013) routed to the plasma membrane after cDNA expression was too small to be detectable by confocal laser scanning microscopy. That means that also after cDNA expression the amount of functional channels in the plasma membrane is very low, like in native cells, but sufficient for current recording in one system but not in the other. May measuring native currents under divalent-free conditions improve this situation?

The measurement of currents through TRP channel is also very much complicated because of the lack of selective and reversible inhibitory tools. Also for Ca^{2+}-activated TRP channels, TRM4 and TRPM5, identification of those currents is possible by selective activation of those channel by Ca^{2+} uncaging (Ullrich et al. 2005). However, in native cells, there are always the problems to dissect these currents through TRP channels from other Ca^{2+}-activated currents. In addition, using "selective" tools like siRNA or comparison of currents in native cells from wild-type and knockout animals is difficult and often based on unreliable readouts. Thus, any functional mechanism involving TRP channels should address this problem of measuring TRP-mediated currents in native cells by the use of identified pharmacological agonist and antagonists which act reversible and selective. The problem of tiny current should be tried to overcome by fluctuation analysis and single-channel measurements, e.g., TRPM4 identification in inside-out patches from sinoatrial nodes (Demion et al. 2007).

17 Species Dependence

There are plenty of examples that activation and modulation of TRP channels are highly species dependent, also within mammalian species. Only in some examples, in the presence of saturating capsaicin, rat TRPV1 reaches full activation, with no further stimulation by protons. In contrast, human TRPV1 (hTRPV1) is potentiated by extracellular protons and magnesium, even at saturating capsaicin (probably due to differences in negatively charged residues in the TM3–TM4 linker) (Wang et al. 2010). Also in the search for effective TRPV1 antagonists, the significant lack of consistency of the pharmacology of many TRPV1 antagonists across different species has created a substantial obstacle (probably due to differences in the pore) (Papakosta et al. 2011). In the search for TRPA1 antagonists, trichloro (sulfanyl)ethyl benzamides were discovered to be active in humans but had no or even activating activity in rodents (Klionsky et al. 2007). Both rodent and rhesus monkey TRPA1 do not respond to extracellular acidosis, and protons even inhibited rodent TRPA1, but human TRPA1 is activated by protons (an effect probably caused by differences in TM5, TM6) (de la Roche et al. 2013). Electrophilic thiamines block human TRPA1 but activate rat TRPA1 (Chen et al. 2008). Non-electrophilic menthol is an activator of mouse TRPA1 at low concentrations and a blocker at high concentrations; but it only activates human TRPA1 (differences in the TM5–TM6 pore region) (Xiao et al. 2008a). Caffeine activates mouse TRPA1 but blocks human TRPA1 (Met268Pro point mutation in the mouse TRPA1 N-terminus changes activation into inhibition) (Nagatomo et al. 2010). Many more of such examples could be mentioned. Even more important, in invertebrates and ancestral vertebrates, TRPA1 serves as a heat receptor. In mammals, TRPA1 is probably a receptor for noxious cold. This is, however, controversial. Cold activates rat and mouse TRPA1 but not human or rhesus monkey TRPA1. At the molecular level, a single residue within the S5 transmembrane domain (G878 in rodent but V875 in primate) seems to account for the observed difference in cold sensitivity (Chen et al. 2013).

It might be important to consider that such species-dependent effects may give some insight into structural requirements of channel activation and modulation. It should also be mentioned that of course basic functional differences in organ function, e.g., the beat frequency in rodents as compared to human hearts, might substantially alter effects of TRP activation raising, e.g., the problem whether rodent models are in general appropriate for an extrapolation to humans (e.g., the TRPM4 effect in mice might be underestimated as compared to humans).

18 Are TRPs Really the Aristotle's Main Sensory Players? Surprises?

Since their discovery, TRP channels have been always considered as the main players in the sensory system and were even considered as the main signal transducers providing via Aristotle's classical five-sense recognition of the outer world (Damann et al. 2008). Vision is in Aristotle's view the noblest of all five senses. So far, TRPs were only considered as photoactive channel in invertebrates. In fact, TRP and TRPL in Drosophila are main players in vision as many times excellently reviewed (Hardie and Postma 2008; Montell 2012). A more recently evolving topic in mammals concerns intrinsically photosensitive retinal ganglion cells (ipRGCs) which are discovered photoreceptors in the mammalian eye which mediate primarily nonimage visual functions, such as pupillary light reflex and circadian photo-entrainment, and are expected to respond to the absolute light intensity (for a recent reviews see Lucas 2013; Pickard and Sollars 2012). It is intriguing to unravel the functioning of this photo-sensing including possible TRP channels and its modulation by different chromophores (Schmidt et al. 2011). A first surprise, melanopsin-expressing photosensitive retinal ganglion cells (pRGCs) express TRPM1 and TRPM3 which are involved in the nonimage-forming responses to light (Hughes et al. 2012). Melanopsin signaling depends also on TRPC6 and TRPC7. Iris muscles isolated from nocturnal mammals such as mice contract when exposed to light through the action of a melanopsin-based signaling pathway. Ablating in mouse the expression of both TRPC6 and TRPC7 also eliminated the light response in the M1 subtype of melanopsin-expressing, intrinsically photosensitive retinal ganglion cells (M1 ipRGCs) (Xue et al. 2011).

At note, very surprisingly, it was recently shown that gating of Drosophila TRP involved not only a photosensitive step coupled to light sensing by rhodopsin but also a mechano-sensitive step (Hardie and Franze 2012). In general, we have to expect that polymodal stimuli, such as light, heat, and force, converge on chromophores which in turn modulate various TRP channels, as shown now in Drosophila (Kirkwood and Albert 2013).

Highly hypothetical but intriguing to consider is the possible existence of "deep brain photoreception" by extraocular light detectors also in mammals, e.g., in dopaminergic amacrine neurons, which may again settled by a melanopsin–TRP interaction. Such sensing may still form a connection with neuroendocrine cells for the control of mood (Fernandes et al. 2012, 2013).

TRPM1 has now been also described as a direct player in image-visual function. It is TRPM1 expressed on ON bipolar cells in the retina which invert the sign of light responses from hyperpolarizing to depolarizing before passing them on to ganglion cells (ON response). TRPM1 colocalizes with metabotropic glutamate receptor mGluR6 (GRM6) coupled to $G\alpha_o$ and $G\beta3$ proteins. Light responses are generated when TRPM1 is activated. The gating mode of this channel is still obscure: Unbinding of glutamate, which is released from adjacent rod photoreceptor cells, from mGluR6 activates TRPM1. Binding of glutamate to mGluR6 closes the channel. The signaling cascade is still not known. After activation, TRPM1 desensitizes and the light response decays (Koike et al. 2010; Morgans et al. 2009). Underlining the important role of this novel TRP-mediated vision pathway, TRPM1 is linked to the autosomal-recessive congenital stationary night blindness type 1C (CSNB1C), a retinal disorder characterized by nonprogressive impaired night vision and decreased visual activity due to ON bipolar cell dysfunction (Audo et al. 2009; Nakamura et al. 2010).

The question remains: Are other TRP channels involved in vision (Gilliam and Wense 2011)? TRPC1 has been detected in cones and rods which may hint to still unknown function beyond light sensing but regulation of Ca^{2+} homeostasis in photoreceptors (Molnar et al. 2012).

TRPs have now been more and more involved in light sensing in cells outside the visual system. Melanocytes sense UV light (mainly UVB light) via TRPM1-mediating melanogenesis (Devi et al. 2009). TRPA1 is also involved in UVB sensing and early melanogenesis; a retinal-dependent G protein-coupled signaling pathway is involved in TRPA1 activation (Bellono et al. 2013; Bellono and Oancea 2013). UVB light is also supposed to activate TRPV4 (Moore et al. 2013). Obviously, the mechanism underlying activation of TRP channels by light is of high interest.

It came also as a surprise that at least in rattlesnake and in bats TRPA1 can be activated by infrared light (Gracheva et al. 2010, 2011; Panzano et al. 2010; Geng et al. 2012; Yokoyama et al. 2011). Activation is reportedly due to heat response of TRPA1 above a temperature threshold for activation at ~28 °C. TRPA1 is a heat-activated channel in these species. Again, the underlying activation mechanism of TRPA1 is highly interesting. If snakes would detect temperature differences of 0.01 °C, which is probably even an underestimation (Ebert and Westhoff 2006), currents through TRPA1 would exhibit a Q^{10} of $>10^{15}$. Obviously, this temperature sensing must involve a nonlinear signal amplification process upstream of channel opening. Vice versa, considering a Q_{10} of ~10, as typical for thermoTRPs, the increase in open probability induced by a warming of 0.003 °C would be only ~0.2 %! It is very unlikely that this would be enough to generate sufficient depolarization to trigger action potentials in the snake trigeminal neurons. One has also to consider that the radiation energy decays with the fourth power of the distance (Stefan–Boltzmann law), which would predict very small activation energies for TRPA1 in the snake's pit organ. Of note, it has never been shown that TRPA1 antagonists impair the infrared sensing of snake or bats! However, infrared radiation sensing by TRP channels is a highly exciting, yet unsolved, topic.

In Aristotle's view, touch, or now more generally described as "somato-sensitivity," was the most primitive sense. However, somato-sensitivity is the most essential and necessary sense for survival, far beyond the other four senses.

This has now become one of the most exciting topics in the TRP field. We will only refer to two examples: mechano- and temperature sensing.

How are TRP channels involved in mechano-sensation? There are several examples which are all reviewed in detail (described in the issue of HEP Plant and in Eijkelkamp et al. 2013). However, with the advent of completely unexpected mechano-sensing channels, Piezo1 and Piezo2, some of the earlier mechano-TRP concepts have certainly to be revised (for a critical discussion, see Nilius and Honore 2013). One recent finding might be mentioned to open our interest novel forms of mechano-sensing. Probably TRPV4 in glia cells can mechanically sense infrasound (16 Hz) which can even be involved in infrasound-caused neurodegeneration (Shi et al. 2013b).

Another important part of somato-sensing is temperature (less correct "thermo-") sensing. It has been one of the most accomplishments in the TRP field to identify the so-called thermoTRPs which are supposed as the main temperature sensors. Without any doubt, some TRPs have a very high temperature sensitivity as expressed in Q_{10} values >10. Mechanistic aspects of thermosensing by TRPs have been addressed in many excellent reviews and will not be further mentioned (Dhaka et al. 2006; Ferrer-Montiel et al. 2012; Patapoutian et al. 2003; Vay et al. 2011; Voets 2012; Wetsel 2011) (also described in this volume by Voets, T. "TRP channels and thermosensation"). The list of thermoTRPs is constantly changing, e.g., TRPV2, TRPV3, and TRPV4 (reviewed by Voets 2012; Nilius et al. 2014) are not anymore considered as functionally involved in temperature sensing, but TRPM3 must be added (Vriens et al. 2011). Several "thermo"TRPs are located in thermosensitive regions such as in skin keratinocytes, in thermosensory nerve endings, and in blood vessels. However, many other cell types are of course also thermosensitive and may not exploit TRP channels. As an important surprise for the coupling of temperature and metabolism, fat cells sense temperature TRP independent (Ye et al. 2013). Temperature sensing has probably many other faces such as metabolic control of many cell functions and homeostasis. At note, TRP channels must be considered as only one player in temperature sensing, as many other channels have been recognized to play an important role in determining the action potential frequency and pattern upon heat or cold stimulation in sensory nerve fibers, in mediating thermoregulatory response, or in triggering temperature-mediated response from the cellular to the behavioral level. Many other proteins, not only ion channel, are involved in orchestrating temperature sensing in temperature-mediated responses. Rhodopsin can function as a temperature sensor and might be involved as an amplifier for thermoTRP activation (Shen et al. 2011). The melatonin-related orphan receptor GPR50 is involved in determining the basal metabolic rate and is a temperature sensor (Swoap 2013). The Ca^{2+} sensor STIM1 is a thermosensor (Xiao et al. 2011). Nav1.7, Nav1.8, IK_D, Kv1, Kv7 (M-current), K_2P channels, the proton channel Hv1 and TMEM16A (Ano1, a Ca^{2+}-activated Cl^- channel), HCN channels, and probably several more function as temperature sensors (see recent reviews Cho et al. 2012; Fujiwara et al. 2012; Madrid et al. 2009; Orio et al. 2013; Voets 2012). This all indicates that TRP will not be the only clue for understanding temperature sensing. Interestingly and illustrating the still mediocre understanding of temperature sensing as a whole physiopsychological event, we refer to the fact that classical phenomena related to thermosensation, including "Weber's silver *Thaler* illusion," the "thermal grill illusion," and "Weber's three-bowl experiment," are not yet understood and cannot be explained at the level of TRPs. It is even difficult to find a quantitative model involving several classes of ion channels to describe the fast and slow adaptation in thermosensory fibers and the mostly unimodal (bell-shaped) steady state discharge.

We will not refer to many unsolved problems in taste, olfaction, and hearing. Many open questions have been addressed in this volume of HEP, e.g., the controversial role of PKD2L1 (TRPP3), the existence of a salt receptor in the TRP family (TRPML3, filed as Senomyx Patent, 2009[1]), the role of TRPV1 in taste rather than chemesthesis, the real function of the many TRP channels in the inner ear [e.g., TRPC3, TRPC5; TRPV1, TRPV4, TRPV5; TRPA1; TRPML3; TRPP3; etc. (Asai et al. 2010)] beyond hearing after identification of the probably main mechano-transducer channels TMC1 and TMC2 (Pan et al. 2013), and a still somewhat elusive role of TRPs in olfaction (Dong et al. 2012).

19 TRPs Go Metabolic

We all agree that TRPs have a huge impact on many sensory functions and behave like unique cellular sensors which are activated by polymodal stimuli. However, more and more evidence defines a role of TRP channels in the regulation of metabolic networks and metabolic diseases, e.g., type 2 diabetes, obesity, dyslipidemia, metabolic syndrome, atherosclerosis, metabolic bone diseases, and electrolyte disturbances. This is certainly a line which deserves a high consideration. Intriguing examples are as follows: (1) The weight of *Trpv1*$^{-/-}$mice heavily exceeds that of controls, which is coupled with a predisposition to age-associated overweight (Garami et al. 2010, 2011); (2) dietary TRPV1 activation improves energy metabolism, muscle growth, and exercise endurance by upregulating PGC-1α and nNOS/mTOR in skeletal muscles (Ito et al. 2012; Luo et al. 2012); (3) TRPV4 antagonists elevate thermogenesis in adipose tissues and protect from diet-induced obesity, adipose inflammation, and insulin resistance (Ye et al. 2012); (4) TRPC1, TRPC5, and TRPC6 expression is altered in a more humanlike metabolic syndrome model, the *Ossabaw miniature pigs* (Hu et al. 2009a); (5) brown fat activity is, besides β3-adrenergic stimulation, modulated by peroxisome proliferator-activated receptor γ (PPARγ) and the connected stimulation of TRPV1 by dietary capsaicin and monoacylglycerols (Birerdinc et al. 2012); (6) TRPV1 activation prevents nonalcoholic fatty liver disease (Li et al. 2012,

[1] Moyer B, Zlotnik A, Hevezi P, Soto H, Lu M, Gao N, Servant G, Brust P, Williams M, Kalabat D, et al. (2009) Identification of Trpml3 (Mcoln3) as a salty taste receptor and use in assays for identifying taste (salty) modulators and/or therapeutics that modulate sodium transport, absorption or excretion and/or aldosterone and/or vasopressin production or release. SENOMYX patent.

2013); and much more published evidence (for detailed reviews see Nilius and Appendino 2011, 2013; Zhu et al. 2010). Concerning these obviously exciting novel approaches to metabolic diseases, some pertinent questions have to be answered: What are the precise signaling pathways up- and downstream of TRPs in adipocytes and in the hypothalamus?

How do TRPs such as TRPV1 exert tissue-specific, positive, and negative effects on metabolism? Do TRPs participate in metabolic diseases, and do known genetic variants (SNPs) in TRP genes influence disease prevalence? Are TRPs useful pharmacological targets for treating obesity and insulin resistance (see for reviews Liu et al. 2008; Nilius and Appendino 2011, 2013; Zhu et al. 2010)?

20 TRPs and Disease

We have to accept a general scheme: TRP channels are not highly expressed on most native cells, and the current density will be low. Therefore, a pertinent question is whether these obviously more "modulatory" ion channels will have such a striking functional impact to cause diseases. Importantly, although many knockout models do not show per se very striking phenotypes, the number of channelopathies connected to dysfunctional TRP channels is amazingly high (for reviews see Cornell et al. 2008; Kiselyov et al. 2007; Nilius 2007; Nilius and Owsianik 2010b; Nilius et al. 2007). One of the probably most exciting puzzles that we have to solve in understanding TRP channelopathies is related to TRPV4 (Nilius and Owsianik 2010a). The Trpv4 gene is prone to many mutations a surprising number of different diseases. Familial digital arthropathy–brachydactyly only affects fingers and toes. Mutations in Trpv4 have also been identified in motor/sensory neuropathies, such as the congenital distal spinal muscle atrophy, scapuloperoneal spinal muscle atrophy, and Charcot–Marie–Tooth disease type 2C,with highly variable phenotypes, i.e., skeletal dysplasias (*brachyolmia, spondylometaphyseal dysplasia Kozlowski, spondyloepimetaphyseal dysplasia Maroteaux pseudo-Morquio type 2, metatropic dysplasia,* and *parastremmatic dysplasia* with a unifying feature short trunk with vertebral platyspondyly and scoliosis), skeletal diseases (*familial digital arthropathy–brachydactyly* only affecting fingers and toes), and motor/sensory neuropathies (*congenital distal spinal muscle atrophy, scapuloperoneal spinal muscle atrophy*, and *Charcot–Marie–Tooth disease type 2C* characterized by degeneration of motor and sensory neurons and peripheral nerves) (for a review, see Nilius and Voets 2013). More than 50 disease-causing mutants have been so far identified. They are distributed over the whole channel protein, with only two hot spots, the ADR (mainly neuropathies) and a region around the MAP-binding site in the C-terminus (skeletal dysplasias) (Nilius and Voets 2013). Another puzzle appeared concerning TRPV4 channelopathies: It is widely discussed that TRPV4 plays an important role in endothelium, contributes to the EDHF mechanism, and subsequently is an endothelial player for the regulation of blood pressure (Earley et al. 2009; Filosa et al. 2013; Kassmann et al. 2013; Kohler et al. 2006; Loot et al. 2008; Rath et al. 2009; Saliez et al. 2008; Vriens et al. 2005; Zhang and Gutterman 2010). However, any vascular phenotype in the TRPV4 patients is missing. The striking challenge is that mutations which are often located in the same domains of the channel proteins and even three identical mutations cause different diseases.

Most TRP-caused diseases have a dominant inheritance pattern, which implies that the patients carry both a mutant and a wild-type allele, which creates a general often overlooked problem: TRPs function probably as tetramers; thus, patients are expected to express TRP channels with variable stoichiometry of WT and mutant subunits. Assuming that wild-type and mutant subunits are equally expressed and assemble into tetrameric channels in a random fashion, one would expect that the TRP channels in a patient's cell consist of 1/16 (6.25 %) pure wild-type channels and 1/16 pure mutant channels and that the 14/16 (87.5 %) of functional channels are heteromultimeric channels consisting of a mixture of wild-type and mutant subunits. This obviously requires the understanding of the properties of such heteromultimeric channels, which have often even not been studied. Understanding a disease needs the functional analysis of such heteromers! It is very likely that they exhibit properties that differ strongly from either pure wild-type or pure mutant homotetramers.

Understanding of the mechanistic background of these channel malfunctions not only is essential for a treatment of these diseases but may also help to understand the functioning of these exciting proteins and their involvement in different signaling cascades.

Obviously, TRP channels are connected to acquired diseases. In these cases, a critical mechanistic understanding is often missing, and far-reaching extrapolations and hopes are focused on TRP channels. We will also refer to a puzzle, seen the huge social impact: TRP causes diseases in the brain. Many discussions have been started about *age as a drug target*, and TRPs are considered as main actors. TRPM7 and TRPM2 have been put forward as potential factors in neurodegenerative diseases including Alzheimer's disease, Parkinson's disease, stroke, and diseases associated with oxidant-mediated neuronal damage (McNulty and Fonfria 2005). Especially TRPM2, which is highly expressed in the striatum (caudate nucleus and putamen), is supposed to play a key role in bipolar disorders (Uemura et al. 2005; Xu et al. 2006). Recent case–control studies implicate TRPM2 conferring risk for bipolar disorder (BD), and genetic variants of TRPM2 have been identified to be coupled with BD supporting a role for this channel in the pathogenesis of this disorder (see for a review Chahl 2007; Xu et al. 2009). Depletion of intracellular Mg^{2+}, a symptom of ischemic brain injury, and a reduction of extracellular Ca^{2+} are both associated with poor neurological outcome and are both conditions which activate TRPM7 and possibly increase the Ca^{2+} load of neuronal cells. This leads to a secondary injury processes and to cell death (Cook et al. 2009). Downregulation of TRPM2 and TRPM7 causes probably neuroprotective effect in the cerebral cortex and hippocampus after brain injury, oxidative stress, inflammation, and neuronal death (Cook et al. 2010). Therefore, TRPM7 might be a potential target for neuroprotection after brain injury. Suppressing the expression of TRPM7 in hippocampal CA1, neurons conferred resistance to ischemic cell death, preserved cell function, and prevented ischemia-induced deficits in memory (Rempe et al. 2009; Sun et al. 2009). Is this an antiaging strategy? On the other hand, mutations in TRPM2 have been discovered which result in rapidly inactivated channels which are unable to maintain sustained Mg^{2+} influx (Hermosura et al. 2002, 2005, 2008; Hermosura and Garruto 2007). Unfortunately, although quite a lot of evidence has been published, we face now clinical studies which completely refuse this even mechanistically underpinned view (Hara et al. 2010). However, still, TRPM7 might be involved in the *familial Alzheimer's disease (FAD)* but mechanistically linked to aberrant $PI(4,5)P_2$ causing TRPM7 dysfunction (Landman et al. 2006). In another view as the abovementioned role for "switching on" of light responses or as tumor suppressor, TRPM1 could be linked to a complex neurodevelopmental disorder characterized by severe visual impairment, intellectual disability, and refractory epilepsy. This disease is caused by a microdeletion in chromosome 15q13.3 carrying the *trpm1* gene (Lepichon et al. 2010). Aging increases also a risk for ischemic brain injury by oxidative stress, edema, inflammation, and excitotoxicity. TRPM4 is reportedly involved in the damaging secondary events that accompany brain injury. Inhibition of TRPM4 channels provides beneficial effects in ischemia-associated cerebral edema and secondary hemorrhage (Simard et al. 2010) and has been considered as a neuroprotective target in the brain or spinal cord lesion (Gerzanich et al. 2009). However, the pharmacological evidence that TRPM4 is really blocked by K_{ATP} (SUR1 bound) inhibitors is still controversial. Also the question arises whether SUR1 is indeed a TRPM4 subunit as suggested (Woo et al. 2013).

This is just to say that although brain injury confers a major burden to Western society and effective treatments are urgently required, the enthusiasm that TRP channel may solve many of such problems should be qualified by a more rigorous investigation of TRP channel functions in more refined and suitable models! But which mechanisms are really involved? How can we really understand that TRP channels are causative in all the sketched diseases?

Obviously, we have to admit that in many diseases the role of TRPs might be somewhat overinterpreted/overstressed. Better pathophysiological models which consider the difficulties to handle TRP must be developed. We are still in the very beginning of understanding disease-causing dysfunctions of TRP channels.

Because of all the controversies and gaps in our understanding on TRP, we have to confess that we have opened an exciting field, but the current state of TRP channel research, as described in this book, is still in its infancy!

References

Almers W (1978) Gating currents and charge movements in excitable membranes. Rev Physiol Biochem Pharmacol 82:96–190

Anderson NL, Anderson NG (1998) Proteome and proteomics: new technologies, new concepts, and new words. Electrophoresis 19:1853–1861

Anderson L, Seilhamer J (1997) A comparison of selected mRNA and protein abundances in human liver. Electrophoresis 18:533–537

Andersson DA, Gentry C, Moss S, Bevan S (2009) Clioquinol and pyrithione activate TRPA1 by increasing intracellular Zn2+. Proc Natl Acad Sci U S A 106:8374–8379

Asai Y, Holt JR, Géléoc GS (2010) A quantitative analysis of the spatiotemporal pattern of transient receptor potential gene expression in the developing mouse cochlea. J Assoc Res Otolaryngol 11:27–37

Asakawa M, Yoshioka T, Matsutani T, Hikita I, Suzuki M, Oshima I, Tsukahara K, Arimura A, Horikawa T, Hirasawa T, Sakata T (2006) Association of a mutation in TRPV3 with defective hair growth in rodents. J Invest Dermatol 126:2664–2672

Audo I, Kohl S, Leroy BP, Munier FL, Guillonneau X, Mohand-Said S, Bujakowska K, Nandrot EF, Lorenz B, Preising M et al (2009) TRPM1 Is mutated in patients with autosomal-recessive complete congenital stationary night blindness. Am J Hum Genet 85:720–729

Bai CX, Giamarchi A, Rodat-Despoix L, Padilla F, Downs T, Tsiokas L, Delmas P (2008) Formation of a new receptor-operated channel by heteromeric assembly of TRPP2 and TRPC1 subunits. EMBO Rep 9:472–479

Banke TG, Wickenden AD (2009) Intracellular zinc irritates TRPA1. Nat Chem Biol 5:141–142

Bavassano C, Marvaldi L, Langeslag M, Sarg B, Lindner H, Klimaschewski L, Kress M, Ferrer-Montiel A, Knaus HG (2013) Identification of voltage-gated K channel beta 2 (Kvbeta2) subunit as a novel interaction partner of the pain transducer Transient Receptor Potential Vanilloid 1 channel (TRPV1). Biochimica et biophysica acta 1833:3166–3175

Beck A, Speicher T, Stoerger C, Sell T, Dettmer V, Jusoh SA, Abdulmughni A, Cavali Eacute A, Philipp SE, Zhu MX et al (2013) Conserved gating elements in TRPC4 and TRPC5 channels. J Biol Chem 288:19471–19483

Bellono NW, Oancea E (2013) UV light phototransduction depolarizes human melanocytes. Channels (Austin) 7(3), ahead of pring, PMID 23764911

Bellono NW, Kammel LG, Zimmerman AL, Oancea E (2013) UV light phototransduction activates transient receptor potential A1 ion channels in human melanocytes. Proc Natl Acad Sci U S A 110:2383–2388

Benedikt J, Samad A, Ettrich R, Teisinger J, Vlachova V (2009) Essential role for the putative S6 inner pore region in the activation gating of the human TRPA1 channel. Biochim Biophys Acta 1793:1279–1288

Berbey C, Weiss N, Legrand C, Allard B (2009) Transient receptor potential canonical type 1 (TRPC1) operates as a sarcoplasmic reticulum calcium leak channel in skeletal muscle. J Biol Chem 284:36387–36394

Birerdinc A, Jarrar M, Stotish T, Randhawa M, Baranova A (2012) Manipulating molecular switches in brown adipocytes and their precursors: a therapeutic potential. Prog Lipid Res 52:51–61

Bohlen CJ, Priel A, Zhou S, King D, Siemens J, Julius D (2010) A bivalent Tarantula toxin activates the capsaicin receptor, TRPV1, by targeting the outer pore domain. Cell 141:834–845

Bright JN, Sansom MS (2004) Kv channel S6 helix as a molecular switch: simulation studies. IEE Proc Nanobiotechnol 151:17–27

Cai X (2008) Unicellular Ca2+ signaling 'Toolkit' at the origin of metazoa. Mol Biol Evol 25:1357–1361

Cao E, Cordero-Morales JF, Liu B, Qin F, Julius D (2013) TRPV1 channels are intrinsically heat sensitive and negatively regulated by phosphoinositide lipids. Neuron 77:667–679

Cavanaugh EJ, Simkin D, Kim D (2008) Activation of TRPA1 by mustard oil, tetrahydrocannabinol and Ca2+ reveals different functional states of TRPA1. Neuroscience 154:1467–1476

Cavanaugh DJ, Chesler AT, Braz JM, Shah NM, Julius D, Basbaum AI (2011a) Restriction of transient receptor potential vanilloid-1 to the peptidergic subset of primary afferent neurons follows its developmental downregulation in nonpeptidergic neurons. J Neurosci 31:10119–10127

Cavanaugh DJ, Chesler AT, Jackson AC, Sigal YM, Yamanaka H, Grant R, O'Donnell D, Nicoll RA, Shah NM, Julius D, Basbaum AI (2011b) Trpv1 reporter mice reveal highly restricted

brain distribution and functional expression in arteriolar smooth muscle cells. J Neurosci 31:5067–5077
Chahl LA (2007) TRP's: links to schizophrenia? Biochim Biophys Acta 1772:968–977
Chen J, Zhang XF, Kort ME, Huth JR, Sun C, Miesbauer LJ, Cassar SC, Neelands T, Scott VE, Moreland RB et al (2008) Molecular determinants of species-specific activation or blockade of TRPA1 channels. J Neurosci 28:5063–5071
Chen J, Kang D, Xu J, Lake M, Hogan JO, Sun C, Walter K, Yao B, Kim D (2013) Species differences and molecular determinant of TRPA1 cold sensitivity. Nat Commun 4:2501
Cheng X, Shen D, Samie M, Xu H (2010) Mucolipins: intracellular TRPML1-3 channels. FEBS Lett 584:2013–2021
Cho H, Yang YD, Lee J, Lee B, Kim T, Jang Y, Back SK, Na HS, Harfe BD, Wang F et al (2012) The calcium-activated chloride channel anoctamin 1 acts as a heat sensor in nociceptive neurons. Nat Neurosci 15:1015–1021
Chu B, Postma M, Hardie RC (2013) Fractional Ca2+ currents through TRP and TRPL channels in Drosophila photoreceptors. Biophys J 104:1905–1916
Chuang HH, Prescott ED, Kong H, Shields S, Jordt SE, Basbaum AI, Chao MV, Julius D (2001) Bradykinin and nerve growth factor release the capsaicin receptor from PtdIns(4,5)P2-mediated inhibition. Nature 411:957–962
Chun JN, Lim JM, Kang Y, Kim EH, Shin YC, Kim HG, Jang D, Kwon D, Shin SY, So I, Jeon JH (2013) A network perspective on unraveling the role of TRP channels in biology and disease. Pflugers Arch 466:173–182
Chung MK, Guler AD, Caterina MJ (2008) TRPV1 shows dynamic ionic selectivity during agonist stimulation. Nat Neurosci 11:555–564
Cook NL, Van Den Heuvel C, Vink R (2009) Are the transient receptor potential melastatin (TRPM) channels important in magnesium homeostasis following traumatic brain injury? Magnes Res 22:225–234
Cook NL, Vink R, Helps SC, Manavis J, van den Heuvel C (2010) Transient receptor potential melastatin 2 expression is increased following experimental traumatic brain injury in rats. J Mol Neurosci 42:192–199
Cornell RA, Aarts M, Bautista D, Garcia-Anoveros J, Kiselyov K, Liman ER (2008) A double TRPtych: six views of transient receptor potential channels in disease and health. J Neurosci 28:11778–11784
Cosens DJ, Manning A (1969) Abnormal electroretinogram from a Drosophila mutant. Nature 224:285–287
Courjaret R, Hubrack S, Daalis A, Dib M, Machaca K (2013) The Xenopus TRPV6 homolog encodes a Mg-permeant channel that is inhibited by interaction with TRPC1. J Cell Physiol 228:2386–2398
Cuajungco MP, Grimm C, Oshima K, D'Hoedt D, Nilius B, Mensenkamp AR, Bindels RJ, Plomann M, Heller S (2006) PACSINs bind to the TRPV4 cation channel: PACSIN 3 modulates the subcellular localization of TRPV4. J Biol Chem 281:18753–18762
Cvetkov TL, Huynh KW, Cohen MR, Moiseenkova-Bell VY (2011) Molecular architecture and subunit organization of TRPA1 channel revealed by electron microscopy. J Biol Chem 286:38168–38176
D'hoedt D, Owsianik G, Prenen P, Cuajungco MP, Grimm G, Heller S, Voets T, Nilius B (2007) Stimulus-specific modulation of the cation channel TRPV4 by PACSIN 3. J Biol Chem 283:6272–6280
Damann N, Voets T, Nilius B (2008) TRPs in our senses. Curr Biol 18:R880–889
de la Roche J, Eberhardt MJ, Klinger AB, Stanslowksy N, Wegner F, Koppert W, Reeh PW, Lampert A, Fischer MJ, Leffler A (2013) The molecular basis for species-specific activation of human TRPA1 by protons involves poorly conserved residues within transmembrane domains 5 and 6. J Biol Chem 288:20280–20292
Demion M, Bois P, Launay P, Guinamard R (2007) TRPM4, a Ca(2+)-activated nonselective cation channel in mouse sino-atrial node cells. Cardiovasc Res 73:531–538

den Dekker E, Schoeber J, Topala CN, van de Graaf SF, Hoenderop JG, Bindels RJ (2005) Characterization of a Madin-Darby canine kidney cell line stably expressing TRPV5. Pflugers Arch 450:236–244
Devi S, Kedlaya R, Maddodi N, Bhat KM, Weber CS, Valdivia HH, Setaluri V (2009) Calcium homeostasis in human melanocytes: role of transient receptor potential melastatin 1 (TRPM1) and its regulation by ultraviolet light. Am J Physiol Cell Physiol 297:C679–C687
Dhaka A, Viswanath V, Patapoutian A (2006) TRP ion channels and temperature sensation. Annu Rev Neurosci 29:135–161
Dong XP, Shen D, Wang X, Dawson T, Li X, Zhang Q, Cheng X, Zhang Y, Weisman LS, Delling M, Xu H (2010) PI(3,5)P(2) controls membrane traffic by direct activation of mucolipin Ca release channels in the endolysosome. Nat Commun 1:38
Dong HW, Davis JC, Ding SY, Nai Q, Zhou FM, Ennis M (2012) Expression of transient receptor potential (TRP) channel mRNAs in the mouse olfactory bulb. Neurosci Lett 524:49–54
Earley S, Pauy T, Drapp R, Tavares MJ, Liedtke W, Brayden JE (2009) TRPV4-dependent dilation of peripheral resistance arteries influences arterial pressure. Am J Physiol 297:H1096–H1102
Ebert J, Westhoff G (2006) Behavioural examination of the infrared sensitivity of rattlesnakes (Crotalus atrox). J Comp Physiol A 192:941–947
Eijkelkamp N, Quick K, Wood JN (2013) Transient receptor potential channels and mechanosensation. Annu Rev Neurosci 36:519–546
Everaerts W, Nilius B, Owsianik G (2010) The vanilloid transient receptor potential channel Trpv4: From structure to disease. Prog Biophys Mol Biol 103:2–17
Fecher-Trost C, Wissenbach U, Beck A, Schalkowsky P, Stoerger C, Doerr J, Dembek A, Simon-Thomas M, Weber A, Wollenberg P et al (2013) The in vivo TRPV6 protein starts at a non-AUG triplet decoded as methionine upstream the canonical initiation at AUG. J Biol Chem 288:16629–16644
Feng S, Li H, Tai Y, Huang J, Su Y, Abramowitz J, Zhu MX, Birnbaumer L, Wang Y (2013) Canonical transient receptor potential 3 channels regulate mitochondrial calcium uptake. Proc Natl Acad Sci U S A 110:11011–11016
Fernandes AM, Fero K, Arrenberg AB, Bergeron SA, Driever W, Burgess HA (2012) Deep brain photoreceptors control light-seeking behavior in zebrafish larvae. Curr Biol 22:2042–2047
Fernandes AM, Fero K, Driever W, Burgess HA (2013) Enlightening the brain: linking deep brain photoreception with behavior and physiology. BioEssays 35:775–779
Ferrer-Montiel A, Fernández-Carvajal A, Planells-Cases R, Fernández-Ballester G, González-Ros JM, Messeguer A, González-Muñiz R (2012) Advances in modulating thermosensory TRP channels. Expert Opin Ther Pat 22:999–1017
Filosa JA, Yao X, Rath G (2013) TRPV4 and the regulation of vascular tone. J Cardiovasc Pharmacol 61:113–119
Flockerzi V, Jung C, Aberle T, Meissner M, Freichel M, Philipp SE, Nastainczyk W, Maurer P, Zimmermann R (2005) Specific detection and semi-quantitative analysis of TRPC4 protein expression by antibodies. Pflugers Arch 451:81–86
Fujiwara Y, Minor DL Jr (2008) X-ray crystal structure of a TRPM assembly domain reveals an antiparallel four-stranded coiled-coil. J Mol Biol 383:854–870
Fujiwara Y, Kurokawa T, Takeshita K, Kobayashi M, Okochi Y, Nakagawa A, Okamura Y (2012) The cytoplasmic coiled-coil mediates cooperative gating temperature sensitivity in the voltage-gated H(+) channel Hv1. Nat Commun 3:816
Gamal El-Din TM, Heldstab H, Lehmann C, Greeff NG (2011) Double gaps along Shaker S4 demonstrate omega currents at three different closed states. Channels (Austin) 4:93–100
Garami A, Shimansky YP, Pakai E, Oliveira DL, Gavva NR, Romanovsky AA (2010) Contributions of different modes of TRPV1 activation to TRPV1 antagonist-induced hyperthermia. J Neurosci 30:1435–1440
Garami A, Pakai E, Oliveira DL, Steiner AA, Wanner SP, Almeida MC, Lesnikov VA, Gavva NR, Romanovsky AA (2011) Thermoregulatory phenotype of the Trpv1 knockout mouse: thermoeffector dysbalance with hyperkinesis. J Neurosci 31:721–173

Garcia-Sanz N, Fernandez-Carvajal A, Morenilla-Palao C, Planells-Cases R, Fajardo-Sanchez E, Fernandez-Ballester G, Ferrer-Montiel A (2004) Identification of a tetramerization domain in the C terminus of the vanilloid receptor. J Neurosci 24:5307–5314

Gaudet R (2008a) A primer on ankyrin repeat function in TRP channels and beyond. Mol Biosyst 4:372–379

Gaudet R (2008b) TRP channels entering the structural era. J Physiol 586:3565–3575

Gees M, Colsoul B, Nilius B (2010) The role of transient receptor potential cation channels in Ca2+ signaling. Cold Spring Harb Perspect Biol 2:a003962

Geng J, Liang D, Jiang K, Zhang P (2012) Molecular evolution of the infrared sensory gene TRPA1 in snakes and implications for functional studies. PLoS One 6:e28644

Gerzanich V, Woo KS, Vennekens R, Orest Tsymbalyuk SO, Ivanova S, Ivanov A, Geng Z, Chen Z, Nilius B, Flockerzi V et al (2009) De novo expression of TRPM4 initiates secondary hemorrhage in spinal cord injury. Nat Med 15:185–191

Gilliam JC, Wense TG (2011) TRP channel gene expression in the mouse retina. Vision Research 51, 2440-2452.x. Vision Res 51:2440–2452

Gordon-Shaag A, Zagotta WN, Gordon SE (2008) Mechanism of Ca^{2+}-dependent desensitization of TRP channels. Channels 2:125–129

Gracheva EO, Ingolia NT, Kelly YM, Cordero-Morales JF, Hollopeter G, Chesler AT, Sanchez EE, Perez JC, Weissman JS, Julius D (2010) Molecular basis of infrared detection by snakes. Nature 464:1006–1012

Gracheva EO, Cordero-Morales JF, Gonzalez-Carcacia JA, Ingolia NT, Manno C, Aranguren CI, Weissman JS, Julius D (2011) Ganglion-specific splicing of TRPV1 underlies infrared sensation in vampire bats. Nature 476:88–91

Griffith TN, Varela-Nallar L, Dinamarca MC, Inestrosa NC (2010) Neurobiological effects of hyperforin and its potential in Alzheimer's disease therapy. Curr Med Chem 17:391–406

Gu Q, Lin RL (2010) Heavy metals zinc, cadmium and copper stimulate pulmonary sensory neurons via direct activation of TRPA1. J Appl Physiol 108:891–897

Gygi SP, Rochon Y, Franza BR, Aebersold R (1999) Correlation between protein and mRNA abundance in yeast. Mol Cell Biol 19:1720–1730

Hara K, Kokubo Y, Ishiura H, Fukuda Y, Miyashita A, Kuwano R, Sasaki R, Goto J, Nishizawa M, Kuzuhara S, Tsuji S (2010) TRPM7 is not associated with amyotrophic lateral sclerosis-parkinsonism dementia complex in the Kii peninsula of Japan. Am J Med Genet B Neuropsychiatr Genet 153B:310–313

Hardie RC, Franze K (2012) Photomechanical responses in Drosophila photoreceptors. Science 338:260–263

Hardie RC, Minke B (1992) The trp gene is essential for a light-activated Ca^{2+} channel in Drosophila photoreceptors. Neuron 8:643–651

Hardie RC, Minke B (1993) Novel Ca2+ channels underlying transduction in Drosophila photoreceptors: implications for phosphoinositide-mediated Ca2+ mobilization. Trends Neurosci 16:371–376

Hardie RC, Postma M (2008) Phototransduction in microvillar photoreceptors of Drosophila and other invertebrates. In: Basbaum AI, Shepherd GM, Westheimer G (eds) The senses: a comprehensive reference, Vision I. Academic, San Diego, pp 77–130

Hermosura MC, Garruto RM (2007) TRPM7 and TRPM2-CANDIDATE susceptibility genes for Western Pacific ALS and PD? Biochim Biophys Acta 1772:822–835

Hermosura MC, Monteilh-Zoller MK, Scharenberg AM, Penner R, Fleig A (2002) Dissociation of the store-operated calcium current ICRAC and the Mg-nucleotide-regulated metal ion current MagNuM. J Physiol (Lond) 539:445–458

Hermosura MC, Nayakanti H, Dorovkov MV, Calderon FR, Ryazanov AG, Haymer DS, Garruto RM (2005) A TRPM7 variant shows altered sensitivity to magnesium that may contribute to the pathogenesis of two Guamanian neurodegenerative disorders. Proc Natl Acad Sci U S A 102:11510–11515

Hermosura M, Cui AM, Go G, Davenport B, Shetler C, Heizer J, Schmitz C, Mocz G, Garruto R, Perraud A-L (2008) Altered functional properties of a TRPM2 variant in Guamanian ALS and PD. Proc Natl Acad Sci U S A 105:18029–18034

Hoenderop JG, Voets T, Hoefs S, Weidema F, Prenen J, Nilius B, Bindels RJ (2003) Homo- and heterotetrameric architecture of the epithelial Ca2+ channels TRPV5 and TRPV6. EMBO J 22:776–785

Hofmann T, Schaefer M, Schultz G, Gudermann T (2002) Subunit composition of mammalian transient receptor potential channels in living cells. Proc Natl Acad Sci U S A 99:7461–7466

Hofmann F, Flockerz IV, Kahl S, Wegener JW (2014) L-type CaV1.2 calcium channels: from in vitro findings to in vivo function. Physiol Rev 94:303–26

Hu G, Oboukhova EA, Kumar S, Sturek M, Obukhov AG (2009a) Canonical transient receptor potential channels expression is elevated in a porcine model of metabolic syndrome. Mol Endocrinol 23:689–699

Hu H, Bandell M, Petrus MJ, Zhu MX, Patapoutian A (2009b) Zinc activates damage-sensing TRPA1 ion channels. Nat Chem Biol 5:183–190

Hughes S, Pothecary CA, Jagannath A, Foster RG, Hankins MW, Peirson SN (2012) Profound defects in pupillary responses to light in TRPM-channel null mice: a role for TRPM channels in non-image-forming photoreception. Eur J Neurosci 35:34–43

Ihara M, Hamamoto S, Miyanoiri Y, Takeda M, Kainosho M, Yabe I, Uozumi N, Yamashita A (2013) Molecular bases of multimodal regulation of a fungal transient receptor potential (TRP) channel. J Biol Chem 288

Inada H, Procko E, Sotomayor M, Gaudet R (2012) Structural and biochemical consequences of disease-causing mutations in the ankyrin repeat domain of the human TRPV4 channel. Biochemistry 51:6195–6206

Ito N, Ruegg UT, Kudo A, Miyagoe-Suzuki Y, Takeda S (2012) Activation of calcium signaling through Trpv1 by nNOS and peroxynitrite as a key trigger of skeletal muscle hypertrophy. Nat Med 19:101–106

Jansson ET, Trkulja CL, Ahemaiti A, Millingen M, Jeffries GD, Jardemark K, Orwar O (2013) Effect of cholesterol depletion on the pore dilation of TRPV1. Mol Pain 9:1

Jeon JP, Lee KP, Park EJ, Sung TS, Kim BJ, Jeon JH, So I (2008) The specific activation of TRPC4 by Gi protein subtype. Biochem Biophys Res Commun 377:538–543

Jeon JP, Hong C, Park EJ, Jeon JH, Cho NH, Kim IG, Choe H, Muallem S, Kim HJ, So I (2012) Selective Galphai subunits as novel direct activators of TRPC4 and TRPC5 channels. J Biol Chem 287:17029–17039

Jeon JP, Roh SE, Wie J, Kim J, Kim H, Lee KP, Yang D, Jeon JH, Cho NH, Kim IG et al (2013) Activation of TRPC4beta by Galpha subunit increases Ca selectivity and controls neurite morphogenesis in cultured hippocampal neuron. Cell Calcium 54:307–319

Jin X, Touhey J, Gaudet R (2006) Structure of the N-terminal ankyrin repeat domain of the TRPV2 ion channel. J Biol Chem 281:25006–25010

Jordt SE, Julius D (2002) Molecular basis for species-specific sensitivity to "hot" chili peppers. Cell 108:421–430

Jurkat-Rott K, Holzherr B, Fauler M, Lehmann-Horn F (2010) Sodium channelopathies of skeletal muscle result from gain or loss of function. Pflugers Archiv 460:239–248

Kalia J, Swartz KJ (2013) Exploring structure-function relationships between TRP and Kv channels. Sci Rep 3:1523

Karashima Y, Prenen J, Meseguer V, Owsianik G, Voets T, Nilius B (2008) Modulation of the transient receptor potential channel TRPA1 by phosphatidylinositol 4,5-biphosphate manipulators. Pflugers Arch 457:77–89

Karashima Y, Prenen J, Talavera K, Janssens A, Voets T, Nilius B (2010) Agonist-induced changes in Ca(2+) permeation through the nociceptor cation channel TRPA1. Biophys J 98:773–783

Kassmann M, Harteneck C, Zhu Z, Nurnberg B, Tepel M, Gollasch M (2013) TRPV1, TRPV4, and the kidney. Acta Physiol (Oxf) 207:546–564

Kedei N, Szabo T, Lile JD, Treanor JJ, Olah Z, Iadarola MJ, Blumberg PM (2001) Analysis of the native quaternary structure of vanilloid receptor 1. J Biol Chem 276:28613–28619
Kim D, Cavanaugh EJ (2007) Requirement of a soluble intracellular factor for activation of transient receptor potential A1 by pungent chemicals: role of inorganic polyphosphates. J Neurosci 27:6500–6509
Kim AY, Tang Z, Liu Q, Patel KN, Maag D, Geng Y, Dong X (2008) Pirt, a phosphoinositide-binding protein, functions as a regulatory subunit of TRPV1. Cell 133:475–485
Kim D, Cavanaugh E, Simkin D (2009) Inhibition of transient receptor potential A1 by phosphatidylinositol-4,5-bisphosphate. Am J Physiol Cell Physiol 295:C92–99
Kirkwood NK, Albert JT (2013) Sensory transduction: confusing the senses. Curr Biol 23:R22–23
Kiselyov K, Soyombo A, Muallem S (2007) TRPpathias. J Physiol 578:641–653
Klein RM, Ufret-Vincenty CA, Hua L, Gordon SE (2008) Determinants of molecular specificity in phosphoinositide regulation: PI(4,5)P2 is the endogenous lipid regulating TRPV1. J Biol Chem 283:26208–26216
Klionsky L, Tamir R, Gao B, Wang W, Immke DC, Nishimura N, Gavva NR (2007) Species-specific pharmacology of Trichloro(sulfanyl)ethyl benzamides as transient receptor potential ankyrin 1 (TRPA1) antagonists. Mol Pain 3:39
Kohler R, Heyken WT, Heinau P, Schubert R, Si H, Kacik M, Busch C, Grgic I, Maier T, Hoyer J (2006) Evidence for a functional role of endothelial transient receptor potential V4 in shear stress-induced vasodilatation. Arterioscler Thromb Vasc Biol 26:1495–1502
Koike C, Obara T, Uriu Y, Numata T, Sanuki R, Miyata K, Koyasu T, Ueno S, Funabiki K, Tani A et al (2010) TRPM1 is a component of the retinal ON bipolar cell transduction channel in the mGluR6 cascade. Proc Natl Acad Sci U S A 107:332–337
Kremeyer B, Lopera F, Cox JJ, Momin A, Rugiero F, Marsh S, Woods CG, Jones NG, Paterson KJ, Fricker FR et al (2010) A gain-of-function mutation in TRPA1 causes familial episodic pain syndrome. Neuron 66:671–680
Landman N, Jeong SY, Shin SY, Voronov SV, Serban G, Kang MS, Park MK, Di Paolo G, Chung S, Kim T-W (2006) Presenilin mutations linked to familial Alzheimer's disease cause an imbalance in phosphatidylinositol 4,5-bisphosphate metabolism. Proc Natl Acad Sci U S A 103:19524–19529
Lee JH, Lee Y, Ryu H, Kang DW, Lee J, Lazar J, Pearce LV, Pavlyukovets VA, Blumberg PM, Choi S (2011) Structural insights into transient receptor potential vanilloid type 1 (TRPV1) from homology modeling, flexible docking, and mutational studies. J Comput Aided Mol Des 25:317–327
Lepichon JB, Bittel DC, Graf WD, Yu S (2010) A 15q13.3 homozygous microdeletion associated with a severe neurodevelopmental disorder suggests putative functions of the TRPM1, CHRNA7, and other homozygously deleted genes. Am J Med Genet A 152A:1300–1304
Leuner K, Kazanski V, Muller M, Essin K, Henke B, Gollasch M, Harteneck C, Muller WE (2007) Hyperforin-a key constituent of St. John's wort specifically activates TRPC6 channels. FASEB J 21:4101–4111
Leuner K, Heiser JH, Derksen S, Mladenov MI, Fehske CJ, Schubert R, Gollasch M, Schneider G, Harteneck C, Chatterjee SS, Muller WE (2010) Simple 2,4 diacylphloroglucinols as TRPC6 activators - identification of a novel pharmacophore. Mol Pharmacol 77:368–377
Li L, Chen J, Ni Y, Feng X, Zhao Z, Wang P, Sun J, Yu H, Yan Z, Liu D et al (2012) TRPV1 activation prevents nonalcoholic fatty liver through UCP2 upregulation in mice. Pflugers Archiv 463:727–732
Li Q, Li L, Wang F, Chen J, Zhao Y, Wang P, Nilius B, Liu D, Zhu Z (2013) Dietary capsaicin prevents nonalcoholic fatty liver disease through transient receptor potential vanilloid 1-mediated peroxisome proliferator-activated receptor delta activation. Pflugers Archiv 465:1303–1316
Liao Y, Erxleben C, Abramowitz J, Flockerzi V, Zhu MX, Armstrong DL, Birnbaumer L (2008) Functional interactions among Orai1, TRPCs, and STIM1 suggest a STIM-regulated

heteromeric Orai/TRPC model for SOCE/Icrac channels. Proc Natl Acad Sci U S A 105:2895–2900
Liscombe DK, Facchini PJ (2008) Evolutionary and cellular webs in benzylisoquinoline alkaloid biosynthesis. Curr Opin Biotechnol 19:173–180
Lishko PV, Procko E, Jin X, Phelps CB, Gaudet R (2007) The Ankyrin repeats of TRPV1 bind multiple ligands and modulate channel sensitivity. Neuron 54:905–918
Liu D, Zhu Z, Tepel M (2008) The role of transient receptor potential channels in metabolic syndrome. Hypertens Res 31:1989–1995
Long SB, Campbell EB, Mackinnon R (2005) Crystal structure of a mammalian voltage-dependent Shaker family K+ channel. Science 309:897–903
Loot AE, Popp R, Fisslthaler B, Vriens J, Nilius B, Fleming I (2008) Role of cytochrome P450-dependent TRPV4 activation in flow-induced vasodilatation. Cardiovasc Res 80:445–452
Lucas RJ (2013) Mammalian inner retinal photoreception. Curr Biol 23:R125–133
Lukacs V, Yudin Y, Hammond GR, Sharma E, Fukami K, Rohacs T (2013) Distinctive changes in plasma membrane phosphoinositides underlie differential regulation of TRPV1 in nociceptive neurons. J Neurosci 33:11451–11463
Luo Z, Ma L, Zhao Z, He H, Yang D, Feng X, Ma S, Chen X, Zhu T, Cao T et al (2012) TRPV1 activation improves exercise endurance and energy metabolism through PGC-1alpha upregulation in mice. Cell Res 22:551–564
Ma X, Qiu S, Luo J, Ma Y, Ngai CY, Shen B, Wong CO, Huang Y, Yao X (2010) Functional role of TRPV4-TRPC1 complex in flow-induced Ca2+ influx. Arterioscler Thromb Vasc Biol 30:851–858
Ma X, Nilius B, Wong JW, Huang Y, Yao X (2011) Electrophysiological properties of heteromeric TRPV4-C1 channels. Biochim Biophys Acta 1808:2807–2818
Madrid R, de la Pen E, Donovan-Rodriguez T, Belmonte C, Viana F (2009) Variable threshold of trigeminal cold-thermosensitive neurons is determined by a balance between TRPM8 and Kv1 potassium channels. J Physiol (Lond) 587:1961–1976
Matsuura H, Sokabe T, Kohno K, Tominaga M, Kadowaki T (2009) Evolutionary conservation and changes in insect TRP channels. BMC Evol Biol 9:228
McNulty S, Fonfria E (2005) The role of TRPM channels in cell death. Pflügers Arch 451:235–242
Meissner M, Obmann VC, Hoschke M, Link S, Jung M, Held G, Philipp SE, Zimmermann R, Flockerzi V (2011) Lessons of studying TRP channels with antibodies. In: Zhu MX (ed) TRP channels. CRC Press, Boca Raton, FL, pp 135–148
Moiseenkova-Bell VY, Stanciu LA, Serysheva II, Tobe BJ, Wensel TG (2008) Structure of TRPV1 channel revealed by electron cryomicroscopy. Proc Natl Acad Sci U S A 105:7451–7455
Molnar T, Barabas P, Birnbaumer L, Punzo C, Kefalov V, Krizaj D (2012) Store-operated channels regulate intracellular calcium in mammalian rods. J Physiol 590:3465–3481
Montell C (2012) Drosophila visual transduction. Trends Neurosci 35:356–363
Montell C, Rubin GM (1989) Molecular characterization of the Drosophila trp locus: a putative integral membrane protein required for phototransduction. Neuron 2:1313–1323
Montell C, Birnbaumer L, Flockerzi V, Bindels RJ, Bruford EA, Caterina MJ, Clapham D, Harteneck C, Heller S, Julius D et al (2002) A unified nomenclature for the superfamily of TRP cation channels. Mol Cell 9:229–231
Moore C, Cevikbas F, Pasolli HA, Chen Y, Kong W, Kempkes C, Parekh P, Lee SH, Kontchoua N-A, Ye I et al (2013) UVB radiation generates sunburn pain and affects skin by activating epidermal TRPV4 ion channels and triggering endothelin-1 signaling. Proc Natl Acad Sci U S A 110:E3225–3234
Morgans CW, Zhang J, Jeffrey BG, Nelson SM, Burke NS, Duvoisin RM, Brown RL (2009) TRPM1 is required for the depolarizing light response in retinal ON-bipolar cells. Proc Natl Acad Sci U S A 106:19174–19178
Mori Y, Wakamori M, Miyakawa T, Hermosura M, Hara Y, Nishida M, Hirose K, Mizushima A, Kurosaki M, Mori E et al (2002) Transient receptor potential 1 regulates capacitative Ca(2+)

entry and Ca(2+) release from endoplasmic reticulum in B lymphocytes. J Exp Med 195:673–681
Moussaieff A, Mechoulam R (2009) Boswellia resin: from religious ceremonies to medical uses; a review of in-vitro, in-vivo and clinical trials. J Pharm Pharmacol 61:1281–1293
Moussaieff A, Rimmerman N, Bregman T, Straiker A, Felder CC, Shoham S, Kashman Y, Huang SM, Lee H, Shohami E et al (2008) Incensole acetate, an incense component, elicits psychoactivity by activating TRPV3 channels in the brain. FASEB J 22:3024–3034
Muller CS, Haupt A, Bildl W, Schindler J, Knaus HG, Meissner M, Rammner B, Striessnig J, Flockerzi V, Fakler B, Schulte U (2010) Quantitative proteomics of the Cav2 channel nano-environments in the mammalian brain. Proc Natl Acad Sci U S A 107:14950–14957
Nagatomo K, Ishii H, Yamamoto T, Nakajo K, Kubo Y (2010) The Met268Pro mutation of mouse TRPA1 changes the effect of caffeine from activation to suppression. Biophys J 99:3609–3618
Nakamura M, Sanuki R, Yasuma TR, Onishi A, Nishiguchi KM, Koike C, Kadowaki M, Kondo M, Miyake Y, Furukawa T (2010) TRPM1 mutations are associated with the complete form of congenital stationary night blindness. Mol Vis 16:425–437
Nieto-Posadas A, Jara-Oseguera A, Rosenbaum T (2011) TRP channel gating physiology. Curr Top Med Chem 11:2131–2150
Nilius B (2007) TRP channels in disease. Biochim Biophys Acta 1772:805–812
Nilius B, Appendino G (2011) Tasty and healthy TR(i)PS: the human quest for culinary pungency. EMBO Rep 12:1094–1101
Nilius B, Appendino G (2013) Spices: the savory and beneficial science of pungency. Rev Physiol Biochem Pharmacol 164:1–76
Nilius B, Honore E (2013) Sensing pressure with ion channels. Trends Neurosci 35:477–486
Nilius B, Owsianik G (2010a) Channelopathies converge on TRPV4. Nat Genet 42:98–100
Nilius B, Owsianik G (2010b) Transient receptor potential channelopathies. Pflugers Archiv 460:437–450
Nilius B, Owsianik G (2011) The transient receptor potential family of ion channels. Genome Biol 12:218–229
Nilius B, Voets T (2013) The puzzle of TRPV4 channelopathies. EMBO Rep 14:152–163
Nilius B, Vennekens R, Prenen J, Hoenderop JG, Bindels RJ, Droogmans G (2000) Whole-cell and single channel monovalent cation currents through the novel rabbit epithelial Ca^{2+} channel ECaC. J Physiol (London) 527:239–248
Nilius B, Prenen J, Tang J, Wang C, Owsianik G, Janssens A, Voets T, Zhu MX (2005a) Regulation of the Ca^{2+} sensitivity of the nonselective cation channel TRPM4. J Biol Chem 280:6423–6433
Nilius B, Talavera K, Owsianik G, Prenen J, Droogmans G, Voets T (2005b) Gating of TRP channels: a voltage connection? J Physiol (Lond) 567:33–44
Nilius B, Mahieu F, Prenen J, Janssens A, Owsianik G, Vennekens R, Voets T (2006) The Ca^{2+}-activated cation channel TRPM4 is regulated by phosphatidylinositol 4,5-biphosphate. EMBO J 25:467–478
Nilius B, Owsianik G, Voets T, Peters JA (2007) Transient receptor potential channels in disease. Physiol Rev 87:165–217
Nilius B, Owsianik G, Voets T (2008) Transient receptor potential channels meet phosphoinositides. EMBO J 27:2809–2816
Nilius B, Prenen J, Owsianik G (2011) Irritating channels: the case of TRPA1. J Physiol 589:1543–1549
Nilius B, Appendino G, Owsianik G (2012) The transient receptor potential channel TRPA1: from gene to pathophysiology. Pflugers Archiv 464:425–458
Nilius B, Biro T, Owsianik G (2014) TRPV3: time to decipher a poorly understood family member! J Physiol 592:295–304
Ong HL, Chen J, Chataway T, Brereton H, Zhang L, Downs T, Tsiokas L, Barritt G (2002) Specific detection of the endogenous transient receptor potential (TRP)-1 protein in liver and airway

smooth muscle cells using immunoprecipitation and Western-blot analysis. Biochem J 364:641–648
Ong EC, Nesin V, Long CL, Bai CX, Guz JL, Ivanov IP, Abramowitz J, Birnbaumer L, Humphrey MB, Tsiokas L (2013) A TRPC1-dependent pathway regulates osteoclast formation and function. J Biol Chem 288:22219–22232
Orio P, Parra A, Madrid R, Gonzalez O, Belmonte C, Viana F (2013) Role of Ih in the firing pattern of mammalian cold thermoreceptor endings. J Neurophysiol 108:3009–3023
Otsuguro KI, Tang J, Tang Y, Xiao R, Freichel M, Tsvilovskyy V, Ito S, Flockerzi V, Zhu MX, Zholos AV (2008) Isoform-specific inhibition of TRPC4 channel by phosphatidylinositol 4,5-bisphosphate. J Biol Chem 283:10026–10036
Owsianik G, Talavera K, Voets T, Nilius B (2006) Permeation and selectivity of trp channels. Annu Rev Physiol 68:685–717
Pan B, Géléoc GS, Asai Y, Horwitz GC, Kurima K, Ishikawa K, Kawashima Y, Griffith AJ, Holt JR (2013) TMC1 and TMC2 are components of the mechanotransduction channel in hair cells of the mammalian inner ear. Neuron 79:504–515
Panzano VC, Kang K, Garrity PA (2010) Infrared snake eyes: TRPA1 and the thermal sensitivity of the snake pit organ. Sci Signal 3:22
Papakosta M, Dalle C, Haythornthwaite A, Cao L, Stevens EB, Burgess G, Russell R, Cox PJ, Phillips SC, Grimm C (2011) Chimeric approach reveals that differences in the TRPV1 pore domain determine species-specific sensitivity to block of heat activation. J Biol Chem 286:39663–33967
Patapoutian A, Peier AM, Story GM, Viswanath V (2003) ThermoTRP channels and beyond: mechanisms of temperature sensation. Nat Rev Neurosci 4:529–539
Phelps CB, Procko E, Lishko PV, Wang RR, Gaudet R (2007) Insights into the roles of conserved and divergent residues in the ankyrin repeats of TRPV ion channels. Channels (Austin) 1:148–151
Phelps CB, Huang RJ, Lishko PV, Wang RR, Gaudet R (2008) Structural analyses of the ankyrin repeat domain of TRPV6 and related TRPV ion channels. Biochemistry 47:2476–2484
Phelps CB, Wang RR, Choo SS, Gaudet R (2010) Differential regulation of TRPV1, TRPV3 AND TRPV4 sensitivity through a conserved binding site on the ankyrin repeat domain. J Biol Chem 285:731–740
Pickard GE, Sollars PJ (2012) Intrinsically photosensitive retinal ganglion cells. Rev Physiol Biochem Pharmacol 162:59–90
Pradet-Balade B, Boulme F, Beug H, Mullner EW, Garcia-Sanz JA (2001) Translation control: bridging the gap between genomics and proteomics? Trends Biochem Sci 26:225–229
Prescott ED, Julius D (2003) A modular PIP_2 binding site as a determinant of capsaicin receptor sensitivity. Science 300:1284–1288
Prole DL, Taylor CW (2011) Identification of intracellular and plasma membrane calcium channel homologues in pathogenic parasites. PLoS One 6:e26218
Prole DL, Taylor CW (2012) Identification and analysis of cation channel homologues in human pathogenic fungi. PLoS One 7:e42404
Prütting S, Grissmer S (2011) A novel current pathway parallel to the central pore in a mutant voltage-gated potassium channel. J Biol Chem 286:20031–20042
Ramsey IS, Delling M, Clapham DE (2006) An introduction to TRP channels. Annu Rev Physiol 68:619–647
Rath G, Dessy C, Feron O (2009) Caveolae, caveolin and control of vascular tone: nitric oxide (NO) and endothelium derived hyperpolarizing factor (EDHF) regulation. J Physiol Pharmacol 60(Suppl 4):105–109
Rempe DA, Takano T, Nedergaard M (2009) TR(I)Pping towards treatment for ischemia. Nat Neurosci 12:1215–1216
Richards MW, Butcher AJ, Dolphin AC (2004) Ca2+ channel beta-subunits: structural insights AID our understanding. Trends Pharmacol Sci 25:626–632

Rohacs T (2013) Regulation of transient receptor potential channels by the phospholipase C pathway. Adv Biol Regul 53:341–355

Rohacs T, Nilius B (2007) Regulation of transient receptor potential (trp) channels by phosphoinositides. Pflugers Archiv 455:157–168

Ryu S, Liu B, Yao J, Fu Q, Qin F (2007) Uncoupling proton activation of vanilloid receptor TRPV1. J Neurosci 27:12797–12807

Sala-Rabanal M, Wang S, Nichols CG (2012) On the potential interactions between non-selective cation channel TRPM4 and the sulfonylurea receptor SUR1. J Biol Chem 287:8746–8756

Saliez J, Bouzin C, Rath G, Ghisdal P, Desjardins F, Rezzani R, Rodella LF, Vriens J, Nilius B, Feron O et al (2008) Role of caveolar compartmentation in endothelium-derived hyperpolarizing factor-mediated relaxation: Ca2+ signals and gap junction function are regulated by caveolin in endothelial cells. Circulation 117:1065–1074

Samways DS, Egan TM (2011) Calcium-dependent decrease in the single-channel conductance of TRPV1. Pflugers Archiv 462:681–691

Schindl R, Fritsch R, Jardin I, Frischauf I, Kahr H, Muik M, Riedl MC, Groschner K, Romanin C (2012) Canonical transient receptor potential (TRPC) 1 acts as a negative regulator for vanilloid TRPV6 mediated Ca2+ influx. J Biol Chem 287:35612–33520

Schmidt TM, Chen SK, Hattar S (2011) Intrinsically photosensitive retinal ganglion cells: many subtypes, diverse functions. Trends Neurosci 34:572–580

Schwenk J, Harmel N, Brechet A, Zolles G, Berkefeld H, Muller CS, Bildl W, Baehrens D, Huber B, Kulik A et al (2012) High-resolution proteomics unravel architecture and molecular diversity of native AMPA receptor complexes. Neuron 74:621–633

Sedgwick SG, Smerdon SJ (1999) The ankyrin repeat: a diversity of interactions on a common structural framework. Trends Biochem Sci 24:311–316

Shen WL, Kwon Y, Adegbola AA, Luo J, Chess A, Montell C (2011) Function of rhodopsin in temperature discrimination in Drosophila. Science 331:1333–1336

Shi J, Birnbaumer L, Large WA, Albert AP (2013a) Myristoylated alanine-rich C kinase substrate coordinates native TRPC1 channel activation by phosphatidylinositol 4,5-bisphosphate and protein kinase C in vascular smooth muscle. FASEB J 28:244–255

Shi M, Du F, Liu Y, Li L, Cai J, Zhang GF, Xu XF, Lin T, Cheng HR, Liu XD et al (2013b) Glial cell-expressed mechanosensitive channel TRPV4 mediates infrasound-induced neuronal impairment. Acta neuropathologica 126:725–739

Shibasaki K, Suzuki M, Mizuno A, Tominaga M (2007) Effects of body temperature on neural activity in the hippocampus: regulation of resting membrane potentials by transient receptor potential vanilloid 4. J Neurosci 27:1566–1575

Sigg D, Bezanilla F (1997) Total charge movement per channel. The relation between gating charge displacement and the voltage sensitivity of activation. J Gen Physiol 109:27–39

Simard JM, Kahle KT, Gerzanich V (2010) Molecular mechanisms of microvascular failure in central nervous system injury-synergistic roles of NKCC1 and SUR1/TRPM4. J Neurosurg 113:622–629

Singh BB, Liu X, Tang J, Zhu MX, Ambudkar IS (2002) Calmodulin regulates Ca(2+)-dependent feedback inhibition of store-operated Ca(2+) influx by interaction with a site in the C terminus of TrpC1. Mol Cell 9:739–750

Starace DM, Bezanilla F (2004) A proton pore in a potassium channel voltage sensor reveals a focused electric field. Nature 427:548–553

Storch U, Forst AL, Philipp M, Gudermann T, Mederos YSM (2012) TRPC1 reduces the calcium permeability in heteromeric channel complexes. J Biol Chem 287:3530–3540

Straub I, Krugel U, Mohr F, Teichert J, Rizun O, Konrad M, Oberwinkler J, Schaefer M (2013a) Flavanones that selectively inhibit TRPM3 attenuate thermal nociception in vivo. Mol Pharmacol 84:736–750

Straub I, Mohr F, Stab J, Konrad M, Philipp S, Oberwinkler J, Schaefer M (2013b) Citrus fruit and fabacea secondary metabolites potently and selectively block TRPM3. Br J Pharmacol 168:1835–1850

Strotmann R, Schultz G, Plant TD (2003) Ca^{2+}-dependent potentiation of the nonselective cation channel TRPV4 is mediated by a carboxy terminal calmodulin binding site. J Biol Chem 278:26541–26549

Strotmann R, Semtner M, Kepura F, Plant TD, Schoneberg T (2010) Interdomain interactions control Ca-dependent potentiation in the cation channel TRPV4. PLoS One 5:e10580

Strubing C, Krapivinsky G, Krapivinsky L, Clapham DE (2001) TRPC1 and TRPC5 form a novel cation channel in mammalian brain. Neuron 29:645–655

Sun HS, Jackson MF, Martin LJ, Jansen K, Teves L, Cui H, Kiyonaka S, Mori Y, Jones M, Forder JP et al (2009) Suppression of hippocampal TRPM7 protein prevents delayed neuronal death in brain ischemia. Nat Neurosci 12:1300–1307

Swoap SJ (2013) Thermoregulation: an orphan receptor finds its way in the cold. Curr Biol 22: R17–18

Tang Z, Kim A, Masuch T, Park K, Weng H, Wetzel C, Dong X (2013) Pirt functions as an endogenous regulator of TRPM8. Nat Commun 4:2179

Tew KD, Monks A, Barone L, Rosser D, Akerman G, Montali JA, Wheatley JB, Schmidt DE Jr (1996) Glutathione-associated enzymes in the human cell lines of the National Cancer Institute Drug Screening Program. Mol Pharmacol 50:149–159

Tombola F, Pathak MM, Gorostiza P, Isacoff EY (2007) The twisted ion-permeation pathway of a resting voltage-sensing domain. Nature 445:546–549

Tsiokas L, Arnould T, Zhu C, Kim E, Walz G, Sukhatme V (1999) Specific association of the gene product of PKD2 with the TRPC1 channel. Proc Natl Acad Sci U S A 96:3934–3939

Tsuruda PR, Julius D, Minor DL Jr (2006) Coiled coils direct assembly of a cold-activated TRP channel. Neuron 51:201–212

Tsvilovskyy VV, Zholos AV, Aberle T, Philipp SE, Dietrich A, Zhu MX, Birnbaumer L, Freichel M, Flockerzi V (2009) Deletion of TRPC4 and TRPC6 in mice impairs smooth muscle contraction and intestinal motility in vivo. Gastroenterology 137:1415–1424

Uemura T, Kudoh J, Noda S, Kanba S, Shimizu N (2005) Characterization of human and mouse TRPM2 genes: identification of a novel N-terminal truncated protein specifically expressed in human striatum. Biochem Biophys Res Commun 328:1232–1243

Ufret-Vincenty CA, Klein RM, Hua L, Angueyra J, Gordon SE (2011) Localization of the PIP2 sensor of TRPV1 ion channels. J Biol Chem 283:26208–26216

Ullrich ND, Voets T, Prenen J, Vennekens R, Talavera K, Droogmans G, Nilius B (2005) Comparison of functional properties of the Ca2+-activated cation channels TRPM4 and TRPM5 from mice. Cell Calcium 37:267–278

Vay L, Gu C, McNaughton PA (2011) The thermo-TRP ion channel family: properties and therapeutic implications. Br J Pharmacol 165:787–801

Vennekens R, Hoenderop JG, Prenen J, Stuiver M, Willems PH, Droogmans G, Nilius B, Bindels RJ (2000) Permeation and gating properties of the novel epithelial Ca^{2+} channel. J Biol Chem 275:3963–3969

Vennekens R, Menigoz A, Nilius B (2012) TRPs in the brain. Rev Physiol Biochem Pharmacol 163:27–64

Voets T (2012) Quantifying and modeling the temperature-dependent gating of TRP channels. Rev Physiol Biochem Pharmacol 162:91–119

Voets T, Droogmans G, Wissenbach U, Janssens A, Flockerzi V, Nilius B (2004) The principle of temperature-dependent gating in cold- and heat-sensitive TRP channels. Nature 430:748–754

Voets T, Owsianik G, Janssens A, Talavera K, Nilius B (2007) TRPM8 voltage sensor mutants reveal a mechanism for integrating thermal and chemical stimuli. Nat Chem Biol 3:174–182

Vriens J, Owsianik G, Fisslthaler B, Suzuki M, Janssens A, Voets T, Morisseau C, Hammock BD, Fleming I, Busse R, Nilius B (2005) Modulation of the Ca2 permeable cation channel TRPV4 by cytochrome P450 epoxygenases in vascular endothelium. Circ Res 97:908–915

Vriens J, Owsianik G, Hofmann T, Philipp SE, Stab J, Chen X, Benoit M, Xue F, Janssens A, Kerselaers S et al (2011) TRPM3 is a nociceptor channel INVOLVED in the detection of noxious heat. Neuron 70:482–494

Vriens J, Held K, Janssens A, Tóth BI, Kerselaers S, Nilius B, Vennekens R, Voets T (2014) Opening of an alternative ion permeation pathway in a nociceptor TRP channel. Nature Chem Biol 10:188–195
Wang YY, Chang RB, Waters HN, McKemy DD, Liman ER (2008) The nociceptor ion channel TRPA1 is potentiated and inactivated by permeating calcium ions. J Biol Chem 283:32691–32703
Wang S, Poon K, Oswald RE, Chuang HH (2010) Distinct modulations of human capsaicin receptor by protons and magnesium through different domains. J Biol Chem 285:11547–11156
Wang L, Cvetkov TL, Chance MR, Moiseenkova-Bell VY (2012) Identification of in vivo disulfide conformation of the TRPA1 ion channel. J Biol Chem 287:6169–6176
Watanabe H, Davis JB, Smart D, Jerman JC, Smith GD, Hayes P, Vriens J, Cairns W, Wissenbach U, Prenen J et al (2002) Activation of TRPV4 channels (hVRL-2/mTRP12) by phorbol derivatives. J Biol Chem 277:13569–13577
Wetsel WC (2011) Sensing hot and cold with TRP channels. Int J Hyperthermia 27:388–398
Wheeler GL, Brownlee C (2008) Ca(2+) signalling in plants and green algae - changing channels. Trends Plant Sci 13:506–514
Whorton MR, MacKinnon R (2011) Crystal structure of the mammalian GIRK2 K+ channel and gating regulation by G proteins, PIP2, and sodium. Cell 147:199–208
Woo SK, Kwon MS, Ivanov A, Gerzanich V, Simard JM (2013) The sulfonylurea receptor 1 (Sur1) - transient receptor potential melastatin 4 (Trpm4) channel. J Biol Chem 288:3655–3667
Wu LJ, Sweet TB, Clapham DE (2010) International Union of Basic and Clinical Pharmacology. LXXVI. Current progress in the mammalian TRP ion channel family. Pharmacol Rev 62:381–404
Xiao B, Dubin AE, Bursulaya B, Viswanath V, Jegla TJ, Patapoutian A (2008a) Identification of the transmembrane domain five as a critical molecular determinant of menthol sensitivity in mammalian TRPA1 channels. J Neurosci 28:9640–9651
Xiao R, Tang J, Wang C, Colton CK, Tian J, Zhu MX (2008b) Calcium plays a central role in the sensitization of TRPV3 channel to repetitive stimulations. J Biol Chem 283:6162–6174
Xiao R, Tian J, Tang J, Zhu MX (2008c) The TRPV3 mutation associated with the hairless phenotype in rodents is constitutively active. Cell Calcium 43:334–343
Xiao B, Coste B, Mathur J, Patapoutian A (2011) Temperature-dependent STIM1 activation induces Ca(2+) influx and modulates gene expression. Nat Chem Biol 7:351–358
Xu C, Macciardi F, Li PP, Yoon IS, Cooke RG, Hughes B, Parikh SV, McIntyre RS, Kennedy JL, Warsh JJ (2006) Association of the putative susceptibility gene, transient receptor potential protein melastatin type 2, with bipolar disorder. Am J Med Genet B Neuropsychiatr Genet 141:36–43
Xu C, Li PP, Cooke RG, Parikh SV, Wang K, Kennedy JL, Warsh JJ (2009) TRPM2 variants and bipolar disorder risk: confirmation in a family-based association study. Bipolar Disord 11:1–10
Xue T, Do MT, Riccio A, Jiang Z, Hsieh J, Wang HC, Merbs SL, Welsbie DS, Yoshioka T, Weissgerber P et al (2011) Melanopsin signalling in mammalian iris and retina. Nature 479:67–73
Yamaguchi H, Matsushita M, Nairn AC, Kuriyan J (2001) Crystal structure of the atypical protein kinase domain of a TRP channel with phosphotransferase activity. Mol Cell 7:1047–1057
Ye L, Kleiner S, Wu J, Sah R, Gupta RK, Banks AS, Cohen P, Khandekar MJ, Bostrom P, Mepani RJ et al (2012) TRPV4 is a regulator of adipose oxidative metabolism, inflammation, and energy homeostasis. Cell 151:96–110
Ye L, Wu J, Cohen P, Kazak L, Khandekar MJ, Jedrychowski MP, Zeng X, Gygi SP, Spiegelman BM (2013) Fat cells directly sense temperature to activate thermogenesis. Proc Natl Acad Sci U S A 110:12480–12485
Yokoyama S, Altun A, Denardo DF (2011) Molecular convergence of infrared vision in snakes. Mol Biol Evol 28:45–48

Yu Y, Ulbrich MH, Li MH, Buraei Z, Chen XZ, Ong AC, Tong L, Isacoff EY, Yang J (2009) Structural and molecular basis of the assembly of the TRPP2/PKD1 complex. Proc Natl Acad Sci USA 106:11558–11563

Yuan JP, Zeng W, Huang GN, Worley PF, Muallem S (2007) STIM1 heteromultimerizes TRPC channels to determine their function as store-operated channels. Nat Cell Biol 9:636–645

Zagranichnaya TK, Wu X, Villereal ML (2005) Endogenous TRPC1, TRPC3, and TRPC7 proteins combine to form native store-operated channels in HEK-293 cells. J Biol Chem 280:29559–29569

Zhang DX, Gutterman DD (2010) TRP channel activation and endothelium-dependent dilation in the systemic circulation. J Cardiovasc Pharmacol 52:133–139

Zhang F, Liu S, Yang F, Zheng J, Wang K (2011) Identification of a tetrameric assembly domain in the C-terminus of heat-activated TRPV1 channels. J Biol Chem 286:15308–15316

Zhang X, Li X, Xu H (2012) Phosphoinositide isoforms determine compartment-specific ion channel activity. Proc Natl Acad Sci U S A 109:11384–11389

Zheng J (2013) Molecular mechanism of TRP channels. Compr Physiol 3:221–242

Zholos AV, Zholos AA, Bolton TB (2004) G-protein-gated TRP-like cationic channel activated by muscarinic receptors: effect of potential on single-channel gating. J Gen Physiol 123:581–598

Zhou Y, Suzuki Y, Uchida K, Tominaga M (2013) Identification of a splice variant of mouse TRPA1 that regulates TRPA1 activity. Nat Commun 4:2408

Zhu MX (2005) Multiple roles of calmodulin and other Ca(2+)-binding proteins in the functional regulation of TRP channels. Pflugers Archiv 451:105–115

Zhu X, Chu PB, Peyton M, Birnbaumer L (1995) Molecular cloning of a widely expressed human homologue for the Drosophila trp gene. FEBS Lett 373:193–198

Zhu X, Jiang M, Peyton M, Boulay G, Hurst R, Stefani E, Birnbaumer L (1996) trp, a novel mammalian gene family essential for agonist-activated capacitative Ca^{2+} entry. Cell 85:661–671

Zhu Z, Luo Z, Ma S, Liu D (2010) TRP channels and their implications in metabolic diseases. Pflugers Arch 461:211–223

Zhu J, Yu Y, Ulbrich MH, Li MH, Isacoff EY, Honig B, Yang J (2011) Structural model of the TRPP2/PKD1 C-terminal coiled-coil complex produced by a combined computational and experimental approach. Proc Natl Acad Sci U S A 108:10133–10138

Zitt C, Zobel A, Obukhov AG, Harteneck C, Kalkbrenner F, Lückhoff A, Schultz G (1996) Cloning and functional expression of a human Ca^{2+}-permeable cation channel activated by calcium store depletion. Neuron 16:1189–1196

Index

B. Nilius and V. Flockerzi (eds.), *Mammalian Transient Receptor Potential (TRP) Cation Channels*, Handbook of Experimental Pharmacology 223, DOI 10.1007/978-3-319-05161-1,

N

O

MIX
Papier aus verantwortungsvollen Quellen
Paper from responsible sources
FSC® C105338

If you have any concerns about our products,
you can contact us on
ProductSafety@springernature.com

In case Publisher is established outside the EU,
the EU authorized representative is:
Springer Nature Customer Service Center GmbH
Europaplatz 3, 69115 Heidelberg, Germany

Printed by Libri Plureos GmbH
in Hamburg, Germany